Lecture Notes in Electrical Engineering

Volume 701

The book series *Lecture Notes in Electrical Engineering* (LNEE) publishes the latest developments in Electrical Engineering - quickly, informally and in high quality. While original research reported in proceedings and monographs has traditionally formed the core of LNEE, we also encourage authors to submit books devoted to supporting student education and professional training in the various fields and applications areas of electrical engineering. The series cover classical and emerging topics concerning:

- Communication Engineering, Information Theory and Networks
- Electronics Engineering and Microelectronics
- Signal, Image and Speech Processing
- Wireless and Mobile Communication
- Circuits and Systems
- Energy Systems, Power Electronics and Electrical Machines
- Electro-optical Engineering
- Instrumentation Engineering
- Avionics Engineering
- Control Systems
- Internet-of-Things and Cybersecurity
- Biomedical Devices, MEMS and NEMS

For general information about this book series, comments or suggestions, please contact leontina.dicecco@springer.com.

To submit a proposal or request further information, please contact the Publishing Editor in your country:

China

Jasmine Dou, Associate Editor (jasmine.dou@springer.com)

India, Japan, Rest of Asia

Swati Meherishi, Executive Editor (Swati.Meherishi@springer.com)

Southeast Asia, Australia, New Zealand

Ramesh Nath Premnath, Editor (ramesh.premnath@springernature.com)

USA, Canada:

Michael Luby, Senior Editor (michael.luby@springer.com)

All other Countries:

Leontina Di Cecco, Senior Editor (leontina.dicecco@springer.com)

**** Indexing: Indexed by Scopus. ****

More information about this series at http://www.springer.com/series/7818

Pradeep Kumar Singh · Yashwant Singh ·
Maheshkumar H. Kolekar · Arpan Kumar Kar ·
Jitender Kumar Chhabra · Abhijit Sen
Editors

Recent Innovations in Computing

Proceedings of ICRIC 2020

Springer

Editors
Pradeep Kumar Singh
Department of Computer Science & Engineering
ABES Engineering College
Ghaziabad, Uttar Pradesh, India

Maheshkumar H. Kolekar
Department of Electrical Engineering
Indian Institute of Technology Patna
Patna, Bihar, India

Jitender Kumar Chhabra
Department of Computer Engineering
National Institute of Technology Kurukshetra
Kurukshetra, Haryana, India

Yashwant Singh
Central University of Jammu
Jammu and Kashmir, India

Arpan Kumar Kar
Department of Management Sciences
Indian Institute of Technology Delhi
New Delhi, Delhi, India

Abhijit Sen
Department of Computer Science and Information Technology
Kwantlen Polytechnic University
Surrey, BC, Canada

ISSN 1876-1100 ISSN 1876-1119 (electronic)
Lecture Notes in Electrical Engineering
ISBN 978-981-15-8296-7 ISBN 978-981-15-8297-4 (eBook)
https://doi.org/10.1007/978-981-15-8297-4

This Springer imprint is published by the registered company Springer Nature Singapore Pte Ltd.
The registered company address is: 152 Beach Road, #21-01/04 Gateway East, Singapore 189721, Singapore

Organising Committee

Committee Members

Chief Patron

Shri G. Parthasarthi, Chancellor, Central University of Jammu, Jammu and Kashmir, India

Patron

Prof. Ashok Aima, Vice-Chancellor, Central University of Jammu, Jammu and Kashmir, India

Co-patron

Prof. Devanand, Dean, School of Applied and Basic Sciences, Central University of Jammu, Jammu and Kashmir, India

Honorary Chair

Dr. Bharat Bhargava, Purdue University, USA
Dr. Marcin Paprzycki, Systems Research Institute, Polish Academy of Sciences, Warsaw, Poland
Dr. Abhijit Sen, Computer Science and Information Technology, Kwantlen Polytechnic University, Canada
Dr. Subhas Chandra Mukhopadhyay, Macquarie University, NSW, 2109, Australia

Conference General Chair

Prof. (Dr.) Wei-Chiang Hong, School of Education Intelligent Technology, Jiangsu Normal University, China
Dr. Yashwant Singh, Head and Associate Professor, Department of CS and IT, Central University of Jammu, Jammu and Kashmir, India

Conference Co-general Chair

Prof. (Dr.) Pao-Ann Hsiung, Professor of Computer Science and Information Engineering, National Chung Cheng University, Taiwan
Dr. Sanjay Sood, Joint Director, C-DAC Mohali, India

Conference Publication Chair

Dr. Arpan K. Kar, Associate Professor, IT Area, DMS, IIT Delhi, India
Dr. Pradeep Kumar Singh, Assistant Professor, Jaypee University of Information Technology, Waknaghat, India
Dr. Zoltán Vámossy, Óbuda University, Budapest, Hungary
Prof. (Dr.) Habil. Levente Adalbert Kovács, Óbuda University, Budapest, Hungary

Conference Program Chair

Dr. Maheshkumar H. Kolekar, Associate Professor, Department of Electrical Engineering, IIT Patna, India
Dr. Pljonkin Anton, Institute of Computer Technologies and Information Security, Southern Federal University, Russia
Dr. Arvind Selwal, Assistant Professor, Department of CS and IT, Central University of Jammu, Jammu and Kashmir, India
Dr. Arti Noor, Joint Director, CDAC Noida, India
Dr. Ioan-Cosmin MIHAI, "Alexandru Ioan Cuza" Police Academy, Romania

Technical Program Committee Chair

Dr. Bhavna Arora, Assistant Professor, Department of CS and IT, Central University of Jammu, Jammu and Kashmir, India
Dr. Nagender Kumar Suryadevara, School of Computer and Information Sciences, University of Hyderabad, Hyderabad, Telangana, India
Dr. Zdzislaw Polkowski, Rector's Representative for International Cooperation and Erasmus+Programme, Jan Wyzykowski University, Polkowice, Poland

Publicity Committee Chair

Dr. Deepti Malhotra, Assistant Professor, Department of CS and IT, Central University of Jammu, Jammu and Kashmir, India
Mr. Neerendra Kumar, Assistant Professor, Department of CS and IT, Central University of Jammu, Jammu and Kashmir, India

Technical Program Committee

Dr. Abhijit Sen, Computer Science and Information Technology, Kwantlen Polytechnic University, Canada
Dr. Anil Sharma, Associate Professor, School of Computer Applications, Lovely Professional University, Phagwara, Punjab, India

Dr. Zdzislaw Polkowski, Rector's Representative for International Cooperation and Erasmus+Programme, Jan Wyzykowski University, Polkowice, Poland
Dr. Zoltán Vámossy, Óbuda University, Budapest, Hungary
Mr. Neerendra Kumar, Assistant Professor, Department of CS and IT, Central University of Jammu, Jammu and Kashmir, India
Prof K. N. Mishra, Birla Institute of Technology, Ranchi
Prof. (Dr.) Habil. Levente Adalbert Kovács, Óbuda University, Budapest, Hungary
Prof. (Dr.) Wei-Chiang Hong, School of Education Intelligent Technology, Jiangsu Normal University, China
Prof. Amay Kumar Rath, DRIEMS, Cuttack
Prof. Amit Prakash Singh, USIT, GGSIPU, Delhi, India
Prof. Arti Noor, Joint Director, CDAC, Noida, India
Prof. C. S. Rai, USIT, GGSIPU, Delhi, India
Prof. D. K. Lobiyal, School of Computer and Information Sciences, Jawaharlal Nehru University, Delhi, India
Prof. Jitender Kumar Chhabra, National Institute of Technology, Kurukshetra, India
Prof. Lalit K. Awasthi, National Institute of Technology, Hamirpur
Prof. M. P. S. Bhatia, NSIT Delhi, India
Prof. M. N. Doja, Department of Computer Engineering, Jamia Millia Islamia, New Delhi, India
Prof. Manu Sood, Director, UIIT, HPU, Shimla, India
Prof. Manu Sood, Himachal Pradesh University, Shimla, Himachal Pradesh
Prof. O. P. Rishi, University of Kota, Rajasthan
Prof. Om Prakash Sangwan, Guru Jambheshwar University of Science and Technology, Hisar, India
Prof. Rakesh Kumar, Kurukshetra University, Haryana
Prof. Vibhakar Mansotra, University of Jammu, Jammu
Prof. (Dr.) Pao-Ann Hsiung, Professor of Computer Science and Information Engineering, National Chung Cheng University, Taiwan

Preface

The Third International Conference on Recent Innovations in Computing (ICRIC 2020) targeted researchers from different domains of advanced computing, intelligent networking, image processing and computer vision, e-learning, cloud and big data, security and privacy, and Digital India on a single platform to showcase their research ideas. This is the ongoing event in continuation of its previous version of ICRIC 2019 from Springer. The conference aims to be an annual ongoing event inviting researchers to exchange their ideas and thoughts. We hope that it will continue evolving and contributing in the field of computing technologies. The Third International Conference on Recent Innovations in Computing (ICRIC 2020) was hosted by Central University of Jammu, Jammu and Kashmir, India, during March 20–21, 2020. We are thankful to our valuable authors for their contribution and our Technical Program Committee for their immense support and motivation toward making the 3rd ICRIC 2020 a successful event. We would like to express our sincere gratitude to our keynote speakers—Dr. Zdzislaw Polkowski, Jan Wyzykowski University, Polkowice, Poland; Prof. Manu Sood, HPU, Shimla; Prof. Neeraj Kumar, Thapar, Patiala; Prof. Vinod Sharma, University of Jammu, Jammu and Kashmir; Prof. Meenakshi Sood, NITTTR Chandigarh; and Prof. (Dr.) Sudeep Tanwar, Nirma University, India. We are also thankful to the vice-chancellor of the university, Prof. Ashok Aima, for extending his continuous support to make this event happen. We express our special thanks to Prof. Devanand from CUJ, Jammu and Kashmir, for his valuable suggestions and help during the technical program schedule. We are also thankful to our various session chairs for sharing their technical sessions and enlightening the delegates of the conference. We want to express our thanks to Dr. Arvind Selwal, Dr. Deepti Malhotra, Dr. Bhavna Arora, Dr. Sudhanshu Tyagi, Prof. Devanand and many more professors for spending their valuable time during the paper presentations. Selected papers were presented in various parallel tracks in six sessions during two days of conference. We are deeply indebted to our sponsor SERB, DST, Government of India, for providing fund and supporting this conference. We express our sincere gratitude to our publication partner, Springer, LNEE Series, for believing in us.

Central University of Jammu
Jammu and Kashmir, India
March 2020

Yashwant Singh
Pradeep Kumar Singh
Maheshkumar H. Kolekar
Arpan Kumar Kar
Jitender Kumar Chhabra
Abhijit Sen

Contents

About the Editors

Dr. Pradeep Kumar Singh is currently Professor and Head at the Department of CSE at ABES Engineering College, Ghaziabad, Uttar Pradesh, India. He has completed his Ph.D. in Computer Science & Engineering from Gautam Buddha University (State Government University), Greater Noida, UP, India. He received his M.Tech. (CSE) with Distinction from GGSIPU, New Delhi, India. He is a senior member of the CSI and ACM, and is an associate editor of IJISMD, IJAEC, IGI Global USA, SPY, Wiley & IJISC journals. He has published 90 research papers and edited 10 books for leading publishers and several special issues for SCI and SCIE journals. He has received three research grants from Govt. of India and Govt. of HP worth Rs 25 Lakhs. He has edited total 8 books from Springer and Elsevier and also edited several special issues for SCI and SCIE Journals from Elsevier and IGI Global. He has Google scholar citations 389, H-index 11 and i-10 Index 15 in his account.

Dr. Yashwant Singh is currently an Associate Professor and Head of the Computer Science and IT Department at the Central University of Jammu, India. He has nearly 15 years of teaching and research experience, and has published about 60 papers in various respected journals and conferences. He has served as general chair or publication chair for several conferences.

Dr. Maheshkumar H. Kolekar is an Associate Professor at the Department of Electrical Engineering at the Indian Institute of Technology Patna, India. From 2008 to 2009, he was a Postdoctoral Research Fellow at the Department of Computer Science, University of Missouri, Columbia, USA.

Dr. Arpan Kumar Kar is an Associate Professor of Information Systems at DMS, IIT, Delhi. He has published over 120 high impact articles, and 6 books. Prior to joining academia, he worked at the IBM Research Lab and Cognizant Business

Consulting. He has provided advisory and consultancy services, and has received multiple awards and recognitions from leading organizations like Elsevier, IFIP, I3E, IIT Delhi, IIM Rohtak, PMI, AIMS, TCS, and Jadavpur University.

Dr. Jitender Kumar Chhabra is a Professor at the Computer Engineering Department at the National Institute of Technology, India. He has published 120 papers in respected international and national journals and conferences, including more than 40 publications from IEEE, ACM, Elsevier, and Springer.

Dr. Abhijit Sen is a Professor at Kwantlen Polytechnic University, Canada. He received his Ph.D. from McMaster University, Canada, in 1976 and his M.S. in Electrical and Computer Science from the University of California, USA, in 1970. He graduated from IIT, Kharagpur, in 1968. His research interests include emerging technologies, networking, wireless and distributed computing, and Internet application development.

Advanced Computing

Performance Analysis of Commodity Server with Freeware Remote Terminal Application in Homogeneous and Heterogeneous Mutli-computing Environments

Shamneesh Sharma, Manoj Manuja, and Digvijay Puri

Abstract The technology is bringing new changes rapidly in terms of hardware and software which leads to the increase in the computing cost. One of the prevalent challenges involved in the research endeavors of researchers is to diminish the cost involved in possession of hardware and network technologies. Open source technologies have tried to cut down costs involve at software level, but learning on the software technologies is still a challenge. Commodity hardware utilization is still an option to decrease the hardware level cost. This paper presents the study of the performance of a commodity hardware server in heterogeneous multi-computing environment. Using the commodity hardware of our university and on a client–server-based model, we have developed a system of remote terminal application on a commodity server in homogeneous and heterogeneous multi-computing environments. The nodes used in the current experimental environment are independent of system architecture.

Keywords Hardware cost reduction · Multi-computing environment · Remote terminal · Commodity hardware

1 Introduction

Business continuity is the success mantra of most of the organizations. In the educational institutions, students' satisfaction is the ultimate goal. Moreover, most of these institutions adopted the technology and growing at very fast pace. Information technology is also paying a vital role in the student's satisfaction by setting up a strong

S. Sharma (✉)
School of Computer Science & Engineering, Poornima University, Jaipur, Rajasthan, India
e-mail: shamneesh.sharma@gmail.com

M. Manuja
Rayat Bahra University, Mohali, Punjab, India
e-mail: manoj_manuja@yahoo.in

D. Puri
iNurture Education Solutions, Bangalore, India
e-mail: digvijaypuri@gmail.com

P. K. Singh et al. (eds.), *Recent Innovations in Computing*, Lecture Notes in Electrical Engineering 701, https://doi.org/10.1007/978-981-15-8297-4_1

feedback system. The past decade is the eyewitness of the rapid growth and change in the technology. Field of information technology has come up with exciting new countenances, soothes and well-being to the society. The use of dedicated servers for every application is not an old story. Some decades back dedicated DR servers were increasing the hardware cost in the computing environments. Hardware costs were not optimized due to low utilization rates of these dedicated servers. Technology changed and dedicated servers were replaced by shared hosting techniques. Few years back the strategy of running applications on economical devices became very popular. It has started with the creation of powerful computing nodes that could be linked together and perform as equal an expansive server [1]. There is a challenge for the organizations to replace the existing IT hardware in case of technology changes. The computer nodes with exhausted storage capacity need to be upscale, but this is possible when we have compatible motherboard with the available storage device. We have discussed a case study for the use of commodity server in the laboratory of an educational institution. The rationale behind the commodity hardware utilization in the cluster computing is to employ large numbers of non-accessible computing components to obtain the computation power at an economical cost. The heterogeneity in a distributed system comes at a cost as distributed system engrosses different types of nodes with diverse software systems. Commodity hardware utilization is one of the methods to cut this cost.

2 Related Work

In the paper [2], authors discussed that cloud services using commodity hardware are the idea with the help of which a huge idle computational power can be made accessible through the network which remains unused in the ideal system. An ideal processing power of a workstation in any organization commonly remains vacant 50–70% of time [3]. The computing capacity of desktops is increasing at a good pace these days, and this growth is still not ended. Most of the researchers have focused on the cloud services using commodity hardware [4]. A lot of research has been done on the unstable computational environments. The resilience of the algorithms on the unstable computational environments [5] shows that these environments used intricate methods to solve the complex patterns. The cluster-based technologies are also useful in the field of multi-computing environments; the cluster-based approach is used by researchers in news technologies like WSN and IoT also [6, 7]. In the paper [8], authors discussed the factors affecting the selecting of a cluster interconnection network technology. They have also discussed the scalable coherent interface in the field of cluster computing. In the year 1999, authors of [9] have coined a term Commodity-Off-The-Shelf (COTS) which was for free hardware components. This research study also brought the new development techniques for supercomputing. Clustering with commodity hardware can change the concept of use of high-performance processors and memory components for computing. The applications of the commodity hardware are up to the use and intentions of the researchers and users. Some of the researchers are using it for cloud solutions [10], some are using it for low cost clustering [11], and some are using it for cyber-attacks [12]. The business continuity of prevailing enterprise is dependent on some essential processes like data storage management, virtual machine management, network traffic management and backup management [13]. For small organizations, the commodity IT infrastructure is beneficial, but when it comes to the large organizations, it cannot be opted for the applications, where we need monitoring, managed services, security, performance and visibility to the future [14]. Not only is the hardware but commodity software industry also making an impact in this field. Infrastructure can be seen as the vertebral column of any organization. The work has been done on the network performance using the commodity hardware.

The paper [15] presents a commodity hardware solution for the improvement of packet capturing solution. The authors of [16] have identified the principles for the edifice of elevated functioning of software router organisms on commodity hardware. The level of commodity platforms has reached at a place where the production of low-cost hardware with viable alternatives is very much possible with their implementation to the real-time network functions without forfeiting the performance [17]. Further development of these systems needs secure environment where old hardware can be used as effective infrastructure. In the paper [18], authors have presented a comprehensive study of the features of commodity hardware for its use in the various applicable environments. Low-cost and potent embedded microprocessors have also facilitated an innovative generation of economical six-degree-of-freedom (6DOF) [19] which are suitable for molecular apparition in desktop environments. When most of the systems are moving toward cluster-based solution in cloud environment [20], there is a need of server-based computing in educational institutions.

Findings and Research Gaps

- To repair and replacement of the faulty IT infrastructure is a challenge to the educational organization as an additional cost to be spent on infrastructure needs.
- The proper utilization of the hardware in the educational institutions is still a challenge as technology change can ask the organization to replace the existing IT infrastructure at any time.
- Commodity hardware utilization techniques can overcome this overburden from the organizations.

3 Experimental Setup and Methodology

On a client server model, we have installed 16 nodes of different processing capabilities with a server. These nodes come from different vendors and have different specifications. The main thing that is kept under the consideration these nodes has low computing capabilities. These nodes come from different vendors, have different configurations and generally have low computing power. We have used one 24 port

D-Link nonmanageable switches of 100 Mbps bandwidth to connect these nodes with commodity server with cat-6 cabling structure (Tables 1 and 2).

We have installed a freeware remote terminal application (WTWare) on our server with OS (Windows server 2008) to run critical applications on the heterogeneous nodes. All the nodes in present system will act as a thin client and WTware as remote operating system which helps to make the whole system as network boot operating system. Then, we have performed the performance analysis tests on our commodity server. We have performed the experiment on real test beds with the setup of 16 nodes and one server. All the hardware used in this experimental setup was taken from the waste hardware section of Alakh Prakash Goyal (APG) Shimla University, Shimla (H.P.), India. All the analysis of commodity server was taken into consideration with following methodology:

1. Check the analysis of commodity server; we have put all the nodes of type A, B C and D in the passive mode.
2. Analysis of commodity server when all the nodes of type A are in the active mode and all other types under passive mode (homogeneous environment).
3. Analysis of commodity server when all the nodes of type B are in the active mode and all other types under dead mode (homogeneous environment).

Table 1 Configuration of the server

RAM type	RAM size	HDD type	HDD size	Processor name	Processor speed	Motherboard
DDR3	4 GB	SATA	640 GB	Intel® Core™ i3 3210	3.20 GHz	Simmtronics H61-MX

Table 2 Configuration of the nodes

No. of nodes	Node no.	RAM type	RAM size	Processor name	Core	Processor speed	Motherboard	Node type
4	1, 3, 7 and 10	DDR3	2 GB	Intel Atom™ CPU D425	Single Core	1.80 GHz	Gigabyte GA-D425TUD	A
10	2, 4, 5, 6, 8, 9, 12, 13, 14 and 16	DDR2	1 GB	VIA C7-D	Single Core	1600 MHz	Simmtronics PC2000E+	B
1	11	DDR3	1 GB	Intel Atom™ CPU D2550	Dual Core	1.86 GHz	Digilite Fast X Fast LAN	C
1	15	DDR2	2 GB	VIA C7-D	Single Core	1600 MHz	Simmtronics VX900I	D

4. Analysis of commodity server when all the nodes of type C are in the active mode and all other types under dead mode (homogeneous environment).
5. Analysis of commodity server when all the nodes of type D are in the active mode and all other types under dead mode (homogeneous environment).
6. Analysis of commodity server when all the nodes of type A and B are in the active mode, whereas nodes of type C and D are kept in passive state (heterogeneous environment).
7. Check the analysis of commodity server; we have put all the nodes of type C and D in the active mode, whereas nodes of type A and B are kept in passive state (heterogeneous environment).
8. Check the analysis of commodity server; we have put all the nodes of type A and C in the active mode, whereas nodes of type B and D are kept in passive state (heterogeneous environment).
9. Check the analysis of commodity server; we have put all the nodes of type B and D in the active mode, whereas nodes of type A and C are kept in passive state (heterogeneous environment).
10. Check the analysis of commodity server; we have put all the nodes of type A, B, C and D in the active mode (Table 3).

In most of the infrastructure-oriented networks, data dissemination can be considered as most important parameter [21], while experimenting in the same way, the above set of experiments has been carried out in such a way that data will be rendered from master system to slave ones. About 25% energy in computer systems is consumed by the HDDs [22] in the operations like seeking, rotation, reading, writing and many more. This research carried out a model where systems are without HDDs.

Table 3 Steps used in methodology

Step	Node (type A)	Node (type B)	Node (type C)	Node (type D)	Computing environment
Step 1	Passive	Passive	Passive	Passive	NA
Step 2	Active	Passive	Passive	Passive	Homogeneous
Step 3	Passive	Active	Passive	Passive	Homogeneous
Step 4	Passive	Passive	Active	Passive	Homogeneous
Step 5	Passive	Passive	Passive	Active	Homogeneous
Step 6	Active	Active	Passive	Passive	Heterogeneous
Step 7	Passive	Passive	Active	Active	Heterogeneous
Step 8	Active	Passive	Active	Passive	Heterogeneous
Step 9	Passive	Active	Passive	Active	Heterogeneous
Step 10	Active	Active	Active	Active	Heterogeneous

4 Results and Discussions

After experimenting on the above methodology in the Step 1, by keeping all nodes in the passive state, we have checked the performance server based on the CPU and memory usages. In this case, the maximum utilization of processor was 3%, and memory was 22%.

After this, we have shifter to the second step by logging on with one node of type A and checked the performance of the server on the processor, memory and network usages with and without running any application on it (Fig. 1).

Without running any application, the performance of the server was same, but after running the application on it, CPU usage increased by 2%, and memory usage reached from 22 to 40%, whereas the network usage was stable on 1% (Fig. 2).

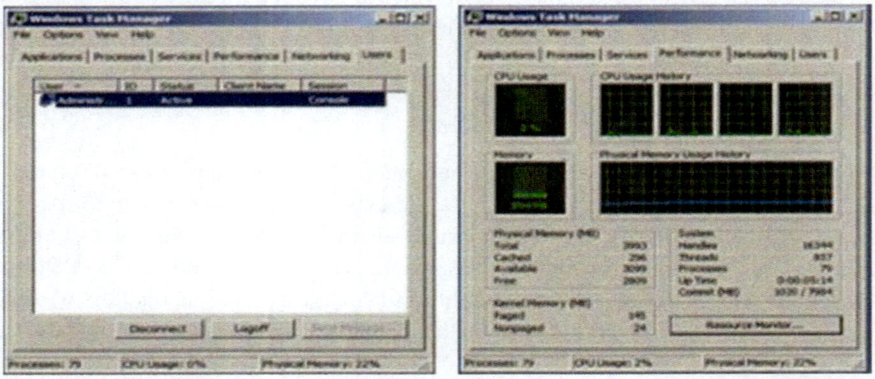

Fig. 1 Server performance when no active user on the network

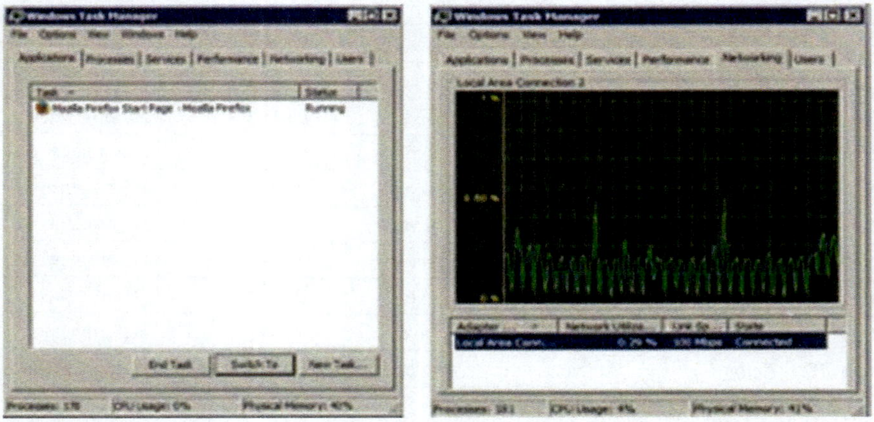

Fig. 2 Server performance when one active user on the network with the execution of one critical application

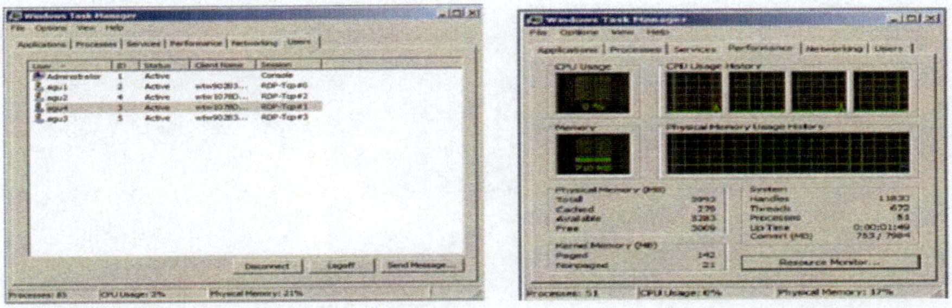

Fig. 3 Server performance in homogeneous environment without any critical application

To create the homogeneous setup, we have examined the four systems of same configuration (4, 3, 7 and 10) by logging on them with and without one critical application. While experimenting, we have found that without running any application, the use of memory was 17% (Fig. 3).

The memory usage has reached to 34% when we executed a critical application on all the nodes (Fig. 4).

On the logging on of all the 16 nodes on the network without running any critical applications on it, we have checked the performance of the server in terms of processor, memory and network usage. The processor and memory usages were stable, whereas network usage reached up to 15% (Fig. 5).

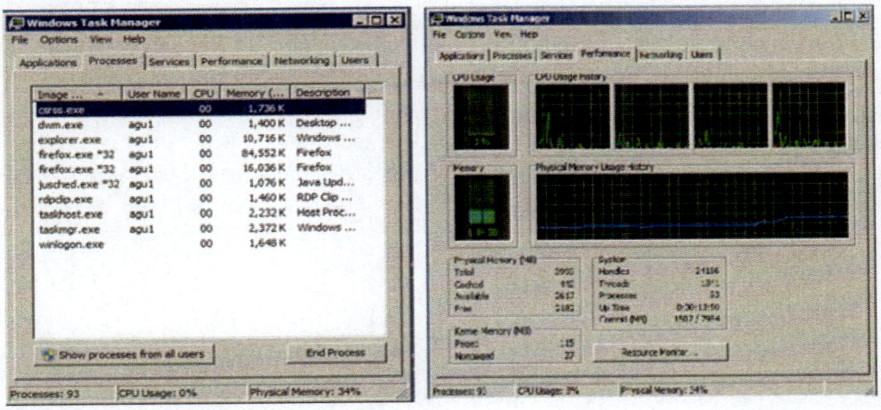

Fig. 4 Server performance in homogeneous environment with the execution of critical application

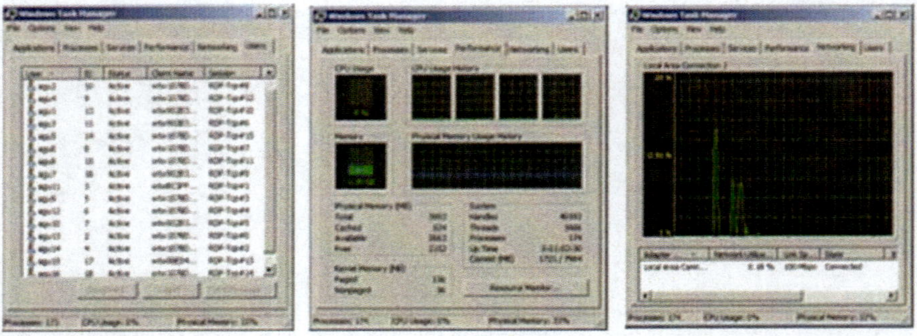

Fig. 5 Server performance when all active user on the network without executing any critical application

5 Conclusion

Today, the main demand in the field of computing is to reduce the setup cost, so in this present research, we have tried to check the performance of a commodity hardware platform. We conclude that commodity hardware is very useful in the organizations to fulfill the computing demands. It not only performs the computing tasks efficiently but also reduced the burden of the hardware setup cost. Moreover, the commodity servers can perform well in the both homogeneous and heterogeneous multi-computing environments in the small organizations. We recommend the use of this model for the setup of small labs with 25 nodes (maximum) with one server of same configuration. After the addition of more nodes, the performance of the server gets reduced to 58%. The best results can be achieved with the node size 15–20 with current configuration. So, a low cost and small setup of IT laboratories can be implemented in the organizations for providing the students with better computing skills. It will not only reduce the cost but also prepare a center for waste IT accessories management.

6 Future Scope

There is a need to do a lot on commodity hardware. The need of computing is growing day by day so the resources, but the cost of hardware and network setup is also increasing at a parallel pace. Researcher needs to focus on the reduction of this setup cost. The present work focuses on a client server computing model setup on a commodity hardware platform. Use of different networking topologies in the computing environments can lead to more experimental setups and exciting results. The distributed computing model can be tested on the same platform. With the uplift in the server configuration, more experiments can be done in homogeneous and heterogeneous environments. Single point of failure is the problem with the current setup, so one more step toward this problem can lead to a research problem.

Acknowledgements This work is supported by Alakh Prakash Goyal (APG) Shimla University, Shimla (H.P.), India. One of our authors would like to thank Dr. Ramesh Chaudhary, Vice Chancellor, Alakh Prakash Goyal (APG) Shimla University to give us this problem which has ultimately twirl to a research challenge for our team. Our team is also grateful to Dr. Purnima Bali, Shoolini University, Solan (H.P.), India, for proofreading this manuscript.

References

1. Bob, R.: 12 ways to reduce your IT costs. IDG Communications (7 Mar 2017). Available at https://www.cio.com/article/3176836/best-practices/12-ways-to-reduce-your-it-costs.html on 11th Oct 2018 15:01 PM
2. Dongre, D., Sharma, G., Kurhekar, M.P., Keskar, R.B., Radke, M.A.: Scalable cloud deployment on commodity hardware using OpenStack. In: Advanced Computing, Networking and Informatics, vol. 2. Smart Innovation, Systems and Technologies 28. Springer International Publishing Switzerland (2014) (Bob)
3. Mutual, M.W., Livny, M.: The available capacity of a privately owned workstation environment. J. Perform. Eval. Arch. **12**(4) (1991)
4. Meena, J., Kumar, M., Vardhan, M.: Efficient utilization of commodity computers in academic institutes: a cloud computing approach. World Acad. Sci. Eng. Technol. Int. J. Comput. Inf. Eng. **9**(2) (2015)
5. Nogueras, R., Cotta, C.: On the use of self-island-based evolutionary computation methods on complex environments. Comput. Sci. Inf. Syst. **15**(3), 733–750 (2018). https://doi.org/10.2298/CSIS180115032N
6. Sharma, S., Singh, S., Chaudhary, V.: Implementation of enhanced reliable distributed energy efficient protocol for WSN using MATLAB. Int. J. Electron. Commun. Technol. **4**(2) (2013)
7. Sharma, S., Manuja, M., Kishore, K.: Node-level self-adaptive network path restructuring technique for internet of things (IoT). In: Choudhury, S., Mishra, R., Mishra, R., Kumar, A. (eds.) Intelligent Communication, Control and Devices. Advances in Intelligent Systems and Computing, vol. 989. Springer, Singapore (2020)
8. Yeo, C.S., Buyya, R., Pourreza, H., Eskicioglu, R., Graham, P., Sommers, F.: Cluster computing: high-performance, high-availability, and high-throughput processing on a network of computers. In: Zomaya, A.Y. (ed.) Handbook of Nature-Inspired and Innovative Computing: Integrating Classical Models with Emerging Technologies, Chap. 16, pp. 521–551, Springer, Berlin (2006)
9. Baker, M., Buyya, R.: Cluster computing: the commodity supercomputer. Softw. Pract. Experience **29**(6), 551–576 (1999)
10. Mittal, P.: Building your own cloud with commodity hardware and "Ganeti". PyCon India (2013). Available at https://in.pycon.org/funnel/2013/63-building-your-own-cloud-with-commodity-hardware-and-ganeti/ on 23rd Oct 2018 14:49 PM
11. Dorband John, E., Raytheon, J.P., Ranawake, U.: Commodity computing clusters at Goddard space flight center. Online J. Space Commun. (2010)
12. Vanhoef, M., Piessens, F.: Advanced Wi-Fi attacks using commodity hardware. In: ACSAC'14, New Orleans, LA, USA, 08–12 Dec 2014
13. Chen, C., Gurganus, J.: Statistical anomaly detection on metadata streams via commodity software to protect company infrastructure: a case study. In: IEEE 37th International Conference on Distributed Computing Systems Workshops, Atlanta, GA, USA (2017)

14. Swanson, G.: Commodity hardware-modernize your infrastructure with Avi networks. White Paper, ANI Networks, India (18 Dec 2015). Available at https://blog.avinetworks.com/modernizing-your-infrastructure-with-open-source-commodity-hardware on 28th Oct 2018 10:56 AM
15. Braun, L., Didebulidze, A., Kammenhuber, N., Carle, G.: Comparing and improving current packet capturing solutions based on commodity hardware. In: IMC'10, Melbourne, Australia, ACM, 1–3 Nov 2010
16. Egi, N., Greenhalgh, A., Handley, M.: Towards High Performance Virtual Routers on Commodity Hardware. ACM CoNEXT, Madrid, Spain (2008)
17. Bonelli, N., Giordano, S., Procissi, G.: Enif-Lang: a specialized language for programming network functions on commodity hardware. J. Sens. Actuator Netw. **7**, 34 (2018)
18. Koning, K., Chen, X., Bos, H., Giuffrida, C., Athanasopoulos, E.: No need to hide: protecting safe regions on commodity hardware. In: Proceedings of the 12th European Conference on Computer Systems, EuroSys, pp. 437–452. Association for Computing Machinery (2017)
19. Stone, J.E., Kohlmeyer, A., Vandivort, K.L., Schulten, K.: Immersive molecular visualization and interactive modeling with commodity hardware. In: Bebis, G., et al. (eds.) Advances in Visual Computing, vol. 6454. Springer, Berlin, Heidelberg (2010)
20. Nikita, G., Sharma, L.S.: Performance evaluation and modelling of the Linux firewall under stress test. In: Singh, P.K., Kar, A.K., Singh, Y., Kolekar, M.H., Tanwar, S., et al. (eds.) Proceedings of ICRIC 2019, Recent Innovations in Computing, 2020. Lecture Notes in Electrical Engineering, vol. 597, pp. 3–920. Springer, Cham, Switzerland (2019)
21. Sharma, S., Kishore, K.: Data dissemination algorithm using cloud services: a proposed integrated architecture using IoT. In: 2nd International Conference on Innovative Research in Engineering Science and Technology (IREST-2017), Eternal University, Baru Sahib, Sirmour (H.P.), India, 7–8 Apr 2017
22. Baghel, A., Srivastava, A., Tyagi, A., Goel, S., Nagrath, P.: Analysis of Ex-YOLO algorithm with other real-time algorithms for emergency vehicle detection. In: Singh, P., Pawłowski, W., Tanwar, S., Kumar, N., Rodrigues, J., Obaidat, M. (eds.) Proceedings of First International Conference on Computing, Communications, and Cyber-Security (IC4S 2019). Lecture Notes in Networks and Systems, vol. 121. Springer, Singapore (2020)

Machine Learning-Based Information Retrieval System

Manpreet Singh Bajwa, Ravi Rana, and Geetanshi Bagga

Abstract The machine learning-based information retrieval model would encourage the user(s) to register on a user platform by authentication of identity information by assigning them a unique membership number. The platform would register the user(s) for a paid membership. The user would be able to search for da keyword(s) or phrase(s) on which the platform would apply auto-correction and clustering of the keyword into the databases would be done. The user would be alerted on his search, and the information would be displayed to the logged-in user, using a plagiarism detection algorithm. This system would come on handy as a tool for efficient search, the results displayed are more to the point and of significant relevance of the keyword or phrase entered by the user. The platform would integrate machine learning-based search giving benefits to students, teachers and scholars as a way of efficient searching protocol.

Keywords Machine learning · Information · Retrieval · Keyword · Auto-correction · Clustering · Plagiarism

1 Introduction

Information retrieval is as much essential as information storage. An information retrieval software (IRS) is not where the data is stored, but it acts as a platform to access the data stored in another remote database. The data is linked to the IRS using metadata. The said platform would require the user to register on it with a user Id

M. S. Bajwa
Department of Computer Science and Engineering, Faculty of Engineering and Technology, SGT University, Gurugram, Haryana 122505, India
e-mail: manpreet_feat@sgtuniversity.org

R. Rana · G. Bagga (✉)
CGC College of Engineering, Landran, Mohali, Punjab 140307, India
e-mail: geetanshibagga23@gmail.com

R. Rana
e-mail: ravirana2319@gmail.com

P. K. Singh et al. (eds.), *Recent Innovations in Computing*, Lecture Notes in Electrical Engineering 701, https://doi.org/10.1007/978-981-15-8297-4_2

and password which would be verified by an authentication mechanism. The user would search the IRS using keyword(s) or phrases. The above-mentioned platform would use various machine learning algorithms to cluster the keyword, suggest auto-correction and implement them to search the database and would provide user(s) with a most suitable match.

Keyword-based searching is the basis of searching today. Many times, it happens the right keyword is not found in the search query, due to this, the conventional systems may display mismatched or no results at all. In situations like this, a machine learning-based IRS would come in handy. A machine learning-based IRS would extract keywords and would work on sequence prediction to predict the most relatable keyword from a cluster of similar queries and would display the most accurate/precise results.

2 Objective

The main objective of implementing a machine learning-based information retrieval system is to make searching easier in a large pool of data. Another objective of implementing this system is for online science journals, and this system could be used as a better approach for a user to get the most accurate result.

3 Information Retrieval System

A traditional information retrieval system (IRS) is a method to retrieve/search information from a database.

A database is a collection of images, videos, data and spreadsheets. In general, a database is collection of information units containing similar and related information.

An IRS does all the difficult work for us, i.e., collecting relevant and desired information from the greater pool of data, i.e., a database. An IRS reduces the time for searching a query in the database by gathering the closest data matching the query [1]. An IRS works on "attributes" closest to the search query. There are different kinds of attributes on which the IRS works, few are listed below:

1. Use Attributes—Field Retrieval
2. Relation Attributes—Boolean Retrieval
3. Truncation Attributes—Truncation Retrieval
4. Completeness Attributes—Exact and Fuzzy Matching
5. Structure Attributes—Search terms types
6. Position Attributes—Whether search terms exist in the fields or subfields.

These attributes work in a specific way which is mentioned against them.

For large-scale information retrieval, we use "Distributed Information Retrieval" [2]. This method is generally employed in distributed environments, and a primary example of a distributed environment is on a multiclass database model. A multiclass model is a model where the contents of the database are copyrighted or owned by a group.

By calculating mutual independence through constructing class, output received is classes of independence which will help in need of information and its easy access, but this system will have low efficiency with the increasing amount of data [6].

For the sake of clarity in this paper, we would exclusively talk about a set of objects in a relational database schema format called interactive book format (IBF) and the system software which is based on machine learning approaches controlling it. This format helps in viewing books in a manner that organizes additional information and provides interaction with the original book, also not hindering with the original published book and would protect the copyrighted publication from privacy by using industry standard digital rights management security features.

4 Machine Learning

Machine learning is a buzz word today in the field of computer and data science. It is a concept that opened the portal of mass opportunities and has a wide range of applications around the globe.

Machine learning is the direct application of artificial intelligence (AI). It allows the machine/program to learn on its own. It eliminates the need of explicitly or exclusive programming. All you have to do is provide data to your machine learning model, and it would train on the patterns present in your data.

The most common models of machine learning applications are prediction, classification, clustering and regression. The need for a model to be applied to a problem is purely a use case scenario. You can use a single model on a problem or work upon a hybrid approach.

A machine learning model is capable of learning its own. We just need to find out which model is to be used that would fit into our problem. In the case of IRS, we would use clustering and classification.

Clustering would be used in case of similar queries matching a dataset. For example, we need to search for the query "data science" in our database, and our clustering algorithm would cluster all the queries which would contain the term "data science" and would show results accordingly [9]. Whereas in our case of clustering of the common words, we use unsupervised machine learning, we would use Elman recurrent neural network, in which there are hidden node activators that are latched and are connected to the input sequence, by applying this technique, we make our model ready to use the current data for clustering by identifying temporal features in the input query.

Classification is used for auto-correction and predictive keyword. Classification is a general way of implementing several techniques that would help us in auto-correction and prediction. Techniques such as the KMP algorithm, Hash map and Trie (digital tree) would be applied for the above-mentioned practices.

If used together, these two powerful algorithms would work wonders in our favor. Machine learning is the perfect choice for the implementation of searching and integrating relevant information from a wide pool of information bucket.

5 Need to Implement Machine Learning Based IRS

The primary purpose of this Web portal is to reduce search time and provide better results, content to the user. Wide range of keywords in the database allows user to have an efficient search which leads to better search time. More precise auto-corrections and suggestions. Search results are in the form of videos, images and news.

6 Conventional IRS Versus Machine Learning Based IRS

As mentioned above, machine learning algorithms would be a powerful optimization in searching. In conventional IRS, if one individual searches for a query, the most recently updated data is shown on the result page. Indexing of searches in a conventional IRS is usually based on alphabetical sorting or date wise sorting. The problem which arises in situations like these is the most relevant search the result is hidden from the user just because it was uploaded a long time ago or alphabetically its precedence is low. These two problems could be combined, and the user would be deprived of the best possible match for his query just because he does not look for it through an enormous figure of pages.

Figure 1 is the pictographic representation of the conventional IRS model used in basic Web search engine [3].

Many techniques are used for optimizing the search results for a user the most common being search engine optimization (SEO), but the main drawback of SEO-based optimization in Internet-based search engines is its use by many advertising agencies to claim traffic to their Web sites which may often lead to false information.

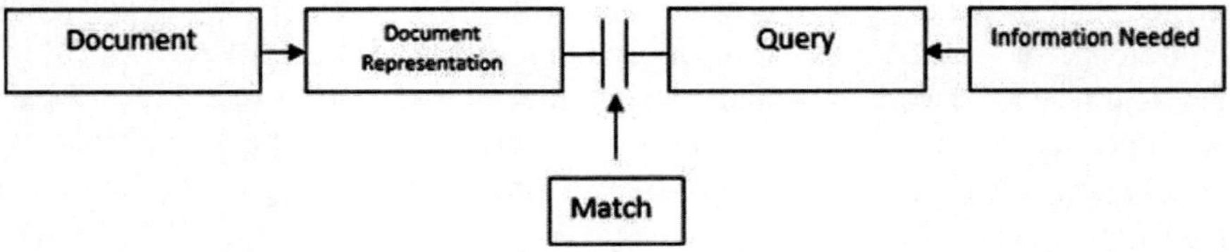

Fig. 1 Representation of a conventional information retrieval system

Machine learning is the next level approach for optimizing the results given by an IRS. As mentioned above, we would talk about the implementation of machine learning exclusively on IBF. In addition to that, we would talk about a platform on which the user would search. Implementation of machine learning on IBF is a three-step process.

1. Entering of the search query by the user—The user would enter the desired query in the search bar, on which auto correct would be applied (if any grammatical or spelling error is present).
2. Application of machine learning algorithms—The search query would be classified using classification algorithm, then it is clustered to the most suitable category of previously indexed searches.
3. Display of results to the user—Most suitable result(s) would be displayed to the user, and in case if the query is not found in the database, the most suitable match would be displayed by the application of KNN algorithm.

7 Rank Power of the Search Results

As suggested in the paper on performance measure for information retrieval systems by Meng [4], expected search length ESL and average search length are very affective key features of single-valued search in information retrieval systems [5]. A hybrid approach with a comparative study using SVM, PSO would be used which focuses on ranking with the new methodology, with appropriate parameters for finding potential solutions. Comparison between different models of Boolean, vector and probabilistic for query search is done.

8 Implementation of Machine Learning in IRS

As mentioned above, the implementation of machine learning in IRS is a three-step process.

- When a user enters a search query onto the platform, the auto checking algorithm(s) present in the platform would first of all search the query for any possible grammatical or spelling mistakes which would be corrected automatically. In addition to auto checking, the platform would also suggest the related search queries to the user, the algorithms used for auto checking and suggestions would be KMP, Hash maps would also be used along with TRIE.
- When the search query is entered onto the platform and various checks and suggestions are made, the classification algorithm(s) would classify the query according to its nature, for instance if a query named "Data Science" is present, the algorithm could classify the query into various previously defined indices like "Deep Learning", "Machine Learning", "Economics", etc. This would help the platform

to produce a variety of results and that too with the closest resemblance and similarity to the search query. When the classification would be completed, the clustering algorithm(s) would first of all compare the query with all of the indices derives by the classification algorithm(s), then it would cluster the query with the most suitable index. The algorithm(s) which can be employed for clustering and classification are DBSCAN, decision tress and Bayesian classification.
- The search results would be displayed by the platform in accordance with the priority index assigned by the clustering and classifications algorithms explained previously. Priority index would be assigned in a manner that the most similar cluster would be given the most priority and the priority decreases for the next cluster. With the use of priority indices, the index with the most priority is displayed first, and the next iteration would have priority less than the first one and so on.

Our platform would use supervised machine learning, i.e., our platform would be trained by a sample set of data, and then, it would learn by itself using the data provided to it by its users. This would help it to increase its accuracy in every search conducted by a user [12]. The machine learning model would be highly optimized by using PCA such that it results in highly efficient search results with optimized time.

User needs to enter the keyword or phrase in the search bar to open content-based pages regarding books, conferences, courses, generals, magazines, authors, standards, etc. User can also opt for advanced search if unsatisfactory search results are displayed. Keyword-related content with authors, publishing year, size of the document, type of document and its copyright terms will be displayed. There will be an inbuilt plagiarism software to check the originality of the paper that is to be published. Users can interact with the expert for any kind of advice through a chat-bot. It will also include personal sign up, account login, registration for membership and payment options. Search can be more defined either by year or by author.

9 Methodology

The methodology behind the implementation of this system could be summed up as:

1. Query evaluation.
2. Query formulation relevance feedback will be based with synonym, thesaurus, etc.
3. Feature extraction for extracting the keyword from the search query which would be used for searching the database, the extracted features would be based on priority, i.e., low or high.

10 Literature Survey

We have done a thorough survey on how to implement various algorithms onto the platform.

As we enter the query into the search box, the first step would be to extract the most meaningful words from the query, for this, natural language processing (NLP) would be used with the help of natural language toolkit.

After taking meaningful observations from the query, the platform would cluster most common words together using the Elman [7] approach in the machine learning model, we would make an Elman recurrent neural network in which there are hidden node activators which are latched to the input so that our model could extract temporal features from the query and cluster the keywords accordingly.

As suggested in Chi-Chun Huang paper on SVM bases Reduced NN Classification, we would use instance-based learning (IBL), to classify data in accordance with the search query and hence display the results in a more organized form [8]. A set of queries would be collected in accordance with its nearest neighbor or instances, and this is also known as nearest neighbor rule. A set of search queries would be initially provided to the model which would be used as the original dataset, the further search queries would be classified in accordance with the training dataset, the dataset would be trained in accordance with the Elman approach so that it could learn from different data and hence provide better classification after every search.

The fault would be minimized in the search results by the implementation of CK and PCA metrics [11].

For suggestive algorithms, we would again use SVM approach to string together different words and form a word sequence from the query which would look up the search query against the databases with respect to the sentences having similar meaning. Two kernel versions would be used: all common subsequences [9].

(ACS) for counting the number of common subsequences between two queries, Sequence Kernel (SK) for aggregating the frequency of matching subsequences between two subsequences [10]. The searching can be done as suggested in the paper on ternary search, and the query could be implemented as an array and can be manipulated into three parts for efficient searching, which would take less time.

It is the MLBT architecture which can be used to design and deploy data based on the MLBT system of research [13]. A conversation and a discussion of various existing survey are made available. Then, we incorporated ML-BT Taxonomic approach, countermeasures and aspects of smart systems. A benchmark analyzing the methodologies and methods necessary is seen in every plane.

Several approaches targeted to machine learning were used for the same reason in the industry, but small changes that also affect efficiency were not given importance [14]. In the analysis, this was taken care after. Focus was put on the significance of preprocessing the data for correct calculation with the accuracy of the dataset being retained.

11 The Platform

The platform would be a Web-based application, a typical Web site or a mobile application which would require an Internet connection to access the data from the database on which the data in the form of IBF is stored.

A user would be required to register beforehand, the registration would be email and phone number-based, after successful completion of the registration the user would be assigned a unique user ID, which would reflect the membership number of the user. The process of registration would make sure all the searches of a user are logged, and this would also help the platform to optimize and display recommendations to a user in accordance with the previous search history.

The platform would be designed to contain a search box, to take queries from the user. Every user would have his profile on which he could keep track of his previous searches. The user would also be allowed to analyze the time spent by him on the platform and his reading habits, and the profile section would also display downloads made by the user. The user would also be notified with system notifications regarding download process and completed downloads.

The platform would also help to display recommendations to a user based upon the data collected by his previous searches. It would also help in improving the accuracy of the searches by using autofill and predictive text-based algorithm(s). The search results would also be incorporated with a plagiarism check feature which would display the amount of plagiarism present in a search result (if any).

Portal will display all the necessary information such as notices, FAQ(s) and Help. There is also an option of assistant and live chat for the user. Natural language processing will help you convert your voice commands in queries.

The platform would also be used for the publishing of research papers, and for the registration on the platform, the user would be directed to a payments page where he would complete the payment and would be given a unique membership number which would be his user ID.

The user does not have to jump or hop to different Web sites. Many options are available for search refining. Paper publishing with plagiarism software ensures the originality of the paper. The user will put the related keyword in the search box and select the related checkbox for a refined search. Keyword-related content with authors, publishing year, size of the document, type of document and its copyright terms will be displayed. Just to help the user to catch up from where he left, there will be an option of search history, and to keep him updated, there will be search alerts. Guidelines will be provided for writing a thesis, submission of thesis, modification and annexure publication. Search results will be in the form of news, images, videos, books, etc. (Fig. 2).

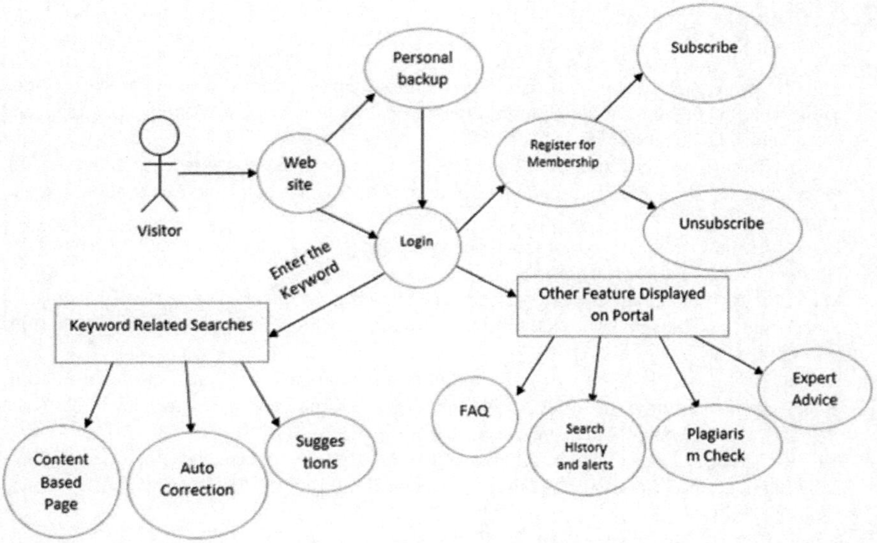

Fig. 2 Representation of machine learning-based information retrieval system

12 Possible Applications

In this paper, we have exclusively talked about the implementation of machine learning-based IRS on a set of objects in a relational database schema format called interactive book format (IBF). But this approach could be implemented on any search query-based system.

The most popular example of where this approach could be used in real world scenario is the web based on search engines. The retrieval of information from Internet would become more accurate and optimized by the implementation of machine learning-based IRS.

This technique could also be used in various organizations like hospitals, schools and colleges where data is stored and retrieval of data is needed. Banking is another sector which would be benefitted from the usage of machine learning-based IRS.

13 Conclusion

Machine learning-based IRS is the next step in the field of information retrieval. The use of this technology would help in optimizing time and provide better search results for the user. This technology would also help to provide an enhanced user experience to the user.

References

1. Zhu, Z., Wang, J.-Y., Yang, Z., Lei, F.: Internet information retrieval system based on mobile agent. In: Proceedings of the Fourth International Conference on Machine Learning and Cybernetics, Guangzhou, 18–21 Aug 2005
2. Li, M., Cao, S.: A serie method of massive information storage, retrieval and sharing. In: Proceedings of 2014 IEEE International Conference on Mechatronics and Automation, Tianjin, China, 3–6 Aug 2014
3. Bates, M.J.: The design of browsing and berrypicking techniques for the online search interface. Online Rev. **13**(5), 407–424 (1989)
4. Meng, X.: A comparative study of performance measures for information retrieval systems. In: Proceedings of Third International Conference on Information Technology: New Generations (ITNG'06), Las Vegas, NV, USA
5. Pandey, S., Mathur, I., Joshi, N.: Information retrieval ranking using machine learning techniques. In: Proceedings of 2019 Amity International Conference on Artificial Intelligence (AICAI), Dubai, United Arab Emirates (2019)
6. Luo, R., Xue, Q.: The model of information retrieval based on independence. In: Proceedings of 2009 International Conference on Future BioMedical Information Engineering (FBIE), Sanya, China (2009)
7. Elman, J.L.: Finding structure in time. Cogn. Sci. **14**(2), 179–211 (1990)
8. Huang, C.-C., Chang, H.-Y.: A novel SVM-based reduced NN classification method. In: Proceedings of 2015 11th International Conference on Computational Intelligence and Security (CIS), Shenzhen, China (2015)
9. Trindade, L.A., Wang, H., Blackburn, W., Rooney, N.: Proceedings of the 2011 International Conference on Machine Learning and Cybernetics, Guilin, 10–13 July 2011
10. Bajwa, M.S., Agarwal, A.P., Manchanda, S.: Ternary search algorithm: improvement of binary search. In: 2015 2nd International Conference on Computing for Sustainable Global Development (INDIACom), New Delhi, pp. 1723–1725 (2015)
11. Bajwa, M.S., Singh, P.K., Agarwal, A.P.: Analyzing and interpreting the fault localized using PCA with CK metrics. In: 2016 Fourth International Conference on Parallel, Distributed and Grid Computing (PDGC), Waknaghat, pp. 575–580 (2016)
12. Bajwa, M.S., Agarwal, A.P., Gupta, N.: Code optimization as a tool for testing software. In: 2016 3rd International Conference on Computing for Sustainable Global Development (INDIACom), New Delhi, pp. 961–967 (2016)
13. Tanwar, S., Bhatia, Q., Patel, P., Kumari, A., Singh, P.K., Hong, W.: Machine learning adoption in blockchain-based smart applications: the challenges, and a way forward. IEEE Access **8**, 474–488 (2020)
14. Polkowski, Z., Vora, J., Tanwar, S., Tyagi, S., Singh, P.K., Singh, Y.: Machine learning-based software effort estimation: an analysis. In: 2019 11th International Conference on Electronics, Computers and Artificial Intelligence (ECAI), Pitesti, Romania, pp. 1–6 (2019)

Web Service Clustering Approaches to Enhance Service Discovery: A Review

Neha Agarwal, Geeta Sikka, and Lalit Kumar Awasthi

Abstract Due to the emergence of Internet technologies and service-oriented computing, there is a rapid growth in the quantity and variety of services on the web. Discovering the web services as per the request is not an easy task because of the advancement of service-oriented computing which includes web services, cloud services, mobile services, etc. These services are dynamic and published according to the emerging standards in repositories. Web service clustering plays a crucial role in web service discovery. When services are grouped according to the similarity, then it reduces the search space and time, so that services can be discovered efficiently. Many eminent researchers have proposed approaches for efficient web service discovery by incorporating web service clustering. In this paper, we review different approaches that are proposed for web service clustering to enhance the discovery process. A comparison among existing approaches is carried out based on the vector representation approach, dimensionality reduction technique, method to capture semantic relationship among features, clustering technique, type of input dataset, number of web services in dataset and service repository. This review will help the researchers to understand the existing techniques to group services in similar clusters to improvise service discovery and the scope for improvement by future directions.

Keywords Service-oriented architecture · Web services/Web API · Web service clustering · Web service discovery

N. Agarwal (✉) · G. Sikka · L. K. Awasthi
Department of Computer Science and Engineering, Dr B R Ambedkar National Institute of Technology, Jalandhar, Punjab, India
e-mail: nehaa.cs.18@nitj.ac.in

G. Sikka
e-mail: sikkag@nitj.ac.in

L. K. Awasthi
e-mail: director@nitj.ac.in

P. K. Singh et al. (eds.), *Recent Innovations in Computing*, Lecture Notes in Electrical Engineering 701, https://doi.org/10.1007/978-981-15-8297-4_3

1 Introduction

With the vast development of Internet technologies and service-oriented computing, the demand for web services is touching the sky because of its numerous benefits like reusability, anywhere accessibility, information exchange, data integration, versatility, etc. Various vendors such that Amazon, IBM, Microsoft, etc., are utilizing the profits of services and delivering services or APIs according to the need for customers by following the global web service standards. Services are the basic building blocks of a software system. A service can be described by its functionality, which is provided by the vendor who is specialized in performing that specific operation. Web services can be categorized into two segments: Simple Object Access Protocol (SOAP)-based services and REpresentational State Transfer (REST)-based services (Web API).

The potential of SOAP-based web service lies in XML and three core technologies: Universal Description Discovery and Integration (UDDI), Web Service Description Language (WSDL) and Simple Object Access Protocol (SOAP). Firstly, the vendor develops service and publishes the description file of service in WSDL format in UDDI, which is a repository to store web services. WSDL file contains documentation for the functionality of service, its name, bindings, type of messages, etc., which are generally desired by a customer to get information about that service. The customer searches web service according to its requirements from UDDI and contacts to the vendor to utilize that web service by using SOAP messages [1].

REST-based services are mainly depended on URI for resource identification and interaction, and HTTP for message transmission. These services are described by XML based languages such as WSDL and Web Application Description Language (WADL), and often service providers use simple natural language text to explain the functionality of service [2]. In general, web service discovery under service-oriented architecture is depicted in Fig. 1. Web service discovery from repository works as a crucial central point to establish a path between the service provider and consumer.

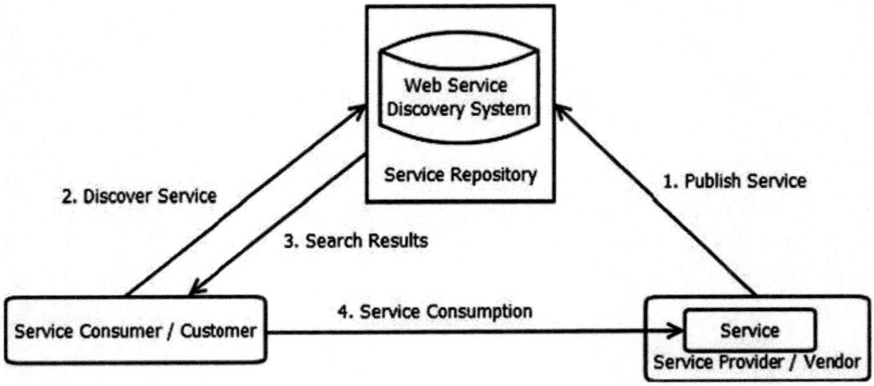

Fig. 1 Web service discovery

In today's world, as people are mainly relying on services for day-to-day work like fetching geographical location, ordering goods and food online, picture enhancement, video calling, etc., there is a vast proliferation in quantity, quality and variety of services on the Internet. With the expediting evolution of Web 2.0, many developers are adopting multifunctional and Internet-based applications known as Mashups by aggregating existing REST-based services [3–5]. Basically; web services are accessible in repositories like UDDI, search engines, web portals, etc. However, it is not a piece of cake to discover appropriate and desired service from repositories as the searching process of services mainly relies on keyword-based matching [6]. With the syntactic analysis of service description files, the discovery process of the desired service suffers from vocabulary problems, i.e., polysemy, synonymy, quasi-synonyms [7, 8].

Semantic web services have also evolved for automatic discovery of services by annotating services [1, 9]. Many eminent researchers have proposed models for semantically describing web services like Web Service Modeling Language (WSML) [10], Web Service Modeling Ontology (WSMO) [11], Web Service Modeling Ontology for Semantics (OWLS) [12]. However, it is not an easy task to annotate web services manually, and extensive endeavor is demanded to describe, share and manage ontologies [13–15]. By considering these constraints, mainly web service description files represented in WSDL or natural language short text are mainly adopted for text mining techniques and for extracting semantic meaning.

For web service discovery, current repositories are still relying on keyword-based searching techniques, which results in low recall and is not an appropriate technique to discover suitable services. Grouping the service according to their functionality is a proficient way to improve web service discovery. Web service clustering is the most adoptive way to enhance web service discovery because by generating groups of similar services according to their functionality, search space and time to discover service are reduced [16–18].

2 Literature Survey

In this section, literature related to web service clustering to improve web service discovery is segregated into four subsections. The first subsection illustrates various methods to represent service in vector space and how similarity among services is computed to cluster them. The second subsection describes different models that are used for dimensionality reduction of features and to group similar services in a cluster. In the third subsection, methods to discover semantic meaning from services are explored. The last subsection shows the meta-heuristic techniques that are utilized in the domain of web service clustering to enhance the performance of clustering.

2.1 Vector Space Representation Methods for Similarity Measurement and Clustering

Various methods have been proposed for enhancing web service clustering based on service representation in vector space. Quality threshold (QT) clustering method is proposed to group up similar service into a cluster by measuring similarity among features extracted from the WSDL documents [19]. The main shortcoming of QT clustering method is that it is computationally expensive. In a paper [20], extracted features are represented in vector space using term-frequency (TF) method. By computing similarity based on Normalized Google Distance (NGD), K-means clustering is applied to enhance web service discovery. For automated classification of services, the maximum entropy method is proposed, and features of WSDL documents are represented using TF method [14]. The limitation of this classification approach is that the maximum entropy suffers from overfitting issues. TF method for vector space representation is not an efficient one because this representation method is unable to find the importance of the term in a collection of documents.

A lexical semantic network is used for automated categorization of services with TF-IDF method for service representation [21]. In a paper [22], neural network and K-Nearest Neighbor (KNN) techniques are incorporated with TF-IDF for the classification of web services.

2.2 Model-Based Clustering

When services are represented in vector space by the TF-IDF method, then the generated matrix is sparse, and there is a requirement to have only relevant features. So to deal with this problem, various models are proposed for dimensionality reduction and extracting semantic meaning from features of services. For dimensionality reduction, principal component analysis (PCA) method is used in literature for feature reduction and to map services from high-dimensional space to low-dimensional space [23, 24].

An efficient search engine is proposed by utilizing the benefits of latent Dirichlet allocation (LDA) as dimensionality reduction, and the K-Means algorithm is used for clustering [25]. An approach for clustering of Mashup services is proposed in which the representation of services in topic distribution form is performed by using LDA, and a combination of K-Means and Agnes algorithms is used for clustering [26].

2.3 Extraction of Semantically Relevant Words

Literature shows that LDA and its amended methods are widely used to enhance the clustering performance. However, LDA suffers from the sparsity problem, and it is unable to detect noise words. WT-LDA method is proposed to deal with this issue in which user tags are incorporated with LDA to improve clustering performance [16]. An approach named as Similar Words and TF-IDF Augmented Latent Dirichlet Allocation (ST-LDA) is proposed to learn similar words and to filter noise due to unnecessary words [27].

In traditional topic modeling (PLSA, LDA) models, it is observed that these models provide a discrete topic distributed for every service. In word embedding, vectors are continuous. So results of topic modeling techniques are not improved with word embedding. To cope with this situation, Gaussian LDA is proposed with word embedding techniques in which input words are transformed into continuous vectors [28, 29]. Another word embedding technique GLoVe (Global Vectors for Word Representation) is proposed for improving web service discovery by retrieving relevant services [8].

2.4 Meta-heuristic Algorithms for Web Service Clustering

In recent years, meta-heuristic algorithms are widely adopted for improving the performance of clustering. Two different hybrid nature-inspired algorithms are presented for web service clustering in which one is inspired by a bird's behavior, and another is based on the practice of ants [30]. For enhancing web service discovery and for automatic categorization of web services, the Meta-heuristic Cat Swarm Optimization (CSO) algorithm is adopted, which is further optimized by PCA for dimensionality reduction [24, 31]. In this approach, it is proved that CSO is performing better than the K-Means clustering algorithm. In a paper [32], bio-inspired algorithms are investigated in the domain of web service clustering, and a hybrid approach using Artificial Bee Colony (ABC) is proposed for web service clustering.

A comparison of related work based on web service clustering is shown in Table 1. This table refers vector representation approach, dimensionality reduction technique, and methods to capture semantic relationship among features, clustering technique, type of input dataset, number of web services in datasets, service repository and crisp description of the proposed approach.

3 Analysis of Literature

To study the existing work in the domain of web service clustering, the papers published over ten years are reviewed. After reviewing work, information extraction and rigorous scanning, total 45 papers in this domain are referenced. In Fig. 2, the distribution of paper published over ten years is shown from which it is observed that in the last three years, lot of work is done in this field for improving the discovery process of services.

Figure 3 demonstrates the types of services used in these referenced papers. From

Table 1 Comparison of related work based on web service clustering

Work	Vector representation approach	Dimensionality reduction	Semantic relationship	Clustering technique	Input dataset	Number of web services in dataset	Service web repository
[19]	TF	No	No	Quality threshold (QT) clustering	Non-semantic services	400	WebserviceList, WebserviceX, xMethods
[33]	TF-IDF	PLSA and LDA	Yes	PLSA and LDA	Semantic web services	1007	OWLS-TC4
[16]	LDA	LDA	No	LDA	Non-semantic services	185	Seekda
[34]	TF-IDF	LDA and CTM	No	LDA and CTM	Non-semantic services	1051	OWLS-TC4
[35]	TF-IDF	PLSA and LDA	Yes	PLSA and LDA	Semantic web services	1007	OWLS-TC4
[36]	TF-IDF	By computing similarity of feature words	WordNet	Hierarchical clustering	Semantic web services	1083	OWLS-TC4
[37]	TF-IDF	No	No	*K*-means	Non-semantic services	15,968	Seekda
[14]	TF	No	No	Maximum Entropy	Non-semantic services	600	WebserviceList, WebserviceX, xMethods
[38]	TF-IDF	No	No	*K*-nearest neighbor	Non-semantic services	3176	Web service benchmark
[39]	TF-IDF	Singular value decomposition	WordNet	Artificial neural network	Non-semantic services	1083	OWLS-TC4

(continued)

Table 1 (continued)

Work	Vector representation approach	Dimensionality reduction	Semantic relationship	Clustering technique	Input dataset	Number of web services in dataset	Service web repository
[31]	TF-IDF	No	No	CSO	Non-semantic services	684	OWLS-TC4
[9]	TF	No	WordNet	NA	Semantic web services	1007	OWLS-TC4
[5]	LDA	LDA	Wor2Vec	*K*-means++ clustering	Non-semantic services	12,920	PW
[40]	LDA	Word2vec	Wor2Vec	LDA	Non-semantic services	3412	PW
[41]	HDP	HDP	No	Affinity propagation clustering	Non-semantic services	12,879	PW
[27]	Word2Vec	LDA	Word2Vec	LDA	Non-semantic services	3660	PW
[24]	TF-IDF	PCA	No	CSO	Non-semantic services	1083	OWLS-TC4
[25]	TF-IDF	LDA	No	*K* means	Non-semantic services	7560	PW
[42]	Mashup API network	No	No	GA-based clustering algorithm	Non-semantic services	6270	PW
[43]	TF-IDF	LDA	No	*K*-medoids	Non-semantic services	1000	PW
[44]	LDA	Degree of representation computation	Word2Vec	LDA	Non-semantic services	4160	PW

(continued)

Table 1 (continued)

Work	Vector representation approach	Dimensionality reduction	Semantic relationship	Clustering technique	Input dataset	Number of web services in dataset	Service web repository
[2]	LDA	LDA	No	LDA	Non-semantic services	1249	PW
[45]	LF-LDA	No	Word2Vec	*K*-means	Non-semantic services	12,000	PW
[32]	No	No	WordNet	Artificial Bee Colony	Non-semantic services	1027	NA

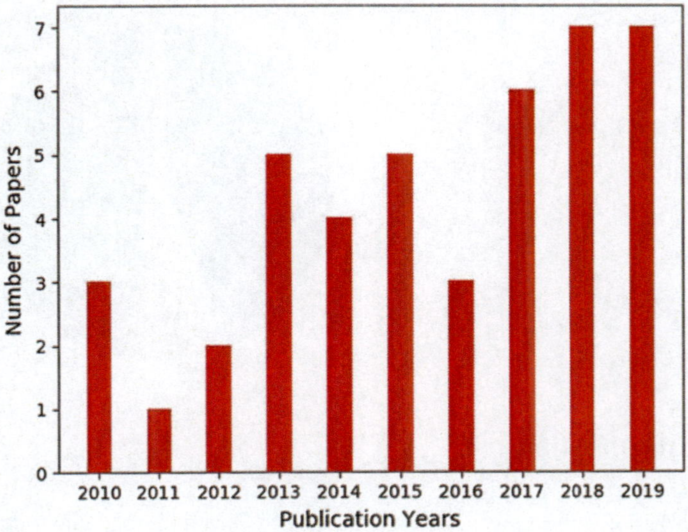

Fig. 2 Distribution of published papers over years

Fig. 3 Types of services used in literature

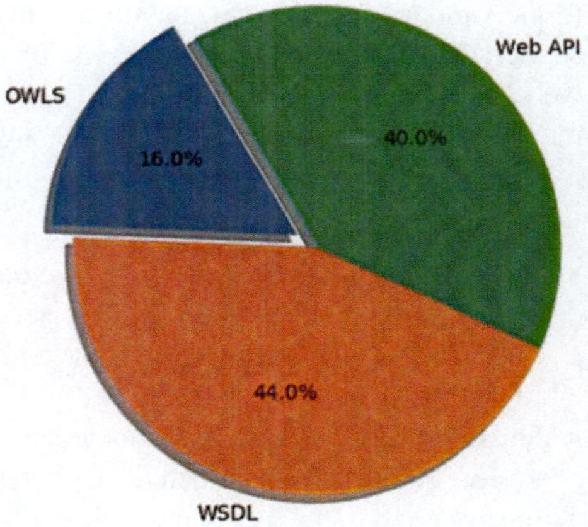

the content of papers, it is noticed that mainly non-semantic web services are preferred by the researchers which are in the form of WSDL (44%) and Web API (40%). Semantic web services, i.e., OWLS, are less preferred because it is not a piece of cake to annotate services. Different datasets like ProgrammableWeb.com (PW), Seedka, OWLS-TC4, SAWSDL, etc., are utilized for performing experiments. In 43.5% papers, PW dataset is preferred which is shown in Fig. 4.

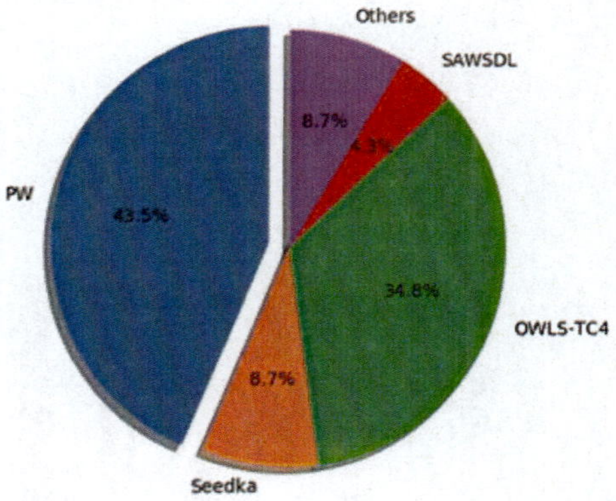

Fig. 4 Datasets used in literature

4 Future Directions

As the number of web services is increasing day by day, it has become a challenging task for users to discover appropriate web services. There is a demand for an expandable, efficient and expeditious search engine that can deal with the large volume of web services. When service repositories are appropriately organized according to the similarity of services, then it automatically boosts up the efficiency of web service discovery, selection, searching, recommendation, ranking, etc. [46, 47]. In web service recommendation, from a large pool of the repository of services, the requested service is selected to be shown in the result of a query on the basis of user requirement. Service ranking also plays a major role in recommendation of services. Future work that can be accomplished to enhance service discovery is as follows:

- In web service discovery, services are discovered according to the matchmaking process. If techniques regarding the service recommendations are also incorporated with service clustering, then the model will be able to locate the right services to the user.
- Quality of Service (QoS) parameters are very essential as non-functional qualities of services are described through these parameters. QoS parameters must be incorporated in the web service clustering process.
- As various vendors are providing services with similar functionality. So in the response to customer's query, the feedback of the previous customer can also be considered to provide results to the present customer.
- Some malfunctions, viruses or malware can be attached with the service to harm the customer's data or to gain some useful information from the customer. So security factors can be integrated into the clustering mechanism.
- Multi-Criteria Decision-Making (MCDM) techniques, for ranking of web services, are needed to be more explored in order to arrange the services as per customer's requirements.
- The combination of big data analytics, machine learning and deep learning (hybrid approaches) can be utilized for efficient service clustering.

- The approaches which are used to find the similarity between customer's query and service need to be more efficient and accurate.
- Semantic-based service discovery is more efficient than a syntactic-based approach. A model can be proposed to automatically annotate web services to overcome the limitations of semantic web service.
- Trust, risk and reputation factors can be evaluated before providing the results of web service discovery to deliver the right service.

5 Conclusion

In this paper, a review of web service clustering approaches to improve web service discovery is provided. In the presence of various vendors, dynamic nature of service, service's functionality description in short natural text, Mashup services, etc., web service discovery is a prominent field for research. Tremendous work has been carried out to enhance web service discovery by grouping services in clusters. In this paper, we have analyzed the different approaches proposed by eminent researchers in the field of web service clustering by focusing on types of used services, clustering method, semantic similarity approaches, dimensionality reduction techniques, datasets, etc. From the study, it is observed that much work remains in the area of web service clustering to improve the efficiency and accuracy of clustering algorithms so that other processes of service like discovery, selection, recommendation, ranking, etc., can be enhanced. We firmly believe that this paper will not only help the researchers to study existing techniques in this domain but will also motivate them to deliver and propose efficient techniques to enhance service discovery.

References

1. Bhardwaj, K.C., Sharma, R.: Machine learning in efficient and effective web service discovery. J. Web Eng. **14**(3 & 4), 196–214 (2015)
2. Zhang, N., Wang, J., He, K., Li, Z., Huang, Y.: Mining and clustering service goals for restful service discovery. Knowl. Inf. Syst. **58**(3), 669–700 (2019)
3. Blake, M.B., Nowlan, M.E.: Knowledge discovery in services (KDS): aggregating software services to discover enterprise mashups. IEEE Trans. Knowl. Data Eng. **23**(6), 889–901 (2010)
4. Cao, B., Liu, X.F., Liu, J., Tang, M.: Effective mashup service clustering method by exploiting LDA topic model from multiple data sources. In: Asia-Pacific Services Computing Conference, pp. 165–180. Springer, Berlin (2015)
5. Shi, M., Liu, J., Zhou, D., Tang, M., Cao, B.: WE-LDA: a word embeddings augmented LDA model for web services clustering. In: 2017 IEEE International Conference on Web Services (ICWS), pp. 9–16. IEEE (2017)
6. Sharma, S., Rana, V.: Web search personalization using semantic similarity measure. In: Proceedings of ICRIC 2019, pp. 273–288. Springer, Cham (2020)
7. Chen, F., Lu, C., Wu, H., Li, M.: A semantic similarity measure integrating multiple conceptual relationships for web service discovery. Expert Syst. Appl. **67**, 19–31 (2017)

8. Lizarralde, I., Rodriguez, J.M., Mateos, C., Zunino, A.: Word embeddings for improving rest services discoverability. In: 2017 XLIII Latin American Computer Conference (CLEI), pp. 1–8. IEEE (2017)
9. Chen, F., Li, M., Wu, H., Xie, L.: Web service discovery among large service pools utilizing semantic similarity and clustering. Enterp. Inf. Syst. **11**(3), 452–469 (2017)
10. De Bruijn, J., Lausen, H., Polleres, A., Fensel, D.: The web service modeling language WSML: an overview. In: European Semantic Web Conference, pp. 590–604. Springer, Berlin (2006)
11. Fensel, D., Lausen, H., Polleres, A., De Bruijn, J., Stollberg, M., Roman, D., Domingue, J.: Enabling semantic web services: the web service modeling ontology. Springer Science & Business Media, Berlin (2006)
12. Martin, D., Burstein, M., Hobbs, J., Lassila, O., McDermott, D., McIlraith, S., Narayanan, S., Paolucci, M., Parsia, B., Payne, T., et al.: Owl-S: semantic markup for web services. W3C Member Submission **22**(4) (2004)
13. Crasso, M., Zunino, A., Campo, M.: A survey of approaches to web service discovery in service-oriented architectures. J. Database Manage. (JDM) **22**(1), 102–132 (2011)
14. Nisa, R., Qamar, U.: A text mining based approach for web service classification. IseB **13**(4), 751–768 (2015)
15. Wang, J., Gao, P., Ma, Y., He, K.: Common topic group mining for web service discovery. In: Asia-Pacific Services Computing Conference, pp. 92–107. Springer, Berlin (2015)
16. Chen, L., Wang, Y., Yu, Q., Zheng, Z., Wu, J.: WT-LDA: user tagging augmented LDA for web service clustering. In: International Conference on Service-Oriented Computing, pp. 162–176. Springer, Berlin (2013)
17. Kumara, B.T., Paik, I., Koswatte, K.R., Chen, W.: Improving web service clustering through post filtering to bootstrap the service discovery. Int. J. Serv. Comput. **2**(3), 1–13 (2014)
18. Kathuria, A., Mukhopadhyay, D., Thakur, N.: Evaluating cohesion score with email clustering. In: Proceedings of First International Conference on Computing, Communications, and Cyber-Security (IC4S 2019), pp. 107–119. Springer, Singapore (2020)
19. Elgazzar, K., Hassan, A.E., Martin, P.: Clustering WSDL documents to bootstrap the discovery of web services. In: 2010 IEEE International Conference on Web Services, pp. 147–154. IEEE (2010)
20. Vijayan, A.S., Balasundaram, S.: Effective web-service discovery using k-means clustering. In: International Conference on Distributed Computing and Internet Technology, pp. 455–464. Springer, Berlin (2013)
21. Sharma, S., Lather, J., Dave, M.: Semantic approach for classification of web services using unsupervised normalized similarity measure. J. Emerg. Technol. Web Intell. **6**(3), 364–372 (2014)
22. Wang, X., Chen, F., Li, M.: Web service classification approach with an integrated similarity measure. In: Proceedings of the 23rd International Conference on Industrial Engineering and Engineering Management 2016, pp. 251–255. Springer, Berlin (2017)
23. Kang, G., Liu, J., Tang, M., Cao, B.: Web service selection algorithm based on principal component analysis. J. Electron. **30**(2), 204–212 (2013)
24. Kotekar, S., Kamath, S.S.: Enhancing web service discovery using meta-heuristic CSO and PCA based clustering. In: Progress in Intelligent Computing Techniques: Theory, Practice, and Applications, pp. 393–403. Springer, Berlin (2018)
25. Bukhari, A., Liu, X.: A web service search engine for large-scale web service discovery based on the probabilistic topic modeling and clustering. SOCA **12**(2), 169–182 (2018)
26. Cao, B., Liu, X.F., Liu, J., Tang, M.: Domain-aware mashup service clustering based on LDA topic model from multiple data sources. Inf. Softw. Technol. **90**, 40–54 (2017)
27. Zhao, Y., He, K., Qiao, Y.: ST-LDA: high quality similar words augmented LDA for service clustering. In: International Conference on Algorithms and Architectures for Parallel Processing, pp. 46–59. Springer, Berlin (2018)
28. Tian, G., Wang, J., Zhao, Z., Liu, J.: Gaussian LDA and word embedding for semantic sparse web service discovery. In: International Conference on Collaborative Computing: Networking, Applications and Worksharing, pp. 48–59. Springer, Berlin (2016)

29. Tian, G., Zhao, S., Wang, J., Zhao, Z., Liu, J., Guo, L.: Semantic sparse service discovery using word embedding and Gaussian LDA. IEEE Access **7**, 88231–88242 (2019)
30. Pop, C.B., Chifu, V.R., Salomie, I., Dinsoreanu, M., David, T., Acretoaie, V., Nagy, A., Oprisa, C.: Biologically-inspired clustering of semantic web services. Birds or ants intelligence? Concurrency Comput. Pract. Experience **24**(6), 619–633 (2012)
31. Kotekar, S., Kamath, S.S.: Enhancing service discovery using cat swarm optimisation based web service clustering. Perspect. Sci. **8**, 715–717 (2016)
32. Bravo, M., Mora-Gutierrez, R.A., Hoyos-Reyes, L.F.: Bio-inspired hybrid algorithm for web services clustering. In: Advanced Analytics and Artificial Intelligence Applications. IntechOpen (2019)
33. Cassar, G., Barnaghi, P.M., Moessner, K.: Probabilistic methods for service clustering. In: SMRR@ ISWC (2010)
34. Aznag, M., Quafafou, M., Rochd, E.M., Jarir, Z.: Probabilistic topic models for web services clustering and discovery. In: European Conference on Service-Oriented and Cloud Computing, pp. 19–33. Springer, Berlin (2013)
35. Cassar, G., Barnaghi, P., Moessner, K.: Probabilistic matchmaking methods for automated service discovery. IEEE Trans. Serv. Comput. **7**(4), 654–666 (2013)
36. Gao, H., Wang, S., Sun, L., Nian, F.: Hierarchical clustering based web service discovery. In: International Conference on Informatics and Semiotics in Organisations, pp. 281–291. Springer, Berlin (2014)
37. Wu, J., Chen, L., Zheng, Z., Lyu, M.R., Wu, Z.: Clustering web services to facilitate service discovery. Knowl. Inf. Syst. **38**(1), 207–229 (2014)
38. Elshater, Y., Elgazzar, K., Martin, P.: goDiscovery: web service discovery made efficient. In: 2015 IEEE International Conference on Web Services, pp. 711–716. IEEE (2015)
39. Kamath, S., Ananthanarayama, V.: Semantic similarity based context-aware web service discovery using NLP techniques. J. Web Eng. **15**(1 & 2), 110–139 (2016)
40. Zhao, Y., Wang, C., Wang, J., He, K.: Incorporating LDA with word embedding for web service clustering. Int. J. Web Serv. Res. (IJWSR) **15**(4), 29–44 (2018)
41. Fletcher, K.K.: A quality-based web API selection for mashup development using affinity propagation. In: International Conference on Services Computing, pp. 153–165. Springer, Berlin (2018)
42. Pan, W., Chai, C.: Structure-aware mashup service clustering for cloud-based Internet of Things using genetic algorithm based clustering algorithm. Future Gener. Comput. Syst. **87**, 267–277 (2018)
43. Jalal, S., Yadav, D.K., Negi, C.S.: Web service discovery with incorporation of web services clustering. Int. J. Comput. Appl. 1–12 (2019)
44. Zhao, Y., Qiao, Y., He, K.: A novel tagging augmented LDA model for clustering. Int. J. Web Serv. Res. (IJWSR) **16**(3), 59–77 (2019)
45. Chen, Y., Wang, X., Xia, H., Wang, Z., Yv, Z.: Research on web service clustering method based on word embedding and topic model. In: The International Conference on Natural Computation, Fuzzy Systems and Knowledge Discovery, pp. 980–987. Springer, Berlin (2019)
46. Chen, L., Zheng, Z., Feng, Y., Wu, J., Lyu, M.R.: WSTRank: ranking tags to facilitate web service mining. In: International Conference on Service-Oriented Computing, pp. 574–581. Springer, Berlin (2012)
47. Obidallah, W.J., Raahemi, B., Ruhi, U.: Clustering and association rules for web service discovery and recommendation: a systematic literature review. SN Comput. Sci. **1**(1), 27 (2020)

Path Finding Using PSO Cooperated with Randomized Noise Functions

Aridaman Singh Nandan and Geeta Sikka

Abstract Path planning is a crucial navigation technology for routing and shortest path problems. The paper discusses a modified approach to the collision-free searching problem using particle swarm optimization incorporated with noise functions like Gaussian and Perlin noise. Aiming at PSO's shortcoming of quickly diving into local minima, the added random noise functions escape the local minima in the convergence process; hence, look for global convergence maintaining fast speed in the early phase. The random sampling improves the particle update procedure to look for broader search space, which is otherwise constraint to global best of population. This variation guarantees solution space exploitation. Particle position vector precision is improved by adding noise (by some predefined factor), and the PSO algorithm is run to get the best particle as a candidate solution. Particle swarm optimization is a low-overhead and easy to implement the technique. Obstacles are incorporated into the algorithm to improve effectiveness. Particles need to reach the destination without colliding with any of the obstacles. Finally, simulation scenarios demonstrate effectiveness considering multiple target positions.

Keywords Enhanced PSO · Path planning · Noise functions · Randomized search · Gaussian and Perlin noise

1 Introduction

One of the fundamental requirements for planning a search path is to plan a collision-free path [1]. Path planning can be global, i.e., with known environment or can be local, i.e., totally unknown environment [2]. Researchers suggest many approaches, which include the widespread use of genetic and evolutionary algorithms [3–5].

A. S. Nandan (✉) · G. Sikka
Dr B R Ambedkar National Institute of Technology Jalandhar, Jalandhar, Punjab, India
e-mail: aridamansn.cs.18@nitj.ac.in

G. Sikka
e-mail: sikkag@nitj.ac.in

P. K. Singh et al. (eds.), *Recent Innovations in Computing*, Lecture Notes in Electrical Engineering 701, https://doi.org/10.1007/978-981-15-8297-4_4

Compared to GA, PSO is easier to implement with fewer parameters to be adjusted. It is a simple algorithm with quick convergence. Particle swarm optimization is a computational method that tries to improve the candidate solution following a given measurable quantity. PSO has been applied to continuous optimization problems because of its specific algorithmic structure (updating velocity and position in every iteration) [6]. This paper studies the problem of path finding where obstacles have uncertain position [7–10], based on improved PSO algorithm.

First, a workspace is modeled to test the implementation of the algorithm. Obstacles are set at discrete points. A basic particle swarm optimization algorithm is implemented with three different target points, and solution precision is checked. This is followed by modifying PSO by adding Perlin and Gaussian noise to explore the gradient position corresponding to each particle position.

2 Particle Swarm Optimization (PSO)

PSO is a stochastic technique for optimization animating social behavior of bird flock, as developed by Kennedy and Eberhart, very effective in solving optimization problems in multi-dimensions. It starts with particles, with random positions and velocities initialized in search space, which potentially represents an initial candidate solution for the problem. To reach an optimal solution, particle velocities are updated [11], hence positions, in each iteration/generation in a specified manner, which follows behavior to reach the solution [6]. A fitness measure determines the fitness of each particle based on its position. The velocity of each particle is updated by keeping track of two *best* positions.

1. Best position a particle has traversed so far—This is called pBest value. pBest value is associated with each particle describing its personal best solution.
2. Best position from the whole population—This is called gBest value. This is single value describing the best candidate solution so far.

Updation of k'th particle's position and velocity is done as follows:

$$\mathrm{vel}_{kd} = \mathrm{vel}_{kd} + c_1 r_1(\mathrm{gBest}_{kd} - x_{kd}) + c_2 r_2(\mathrm{pBest}_{kd} - x_{kd}); \quad k = 1, \ldots, N \qquad (1)$$

$$x_{kd} = x_{kd} + v_{kd}; \quad d = 1, \ldots, D \qquad (2)$$

where c_1 and c_2 are positive constants, termed as acceleration coefficients, N is swarm population and D is dimensions of problem space. Dimension is the number of parameters of the function being optimized, r_1 and r_2 are random numbers in range of 0–1.

Current velocities are calculated using Eq. (1) based on its previous velocity, the personal best position of the particle achieved so far, and the best from the entire population achieved. Equation (2) updates particle's position in solution space. Learning factors c_1 and c_2 control the influence of pBest and gBest on search procedure. Both c_1 and c_2 are taken to be 2 yielding good results. In most of the cases, velocities quickly attain large values which leaves the particle out of the boundary of search space. To control this massive increase in velocity, bound velocity check is used in Eq. (1). Thus, if velocity exceeds a specified maximum value V^{max}, then the velocity is clamped to V^{max}. A review of modifications into this basic algorithm can be seen in 27.

Chaotic Particle Swarm Optimization [1, 12] is one of the modifications adopting logistic mapping. in chaotic particle swarm optimization (CPSO) approach, the algorithm first finds the gBest position as a candidate solution, and then chaotic space (0, 1) is mapped into neighborhood of this candidate solution. This implies that a better solution is searched by means of chaos.

3 Noise Functions

Noise algorithms have graphical applications to produce procedural textures, natural motion, shapes, terrains, etc. Noise is an algorithm for generating wavelets that are seemingly pseudorandom but mathematically predictable and used for procedural generation of textures and terrains. What so great about using the noise algorithm to create underlying data is that it has smoothness and predictability that is not provided by the random number generation. As you can see from Fig. 1 generated with both random number and noise, the line being created by the noise algorithm is far more superior representation of landscape, terrain or valleys.

Noise can be generated in any number of dimensions. 1-D noise appears as a line graph and is generated using one value as input, i.e., x value. 2-D noise can be used to generate textures or height values for landscapes.

Fig. 1 Random noise (yellow) and Perlin noise (red)

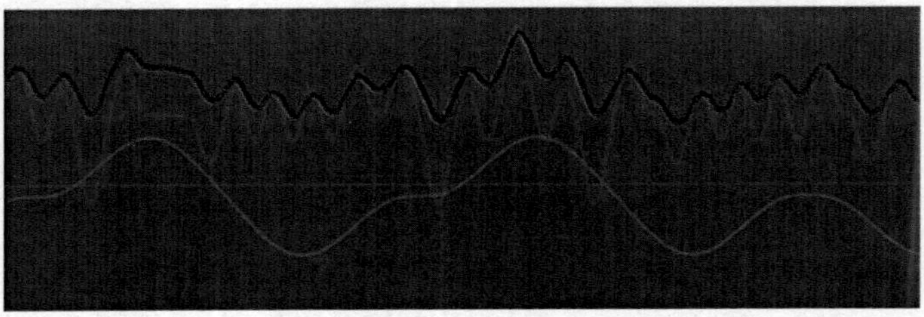

Fig. 2 Perlin noise with different increment values deciding curve smoothness

Gaussian noise is statistical noise fitting on Gaussian distribution. There is no minimum or maximum value that Gaussian noise might return. The noise is based on mean and standard deviation, and there is a very low probability that of returning a number far from mean, and a higher probability that numbers near mean is returned. Perlin noise returns a continuous waveform with values between 0 and 1. Because it is mathematically calculated, it can also be zoomed in. The smaller the increments sent to the function, the smoother it is. If Perlin noise is generated with an increment of 0.01, it returns smooth data. If increment is increased, it's like zooming out, and returned values give much sharper appearance. The decreasing increment is like zooming in a particular portion of data. With enormous increment value, minute details in data are missed, and taking too much smaller value can end up looking for invalid patterns. From Fig. 2, the Perlin noise wavelet with yellow color looks more natural.

4 PSO with Noise Functions

The path planning procedure for particles to reach the destination tackling obstacles is incorporated with the addition of noise values. The velocity vector of each particle in basic PSO is updated using the acceleration coefficient, random numbers, personal best and global best. A new noise factor is added in velocity vector update formulae based on gBest (in case of Perlin noise) and standard deviation (in case of Gaussian noise). The modified velocity vector formulation searches the environment space over a much more extensive area, looking for new paths that are not explored by basic PSO implementation. Hence, velocity vector is updated as:

$$\begin{aligned}\mathrm{vel}_{kd} &= \mathrm{vel}_{kd} + c_1 r_1(\mathrm{gBest}_{kd} - x_{kd}) + c_2 r_2(\mathrm{pBest}_{kd} - x_{kd}) \\ &\quad + \mathrm{noise}\left(\mathrm{gBest}_{kd}\right); \quad k = 1, \ldots, N\end{aligned} \tag{3}$$

$$\begin{aligned}\mathrm{vel}_{kd} &= \mathrm{vel}_{kd} + c_1 r_1(\mathrm{gBest}_{kd} - x_{kd}) + c_2 r_2(\mathrm{pBest}_{kd} - x_{kd}) \\ &\quad + \text{random Gaussian}(m, sd); \quad i = 1, \ldots, N\end{aligned} \tag{4}$$

Equation (3) introduces Perlin noise to the velocity vector. The global best among the swarm population is chosen for generating noise value. This gives a value that has a minor gradient from gBest position values and might explore an obstacle gap that was otherwise not considered as one of the possible paths. In elementary PSO, particles try to reach the gBest position vector. With the addition of noise, the area around the gBest position vector is explored hence increasing search space without any over-head. The particles try to converge through new paths and find different ways to tackle the same obstacle space.

Equation (4) introduces random Gaussian noise to velocity vectors based on the predefined value of mean and standard deviation. Each particle's velocity vector is updated independently of other particle's position to look for variety in their local search space. The decision of the range that the particle looks upon in its local search space is based on the standard deviation of Gaussian distribution.

The more the standard deviation, the more random the particle can go in its local search space. The divergence from its previous position might be drastic, but new regions of environment space are explored. Algorithm 1 describes the searching procedure approach. Only those particles participate in the procedure, which is not yet hit with any obstacle.

Algorithm 1 Path Planning algorithm using PSO

Step 1: Randomly initialize N particles X_i^d and V_i^d
Step 2: Evaluate fitness and determine best positions: pBest_i and gBest
Step 3: Update velocity and position vectors $V(k)_i^d$ and $X(k)_i^d$ in k'th iteration
Step 4: Detect collisions with obstacles and update particle's hit status
Step 5: Repeat steps 2–4 until maximum iteration number is attained or destination is not reached.

5 Experiment and Simulation Results

5.1 Numerical Simulation

Some simulations are put to illustrate the modified algorithm. A two-dimensional world with size 100 × 100 is created along with some obstacles, target being represented as a black dot. Particles are presented in green color, and gBest is distinguished with blue color. The blue line represents the trace of the gBest during the simulation to reach the destination. The initial population of particles starts from the bottom left corner of the experimental system (see Fig. 3). The aim is to reach the destination without any collision with obstacles by updating the positions as per the PSO algorithm. Three cases with different target positions are considered. Target is placed in obstacles to test crisp turns. Related parameters are set as: $c_1 = c_2 = 2$, $N = 50$, $D = 2$ (x and y). The velocity factor of each particle is updated using basic PSO, PSO with Perlin noise, and PSO with Gaussian noise and observations are recorded. All of the

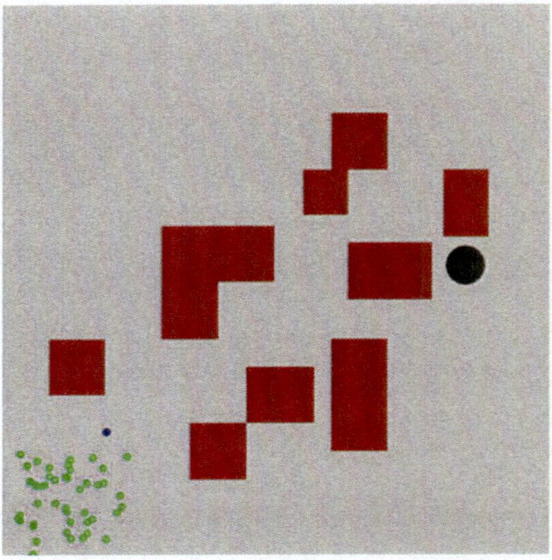

Fig. 3 Beginning scenario: blue color particle indicate best fitness till that iteration

particles either collide with obstacles (see Fig. 4), or some of the particles reach the destination. The more the particle reaches closer to the destination, the more is its fitness. The corresponding fitness function is taken as

$$F(i, k) = \frac{1}{(\text{distance}(\text{particle}(i), \text{target}))} \tag{5}$$

The equation describes the fitness of ith particle in kth iteration. The distance value is squared to distinguish between two particles closer to each other. Two particles closer to each other have an approximately closer fitness value, but squaring the distance expands the fitness gap between the particles; hence, particles become much more distinguished [13].

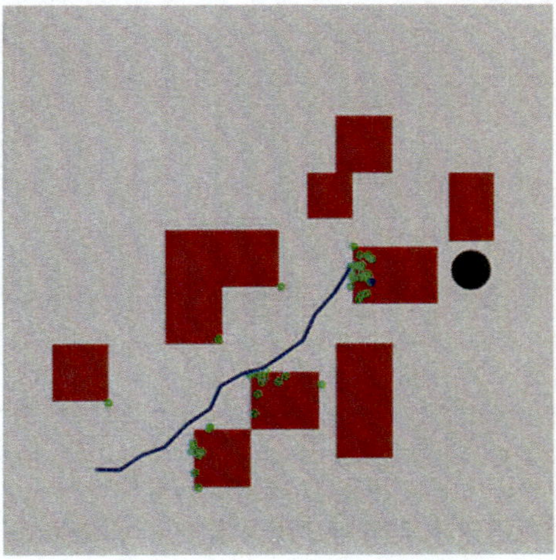

Fig. 4 Collision scenario: all particles hit obstacle

The fitness of a particle needs to be maximized. The particles start moving closer to the destination by updating velocity, followed by updating the position vector as per the particle swarm optimization algorithm. Each particle keeps track of its pBest, and the global best is chosen (blue particle) with maximum pBest fitness value from the population.

5.2 Trajectory Analysis

The process aims to obtain a collision-free shortest path in the 2-D experimental system. The addition of random noise to the particle swarm optimization aims to exploit the solution space to a broader extent. Comparing Figs. 7 and 10, basic PSO finds a longer path as compared to PSO incorporated with Perlin noise. The addition of Perlin noise explores a broader area and look for alternative paths. The exploration results in finding a new shorter path (see Fig. 10). Also, the number of particles reaching the destination in case of Perlin noise updation is more as compared to the basic PSO procedure. Most of the particles in basic PSO collide with the obstacles at crisp turns; hence, the algorithm struggles in taking crisp turns (Figs. 6 and 7). Both basic PSO and PSO with Perlin noise give high solution precision but later outperforms in terms of exploiting new shorter collision-free paths and failure rate. Failure rate is the number of times any one of the particles of the population fails to reach the destination.

5.3 Results

Simulation is carried on three different target positions using basic PSO (see Figs. 5, 6 and 7), PSO incorporated with Gaussian noise (see Figs. 11, 12, 13, 14, 15, 16 and 17), PSO incorporated with Perlin noise (see Figs. 8, 9 and 10).

From Tables 1 and 2, it is clear that the addition of Perlin noise has positively influenced the result. Time to reach the destination is approximately same but, the vast difference could be seen in the number of particles alive and their fitness value. The noise has also handled the crisp turn target position (84, 47) in a better way as compared to basic PSO. The addition of Perlin noise also allows the particle to explore the area around the global best particle. Rather than blindly directing to gBest direction, the particles now make a decision based on gBest position and explore the region surrounding it. Hence, new paths to reach the destination are found with Perlin noise addition that was otherwise not tracked by basic PSO.

The addition of Gaussian noise increased the time to reach the destination. A higher value of standard deviation particle tries to explore regions around its position. Magnitude of region explored directly depends on the standard deviation value. This exploration time increased the time to reach the destination. The particle trajectory is distorted or not smooth because it elegantly tries to exploit the region around it

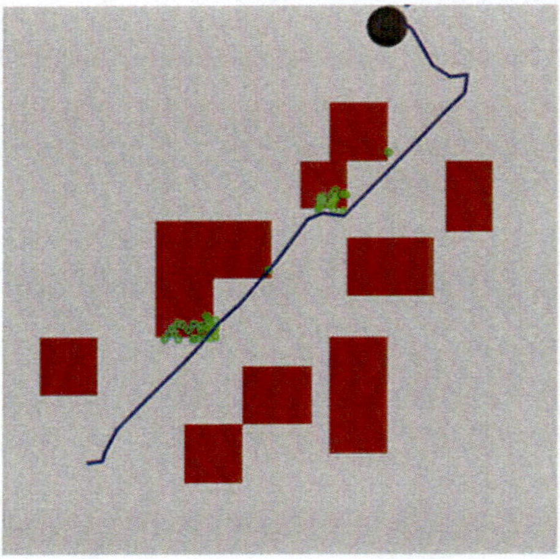

Fig. 5 Basic PSO and target at (70, 7)

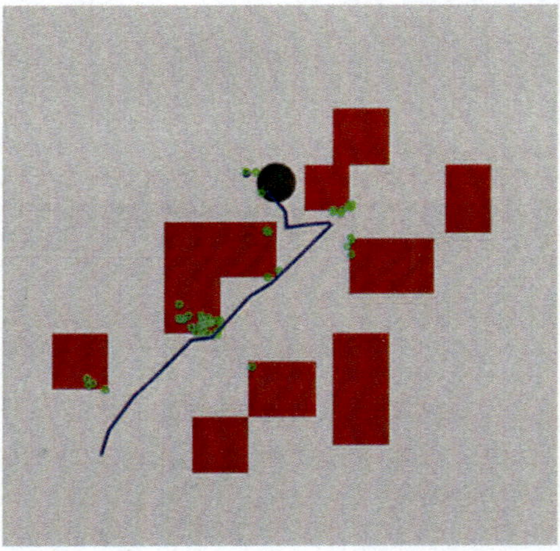

Fig. 6 Basic PSO and target at (50, 33)

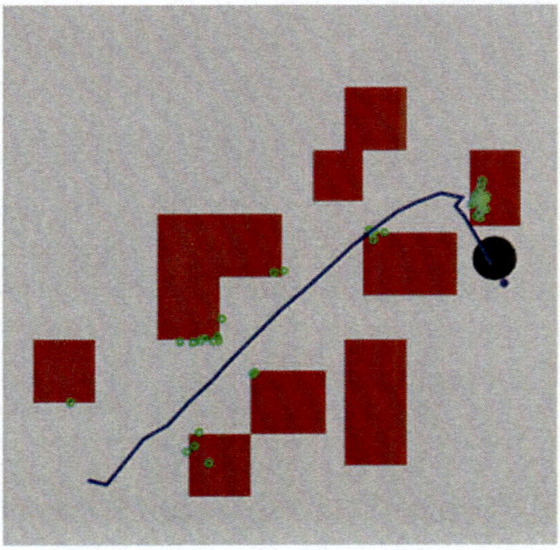

Fig. 7 Basic PSO and target at (84, 47)

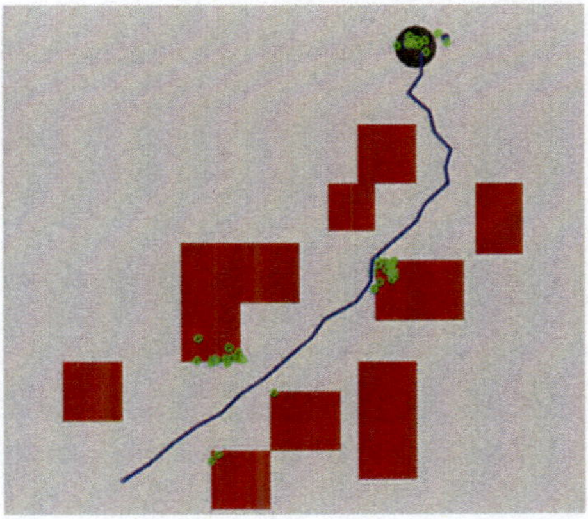

Fig. 8 PSO with Perlin noise and target at (70, 7)

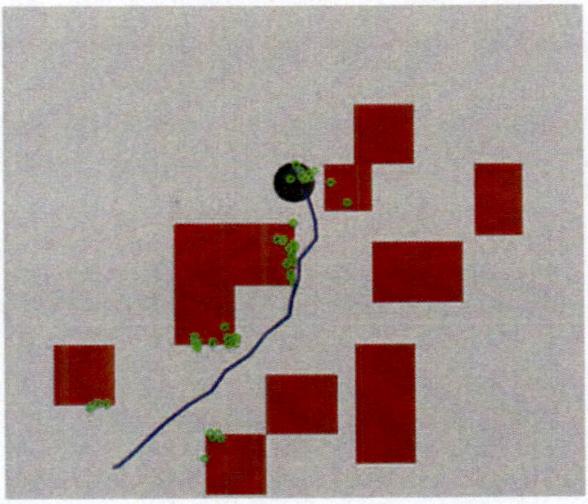

Fig. 9 PSO with Perlin noise and target at (50, 33)

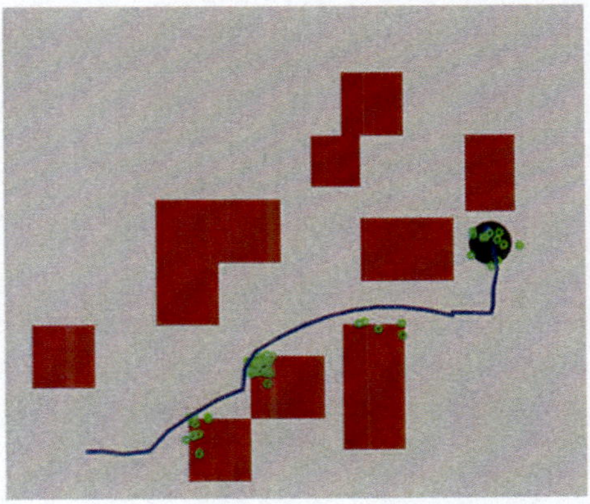

Fig. 10 PSO with Perlin noise and target at (84, 47)

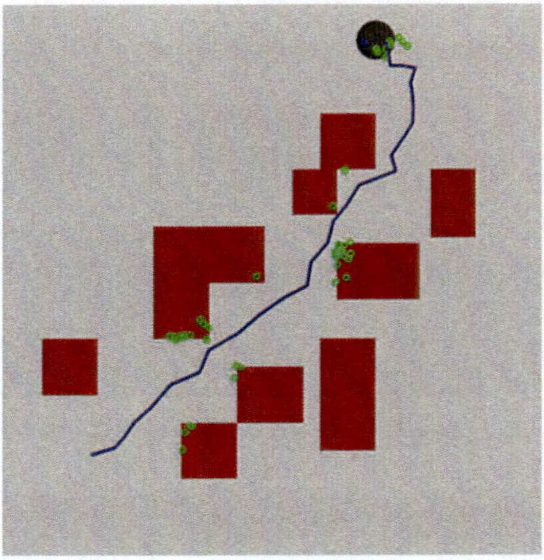

Fig. 11 PSO with Gaussian noise ($m = 1$, sd $= 1$) and target at (70, 7)

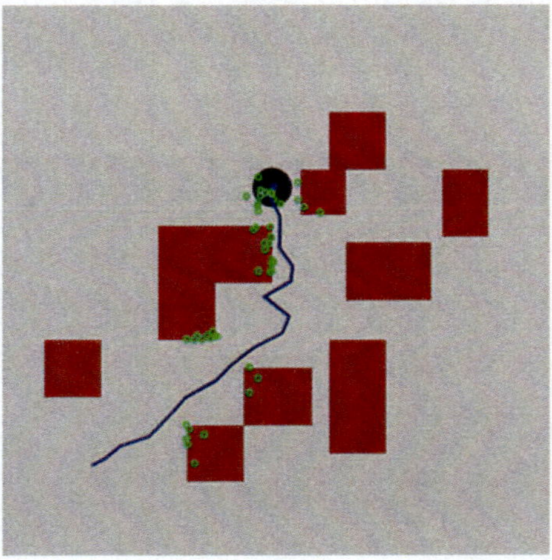

Fig. 12 PSO with Gaussian noise ($m = 1$, sd $= 1$) and target at (50, 33)

(see Fig. 17). Also, sharp turns are taken by the global best particle to reach the destination. From Table 3, it can be seen that failure to reach destination increases. In an attempt to explore regions around its position, a particle hit the obstacle quite often. Hence, the population dies quite early. The number of particles alive also reduces as compared to Perlin noise results.

6 Conclusion

Particle swarm optimization is incorporated with Gaussian and Perlin noise to obtain better result in terms of region exploration and solution precision. Basic PSO and

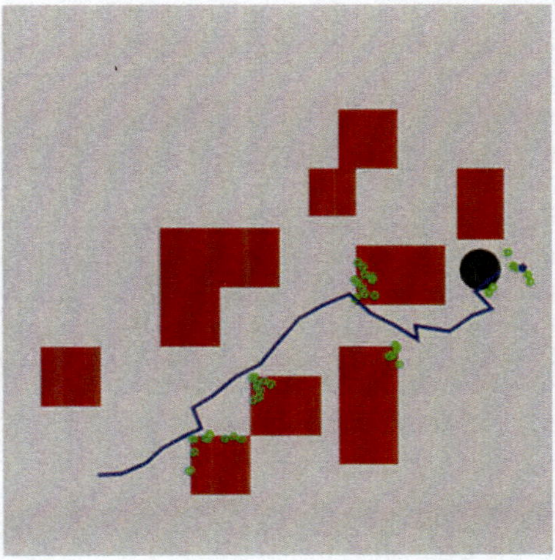

Fig. 13 PSO with Gaussian noise ($m = 1$, sd = 1) and target at (84, 47)

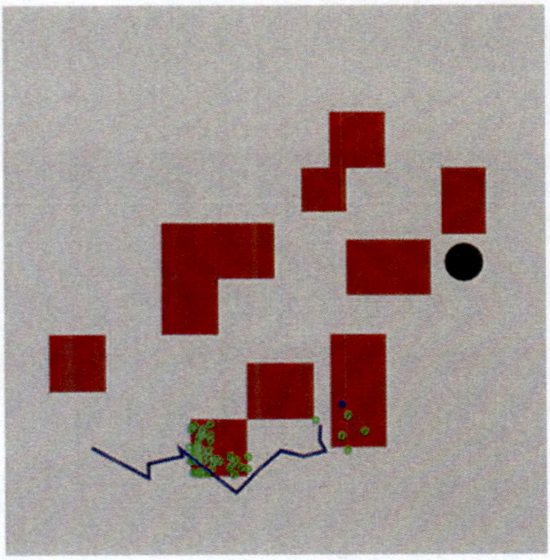

Fig. 14 Gaussian noise (sd = 5) with high variation lead to collision

one with added Perlin noise get a shorter path with higher solution precision. But, the introduction of Perlin noise improves the solution by taking crisp turns smoothly. Obstacle resistance also increases with the introduction of Perlin noise. Compared to Perlin noise results, Gaussian noise explores more vast space, but in this attempt, the path to reach the destination gets distorted. The path becomes more extended, and solution precision decrease by a small factor. But new paths are explored more with the incorporation of Gaussian noise as compared to other results.

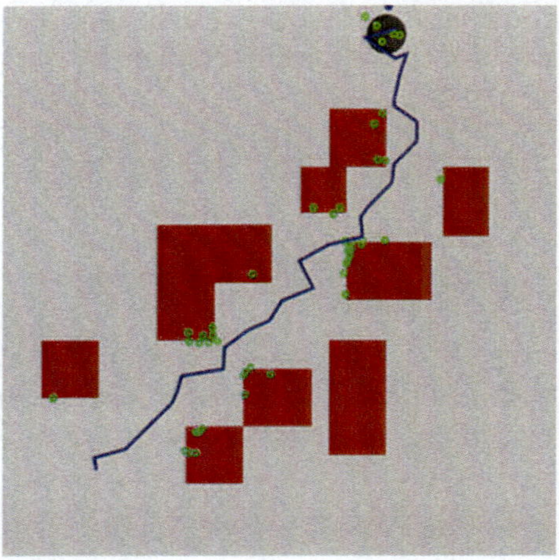

Fig. 15 PSO with Gaussian noise ($m = 1$, sd $= 3$) and target at (70, 7)

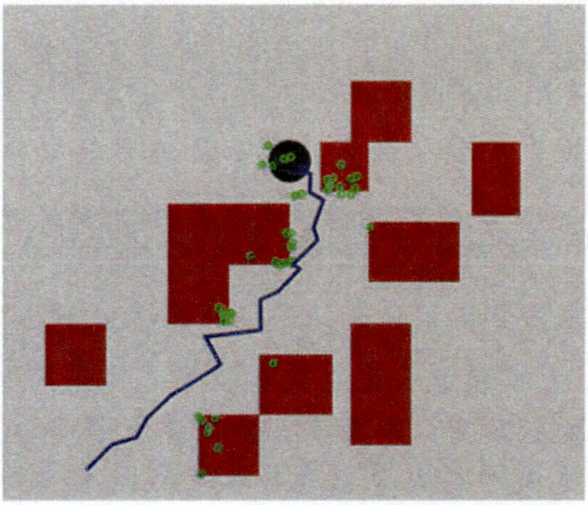

Fig. 16 PSO with Gaussian noise ($m = 1$, sd $= 3$) and target at (50, 33)

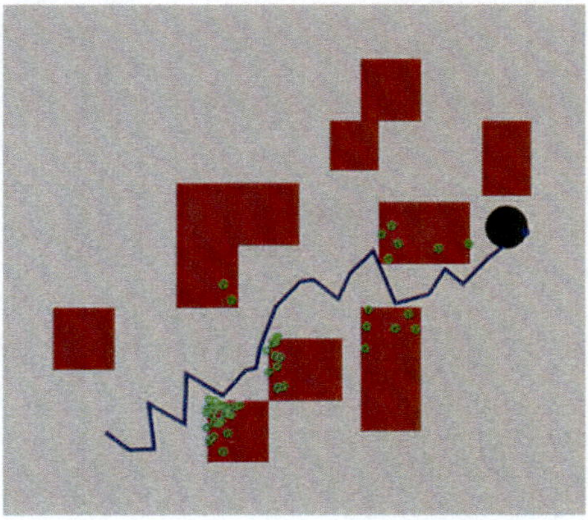

Fig. 17 PSO with Gaussian noise ($m = 1$, sd $= 3$) and target at (84, 47)

Table 1 Simulation results of PSO

Target position	Time to reach	Alive particles range	Fitness	Failure to reach
(70, 7)	26	Variable	0.00234	1/50
(50, 33)	17	[1–5]	0.003234	1/50
(84, 47)	20	[2–7]	0.00215	1/50

Table 2 Simulation results of PSO incorporated with Perlin noise

Target position	Time to reach	Alive particles range	Fitness	Failure to reach
(70, 7)	25	[9–16]	0.0238	1/50
(50, 33)	17	[11–20]	0.0793	1/50
(84, 47)	21	[12–17]	0.0128	1/5

Table 3 Simulation results of PSO incorporated with Gaussian noise

Target position	Time to reach	Alive particles range	Fitness	Failure to reach
Mean = 1, Standard deviation = 1				
(70, 7)	27	Variable	0.0310	1/50
(50, 33)	17	[13–27]	0.1235	1/50
(84, 47)	29	Variable	0.0153	2/7
Mean = 1, Standard deviation = 3				
(70, 7)	29	[3–8]	0.0490	1/50
(50, 33)	19	[9–13]	0.0224	1/5
(84, 47)	24	[2–3]	0.0264	2/5

References

1. Zhao, Q., Yan, S.: Collision-free path planning for mobile robots using chaotic particle swarm optimization. In: International Conference on Natural Computation, pp. 632–635. Springer, Berlin, Heidelberg (2005, Aug)
2. Qin, Y.Q., Sun, D.B., Li, N., Cen, Y.G.: Path planning for mobile robot using the particle swarm optimization with mutation operator. In: Proceedings of 2004 International Conference on Machine Learning and Cybernetics (IEEE Cat. No. 04EX826), vol. 4, pp. 2473–2478. IEEE (2004, Aug)
3. Foo, J.L., Knutzon, J., Kalivarapu, V., Oliver, J., Winer, E.: Path planning of unmanned aerial vehicles using B-splines and particle swarm optimization. J. Aerosp. Comput. Inf. Commun. **6**(4), 271–290 (2009)
4. Ranaweera, D.M., Hemapala, K.U., Buddhika, A.G., Jayasekara, P.: A shortest path planning algorithm for PSO base firefighting robots. In: 2018 Fourth International Conference on Advances in Electrical, Electronics, Information, Communication and Bio-informatics (AEEICB), pp. 1–5. IEEE (2018, Feb)
5. Sadiq, A.T., Hasan, A.H.: Robot path planning based on PSO and D Star algorithms in dynamic environment. In: 2017 International Conference on Current Research in Computer Science and Information Technology (ICCIT), pp. 145–150. IEEE (2017, Apr)

6. Eberhart, R., Kennedy, J.: A new optimizer using particle swarm theory. In: MHS'95 Proceedings of the Sixth International Symposium on Micro Machine and Human Science, pp. 39–43. IEEE (1995, Oct)
7. Pattanayak, S., Agarwal, S., Choudhury, B.B., Sahoo, S.C.: Path planning of mobile robot using PSO algorithm. In: Information and Communication Technology for Intelligent Systems, pp. 515–522. Springer, Singapore (2019)
8. Ayari, A., Bouamama, S.: A new multiple robot path planning algorithm: dynamic distributed particle swarm optimization. Robot. Biomimetics **4**(1), 8 (2017)
9. Solea, R., Cernega, D.: Online path planner for mobile robots using particle swarm optimization. In: 2016 20th International Conference on System Theory, Control and Computing (ICSTCC), pp. 222–227. IEEE (2016, Oct)
10. Kang, H.I., Lee, B., Kim, K.: Path planning algorithm using the particle swarm optimization and the improved Dijkstra algorithm. In: 2008 IEEE Pacific-Asia Workshop on Computational Intelligence and Industrial Application, vol. 2, pp. 1002–1004. IEEE (2008, Dec)
11. Zhang, Y., Gong, D.W., Zhang, J.H.: Robot path planning in uncertain environment using multi-objective particle swarm optimization. Neurocomputing **103**, 172–185 (2013)
12. Tharwat, A., Elhoseny, M., Hassanien, A.E., Gabel, T., Kumar, A.: Intelligent B´ezier curve-based path planning model using Chaotic Particle Swarm Optimization Algorithm. Cluster Comput. 1–22 (2018)
13. Mandava, R.K., Bondada, S., Vundavilli, P.R.: An optimized path planning for the mobile robot using potential field method and PSO algorithm. In: Soft Computing for Problem Solving, pp. 139–150. Springer, Singapore (2019)

A Comparative Study of NoSQL Databases

Simmi Bagga and Anil Sharma

Abstract The need and trend of data record analysis has seen an enormous rise in the past. More and more organizations are realizing the need for a schematic decision making procedure which makes them rely on past data to make future predictions. In this run, the data analysis techniques have also developed along with the advancement of data formats available and now trends are more towards NoSQL (Not Only SQL) type of data stores than the relational ones. This paper explores the types of NoSQL which offer high availability, performance, and eventual concurrency applications but losing the ACID properties of the traditional databases. The authors discuss various data stores in brief and also compare these data stores based on different aspects.

Keywords No SQL database · Sharding · Scalability · Replication · Consistency · Performance

1 Introduction

The business leaders of this dynamic world mostly depend upon the data analysis techniques for the decision making processes of their respective organizations. The data that is being analyzed is continuous and related to customer and market trends which is usually collected from various sources that use different structures to maintain their database. As a result, a heterogeneous collection of structured, semi-structured, and unstructured data is obtained. Due to so much complexity in the format of the data, the traditionally popular RDBMS approach of data analysis is not applicable universally. Rather, this RDBMS approach has very limited applications for well-structured data only.

S. Bagga (✉) · A. Sharma
School of Computer Applications, Lovely Professional University, Phagwara, Punjab, India
e-mail: simmibagga12@gmail.com

A. Sharma
e-mail: anil.19656@lpu.co.in

P. K. Singh et al. (eds.), *Recent Innovations in Computing*, Lecture Notes in Electrical Engineering 701, https://doi.org/10.1007/978-981-15-8297-4_5

To overcome the limitations of the RDBMS approach, the NoSQL approach was developed in 1988, by Carlo Strozzi which can handle structured, semi-structured, and unstructured data. The main aim was to provide a flexible architecture that could handle highly scalable and flexible data. Non-relational database fulfilled this aim because of the provided increased scalability and no rigid schema for inserting data of different types. Because of the architecture being scalable horizontally, even if some nodes fail, the overall reliability and consistency of the cluster will remain high. Although, there can be a problem related to additional storage may occur because data is deformalized but the advantages of processing efficiency over RDBMS provide a comparatively greater advantage. While working with NoSQL, large entities of data are stored using XML (eX-tensible Markup Language) format [1–3]. There are some major advantages of using NoSQL as listed below:

1. NoSQL database don't get attached to the relational model. They can either be Document-oriented, key-value data store, Column-oriented, and Graph Data Stores.
2. It does not strictly follow any schema structure rather supports flexible schema that is known as Dynamic Schema Evolution. It allows the insertion of data without predefines schema, facilitates real-time changes, easy integration and less database administration.
3. It has scale-out architecture instead of the monolithic architecture of a relational database.
4. It supports replication which guarantees high availability, fault tolerance, and disaster recovery.
5. NoSQL does not support ACID property rather BASE (Basically Available, Soft State, Eventually Consistent) which follows application should always be available, not all the sites contain current data but the system is in a consistent state.
6. No SQL does not support join. It does not use Normalization so no need to support join, group by operation [3, 4].

2 Types of NoSQL

No SQL is a non-relational, open-source, distributed database that can scale-out or scale horizontally and can deal with the variety of datasets. A non-relational database is a versatile solution, well adapted for varied data types. It is used to handle Big Data and other real-time web application data. Different No SQL databases use different methods of storing data. These databases can handle semi-structured and unstructured data very easily [4]. Based on the structure of data, No SQL is classified into four different categories which are:

2.1 Key-Value Datastore
2.2 Column-oriented Data Store
2.3 Document Data Store

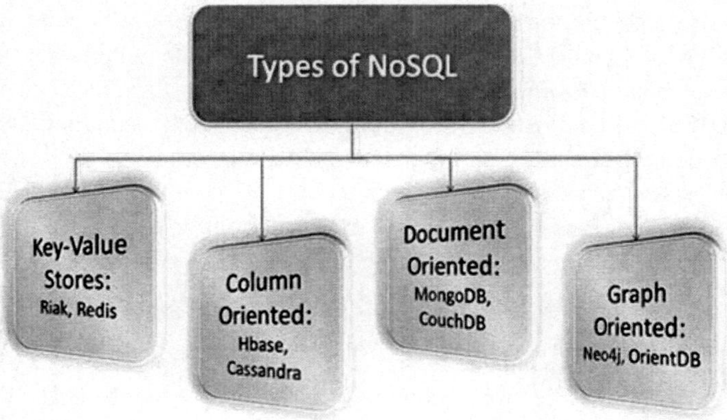

Fig. 1 Types of NoSQL

2.4 Graph Data Store

The key-value data stores share similarities with hash tables. Both use keys as indices, which result in speed improvement over RDBMS. Thus, this data model resembles a dictionary or a map where values can be addressed using the specified keys. Document-oriented database are well suited for the applications in which data available is not uniform-sized fields, rather data is in the form of a document having distinct characteristics. Column-oriented data store stores data in the tables and perform better than traditional row-oriented database systems. The goal of Column-oriented data stores is to efficiently read and write data. The Graph datastore easily handle interconnected large amount of data but does not support the scalability [4, 5] (Fig. 1).

3 Literature Review

Tauro et al. explained the limitations of Relational databases, i.e., SQL when used with data aggregation. In NoSQL, a large volume of data can be aggregated because of an efficient framework. The data query, replication, and consistent models of various NoSQL databases are compared in this paper. NoSQL Databases are useful when there is a need to price a huge volume of data with high scalability. This research work also provides key points that are to be needed to select the database and efficiently using it [1].

Kaur and Rani explained the development of non-relational databases over the traditional model. This includes the enhancement in scalability, flexibility (no rigid schema) to insert data, and easiness in capturing data of dissimilar types. This improvement in scalability, performance, and flexibility is achieved on the tradeoff of additional storage as data is de-normalized. The four classes of NoSQL that are: Column-oriented databases, Document Databases, Key-value databases, and Graph

Databases are also introduced by the authors, however, only document-oriented and graph-based databases are discussed in the paper. The authors also discussed a case study related to data modeling. For their study, they have chosen three main languages that are Postgre SQL, MongoDB and Neo4j as relational databases, document-oriented databases, and Graph database respectively [4].

Patal et al., explains the flexible architecture of NoSQL databases for highly scalable data storage needs. Authors also explained that with the advent of NoSQL databases, RDBMS databases don't come to an end. The Authors have considered the polyglot persistence of NoSQL and RDBMS ad the future of data management [6].

Moniruzzaman et al. explained the present state of NoSQL databases along with the characteristics, classification, evaluation, and comparison of NoSQL. The author gave information about the need for understanding the weaknesses and strengths of different NoSQL databases [7].

Kumar et al. highlighted the need for the NoSQL database to substitute the relational model to fulfill the modern-day database requirements. MongoDB, a new NoSQL platform is mainly highlighted by the authors. A comparative study is made with MySQL and it is justified that why MongoDB is liked over MySQL. Finally, towards the end of the paper, it is suggested that a middleware metadata can be introduced between the database layer and the application layer to integrate these different technologies. An open-source language like PHP can be used for this [8].

Merlin Shalini et al. compared two database models to identify which one has better performance and scalability over the other. The finding was that MongoDB is a better database for complicated queries. But this could be achieved at the cost of data redundancy [9].

Venkatraman et al. explained about different NoSQL data models. Authors also explained the difference between SQL and NoSQL based on their performance and found that NoSQL performs better in various business analysis situations where business demands high performance and scalability in the simplicity manner. However, the authors sum up by saying that NoSQL would still exist with SQL as its complement [10].

Lourenco et al. compared the performance of various NoSQL database engines. The author used the basis like most useful use case situations from the programmer viewpoint, benefits, and disadvantages, etc. to make comparisons. Authors demonstrated that there is no significant difference between a column-based database such as Cassandra and SQL Server, i.e., between the relational database and the document-based database. The conclusion was that every database user's requirement is different and operations systems also differ from one another. These factors have an extreme impact on the database selection [11].

Mohamed et al. gave a brief introduction to the basic concepts of relational and non-relational databases. The contrast between relational data models and the NoSQL type model is shown by highlighting some key points like Transaction Reliability, Data Model, Scalability, Cloud Suitability, Big Data Handling, Crash Recovery, and Security. Also, the paper focused on security because it is an underlying feature.

Finally, the author concludes by discussing the future scope of NoSQL databases [12].

Gupta et al. begin by defining relational databases and how they need NoSQL databases to arise by telling the importance of the NoSQL database in the current industrial scenario. Different types of NoSQL databases with a software example of each are given. These types are further compared with each other based on different application requirements [13].

Engle et al. start by discussing what is the need for NoSQL databases by their basic feature discussion. Then the general aspects of different types of NoSQL are mentioned comprehensively. The NoSQL databases are single boxes evaluated as per the parameters like Storage, Retrieval, Cross-Aggregate consistency, data type enforcement, small and large transaction performance, manipulation, and plasticity [14].

Sharma et al. started with how NoSQL came into existence and how it evolved. Types of NoSQL are explained and the difference between SQL and NoSQL is given based on certain characteristic features. The CAP theorem gave information about consistency, availability, and partial tolerance of distributed systems [15].

Mitreva and Kaloyanova presented the basic terms that should be known to study the big data along with the NoSQL model. Further, the types of NoSQL databases were studied highlighting their features like Indexing, Searching, Usage, and Joins. The research was concluded by giving the note that NoSQL can offer better optimizations over ACID properties of Relational Databases [16].

4 Comparison of Various NoSQL Databases

There are many parameters based on which NoSQL Databases can be studied and compared. These are Replication, Consistency, Complexity, Query Processing, Sharding, Performance, Speed of accessing, etc. But in this study, only the following parameters are taken into consideration.

4.1 Replication

Replication helps to maintain high availability and durability in the case of any failure. On the other hand, when the same data is being stored on different machines it leads to the problem of consistency and is required to perform multiple updates. There is a tradeoff between consistency, latency, and availability. The two main methods of replication used by different databases are Synchronous and Asynchronous. There are other categories of Replications that are: Master-Slave scheme and Multi-master scheme. Various NoSQL databases use different methods of replication.

MongoDB follows an asynchronous replication method where only one node is dedicated for write operation called a primary node and the other is the secondary node that applies operations from the primary node. It provides up-to-date primary node to give local consistency. The replica set provides automatic failover and automatic recovery [2, 17].

CouchDB can replicate synchronously. CouchDB uses two different methods of Replication: Transient and Persistent Replication, Triggering, Stopping, and Monitoring Replication.

HBase uses the Asynchronous replication method where clusters are distributed geo-graphically and the link between the clusters need not be online. The rows inserted on the master cluster are not immediately replicated to the slave clusters. So there is a weak level of consistency known as eventual consistency. HBase replication on each regional server is based on HLogs [18].

Riak supports Multi-datacenter replication that provides two modes of replication: Full sync and Real-time sync. In full sync mode, replication from the primary site to the secondary site is performed on suitable intervals and the default interval is taken six hours. During the replication process, the key block value is computed and compared with replicas. Any missing or updated block of the primary cluster is needed to modify the secondary cluster. In Real-time sync when any updation is done in the primary site, it is followed by sending the same updation in real-time to the secondary sites. Multi data center replication helps to maintain the backup and to meet disaster recovery easily.

Redis replicates all data from the master node to slave node asynchronously, i.e., the slave does not always contain update data. While modifying, the process of replication copies the modification data to all copies. There is no provision of selective replication of data. The slave node always listens to the master node for updation. If any new updation is available in the master node, the slave node also updates automatically.

Neo4j supports replication for more reliability but the cluster replication is synchronous.

4.2 Consistency

Consistency means after performing operations like update or write, the database must be inconsistent form. It means that when users try to access or fetch data from any node then the same data should be displayed and any update performed to one node should be automatically updated to all other nodes. Most NoSQL databases follow BASE properties. There are two main types of Consistency that various databases follow, i.e., Eventual Consistency and Strong Consistency.

Riak supports the Eventual consistency model in former versions but Riak 2.0 uses a strong consistency system. In Riak, there is a tradeoff between availability and consistency. The fine-tuning of consistency sacrifices that availability.

Redis supports object-level atomic updates to achieve strong consistency that guarantees consistent updates to every single object replica that is atomic. Redis is Eventual Consistent when read is performed from the replica node by Client. For strong consistency in a distributed environment, it is required to configure access to the quorum for read and write. Redis does not support the setting to a specific quorum for reads and writes. In Redis, specifying the number of replicas to reads and write is not possible.

Cassandra can be strict consistent data to an eventually consistent system. If it follows strict consistency then it gets the latest data while any read request and old data are sometimes provided in the case of eventual consistency. In the case where read operation and write operations greater than the number of replication then the database provides strong consistency.

CouchDB provides eventual consistency when the replication feature is used in the master-master setup but it provides strong when uses a master-slave setup.

HBase is strictly consistent when data is not replicated. It guarantees eventual consistency when the replication of data is enabled [18].

MongoDB provides eventual consistency. It has no version of concurrency control. Read operations from the client provide consistency control [17].

4.3 Sharding

Sharding refers to the splitting of data across multiple machines. The sharding process helps to reduce the load in the existing machines by adding new machines. NoSQL databases can easily handle scalability by increase the nodes in the cluster. This is because NoSQL follows more of a peer-to-peer architecture with all the nodes being the same which speeds up the read/write operations and don't take overhead costs.

Riak scales with additional nodes and automatically rebalances immediately at the time of node failure. Thus, it guarantees high availability and reliability.

HBase supports the concept of scaling by adding more nodes to the cluster. This helps in reducing the load on the servers and speeds up both read and writes operations on the cluster [18].

MongoDB uses auto sharding. MongoDB supports horizontal scaling and distributes load to thousands of nodes through automatic sharding. A MongoDB shard cluster includes three main components that are shard, mongos, and config servers. MongoDB automatically detects the imbalance among shards and automatically rebalances the load by adding more machines. MongoDB support two sharding strategies one is Hashed Sharding and the other is Ranged Sharding [2, 4, 17].

CouchDB does not support the sharding mechanism inbuilt but can be achieved by using two projects The Lounge and The Cloudant.

Cassandra has a special design feature that is it has the facility to scale incrementally. It scales dynamically when the load increases.

Redis support sharding to reduce load from a single machine. The sharded cluster can be set up to store more that cannot be handled by a single node. The setup sharded cluster is done by dividing the keyspace into the same number of parts.

4.4 Performance

MongoDB Performs well in the cases where data is rarely read but frequently changing data is required. MongoDB also supports dynamic queries. It also retains some properties of SQL so it performs well when data is available in the form of documents. MongoDB uses a replica set to achieve replication for high availability and disaster recovery [19, 20].

Cassandra is a column-oriented data store best suited for the situations where writes are more often than reads. Cassandra provides high availability, partition tolerance, and concurrency control [21].

Redis falls in the category of Key-Value data store and performs well for the statistical rapidly changing data and needs more frequent updates where a read operation is rarely required. Redis performs write operations very fast.

HBase falls in the category of a column-oriented database and can handle and store a huge quantity of data. It works efficiently where random reads and write from large databases are required.

Riak is a key-value database and used where the high availability is most important to apply. Riak is a Fault-tolerant system and has not any single point of failure which guarantees high availability [22, 23].

CouchDB is easy to use and consistent where data changes are done with predefined queries. CouchDB provides high availability, fault tolerance, and concurrency control.

Neo4j is a graph-based data store that is suitable for highly connected data like Facebook. The query time of the graph data store is constant and does not depends upon the size of the database.

5 Contributions

Different NoSQL has different features. They are compared based on some common features like replication, Availability, Scalability. Some data stores like MongoDB are popular for document type data storage where frequent data modification and dynamic querying are often required. On the other hand, some data stores like Cassandra are there that provide benefits like high availability, partial tolerance, concurrency control and write frequently in a column bases database architecture.

Redis is preferable where data is rarely read but frequently updating because it is very fast updating data. Redis replicates all data from the master node to slave node asynchronously, i.e., the slave does not always contain update data. For random read/write from a large Column-oriented database, HBase is preferable. For high availability, Riak is more preferable as it is fault-tolerant and has no single point of failure. On the basis, some common features comparison of some datastore is presented in the table.

6 Conclusion

In the study of different types of NoSQL databases and the review of some major vendors of each type of database, it has been identified that, in general, no database is suitable for all applications because these databases were built to satisfy a particular class of users' needs. As was shown in the comparison table (Table 1) of NoSQL, different data stores are suitable according to different purposes and features demanded in the user's problem; the choice is made in favor of the database that offers the appropriate features. The Non-relational database is yet not as much developed to have an ideal NoSQL database management system. Although some of the applications are more popular and practical among the others (like document type is most popular these days), giving their data stores greater popularity. Also, it is important to note that how these databases achieve the basic features of NoSQL like replication, sharding, consistency, etc. for handling the big data. This study can extend the help to know about the tradeoffs being made while choosing the database. So, it is important to understand the correct need for using the database and then make a wise choice to suit application needs.

Table 1 Comparison of different NoSQL datastores

Features	Datastore						
	MongoDB	CouchDB	HBase	Cassandra	Riak	Redis	Neo4j
Programming language	C++	Java Script	Java	Java	Java Script	C	Java
Platform	Linux, Mac OS, and Windows	Linux, Mac OS, and Windows	Linux, Mac OS, and Windows	Linux, Mac OS, and Windows	Linux, Mac OS, and Windows	Linux, Mac OS, and Windows	Windows, Linux,
Storage type	Document	Document	Column	Column	Key-value	Key-value	Graph
Replication	Yes	Yes (unidirectional)	Yes (using DFS)	Yes	Yes	Yes (unidirectional)	Yes
Consistency	Yes	No	Yes	No	No	Yes	Yes
Partition tolerance	Yes	Yes	Yes	Yes	Yes	Yes	Yes
Sharding	Yes	Using external tools	Yes	Yes	Yes	Using external tools	yes

References

1. Tauro, C.J.M., Rao Patil, B., Prashanth, K.R.: A Comparative Analysis of Different NoSQL Databases on Data Model, Query Model, and Replication Mode. Elsevier Publications (2013)
2. Prasad, A., et al.: A comparative study of NoSQL databases. Int. J. Adv. Res. Comput. Sci. (2014)
3. Kim, W., et al.: Classifying schematic and data heterogeneity in multidatabase NoSQL database: new era of databases for big data analytics—classification, characteristics and comparison se systems. Computer (1991)
4. Kaur, K., Rani, R.: Modeling and querying data in NoSQL databases. In: Big Data (IEEE International Congress) (2013)
5. Sharma, R., et al.: A study of NoSQL databases and working overviews. Int. J. Recent Trends Eng. Res. (2016)
6. Patel, T., Eltaieb, T.: Relational database vs NoSQL. J. Multidiscip. Eng. Sci. Technol. (JMEST) (2015)
7. Moniruzzaman, A.B.M., et al.: NoSQL database: new era of databases for big data analytics—classification, characteristics, and comparison. Int. J. Database Theory Appl. (2013)
8. Kumar, L., et al.: Comparative analysis of NoSQL (MongoDB) with MySQL database. Int. J. Modern Trends Eng. Res. (IJMTER) (2015)
9. Merlin Shalini, D.R., et al.: Performance and scaling comparison study of RDBMS and NoSQL (MongoDB). Elixir Comp. Eng. (2015)
10. Venkatraman, S., et al.: SQL versus NoSQL movement with big data analytics. Int. J. Inf. Technol. Comput. Sci. (2016)
11. Lourenco, J.R., et al.: Comparing NoSQL databases with a relational database: performance and space. Int. J. Big Data (2015)
12. Mohamed, M.A., Altrafi, O.G., Ismail, M.O.: Relational vs. NoSQL databases: a survey. Int. J. Comput. Inf. Technol. (IJCIT) (2014)
13. Gupta, A., Tyagi, S., et al.: NoSQL databases: critical analysis and comparison. In: International Conference on Computing and Communication Technologies for Smart Nation (IC3TSN) (2017)
14. Engle, R.D.L., Langhals, B.T., Grimaila, M.R., Hodson, D.D.: Evaluation criteria for selecting NoSQL databases in a single-box environment. Int. J. Database Manage. Syst. (IJDMS) (2018)
15. Sharma, N., Jain, C., et al.: Comparative study of distributed, scalable & high-performance NoSQL databases. Int. J. IT Eng. (2015)
16. Mitreva, E., Kaloyanova, K.: NoSQL solutions to handle big data. Research Gate (2013)
17. Cattell, R.: Scalable SQL and NoSQL data stores. SIGMOD (2010)
18. Chrimes, D., Zamani, H.: Using distributed data over HBase in big data analytics platform for clinical services. Comput. Math. Methods Med. (2017)
19. Katkar, M.: Performance analysis for NoSQL and SQL. Int. J. Innov. Emerg. Res. Eng. (2015)
20. Damodaran, D., et al.: Performance evaluation of MySQL and MongoDB databases. Int. J. Cybern. Inf. (IJCI) (2016)
21. Aniceto, R., et al.: Evaluating the Cassandra NoSQL database approach for genomic data persistency. Int. J. Genomics (2015)
22. Karande, N.D.: A survey paper on NoSQL databases: key-value data stores and document stores. Int. J. Res. Advent Technol. (2018)
23. Rossel, G., et al.: A modeling methodology for NoSQL key-value databases. Database Syst. J. (2017)

SMT Versus NMT: An Experiment with Punjabi–English

Kamal Deep, Ajit Kumar, and Vishal Goyal

Abstract In this paper, a comparison of the Statistical machine translation (SMT) and neural machine translation (NMT) for Punjabi to English in the fixed domain of health and tourism is provided. We have tried to answer does NMT perform equivalent well or better with respect to the SMT system? We have developed the three base models viz., SMT-based model using the Moses toolkit, followed by long short-term memory (LSTM) model and bidirectional LSTM model using the OpenNMT toolkit. All three models used the Punjabi–English parallel corpus of TDIL health and tourism. Finally, the quality of translation is validated using the automatic parameter like the BLEU score and the TER score. We observed that bidirectional LSTM performs better than simple LSTM an SMT system.

Keywords SMT · NMT · TDIL parallel corpus

1 Introduction

Machine translation (MT) is a computer software-aided translation system. A computer program is used to translate one natural language (like Punjabi) to another natural language (like English). The translation is a complex task for humans as well as for computer machines. Accurate translation needs a syntactical and semantically understanding of both source and target languages. Research in MT started with a rule-based approach [1, 2]. Primarily, for building MT systems, corpus-based approach is used [3]. The corpus-based approach uses two techniques to develop the MT system. One is statistical techniques and the second is artificial neural network-based techniques [4, 5].

K. Deep (✉) · V. Goyal
Punjabi University, Patiala, Punjab, India
e-mail: kamal.1cse@gmail.com

A. Kumar
Multani Mal Modi College, Patiala, India

P. K. Singh et al. (eds.), *Recent Innovations in Computing*, Lecture Notes in Electrical Engineering 701, https://doi.org/10.1007/978-981-15-8297-4_6

MT system developed using the statistical techniques is known as statistical machine translation (SMT) system and developed using the artificial neural network-based techniques is known as neural machine translation (NMT) system. SMT system uses the source-target language parallel corpus and monolingual target language corpus to develop the statistical models [6, 7]. The decoder decodes the input sentences using these statistical models. SMT has provided good outcomes for the pair of the language that has the same sentence structure as Hindi and Punjabi have sentence structure of Subject-Object-Verb (SOV). But in our case of Punjabi and English sentence, structure is different. Punjabi has Subject-Object-Verb (SOV) structure and English has Subject-Verb-Object (SVO) [8]. So reordering is also a major problem in the Punjabi to English SMT system.

NMT is a relatively new way to develop a neural machine translation system. NMT has a limited vocabulary due to the use of softmax function in the last layer of the neural network [9]. There is a problem of out of vocabulary words in the case of the NMT system. Researchers [10, 11] have given global and local approaches to resolve this problem.

In this research, the performance of SMT and NMT is tested by using Punjabi–English corpus provided by TDIL [12]. The whole paper is divided into various sections. Section 2 is about corpus preparation. The corpus used in the training, tuning and testing is discussed. Section 3 explains the experimental setup for the SMT and NMT system. Section 4 describes the evaluation of the output with respect to various metrics and Sect. 5 finally draws the conclusion.

2 Corpus Preparation

To provide an accurate comparison for SMT and NMT system, the same data sets are used in each approach. We have downloaded the Punjabi-English parallel tagged corpus of the health and tourism domain from TDIL [12] website. Tags are removed from a parallel corpus. Corpus is manually checked and incorrect or incomplete sentences are removed from the corpus. Table 1 shows the domain wise count of the parallel sentences.

This parallel corpus is divided into three files: training, tuning, and testing. A tuning file is used as a validation file in the NMT system. The corpus statistics are shown in Table 2.

Table 1 TDIL health and tourism corpus

Corpus	Parallel sentences	Tokens (English)
TDIL health	24,605	421,825
TDIL tourism	26,823	459,433

Table 2 TDIL corpus is divided into training, tuning, and test sets

	Parallel sentences	Tokens (English)	Tokens (Punjabi)
Training	50,323	849,152	883,210
Tuning	969	16,322	16,929
Testing	969	16,638	17,347

3 Experiment Set-Up

There was a need for three experimental models: one for training SMT system and another two for training different NMT system. All set-ups are discussed below one by one.

3.1 SMT

Moses [13] is an SMT system that by use of parallel text allows the training of the translation model. After the SMT model is built, the decoder uses the beam search algorithm to translate one language text to another language. Beam search algorithm finds the highest probability translation among the exponential number of choices [14].

3.2 SMT Training

In the SMT system, there is a need for language model (LM) and translation model(TM). The language model is built using the target language, English is the target language in our proposed model. LM helps to ensure fluent output. To develop LM, KenLM was used. Giza++ was used to learn word alignment. SMT is trained using the Moses toolkit (see Fig. 1) depicts the whole process.

To develop the SMT system, we used Punjabi as the source language and English as the target language. There is a need for pre-processing before to train the SMT system. Various pre-processing steps are performed on the source and target language.

1. Tokenization: it is a process of divided text into a set of meaningful pieces. These pieces are called tokens. Tokens are words, numbers and punctuation marks in case of languages. Tokenization is performed on both English and Punjabi parallel corpus.
2. True casing: it is the problem of finding the proper capitalization of words within a text where such information is unavailable. For English, all text is lowercased.

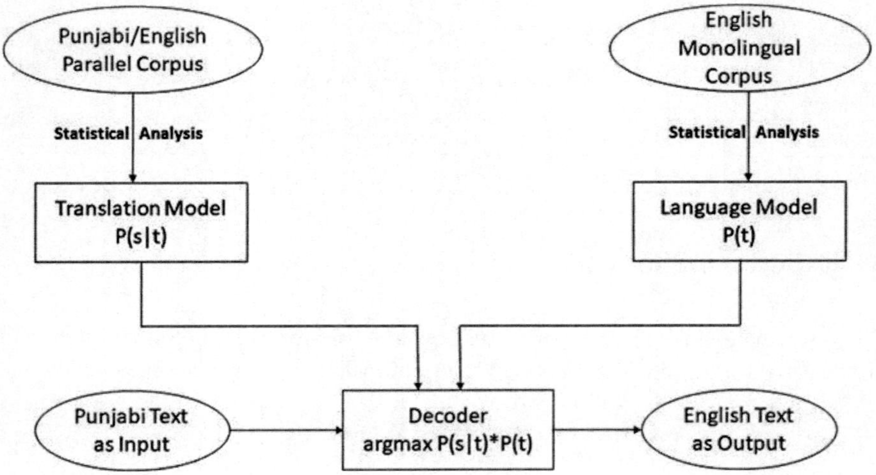

Fig. 1 Statistical machine translation system

3. Cleaning: the length of a sentence also affects the quality of the translation model. So long sentences (# of tokens > 40) were removed from the corpus. After cleaning, # of sentences in the training file is 49,432 from 50,323 sentences.

3.3 NMT

NMT [15] is based on the artificial neural network. NMT system use encoder and decoder architecture to learn the translation. Encoder reads the source text. This text is encoded into a vector of fixed size. The decoder decodes this fixed-size vector into target text which is of variable size and completes the whole translation system. Recurrent neural network (RNN) [16] is used in encoder and decoder of the NMT system.

Let S and T be the source and target sentence pair with S as $s_1, s_2, s_3, \ldots, s_x$ and T as $t_1, t_2, t_3, \ldots, t_y$. Here x and y is the number of words in source and target sentence. An encoder converts the source sentences S into the fixed-size vector. The decoder uses the conditional probability to output one word at a time.

$$P(T|S) = P(T|S_1, S_2, S_3, \ldots, S_M) \quad (1)$$

Here $S_1, S_2, S_3, \ldots, S_M$ is a fixed-size vector encoded by encoder. The chain rule is applied to the above Eq. (1) and it is converted into a new equation.

$$P(T|S) = P(t_i|t_1, t_2, t_3, \ldots, t_{i-1}; S_1, S_2, S_3, \ldots, S_M) \quad (2)$$

Decoder predicts the next word using words predicted till now and by source sentences *S*.

There are two types of encoders used to develop two models: long short-term memory (LSTM) [17] and bidirectional LSTM [18]. Bidirectional LSTM has the advantage of learning the sentence from both directions left to right as well as the right to left [19].

3.4 NMT Training

OpenNMT [20] toolkit is used for training the two NMT model, one by using LSTM and another by bi-LSTM. Both proposed model is shown in (see Figs. 2 and 3). Tokenization, true casing and cleaning were also performed as a pre-processing step to train both NMT models. Vocabulary size of 34,000 words for Punjabi language

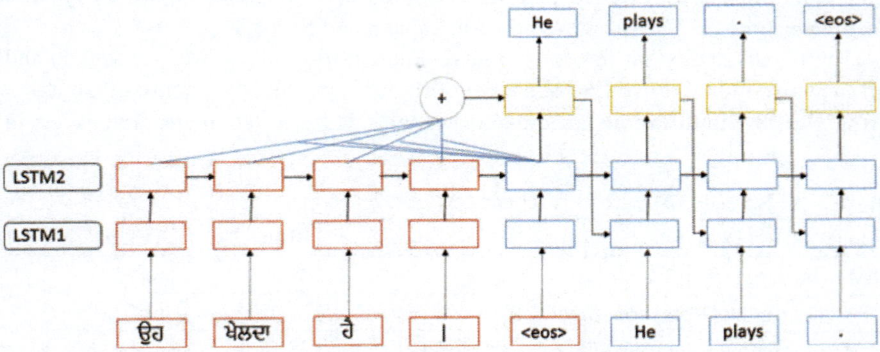

Fig. 2 NMT model by using LSTM as encoder and decoder having two hidden layers

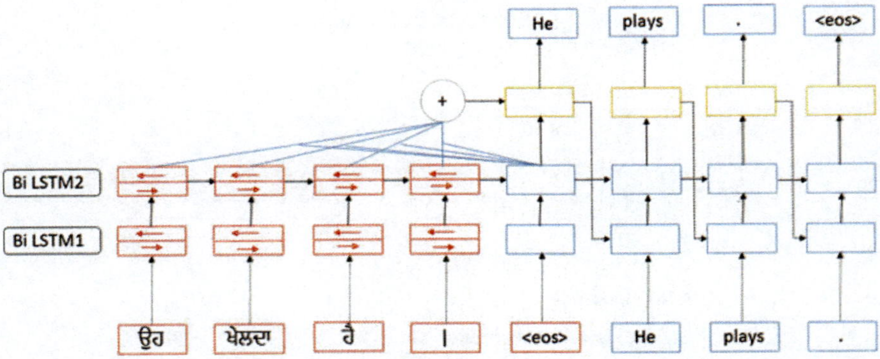

Fig. 3 NMT model by using bidirectional LSTM as encoder and LSTM as decoder having two hidden layers

and 35,000 words for the English language, respectively, is developed. A batch size of 64 and 102 epochs for training is fixed. The activation function used as softmax, sgd as an optimizer and the loss calculation at each step was done using *categorical cross-entropy*. In one model, two layers for LSTM encoder and decoder with 500 cells at each layer are used for training. In the second model, two layers for bidirectional LSTM encoder and LSTM decoder with 500 cells at each layer are used for training. Training time depends upon whether the model is developed using the CPU or GPU. We have used 4 GB NVIDIA GeForce GTX1050Ti GPU to fast the training. LSTM model took the 6.8 h to train and the Bi-LSTM model took the 7.4 h to train full baseline models.

4 Result and Analysis

We have tested all models on the same test set that was discussed earlier in Sect. 2 of corpus preparation. BLEU and TER score is used as the automatic evaluation parameter for MT. The result is shown in Table 3 and Fig. 4.

The results show that for our proposed Punjabi to English SMT (Model 1) and NMT with a bidirectional LSTM encoder (Model 2) produces the approximate same BLEU score. But this score is more than Model 2. The translation error rate is less in

Table 3 BLEU and TER score for SMT and NMT models

Model	BLEU	TER (%)
SMT (Model 1)	38.39	60.75
NMT (2 layer LSTM encoder, Model 2)	29.97	62.59
NMT (2 layer bidirectional LSTM encoder, Model 3)	**38.46**	**53.84**

Bold shows the best value of the BLEU and TER

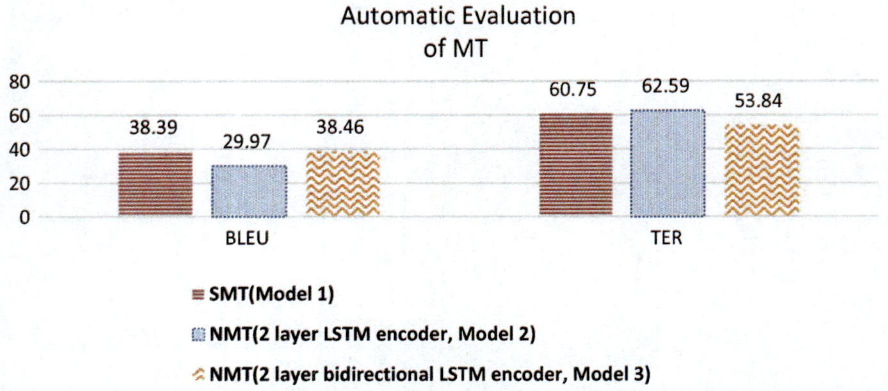

Fig. 4 Bar graph displaying the BLEU and TER score for all the three proposed models

bidirectional LSTM encoder NMT (Model 3) as compared to the other two models. This means that the third proposed model is giving a more accurate output as compare to the other two models.

4.1 BLEU Score on the Sentence Level

To gain insight into the specific difference between all three proposed models for Punjabi to English, sentence-level BLEU score is checked. It is checked on the health and tourism domain separately. First, two sentences are taken from the health domain.

1. Source Text: ਇਹ ਸੇਬ ਖਣਿਜ ਅਤੇ ਵਿਟਾਮਿਨਾਂ ਨਾਲ ਭਰਪੂਰ ਹੈ ।
   English Reference: this apple is full of minerals and vitamins.
   SMT: this *is* the apple is full of minerals and vitamins.
   NMT (LSTM): this apple is rich with minerals and vitamins.
   NMT (bidirectional LSTM): this apple is full of minerals and vitamins.
2. Source Text: ਪਰਿਅੱਡਾੱਨਿਟਸਟ ਗਮਸ ਦਾ ਇਲਾਜ਼ ਕਰਦਾ ਹੈ |
   English Reference: pyrrheodontist does the treatment of gums.
   SMT: the treatment of ਪਰਿਅੱਡਾੱਨਿਟਸਟ gums.
   NMT (LSTM): pyrrheodontist is the treatment for the treatment of tap.
   NMT (bidirectional LSTM): pyrrheodontist does the treatment of gums.

In example 1, NMT with bidirectional LSTM provides the exact output as given in the reference output. SMT system also shows the correct output but it is showing "is" also with "this." The "ਭਰਪੂਰ" is translated into "rich" by the NMT with LSTM encoder. It is also the correct meaning of the word "ਭਰਪੂਰ," but "full" fits more in this context as compared to "rich" for the word "ਭਰਪੂਰ."

In example 2, the SMT system is not able to translate ਪਰਿਅੱਡਾੱਨਿਟਸਟ to English and moreover sentence structure is not correct. A simple LSTM system has given incorrect output. Looking at bidirectional NMT system output, it is completely accurate and matches with reference also. Now, we are taking sentences from the tourism domain to see how each of three systems performs with smaller, medium and longer sentences.

3. Source Text: ਜੰਮੂ ਅਤੇ ਕਸ਼ਮੀਰ ਦਾ ਬਿਲੌਰ ਇਕ ਤੀਰਥ ਸਥਾਨ ਹੈ |
   English Reference: billour of jammu and kashmir is a pilgrimage.
   SMT: ਬਿਲੌਰ of jammu and kashmir is a pilgrimage.
   NMT (LSTM): charar-e-sharif of jammu and kashmir is chalets.
   NMT (bidirectional LSTM): the interiors of jammu and kashmir is a pilgrimage.
4. Source Text: ਔਲੀ ਦਾ ਰੋਪਵੇ ਏਸ਼ੀਆ ਦਾ ਸਭ ਤੋਂ ਲੰਬਾ ਅਤੇ ਉੱਚਾ ਰੋਪਵੇ ਹੈ ।
   English Reference: the ropeway of auli is the longest and highest ropeway of asia.
   SMT: the ropeway of auli is the longest and highest ropeway.
   NMT (LSTM): the ropeway of auli is the longest and highest ropeway of asia.

NMT (bidirectional LSTM): the ropeway of auli is the longest and highest ropeway of asia.

5. Source Text: ਇਸਦੇ ਇਲਾਵਾ ਬ੍ਰਹਮਾ ਮੰਦਿਰ, ਸ੍ਰੀ ਦਾਮੋਦਰ ਮੰਦਿਰ, ਸ੍ਰੀ ਗੋਪਾਲ ਗਣਪਤੀ ਮੰਦਿਰ, ੮੦੦ ਸਾਲ ਪੁਰਾਣਾ ਸ੍ਰੀ ਕਾਲਿਕਾ ਦੇਵੀ ਮੰਦਿਰ ਬੇਹੱਦ ਪ੍ਰਸਿੱਧ ਹਨ।
English Reference: besides this brahma mandir, shri damodar mandir, shri gopal ganapati mandir, 800 years old shri kalika devi mandir are extremely famous.
SMT: besides this brahma mandir, shri damodar mandir, shri gopal ganapati mandir, 800 years old shri kalika devi mandir are extremely famous.
NMT (LSTM): apart from this brahma temple, shri damodar temple, shri gopal mandir, meera devi temple, 800 years old shri kalika devi temple are very famous.
NMT (bidirectional LSTM): besides this brahma temple, shri damodar temple, shri gopal ganpati temple, 800 years old shree devi temple is very famous.

In example 3, all three models are not able to translate the word "ਬਿਲੌਰ." SMT has given "ਬਿਲੌਰ" as such in the output whereas Simple LSTM translated "ਬਿਲੌਰ" into "charar-e-sharif" and bidirectional NMT system translated "ਬਿਲੌਰ" into "interiors." It is an out of vocabulary problem. In example 4, the SMT system has skipped the translation of "ਏਸ਼ੀਆ ਦਾ." Both the LSTM system yields the correct translation. Example 6 is a longer sentence from the tourism domain. The word "ਕਾਲਿਕਾ" is correctly transliterated by the SMT and LSTM encoder but bidirectional LSTM encoder has skipped it. "ਇਸਦੇ ਇਲਾਵਾ" is translated to "apart" by LSTM encoder and into "besides" by other two models. But both are the correct translations of the word "ਇਸਦੇ ਇਲਾਵਾ."

5 Conclusion

This research uncovers that the proposed NMT system with bidirectional LSTM as an encoder produces a better translation quality as compare to the NMT system with simple LSTM as an encoder and SMT system. We have observed that many words are translated incorrectly in the bidirectional NMT system as the translation error rate is 53.84%. In the future, we plan to find methods to incorporate into the bidirectional NMT system so that it will give a high BLEU score and low TER score as possible.

References

1. Garje, G.V., Bansode, A., Gandhi, S., Kulkarni, A.: Marathi to English sentence translator for simple assertive and interrogative sentences. Int. J. Comput. Appl. **138**(5), 42–45 (2016). https://doi.org/10.5120/ijca2016908837
2. Singh, P.K., Kar, A.K., Singh, Y., Kolekar, M.H., Tanwar, S.: Proceedings of ICRIC 2019. Recent Innovations in Computing, 2020. Lecture Notes in Electrical Engineering, vol. 597, pp. 3–920. Springer, Cham, Switzerland

3. Singh, U., Goyal, V., Singh, G.: Urdu to Punjabi machine translation: an incremental training approach. Int. J. Adv. Comput. Sci. Appl. **7**(4), 227–238 (2016). https://doi.org/10.14569/ijacsa.2016.070428
4. Chaudhary, J.R., Patel, A.C.: Machine translation using deep learning: a survey. Int. J. Sci. Res. Sci. Eng. Technol. **4**(2), 145–150 (2018)
5. Mahata, S.K., Mandal, S., Das, D., Bandyopadhyay, S.: SMT vs NMT: a comparison over Hindi & Bengali simple sentences. In: International Conference on Natural Language Processing, Dec 2018, pp. 175–182. [Online]. Available: https://arxiv.org/abs/1812.04898
6. Bombay, T.: Statistical machine translation with rule based re-ordering of source sentences. In: International Conference on Natural Language Processing, Aug 2008
7. Singh, P.K., Pawłowski, W., Tanwar, S., Kumar, N., Rodrigues, J.J.P.C.: Proceedings of First International Conference on Computing, Communications, and Cyber-Security (IC4S 2019), 2020. Lecture Notes in Networks and Systems, vol. 121, pp. 3–917. Springer, Cham, Switzerland
8. Visweswariah, K., Rajkumar, R., Gandhe, A., Ramanathan, A., Navratil, J.: A word reordering model for improved machine translation. In: EMNLP-2011, pp. 486–496. [Online]. Available: https://aclweb.org/anthology-new/D/D11/D11-1045.pdf
9. Anand, A., Bhattacharyya, P.: Literature survey: neural machine translation (2014)
10. He, W., He, Z., Wu, H., Wang, H.: Improved neural machine translation with SMT features. In: 30th AAAI Conference on Artificial Intelligence, AAAI 2016, no. 10, pp. 151–157 (2016)
11. Luong, M.-T., Sutskever, I., Le, Q.V., Vinyals, O., Zaremba, W.: Addressing the rare word problem in neural machine translation (2014). https://doi.org/10.3115/v1/P15-1002
12. TDIL CORPUS Homepage. https://www.tdil-dc.in/index.php?lang=enwww.tdil-dc.in. Last accessed 21 Nov 2018
13. Bhalla, D., Joshi, N., Mathur, I.: Rule based transliteration scheme for English to Punjabi. Int. J. Nat. Lang. Comput. **2**(2), 67–73 (2013). https://doi.org/10.5121/ijnlc.2013.2207
14. Azath, M., Kiros, T.: Statistical machine translator for English to Tigrigna translation. Int. J. Sci. Technol. Res. **9**(1), 2095–2099 (2020)
15. Östling, R., Scherrer, Y., Tiedemann, J., Tang, G., Nieminen, T.: The Helsinki neural machine translation system, vol. 2, pp. 338–347 (2018). https://doi.org/10.18653/v1/w17-4733
16. Cho, K., van Merrienboer, B., Bahdanau, D., Bengio, Y.: On the properties of neural machine translation: encoder–decoder approaches. In: Proceedings of SSST-8, Eighth Workshop on Syntax, Semantics and Structure in Statistical Translation, pp. 103–111 (2014). [Online]. Available: https://arxiv.org/pdf/1409.1259v2.pdf%5Cn; https://arxiv.org/abs/1409.1259
17. Wu, Y., et al.: Google's neural machine translation system: bridging the gap between human and machine translation, pp. 1–23 (2016). [Online]. Available: https://arxiv.org/abs/1609.08144
18. Zaremba, W., Sutskever, I., Vinyals, O.: Recurrent neural network regularization, no. 2013, pp. 1–8 (2014)
19. Dušek, O., Novikova, J., Rieser, V.: Evaluating the state-of-the-art of end-to-end natural language generation: the E2E NLG challenge. Comput. Speech Lang. **59**, 123–156 (2020). https://doi.org/10.1016/j.csl.2019.06.009
20. OPENMT Homepage. https://github.com/OpenNMT/OpenNMT-py. Last accessed 21 Nov 2018

Performance Enhancement of MIMO Configurations in FSO System Under Different Weather Conditions

Arjun Dubey and Harmeet Singh

Abstract Free-space optics (FSO) is an optical wireless communication technique in which free space (air, outer space or vacuum) acts as a medium between transceivers, and for the effective transmission of the optical signal, line of sight (LOS) is necessary between the transceivers. FSO works as same as in optical fiber communication but in this optical beam propagates over free space rather than cores of fiber. FSO communication is also known as terahertz communication and optical wireless communication (OWC). Weather condition (clear air, haze and fog) is one of the main impairments affecting free-space optical communication. In this simulation model under different weather conditions such as clear air, haze and fog, the two systems MIMO-FSO (4 × 4) and MIMO-FSO (8 × 8) have been analyzed. Free-space optics emerged as one of the various merits over the radio spectrum. FSO can achieve high capacity with huge unlicensed optical spectrum and less operational costs and the MIMO technique in wireless communication systems is widely used because it provides huge data throughput and increased link range without the addition of neither bandwidth nor transmitted power. To bring out results under different weather conditions (clear air, haze and fog), simulation model has been analyzed for an array of 1 km at the frequency of 193.1 GHz. The parameters of the MIMO-FSO system, i.e., Q-factor and BER, have been analyzed for MIMO-FSO (4 × 4) and MIMO-FSO (8 × 8). In this, we analyze the difference between the two MIMO-FSO (4 × 4) and MIMO-FSO (8 × 8) system and this shows that 8 × 8 MIMO-FSO has better performance to the 4 × 4 MIMO-FSO system in different weather conditions.

Keywords Free-space optics · MIMO · Bit error rate · Wireless communication · Q-factor · OptiSystem 13.0

A. Dubey (✉) · H. Singh
Department of Electronics and Communication Engineering, Chandigarh University, Mohali, India
e-mail: arjundubey0084@gmail.com

H. Singh
e-mail: harmeetsingh85.hs@gmail.com

P. K. Singh et al. (eds.), *Recent Innovations in Computing*, Lecture Notes in Electrical Engineering 701, https://doi.org/10.1007/978-981-15-8297-4_7

1 Introduction

1.1 FSO

Day by day our modern digital infrastructure is growing rapidly and the demand for huge data rate access is also growing rapidly, and to fulfill this, we need faster communication access [1]. But in these days, we have many communication technologies such as coaxial cable, copper wires and optical fibers. In optical fiber communication, we have attained a very high amount of data rates because of the light beam as a source of propagation [2]. These advances in communication have few constraints, i.e., congested spectrum and range, low information rate, costly authorization, security issues and high cost of establishment and installment [1]. FSO works as same as in optical fiber communication but in this optical beam is propagates over free space rather than cores of fiber [3]. The beauty of this technology is that it utilizes license-free spectrum band, i.e., no need to buy spectrum or other types of authorization for installation [1]. But this technology has demerits as well, in which transmission is affected badly due to attenuation of the signal. Attenuation arises when the atmosphere is cloudy, foggy and rainy outside [4]. Other factors are also affecting the transmission are, flying birds can also block a single beam for a short period of time which arises only very short interruptions and then transmissions of signals are automatically resumed [5]. Beam spreading and roving due to propagation through the air packets of fluctuating temperature, density, and refractive index and scintillation but these effects are temporarily smaller than atmospheric turbulence [6] (Fig. 1).

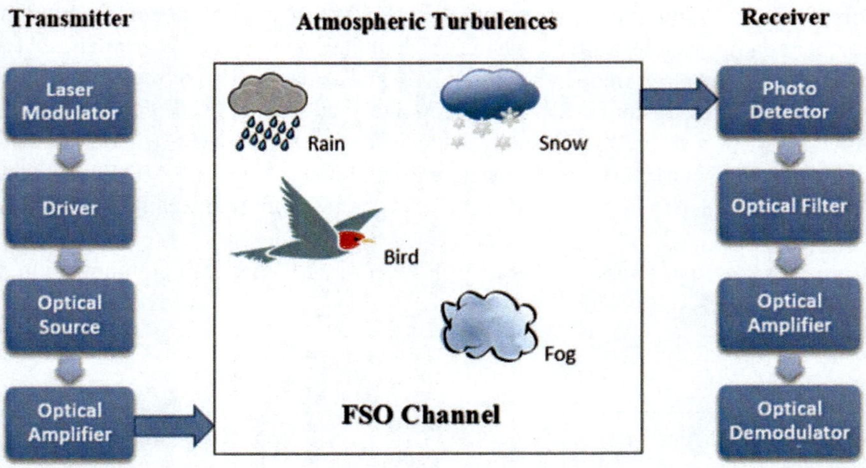

Fig. 1 Block diagram of free-space optical communication (FSO)

1.2 MIMO Technique

MIMO technology is considered as the most widely used technology in wireless communication [7]. MIMO systems are the most favorable strategy to achieve excellent performance and exceptionally high efficiency. The communication system using N_T, i.e., number of transmitting Tx antennas at transmitter and N_R number of receiving Rx antennas at receiver is generally referred to as a multi-input multi-output (MIMO) system [3–5].

- The special case $N_T = N_R = 1$ is called a single input single output (SISO) system.
- $N_T = 1; N_R \geq 2$ is called a single input multiple output (SIMO) system.
- If $N_T \geq 2; N_R = 1$, then system is denoted to as multiple input single output (MISO) system.
- The resulting spatial channel in a MIMO system is termed as MIMO channel [8].
- In a MIMO network with N_T, i.e., transmitter antennas and N_R, i.e., receiver antennas, we denote the corresponding low-pass channel impulse response between the antenna transmitting '*j*th' and the antenna receiving '*i*th' as $h_{ij}(\tau; t)$ (Figs. 2, 3 and 4).

  where 'τ' is called 'delay' variable and 't' is called 'time' variable.

  Hence, the system model for received sample signal is expressed by:

$$y = Hx + n$$

where y symbolize the received and x symbolize the transmitted vectors [3]. H is the channel matrix and n is the noise vector.

2 System Design Model

FSO optimal design in the software (OptiSystem 13.0) is done by two different systems, i.e., SISO and MIMO, firstly we use single input single output (SISO) in which only single transmitter and single receiver are installed in the system model [9]. But in multiple input multiple output (MIMO), there are multiple transmitters and receivers respectively in the system model. In the below MIMO-FSO design model,

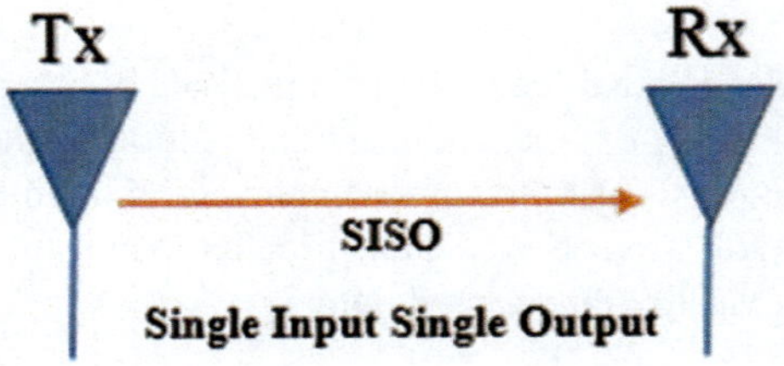

Fig. 2 SISO (existing technology)

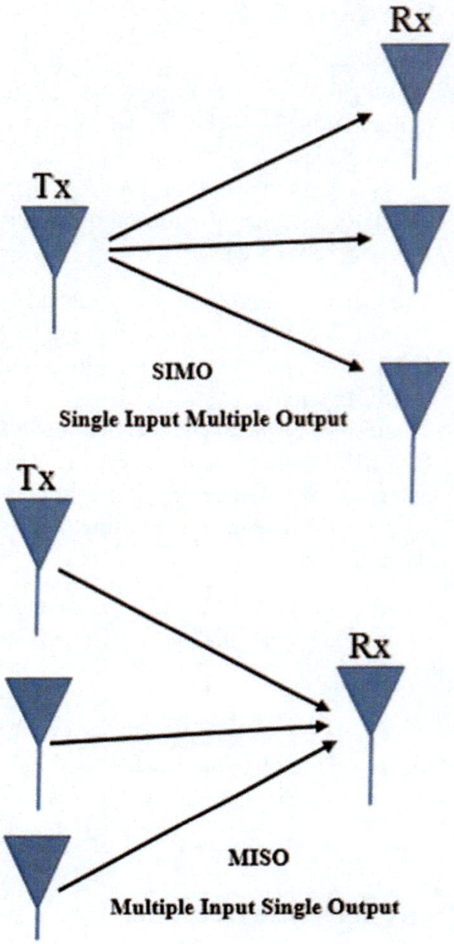

Fig. 3 SIMO and MISO (smart antenna systems)

we use 4 × 4 MIMO-FSO and 8 × 8 MIMO-FSO, i.e., there are four transceivers and eight transceivers, respectively, in the system model [10].

2.1 4 x 4 MIMO-FSO Systems

There are three sections in the FSO design model, i.e., transmitting section, channel mode, i.e., FSO and receiving section. In the system model, the transmitting section consists of following elements, i.e., continuous wave (CW) laser, non-return to zero (NRZ) pulse generator, pseudorandom bit sequence (PRBS generator) and MZM (Mach-Zehnder modulator).

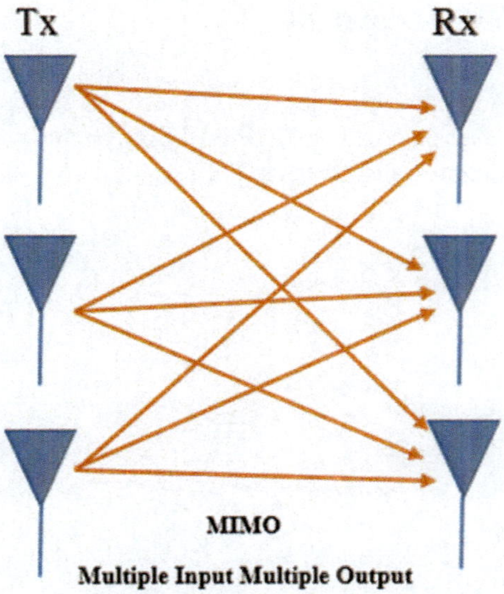

Fig. 4 MIMO systems

In the transmitting side, the signal is modulated by the use of MZM modulator using the most accepted modulation method (intensity modulation). Laser source (CW laser) in the transmitter section converts electrical signal into optical signal. And the fork module in the system model is the number of output ports which came from the previous module with same value. Several laser beams are formed by the fork module and the power combiner through FSO channel combined all these laser beams [11]. Through the avalanche photodetector (APD) and low-pass Bessel filter, the optical signal is received and then the simulation is carried out to analyze 'BER analyzer' in the software. The calculation of Q-factor, BER, eye height, eye diagram, etc. is done by the BER analyzer automatically. Figure 5 is the simulation layout of 4 × 4 MIMO-FSO in the software.

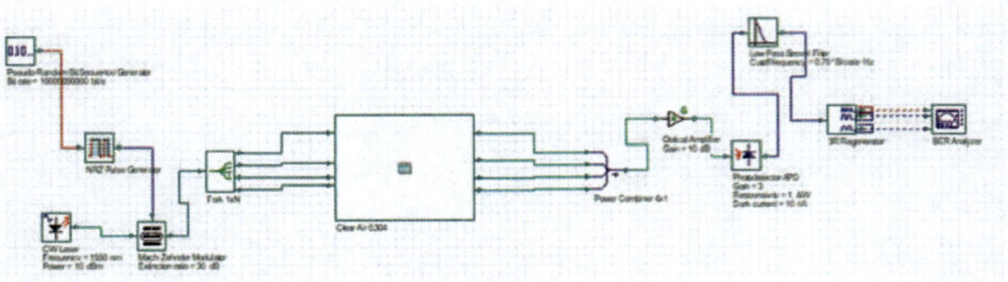

Fig. 5 4 × 4 MIMO-FSO system

2.2 8 × 8 MIMO-FSO Systems

In Fig. 6 system layout, there are eight transceivers. All the components and their parameters are the same as in the 4 × 4 MIMO-FSO system layout. The only difference is that there are eight transceivers in the system model of 8 × 8 MIMO-FSO (Tables 1 and 2).

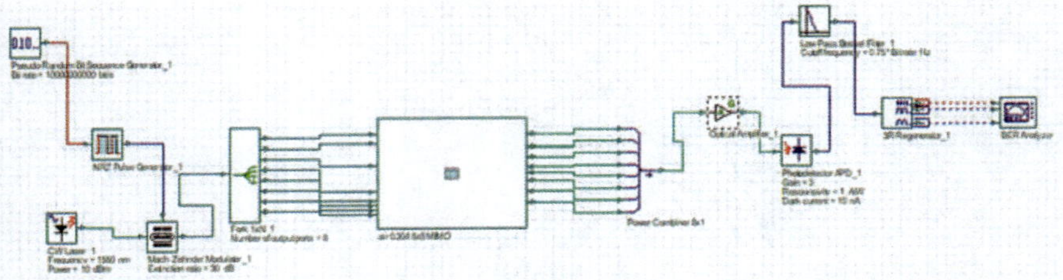

Fig. 6 8 × 8 MIMO-FSO system

Table 1 Parameters for the MIMO-FSO system

Parameters	Value
Data rate (in Gbps)	10
Wavelength (λ)	1550 nm
Extinction ratio	30 dB
Transmitter aperture	2.5 cm
Transmitter loss	1.8 dB
Receiver aperture	5 cm
Receiver loss	1.8 dB
Additional loss	1 dB
APD gain	3 dB

Table 2 Attenuation of weather conditions for different MIMO-FSO configurations

Weather conditions	Attenuation (dB/km)
Clear air	0.304
Haze	4.319
Fog	15.56

3 Simulation Results and Discussions

3.1 Eye Diagram Analysis of MIMO-FSO System in Different Weather Conditions

3.1.1 Clear Air

Under clear air atmospheric conditions, the atmospheric loss associated with visibility is minor. But in the air, we experience other adverse effects such as scintillation and fading of the signal [12]. Due to solar heating, wind, heat from the air conditioning ducts, etc. give rise to the variation in the refractive index of the air along the transmission path which results in a change of amplitude and phase of the received signal, i.e., channel fading.

Under constant distance 1 km and constant data rate of 10 Gbps from Fig. 7a, b we conclude that under attenuation 0.304 dB/km of clear air, the Q-factor and BER of 4 × 4 MIMO-FSO are 28.421 and 4.32914×10^{-178}, respectively, and 8 × 8 MIMO-FSO have 37.9249 and 4.144×10^{-315}. Here, the performance of 8 × 8 MIMO-FSO is better than the 4 × 4 MIMO-FSO system under clear weather conditions. And we see that there is 4.144×10^{-315} BER in 8 × 8 MIMO-FSO system which shows that the efficiency of this system model is better than the other.

3.1.2 Haze

Haze is an atmospheric change in which dust, sand storms, mist, smoke, smog and other types of dry particulates suspended in the air which affects the visibility or clarity of the sky (Fig. 8) [13].

In this, we conclude that under attenuation 4.319 dB/km, the Q-factor and BER of 4 × 4 MIMO-FSO are 16.104 and 1.06167×10^{-058}, respectively, and for 8 × 8 MIMO-FSO is 25.6477 and 1.8618×10^{-145}, respectively. Here also, the performance of 8 × 8 MIMO-FSO is better than the 4 × 4 MIMO-FSO but here the performance of Q-factor under haze is lower than the Q-factor of clear air, i.e., 25.6477 which shows that we have better results in clear air.

3.1.3 Fog

FSO performance's significant downside is fog because the particle size of the cloud is close enough to the wavelength of the transmitted signal used in free-space optical communications. The absorption occurs in near-IR occurs due to the presence of water droplets in fog and this absorption increases as the size of water droplets increases as such in case of rain and snow [14].

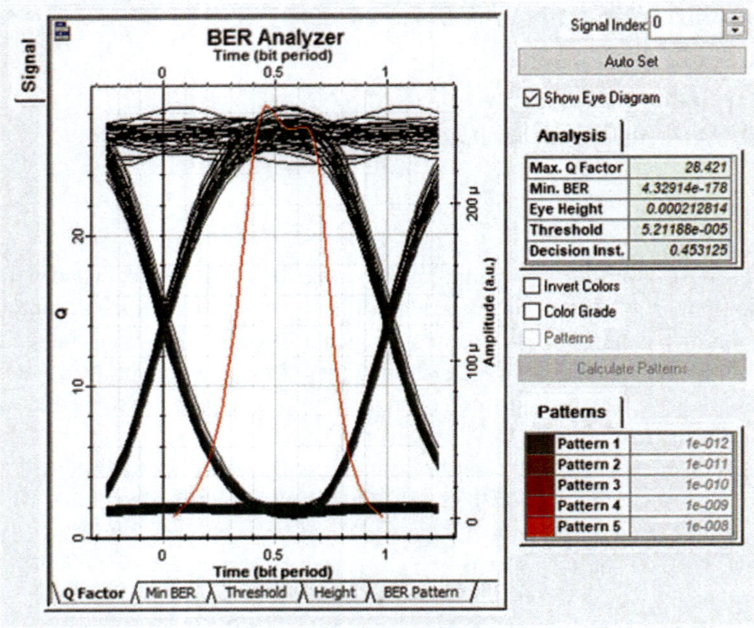

(a)

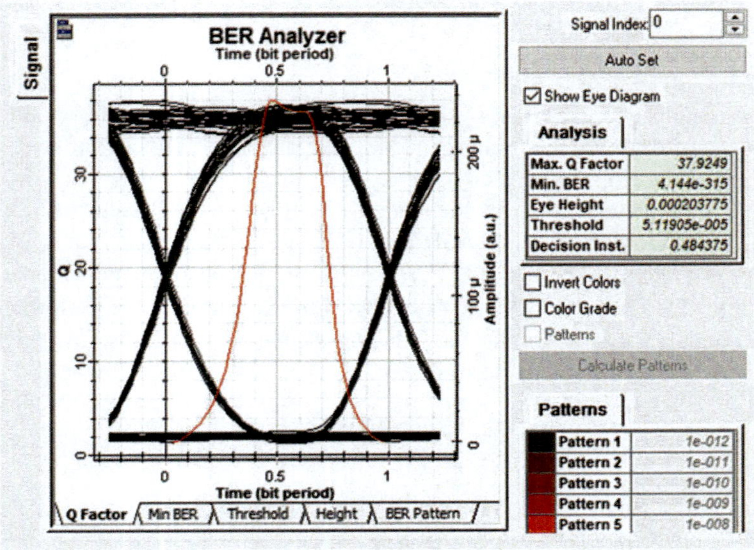

(b)

Fig. 7 **a** Eye diagram of 4 × 4 MIMO-FSO, **b** eye diagram of 8 × 8 MIMO-FSO for clear air 0.304 dB/km

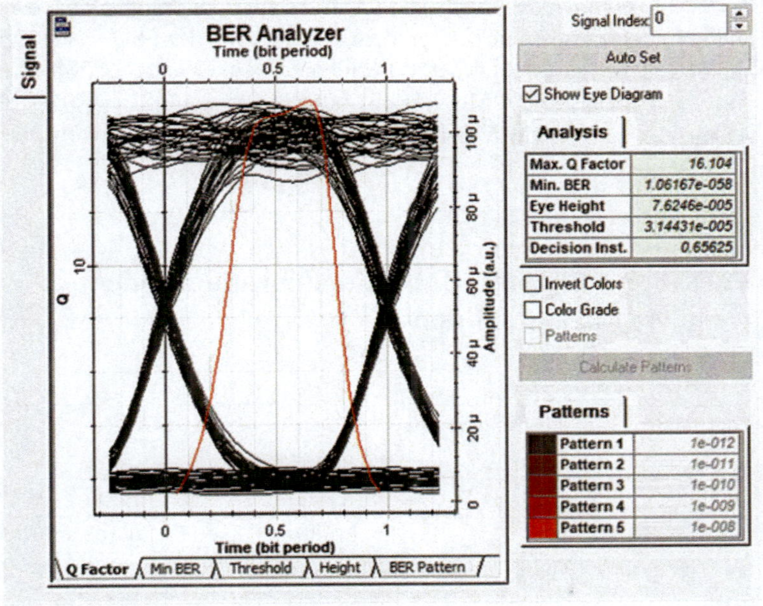

(a)

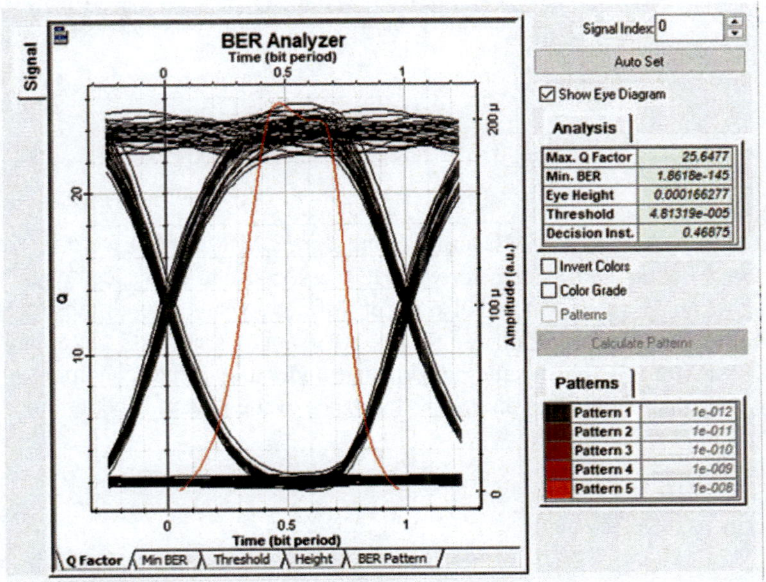

(b)

Fig. 8 **a** Eye diagram of 4 × 4 MIMO-FSO, **b** eye diagram of 8 × 8 MIMO-FSO for haze 4.319 dB/km

In Fig. 9a, b we conclude that during harsh weather condition, i.e., fog having an attenuation of 15.56 dB/km is degraded in 8 × 8 MIMO-FSO system model. The Q-factor and BER of 4 × 4 MIMO-FSO are 2.44677 and 7.02588×10^{-03}, respectively, and for 8 × 8 MIMO-FSO is 4.05579 and 2.46682×10^{-005}. From here, we analyze that 8 × 8 MIMO-FSO system model has better performance than the 4 × 4 MIMO-FSO.

4 Q-Factor and BER of MIMO Configuration Under Diverse Weather Conditions

MIMO configuration	Weather condition	Range = 1 km	
		Q-factor	BER
4 × 4 MIMO-FSO	Clear air	28.421	4.32914×10^{-178}
	Haze	16.104	1.06167×10^{-058}
	Fog	2.44677	7.02588×10^{-03}
8 × 8 MIMO-FSO	Clear air	37.9249	4.144×10^{-315}
	Haze	25.6477	1.8618×10^{-145}
	Fog	4.05579	2.46682×10^{-005}

5 Conclusion

The tests of optical 4 × 4 MIMO-FSO and 8 × 8 MIMO-FSO were inferred in this paper by using OptiSystem software 13.0 at a constant data speed of 10 Gbps, constant distance, i.e., 1 km and also under different weather conditions (clear air, haze and fog). From the simulation results, we examine that the Q-factor and BER of 8 × 8 MIMO-FSO have better performance under all weather conditions and we also analyze that under clear air, the performance of the system is better one.

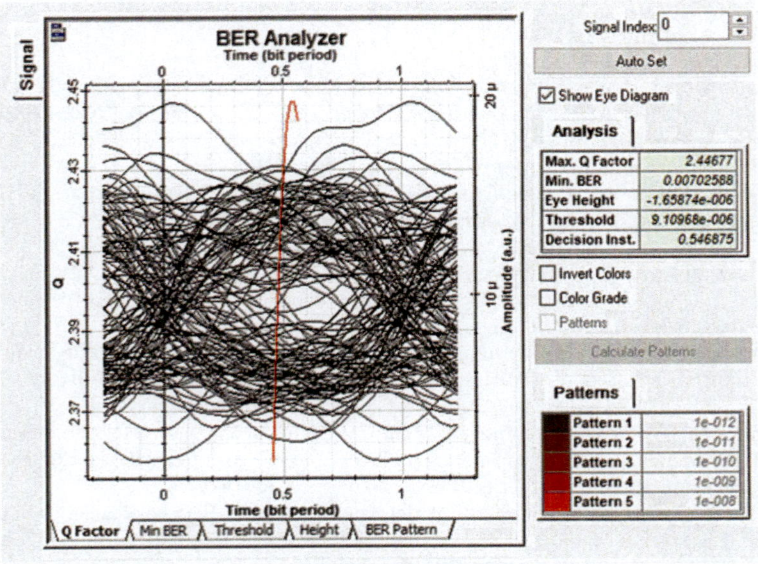

(a)

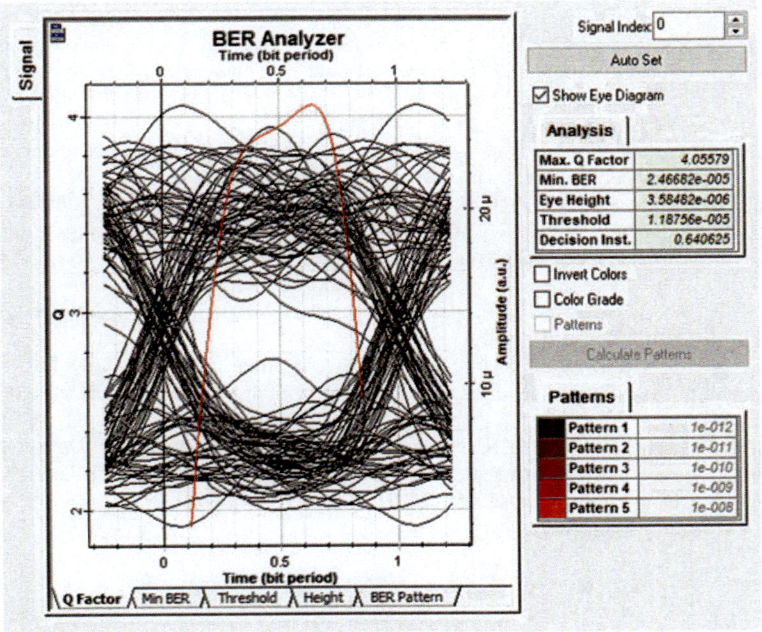

(b)

Fig. 9 **a** Eye diagram of 4 × 4 MIMO-FSO, **b** eye diagram of 8 × 8 MIMO-FSO for fog 15.56 dB/km

References

1. Khalighi, M.A., Uysal, M.: Survey on free space optical communication: a communication theory perspective. IEEE Commun. Surv. Tutorials **16**(4), 2231–2258, 4th Quart. (2014)
2. Sarkar, S., Janyani, V., Singh, G., Ismail, T., Selmy, H.A.: Design of 64 QAM transceiver model and its performance analysis for FSO communication (2018)
3. Chatti, I., Baklouti, F., Chekir, F., Attia, R.: Comparative analysis of MIMO-based FSO and MIMO-based MGDM communications. Opt. Rev. (2019). https://doi.org/10.1007/s10043-019-00537-z
4. Esmail, M.A., Fathallah, H., Alouini, M.S.: Outdoor FSO communications under fog: attenuation modeling and performance evaluation. IEEE Photonics J. **8**(4), 1–22 (2016). https://doi.org/10.1109/jphot.2016.2592705
5. Kaur, P., Kar, S., Jain, V.K.: Performance analysis of free space optical links using multi-input multi-output and aperture averaging in presence of turbulence and various weather conditions. IET Commun. **9**(8), 1104–1109 (2015). https://doi.org/10.1049/iet-com.2014.0926
6. Esmail, M.A., Fathallah, H., Alouini, M.S.: Channel modeling and performance evaluation of FSO communication systems in fog. In: Proceedings of 23rd International Conference on Telecommunications (ICT), IEEE, Thessaloniki, Greece (2016)
7. Prabu, K., Kumar, D.S., Srinivas, T.: Performance analysis of FSO links under strong atmospheric turbulence conditions using various modulation schemes. Optik Int. J. Light Electron. Opt. **125**(19) (2014). https://doi.org/10.1016/j.ijleo.2014.07.028
8. Singh, H., Sappal, A.S.: Analytic and simulative comparison of turbulent FSO system with different modulation techniques. Opt. Laser Technol. **114**, 49–59 (2019). https://doi.org/10.1016/j.optlastec.2019.01.013
9. Mahajan, S., Prakash, D., Singh, H.: Performance analysis of free space optical system under different weather conditions. In: International Conference on Signal Processing & Integrated Networks, SPIN 2019. IEEE Xplore Digital Library. https://doi.org/10.1109/SPIN.2019.8711687
10. Kaushal, H., Kaddoum, G., Jain, V.K., Kar, S.: Experimental investigation of optimum beam size for FSO uplink. Opt. Commun. **400**, 106–114 (2017)
11. Mahajan, S., Prakash, D., Singh, H.: Analysis of free space optical system under 4-channel spectrum slicing wavelength division multiplexing (SS-WDM). In: Proceedings of the Third International Conference on Advanced Informatics for Computing Research, p. 20. ACM (2019)
12. Son, I.K., Mao, S.: A survey of free space optical networks. Digital Commun. Networks. https://doi.org/10.1016/j.dcan.2016.11.00
13. Mahajan, S., Parkash, D., Singh, H.: Design and investigation of multiple TX/RX FSO systems under different weather conditions. In: Proceedings of ICRIC 2019, 2020, pp. 377–388. Springer, Cham (2019)
14. Chechi, D., Singh, S., Sharma, S.: Analysis and design of WDM optical OFDM system with coherent detection using different channel spacing. In: Proceedings of ICRIC 2019, 2020. Springer, Cham (2019). https://doi.org/10.1007/978-3-030-29407-6_27

A Low-Power Hara Inductor-Based Differential Ring Voltage-Controlled Oscillator

Misbah Manzoor Kiloo, Vikram Singh, and Mrinalini Gupta

Abstract The most important component needed for all wireless and communication systems is the voltage-controlled oscillator (VCO). In this paper, a four-stage low-power differential ring voltage-controlled oscillator (DRVCO) is presented. The proposed DRVCO is designed using new differential delay cell with dual delay path and Hara inductor to obtain a high frequency VCO with low-power consumption. Results have been obtained at supply voltage of 1.8 V using 0.18 μm TSMC complementary metal oxide semiconductor (CMOS) process. The tuning range for the proposed VCO varies from 4.6 to 5.5 GHz. This low-power VCO has a power consumption of about 5–10 mW over a control voltage variation of 0.1–1.0 V. The proposed VCO circuit at an offset frequency of 1 MHz achieves a phase noise of −67.9966 dBc/Hz. The figure of merit of proposed circuit is −135 dBc/Hz.

Keywords Active inductor · Differential ring voltage-controlled oscillator (DRVCO) · Low power · Voltage-controlled oscillator (VCO) · Wireless communication

1 Introduction

The integrated circuits using CMOS technology have enacted an active role in the rapid expansion of the wireless communication field and other diverse applications. Fast speed, cheaper analog signal processing, low production cost, etc. are the notable features that popularized CMOS technology [1–7]. The multi-purpose radio frequency block of modern communication network is the VCO. It is a special type of oscillating circuit which varies output signal over a dynamic range which is driven by input voltage and presents a linear relation of output frequency to the input voltage [4]. VCO finds application in clock recovery circuits [8], disk drives [9], frequency synthesizer [4], clock synchronization [10], phase-locked loop [11].

M. M. Kiloo (✉) · V. Singh · M. Gupta
Department of Electronics and Communication Engineering, Shri Mata Vaishno Devi University, Katra, Jammu & Kashmir, India
e-mail: 18mmc011@smvdu.ac.in

P. K. Singh et al. (eds.), *Recent Innovations in Computing*, Lecture Notes in Electrical Engineering 701, https://doi.org/10.1007/978-981-15-8297-4_8

The two primary ways of modeling VCO are: resistor–capacitor (RC) network and inductor–capacitor (LC) network [12]. Ring oscillators, a type of RC oscillators, have become a popular building block in many digital and communication systems due to their integrated design, wide tuning range, ease of manufacturing process [13, 14].

Ring VCOs output voltage should be configured to have quick rising and falling edges so that the transition period during which circuit noise contributes the most is minimized. The reduced delay can be achieved by employing active inductor. The active inductor network shows inductive characteristic within a narrow frequency range under certain dc biasing conditions and constraints of signal swing [15]. Compared to their passive equivalents, the key advantages of active devices are reduced layout area, adjustable self-resonant frequency, increased gain, increased inductance, enhanced bandwidth and quality factor with the ability to adjust all parameters and full compliance with digitally operated CMOS technologies [16]. The ignited interest in the design of VCO using active inductor over last few years is due to enhanced achievable tuning range and reduced power consumption [17, 18]. This paper illustrates a VCO design consisting of following sections. Section 2 illustrates VCO using active inductor and presents the proposed VCO design along with corresponding results. This section finally presents the comparisons with existing VCOs and concludes the design.

2 VCO Using Active Inductor

Hara inductor is a gyrator-C active inductor [16, 19]. The Hara inductor and its equivalent small-signal model are depicted in Figs. 1 and 2, respectively.

Due to the feedback used in Hara inductor, increasing input current results in an increase in voltage at the input node. The voltage at the gate is maintained at V_{dd} and V_{gs} is reduced, which tends to lower the current flow from the active inductor. The small-signal model gives an expression for calculation of input impedance as:

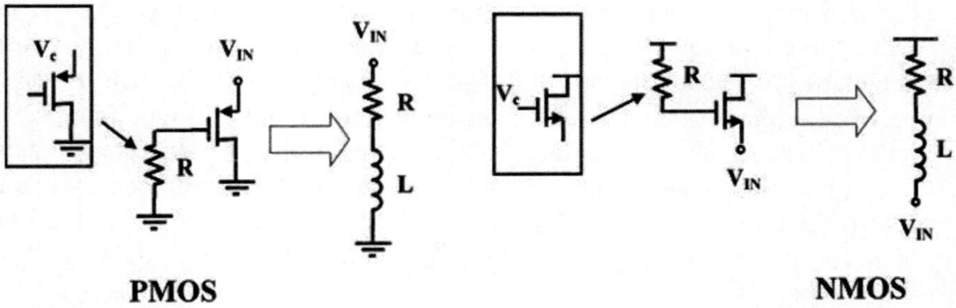

Fig. 1 Hara inductor

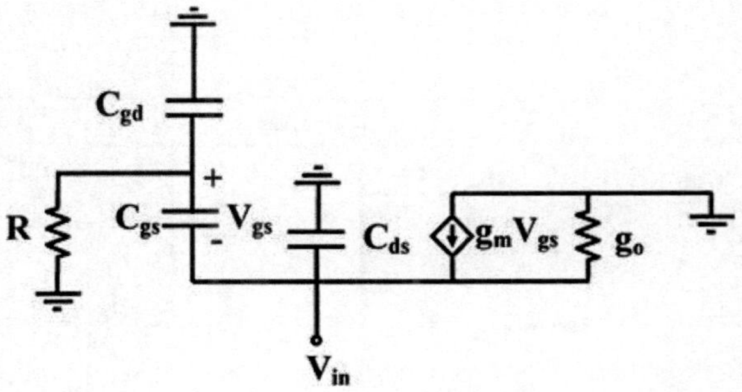

Fig. 2 Small-signal model of Hara inductor

$$Z \approx \left(\frac{1}{RC_{\rm gs}C_{\rm gd}}\right)\frac{sRC_{\rm gd}+1}{s^2+s\frac{g_m}{C_{\rm gs}}+\frac{g_m}{RC_{\rm gs}C_{\rm gd}}} \tag{1}$$

Considering $C_{\rm gs} \gg C_{\rm gd}$ and $g_m \gg g_o$ further simplifies the expression of impedance. The self-resonant frequency can be described as:

$$\omega_o = \sqrt{\omega_t \omega_z} \tag{2}$$

where $\omega_t = g_m/C_{\rm gs}$ and $\omega_Z = 1/RC_{\rm gd}$ The value of resistance is expressed in Eq. (3)

$$R = (g_m + \omega^2 C_{\rm gs}^2 R)/(g_m^2 + \omega^2 C_{\rm gs}^2) \tag{3}$$

The inductance of Hara inductor is

$$L = C_{\rm gs}(g_m R - 1)/(g_m^2 + \omega^2 C_{\rm gs}^2) \tag{4}$$

Proposed VCO Design This paper implements a four-stage differential ring voltage-controlled oscillator by employing Hara inductor. The schematic of the proposed VCO delay cell is shown in Fig. 3.

The control voltage ($V_{\rm c}$) is applied across a pair of PMOS transistors (M_9 and M_{10}) and drain of M_7 and M_8. The variation of $V_{\rm C}$ enables us to obtain a tuning range of frequency. The expression for calculating oscillation frequency of ring oscillator $f_{\rm osc} = (2Nt_{\rm d})^{-1}$ depicts that the frequency is limited by number of delay cells (N) and propagation delay ($t_{\rm d}$) [12]. The drawback of conventional VCO can be overcome by employing negative skewed delay technique, which in this case is provided by ($V_{\rm IN2-}$ and $V_{\rm IN2+}$) to M_{11} and M_{12} [18]. This ensures that PMOS is turned on preemptively and thereby improves the performance. The PMOS transistors M_7 and M_8 act as inversion-MOS (IMOS) varactor capacitances at the differential output terminals and are controlled using the control voltage ($V_{\rm c}$). This series configuration tends to decrease the overall capacitance and thereby tends to increase tuning range. The transistors M_9 and M_{10} are acting as load. The active inductor biasing is provided by Hara inductor. The transistors M_{11}, M_{12} act as a series resistance of the inductors formed by M_5 and M_6.

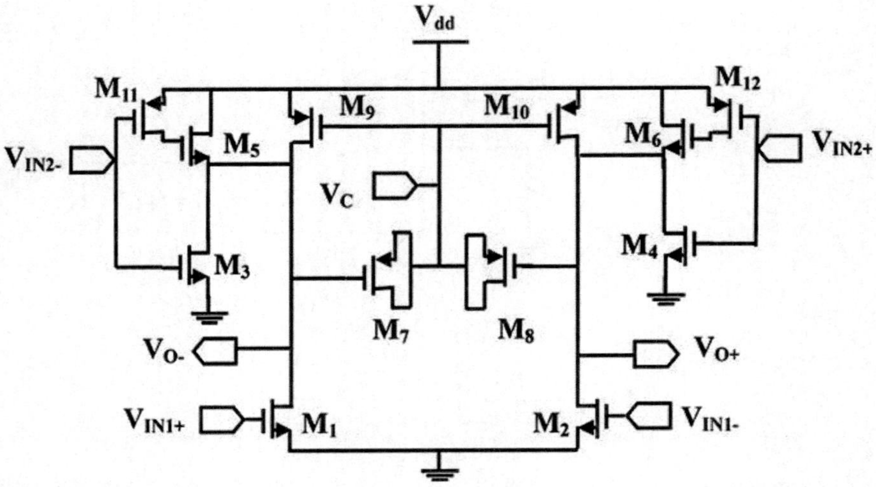

Fig. 3 Proposed delay cell

The four-stage schematic of designed VCO is shown in Fig. 4. The dotted lines in Fig. 4 represent the skewed signal path. Table 1 gives the aspect ratio of the transistors used in the proposed VCO delay cell.

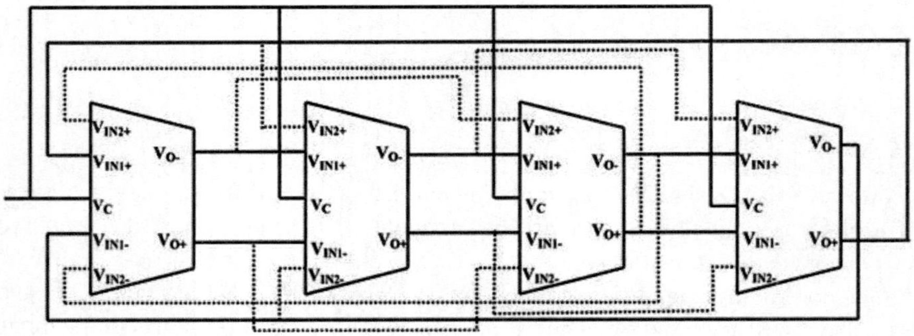

Fig. 4 Four-stage skewed delay VCO

Table 1 Size of transistors

Transistors	W (μm)/L (μm)
M_1, M_2	5/0.18
M_3, M_4	1.5/0.18
M_5, M_6	1/0.18
M_7, M_8	2/0.18
M_9, M_{10}	4/0.18
M_{11}, M_{12}	1/0.18

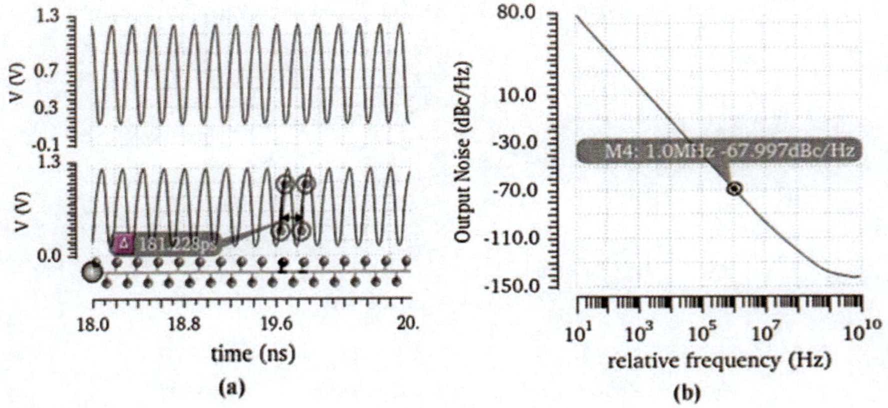

Fig. 5 **a** Transient analysis of proposed circuit. **b** Phase noise of proposed circuit

Results and Discussion The proposed DRVCO is designed using TSMC 0.18 μm CMOS technology. The transient analysis and phase noise of proposed DRVCO are depicted by Fig. 5a, b, respectively. The figure of merit (FOM) can be calculated from the expression [20]:

$$\mathrm{FOM} = L(f_0, \Delta f) - 20\log\left(\frac{f_0}{\Delta f}\right) + 10\log\left(\frac{P_{\mathrm{Supply}}}{[\mathrm{mW}]}\right) \tag{5}$$

where the phase noise is represented by $L(f_0, \Delta f)$ in dBc/Hz, f_0 is the carrier frequency and P_{Supply} is the power dissipation in mW. The simulation is carried out at $V_{\mathrm{dd}} = 1.8$ V and over control voltage (V_c) varied from 0.1 to 1.0 V.

The power dissipation over control voltage ranging from 0.1 to 0.8 V at width of 1 μm, 2 μm and 3 μm for NMOS active inductors M_{11} and M_{12} is depicted in Fig. 6a. The result presents a maximum power dissipation of 7.91 mW at width M_5 and M_6 = 3 μm for control voltage 0.1 V. The minimum power dissipation of 4.78 mW is obtained at width M_5 and M_6 = 1 μm for control voltage of 0.8 V. The result depicts an increase in power dissipation with increase in width of NMOS transistors M_5 and M_6. The increased power dissipation tends to decrease as the control voltage is increased for individual width. This effect can be understood by expression (4). The increase in width of NMOS transistors M_5 and M_6 decreases the inductance value and this eventually leads to decreased power dissipation with increased control voltage.

Figure 6b presents frequency variation over control voltage ranging from 0.1 to 0.8 V at width of 1 μm, 2 μm and 3 μm for NMOS active inductors M_5 and M_6. The result depicts a decrease in tuning range with the increase in width of NMOS transistors M_5 and M_6. The maximum tuning range of 4.6–5.5 GHz is obtained at a width of 1 μm for NMOS transistors M_5 and M_6. The decreased tuning range can be explained as: The increase in width tends to increase the transconductance

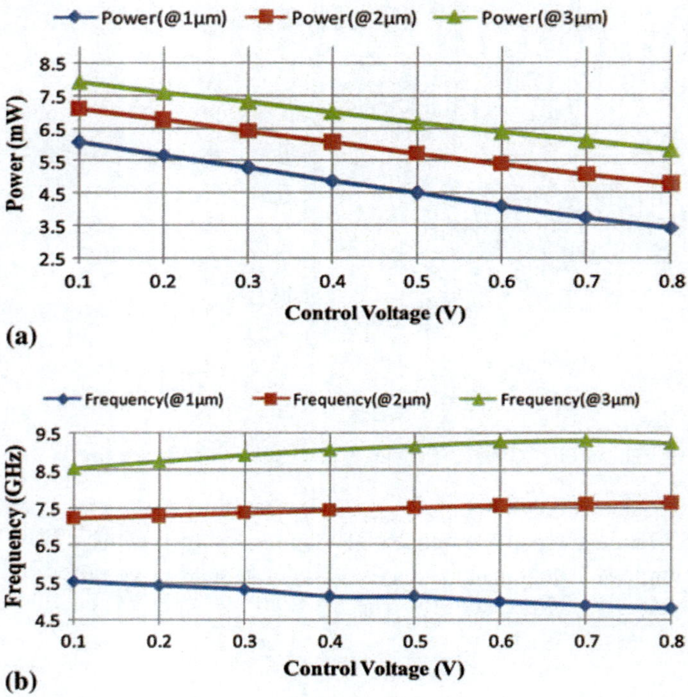

Fig. 6 Effect of variation in width of $M_5 = M_6$ at 1, 2 and 3 μm: **a** power versus control voltage, **b** output frequency versus control voltage

which tends to increase the inductance which is evident from Eq. (4). This increased inductance will tend to decrease the frequency tuning range.

The power dissipation over control voltage ranging from 0.1 to 0.8 V at width of 1 μm, 2 μm and 3 μm for PMOS active inductors M_5 and M_6 is depicted in Fig. 7a. The result presents constant power dissipation at different widths of M_5 and M_6 which can be understood from small-signal model of active inductor. The result presents a maximum power dissipation of 6.06 mW for control voltage 0.1 V. The minimum power dissipation of 3.40 mW is obtained at control voltage of 0.8 V.

Figure 7b presents the frequency variation over control voltage ranging from 0.1 to 0.8 V at width of 1 μm, 2 μm and 3 μm for PMOS active inductors M_{11} and M_{12}. The result depicts a decrease in tuning range with the increasing width. The maximum tuning range of 4.6–5.5 GHz is obtained at a width of 1 μm for PMOS transistors M_{11} and M_{12}. The decreased tuning range can be explained as: The increase in width tends to increase the transconductance which tends to increase the inductance which is evident from Eq. (4). This increased inductance will tend to decrease the frequency tuning range.

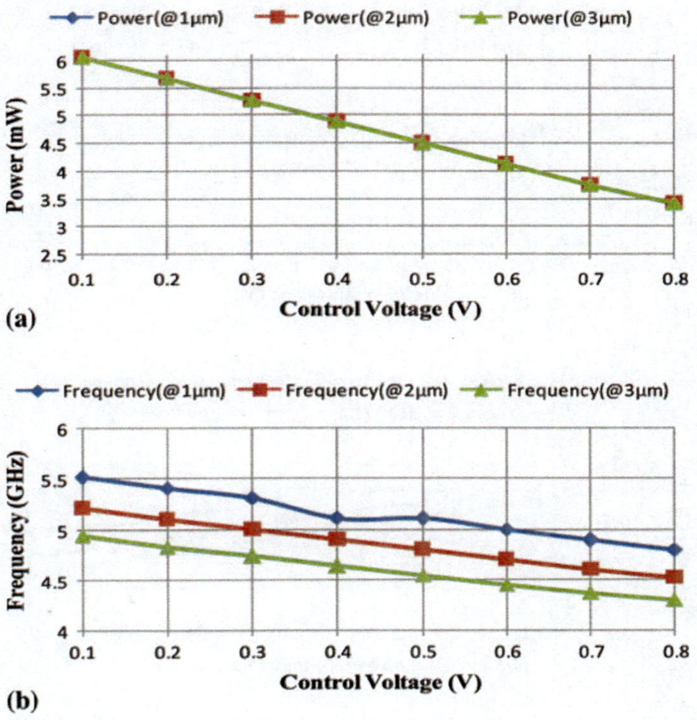

Fig. 7 Effect of variation in width of $M_{11} = M_{12}$ at 1, 2 and 3 µm: **a** power versus control voltage, **b** output frequency versus control voltage

The effect of increase in temperature (−40 to 50 °C) for control voltage range of 0.1–0.8 V over the power dissipation is shown in Fig. 8a. The result depicts constant power dissipation over different temperatures. The result presents a maximum power dissipation of 6.06 mW at 0.1 V and minimum power dissipation of 3.40 mW at 0.8 V.

The frequency variation over range of temperature (−40 to 50 °C) with control voltage spread from 0.1 to 0.8 V is presented in Fig. 8b. The maximum value of frequency is 5.78 GHz for −40 °C at a control voltage of 0.1 V. The minimum value is 4.80 GHz for 50 °C at control voltage of 0.8 V. The transient response across different temperatures is presented in Fig. 9.

A description of the performance of the active inductor-based DRVCO presented in this paper is compared to the recently reported VCOs and the comparison is given in Table 2. The proposed VCO design presents a least power dissipation in comparison to [21–24]. It presents a considerable phase noise and FOM. It presents a wider tuning range in comparison to [25, 26].

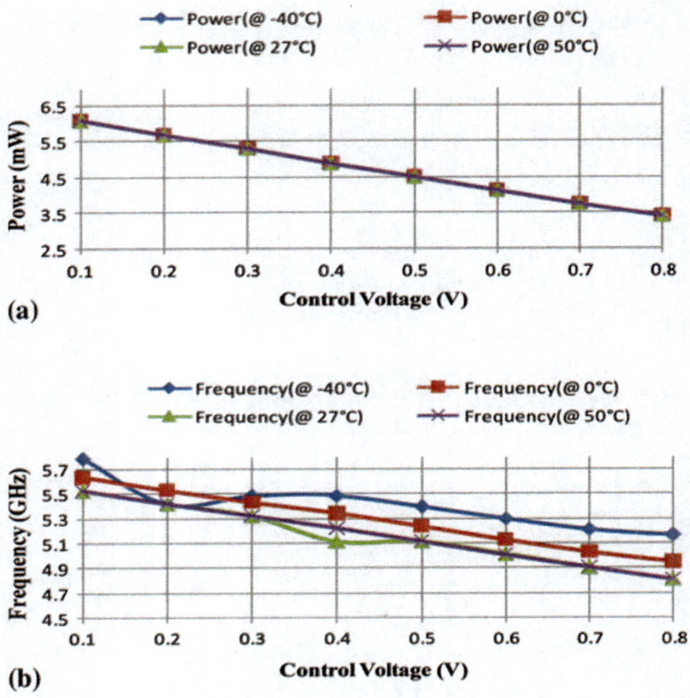

Fig. 8 Effect of variation in temperature at −40, 0, 27 and 50 °C: **a** power versus control voltage, **b** output frequency versus control voltage

3 Conclusion

This paper presents a Hara inductor-based dual delay cell for four-stage DRVCO. A low-power VCO design with considerable phase noise and FOM is presented. It has two input voltages. The skewed voltage ensures early turn on of PMOS transistor to enhance the performance. Active inductor has been employed to achieve a reduced chip area. The proposed VCO is simulated in 180-nm TSMC CMOS technology in Cadence Specter RF software under 1.8-V supply voltage. This VCO has a frequency range of 4.6–5.5 GHz, with phase noise of −67.99 dBc/Hz at 1 MHz offset frequency and FOM of −135 dBc/Hz. The circuit achieves a low-power dissipation of 2.83–6.06 mW. The proposed VCO finds application in areas with requirements of high frequency, low power and low area [27].

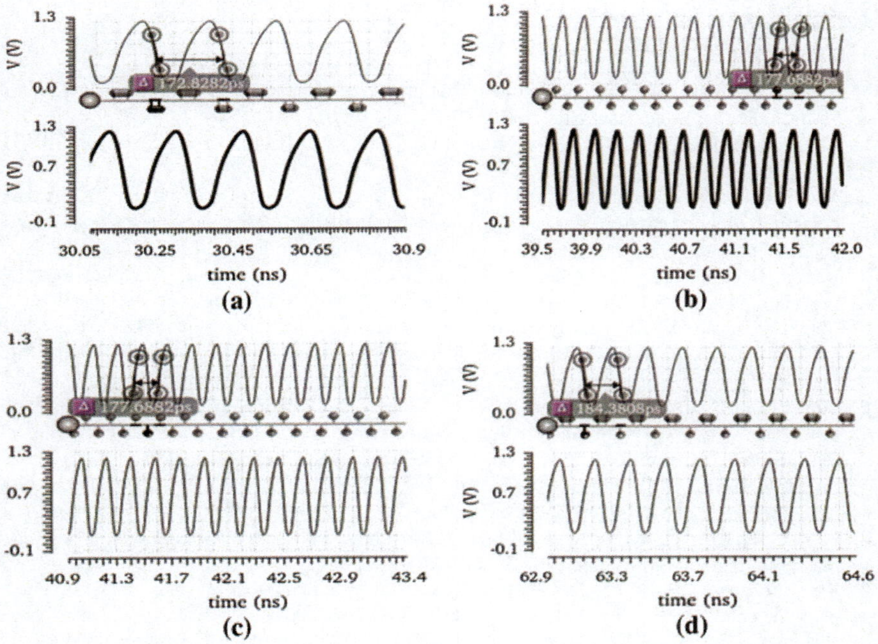

Fig. 9 Transient analysis **a** at −40 °C **b** at 0 °C **c** at 27 °C **d** at 50 °C

Table 2 Comparison of various active inductor-based DRVCO

References	Tech. (μm)	Topology	Supply voltage (mV)	No. of stages	Tuning range (GHz)	Phase noise (dBc/Hz)	Power diss. (mW)	FOM (dBc/Hz)
[21]	0.18	SC3A	2.0	5	4.3–6.1	−85	80	−146.9
[22]	0.18	CO	1.8	2	2.5–9	−82	50.9	−145.9
[23]	0.18	HAI	1.8	4	4.9–5.9	−86.7	8.1	−149.7
[24]	0.18	DI	1.5	3	2.7–4.1	−70	9.5	−135.6
[25]	0.45	FAI	1	–	1.1–1.8	−98.37	1.1	−176.69
[26]	0.35	XG and NV	1.8–3.3	3	0.34–0.62	–	0.16–1.12	–
This work	0.18	HAI	1.8	4	4.6–5.5	−67.99	6.06	−135

SC3A—source capacitively coupled current amplifier; DI—differential inverter; CO—coupled oscillators; HAI—Hara active inductor; FAI—floating active inductor; XG and NV—XOR gates and NMOS varactor

References

1. Tiebout, M.: Low power low phase noise differentially tuned quadrature VCO design in standard CMOS. IEEE J. Solid-State Circ. **36**(7), 1018–1024 (2001)
2. Lee, T.H., Hajimiri, A.: Oscillator phase noise: a tutorial. IEEE J. Solid-State Circ. **35**(3), 326–336 (2000)
3. McCorquodale, M.S., Gupta, V.: A history of the development of CMOS oscillators: the dark horse in frequency control. In: Joint Conference of the IEEE International Frequency Control and the European Frequency and Time Forum (FCS) Proceedings, pp. 437–442 (2011)
4. Razavi, B.: Challenges in the design of frequency synthesizers for wireless applications. In: Proceedings of IEEE CICC 97—Custom Integrated Circuits Conference, pp. 1–8 (1997)
5. Singh, V., Arya, S.K., Kumar, M.: Gm-boosted current-reuse inductive-peaking common source LNA for 3.1–10.6 GHz UWB wireless applications in 32 nm CMOS. Analog Integr. Circ. Sig. Process. **97**(2), 351–363 (2018)
6. Singh, P.K., Bhargava, B.K., Paprzycki, M., Kaushal, N.C., Hong, W.C.: Handbook of wireless sensor networks: issues and challenges in current scenario's. In: Advances in Intelligent Systems and Computing, vol. 1132, pp. 155–437. Springer, Cham, Switzerland (2020)
7. Singh, V., Arya, S.K., Kumar, M.: A 5.7 mw, UWB LNA for wireless applications using noise canceling technique in 90 nm CMOS. Frequenz **74**(1–2), 83–93 (2020)
8. Park, C.H., Kim, O., Kim, B.: A 1.8-GHz self-calibrated phase locked loop with precise I/Q matching. IEEE J. Solid-State Circ. **36**(5), 777–783 (2001)
9. Savoj, J., Razavi, B.: A 10-Gb/s CMOS clock and data recovery circuit with a half-rate linear phase detector. IEEE J. Solid-State Circ. **36**(5), 761–767 (2001)
10. Weigandt, T.C., Kim, B.K., Gray, P.: Analysis of timing jitter in CMOS ring oscillators. In: Proceedings of IEEE International Symposium on Circuits and Systems—ISCAS, pp. 27–30 (1994)
11. Gardner, F.M.: Phaselock Techniques, 2nd edn. Wiley, New York (1979)
12. Vaucher, C.S., Leenaerts, D., Tang J.V.D.: Circuit Design for RF Transceivers USA, pp. 185–238. Kluwer Academic Publishers (2003)
13. Abidi, A.A.: Phase noise and jitter in CMOS ring oscillators. IEEE J. Solid-State Circ. **41**(8), (2006)
14. Mandal, M.K., Sarkar, B.C.: Ring oscillators: characteristics and applications. Indian J. Pure Appl. Phys. **48**, 136–145 (2010)
15. Fillaud, M., Barthelemy, H.: Design of a wide tuning range VCO using an active inductor. In: Joint 6th International IEEE Northeast Workshop on Circuits and Systems and TAISA Conference, pp. 13–16 (2008)
16. Yuan, F.: CMOS Active Inductors and Transformers: Principle, Implementation, and Applications. Springer, Toronto, Ontario, Canada (2008)
17. Laskar, J., Mukhopadhyay, R., Lee, C.H.: Active inductor-based oscillator: a promising candidate for low-cost low-power multi-standard signal generation. In: Proceedings of IEEE Radio Wireless Symposium, pp. 31–34 (2007)
18. Lu, L.H., Hsieh, H.H., Liao, Y.T.: A wide tuning-range CMOS VCO with a differential tunable active inductor. IEEE Trans. Microw. Theory Tech. **54**(9), 3462–3468 (2006)
19. Hara, S., Tokumitsu, T., Tanaka, T., Aikawa, M.: Broadband monolithic microwave active inductor and its application to miniaturized wideband amplifiers. IEEE Trans. Microw. Theory Appl. **36**(12), 1920–1924 (1988)
20. Lingala, S., Pokharel, R.K., Tomar, A., Kanaya, H., Yoshida, K.: A wide tuning range –163 FOM CMOS quadrature ring oscillator for inductorless reconfigurable PLL. In: International Symposium on Signals, Systems and Electronics (ISSSE2010), pp. 1–4 (2010)
21. Tao, R., Berroth, M.: The design of 5 GHz voltage controlled ring oscillator using source capacitively coupled current amplifier. In: IEEE MTTS International Microwave Symposium Digest A, pp. 109–A112 (2003)
22. Rezayee, A., Martin, K.: A coupled two-stage ring oscillator. In: IEEE Midwest Symposium on Circuits and Systems, pp. 878–881 (2011)

23. Zhang, C., Li, Z., Fang, J., Zhao, J., Guo, Y., Chen, J.: A novel high-speed CMOS fully-differential ring VCO. In: IEEE International Conference on Solid-State and Integrated Circuit Technology (ICSICT) (2014)
24. Wenhua, Z., Hvolgaard, J.M., Larsen, T.: A 0.18 μm CMOS low power ring VCO with 1 GHz tuning range for 3–5 GHz FM-UWB applications. In: IEEE 10th International Conference Communication Systems, pp. 1–5 (2006)
25. Mehra, R., Kumar, V., Islam, A.: Floating active inductor based Class-C VCO with 8 digitally tuned sub-bands. AEU Int. J. Electron. Commun. **83**, 1–10 (2018)
26. Kumar, M.: Voltage-controlled oscillator design using MOS varactor. J. Inst. Eng. (India) Ser. B (2019)
27. Park, C.H., Kim, B.: A low-noise 900-MHz VCO in 0.6-μm CMOS. IEEE J. Solid-State Circ. **34**(5), 586–591 (1999)

Design and Simulation of Optimum Digital Filter for Removal of Power Line Interference Noise

Shruti Jain, Manasvi Kashyap, Mohit Garg, and Shailu Srivastava

Abstract The optimal digital filter design emerged as one of the key research issues from the past few years in biomedical engineering field. Digital filters are designed to preprocess the raw biomedical signal and extract useful information from them. Generally, all the biomedical signals are corrupted from the power line interference (PLI) noise. In this research work, an optimal and stable notch filter is designed for the elimination of PLI noise. The entire simulations were carried out using LabVIEW, and the results are validated using MATLAB. After analyzing theoretically and experimentally, it is inferred that window techniques perform better as compared to the existing filter designing techniques. The Kaiser window function gives better performance as compared to other traditional window techniques due to its sharp lobe that helps in reduction of PLI noise. Additionally, to mark the effectiveness of the selected window function, various parameters like effect of order, threshold frequencies, and side lobe attenuation are analyzed and compared.

Keywords Digital filters · FIR filters · Window technique · LabVIEW software

1 Introduction

Signal processing in electrical and electronics engineering is based on modifying and analyzing different signals such as images, sounds, and biological measurements. To an electronics engineer, it can be restricted to sampling, digitization, filtering, and

S. Jain · M. Kashyap · M. Garg · S. Srivastava (✉)
Jaypee University of Information Technology, Solan, India
e-mail: shailusrivastava5678@gmail.com

S. Jain
e-mail: jain.shruti15@gmail.com

M. Kashyap
e-mail: manasvi912@gmail.com

M. Garg
e-mail: mohit201097@gmail.com

P. K. Singh et al. (eds.), *Recent Innovations in Computing*, Lecture Notes in Electrical Engineering 701, https://doi.org/10.1007/978-981-15-8297-4_9

other spectral analysis [1, 2]. In filters, the finite impulse response (FIR) filters are considered to be the popular that can be used in the biomedical signal for removing noise and spectral analysis [3, 4]. The FIR filter is used to select the desired frequency spectrum that can permit adaptable filtering with better rejection property without any change in the hardware structure [5]. An FIR filter is a filter that is used to create frequency response digitally. The filter can be designed using various methods, but many of them rely on filter approximation. The FIR filter transfer function approaches to the ideal when the order of the filter increases, but that will lead to an increase in the complexity and total time needed to process the given input signal.

In this paper, the authors have studied different research papers on FIR filter design. Park and Meher [6] suggested the efficient FPGA and ASIC realizations of DA-based reconfigurable FIR digital filter. Ray et al. [7] suggested an efficient shift-add implementation of FIR filters using variable partition hybrid form structures. Pun et al. [8] suggested the design and implementation of FIR and IIR digital filters with variable frequency characteristics. Barnela et al. [9] implemented FIR filter using MATLAB and evaluated various parameters. The authors in [10] implemented FIR filter using FPGA. In [11], the author proposed a novel neural network-based approach for designing FIR filters. The authors of [12] implemented an improved back-propagation neural network algorithm on classification problems.

As the society progresses, the infrastructure along with technology enhancement is much important. Nowadays, a technology does not even last few weeks and a replacement of it in the market is there. Everyone wants things to be done at fast rate along with best results. This paper mainly focuses on the filtering of power line interference (PLI) noise from any biomedical signal using window functions. The research work highlights on the design of notch filter using various window techniques for an upgraded framework of a FIR filter using LabVIEW software. The results are validated using MATLAB software. The main requirement for the selection of window function is as follows: The main lobe width should be as narrow as possible, attenuation (spectral leakage) of the lobes present inside must be minimum and how fast does the side lobes die out. Later, in this paper, the effect of order, threshold frequencies, and side lobe attenuation is also analyzed on the output.

The rest of this paper is structured as: In Sect. 2, methodology is discussed; in Sect. 3, the obtained results using LabVIEW and MATLAB are discussed followed by a conclusion and future scope in Sect. 4.

2 Methodology

Various types of filters are categorized on the basis of signal processing, elements, construction filters, impulse response, and frequency range. Figure 1 shows the block diagram of different filters. Depending upon different *techniques used in signal processing, filters* are classified as analog and digital. The filter that works on continuous time signals is known as analog filter. On the other hand, the filter which is capable of doing different mathematical operations on a discrete time signal is

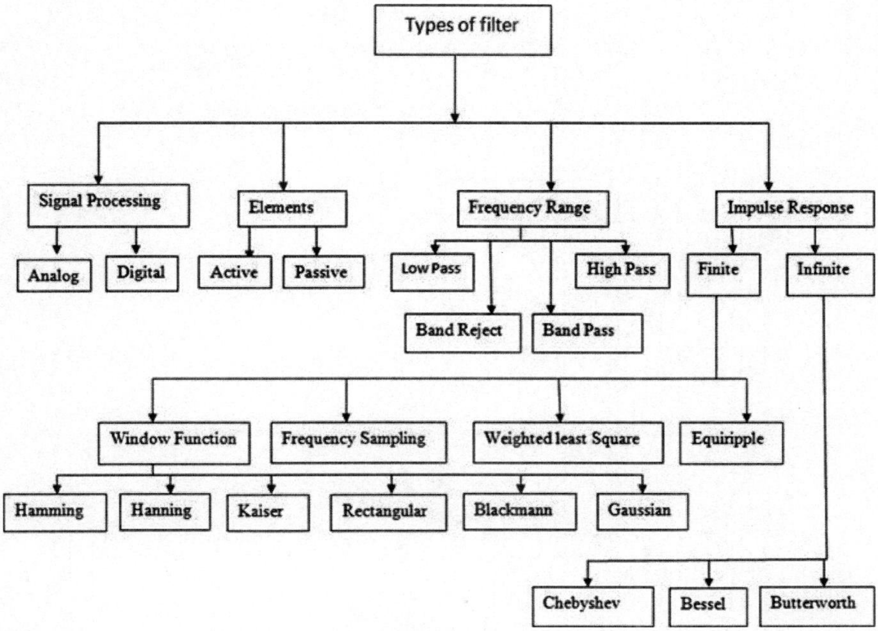

Fig. 1 Block diagram of different types of filters

called digital filters. Depending on the type of *elements*, further filters are classified into active and passive filters. An active filter comprises with transistors and operational amplifiers, whereas passive filters utilizes passive elements like resistors and capacitors.

According to different *operating frequency ranges*, filters are categorized as low-pass filter (LPF), high-pass filter (HPF), band-pass (BP), band reject (BR) and all-pass filter (APF). An LPF is that type of filter which allows passing signals with a frequency less than cutoff frequency (f_c) and truncating signals with frequencies more than that of f_c [13, 14]. A HPF is a filter type that passes signals with a range of frequencies higher than that of f_c and not allows the signals to pass with frequencies lower than f_c. A BPF is a filter that allows certain frequencies within a specified range and rejects frequencies other than the selected range. An APF is a type of filter that passes all frequency ranges equally. According to the *Impulse Responses*, we have two different types of filters that are IIR and FIR [15, 16].

The computational time and memory consumption are more in FIR as compared to IIR. Lower order of filter can be used for IIR filter design which needs less computation to achieve the same results. However, an FIR has linear phase property and is quite stable, which makes it advantageous over IIR filtering. The linear phase means that the filter has no phase shift across the band of frequencies.

FIR filter design steps: Various filter design steps are illustrated in Fig. 2 which includes specification of filter, selection of window, calculation of filter coefficient,

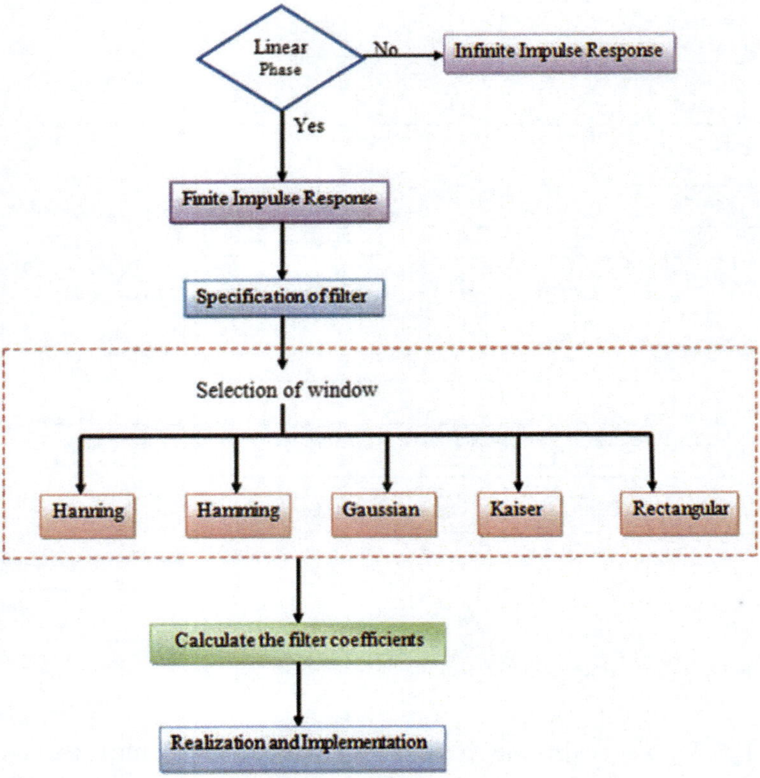

Fig. 2 Filter design steps

and implementation. To design a FIR digital filters using windowing techniques, it is important to select the filter order depending upon filter specifications (attenuation factors and selectivity). The design should be such that the filter should have higher suppression of unwanted spectrum (higher stopband attenuation) and narrow suppression in the transition region. The filter coefficients depend on the input conditions (such as type of filter). Later, Fourier analysis is performed to get the phase response and the magnitude of the filter. In this study, LabVIEW software is used to implement notch FIR filter with the help of different window functions such as rectangular, Hamming, and Hanning. [14].

3 Results and Discussion

FIR filter design can be done in different ways, namely window method, frequency sampling, weighted least squares, and equi-ripple filters in the time and frequency domain. The impulse response of an ideal digital filter can be calculated using the

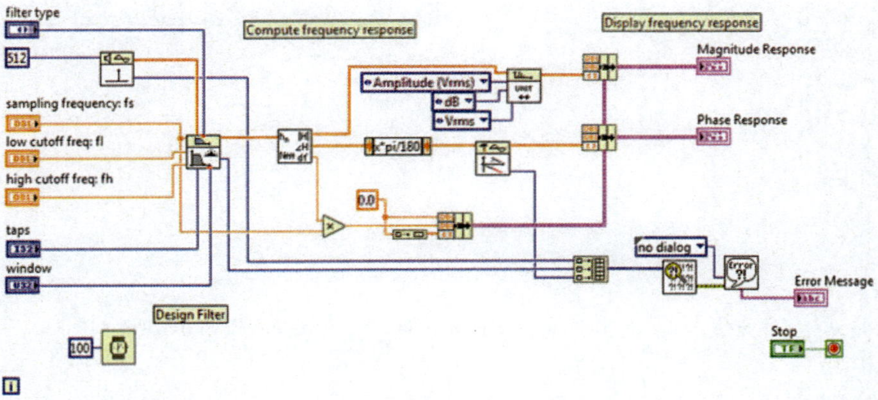

Fig. 3 Implementation of notch filter using LabVIEW

windowing method. The main reason to choose a windowing technique for analyzing any filter is the simplicity in the design. The window functions are widely used because of its small stopband attenuation for the same number of taps. The characteristic of the FIR filter depends on the chosen window. There are number of windows to achieve the required stopband attenuation. The transition region from passband to stopband increases as the attenuation increases. So, to shape the response of the filter, the windows are used.

In this research article, FIR notch filter is designed to remove the 50 Hz/60 Hz PLI noise from any biomedical signal. The filter is designed on LabVIEW considering the order of 100, lower cutoff frequency (f_{c1}) as 49 Hz, upper cutoff frequency (f_{c2}) as 51 Hz, and sampling frequency (f_S) as 1000 Hz. The evaluation environment consists of a PC with an Intel Core i5 processor and 16 GB RAM. Initially, the results were simulated in LabVIEW; then, the results are validated using MATLAB 2018b software. Figure 3 shows the schematic diagram of the notch filter design in the LabVIEW software.

To design notch filter, different windows are used, namely Hamming, Kaiser, Hanning, Gaussian, Blackmann, and rectangular windows.

A. *Hamming window* is developed by Richard Hamming, who was a member of the Manhattan Project. Equation (1) shows the Hamming window function:

$$w(n) = 0.54 - 0.46\cos\left\{\frac{2\pi n}{n-1}\right\}, \quad 0 \le n \le N-1 \tag{1}$$

where n is the value between 0 and $N - 1$ and N is the order of the filter. This filter has a narrow transition zone, smaller ripples than Hanning window. Figure 4 shows the LabVIEW and MATLAB results for notch filter using Hamming window.

B. *Hanning window* has the narrowest transition band, but a large ripple in stopband. The window function is represented by Eq. (2). Figure 5 shows the LabVIEW

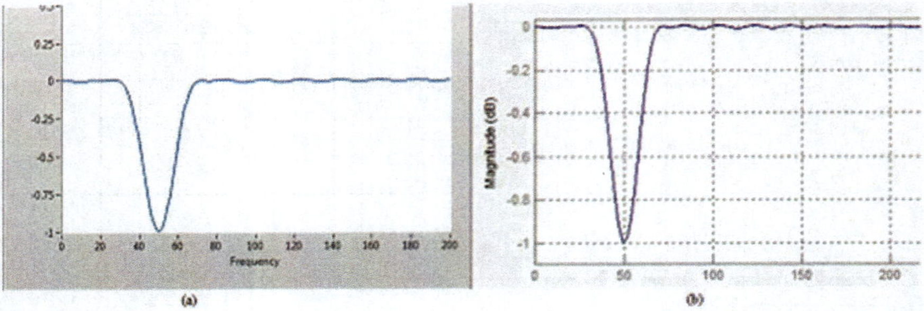

Fig. 4 Notch filter design at 50 Hz using a Hamming window **a** LabVIEW, **b** MATLAB

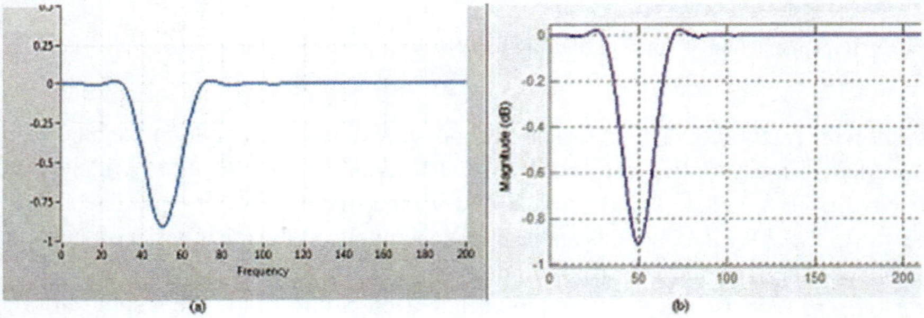

Fig. 5 Notch filter design at 50 Hz using a Hanning window **a** LabVIEW, **b** MATLAB

and MATLAB results.

$$w_{Ha}(n) = 0.5\,(1 - \cos(2\pi nN)), \quad 0 \leq n \leq N \tag{2}$$

where n is the value between 0 and $N - 1$ and N is the order of the filter.

C. The *Gaussian window* is a nonzero function having a bell-liked shape in closed form that transforms itself. For a given temporal width, it produces the smallest RMS frequency. Equation (3) represents the function for Gaussian window, and the simulation results are shown in Fig. 6

$$w_G(n) = e^{-\frac{1}{2}\left(\frac{n-N/2}{\sigma N/2}\right)^2} \quad 0 \leq n \leq N,\ \sigma \leq 0.5 \tag{3}$$

where σ is the standard deviation.

D. *Kaiser window* is developed by James Kaiser at Bell Laboratories. Equation (4) illustrates the Kaiser window function. In the stopband region, this window has small-amplitude ripple and has a wide transition width. Figure 7 shows the simulation results of Kaiser window.

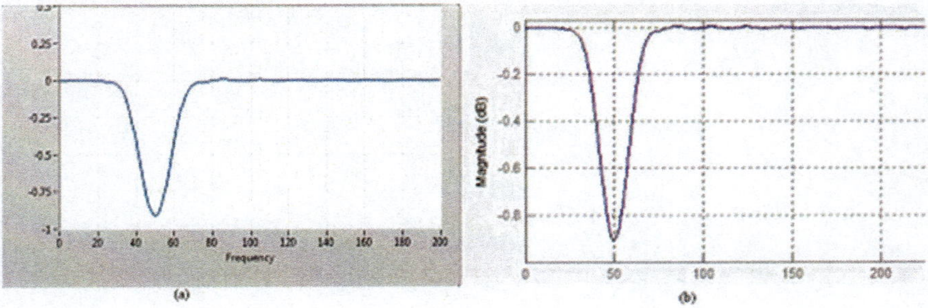

Fig. 6 Notch filter design using Gaussian window **a** LabVIEW, **b** MATLAB

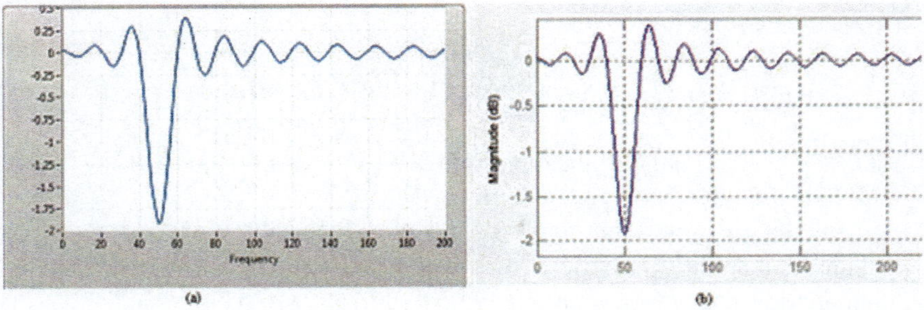

Fig. 7 Notch filter design using Kaiser window ($\beta = 0.5$) **a** LabVIEW, **b** MATLAB

$$w_k(n) = \begin{cases} \dfrac{I_o\left(\beta\sqrt{1-\left(\frac{n}{M/2}\right)^2}\right)}{I_o(\beta)} & -\frac{M-1}{2} \le n \le \frac{M-1}{2} \\ 0 & \text{elsewhere} \end{cases} \tag{4}$$

where M is the order, β is the side lobe attenuation.

E. *Blackmann window* design has a minimal leakage and is close to optimal. Figure 8 shows the LabVIEW and MATLAB results. The window function is

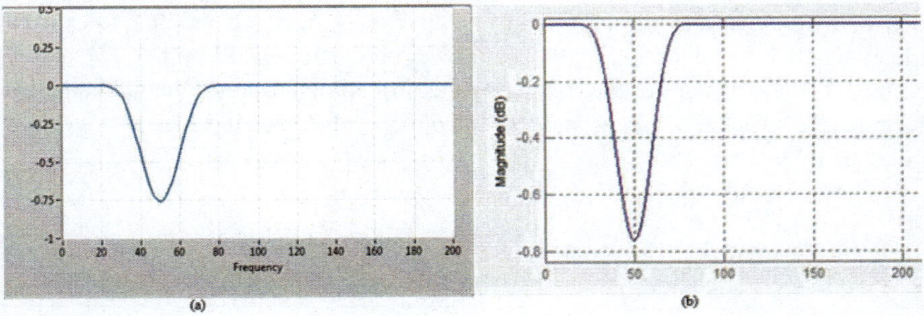

Fig. 8 Notch filter design at 50 Hz using Blackmann window **a** LabVIEW, **b** MATLAB

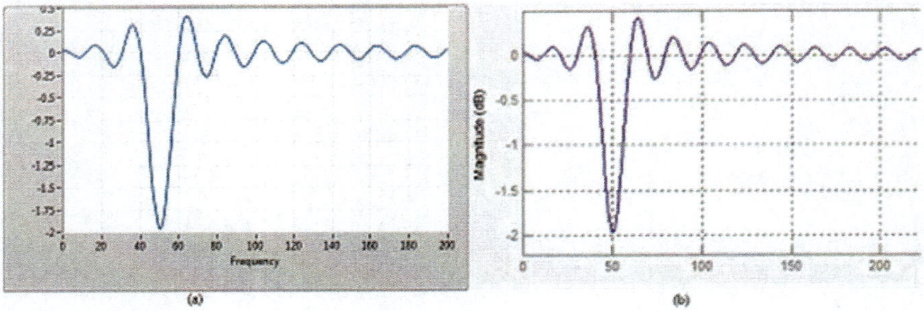

Fig. 9 Notch filter design at 50 Hz using rectangular window **a** LabVIEW, **b** MATLAB

expressed by Eq. (5)

$$w_B(n) = 0.42 - 0.5\cos(2\pi n/N - 1) + 0.08\ \cos(4\pi n/N - 1) \tag{5}$$

for $0 \leq n \leq M$, where N is the order, M is $N/2$ (N = even) and $(N + 1)/2$ (N = odd), n is the value between 0 and M.

F. *Rectangular window* is the simplest window and the entire interval has a value of unity. Equation (6) represents the function for rectangular window, and the simulation results are shown in Fig. 9

$$w_R(n) = \begin{Bmatrix} 1 - \frac{M-1}{2} \leq n \leq \frac{M-1}{2} \\ 0 \text{ Otherwise} \end{Bmatrix} \tag{6}$$

The Hanning, Hamming, and Blackmann give a smoother trimming of the impulse response and do not have a sharp lobe which can be used for the removal of PLI. The Blackmann window has a wider lobe but comparatively is has lesser ripple content. It is observed that for side lobe attenuation (β) 0.5, Kaiser window has the sharp notch on the expense of higher content of ripples which are present in it. Other filters have no ripple content, but the notch width is much wider as compared to Kaiser window. After studying and simulating different window functions, Kaiser window results better because of its sharp lobe that helps in reduction of PLI noise which occurs in biomedical signals. From the results (Figs. 4, 5, 6, 7, 8 and 9), it is seen that the gain of notch is not crossing the half power point (the −3 dB point). The quality factor (Q) is calculated for the designed filter. Equation (7) shows the mathematical formulation of quality factor, and the calculated values for the Kaiser window is given in Table 1.

Table 1 Calculation of notch frequency and Q value for the Kaiser window

f_{c1} (Hz)	f_{c2} (Hz)	f_0 (Hz)	Q
49	51	50	0.04
59	61	60	0.033

$$Q = \frac{f_{c2} - f_{c1}}{f_0} \tag{7}$$

where notch frequency $f_0 = \frac{f_{c1}+f_{c2}}{2}$.

Different simulations were performed to analyze the effect of the order of the filter and threshold values on the designed Kaiser window. Also, the effect of β is analyzed so as to reduce the ripple content.

Effect of order on the gain of Kaiser window: In this paper, the authors have studied the effect of order on the designed Kaiser window filter, and it is found that as the order of the filter increases, the notch is sharper and the gain value will be more as given in Table 2.

From our results, it is interpreted that for the higher order, gain becomes better.

Effect of threshold frequencies on gain of Kaiser Window: By varying the threshold frequencies of the Kaiser window, it is observed that as the frequency increases, the notch becomes sharper which results in better gain and reduces the ripples. Results for variable threshold frequencies are given in Table 3.

From our results, it is interpreted that for f_{c1} = 45 Hz and f_{c2} = 55 Hz, the gain and the notch of the filter show remarkable result as shown in Fig. 10. To further reduce the ripple content, the authors have also seen the effect of β on the gain.

Effect of β on the gain of Kaiser window: In this research article, the authors have also seen the effect of β on the designed Kaiser window filter. It is observed that the ripples are dominating factor as the value of β increases the ripples decreases but the notch width increases. The authors also studied the effect of β on the gain of Kaiser window output by keeping the threshold frequency constant as f_{c1} = 45 Hz and f_{c2} = 55 Hz. Table 4 tabulates the effect of β on the gain.

Table 2 Effect of order on gain of Kaiser window for f_{c1} = 49 Hz and f_{c2} = 51 Hz

Order	Gain(dB)
50	−0.899
100	−1.92
150	−2.98
200	−4.25
300	−7.19

Table 3 Effect of threshold frequencies on gain of Kaiser window for $\beta = 0.5$

f_{c1} (Hz)	f_{c2} (Hz)	Gain (dB)
45	55	−17.65
46	54	−11.47
47	53	−7.40
48	52	−5.77

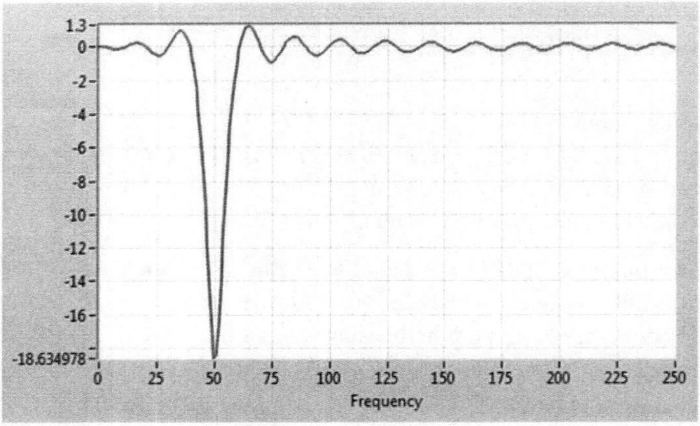

Fig. 10 Notch filter for $f_{c1} = 45$ Hz and $f_{c2} = 55$ Hz

Table 4 Effect of on gain the output of Kaiser window for $f_{c1} = 45$ Hz and $f_{c2} = 55$ Hz

β	Gain (dB)
0.5	−17.63
1	−15.24
2	−10.98
3	−8.53
4	−7.10
5	−6.19
6	−5.54

It is interpreted that the best results for Kaiser window are shown when $\beta = 0.5$. If the β value increases to 6, then no ripples were observed. From all the observations, it is observed that for better gain and notch, only one parameter β value or threshold value can be changed.

4 Conclusion and Future Scope

In the processing of signals, the main task of a filter is to discard unnecessary signal components. Another task is to obtain much important data from the input signal, such as the information which is present in a specified frequency range. In this paper, an optimum filter is designed to remove the power line interference noise of any biomedical signal. This research paper carried out the analysis of various FIR filter types by simulating in LabVIEW and validates the result using MATLAB. After simulations, it is observed that Kaiser window results in better performance because of its sharper notch and β factor which plays major role in reducing the ripples than any other window method. To reduce the effect of ripples which are present in the passband of notch filter, β factor is analyzed. It is observed that on increasing the order of the filter for the same parameter, the notch is sharper and results in better gain. The proposed system is reliable and sustainable for the future use. In the future, the authors will try to have a hardware implementation of the different digital filter design using field-programmable gate array (FPGA).

References

1. Proakis, J.G.: Digital Signal Processing Principles, Algorithms, and Applications, 3rd edn. Northeastern University Dimitris G. Manolakis Boston College
2. Jackson, L.B.: Digital Filters and Signal Processing: With MATLAB Exercises. Springer Science & Business Media (2013)
3. Prashar, N., Dogra, J., Sood, M., Jain, S.: Removal of Electromyography Noise from ECG for High Performance Biomedical Systems. Network Biology. **8**(1), 12–24 (2018)
4. Kirti, S.H., Jain, S.: FPGA implementation of Power-Efficient ECG pre-processing block. Int. J. Recent Technol. Eng. 8(1):2899–2904
5. Schaumann, R., Xiao, H., Mac, V.V.: Design of analog filters, 2nd edn. The Oxford Series in Electrical and Computer Engineering (2009)
6. Park, S.Y., Meher, P.K.: Efficient FPGA and ASIC realizations of DA-based reconfigurable FIR DIGITAL Filter. IEEE Trans. Circ. Syst.-II: Express Briefs, 1–5
7. Ray, D., George, N.V., Meher, P.K.: Efficient shift-add implementation of FIR filters using variable partition hybrid form structures. IEEE Trans. Circ. Syst.–I: Regular Papers **65**(12), 4247–4257 (2018)
8. Pun, C.K.S, Chan, S.C., Yeung, K.S., Ho, K.L.: On the design and implementation of FIR and IIR digital filters with variable frequency characteristics. IEEE Trans. Circ. Syst.—II: Analog Digit Signal Process. **49**(11), 689–703 (2002)
9. Barnela, M., Kumar, S., Kaushik, A.: Satvika, "Implementation and performance estimation of FIR digital filters using MATLAB Simulink. Int. J. Eng. Adv. Technol. **3**(5), 62–65 (2014)
10. Joshi, S., Ainapure, B.: FPGA based FIR filter. Int. J. Eng. Sci. Technol. **2**(12), 7320–7323 (2010)
11. Wang, X., Meng, X., He, Y.: A novel neural networks-based approach for designing FIR filters. In: Proceedings of the 6th World Congress on Intelligent Control and Automation, June 21–23, 2006, Dalian, China, pp. 4029–4032
12. Nawi, N.M., Ransing, R.S., Najib, M., Salleh, M., Ghazali, R., Hamid, N.A.: An Improved Back Propagation Neural Network Algorithm on Classification Problems" et al. (Eds.): DTA/BSBT 2010, CCIS 118, , pp. 177–188 (2010). © Springer-Verlag Berlin Heidelberg 2010
13. Prashar, N., Jain, S., Sood, M., Dogra, J.: Review of biomedical system for high performance applications. In: 4th IEEE International Conference on signal processing and control (ISPCC 2017), Jaypee University of Information technology, Waknaghat, Solan, H.P, India, pp. 300–304, 21–23 September 2017
14. Prashar, N., Sood, M., Jain, S.: Design and performance analysis of cascade digital filter for ECG signal processing. Int. J. Innov. Technol. Explor. Eng. (IJITEE) **8**(8), 2659–2665 (2019)
15. Aparna, R., Chithra, P.L.: An effective method for continuous speech segmentation using filters. Natl. Conf. Comput. Intell. Syst. **1**(1), 17–23 (2012)
16. Subhadeep, C.: Advantages of Blackman window over hamming window method for designing FIR filter. Int. J. Comput. Sci. Eng. Technol. **4**(8), 1181–1189 (2013)

Centralized Blood Bank Database and Management System

Osunlana Ismail, Sanjay Misra, Jonathan Oluranti, and Ravin Ahuja

Abstract A blood bank is a place where blood is collected and stored to be used by other individuals who need them either due to health emergencies or blood shortages. Blood banks are scattered all over places and not easily assessable to donors and patients who need them. So, it is important to have an organized database to help in allowing donors easily locate the nearest blood banks and donate blood, and also to make patients easily access blood when they need them within the shortest possible time. The aim of this research is to build a feasible system to help in the efficient management of blood bank activities and also provide easy platforms for patients to easily access blood during emergencies. This app would be built on the android platform connected with a secured online cloud-based database to keep the patients, donors and blood banks' details safe. This is an efficient management system for blood banks as their strenuous process is now being made easy using technology.

Keywords Blood · Blood banks · Hospitals · Donors

1 Introduction

Blood is the one of the most important elements in life, and it is often referred to as the "Essence of Life" [1]. Easy access to blood by patients is a major challenge in Nigeria. During emergencies, a patient searches through his family members first for

---

O. Ismail · S. Misra (✉) · J. Oluranti
Covenant University, Ota, Nigeria
e-mail: Sanjay.misra@covenantuniversity.edu.ng

O. Ismail
e-mail: ismailosunlana@yahoo.com

J. Oluranti
e-mail: jonathan.oluranti@covenantuniversity.edu.ng

R. Ahuja
Shri Vishwakarma Skill University, Gurgaon, India
e-mail: ravinahujadce@gmail.com

P. K. Singh et al. (eds.), *Recent Innovations in Computing*, Lecture Notes in Electrical Engineering 701, https://doi.org/10.1007/978-981-15-8297-4_10

matching blood type; if he is not able to find, he then starts contacting different blood banks. It is a strenuous and time-consuming process, and the patient may not get the blood within the shortest possible time. The task of a blood bank is to organize and manage blood received from blood donors and ensure that blood is properly kept and then distributed efficiently to patients who need them. Nigeria currently has about 1.7 million pints of blood as yearly deficits. Out of about 1.8 million pints of blood being required every year, only 66,000 pints of blood are being met which makes the deficit about 1.7 million pints [2]. This means a lot of patients are dying daily due to the unavailability of blood. The blood is majorly used for hemostatic resuscitation [3]. In developing countries especially Nigeria, it is very difficult getting access to blood from blood banks as most of their processes are manual. The blood banks do not have an organized database, they need to manually check all the blood they have on request, and this may be time consuming for patients who are on emergencies [4].

A lot of research papers have been written about centralized blood bank systems, with each of them suggesting different models. Most of the suggested models have been focused on other countries with none focused on Nigeria [5]. Also, the existing papers have only been proposing models that are not real time and not focused on security. Also in Nigeria, a major factor that needs to be considered is logistics, and we have poor transportation and logistics systems that could impede on blood delivery time and may be very dangerous for patients who require blood instantly such as cases of accident victims or bleeding of delivery women [6]. No existing model is robust enough to cover the challenges peculiar to the Nigerian environment.

Our new system is an android app with a cloud-based database to efficiently manage all collected data, especially client and donor data. Through this system, a blood donor can easily create an account as a donor and locate the nearest government-approved blood bank to go for screening and if successful, he would donate his blood [7]. Also, the patients can also find matching blood types on the app and locate the nearest blood banks to get blood [8]. The system is a real-time platform such that when new blood data are being uploaded, it gets updated immediately in the database and can be accessed by the donors, patients and blood banks. All blood banks on this system are government-approved which means that no illegal purchase of blood would be made. The system is also hosted on a secured online cloud system to ensure it is not being infiltrated [9].

There are four classifications of blood with each having both positive and negative variations. All these variations and other relevant blood data like sugar content, packed cell volume, disease conditions, antibodies are considered when matching donors to patients [10]. There need to be proper frameworks in place for the storage of all this information, and there also need to be proper structures in place to ensure these data can be easily searched and sorted especially by those who are in need of blood. This project would make blood available for patients in emergencies and would be of great benefits in saving lives.

The work is sectioned as follows. Section one contains the introduction and provides justification for the paper. Section 2 contains the literature review and motivation for the study. Section 3 covers the methodologies and materials used in the work. Section 4 talks about the design of the work. Section 5 shows the demonstration and results. Section 6 covers the comparison with other systems, while Section seven is the conclusion.

2 Literature Review

A lot of research papers have been written about blood bank management systems with each of them proposing a model that is entirely different from the Nigerian environment.

Ali et al. [11] developed a Web application and an android application that allows blood recipients to be able to search for donors around them. The system makes use of Google Maps to detect the nearest donor around the blood recipients. The system gives the recipients all the information about the donors including their phone numbers and email address. This is a system that cannot be applied in a developing country like Nigeria where Google Maps is not efficient due to irregularities in our housing planning [11]. Most places in Nigeria have not been mapped to Google Maps yet. Also, there are a lot of privacy and security issues with this model as private information is shared. Nigeria is known for its highest numbers of frauds; fraudsters can use such personal information in perpetuating their evil acts.

Makau et al. [12] proposed a Web-based model for the implementation of a blood bank repository system in Kenya. Ibrahim et al. [10] developed an android-based system that helps in coordinating the activities of blood banks and donation centers. Their android app allows blood banks and hospitals to be able to manage records of blood donors and also to send them requests should in case their blood type is being needed. This system is not as efficient as there is a delay in getting blood across the recipients. Ashita et al. [13] built an inventory management system for blood banks to be able to easily gather and manage all blood donor's information. Their system was majorly based on information submitted by the donors, and it is divided into three modules: the patient module, the donor module and blood bank module. Their system lacks a sustainable administration system. Muddu et al. [14] created a system to aggregate all blood donor's information in a database and also to consistently inform them of patient who needs matching blood via SMS. There is no administration module to completely monitor the entire system and its purely autopilot. Morrisa et al. [3] conducted a study to investigate the major causes of blood usage during emergencies in South Africa. They were able to come to a conclusion that trauma accounts for the highest reason for blood transfusion. The study was conducted on three secondary emergency centers in Cape Town, and the data were recorded over three months. A total of 210 emergency situations was recorded, and 329 blood units were used. Trauma makes up a total of 39% (81), surgical conditions account for 22% (47), upper gastrointestinal diseases account for 11% (24), while bleeding accounts for 8% (16). Medical conditions account for 15% (31). The majority of the emergency blood was used during emergency surgical operating theaters; this was about 77% ($n = 253$). Anitha et al. [15] designed and developed an embedded system that brings blood donors and blood recipients together. The system allows blood patients to easily source for blood all over the country. They designed a low-cost GSM-based Raspberry Pi that allows blood donors and recipients to be able to communicate via SMS. Their platform aggregates blood donors together and then allows blood recipients to access matching blood donors and communicate with them via SMS.

Selvamani et al. [16] proposed a novel approach for the management of online blood bank databases. Their system connects blood donors and recipients in real time during the urgent need of blood. They created an online database that aggregates details of blood banks and donors from a variety of sources including NGOs, medical centers, online mediums, etc. Their platform was created such that the communication barriers between the blood banks and donors were minimized. Their blood bank central database platform is being managed by appointed online administrators. Their system is also being embedded with an algorithm that searches the list of blood donors and recommends the most suitable one to the patients, thereby reducing time spent searching for donors. The variables considered for recommending suitable blood donors include nearness to patient's locations, blood compatibility, availability of donors. After the selection is being made, the patients can also make a call to the donor from the app. Clemen et al. [17] designed an information management system that assists blood banks in effectively managing and optimizing their operations. It is a central database system with a powerful search and sorting algorithm that enables you filter through blood bank database records easily. Muhammad et al. [18] created a blood donor recruitment platform that allows blood banks to easily recruit donor. In some other related works, the authors have presented electronic health record and e-healthcare access [19, 20].

Based on the above, the motivation for this project is due to the fact that there are no existing blood bank management systems in Nigeria, and this system if implemented would improve existing blood bank efficiency and optimization. This would also make blood banks more accessible especially during emergencies.

3 Methodologies

Our system would be a mobile app that allows hospitals and patients to be able to get access to blood in the shortest possible time. The front end which is being accessed by the hospitals, blood banks and patients is an android-based application. While the back end which allows for the storage of data and information is being implemented with Firebase Real-time Database. This system is designed to ensure seamless searching and sorting of blood information. It is also designed to maintain privacy and data security.

Firebase Database: Firebase is a real-time online database hosting platform created and maintained by Google. This database is targeted at mobile platforms and apps. Firebase is a back-end platform that allows data to be synced in real-time across multiple devices. Unlike MySQL database, Firebase is a NoSQL database that makes real-time syncing of data possible across all users, thereby making collaboration easier. Also, it has an offline cache capability which means users can still access information without Internet connection [21]. Firebase has a lot of modules or database, but we are making use of three of its modules as follows.

Firebase Authentication: The Firebase Authentication modules allow us to authenticate users who make use of our app. It allows the implementation of sign-up and login for our users. The users are being authenticated uniquely with their email address and password.

Firebase Database: The Firebase Database allows us to store data to be accessed in real time. We use it to store blood donors, hospitals and blood banks' details. It is used to store and sync all our app data.

Firebase Cloud Messaging: We make use of the Firebase Cloud Messaging module to message blood donors of their blood testing appointments, pass information to donors and blood banks, etc. [22].

PHP: PHP is one of the most popular server-side scripting languages mostly used for Web application development. It is designed as a general-purpose programming language and widely used all over the world. PHP powers most of the online Web sites, and it is currently active on about 240 million Web sites and 2.1 million Web servers.

While PHP is a server site programming language, it is commonly used with other programming and markup languages like HTML, JavaScript, XML, CSS, etc. In this app, we would use PHP to create a JSON API output to enable our mobile app to communicate with our online Firebase Database. Our android app needs an API gateway to be able to communicate with the online database to perform functions like adding data to the database, fetching data, updating data, etc. The JSON API is the most efficient API gateway to use, and PHP is a programming language that can be used to create such JSON API [23].

Android Java: Android is a mobile operating system (OS) used majorly for touch screen mobile devices. It is based on the Linux kernel and currently developed and maintained by Google Engineering Team. We would be designing the front end of this app on the android platform as mobile phones are the easily assessable platforms. This would allow for patient and hospitals to get access to blood banks and blood information in real time on their mobile devices [24, 25].

4 Design

The design of the proposed system and flowchart of the blood management application are given in Figs. 1 and 2, respectively.

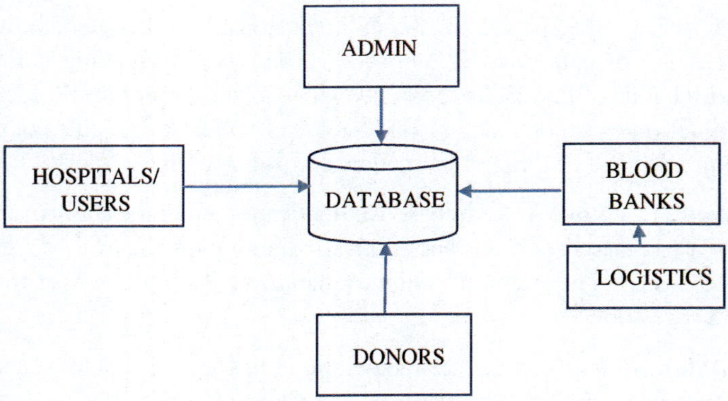

Fig. 1 Database design for the blood bank management app

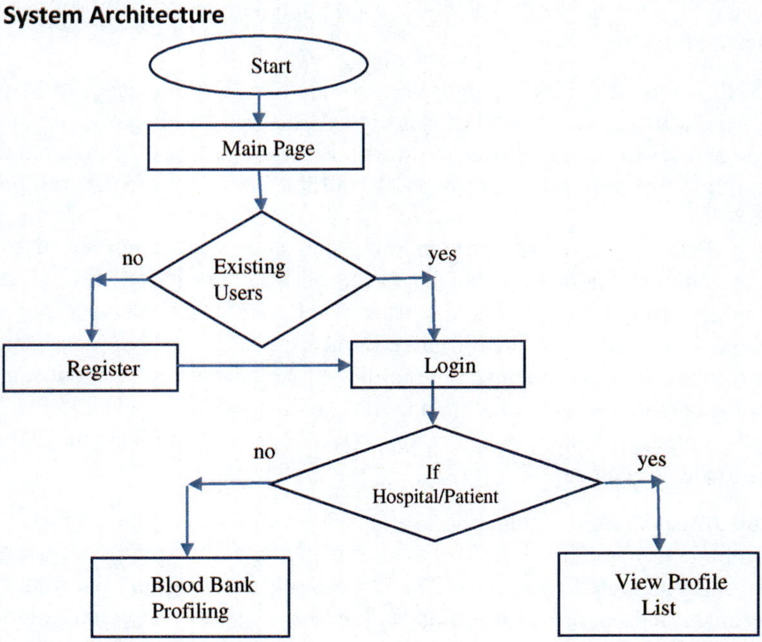

Fig. 2 Flowchart of the blood management application

4.1 System Architecture

Admin Module—The admin module is the centralized module that is in charge of all admin activities which includes: blood bank approval, blood donor approval, logistic company approval, data management and analysis, blood banks and user management.

Blood Donor Module—The blood donor's module is an important aspect and allows for—user's authentication (registration and login), data and information upload, and blood screening appointment booking.

Hospital/User Module—The hospitals or patients who need blood would be able to perform the following processes through their module—user authentication, data and information upload, search blood banks, search donors and request for blood.

Blood Bank Module—The major function of the blood banks is to be able to manage and organize all blood donors' data. The blood bank is also in charge of scheduling and conducting tests for blood donors to ensure they meet blood donation requirement and are fit to donate blood. The blood bank module would contain the following—blood bank authentication, schedule of blood test date for donors, blood data and information upload and blood donor information update.

Logistics Module—The logistics company is in charge of delivering blood in the shortest possible time. The logistics company would get requests from the blood banks, and then they pick the blood up and deliver to the hospitals or patients who need the blood. The logistics company would be able to perform the following functions through their module—authentication (registration and login), receive delivery requests from blood banks, manage deliveries in real time.

Table 1 shows the database table structure for the blood donors. This table contains all blood donor information. It also keeps tracks of donor login seasons, donation days and their personal details. Table 2 shows the database table structure for all registered blood banks. This table majorly contains the contact details of the blood banks and also keeps records of the blood banks' available bloods. Table 3 shows the database table structure for blood screening and donation appointment bookings. It

Table 1 Database design for blood donors

Name	Data type	Constraints
Email_id	varchar(100)	Primary key
Name	varchar(100)	
Phone number	varchar(100)	Allow null
Sex	varchar(100)	Allow null
Religion	varchar(100)	Allow null
Location	varchar(100)	Allow null
Picture	Image	Allow null
Blood group	varchar(100)	Allow null
Blood type	varchar(100)	Allow null
Username	varchar(100)	
Password	varchar(100)	
Blood_Donated	varchar(100)	Allow null
Date donated	varchar(100)	Allow null

Table 2 Database design for blood banks

Name	Data type	Constraints
Email_Id	varchar(100)	Primary key
Blood_bank_name	varchar(100)	
Approval_No	varchar(100)	
Approval_Details	varchar(100)	Allow null
Phone_Number	varchar(100)	Allow null
Location	varchar(100)	
State	varchar(100)	Allow null

Table 3 Database design for blood donation appointment

Name	Data type	Constraints
Email_id	Int	Primary key
Donor_name	varchar(100)	
Phone_Number	varchar(100)	
Location	varchar(100)	
Blood_Group	varchar(100)	
Amount_Donated	varchar(100)	
Appointment_Date	varchar(100)	
Date donated	varchar(100)	Allow null

keeps records of all appointment dates. This table is used by the blood banks to send reminders to blood donors.

The software packages and frameworks used in this system are Android Studio, SQL Studio, PhpStorm, Postman, Firebase.

5 Demonstration and Results

After testing, the blood bank application was able to achieve all its intended purposes. Figure 3 shows the authentication system that allows users to register. Blood donors, hospitals and patients can create account after which they can log in to access the dashboard.

Figure 4 shows the blood donor page where patients can search for matching donors around them. The patients can make searches selecting his/her nearest local government area (LGA). All listed donors are verified donors that have gone through the necessary medical checkups and are deemed fit by government-approved hospitals. Figure 5 is the Blood Request Page where patients can confirm matching donors and contact them. The patients or hospitals can either contact the donor through phone calls, SMS or email. After putting a call through to the donor, the patients can

Fig. 3 Authentication page for blood donor/patients

Fig. 4 Search page

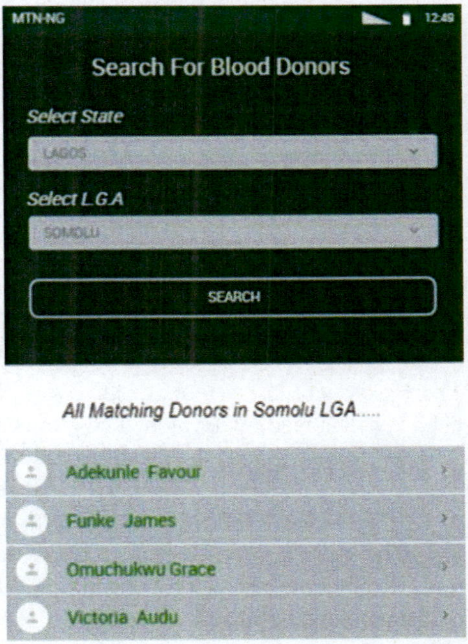

Fig. 5 Blood Request Page

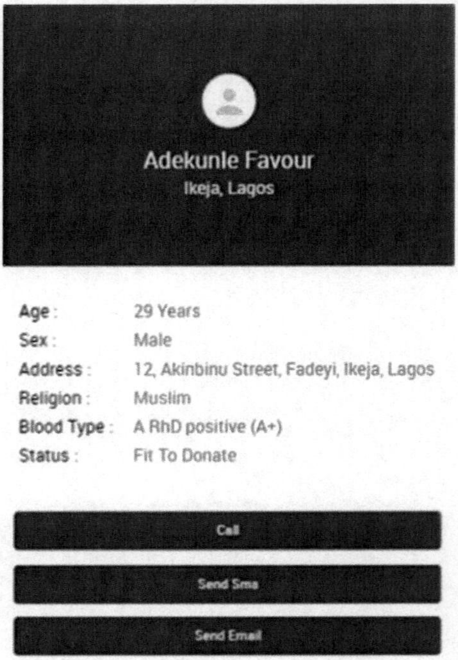

then make negotiations with the donor to come to the hospital for donation or locate the nearest blood banks for the donation.

Figure 6 shows the appointment booking page for interested donors to book appointments with the nearest blood banks so as to confirm their eligibility and also donate blood. When appointments are booked, the respective blood banks receive a notification via mail and can prepare for the donors' visit. An interested donor would need to first book a screening appointment date, and then subsequently he can start booking donation dates.

6 Discussion and Comparison

There exist similar systems to this work, but most of the proposed systems have been implemented by developed countries and none has been tailored to developing African nations. This system tends to put into considerations African local challenges like logistics challenges, poor network coverage, etc.

Ibrahim et al. [10] developed a similar system, and their project was an android-based app that allows patients to request the blood solely online with GPS functionality. This would not be efficient in the African environment due to poor Internet connection and unreliable GPS data.

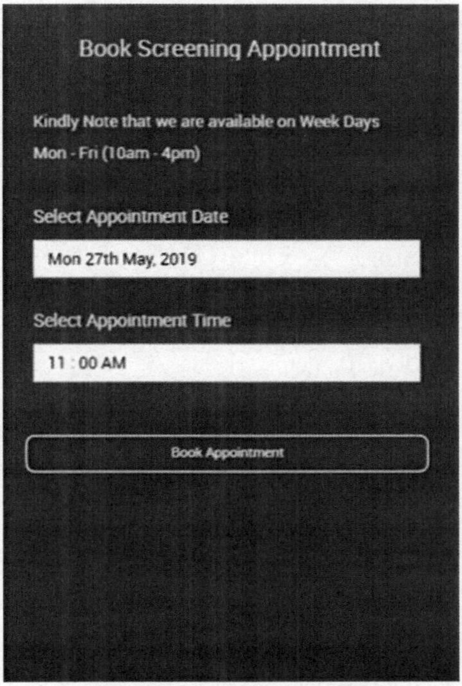

Fig. 6 Appointment Scheduling Page for interested donors

Muhammad et al. [18] developed a similar system, but their platform is only based on SMS functionality. Their app only sends SMS messages alone to donors, and this is not always a reliable means of communication in Africa. This project solves this challenge by providing diverse avenues for patients to contact donors. The avenues this project offers include SMS, phone calls and email.

7 Conclusion

A lot of research proposals have been made for the development of online database to help patients get easy access to blood. All the proposals have been focused on advanced countries which have different environmental terrains with developing countries like Nigeria. In this paper, we have discussed the major issues and different challenges peculiar to implementing online blood bank solutions in Nigeria, and we have proposed suitable measures to overcome those challenges. This system allows easy communication between blood banks, hospitals, patients and blood donors.

On a long-term basis, this project can easily be scaled to integrate a lot of functionalities that would make it better. Since this project was about the development of the android app, in the future, the iOS app can be developed to cater for iPhone users. Also, GPS tracking can be incorporated into the app to make location reporting possible. This would allow blood banks to see blood donor's location in real time and can easily make fast decisions based on the donor's location.

References

1. Sunita, B., Kajal, J., Snehal, K., Varsha, P.: Blood bank management system using android app. Int. Eng. Res. J. (IERJ), 4467–4471 (2019)
2. Christiana, N., Patience, I.: 1.7 million pints of blood as yearly deficit hits Nigerian hospitals. Available: https://leadership.ng/2018/12/23/1-7m-pints-of-blood-yearly-deficit-hits-nigerian-hospitals/ (2019)
3. David, D. Morrisa, H., Melanie, S., Stevan, B.: Utilisation of emergency blood in a cohort of South African emergency. Afr. J. Emerg. Med. (2019)
4. Abhijeet, G., Nilofar, M., Tejashri, W., Raviraj, I., Brijendra, G., Kama, R.: Smart blood finder. Int. J. Trend Sci. Res. Dev. **2**, 1027–1032 (2017)
5. Vikas, K., Sharad, M.: Blood bank management information system in India. Int. J. Eng. Res. Appl. (IJERA) **1**, 260–263 (2011)
6. Kazeem, Y.: QZ. [Online] This Lagos startup will save lives by making it easier to store and deliver blood for hospitals. https://qz.com/africa/708435/in-lagos-delivering-donated-blood-to-patients-is-tougher-than-finding-blood-donors/. Accessed 6 April 2019
7. Alagbe, J.: Punch Nigeria Newspaper. How a network of young Nigerians is solving blood shortage in hospitals. https://punchng.com/how-a-network-of-young-nigerians-is-solving-blood-shortage-in-hospitals/. Accessed 6 June 2019.
8. Amarjeet, S., Siddharth, S., Srivastava, P., Murthy, B.: A standard compliant Blood Bank Management System with enforcing mechanism. In: 2015 International Conference on Computing, Communication and Security (ICCCS), Pamplemousses, Mauritius (2015)
9. Abdulrahman, A., Fatma, E., Altaf, A.: Blood bank smart phone application for managing and organizing the blood donation. Int. J. New Comput. Archit. Appl. (IJNCAA), 86–91 (2016)
10. Ibrahim, F., Tukur, A., Mohamed, I.: CBBR centralized blood bank repository. Int. J. Inf. Syst. Eng. **3**, 85–97 (2015)
11. Akkas, K., Israt, A., Arifu, M.: Blood donation management system. Am. J. Eng. Res. (AJER) **4**, 123–136 (2015)
12. Makau, N., Fanon, A.: Blood bank management information system.. A Case Study of the Kenya National Blood Transfusion Services (2013)
13. Ashita, J., Amit, N., Nitish, S., Shubhada, M.: Online blood bank management system using android. Int. J. Innov. Stud. Sci. Eng. Technol. **2** (2012)
14. Muddu, G., Nagaraju, S.: Design and implementation of short message service (SMS) based blood bank. In: International Conference on Inventive Computation Technologies (ICICT), Coimbatore, India (2016)
15. Anitha, J., Bala, L., Senthil, M.: Design and implementation of automated blood bank using embedded systems. Research Gate (2015)
16. Selvamani, K., Ashok, K.: A novel technique for online blood bank management. Procedia Comput. Sci. **48**, 568–573 (2015)
17. Clemen, T., Teena, S., Sankar, K.: A study on blood bank management. Middle-East J. Sci. Res., 1123–1126 (2014)
18. Muhammad, S., Khondoker, A., Shakil, H., Anjon, B., Syed, I.: Smart Blood Query: A Novel Mobile Phone Based Privacy-aware Blood Donor Recruitment and Management System for Developing Regions. Department of Computer Science & Engineering, Bangladesh University of Engineering & Technology (BUET) (2015)
19. Ayeni, F., Misra, S.: Overcoming barriers of effective health care delivery and electronic health records in Nigeria using socialized medicine. In: 2014 11th International Conference on Electronics, Computer and Computation (ICECCO), pp. 1–4. IEEE (2014)

20. Ayeni, F., Omogbadegun, Z., Omoregbe, N., Misra, S., Garg, L.: Overcoming barriers to healthcare access and delivery. EAI Endorsed Trans Pervasive Health Technol **s**(15) (2018)
21. Chunnu, K., Pritam, S.: Application of firebase in android app development—a study. Int. J. Comput. Appl. **179** (2018)
22. Nilanjan, C., Souvik, C., Asoke, N., Decosta, A.: Real-time communication application based on android using google firebase. Int. J. Adv. Res. Comput. Sci. Manag. Stud. (2018)
23. Punam, K., Rainu, N.: A Research Paper On Website Development OptimizationUsing Xampp/PHP, Vol. 8. International Journal of Advanced Research in Computer Science (2017).
24. Ma, L., Gu, L., Wang, J.: Research and development of mobile application for android platform. Int. J. Multimedia Ubiquitous Eng., 187–198 (2014)
25. Şenay, K.: Developing of android mobile application using Java and Eclipse: an application. Int. J. Electron. Mech. Mechatron. Eng., 1335–1354 (2017)

Effect of Supercapacitor on Power Supply for Rechargeable Implanted Medical Devices

Attah Amarachi Rita, Sanjay Misra, Ravin Ahuja, and Jonathan Oluranti

Abstract The need for medical devices to be planted into living organisms to perform the function of a dysfunctional body part is increasing by the day. Most of these devices require power supply of some sort to function appropriately. The supply can be taken care of by batteries but the batteries have a life span which will never be long enough, especially if the implant is in a human. This will mean that every time the battery dies the device will have to be brought out and the batteries changed. This paper seeks to explore the existing energy storage capacities for a wireless setup. The addition of a supercapacitor to the battery or replacement in the power pack was simulated and analyzed. Then, a proffered solution which is introducing a microcontroller to determine the switching between battery and super capacitor was proposed. Also some level of communication and control of the implant by the external circuit through the capacitor.

Keywords Implanted medical devices · Supercapacitor · Batteries

1 Introduction

Devices like implanted medical devices can be very crucial for the survival of its host. At the same time can cause more serious complications if not handled appropriately. Many times as living organisms go about their various activities they are exposed to

A. A. Rita · S. Misra (✉) · J. Oluranti
Covenant University, Ota, Ogun State, Nigeria
e-mail: Sanjay.misra@covenantuniversity.edu.ng

A. A. Rita
e-mail: Amarachi.attah.rita@gmail.com

J. Oluranti
e-mail: jonathan.oluranti@covenantuniversity.edu.ng

R. Ahuja
ShriVishwakarma Skill University, Gurgaon, India
e-mail: ravinahujadce@gmail.com

P. K. Singh et al. (eds.), *Recent Innovations in Computing*, Lecture Notes in Electrical Engineering 701, https://doi.org/10.1007/978-981-15-8297-4_11

things that could be detrimental to their health and existence. Some of these harmful circumstances may not necessarily eliminate the organism but could weaken part of the organism and cause that part to stop functioning. Due to medical advancements, the dysfunctional part can be assisted or replaced by a device we call implanted medical device (IMD) [1]. The use of these devices has gone from being an option to the only reasonable option. Hence, the use has become more popular over the last years.

Devices like this exist in various forms for different functions. The devices are sometimes placed inside the organism or attached to a part of the organism. Some of these devices do not need power or an internal drive to perform its function while others do. Cochlear implant is a typical example of an IMD that needs power supply, and these devices are referred to as active devices; others examples are pacemakers and insulin pumps [1]. While vascular grafts and retinal implants are passive and do not require energy of any sort. The devices which require power supply usually require power supply of between milli and micro watts. This power is readily made available using batteries.

The life span of the IMDs and the batteries is never the same. Usually, the IMDs are more durable than the batteries in them. The average life span for the batteries in implanted medical devices in general is between 5 and 10 years [2]. Such batteries will need to be changed to maintain the function of the IMDs. Most of the active devices are usually placed inside the living organisms. This means the devices will need to be brought out for the batteries to be changed. This will require surgery and expenses all over again every time the battery is to be changed [3]. The employed solution over the last years to this is providing the IMD with its power supply while it remains inside the organism [4]. This can be done by either energy harvesting from body motions or wirelessly transmitting power through the skin to the device in the body [5, 6]. This will mean the employment of rechargeable batteries [7]. These batteries must be able to charge fast and discharge very slowly to avoid dependence on charging unit. The charging has to be both efficient and effective as any losses will cause a direct harm to the organism. In this paper, pacemakers are used as case study, the modern day pacemaker is programmable, and depending on its configuration the energy consumptions are different [8].

In the next section, we will analyze various existing wireless systems and their various energy storage capacities. Then, we will explore powering the IMD with a lithium polymer battery, a supercapacitor, and then both with the help of a microcontroller. The various voltage levels and energy densities were explored together with time taken for charging to be complete. This paper is structured as follows. Second section provides a brief introduction of background and the literature survey. Section 3 presents the proposed system while in Sect. 4 we compare the performance of a supercapcitor and battery using simulink. In Sect. 5, we discuss the outcome of the simulation.

2 Background and Literature Survey

2.1 Wireless Medium

To provide the internal charging of the IMD, many propositions have been made. The most popular is the use of inductive charging system. This is where a primary coil outside the body induces voltage in a secondary coil inside a recharging circuit inside the body [9]. Just like a transformer this work employs Faraday's law of electromagnetic induction [10]. A primary coil induces voltage in a secondary coil. The primary coil will be connected to a power circuit which will be in a recharging device. This device will be external or outside the patient or organism. The secondary coil will also be connected to a recharging circuit. This circuit together with the battery and the medical component will make up the implanted device. Through the magnetic connection between the coils, power can be transferred from the primary coil (outside) to the secondary coil (inside) without wires for electric energy conduction. Another is the use of the optimum power transfer method using impedance matching external and internal circuits. The external circuit makes use of both a directional coupler and an impedance matching circuit [11]. Also the use of laser diode and an array of diodes has been employed in recent times [12].

2.2 Existing Systems

In general, there are three major divisions when it comes to wireless energy transfer: far field, non-resonant coupling to near field, and resonant coupling to the near field [10]. The following reviews will reflect the various classifications. Himanshu Joshi and his team proposed that the circuit be operated at its resonant frequency. The method used for determining the resonant frequency in this paper was by having a sensing or monitoring unit or device in the IMD to indicate when the circuit is in its resonant frequency. This was done based on the fact that during a duty cycle, the voltage applied by the recharging circuit to the battery will be the highest when the signal is at the resonant frequency of the recharging circuit. The sensing unit will then detect and indicate once there is an alignment. Once the alignment is established, the charging of the battery begins [9]. However, they had issues with time taken for alignment, heating involved, initial danger of the circuit to the patient, changing frequency of the signal coupled may create regulatory problems. Instead of systems with just two coils, some systems for better efficiency used four coils [13, 14].

In another paper, Mahammad employed clean energy. This is energy was harvested from human energy or energy in the environment then transduced using piezoelectric transducer [15] to energy usable in IMDs. This form of energy harvesting will take care of wireless power transfer that will reach deep into the device and still maintain very high efficiency. It also makes use of the impedance matching for maximum power transfer then a rectifier to get a steady DC output. The presence of a capacitor in the transducer will limit the output of the transducer and the impedance matching capability of the circuit. To avoid this, an inductor was added in place of the existing transducer reactance. This will also boost the input AC voltage to the rectifier. The inductor size and resonant frequency suitable for an IMD were chosen to reduce losses and attenuation during energy transfer [3]. The drawbacks here were long charging time as a result of large time constant external circuit will always need to be handy, and very low energy is usually gotten from the energy harvesting.

Another system was designed by Pablo Mendoza and his team both for communication with and powering of the device. The IMD setup was designed in such a way that it will operate in two modes. In the first, which is the idle mode the power consumption will be at the barest minimum. During this mode, the device will be powered by the capacitor so the communication unit will be in shutdown mode. The low voltage consumption of the device is achieved using a voltage regulator. Then, during the charging or active mode which lasts for only about 7 ms, the device will operate normally and consume at its full capacity. This setup provides fast charging time and long-independent operation time. Although the idle time will not allow for the maximum output of the device at all times [6].

This paper makes use of planar inverted F-antenna to explore the possibility of using radiation instead of inductive power harvesting. This was used for IMDs that are implanted deeper than 10 mm. The case study was a device placed in the muscle tissue. So a three layered phantom was used to represent the arm. This setup used lower frequency transmission to achieve higher power transmission capacity. A harvesting system is also designed for the systems using off-the-shelf components. The inductive coupling is used while the antenna is used to achieve optimum radiation and DC power harvesting [16]. Areas for improvement are in the long charging time, and the safety of patient is not at its best. There is also an existing system for near field charging using round wire coils. This charges a rechargeable battery through a rectifier [17].

This paper makes use of transferring energy or power through resonance. Using smaller resonators, they designed a magnetic field for transferring power wirelessly. There was also the availability of monitoring and displaying the state of the device inside the body. The internal and external circuits are connected to each other and can transfer power to each other through the resonators in a strong magnetic circuit. Software was used to simulate and pick the frequency at which the transmission will be done (resonant frequency). The receiver unit was tried in three conditions, and comparisons were made [18]. Temperature rise of battery in the implanted device will be a problem in this setup, and it may not be so effective for deep implanted devices.

This setup uses a four-coil system to provide ultra-performance implants with power supply that needs frequent, fast battery recharging. This design was able to achieve power transfer wirelessly beyond 1 cm. Within 60 min, a 200 mAh battery (medical grade battery) will be up to 84% of its capacity. The temperature rise of the battery is 2.1 °C from simulation results [19]. This setup will need frequent recharging.

In this study, resonant coupling technique was also employed but for a brain implant. The two coils were coupled together using a magnetic field. The power supply in this setup is very small compared to the one needed in a pacemaker. The system was able to transfer 160 nW at a frequency of 13.56 MHz with a supply of 1 W from external circuit. These results were found after simulation using orCad software. In this setup, there could be possible circuit interference from external circuit [20].

Hafzaliza and others proposed using energy harvesting to acquire energy from body energy sources and stored the energy using supercapacitors. A simulation was done with coventorware software, and the supercapacitor was modeled with capacitors with capacitance levels of between 0.22 mF/cm^{-2} and 0.48 cm^{-2} representing the planer and double-stacked structure, respectively [21]. The future for implanted medical devices is the injectable and leadless devices [22].

2.3 Power Storage—Rechargeable Batteries

The type of storage that is applicable to a system like this is one with both high charging speed and substantial energy discharge time. There are many available batteries that can discharge at a very slow rate which will be very appropriate for this application but another consideration is the available size of the devices; for example, one can get a lithium ion battery with up to 2000 mAh but the size of the rechargeable battery of this size is too enormous; hence, we maintain below 1000 mAh [23]. Capacitors usually provide very fast charging times. This will make for more comfort as the patient or organism will not need to be too long while charging [24]. Batteries on the other hand stay longer after they have been charged.

There are various implantable medical devices with varying power demands (Table 1). There are also varying power supplies designed to suit the various needs. Table 2 reflects differents types of power supply and their various capacities. Though the power supply needed by these devices is small, there a great need for high energy density. Also the temperature rise of the battery during operations is very critical. Using pacemakers and defibrillators as a case study for the purpose of this design.

A very important aspect to be considered is the effect of the charging on the human skin. The effect of temperature rise rays from wireless medium and the time span of

Table 1 Various power demands from implantable medical devices

IMD	Power demand
Cochlear implants	10 mW
Defibrillator and pacemakers	30–100 μW
Drug pump	100 μW–2 mW
Neurological stimulator	30 μW–10 mW
Retinal prosthesis	40 mW–250 mW

Table 2 Rechargeable battery types available and properties

Rechargeable battery types available	Battery capacity	Charging time	Discharge time
Lithium ion cell	1000 mAH	2–3 h	
Lithium polymer	150mAH	60 min	85 h
Supercapacitor	88 mF	81 s	1 day +

Rechargeable battery types available	Voltage range	Temperature rise (°C)	IMD application
Lithium ion cell	1.5–4.2	26–40	Pacemaker, defibrillators
Lithium polymer	3.7	20–60	Pacemaker, defrillators
Supercapacitor	3.3	10–60	Pacemaker

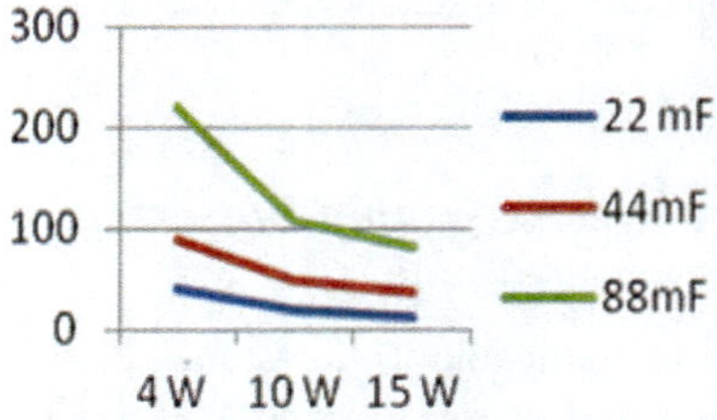

Fig. 1 The graph of the different charging times (in seconds) for different power input and different capacity values

the charging on the skin. Hence, wireless mediums with low-power densities are the most appropriate choice [25–27].

This leads to another lapse which is the fact that the power gotten from the wireless communication is usually very small hences longer charging times. As shown from the chat using supercapacitor as a case study (Fig. 1).

3 Proposed Design

The design (Fig. 2) of circuit operation and circuit components and schematic diagram for proposed system (Fig. 3) are provided in below subsections.

3.1 Circuit Operation

In this proposed setup, a microcontroller is used to alternate the power supply between a supercapacitor and battery. Adding supercapacitors will allow for longer operating time. Hence, this setup has two modes: battery powered and capacitor powered. When

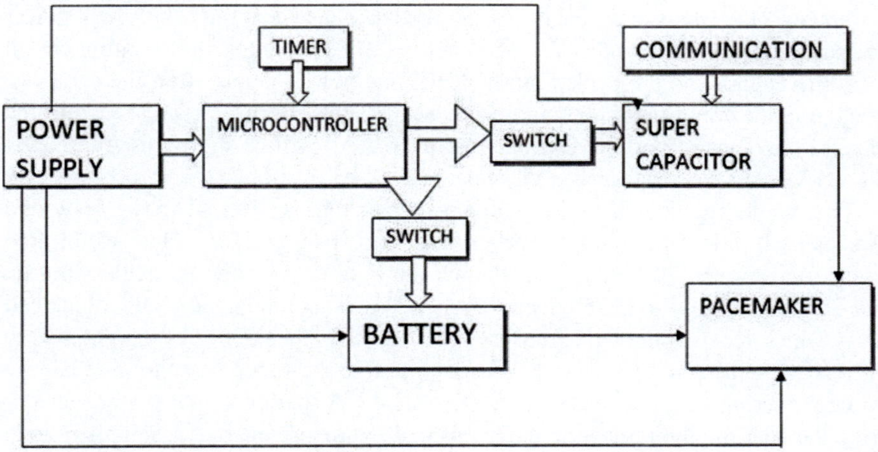

Fig. 2 Proposed design

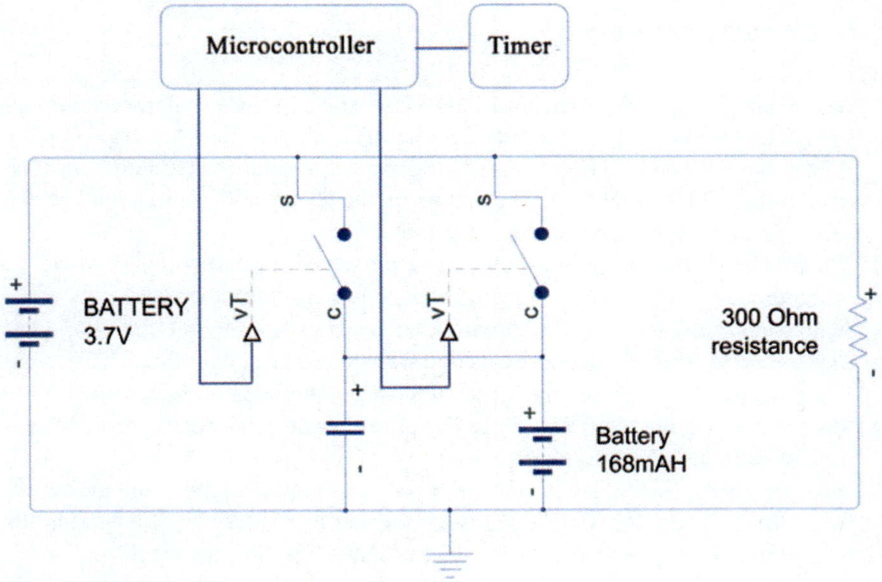

Fig. 3 Schematic diagram for proposed system

the power supply is available, it powers the microcontroller, battery, and supercapacitor. The power supply will charge the battery and capacitor simultaneously. When the power supply is removed, the super capacitor will power both microcontroller and pacemaker. Once the capacitor is almost fully discharged, the timer will send a signal to the microcontroller which will close the switch that connects the battery. The microcontroller will immediately turn off the capacitor. This will switch the microcontroller off as the capacitor is its only back up. Then, the battery will power the pacemaker till it fully discharges. When it fully discharges, the charging circuit is applied again and the cycle repeats itself. The power supply from the capacitor could provide opportunity for communication with the external circuit as regards the state of the device. While the power from the battery will be strictly used to power the device as it lasts longer then the capacitor.

The whole circuit will be in two parts: a part that is external to the body and the internal part of the circuit. The external part of it will be a close-to-infrared light (transmitter). To be able concentrate the light rays on the particular spot on the skin, a lens is used. On the receiver side, there is an implanted photovoltaic cell array which receives the light from the lens. The cells then convert the light rays to electrical energy (DC) which in turn used to charge the super capacitor and battery while powering the pacemaker in charging mode. A unidirectional flow of energy from the charging point is desired and ensured by the presence of a Schottky diode [28].

3.2 Circuit Components

1. Power supply: The light produced from a laser source (diode) will produce about 5 mW and a wavelength of about 750 nm. Modeled as a 3.7 V battery
2. Lens: The light rays if not converged properly will not have maximum impact on the PV cells behind the skin. Depending on the shape and size of the PV cells, the light is usually shaped using the lens and filter.
3. Photovoltaic cells: These are simply an arrangement of photodiodes. About 12 photodiodes (cells) 6 cells in series connection and the other 6 also in series with each other, then an overall parallel connection to combine the two. After the cell arrangement, the photosensitive area should be about 90 mm^2. To avoid temperature increase, the other parts of the array that are not photosensitive are covered with shiny white covering. The area of photosensitivity determines the output from the PV cells.
4. Switch: This is used to connect the microcontroller to the battery and the capacitor. When the device is not charging to prevent the battery from powering the charging unit and dissipating the energy faster, we use this device [29].
5. Microcontroller and timer: This device counts and sends signals to the microcontroller to determine when the capacitor goes off and the battery comes on.
6. Supercapacitor: A large capacitor of about 88 mF connected in parallel with the battery of the device will provide fast charging and some hours of use without the transmitter. A capacitor was used to model the supercapacitor.
7. Battery: A 168 mAh lithium polymer battery was used. These batteries have an outstanding life cycle and acceptable temperature rise during operation. With the addition of the capacitor when the transmitter (power supply) is off, the capacitor

will power the battery and give a longer lasting power supply to the pacemaker. A 300 Ω purely resistive load was used to model the pacemaker.

4 Experimentation and Validation and Results

The energy and power density for the supercapacitor and small medical batteries were compared and analyzed. Using Simulink, the power supply from charging unit was modeled as a simple voltage source. So with equal input power the state of charge, current, output voltage was compared. The voltage source was modeled using a DC voltage source (3.7 V battery). The voltage of 3.7 V was used to charge the rechargeable power supplies to their maximum energy storage capacity since the medical device to be powered needs an input voltage of only about 3–4 V. The micro supercapacitor used for this experiment was 100 mF. The battery was a lithium polymer battery of about 168 mAh. The physical measurement of this battery is a measurement of 24 mm, 18 mm, and 3 mm as the length, breadth, and thickness, respectively.

The control between the battery supply and supercapacitor was modeled by the use of a microcontroller and timer controlling a pair of switches. When the power supply is applied, it will charge both of the storage devices for 81 s; then, the microcontroller opens the switch to the capacitor, while the battery continues the charge for the next 1 h. After the charging, the battery is disconnected and the supercapacitor is reconnected. The power supply is then disconnected, and the supercapacitor maintains supply to the controller and pacemaker. This was modeled by measuring the power storage capabilities of the two storage devices individually on simulink. Then, a combined display of their individual properties was displayed. The pacemaker was modeled using a 300 ohms resistor (Fig. 4).

From the simulations done, a comparison was done between the performance of the supercapacitor and the battery. Inserting a supercapacitor into the system will increase the operating time to approximately four days which will make for less dependence on the charging unit. It will begin the cycle with one day in full circuit

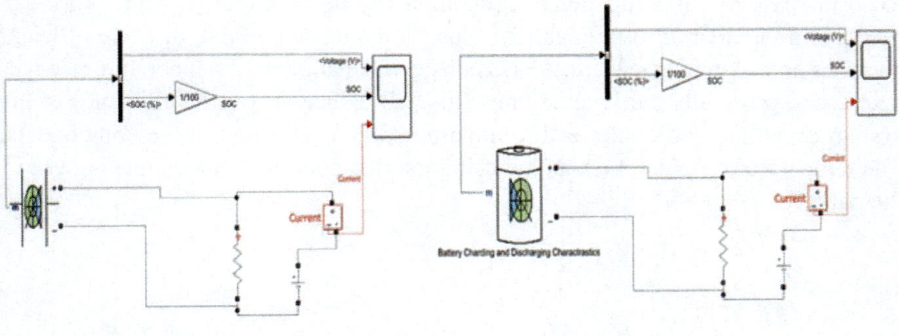

Fig. 4 The parallel between the energy densities of supercapacitor and battery

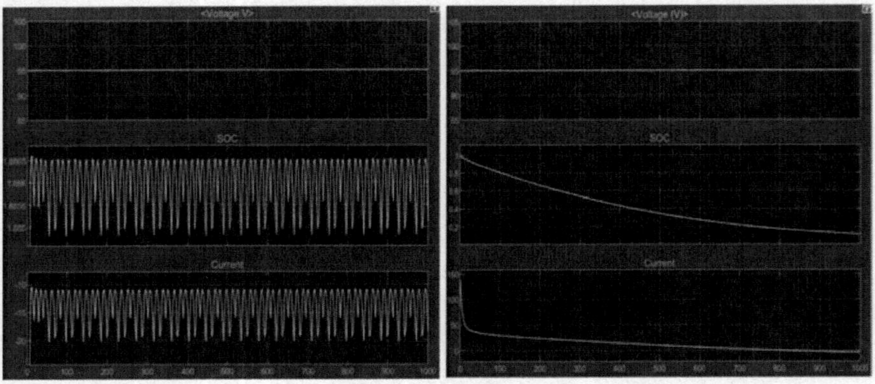

Fig. 5 The results of the supercapacitor and battery discharge characteristics

operation and the remaining three days in power saving mode. Also the charging time increases from 1 h to about 1 h and 2 min with the supercapacitor in the system. The voltage supplied to the pacemaker throughout the standby time will remain constant from the graphs (Fig. 5).

5 Conclusion and Future Work

This paper presents a system with two energy storage devices. This system switches from a supercapacitor to a battery in a power saving mode to make for longer stay on time without the power pack. The simulation demonstrated the effect of the addition of a supercapacitor to the battery for use as a power supply for the pacemaker. With the addition of the capacitor, the energy discharge time was increased to 96 h. The inclusion of the supercapacitor will lead to a slight increase in charging time for the device from 1 to 1 h 2 min assuming a loss free setup. The pacemaker supply voltage of 3.7 V is maintained throughout the discharge time.

The switching between the devices will also lead to significant losses. This is due to an increase in the components and complexity of the circuitry. More work can be done in the area of loss reduction. This can be made possible by more efficient switches and switching system. Also a considerable amount of temperature rise will be present, especially during charging. This will reduce the overall efficiency of the system over time and could lead to a malfunctioning of some of the components. The temperature rise can be handled using internal cooling methods making use of the natural flow of body fluids.

References

1. Jing, L, Xiaojuan, W.: Power Sources and Electrical Recharging Strategies for Implanted Medical Devices, pp. 1–3. Higher Education Press and Springer-Verlag (2008)
2. Chan, V, Illumin, G.O.: Engineering heartbeats, 6 may 2013. [Online]. Available: https://illumin.usc.edu/engineering-heartbeats-the-evolution-of-artificial-pacemakers/. Accessed 15 April 2019.
3. Lee, J.-S, Lee, S.-G, Hoang, N. K.: maximum power transfer considering limited available input power in ultrasonic wireless power transfer for implanted medical devices. In: 2014 IEEE Fourth International Conference on Consumer Electronics Berlin (ICCE-Berlin), Berlin Germany (2014)
4. Tudor, M. J, White, N. M, Beeby, S.P.: Energy Harvesting Vibration Sources for Microsystems Applications, pp. 175–195, 19 July (2006)
5. Ochoa, M., Rahimi, R., Ziaie, B., Kim, A.: New and emerging energy sources for implantable wireless microdevices. Special Section on Nanobiosensors **3**, 89–98 (2015)
6. Bibin, J., Dietmar, S., Wolfgang, H. K, Pablo, M.-P.: Super-capacitors for implantable medical devices with wireless power transmission. In: 2018 14th Conference on Ph.D. Research in Microelectronics and Electronics (PRIME), Prague, Czech Republic (2018)
7. Acharf, B., Ammar, A.B, Hung, K., Cao.: Power approaches for implantable medical devices, 15 November (2015).
8. Dennis, D., Fitzpatrick.: Pacemakers and implanted cardioverter defibrilllators. In: Implantable Electronic Medical Devices, pp. 75–97 (2015)
9. Norman, O., Joshi, H.: United States of America Patent US 8,843,207 B2 (2014)
10. Rahman, S., Al-Haddad, K., Fadhel, Y.B.: General Principles of wireless power transmission and its applications in implantable medical devices, pp. 5216–5221. IEEE (2016)
11. Griffith, A.G, Newbury, P., Tae, W.H.: United States of America Patent US 6,212,431 B1, (2001)
12. K.awata, K., Murakawar, M., Kobayashi, O., Nakamuka, S.: A Wireless near-infrared energy system for medical implants. IEEE Eng. Med. **18**(6), 70–72 (1999)
13. Kavyashue, p., Qaroot, A., Perez, S., Thomas, S., Tan, Y.K., Ababneh, M.M.: Optimized Power Management Circuit for Implantable Rectenna for in-body Medical Devices, pp. 30–34, 12–15 December (2017).
14. Babu, J.V, Bobba, B.P.: Design and analysis of a robust system for wirelessly powering implantable devices. In: 1st IEEE International conference on Power Electronics, Inteligent Control and Energy Systems (2016)
15. Mutashar, S., Asamad, S., Hussain, A., Hannan, M. A.: "Energy harvesting for the implantable biomedical devices issues and chaleges," p. 23, (2014).
16. Bakogianni, S., Diet, A., Bihan, Y.L., Pichon, L., Kouloundis, S.: Investigation of Efficient Wireless Charging for Deep Implanted Medical Devices, Paris (2016)
17. Liang, S-y., Lee, C.-J., Cheng, M.-C.: A low-power bidirectional telemetry device with a near-field charging feature for a cardiac microstimulator. IEEE Trans. Biomed. Circuits Syst., 357–367 (2011)
18. Yang, X.S., Qingxin, Y., Jun, Z.Y., Xu, G.: Design on Magnetic Coupling Resonance Wireless Energy Transmission and Monitoring System for Implanted Devices, p. 4 (2015)
19. Nurmikko, J.L., Arto. V.: Multi-coil high efficiency wireless charger system for hermetically sealed biomedical implants, p. 4 (2018)
20. Ktata, S., Rahmani, S., Al-Haddad, K., Fadhel, Y.B.: Design and simulation of wireless power transfer system for brain implant. In: 15th International Multi-CONFERENCE on Systems, Signals and Devices (SSD), Canada (2018)
21. Azrul,. A., Hamzah, J.Y., Mohamed, A.M., Burhanuddin,Y.M., Hafzaliza, E., Zainal, A.: Interdigitated MEMS supercapacitor for powering heart pacemaker. INTECH, pp. 145–163 (2016)
22. Sanghani, P. S., Morris, M.M., Chow, E.Y.: Wireless MEMS-based implantable medical devices for cardiology. Wirel. MEMS Netw. Appl., 78–100 (2017)

23. Weerasinghe, D.P., Chandima, K.D.U.I., Dayarathna, H.G. D.A., Jayasinghe, W. K. K. R., Dharmasin, H.M.: Inductive power transmission charging and communication for implanted devices. In: ICIIS'2017 1570371694 (2017)
24. Gogotsi, Y., Dunn, B., Simon, P.: Where do batteries end and supercapacitors begin. In: AAAS, pp 1210–1211 (2014)
25. Battery University, 24 April 2018. [Online]. Available: https://batteryuniversity.com. Accessed 30 May 2019
26. Takeuechi, D.M., Spillman, E.S.: Lithium ion Battries for Medical Devices. Wilson Greatbatch ltd, New York (1999)
27. Gan, H., Takeuchi, E.S., Rubino, R.S.: Implantable Medical Applications of Lithium-ion Technology. IEEE, New York (2002)
28. Iqbal, S., Karmaker, M., Zinnat, S.F., Ali, M.T., Saha, A.: A wireless optical power system for medical implants using low power near-IR laser, pp. 1978–1981 (2017)
29. Arnaud, A., Szollosy, G.M.: Integrated switch for implantable medical devices in HV-MOS technology, pp. 1–95 (2010)

Intelligent Networking

Particle Swarm Optimization and Genetic Mutation Based Routing Technique for IoT-Based Homogeneous Software-Defined WSNs

Rohit Ramteke and Samayveer Singh

Abstract Recent advancement in wireless sensor network has evolved as an open system which can be reconfigured dynamically. Generally, these networks have different limitations and challenges such as energy consumption in data collection, control node election, load balancing, etc. An efficient load balancing in terms of data collection and forwarding is depended on the routing techniques which are responsible to provide an effective path to transmit the collected data such that the minimum amount of energy should be consumed in the process. The control nodes are responsible for assigning the task and data transmission in the cluster-based routing techniques. The selection of the control node is an NP-hard problem. In this paper, an adaptive particle swarm optimization (PSO) ensemble with genetic mutation-based routing is introduced to select control nodes for IOT-based software-defined WSN. The proposed method plays a significant role in selecting the control nodes based on the fitness value. Fitness value takes energy and distance parameters into consideration. First, the proposed work is implemented for homogeneous nodes that can be deployed with single and multiple sinks. Further, the proposed work can also implement for the heterogeneous sensor nodes having different computing power accompanied by single and multiple sinks. The simulation result of the proposed method outperforms over some other existing algorithms under the different arrangements of the network.

Keywords Adaptive PSO · SDWSN · IoT · Homogeneous nodes · Single and multi-sink

---

R. Ramteke · S. Singh (✉)
Department of Computer Science & Engineering, Dr. B R Ambedkar National Institute of Technology Jalandhar, Jalandhar, Punjab, India
e-mail: samays@nitj.ac.in

R. Ramteke
e-mail: ramteke.rohit3@gmail.com

P. K. Singh et al. (eds.), *Recent Innovations in Computing*, Lecture Notes in Electrical Engineering 701, https://doi.org/10.1007/978-981-15-8297-4_12

1 Introduction

In the time of digitalization, everyone is digitally engaged because of the huge advancement in the technological. It starts from a small programmable chip to a large space shuttle for space expeditions since science is always laid on the principle of creating, monitor, and development. The huge deployment of the sensors in wireless modules, embedded controllers, and the cheap sensors have forced the emergence of the Industrial Internet of Things (IIoT) which is introduced as the fourth industrialization revolution called Industry 4.0 (e.g., Industrial intelligence robot) [1]. The sensor node is considered as a complete module, machine, or subsystem whose intention is to recognize, trace the changes in surrounding conditions. The complexity of the operations changed when the monitoring area changed. Thus, there is a need to adapt different deployment technologies of sensors which should be a more flexible and adaptable system. The wireless networks always come into the picture for the flexible systems in the priority. The WSN is primarily deployed in a harsh environment. The sensor nodes are neither manageable by the direct human interference nor with the wired interconnection because, in a harsh environment, it's not possible to operate the system manually.

With the evolution of 5G technology and micro-electronics, most of the technological utilities are shrinking in small devices like smartphones, Notepad, etc. Integration of the new technology like cloud computing, fog computing, Big-Data, micro-electronics has led to the new domain for the deployment of wireless sensor networks which comprises the internet module along with WSN architecture known as the IoT [2]. These IoT infrastructures have a base of software-defined network and core module of wireless sensor networks which are broadly required to connect with the internet so that it can be synchronized with the current technologies. They are capable of effectively detecting the failure and initiating maintenance and reconfiguration process. Here devices can communicate, share, process the collected information, and perform decision making. This infrastructure is the service provider in three phases, like data gathering service, data transfer service, and processing service. The sensor nodes can be used as a dedicated task monitoring system. Some regular engagements of sensor networks are military applications, environmental applications, health applications, commercial applications. The new advancement in the field of software-defined wireless sensor networks has evolved wireless sensor technology to a very extent. Wireless sensors, Cloud computing, artificial intelligence, and other supporting technologies have set-up collectively a new paradigm of information processing as shown in Fig. 1 because of the real-time integration of the data this transformation has increased the demand for this prototype. IoT structure was comprised base of software-defined WSNs. It's an arrangement consisting of interrelated computing and processing devices, mechanical units, computerized equipment, that can move and process the information over the system without human intervention [4].

It includes artificial intelligence as well as the decision making from the aggregation of the collected data. IoT is the innovation that can lead the communication of

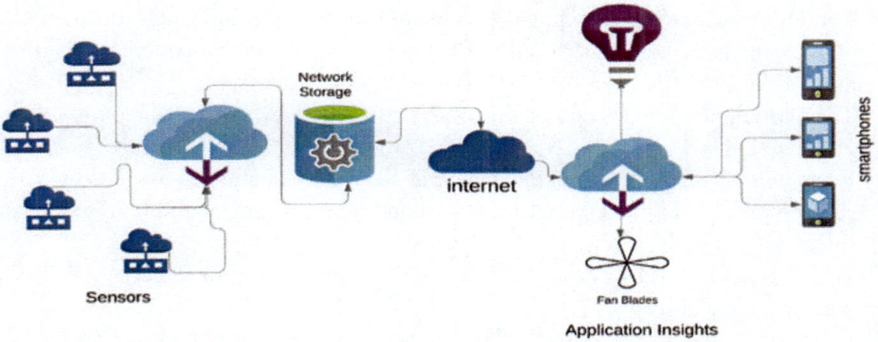

Fig. 1 IoT supported SDSWSN architecture

the devices to the next level and it doesn't limit only for senor network. It includes smart systems having intelligence, for example, industrial robots, and home automation systems. Take your smart fitness band, having some sensor to check your heart rate once you come from the morning walk. It will automatically trace and sends the information to all household devices. They can adjust themselves according to the current biological condition of the person like to drop-down or raise the room temperature.

In this paper, the genetic algorithm is used for the selection of control nodes along with an effective fitness function. In the proposed method the control nodes are selected dynamically, and the genetic algorithm helps to remove the uneven clustering problem. An adaptive inertia weight tuner is used for effective convergence. For better load balancing multiple sinks, the model is implemented using the genetic algorithm.

Rest of the paper organized as follows: a system model of the proposed method is discussed in Sect. 2. The literature review is discussed in Sect. 3. The proposed method is discussed in Sect. 4. Section 5 discusses the performance study of the proposed and the existing work. Finally, the paper is concluded in Sect. 6.

2 System Model

An IoT supported software-defined WSN model is observed as an operational WSN architecture as a digraph $G_n = (V, L)$. In the given model V specifies the vertex set, which having Software-defined WSNs or common nodes, control nodes (CNs), and the control server (CS) or sinks node. They are randomly distributed within the specific monitoring area. The L in the Graph specifies the collection of directed transmission or communication link that is dedicated to the transmission of the collected data from common nodes to the control node (CN) and the control server [14].

The assumptions of the SDWSN environment are listed as follows.

- IoT derived SDWSNs committed to detecting various sensing targets like temperature, moistness, etc. are randomly disseminated inside the equivalent geological area of the SDWSN.
- The participating IoT enabled SDWSNs must have a universal identification number (UIN).
- Each unit in the IoT deployment having SDWSN capabilities is dedicated to sensing the collect the data from the surrounding environment and send that data to the control node (CN), control server (CS).
- Whether it's a normal sensor unit or the dedicated sensor unit, each SDWSNs have equal capability.
- In the unattainable deployment, all units including the control node are equipped with a non-replaceable battery. The energy distribution within all nodes is fairly allocated.
- Traditional network configuration suggests that the control server is having external energy resources since it's the dedicated server to carry out all processing of the network.

2.1 Energy Dissipation

The considered environment of IoT enabled SDWSN is most widely adopted data transmission model which is based on the path loss concept and the model consists of both multipath (E_{mp}), fading (d^4 power loss), and free space (E_{fs}) (d^2 power loss) channel utilization. The energy consumption of the model depends on the distance (d) between two entities. The ith the transmitter is having the (X_i, Y_i) coordinate, and another jth receiver is having the coordinates (X_j, Y_j). The distanced can be calculated by using the Euclidean distance Formula and is formulated as:

$$\sqrt{(Xj - Xi)^2 + (Yj - Yi)^2} \tag{1}$$

In this work, we use the power control mechanism if the calculated d is lower than the threshold (d_0), then the free space model is used else multipath model is used to remunerate the path loss concept [5]. To send a l-bit message over the distanced$d$, energy dissipation to transmit the data can be calculated for the common SDWSN node ($E_{\text{TXN\_ SDWSN}}$) is given as follows:

$$E_{\text{TXN\_ SDWSN}(l,d)} = \begin{cases} k \times E_{\text{elec}} + k \times E_{\text{fs}} \times d^2 d \leq d_0 \\ k \times E_{\text{elec}} + k \times E_{\text{mp}} \times d^4 d > d_0 \end{cases} \tag{2}$$

The energy transmission for control node of l-bit data packet is as follow:

$$E_{\text{TXN CN}}(l, d) = \begin{cases} k \times (E_{\text{elec}} + E_{\text{DA}}) + k \times E_{\text{fs}} \times d^2 d \leq d_0 \\ k \times (E_{\text{elec}} + E_{\text{DA}}) + k \times E_{\text{mp}} \times d^4 d > d_0 \end{cases} \tag{3}$$

The E_{TXN} is known for the energy required for transmission, control node consume energy E_{DA} for data aggregation whereas the distance between two sensor units of SDWSN or between nodes to the control server is defined by d. The energy dissipates per bit to run transmitter, or a receiver circuit is defined by E_{elec}. It depends on various factors such as modulation, digital coding, source coding, filtering, and signal spreading. E_{fs} and E_{mp} depends on the transmitter amplifier model. Here l is the data length to be transmitted and d_0 is the threshold value for transmission distance and usually formulated as below

$$d_0 = \sqrt{\frac{E_{\mathrm{fs}}}{E_{\mathrm{mp}}}} \tag{4}$$

The radio transmitter consumes the following amount of energy to receive a l-bit message:

$$E_{\mathrm{RXN}}(l) = l * E_{\mathrm{elec}} \tag{5}$$

3 Literature Review

The most significant concern in a software-defined sensor network (SDSN) is to build a routing algorithm that manages an effective consumption of the energy. Thus, minimum energy consumption is an important objective in a wireless sensor network (WSN) where one of the possible ways to achieve this objective is efficient routing for data transmission [7]. The effectiveness of a particular routing algorithm mainly depends on the capabilities of the sensor nodes, control node, and the dedicated application requirements. Clustering plays a very vital role in energy balancing between the control server, control nodes, and normal nodes [8]. The clustering-based routing algorithms can be considered as the most sophisticated categories which maintain load balancing among all types of nodes. Heinzelman et al. introduced a classical clustered oriented routing algorithm namely "low energy adaptive clustering hierarchy" abbreviated as LEACH [7]. It's still considered as the base for other evolving advanced clustering algorithms. This technique used a randomization strategy in the distribution of energy between the nodes of SDWSN by the selection of local control nodes rotationally. When the control node is far away from the control server then energy depletion is more, so the distance plays an important role in control node selection. After solving various state of art in the LEACH there are different LEACH versions are introduced by the preceding researchers. Ran et al. discuss LEACH-FL [9] in which they introduced a fuzzy logic-based two-level hierarchical control node selection techniques. In this work, the probability is calculated by considering the fuzzy logic which leads to minimizing energy consumption and it prolongs the lifetime.

Younis et al. introduce "HEED" a hybrid approach that is energy efficient and having a distributed strategy based clustering [7]. It included residual energy as the parameter for control node selection. However, it was introduced for the Ad-Hoc network, so its effectiveness was not tested against the software-defined wireless sensor nodes. Xiaorong et al. propose the "Hausdorff clustering" mechanism for WSN. It considered the node location, network connectivity, and communication effectiveness as the parameter for control node selection [10]. This clustering-based algorithm is generally suited for the Ad-hoc network and Wireless network. Singh et al. propose PSO-C swarm intelligence-based optimization algorithm for control node selection. Their aim is to localize control nodes surrounding the center density [11]. Xiang et al. propose a NWPSO which is a variant of PSO. Notably, it depends on the variance on inertia function as a non-linear weight to select the control nodes from the network [5]. The later work proposed by Kumar et al. proposed integration of the genetic algorithm with the PSO for performance enhancement of network [6]. This paper has a trace of different parameter considerations for the control node selection. The various version of PSO was introduced in recent years since PSO is now prominently used in the different research work carried out for the software-based wireless sensor technology. From the above literature survey, we can observe that the classical techniques for clustering and control node selection are efficient in the domain of WSN with some restrictions. The changing scenario of sensor leads this technology to other extends. This type of improvement in the algorithms and the protocols are necessary with time. Here PSO has played a vital role to provide a scalable, efficient, and modern protocol approach for the dynamic working of the network.

4 Proposed GA-Based PSO

In 1995, Kennedy et al. propose a population-based metaheuristic global search optimization technique known as Particle Swarm optimization. Originally, they worked on a model that describes the social behavior of creatures like a flock of birds and the school of fishes. However, their model was competent to do optimization test so they proposed a new optimization technique. The PSO works in a multidimensional search space were each particle initially consists of two entities, i.e., position vector and velocity vector. The initial position of the particles is considered as the potential solution. All particles move continuously towards the best solution by updating their respective pbest and gbest solution. As the PSO terminated on some defined condition all the particle is reached nearby to an optimal solution. In PSO position and velocity is updated using the following formula which consists of previous communicated learning like pbest, gbest and other factors of particles. The velocity updation rule of PSO is as follows:

$$V(t) = \omega \times V_i(t-1) + c_1 \times r_1(\text{pbest}(t-1) - X_i(t-1))$$

$$+ c_2 \times r_2(\text{gbest} - X_i(t-1)) \tag{6}$$

Position updation in PSO is as follows:

$$X_j^i(t) = X_i(t-1) + V_i(t) \tag{7}$$

4.1 Proposed Method

The SDWSN is equipped with sensor nodes having two functionalities, i.e., nodes can collect and transmit the data to control nodes or base stations for further processing. The control nodes are selected from the common nodes. Starting with a homogeneous network having a single sink or control server the proposed PSO is executed with a predefined number of N particles which is an individual network entity. Particles are generated randomly were X_i represents the position vector and V_i is the velocity vector of ith particle at iteration t in jth dimension. Every ith particle attended with a random number (rvar) of control nodes. $rvar$ is within a range of minimum (min CN) and maximum (max CN) number of control nodes. For ith particle rvar_i elements are considered as control nodes. The randomly generated position vector and a velocity vector of ith particles are of continuous value. The sequence position vector $S_i(t)$ as discrete values are further generated by applying the smallest position value (SPV) rule over corresponding position vector $X_i(t)$. The primary role of SPV rule is to assign indexing for the position vector X_i of each randomize generated particle. Although this sequence position vector is used for the clustering. We pick first rvar_i as control nodes from sequence vector indexes. After cluster nodes selection the control nodes CN_i are enumerated with their corresponding common nodes. This grouping is done by finding the closest control node for each common node. Further selected rvar used in Fitness calculation of each particle is which is performed by an effective fitness function. The fitness function is formulated using different parameters of the network. We have taken distance and the energy into consideration for the fitness calculation. Sequence position vector and rvar_i are evaluated using defined fitness function for each particle of the swarm. The pbest and gbest solution for the swarm is updated based on the calculated fitness. After executing the defined process by the proposed method, it exhibits the gbest solution among all particles of each iteration. The number of iteration is one of the control parameters in the PSO, so after every iteration, PSO checks for the satisfying criteria to discontinue the process.

Further, in consecutive processing, the updation of the position vector is done followed by updation of velocity vector using given Eq. 6. The formulation of the velocity vector in conventional PSO is having fixed inertia weight. Inertia weight plays an important role in velocity updation of the particle, it is one of the control parameters. Large inertia weight facilitates greater global search and small inertia weight is facilitating greater local search therefore proposed method uses adaptively tuned inertia. In this approach, at the beginning of process, inertia weight is comparatively large, which needs to be slow down in later iterations. We carry out this task by iteratively damping the inertia weight using Eq. 9. The newly generated velocity vector further used in Eq. 7 to update the corresponding particle position.

Now fork and merge model is starts to execute where the SPV rule is applied again on the newly updated particle to sustain the individual's restraint. In a fork, each particle induces to respective offspring (t % of initial population IP) accompanied by parent particle using genetic mutation. PSO does not have a selection operator whereas GA has a selection operator which helps other particles to evolve with the fittest particle. The genetic mutation operator GMO uses a two-point random mutation which is the selection feature of a genetic algorithm. The fittest particle will lead the solution toward a global solution. This genetically muted offspring particle is only differentiable over sequence position vector, there corresponding position and velocity vector are the same. Each offspring or sub-particle is having corresponding control nodes sequence. GMO implemented with sequence position value accompanied by rvar number of CNs such that it should propagate the good feature of fittest control node in the offspring. Further, the fitness function is applied over generated offspring to evaluate the fitness of each sub-particles. All the offspring and parent particles are merged into a single particle based on fitness value. This is the iterative process in each iteration the fittest particle is selected, and that particle will help to update the gbest and pbest solution for the given population till the termination criteria do not meet. A single sink network is prone to more energy consumption. The transmission process will halt if the sink gets damaged or any failure related to the sink. Load balancing is not possible in this type of network; congestion is also a major issue. The use of multiple sinks can avoid this problem to some extends [13]. The multiple sinks are deployed over a defined observation area. The workability of the proposed method is extended for multiple sinks. As the pbest and gbest solution depend on fitness value given by fitness function and the proposed fitness function is based on energy and distance parameter of control nodes, control sink, and the common node. Fitness value F is calculated using Eq. 8.

$$F = \alpha f_1 + \beta f_2 = \alpha \frac{E_{\mathrm{CN}\,to\,\mathrm{CS}}}{E_{\mathrm{SDCN}\,to\,\mathrm{CN}}} + \beta \frac{1}{D_{\mathrm{SDCN\ to\ CN}} + D_{\mathrm{CN\ to\ CS}}} \tag{8}$$

Here, $D_{\mathrm{SDCN\ to\ CN}}$ and $D_{\mathrm{CN\ to\ CS}}$ describe the average distance within the common nodes (SDSNs) & the corresponding CN and the average distance within the CNs & the CS respectively. Furthermore, $E_{\mathrm{SDCN\ to\ CN}}$ and $E_{\mathrm{CN\ to\ CS}}$ represent the power dissipated in data communication between the common nodes (SDSNs) and the corresponding CN and the power dissipated in data communication within the CNs & the CS respectively [6]. To make a balanced tradeoff between distance fitness and energy fitness α, β are set to value 0.5. This modification drastically effects on fitness value of the network. Control nodes selection is done by taking sink distance into consideration and clusters are formed. The existing inertia tuner was very complex and increases the computation time of the overall process. Whereas the proposed inertia tuner able to perform effectively for both single sink and multi-sink model. It is having the ability to modify according to the number of iterations. In the proposed method inertia weight is adaptively tuned using Eq. 9.

$$\text{inertia} = \text{inertia} - \delta \times \text{inertia} \tag{9}$$

Here δ denotes a seed value (δ =0.02) that can be adjusted on the network configuration.

Algorithm 1: Proposed algorithm

---

1. Initialize N number of particles with their position and Velocity vector.
2. Initialize the number of sink (single or multiple).
3. Apply SPV rule over position vector and a sequential vector is generated
4. Find fitness of particle accompanied by $rvar$ number of control nodes
5. Update pbest solution for a given population
4. Update gbest solution for the population
5. **While** termination criteria do not meet
6. Damping of inertia weight during each iteration
7 **For** every particle
8. Update velocity vector
9. Update position vector
10. Apply SPV rule over updated vector
11. Apply GA
12. Fork *t* % of the particle of the total population
13. Apply GMO over current sequential vector and generate offspring from the current generation
14. Merge the all offspring and parents into single Particle based on fitness value
15. Update pbest and gbest solution.
16. **End for**
17. Project gbest solution
18. **End while**

---

5 Performance Study

In this section, the proposed method is compared with the other exiting methods on performance matrix of fitness value, stability period, inertia weight, and average residual energy. The proposed method is implemented using a javascript framework. The simulation of different scenarios is taken into consideration based on the various parameters specified in Table 1. Where *N* is the number of particles and each particle is having 100 nodes. The proposed method is tested over a given geographical area, where particles are randomly initialized. The GMO is a costly operation where t% is the particle fork percentage therefore, increasing t results in higher computation time of the proposed method. The experimental result over the different values of t shows

Table 1 Testing parameters

Type	Parameter	Value
Network	Area (m) Location of CS Initial energy Number of nodes	100*100 (50*175) 2 J 100
Application	Data packet size advertisement packet length	100 25
Radio model	E_{elec} E_{fs} E_{mp} E_{textDA} d_0	50 nJ/bit 10 pJ/bit/m2 0.0013 nJ/bit/m4 5 nJ/bit/signal 75 m
Proposed model	IP Number of iteration C1 C2 t The maximum value of ω	50 40 2 2 25% 0.94

that t in range of (20–25%) is having better fitness value for efficient computation time. Interia weight is initially set the maximum to 0.94 as mention in the base paper and afterward in each iteration inertia weight is tuned by using adaptive inertia tuner. This formulation works effectively as compare to other inertia tuners. Further, based on the optimal number of control nodes, the optimal location for the selected control node is decided. The result shows that after completing 75% of total iterations, the network begins to stabilize for a single sink model, but for multiple sink model, it takes about 80% of the total iteration count. Based on the experiments carried out over different test cases it was found that the optimal number of control sink can vary from 4 to 6 for overall convergence over given geographical area. Whereas the number of sink NS = 4 is taken for effective tradeoff.

The location of the control sink also plays a major role in overall network coverage and energy consumption. So, for the moderate effect of location, placing of a control node in between the corner and halfway from edges is suggested. The proposed method is analyzed for the fitness value over varying initial population and number of iterations. Figure 2 describes experimental results which shows that the proposed fitness function leads to better fitness value for the single sink and multi sink model. Selection of the optimal number of control nodes and their location results in better fitness value of the overall network. For a better convergence, an adaptive inertia tuner is used. Due to the adaptive behavior of proposed inertia, it can dynamically be configured by adjusting seed value. Figure 3 shows a comparative analysis of the existing inertia tuner and proposed inertia tuner. Results depict that the proposed inertia tuner can perform quite better without a complex formulation which minimizes the computation time for the proposed method.

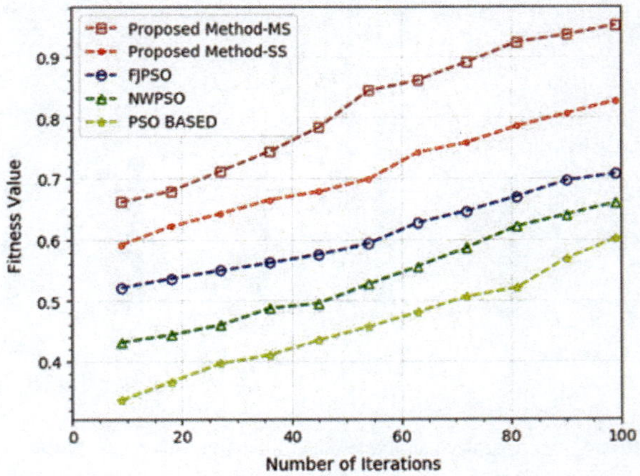

Fig. 2 Comparative analysis of fitness value

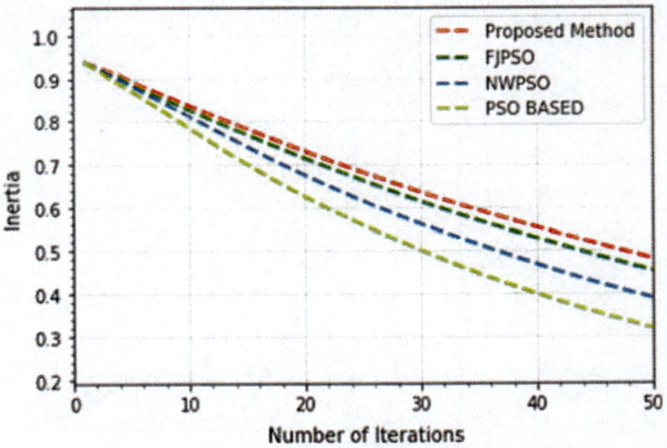

Fig. 3 Comparative analysis of adaptive inertia tuner

In the single sink model, it takes about 725–750 rounds and in the multi sink model, it takes about 954–980 rounds when nodes are starting to drain energy as shown in Fig. 4. The stability period is affected by the number of control nodes and the number of sink nodes. Also, the same number of control node stability periods of the proposed method is greater as compared to existing methods due to two-level optimization of the proposed method using the genetic algorithm.

The comparative analysis is shown in Fig. 5 illustrates that PSO based and NWPSO have soaring power consumption over other methods this is because this method does not consider genetic mutation for selecting fittest control nodes. Here the proposed Method has considerably able to save up to 48.49–51.34% of energy in comparison with PSO based and NWPSO while they save nearly 25.34–30.87% in comparison with FJPSO for single sink model and with slighlty greater proportion for multiple sinks. This improvement is due to better localization of control node using genetic mutation and particle swarm optimization. Here energy consumption depends on so many factors such as the distance of control nodes to common nodes, the distance of

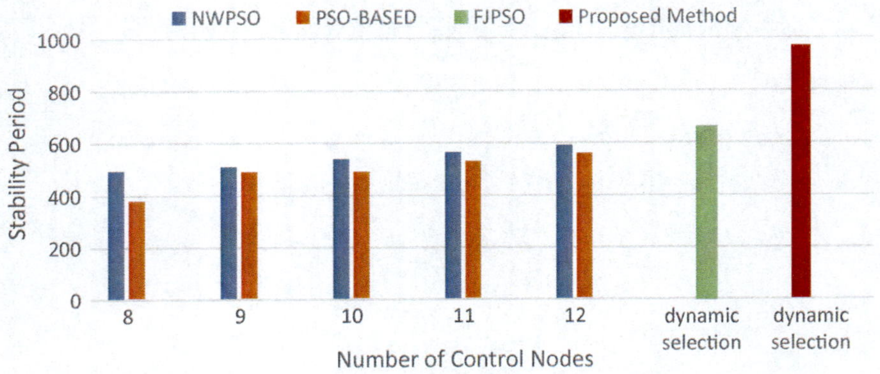

Fig. 4 Comparative analysis—Stability period versus optimal number of control nodes

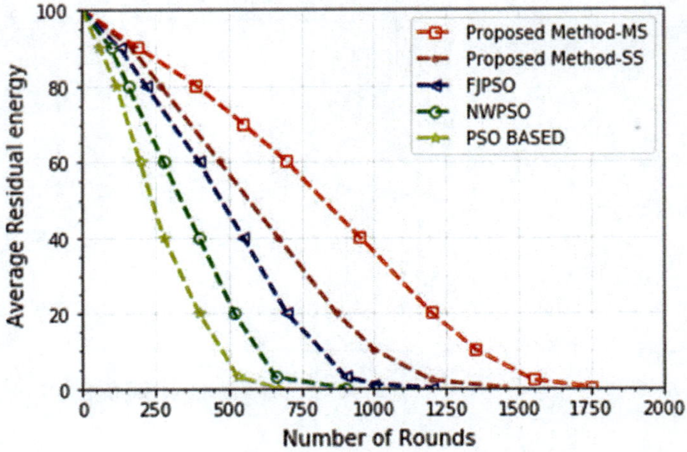

Fig. 5 Comparative analysis of residual energy

control nodes to sink, number of control nodes, load balancing, etc. The multiple sink model helps in load balancing which leads to extending the lifetime of the network.

6 Conclusions

In the proposed method, two-level optimization using GMO is introduced which enables the PSO to work dynamically for control node selection. The proposed method is tested over the different parameters and in various network arrangements and the results show this method outperforms on various state of art of existing methods. The energy consumption and distance tradeoff, inertia tuner, effective fitness function of the proposed method significantly perform better than the existing methods. The multi-sink model has shown the major improvement in terms of energy consumption by the proposed method for homogenous sensor nodes as compared to of single sink model. This model is flexible and extensively modified for different scenarios. Observing the results this model can be seen as the most promising in terms of improving lifespan, energy consumption for the reconfigurable IoT-based SDWSN.

References

1. Zhang, D., Li, G., Zheng, K., Ming, X., Pan, Z.H.: An energy-balanced routing method based on forward-aware factor for wireless sensor networks. IEEE Trans. Ind. Informatics **10**(1), 766–773 (2014). https://doi.org/10.1109/TII.2013.2250910
2. Chang, R.S., Lin, C.F.: A survey of routing algorithms for wireless ad hoc networks. In: Proceedings—2004 Glob. Mob. Congr., pp. 169–174 (2004)
3. Zhou, J., Xu, M., Lu, Y.: Biologically inspired low energy clustering for large scale wireless sensor networks. J. Phys. Conf. Ser. **1267**(1) (2019). https://doi.org/10.1088/1742-6596/1267/1/012004
4. Bounceur, A., Bezoui, M., Lounis, M., Euler, R., Teodorov, C.: A new dominating tree routing algorithm for efficient leader election in IoT networks. In: CCNC 2018—2018 15th IEEE Consumer Communications and Networking Conference, vol. 2018, pp. 1–2 (2018). https://doi.org/10.1109/CCNC.2018.8319292
5. Xiang, W., Wang, N., Zhou, Y.: An energy-efficient routing algorithm for software-defined wireless sensor networks. IEEE Sens. J. **16**(20), 7393–7400 (2016). https://doi.org/10.1109/JSEN.2016.2585019
6. Kumar, N., Vidyarthi, D.P.: A green routing algorithm for IoT-enabled software defined wireless sensor network. IEEE Sens. J. **18**(22), 9449–9460 (2018). https://doi.org/10.1109/JSEN.2018.2869629
7. Heinzelman, W.R., Chandrakasan, A., Balakrishnan, H.: Energy-efficient communication protocol for wireless microsensor networks. Proceedings Hawaii International International Conference on Systems Science, vol. 00, no. c, p. 223 (2000). https://doi.org/10.1109/hicss.2000.926982
8. Misra, S., Kumar, R.: A literature survey on various clustering approaches in wireless sensor network. In: 2nd International Conference on Communication, Control & Intelligent Systems (CCIS 2016), pp. 18–22 (2017). https://doi.org/10.1109/CCIntelS.2016.7878192
9. Ran, G., Zhang, H. and Gong, S.: Improving on LEACH protocol of wireless sensor networks using fuzzy logic. J. Inf. Comput. Sci. **7**(3), 767–775 (2010). https://doi.org/10.1016/j.jradnu.2005.09.004
10. Zhu, X., Shen, L., Yum, T.S.P.: Hausdorff clustering and minimum energy routing for wireless sensor networks. IEEE Trans. Veh. Technol. **58**(2), 990–997 (2009). https://doi.org/10.1109/TVT.2008.926073
11. Singh, B., Lobiyal, D.K.: A novel energy-aware cluster head selection based on particle swarm optimization for wireless sensor networks. Human-Centric Comput. Inf. Sci. **2**(1), 1–18 (2012). https://doi.org/10.1186/2192-1962-2-13
12. Kennedy, J., Eberhart, R.: Particle swarm optimization proceedings. Proceedings of the ICNN'95—International Conference on Neural Networks, vol. 11, no. 1, pp. 111–117 (1995)
13. Huang, Z., Cheng, Y., Liu, W.: A novel energy-efficient routing algorithm in multi-sink wireless sensor networks. In: Proceedings 10th IEEE International Conference on Trust, Security and Privacy in Computing and Communications (2011); 8th IEEE International Conference on

Embedded Software and Systems (ICESS 2011); 6th International Conference on FCST 2011, pp. 1646–1651 (2011). https://doi.org/10.1109/TrustCom.2011.228
14. Sharma, A., Singh, P.K.: Taxonomy on localization issues and challenges in wireless sensor networks. Recent Adv. Electr. Electron. Eng. **13**(2), 193–202 (2020)
15. Singh, P.K., Kar, A.K., Singh, Y., Kolekar, M.H., Tanwar, S.: Proceedings of ICRIC 2019, Recent Innovations in Computing, 2020, Lecture Notes in Electrical Engineering, vol. 597, pp. 3–920. Springer, Cham, Switzerlands

Defining and Evaluating Network Communities Based on Ground-Truth in Online Social Networks

Sanjeev Dhawan, Kulvinder Singh, and Amit Batra

Abstract A social network is a cluster or aggregation of vertices such as persons or social entities, and edges which are used to depict personal relationship between these nodes. Social networks have a noteworthy role in the movement of data, and social network exploration has gained a focus in research. The analysis of these social networks has resulted into uncovering of variety of communities in the network. The main objective of uncovering the structure of a community is to break the network into dense areas of the graph, and these dense areas represent entities which are related closely and hence they belong to a community. Plentiful algorithms have been suggested and recommended, and surveys have been conducted currently. In this manuscript, we will discuss numerous strategies for uncovering the structure of communities and techniques which have been suggested so far. We will divide these algorithms into several categories. These categories correspond to traditional approach of community detection, overlapping community detection, established clustering techniques for uncovering the structure of communities, nonclique-based techniques for uncovering the structure of communities, community detection using genetic algorithms, improved modularity approach for uncovering the structure of communities and so forth. We will start by discussing and understanding several metrics which can be used to ascertain the structure and hence the quality of communities. We will also compare all these community detection algorithms based on approaches used, along with parameters these algorithms depend on.

Keywords Community detection · Social networks analysis · Modularity · Normalized mutual information · Clustering · Overlapping community detection

S. Dhawan · K. Singh · A. Batra (✉)
Department of CSE, University Institute of Engineering and Technology, Kurukshetra University, Kurukshetra, Haryana, India
e-mail: amitbatra2011@gmail.com

S. Dhawan
e-mail: sdhawan2015@kuk.ac.in

K. Singh
e-mail: ksingh2015@kuk.ac.in

P. K. Singh et al. (eds.), *Recent Innovations in Computing*, Lecture Notes in Electrical Engineering 701, https://doi.org/10.1007/978-981-15-8297-4_13

1 Introduction

Since last many years, community discovery has appeared as a milestone in the arena of social graph investigation. The community discovery is performed in such a way that entities or vertices pertaining to a specific community are very alike, analogous, related and similar while they are disparate from vertices or entities that belong to the other communities. Plentiful eminent scientists have conducted a study and stressed on extracting disjoint communities that divide the set of vertices or entities inside a network. Moreover, researchers have found that there is a growth in the intracommunity overlap and suggested and recommended several methods for finding overlapping communities. The networks which present a community arrangement may sometimes show a hierarchical community arrangement also [1]. This paper shows analysis of social networks [2].

2 Community Discovery Algorithms

Fortunato carried out a full evaluation of community discovery in networks in 2010 [3]. Coscia et al. [4] in 2011, Fortunato and Castellano [5] in 2012, Porter et al. [6] in 2009, Danon et al. [7] in 2005 and Plantie and Crampes [8] in 2013 are additional studies in the work.

2.1 Clustering-Based Community Discovery

Out of several most popular clustering-based community structure discovery technique is the Girvan–Newman Community discovery method. It carries out divisions among the vertices. Girvan and Newman [9] proposed an algorithm consisting of following steps:

The edges of the network or graph are removed gradually. The edges to be detached are selected by calculating a metric called betweenness. The betweenness is computed repeatedly for the deletion of each edge. The Girvan–Newman suggested technique successively deletes edges of high betweenness.

Three diverse methods of determining the "edge–betweenness" within nodes of a network have been suggested in the work of Newman and Girvan [10]. Girvan–Newman procedure has been improved by several writers and made practical to several graphs [11–18]. Rattigan et al. [11] suggested the cataloging techniques to decrease the mathematical intricacy of the Girvan–Newman technique considerably. Chen et al. [12] improved the Girvan–Newman technique to divide weighted networks and employed it to recognize serviceable components in the yeast proteome graph. They proposed "weak" and "strong" communities. Moon et al. [19] have suggested and realized the similar form of the Girvan–Newman technique to deal with big data. Newman [20] has applied effort to optimize modularity so that the task of grouping vertices to make communities results in achieving optimal modularity improvement. Newman and Girvan were the first who specified a unit of measurement popularly known as "modularity" to assess the qualitative feature of communities or partitions shaped [10]. AdClust technique [21] can mine components from intricated graphs with noteworthy accuracy and power. Wahl and Sheppard [22] suggested a classified uncertain spectral clustering centered technique. The density-centered network clustering (DENGRAPH) [23] method employs the notion of density-centered progressive clustering of three-dimensional information and is proposed to do the task for big dynamic sets of data containing noise. The Markov clustering technique (MCL) [24] is a network movement simulated method that can be utilized to discover groups in a network and is similar to finding of clusters in the graphs. Nikolaev et al. [25] employed an "entropy centrality metric" centered on the Markovian method to incrementally discover clusters. The technique suggested by Steinhaeuser et al. [26] implements several small random walks and infers vertices which have been visited through the identical walk as alike vertices, which provides a sign that they pertain to the same cluster (Table 1).

2.2 *Modularity Improvement for Community Discovery*

Recall, the definition of modularity.

$$Q = \frac{1}{2m}\sum_{ij}^{n}\left(A_{ij} - \frac{k_i k_j}{2m}\right)\delta(C_i, C_j) \tag{1}$$

Algorithm based on improved modularity [32]:

In the above equation of modularity, $\delta(C_i, C_j)$ will work only in case $C_i = C_j$, meaning that edges inside the communities would be taken into account. One disadvantage of modularity is that it performs poorer in terms of small communities. So, we make changes to the definition of modularity as follows:

$$Q = \frac{1}{2m}\sum_{ij}^{n}\left[\left(A_{ij} - \frac{k_i k_j}{2m}\right)\delta(C_i, C_j) - \beta\left(A_{ij} - \frac{k_i k_j}{2m}\right)^{\alpha}\left(1 - \delta(C_i, C_j)\right)\right]$$

where α and β are the tunable parameters.

$$\left(A_{ij} - \frac{k_i k_j}{2m}\right)\delta(C_i, C_j)$$

is called as intra-factor and

$$\beta\left(A_{ij} - \frac{k_i k_j}{2m}\right)^{\alpha}\left(1 - \delta(C_I, C_j)\right)$$

is called as inter-factor. Simulated annealing is a probabilistic technique for comprehensive maximization. Guimera et al. [29] employed simulated annealing for modularity maximization. By utilizing these two techniques, we expect a quite fast and optimal modularity algorithm. In a latest paper, a expandable form of the technique has been designed employing “MapReduce by Chen et al.” [33]. Newman [34] implemented the “betweenness” method in general for “weighted” graphs. Newman [35] however in another method described the modularity matrix in forms of eigenvectors. Clauset et al. [27] employed a method where modularity can be maximized through

Table 1 Community discovery using clustering

Proposed algorithm	Strategy	Criterion
Clauset et al. [27]	Optimal value of modularity using greedy algorithm	The no. of edges and vertices, modularity value
Girvan and Newman [10]	Disruptive grouping (utilization of modularity as a structure measure)	The value of metric called “edge betweenness”
Newman	Optimization and maximization of modularity	Modularity value, matrix-related eigenvector and eigenvalue
Blondel et al. (Louvain Technique) [28]	Analysis of hierarchical cluster	How many vertices are there, no. of edges, value of “modularity”
Ye et al. (AdClust) [21]	Grouping	Modularity, nodes
Guimera et al. [29], Zhou et al. [30]	Optimal and maximal value of the modularity using probabilistic simulated annealing	Linking probability, no. of relationships in terms of links, no. of partitions, how many modules are there, the value of the modularity, “No. of edges, inter- and intra-aspect”
Wahl and Sheppard [22]	“Hierarchical fuzzy spectral clustering”	“Fuzzy modularity” for fuzzy community structure, Jaccard similarity index or coefficient
Duch et al. [31]	Modularity maximization	How many nodes are there, no. of relationships or links, degree, value of “modularity”
Falkowski et al. [23]	Density-centered grouping	Distance
Steinhauser et al. [26]	Consensus grouping to ascertain community structure, random walk strategy	Resemblance matrix, random walks distance
Dongen et al. (MCL) [24]	Clustering algorithm based on Markovian matrix	How many vertices or nodes are there
Nikolaev et al. [25]	Clustering utilizing efficient use of entropy centrality	A special matrix called transition probability matrix

greedy technique to discover clusters for big graphs. "Blondel et al. [28] implemented an repetitive two stage algorithm called as Louvain algorithm". "Zhou et al. [30] endeavored to maximize modularity employing simulated annealing, presenting the notion of intra and inter edges". Duch et al. [31] suggested a heuristic exploration centered methodology employing an extremal maximization method in order to make best use of the function of modularity having a running time of $O(n^2 \log_2 n)$.

2.3 Community Discovery Using Genetic Algorithms

Pizzuti [36] came with a GA-Net method, which utilizes a graph illustration of the network which was locus based (Table 2). In another flavor of GA, MOGA-Net [37], given by Pizzuti maximizes the values of two objective functions. These are respectively known as score and the fitness of the cluster. The maximum and more the value of the cluster score is, clustering which is obtained is denser [3] and denser and hence good quality community structures are explored. The value of the community fitness can be calculated as the total of fitness of vertices corresponding to a group. As soon as this total achieves its maximal value, the count of exterior relationships across is reduced to its maximum. Hafez [38] proposed an algorithm which takes

Table 2 Community discovery using genetic algorithms

Proposed algorithm	Strategy	Criterion
Pizzuti (GA-Net) [36]	Fitness function using score of the community	Score of the community
Hafez et al. [38]	"Multiobjective", single-objective maximization	How many no. of genes are there, mutation crossover operators
Pizzuti (MOGA-Net) [37]	Multiobjective maximization	Larger the community score denser is the grouping, as the value of community fitness achieves its maximum, no. of external across links get reduced
Liu et al. [40]	GA using clustering	Mass of populace, greatest generation no., optimum no. of "generations" for unaffected "fittest chromosome" portion of extracted centers, how many clusters are there
Zadeh [42]	Multipopulation algorithm	Average dependent on trust space and normative factors
Mazur et al. [39]	Fitness functions utilizing cluster score and modularity	Fitness tasks
Tasgin et al. [41]	Optimal value of modularity	Mass of populace, no. of chromosomes, modularity

both multiobjective optimization and single-objective optimization into consideration for community detection. Mazur et al. [39] have employed the fitness function of modularity along with the score of the community. Liu et al. [40] employed GA along with clustering to determine the cluster arrangements in a graph. "Tasgin et al. [41] have also maximized the graph modularity employing GA". "A multicultural method [42] for cluster discovery uses the fitness function specified by Pizzuti [36] in GA-Net". A genetic algorithm employing maximization of "modularity," suggested by Nicosia et al. [43], was illustrated in the clusters which are "overlapping" in this paper.

3 Community Discovery Techniques

A latest review by Amelio et al. [44] provides a broad survey of main "overlapping cluster" discovery techniques; an effort is made to involve a group of "dynamic" graphs-centered "overlapping community" discovery. Additional work containing comprehensive survey of techniques for detecting overlapping clusters is performed by Xie et al. [45].

3.1 Overlapping Community Detection Using Nonclique-Based Techniques

Few more nonclique techniques to detect clusters which are overlapping are provided in Table 3. A task that employs a technique to overlying cluster discovery is GA-NET+ suggested by Pizzuti [46]. A method [47] to discover communities in graphs which can hold several types of network features, for example, edge weights, hierarchy, edge direction, clusters which are overlapping, and graph dynamics called "order statistics local optimization method (OSLOM)" has been suggested. Baumes et al. [48] has taken into consideration a cluster as a subclass of vertices that persuades a nearby ideal subnetwork with reference to a special function called density function. Chen et al. [49] employ a "game-theoretic" technique to report the problem of clusters which are "overlapping." In other game-theoretic technique, Alvari et al. [50] suggested a technique comprising of two subtechniques, "PSGAME" employing "Pearson correlation and NGGAME" employing adjacent resemblance metric. Alvari et al. [51] suggested a "dynamic game theory" technique "(D-GT)," which considered vertices as "rational agents." A "link-clustering"-centered "GA, GaoCD," suggested by Shi et al. [52] discovers clusters which are overlapping. The "Overlapping Community" Discovery by Local Cluster Extension (OCDLCE) technique [53] for clusters which are overlapping was centered on cluster enlargement. Bhat et al. [54] suggested a novel density-centered cluster discovery method, OCMiner. Zhang et al. [55] have incorporated the conception of implied link predictions in their prototype.

Table 3 Overlapping community detection using nonclique-based techniques

Proposed algorithm	Strategy	Criterion
Pizzuti (GA-NET+) [46]	Genetic algorithm centered	Score of the community
Baumes et al. [48]	Communities of nodes which are overlapping	Internal edge probability, intensity ratio, intensity of external edge, edge ratio, intensity of internal edge
Alvari et al. [50, 51]	Game philosophy centered	Group of "snapshots," nodes and links
Xing et al. [53]	Cluster discovery	Vertices, links, adjacent of vertex
Zhang et al. [55]	"Preference-centered non-negative matrix" categorization	No. of vertices, edges, clusters
Whang et al. [57]	"Seed growth"	Vertices, edges, no. of "seeds, PageRank link" resulting metric
"Nicosia et al." [43]	"Modularity" for clusters which are "overlapping" GA technique	Out degree, belongingness coefficient, in degree
Lancichinetti et al. [47]	Link weights, direction	nodes, links, internal and external subnetwork, degree of subnetwork
Chen et al. [49]	"Game" philosophy centered	Set of clusters, loss and gain function
Shi et al. (GaoCD) [52]	"Partition density" as objective task	Running generation proportion of crossover, size of population, ratio of mutation
Bhat et al. [54]	Density centered	Threshold
Kozdoba et al. [56]	Community Aggregation	Number of components, probability metrics, threshold metric
Rees et al. [58]	Friendship clusters, ego graphs	Number of vertices

In a "clustering-centered" technique, Kozdoba et al. [56] introduced a technique Community Accumulation for Overlying Clusters (CLAGO) in which the task to discover overlapped clusters is partitioned into two stages. "Whang et al. [57] have suggested neighborhood inflated seed expansion (NISE)," a seed-centered growth technique to discover clusters. Rees et al. [58] have employed in their effort, the notion of friendship and ego-networks.

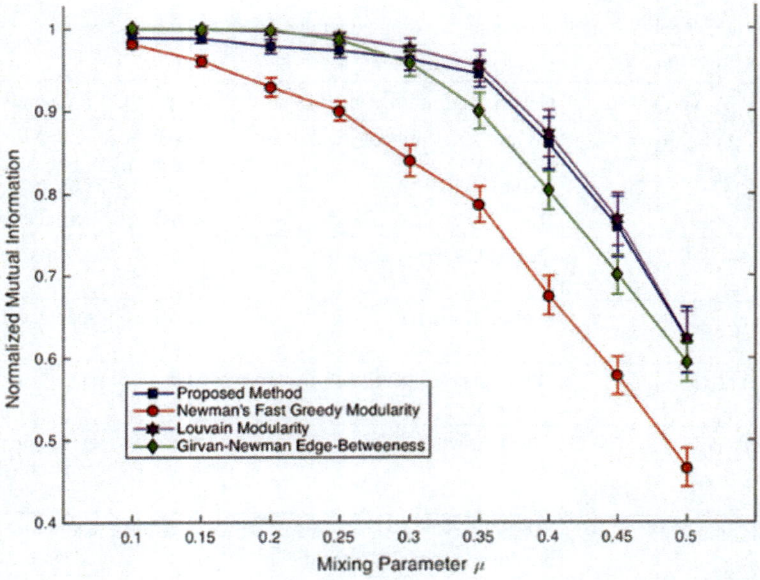

Fig. 1 NMI versus mixing parameter

4 Standard Datasets

Datasets like Zachary [59] Karate club graph which depicts the relationship among 34 people of a karate club come under the category of real-time datasets. Dolphin social graph illustrates the behavior and social collaborations of dolphins for a duration of seven years as examined by Lusseau et al. [60]. There are also few more real-time datasets, for example, the Southern women dataset [61], etc. A technique to produce standard datasets was suggested by Lancichinetti et al., called as benchmark [62]. Another example of real-time input data are American College Football network [9] datasets. On the contrary, example of artificial datasets can be one given by Girvan–Newman, where four communities can be identified starting from 128 vertices. Figure 1 depicts how NMI varies with value of mixing parameter.

5 Critical Applications of Community Detection

Cao et al. [63] have utilized a community discovery technique so as to make the cooperative filtering to do better in the recommendation process. Enhanced fuzzy C-mean-based cluster discovery has been carried out in social graphs employing dynamic parallelism by Mahmoud Al-Ayyoub et al. [64]. A novel trust-centered cluster discovery technique in social graphs has been suggested by Chen et al. [65]. Anew expandable leader-cluster discovery technique for cluster discovery in social graphs has been employed by Ahajjam et al. [66]. User interest cluster discovery on social graphs has been carried out employing collaborative filtering by Jiang et al. [67]. Emotional cluster discovery in social graphs has been suggested by Kanavos et al. [68]. An incremental technique to discover clusters in dynamic evolving social graphs has been suggested by Zhao et al. [69]. Zheng et al. [70] employed privacy-preserved cluster discovery in online social graphs. The area of cluster discovery has huge scope of discovering the good quality clusters in several big complex and online social networks. Furthermore, wireless sensor networks [71] can be integrated with online social networks to provide insights into the communication links and patterns between people. Moreover, smart networks can be built by integrating social and wireless sensor networks as proposed in [72]. The work proposed in [73, 74] can be utilized in near future for detecting high-quality community structures.

6 Conclusion and Future Research Directions

The field of community discovery has massive scope of discovering the good-quality communities in several large complex and online social networks. Several basic concepts like metrics used in community detection and community discovery strategies have been presented in this critical review article. These community detection techniques can be practical to several "real-world" social "networks like Facebook, Twitter, Instagram and LinkedIn," etc. to discover good-quality communities along with substantial amount of information which helps us to understand and visualize these networks. Such vast information can serve a useful purpose in different fields like psychology, sociology, biology and several other fields of science.

References

1. Ozturk, K.: Community Detection in Social Networks. Middle East Technical University, Graduate School of Natural and Applied Sciences (2014)
2. Tang L., Liu H.: Community Detection and Mining in Social Media, vol. 2, no. 1 (2010)
3. Fortunato, S.: Community detection in graphs. Phys. Rep. **486**(3–5), 75–174 (2010)
4. Coscia, M., Giannotti, F., Pedreschi, D.: A classification for community discovery methods in complex networks. Stat. Anal. Data Min. **4**(5), 512–546 (2011)
5. Fortunato S., Castellano C.: Community structure in graphs. Comput. Complex. Theory, Tech. Appl. **9781461418**, 490–512 (2012)
6. Porter, M.A., Onnela, J.P., Mucha, P.J.: Communities in networks. Not. Am. Math Soc. **56**, 1082–1097 (2009)
7. Danon, L., Díaz-Guilera, A., Duch, J., Arenas, A.: Comparing community structure identification. J. Stat. Mech. Theory Exp. **09008**(9), 219–228 (2005)
8. Plantié, M. and Crampes, M.: Survey on social community detection. Soc. Media Retr. Comput. Commun. Networks. London SpringerVerlag, 65–85 (2013)
9. Girvan, M., Newman, M.E.: Community structure in social and biological networks. Proc. Natl. Acad. Sci. **99**, 7821–7826 (2002)

10. Newman M. E. J., Girvan M.: Finding and evaluating community structure in networks. Phys. Rev. E - Stat. Nonlinear, Soft Matter Phys. **69**(22), 1–15 (2004)
11. Rattigan, M.J., Maier, M., Jensen, D.: Graph clustering with network structure indices. ACM Int. Conf. Proc. Ser. **227**, 783–790 (2007)
12. Chen, J., Yuan, B.: Detecting functional modules in the yeast protein-protein interaction network. Bioinformatics **22**(18), 2283–2290 (2006)
13. Holme, P., Huss, M., Jeong, H.: Subnetwork hierarchies of biochemical pathways. Bioinformatics **19**(4), 532–538 (2003)
14. Pinney, J.W., Westhead, D.R.: Betweenness-based decomposition methods for social and biological networks. Soft Matter, 87–90 (2005)
15. Gregory, S.: An algorithm to find overlapping community structure in networks. Lect. Notes Comput. Sci. (including Subser. Lect. Notes Artif. Intell. Lect. Notes Bioinformatics), vol. 4702 LNAI, pp. 91–102, (2007)
16. Guimerà, R., Danon, L., Díaz-Guilera, A., Giralt, F., Arenas, A.: Self-similar community structure in a network of human interactions. Phys. Rev. E - Stat. Physics, Plasmas, Fluids, Relat. Interdiscip. Top. **68**(6), 1–4 (2003)
17. Arenas, A., Danon, L., Díaz-Guilera, A., Gleiser, P.M., Guimerà, R.: Community analysis in social networks. Eur. Phys. J. B **38**(2), 373–380 (2004)
18. Tyler, J.R., Wilkinson, D.M., Huberman, B.A.: E-Mail as spectroscopy: automated discovery of community structure within organizations. Inf. Soc. **21**(2), 133–141 (2005)
19. Moon, S., Lee, J.G., Kang, M., Choy, M., Woo, L.J.: Parallel community detection on large graphs with MapReduce and GraphChi. Data Knowl. Eng. **104**, 17–31 (2016)
20. Newman, M.E.J.: Fast algorithm for detecting community structure in networks. Phys. Rev. E - Stat. Physics, Plasmas, Fluids, Relat. Interdiscip. Top. **69**(6), 5 (2004)
21. Ye, Z., Hu, S., Yu, J.: Adaptive clustering algorithm for community detection in complex networks. Phys. Rev. E - Stat. Nonlinear, Soft Matter Phys. **78**(4), 1–6 (2008)
22. Wahl, S., Sheppard J.: Hierarchical fuzzy spectral clustering in social networks using spectral characterization. In: Proceedings of the 28th International Florida Artificial Intelligence Research Society Conference, FLAIRS 2015, pp. 305–310 (2015)
23. Falkowski, T., Barth, A., Spiliopoulou, M.: DENGRAPH: a density-based community detection algorithm. In: Procedings of the IEEE/WIC/ACM International Conference on Web Intelligence WI 2007, pp. 112–115 (2007)
24. Dongen, S.V.: Graph clustering by flow simulation. University of Utrecht (2000)
25. Nikolaev, A.G., Razib, R., Kucheriya, A.: On efficient use of entropy centrality for social network analysis and community detection. Soc. Netw. **40**, 154–162 (2015)
26. Steinhaeuser, K., Chawla, N.V.: Identifying and evaluating community structure in complex networks. Pattern Recognit. Lett. **31**(5), 413–421 (2010)
27. Clauset, A., Newman, M.E.J., Moore, C.: Finding community structure in very large networks. Phys. Rev. E - Stat. Physics Plasmas Fluids Relat. Interdiscip. Top. **70**(6), 6 (2004)
28. Blondel, V.D., Guillaume, J.L., Lambiotte, R., Lefebvre, E.: Fast unfolding of communities in large networks. J. Stat. Mech. Theory Exp. **10**, 2008 (2008)
29. Guimerà, R., Sales-Pardo, M., Amaral, L.A.N.: Modularity from fluctuations in random graphs and complex networks. Phys. Rev. E - Stat. Physics, Plasmas, Fluids, Relat. Interdiscip. Top. **70**(2), 4 (2004)
30. Zhou, Z., Wang, W., Wang, L.: Community detection based on an improved modularity. Commun. Comput. Inf. Sci. **321 CCIS**, 638–645 (2012)
31. Duch, J., Arenas, A.: Community detection in complex networks using extremal optimization. Phys. Rev. E - Stat. Nonlinear Soft Matter Phys. **72**(2), 1–4 (2005)
32. Zhou, Z., Wang, W., Wang, L.: Community detection based on an improved modularity. In: Liu, C.L., Zhang, C., Wang, L. (eds.) Pattern Recognition. Communications in Computer and Information Science, vol. 321, Springer (2012)
33. Chen, Y., Huang, C., Zhai, K.: Scalable Community Detection Algorithm with MapReduce. Commun. ACM **53**, 359–366 (2009)

34. Newman, M.E.J.: Analysis of weighted networks. Phys. Rev. E - Stat. Physics, Plasmas, Fluids, Relat. Interdiscip. Top. **70**(5), 9 (2004)
35. Newman, M.: Modularity and community structure in networks. Proc. Natl. Acad. Sci. **103**, 8577–8582 (2006)
36. Pizzuti, C.: GA-Net: a genetic algorithm for community detection in social networks. Lect. Notes Comput. Sci. (including Subser. Lect. Notes Artif. Intell. Lect. Notes Bioinformatics), vol. 5199 LNCS, pp. 1081–1090 (2008)
37. Pizzuti, C.: A multiobjective genetic algorithm to find communities in complex networks. IEEE Trans. Evol. Comput. **16**(3), 418–430 (2012)
38. Hafez, A.I., Ghali, N.I., Hassanien, A.E., Fahmy, A.A.: Genetic algorithms for community detection in social networks. Int. Conf. Intell. Syst. Des. Appl. ISDA,. 460–465 (2012)
39. Mazur, P., ZmarzŁowski, K., OrŁowski, A.J.: Genetic algorithms approach to community detection. Acta Phys. Pol. A **117**(4), 703–705 (2010)
40. Liu, X., Li, D., Wang, S., Tao, Z.: Effective algorithm for detecting community structure in complex networks based on GA and clustering. Lect. Notes Comput. Sci. (including Subser. Lect. Notes Artif. Intell. Lect. Notes Bioinformatics), vol. 4488 LNCS, no. PART 2, pp. 657–664 (2007)
41. Tasgin, M., Herdagdelen, A., Bingol, H.: Community detection in complex networks using genetic algorithms. arXiv Prepr. arXiv (2007)
42. Zadeh, P.M., Kobti, Z.: A multi-population cultural algorithm for community detection in social networks. Procedia Comput. Sci. **52**(1), 342–349 (2015)
43. Nicosia, V., Mangioni, G., Carchiolo, V., Malgeri, M.: Extending the definition of modularity to directed graphs with overlapping communities. J. Stat. Mech. Theory Exp. **3**, 2009 (2009)
44. Amelio, A., Pizzuti, C: Overlapping community discovery methods: a survey. In: Social Networks: Analysis and Case Studies. Lecture Notes in Social Networks, pp. 105–125. Weinheim Springer-Verlag (2014)
45. Xie, J., Kelley, S., Szymanski, B.K.: Overlapping community detection in networks: the state-of-the-art and comparative study. ACM Comput. Surv. **45**(4) (2013)
46. Pizzuti, C.: Overlapped community detection in complex networks. Proceedings of the 11th Annual Conference on Genetic and Evolutionary Computation (GECCO-2009), pp. 859–866 (2009)
47. Lancichinetti, A., Radicchi, F., Ramasco, J.J., Fortunato, S.: Finding statistically significant communities in networks. PLoS One 6(4) (2011)
48. Baumes, J., Goldberg M., Krishnamoorthy, M., Magdon-Ismail, M., Preston, N.: Finding communities by clustering a graph into overlapping subgraphs. Int. Conf. Appl. Comput. (IADIS 2005), pp. 97–104, 2005.
49. Chen, W., Liu, Z., Sun, X., Wang, Y.: A game-theoretic framework to identify overlapping communities in social networks. Data Min. Knowl. Discov. **21**(2), 224–240 (2010)
50. Alvari, H., Hashemi, S., Hamzeh, A.: Detecting overlapping communities in social networks by game theory and structural equivalence concept. Lect. Notes Comput. Sci. (including Subser. Lect. Notes Artif. Intell. Lect. Notes Bioinformatics), vol. 7003 LNAI, no. PART 2, pp. 620–630 (2011)
51. Alvari, H., Hajibagheri, A., Sukthankar, G.: Community detection in dynamic social networks: A game-theoretic approach. In: ASONAM 2014—Proceedings of the 2014 IEEE/ACM International Conference on Advances in Social Networks Analysis and Mining, pp. 101–107 (2014)
52. Shi, C., Cai, Y., Fu, D., Dong, Y., Wu, B.: A link clustering based overlapping community detection algorithm. Data Knowl. Eng. **87**, 394–404 (2013)
53. Xing, Y., Fanrong, M., Yong, Z., Ranran, Z.: Overlapping community detection by local community expansion. J. Inf. Sci. Eng. **31**(4), 1213–1232 (2015)
54. Bhat, S.Y., Abulais, M.: OCMiner: A density-based overlapping community detection method for social networks. Intell. Data Anal. **19**(4), 917–974 (2015)
55. Zhang, H., King, I., Lyu, M.R.: Incorporating implicit link preference into overlapping community detection. Proc. Natl. Conf. Artif. Intell. **1**, 396–402 (2015)

56. Kozdoba, M., Mannor, S.: Overlapping Community Detection by Online Cluster Aggregation. arXiv Prepr. arXiv1504.06798, pp. 1–15 (2015)
57. Whang, J.J., Gleich, D.F., Dhillon, I.S.: Overlapping community detection using neighborhood-inflated seed expansion. IEEE Trans. Knowl. Data Eng. **28**(5), 1272–1284 (2016)
58. Rees, B.S., Gallagher, K.B.: Overlapping community detection by collective friendship group inference. In: Proceedings of the 2010 nternational Conference on Advances in Social Networks Analysis and Mining, ASONAM 2010, pp. 375–379 (2010)
59. WW Z.: An information flow model for conflict and fission in small groups. J Anthr. Res, vol. 33, pp. 452–473, 1977.
60. Lusseau, D., Schneider, K., Boisseau, O.J., Haase, P., Slooten, E., Dawson, S.M.: The bottlenose dolphin community of doubtful sound features a large proportion of long-lasting associations: Can geographic isolation explain this unique trait? Behav. Ecol. Sociobiol. **54**(4), 396–405 (2003)
61. Clark, T.D., Davis, A., Gardner, B.B., Gardner, M.R., Warner, W.L.: Deep South: A Social Anthropological Study of Caste and Class. J. South. Hist. **8**(3), 439 (1942)
62. Lancichinetti, A., Fortunato, S.: Benchmarks for testing community detection algorithms on directed and weighted graphs with overlapping communities. Phys. Rev. E - Stat. Nonlinear Soft Matter Phys **80**(1), 1–8 (2009)
63. Cao, C., Ni, Q., Zhai, Y.: An improved collaborative filtering recommendation algorithm based on community detection in social networks. In: Proceedings of the 2015 Annual Conference on Genetic and Evolutionary Computation, ACM, pp. 1–8 (2015)
64. Al-Ayyoub, M., Al-andoli, M., Jararweh, Y., Smadi, M., Gupta, B.: Improving fuzzy C-mean-based community detection in social networks using dynamic parallelism. Comput. Electr. Eng. **74**, 533–546 (2019)
65. Chen, X., Xia, C., Wang, J.: A novel trust-based community detection algorithm used in social networks. Chaos, Solitons Fractals **108**, 57–65 (2018)
66. Ahajjam, S., El Haddad, M., Badir, H.: A new scalable leader-community detection approach for community detection in social networks. Soc. Netw. **54**, 41–49 (2018)
67. Jiang, L., Shi, L., Liu, L., Yao, J., Yousuf, M.A.: User interest community detection on social media using collaborative filtering. Wirel. Netw. **4** (2019)
68. Kanavos, A., Perikos, I., Hatzilygeroudis, I., Tsakalidis, A.: Emotional community detection in social networks. Comput. Electr. Eng. **65**, 449–460 (2018)
69. Zhao, Z., Li, C., Zhang, X., Chiclana, F., Viedma E.H.: An incremental method to detect communities in dynamic evolving social networks. Knowledge-Based Syst. **163**(Chao Li), 404–415 (2019)
70. Zheng, X., Cai, Z., Luo, G., Tian, L., Bai, X.: Privacy-preserved community discovery in online social networks. Futur. Gener. Comput. Syst. **93**, 1002–1009 (2019)
71. Singh, P.K., Paprzycki, M.: Introduction on wireless sensor networks ıssues and challenges in current era. In: Singh, P.K., Bhargava, B.K., Paprzycki, M., Kaushal, N.C., Hong, W.C. (eds.) Handbook of Wireless Sensor Networks: Issues and Challenges in Current Scenario's, Advances in Intelligent Systems and Computing, vol. 1132, pp. 3–12. Springer, Cham, Switzerland (2020)
72. Meet, K., Reecha, S., Anu, S.: A review on hybrid WSN-NGPON2 network for smart world. In: Singh, P.K., Bhargava, B.K., Paprzycki, M., Kaushal, N.C., Hong, W.C. (eds.) Handbook of Wireless Sensor Networks: Issues and Challenges in Current Scenario's, Advances in Intelligent Systems and Computing, vol. 1132, pp. 655–671. Springer, Cham, Switzerland (2020)
73. Chaudhary, L., Singh, B.: Community detection using maximizing modularity and similarity measures in social networks. In: Somani A., Shekhawat R., Mundra A., Srivastava S., Verma V. (eds) Smart Systems and IoT: Innovations in Computing. Smart Innovation, Systems and Technologies, vol. 141, pp. 197–206. Springer (2020)

74. Fatima, S., Badugu, S.: A study on overlapping community detection for multimedia social network. In: Satapathy, S., Raju, K., Shyamala, K., Krishna, D., Favorskaya, M. (eds.) Advances in Decision Sciences, Image Processing, Security and Computer Vision. Learning and Analytics in Intelligent Systems, vol. 4., pp. 572–578. Springer (2020)

Microstrip Patch Antenna with Truncated Edges for Bandwidth Improvement for Wireless Applications

Amandeep Kaur and Praveen Kumar Malik

Abstract In the research article, microstrip patch antenna with truncated edges is designed. Truncated technique is used for bandwidth improvement. Antenna is fabricated using Rogers RT Duroid (5880) substrate with $h = 1.6$ mm and dielectric constant of 2.2, loss tangent 0.0009. Microstrip feed line used to excite patch. Antenna overall size is 12×35 mm. Proposed antenna simulation is carried out using HFSS antenna simulation software, and performance is analyzed using antenna parameter like return loss (S11), VSWR, radiation pattern, Bandwidth, and Gain.

Keywords Gain · Patch antenna · Return loss · VSWR

1 Introduction

The electromagnetic signals are trans-received or broadcast through free space or atmosphere to in-cooperate two or more than two devices is known as wireless communication. The task of radiating and receiving the electromagnetic energy is performed within metallic devices such as a wire, plane sheet or may be a rod is termed as an antenna per IEEE (1969). For more than last seven decades, the antenna in the field of technology is one of the vigorous and essential parts in wireless communication revolution [1]. However, many challenges related to field of antenna are still facing today like design of small size, wideband, and high-gain antennas. Moreover, there is need to make them also compatible with small size communication systems [2]. Nevertheless, as the overall size of communication systems are also becoming smaller day by day due to advancement in VLSI technologies, there is need to further reduce the size of antennas by unchanging existing properties. One of the possible solution to this problem is micro-strip patch antennas (MSPA), which are extremely

A. Kaur · P. K. Malik (✉)
Lovely Professional University, Phagwara, Punjab, India
e-mail: pkmalikmeerut@gmail.com

A. Kaur
e-mail: aman.dhaliwal18@gmail.com

P. K. Singh et al. (eds.), *Recent Innovations in Computing*, Lecture Notes in Electrical Engineering 701, https://doi.org/10.1007/978-981-15-8297-4_14

popular for wireless communication due to their special advantages over other type of antennas [3]. Microstrip patch antenna is in very much demand due to various advantages like compact in size, light weight, easy fabrication, and integration with microwave circuits, which is suitable to use on many wireless communication products like mobile phones, embedded, IoT, aircrafts, satellite, and missile applications [4]. Instead of numerous advantages, microstrip antenna has some shortcomings also like less gain and bandwidth. In the literature, various techniques are used to improve gain and bandwidth like use of EBG structures, shorting pins, air as substrate, slots in patch, defected ground structures, parasitic patch, truncating its corners, using substrates of high thickness and lower dielectric constant can be implemented [5–7]. In this proposed design, to improvement bandwidth of antenna concept of truncated techniques is used. Microstrip patch is considered due to its simplicity in design as base structure. For bandwidth enhancement, all four corners of patch are truncated with 2 mm. Antenna simulation is done using HFSS, and antenna performance is analyzed in terms of bandwidth, gain, VSWR, and return loss. Proposed antenna shows good performance for higher frequencies for wireless applications [8]. Nowadays, there is huge requirement of an antenna which is having small in size, light in weight, high in performance, easy to install, and smooth in shape. So a microstrip antenna is best suited option which fulfills all necessary requirements [9]. Performances of the proposed antenna are compared with that of some already existing papers in terms of their frequency bands, dimensions, structures, and bandwidth. In comparison table, we can found that antennas in [10] are single band antenna with less bandwidth, gain, and large size as compared to proposed design. The antennas in [11] are multi-band antenna but have large geometry, small bandwidth, and low-gain contrast with the presented antenna. The two square patches are placed at an edge to edge distance of 0.012 lambda and center to center distance of 0.027 lambda from each other. The isolation between the two antenna elements is increase from 2.8 db to 20 dB [12]. Two-symmetrical nine shaped rectangular patch MIMO antenna for wireless LAN and X band application designed by author. The geometry of the proposed antenna consists of three slots. First slot at the upper layer and remaining two slots at the ground plane which produce defective ground structure. Traditionally, the technology evolved till 4G was based on hardware improvements and enhancements, which is different when the discussion arises about 5G. The software also has equal importance and significance in the technology enhancement. The technology can be used to increase the connectivity of performance with 100 times of user data rate, better connectivity between devices, and around 1000 mobile data volume. Multi-hop transmission can help be more effective when the communication range to be increased. Half-duplex orthogonal cooperative protocol is used to establish long-range communication in which other nearby vehicles can be used as transponders [13, 14].

2 Antenna Design

In proposed antenna design, microstrip patch is considered due to its simple configuration and ease fabrication [15–17]. For antenna designing, both patch and ground consist of copper. Rogers RT Duroid (5880) substrate with thickness 1.6 mm and dielectric constant of 2.2, loss tangent 0.0009 is used, as bandwidth of antenna highly depends on substrate thickness, dielectric constant, partial or full ground etc. Dimensions for patch considered are 12 mm × 15 mm with edges truncated. In proposed design, partial ground plan is used instead of full ground with dimensions 12 mm × 19 mm, as shown in Table 1. Antenna radiating patch is excited using microstrip feed line with characteristics impedance of 50 Ω. Proposed antenna patch and top view are shown in Fig. 1.

Table 1 Proposed antenna dimensions

Parameters	Dielectric constant	Substrate material	Loss tangent	Height of substrate
Value	2.2	Rogers RT duroid	0.009	1.6 mm
Parameters	Substrate length(Ls)	Substrate width (Ws)	Length of Gnd (Lg)	Width of Gnd (Wg)
Value	35 mm	12 mm	19 mm	12 mm
Parameters	Patch length (Lp)	Patch width (Wp)	Feed line length (Lfl)	Feed line width (Wfl)
Value	15 mm	12 mm	20 mm	2 mm

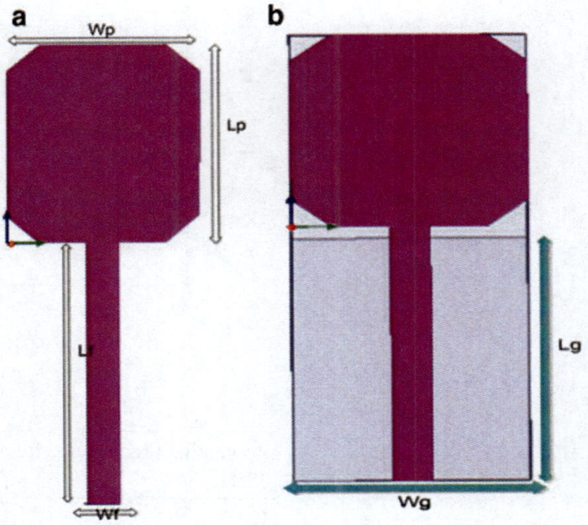

Fig. 1 **a** Patch view. **b** Top View

3 Results and Discussions

To analyzed performance antenna is simulated from 1 to 20 GHz. Proposed antenna performance is simulated using HFSS antenna simulation software in terms of return loss, gain, bandwidth, and VSWR.

3.1 Return Loss

Retun loss is the ratio of the reflected antenna power to the antenna incident power, defined in decibles (dB). Ideally, value of return loss should be low because it indicates the reflected power back [18, 19]. Figure 2 depicts the return loss versus frequency performance of proposed antenna. It is found that antenna resonates at three frequency bands 10.92 GHz, 15.03 GHz, and 18.66 GHz with return loss of −23.43, −24.86, and −17.23 dB, respectively. The wide bandwidth achieved is at 15.03 GHz frequency from 14.27 GHz to 17.99 GHz, i.e., 3720 MHz in comparison to other frequency bands as reflected in Table 2.

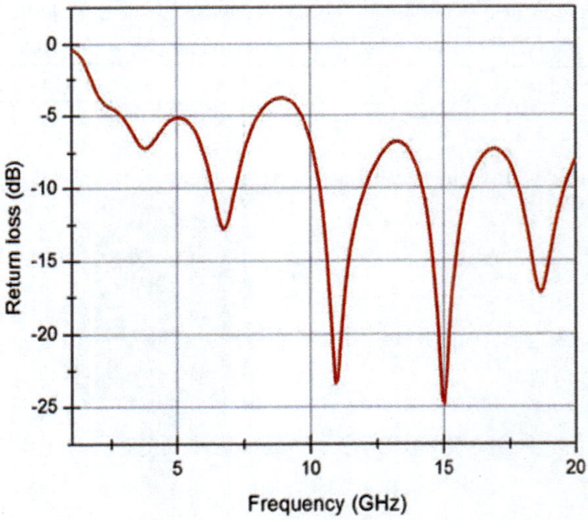

Fig. 2 Proposed antenna return loss performance

Table 2 Bandwidth achieved with return loss

Resonating Frequency (GHz)	Return loss (dB)	Bandwidth (MHz)
10.92	−23.43	1620(10.35–11.97)
15.03	−24.86	3720(14.27–17.99)
18.66	−17.23	1430(18.09–19.52)

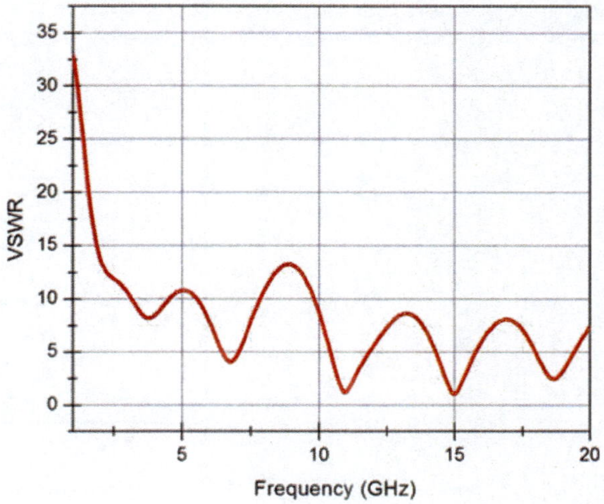

Fig. 3 VSWR versus frequency plot

3.2 Voltage Standing Wave Ratio (VSWR)

VSWR reflects impedance matching between patch and feed line. From ideal conditions, VSWR value lies between 0 and 3 [20, 21]. For proposed antenna, at all resonating frequency bands values of VSWR lies in range of 0 to 2 that is acceptable. Value of VSWR achieved for proposed structure at resonating frequencies 10.92, 15.03, 18.66 GHz is 1.17, 0.99, and 2.40, respectively, as shown in Fig. 3.

3.3 Gain

For good antenna performance, gain of antenna should be more than 3 dB. Antenna gain represents mainly two performance parameters directivity and electrical efficiency. For proposed antenna, figure [4–6] represents the gain versus frequency response for resonating bands 10.92, 15.03, and 18.66 GHz for different values of Phi and Theta.

As depicted from Fig. 4, antenna gain at 10.92 GHz resonating frequency with different values of phi and theta, i.e., Phi = 110° and Theta = 140°, respectively. As illustrated from graph for resonating band 10.35 GHz to 11.97 GHz, gain of proposed antenna varies between 4.44 dB and 4.73 dB, respectively. Maximum gain achieved at frequency 10.92 GHz is 4.68 dB.

Also, as reflected in Fig. 5, for operating bandwidth of 14.27 GHz to 17.99 GHz, antenna gain varies from 4.64 dB to 4.24 dB with maximum values of gain achieved at resonating frequency 15.03 GHz; that is, 4.74 dB at Phi = 300° and Theta = 120°.

Similarly, as illustrated from Fig. 6, at frequency 18.66 GHz, value of gain achieved is 8.12 dB and varies between 8.16 and 7.34 dB for resonating band

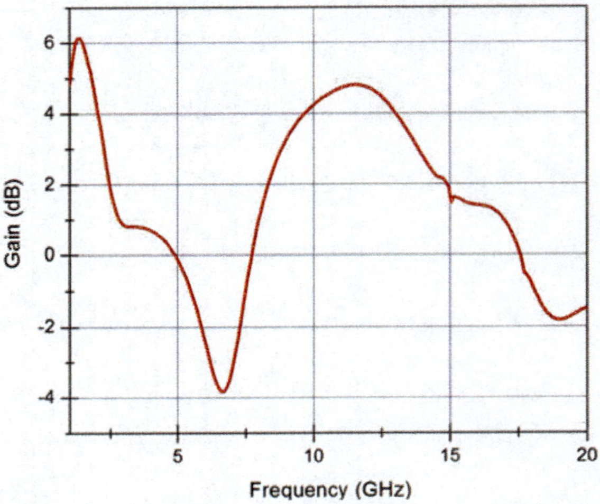

Fig. 4 Gain at 10.92 GHz (Phi = 110°, Theta = 140°)

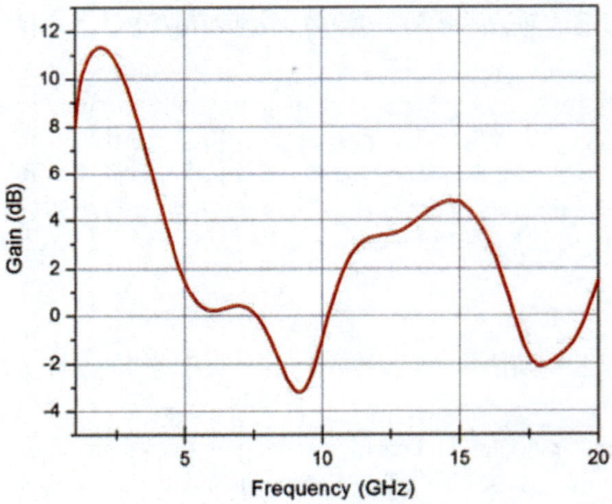

Fig. 5 Gain at 15.03 GHz (Phi = 300°, Theta = 120°)

18.09 GHz to 19.52 GHz at phi = 70° and Theta = 60°. Maximum value of gain achieved is at operating frequency 18.66 GHz, i.e., 8.12 dB.

4 Conclusion

New microstrip patch antenna with truncated edges is designed for bandwidth enhancement for wireless applications. Rectangular patch is truncated to improve the antenna bandwidth for high frequencies. Antenna results are simulated using HFSS simulator, and performance is analyzed in terms of gain, bandwidth, return

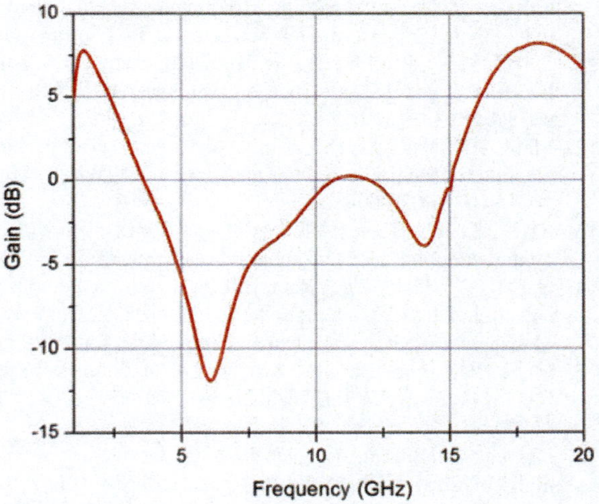

Fig. 6 Gain at 15.03 GHz (Phi = 300°, Theta = 120°)

loss, and VSWR. It is analyzed from simulation results that antenna exhibits multi-band characteristics that is able to cover large frequency bands from 10.92 GHz to 18.66 GHz. Proposed antenna shows good bandwidth of 1620 MHz, 3720 MHz, and 1430 MHz for resonating frequencies 10.92 GHz, 15.03 GHz, and 18.66 GHz, respectively. Antenna shows good performance for these frequency bands for various wireless applications.

References

1. Balanis, C.A.: Antenna Theory, Analysis and Design, 2nd edn. Wiley, New York (1997)
2. Roy, J.S., Thomas, M.: Compact and broadband microstrip antennas for next generation high-speed wireless communication using HIPERLAN/2. Int. J. Microwave Sci. Technol. **2007**, 1–4 (2007)
3. Garg, R., Bhartia, P., Bahl, I., Ittipiboon, A.: Microstrip Antenna Design Handbook. Artech House Inc., Boston (2001)
4. McKinzie, W.E.: 60GHz 2x2 LTCC patch antenna array with an integrated EBG structure for gain enhancement. IEEE J. Antenna Wave Propag. Lett. **15**, 1522–1525 (2016)
5. Zhang, X.: Gain enhanced patch antennas with loading of shorting pins. IEEE Trans. Antennas Propag. **64**, 3310–3318 (2016)
6. Guha, D.: Estimation of gain enhancement replacing PTFE by air substrate in microstrip patch antenna. IEEE Antennas Propag. Mag. **52**, 92–92 (2010)
7. Chen, J.: A dual-band dual-polarization slot patch antenna for GPS and Wi-Fi applications. IEEE Antennas Wirel. Prorog. Lett. **15**, 406–409 (2016)
8. Jilani, S.F.: Millimetre-wave T-shaped MIMI antenna with defected ground structures for 5G cellular network. IET Microwaves J. Antennas Propag. **12**, 672–677 (2018)
9. Mishra, G.P.: Miniaturised microstrip patch design based on highly capacitive defected ground structure with fractal boundary for X-band microwave communications. IET Microwaves J. Antennas Propag. **13**, 1593–1601 (2019)
10. Malik P.K., Parthasarthy H., Tripathi M.P.: Alternative mathematical design of vector potential and radiated fields for parabolic reflector surface. In: Unnikrishnan S., Surve S., Bhoir D.

(eds) Advances in Computing, Communication, and Control. ICAC3 2013. Communications in Computer and Information Science, vol 361. Springer, Berlin, Heidelberg (2013)
11. Surjati, I., Haidi, J.: Increasing bandwidth triangular microstrip antenna using parasitic patch. In: IEEE Topical Conference on Antennas and Propagation in Wireless Communications, pp. 809–812
12. Malik, P.K., Wadhwa, D.S., Khinda, J.S.: A survey of device to device and cooperative communication for the future cellular networks. Int. J. Wireless Inf. Netw. (2020). https://doi.org/10.1007/s10776-020-00482-8
13. Xi, L., Zhai, H., Li, L.: A low profile antenna system with compact new structure for reducing mutual coupling. J. Electromag. Waves Appl. **33**, 71–83 (2019)
14. Gupta, N.P., Malik, P.K., Ram, B.S.: A review on methods and systems for early breast cancer detection. In: 2020 International Conference on Computation, Automation and Knowledge Management (ICCAKM), Dubai, United Arab Emirates, pp. 42–46 (2020)
15. Malik, P., Parthasarthy, H.: Synthesis of randomness in the radiated fields of antenna array. Int. J. Microwave Wirel. Technol. **3**(6), 701–705 (2011). https://doi.org/10.1017/S1759078711000791
16. Dkiouak, A., Zakriti, A., Quahabi, M.E.: Design of a compact dual-band MIMO antenna with high isolation for WLAN and X -band satellite by using orthogonal polarization. J. Electromag. Waves Appl., 1–14 (2019). https://doi.org/10.1080/09205071.2019.1657504
17. Kaur, A., Malik, P.K.: Tri State, T shaped circular cut ground antenna for higher 'X' band frequencies. In: 2020 International Conference on Computation, Automation and Knowledge Management (ICCAKM), Dubai, United Arab Emirates, pp 90–94 (2020)
18. Ghannad, A.A., Khalily, AM., Xiao, P., Tafazolli, R., Kishk, A.A.: Enhanced matching And vialess decoupling of nearby patch antennas for MIMO system. IEEE Antennas Wirel. Propog. Lett. **18**(6) (2019)
19. Tiwari, P., Malik, P.K.: Design of UWB antenna for the 5G mobile communication applications: a review. In: 2020 International Conference on Computation, Automation and Knowledge Management (ICCAKM), Dubai, United Arab Emirates, pp. 24–30 (2020)
20. Huang, C., Zhai, B., Tang, A., Wang, X.: Virtual mesh networking for achieving multi-hop D2D communications in 5G networks. Ad Hoc Netw. **94**, 101936 (2017). ISSN 1570-8705. https://doi.org/10.1016/j.adhoc.2019.101936.
21. Malik, P.K., Parthasarthy, H., Tripathi, M.P.: Axisymmetric excited integral equation using moment method for plane circular disk. Int. J. Sci. Eng. Res. **3**(3), 1–3. ISSN 2229-5518

A Deep Network Model for Paraphrase Detection in Punjabi

Arwinder Singh and Gurpreet Singh Josan

Abstract Paraphrase refers to the text which tells the same meanings but with different expressions. It is important in NLP as it deals with many applications such as information retrieval, information extraction, machine translation, query expansion, question answering, summarization and plagiarism. Paraphrase detection is to find that given two texts are semantically similar or not similar. Though paraphrase detection has wide literature, there is no proper algorithm for paraphrase detection in Punjabi language. A new paraphrase detection model for Punjabi language is developed in this paper. We use two deep learning methods to map sentences as vectors, and these vectors are further used to detect paraphrases. Despite other implementations of paraphrase detection, our model is simple and efficient to detect paraphrases. Qualitative and quantitative evaluations prove the efficiency of the model and can be applied to various NLP applications. The proposed model is trained on Quora's question pair dataset which makes new directions for paraphrasing in Indian languages.

Keywords Word embedding · Sentence embedding · Sentence encoding · Semantic similarity · Paraphrasing

1 Introduction

Paraphrasing means to rephrase the original text without changing its meaning. Paraphrasing can be divided into three categories, i.e. paraphrase detection, extraction and generation. The extraction of paraphrase instances is referred as paraphrase extraction. Paraphrase generation is to generate new text which will be semantically similar to original text. Paraphrase detection refers to find that given two sentences are semantically similar or not similar. If both of the sentences convey the same meaning, it is

A. Singh (✉)
Mata Sundri University Girls College, Mansa, Punjab, India
e-mail: arwindersingh13@gmail.com

G. S. Josan
Department of Computer Science, Punjabi University, Patiala, Punjab, India
e-mail: josangurpreet@pbi.ac.in

P. K. Singh et al. (eds.), *Recent Innovations in Computing*, Lecture Notes in Electrical Engineering 701, https://doi.org/10.1007/978-981-15-8297-4_15

considered as paraphrases. The proposed work will focus on paraphrase detection. The two sentences can be written by changing synonyms or structure. Consider the following two sentences which are written by changing their structure:

ਮੈਂ ਆਪਣਾ ਭਾਰ ਕਿਵੇਂ ਘਟਾਵਾਂ?(main apna bhaar kiven ghatavan?) (how do I reduce my weight?)
ਮੈਂ ਕਿਵੇਂ ਆਪਣਾ ਭਾਰ ਘਟਾਵਾਂ?(main kiven apna bhaar ghatavan?).

These are the paraphrases of each other. These are interrogative sentences and the word ਕਿਵੇਂ (kiven) (how) is making the sentences interrogative. The second sentence is rephrased by changing the position of ਕਿਵੇਂ (kiven) (how) which does not change the semantic meaning of sentence.

We can achieve various significances by detecting paraphrases as it deals with various NLP applications such as information retrieval [27], query expansion [5], question answering [31], summarization [6, 21] and plagiarism detection [11]. The use of Web, smart devices and social media is increasing day by day, and the information is easily available on the Internet. People try to use original text in their work due to easy availability of information by changing synonyms or phrases. So, paraphrase identification is becoming useful in various NLP tasks.

Paraphrasing can be seen as semantic similarity between two texts. The words occurred in the similar context tend to convey the similar meaning [26, 28]. Vector space model (VSM) is a famous approach to deal with semantic similarity from large unlabelled data. Different approaches are available for VSM, i.e. traditional distributed semantic models (DSMs) and novel deep neural network (DNN). The invention of word embeddings [22] leads to major advancements in text processing using deep neural networks. It represents words with vectors (called embeddings), groups the similar words and also preserves the semantic relations between words. The advances in deep learning's generative models help to implement such complex tasks.

Just like word vectors capture the semantic similarity of words, we worked to map sentences as vectors for finding paraphrases from available corpus. In simple terms, vectors are the numerical representation of the words and sentences. The words are represented in multi-dimensional vectors, and word vectors are then combined to form sentence vectors. The two sentences are seen as paraphrases when the distance between their vectors is minimum. There are traditional [28, 34] as well as neural network [15, 17, 19, 22, 23, 35] methods to define sentences as vectors. In this paper, we explored two methods to obtain sentence vectors. The first method is average word vector model which requires word2vec model. Word2vec is the representation of words as vectors. It can group the similar words and also preserve the semantic relations between words. We trained word2vec for Punjabi with continuous bag-of-word (CBOW) architecture defined by Mikolov [22]. The sentence vector is then obtained by averaging vectors of words of the whole sentence. The other method is encoder–decoder with LSTM to generate vectors for sentences. These vectors are then used to find similar sentences for given sentence by finding cosine similarity between vectors.

Research Challenges. So, for a machine, it is difficult to find semantic and syntactic relations between sentences or phrases. It becomes more difficult to detect paraphrases from tweets due to structure of sentences, misspelling and style of writing. It is also challenging to recognize lexical, phrasal or sentential paraphrases in the corpus. There is a small number of datasets in Indian languages especially in Punjabi. Traditional approaches were unable to define similarity between synonyms. These approaches run with data sparseness where feature extraction was time consuming, where deep learning models work with low dimensions, take short time for training, can recognize synonyms and preserve the semantic relation between sentences. To overcome these issues, we propose paraphrase detection model using deep learning techniques.

Scope of the Study. A document can be seen as set of words or phrases or sentences. The representation of a document as sentences is an easy way. The proposed work is also focusing on sentences, and we develop two models to map sentences into vectors. These sentence vectors are then able to perform different similarity tasks on unlabelled data. We applied this approach for detecting paraphrases as explained in Sect. 4. After getting good results in paraphrase detection, we successfully applied this approach for making paraphrasing dataset. There are various news are common in different newspapers. There can be similar headlines or sentences within articles. We collected similar sentences as paraphrases. Sentiment analysis can also be performed with sentence vectors. The approach can also be applied for answer selection.

Section 2 will discuss related work in which different approaches are discussed of word embeddings, sentence embeddings and paraphrase detection. The methodology of this work is discussed in Sect. 3 where two datasets, two methodologies for sentence vector generation and paraphrase detection are explained. Section 4 will show the evaluation and results. The fifth section represents conclusion and future work.

2 Related Work

The earlier implementations of vector models were count-based. Latent semantic analysis was developed [9, 10] in 1980s, and random indexing was proposed by Sahlgren [25] with a great advantage of low dimensionality. Sentence modelling is focused to collect n-gram features [20], syntax features [8, 24]and machine translation features [20]. But the advancement in deep learning changed the scenario from count-based models to predictive models.

Tau Yih et al. [34] have worked with sentence vectors and generated text as tf-idf vectors. Their model combined positive and negative pairs in a logistic loss function. Kiros [18] presented skip-thought algorithm to represent sentences as vectors in a different way. They have used recurrent neural networks (RNN) and focused to consider the word orders. Mikolov's [22] CBOW algorithm was to predict a word from its adjacent words and produced amazing results in semantic similarity tasks.

Kenter [16] implemented this approach for sentences and predicted a sentence from neighbouring sentences.

The weighted average word vectors and parse trees produced word vectors in an order. Matrix vectors are different models for the representation of sentences and phrases as vectors. These models were analysed by Kim [17], Lin et al. [19], Yin et al. [35], Kalchbrenner et al. [15], White et al. [32] who worked on sentence embeddings for the classification of sentences. They tried to find how sentence embeddings capture the meanings. They used three different models, i.e. unfolding recursive autoencoders (UREA), PVDM and PV-DBOW (paragraph vector methods) and sum and mean of word embeddings (SOWE and MOWE) for classification.

Tree-LSTM was proposed by Tai et al. [30] for sentiment classification and semantic relatedness. The sentences were processed by Tree-LSTM instead of skip-thoughts or other standard LSTM. The new model was used by them just as a model which was designed by Kiros et al. [18], but they argued that Tree-LSTM is good for taking necessary information from sentences as it works on syntactic properties of sentences. Achananuparp et al. [1] developed a hybrid approach for finding similar questions. For this, they combined semantic and syntactic similarity to question type similarity. Their approach measured syntactic and semantic similarity from words similarity, word orders and part of speech (POS) information and question type with support vector machine (SVM) classifier.

Periwal [23] proposed RNN-based model to generate similar sentences. He combined variational autoencoders-long short-term memory (VAE-LSTM) to produce semantic sentence. The model is implemented with single and double layer encoders but one decoder with single layer. Zhang et al. [36] worked to encode sentences with common semantic information into similar vector representations. The authors used encoder–decoder model to map sentences as vectors and then used their model for sentence paraphrases and paragraph summarization.

Gharavi et al. [11] followed deep learning's semantic similarity approach for plagiarism detection in Persian. They mapped sentences as vectors with average word vector model and collected similar sentences by performing cosine similarity and further Jaccard similarity measure was performed. Paraphrase detection has many implementations which worked well on clean text but unsuccessful to recognize paraphrases for noisy text. So, Agarwal et al. [2] proposed neural network approach by combining CNN and LSTM. They worked on social media text such as Twitter for detecting paraphrases. The recursive neural network used by Huang [13] to collect phrasal representations. His work mapped phrases as vectors to show that close vectors are similar. The most recent work implemented by [14, 33, 37–40]. Paraphrase detection has wide literature as summarized in Table 1 but is missing in Punjabi language which is implemented in the proposed work.

Table 1 Summary of reviewed literature

S. No.	Model	Dataset	Main objective
1	Similarity learning via Siamese Neural Network (S2Net)	Wikipedia articles	To generate sentence vectors for cross-lingual document retrieval and Ad relevance measures
2	RNN-based skip-thought vectors	Book Corpus	Representation of sentences as vectors and evaluated on eight similarity tasks semantic relatedness and paraphrase detection etc
3	Siamese CBOW	Toronto Book Corpus	To obtain word embeddings to form sentence embeddings for semantic textual similarity
4	CNN sentence classification	Mikolov's pretrained word vector on Google news dataset	Simple CNN model to represent sentences as vectors for sentence classification tasks
5	ABCNN attention-based convolution neural network	WikiQA, Microsoft Research Paraphrase and Sick	For modelling a pair of sentences further applied for answer selection and textual entailment
6	UREA, PV-DM, PV-DBOW and SOWE-MOWE	Microsoft Research Paraphrase and Opinosis	For various classification tasks
7	Tree-LSTM	Pre-trained Glove word vector model	Trained for similarity tasks, i.e. classification on movie review and semantic relatedness on Sick
8	Hybrid question similarity approach	Pre-trained Glove word vector model	To perform similarity tasks for finding similar questions
9	VAE-LSTM for similar sentence generation	Cornell movie dialogue and TV series Friends corpus	To combine VAE and LSTM to semantic similar sentences
10	Encoder–decoder framework for sentence paraphrases	Visual caption dataset for paraphrase and Tacos Multi-Level Corpus	To form sentences into vectors for paraphrasing and summarization
11	Deep learning-based approach for Persian plagiarism detection	Persian PAN2016	Two-step method for detecting paraphrasing in Persian language
12	CNN-LSTM for paraphrase detection	Twitter paraphrase SEmEval 2015 and Microsoft Research Paraphrase	Hybrid approach for paraphrase identification for noisy text from Twitter
13	Recursive autoencoder for paraphrase detection	Microsoft Research Paraphrase and English Gigaword corpus	Target to detect paraphrases in short phrases

3 Methodology

The vectors for sentences were generated with two methods, i.e. average word vectors and encoder–decoder LSTM.

3.1 Dataset

Two datasets are prepared in the proposed work. To generate sentence vectors with average word vector, we require a pre-trained word vector model which is not available in Punjabi language. So, we trained a word vector model for Punjabi. Word2vec requires large amount of data for training, and we have used Dataset-I for training word2vec. Dataset-2 is used for sentence vectors.

Dataset-1. We have collected data from two sources. The first source is Ajit newspaper which has publicly available data about politics, sports and education. The second source is Wikipedia from where Punjabi poems, stories and novels are collected.

Dataset-2. Quora released a paraphrase dataset of duplicate questions in January 2017 which contains 400 K pairs. Each of the pair is associated with binary value 0 or 1. 1 indicates that both of the sentences are similar to each other and 0 means they are not similar. Out of 400 K paraphrasing sentences, we select 36,339 pairs (72,678 sentences) whose binary value is 1. These pairs are in English, so Google translator is used to translate these sentences in Punjabi language. The dataset is divided into two parts for training and testing. 71678 sentences are used for training and 1000 for testing [Tables 2,3].

Table 2 Dataset-I

Item	Value
Total lines	8,95,256
Total words	1,22,33,081
Total vocabulary	285,672

Table 3 Dataset-II

Item	Value
Total lines	72,678
Total words	7,89,563
Total vocabulary	24,398
Max length	58

3.2 *Sentence Vector Generation*

We work on sentence generation with two different models, i.e. SentVecAve and SentVecLSTM.

SentVecAve. SentVecAve model is developed in this study to generate sentence vectors using average word vectors. This model is a two-step process. Initially, we train word2vec model for Punjabi Language which is a famous technique to learn word vectors. Word2vec can be implemented using continuous bag of words (CBOW) and skip-gram. This model is developed by Tomas Mikolov in 2013 at Google [22]. Both of the approaches are able to learn weights which is seen as vector representation of word. For this study, we have opted CBOW which predicts word according to its context. It is fast to train and also has better representations for frequent words. The model is trained with Dataset-I with window size 5 and dimension 100.

The second step is to generate sentence vectors for all sentences. The architecture of sentences generation with SentVecAve is shown in Fig. 1. Our SentVecAve model reads the input text as sentence and split sentence into words (w_1, w_2, …., w_n). Then, model reads the embeddings of each word from the trainedword2Vec model. At the last step, all word's vectors are then averaged to form a single vector which considered as sentence vector D. This simple architecture is recursively called to obtain vectors for all sentences.

$$D = \frac{\sum_{i=1}^{n} w_i}{n} \tag{1}$$

SentVecLSTM. The advancements in neural networks, specially the generative models [4, 7] produced good results for solving complex sequence-related problems. Seq2Seq [3, 29] is a famous technique of recurrent neural network to solve typical

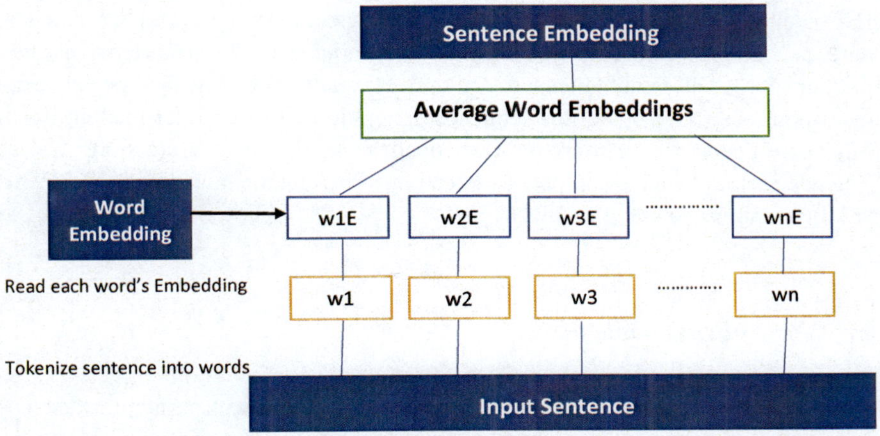

Fig. 1 Sentence embedding with average word vector model

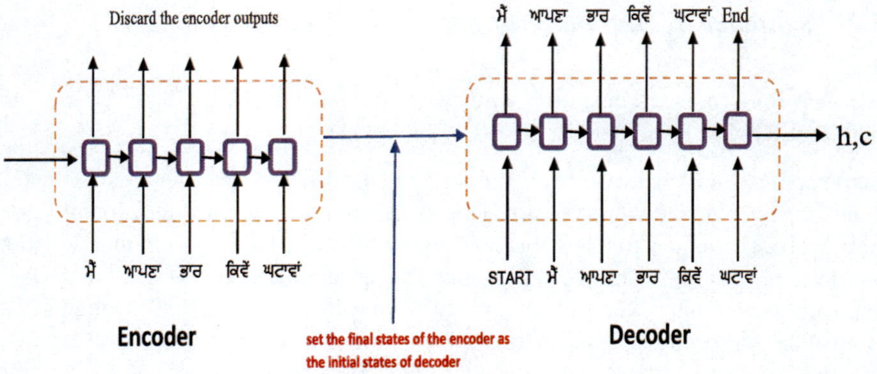

Fig. 2 Sentence embedding with encoder–decoder LSTM

NLP problems such as machine translation, question answering and text summarization. Encoder–decoder architecture is commonly used to implement Seq2Seq models. Both of the encoder and decoder models are LSTM models as LSTMs [12] are able to preserve vector representation as internal states. Encoder reads an input and encodes it into internal state vectors. The decoder takes internal state vector as an input encoded by the encoder.

As shown in Fig. 2, we use neural machine translation model (NMT) in our work. We consider same data for source and target language. The encoder reads some sequence at a time (word in our case). We discard output of each cell and only preserve internal states. NMT models produces output at decoder side after reading the entire sentence. The internal states (ht, ct) of LSTM preserve the vector representation. So, the internal states of the last time step will represent information about whole sentence. The encoded thought vector of encoder becomes the input of decoder. Now, decoder is able to generate one output sequence at a time after reading the complete encoded sentence. For decoder, we add START at start and END at end of each sentence to tell the decoder where to start and end the sentence. Here, decoder sets to predict same sentence as input sentence because we assume that our source and target languages are same. It reads START as the first sequence and produces ਮੈਂ(main), and this sequence generation will stop after reading END. We ignore decoder after training. The encoder's hidden state vectors are read as sentence vectors to make paraphrase detections. We train the model with 20 epochs, 512 dimensions and attention mechanism.

3.3 *Paraphrase Detection*

After generating vectors of all the sentences, the vectors with minimum distance are seen as paraphrases. The input sentence is first vectorized using SentVecAve model. This new vector representation of a sentence is compared with all vectors generated by SentVecAve. This comparison makes with well-known cosine similarity formula (Eq. 2). Two vector representations with minimum distance are read as paraphrases. The input sentence is further used to find its vector representation using SentVecLSTM. Again, it is compared with the vectors created by SentVecLSTM. The cosine angle is calculated to find paraphrases with this model. Our model worked well for finding paraphrases.

$$\text{sim} = \frac{a \cdot b}{|a||b|} \tag{2}$$

4 Evaluation and Results

We performed quantitative and qualitative evaluations on our model. For quantitative evaluation, we use BLEU metrics which is famous for machine translation's evaluations. BLEU is a method to find n-grams between reference sentence and generated sentence. Some of the aspects were not seen by evaluation metrics so that we had to move towards qualitative evaluation. We select 1000 sentences from test dataset and assigned this to three evaluators. The evaluators focus on relevance aspect of the sentences, i.e. whether the detected paraphrase is semantically similar to input sentence.

The results for paraphrase detection are shown in Table 4 for each method SentVecAve and SentVecLSTM. The BLEU score is 0.45 for SentVecLSTM and 0.35 for SentVecAve. The qualitative results show that 70% true paraphrases are detected by our SentVecLSTM model and 35% for SentVecAve. The evaluators marked 1 if the detected sentence is true paraphrase of given sentence, otherwise marked 0. The most of the sentences marked as 1 by all the evaluators. Some of the detected paraphrases are shown in Table 5. The input sentence " "ਨਵਾਂ ਜੀਵਨ ਸ਼ੁਰੂ ਕਰਨ ਲਈ ਮੈਨੂੰ ਕੀ ਕਰਨਾ ਚਾਹੀਦਾ ਹੈ?"" (nvan jeevan shuroo karn lai mainoo ki krna chahida hai?) (What should I do to start a new life?) is detected by SentVecAveas " "ਭਾਰਘਟਾਉਣਲਈਮੈਨੂੰਕੀਕਰਨਾਚਾਹੀਦਾਹੈ?"" (bhaar ghtaun lai mainoo ki krna chahida hai?) (What should I do to reduce weight?), whereas by SentVecLSTM as " ਨਵਾਂ ਜੀਵਨ ਸ਼ੁਰੂ ਕਰਨ ਦਾ ਸਭ ਤੋਂ ਵਧੀਆ ਤਰੀਕਾ ਕੀ ਹੈ?" (nvan jeevan shuroo karn da sabhton vadhia tareeka ki hai?) (What is the best way to start a new life?). As shown in Table 5, most of the paraphrases detected by SentVecLSTM are semantically similar to input sentence, but the paraphrases detected by SentVecAve are not semantically similar to given sentence. The advantage of encoder–decoder

Table 4 Results

Evaluation method	SentVecAve	SentVecLSTM
BLEU score	0.35	0.45
Human evaluation	35%	70%

Table 5 Paraphrases detected by SentVecAve and SentVecLSTM

Input sentence	Paraphrases detected by SentVecAve	Paraphrases detected by SentVecLSTM
ਨਵਾਂ ਜੀਵਨ ਸ਼ੁਰੂ ਕਰਨ ਲਈ ਮੈਨੂੰ ਕੀ ਕਰਨਾ ਚਾਹੀਦਾ ਹੈ? (navan jeevan shuroo karn lai mainoo ki karna chahida hai?) (What should I do to start a new life?)	ਭਾਰ ਘਟਾਉਣ ਲਈ ਮੈਨੂੰ ਕੀ ਕਰਨਾ ਚਾਹੀਦਾ ਹੈ? (bhaar ghataun lai mainoo ki karna chahida hai?) (What should I do to reduce weight?)	ਨਵਾਂ ਜੀਵਨ ਸ਼ੁਰੂ ਕਰਨ ਦਾ ਸਭ ਤੋਂ ਵਧੀਆ ਤਰੀਕਾ ਕੀ ਹੈ? (nvan jeevan shuroo karn da sabhton vadhia tareeka ki hai?) (What is the best way to start anew life?)
ਪੱਤਰਕਾਰੀ ਲਈ ਸਭ ਤੋਂ ਵਧੀਆ ਕਾਲਜ ਕੀ ਹੈ? (pattarkari lai sabh ton vadhia kalajki hai?) (What is the best college for journalism?)	ਕਾਲਜ ਚੁਣਨ ਲਈ ਸਭ ਤੋਂ ਵਧੀਆ ਤਰੀਕਾ ਕੀ ਹੈ? (kalaj chunan lai sabh ton vadhia tareeka ki hai?) (What is the best way to choose a college?)	ਪੱਤਰਕਾਰੀ ਦੇ ਲਈ ਸਭ ਤੋਂ ਵਧੀਆ ਕਾਲਜ ਕਿਹੜਾ ਹੈ? (pattarkari de lai sabh ton vadhia kalaj kihda hai?) (Which is the Best College for Journalism?)
ਚੀਨੀ ਲੋਕਾਂ ਨੂੰ ਤੁਸੀਂ ਕੀ ਸਮਝਦੇ ਹੋ? (cheeni lokan noon tusin ki samjhde ho?) (What do you think of Chinese people?)	ਕੀ ਤੁਸੀਂ ਚੀਨੀ ਲੋਕਾਂ ਨੂੰ ਪਸੰਦ ਕਰਦੇ ਹੋ? (ki tusin cheeni lokan noon pasand karde ho?) (Do you like Chinese people?)	ਚੀਨੀ ਲੋਕਾਂ ਬਾਰੇ ਤੁਸੀਂ ਕੀ ਸੋਚਦੇ ਹੋ? (cheeni lokan bare tusin ki sochde ho?) (What do you think about Chinese people?)
ਮਾਰਕੀਟ ਖੋਜ ਦੀ ਮਹੱਤਤਾ ਕੀ ਹੈ? (markit khoj dee mahatata ki hai?) (What is the importance of market research?)	ਮਾਰਕੀਟ ਦੀ ਖੋਜ ਕੀ ਹੈ ਅਤੇ ਇਸ ਦੀ ਮਹੱਤਤਾ ਕੀ ਹੈ? (markit dee khoj ki hai ate is dee mahatata ki hai?) (What is market research? And what is its importance?)	ਅਸਲ ਵਿੱਚ ਡਿਜਿਟਲ ਕਲਾ ਕੀ ਹੈ? (asal vichch dijital kalaa ki hai?) (What really is digital art?)

model is that it is able to preserve the overall semantic meaning of the sentence, but average word vector model is unable to do this. SentVecAve model wins the battle for some sentences. The last sentence in Table 5 detected by SentVecAve is accurate. So, our SentVecLSTM model is powerful to detect semantically similar and grammatically accurate paraphrases as we get 70% accuracy with human evaluation.

The analysis show that our SentVecLSTM model is better to detect semantically similar sentences. Most of the sentences detected by this model are readable and relevant to given sentences. Most of the detected paraphrases are semantically similar instead of sharing the same words. So, we get low BLUE score as it finds n-grams with the reference sentences. So, results can be improved with clean and large dataset.

5 Conclusion and Future Work

The proposed research focuses on paraphrase detection and its uses in various NLP tasks. We work on two models to generate sentence vectors: first to generate sentence vectors by averaging the word vectors. The next model is trained to generate sentence vectors with encoder–decoder LSTM. Then, these vectors are used to detect paraphrase. We use two datasets in our work, and we evaluate our model with Quora's paraphrase dataset. The model further applies for making paraphrasing corpus. The comparison shows the limitation of averaging word embeddings this approach is unable to preserve word orders. Encoder–decoder LSTM is able to preserve semantic meanings of sentences, and it produces remarkable results. In future, the model will be applied for sentence matching, sentiment analysis and answer selection and also will be fine-tuned for improving accuracy of paraphrase detection.

References

1. Achananuparp, P., Hu, X., Zhou, X., Zhang, X.: Utilizing sentence similarity and question type similarity to response to similar questions in knowledge-sharing community. In: Proceedings of QAWeb 2008 Workshop, Beijing, China (to appear, 2008) (2008).
2. Agarwal, B., Ramampiaro, H., Langseth, H., Ruocco, M.: A deep network model for paraphrase detection in short text messages. Inf. Process. Manag. **54**, 922–937 (2018)
3. Bahdanau, D., Cho, K., Bengio, Y.: Neural machine translation by jointly learning to align and translate. In: 3rd International Conference on Learning Representations, ICLR 2015, San Diego, CA, USA, May 7–9, 2015, Conference Track Proceedings (2015)
4. Bowman, S.R., Vilnis, L., Vinyals, O., Dai, A.M., Józefowicz, R., Bengio, S.: Generating sentences from a continuous space. In: Proceedings of the 20th SIGNLL Conference on Computational Natural Language Learning, pp. 10–21. ACL (2016)
5. Cao, H., Jiang, D., Pei, J., He, Q., Liao, Z., Chen, E., Li, H.: Context-aware query suggestion by mining click-through and session data. In: KDD '08: Proceeding of the 14th ACM SIGKDD international conference on Knowledge discovery and data mining, pp. 875–883 (2008)
6. Chatterjee, N., Mohan, S.: Extraction-based single-document summarization using random indexing. In: Proceedings of the 19th IEEE International Conference on Tools with Artificial Intelligence, vol. 02, pp. 448–455. IEEE Computer Society (2007)

7. Chung, J., Kastner, K., Dinh, L., Goel, K., Courville, A.C., Bengio, Y.: A recurrent latent variable model for sequential data. Adv. Neural. Inf. Process. Syst. **28**, 2980–2988 (2015)
8. Das, D., Smith, N.A.: Paraphrase identification as probabilistic quasi-synchronous recognition. In: Proceedings of the Joint Conference of the 47th Annual Meeting of the ACL and the 4th International Joint Conference on Natural Language Processing of the AFNLP, pp. 468–476. Association for Computational Linguistics, Suntec, Singapore (2009)
9. Deerwester, S.: Improving information retrieval with latent semantic indexing. In: Proceedings of the 51st Annual Meeting of the American Society for Information Science, vol. 25, pp. 36–40 (1988)
10. Deerwester, S., Dumais, S., Furnas, G., Landauer, T., Harshman, R.: Indexing by latent semantic analysis. J. Am. Soc. Inf. Sci. **41**(41), 391–407 (1990)
11. Gharavi, E., Bijari, K., Zahirnia, K., Veisi, H.: A deep learning approach to Persian plagiarism detection. In: Working notes of FIRE 2016—Forum for Information Retrieval Evaluation, Kolkata, India, December 7–10, 2016, vol. 1737, pp. 154–159 (2016)
12. Hochreiter, S., Schmidhuber, J.: Long short-term memory. Neural Comput. **9**(8), 1735–1780 (1997)
13. Huang, E.: Paraphrase detection using recursive autoencoder. In: Stanford NLP Group, Natural Language Processing, Final Projects Reports (Stanford University, Stanford, CA, 2011) (2011); Huang, E.: Paraphrase detection using recursive autoencoder. In: Stanford NLP Group, Natural Language Processing, Final Projects Reports (Stanford University, Stanford, CA, 2011) (2011)
14. El Desouki, M.I., Gomaa, W.H.: Exploring the recent trends of paraphrase detection. Int. J. Comput. Appl. **182**, 1–5 (2019). https://doi.org/10.5120/ijca2019918317
15. Kalchbrenner, N., Grefenstette, E., Blunsom, P.: A convolutional neural network for modelling sentences. In: Proceedings of the 52nd Annual Meeting of the Association for Computational Linguistics (Volume 1: Long Papers), pp. 655–665. Association for Computational Linguistics (2014)
16. Kenter, T., Borisov, A., de Rijke, M.: Siamese CBOW: optimizing word embeddings for sentence representations. In: Proceedings of the 54th Annual Meeting of the Association for Computational Linguistics (Volume 1: Long Papers), pp. 941–951. Association for Computational Linguistics, Berlin, Germany (2016)
17. Kim, Y.: Convolutional neural networks for sentence classification. In: Proceedings of the 2014 Conference on Empirical Methods in Natural Language Processing, EMNLP 2014, October 25–29, 2014, Doha, Qatar, A meeting of SIGDAT, a Special Interest Group of the ACL, pp. 1746–1751 (2014)
18. Kiros, R., Zhu, Y., Salakhutdinov, R., Zemel, R.S., Urtasun, R., Torralba, A., Fidler, S.: Skip-thought vectors. In: NIPS, pp. 3294–3302 (2015)
19. Lin, R., Liu, S., Yang, M., Li, M., Zhou, M., Li, S.: Hierarchical recurrent neural network for document modeling. In: Proceedings of the 2015 Conference on Empirical Methods in Natural Language Processing, pp. 899–907. The Association for Computational Linguistics (2015)
20. Madnani, N., Tetreault, J., Chodorow, M.: Re-examining machine translation metrics for paraphrase identification. In: Proceedings of the 2012 Conference of the North American Chapter of the Association for Computational Linguistics: Human Language Technologies, pp. 182–190. Association for Computational Linguistics (2012)
21. Mani, I.: Summarization evaluation: An overview. In: In Proceedings of the North American chapter of the association for computational linguistics (NAACL). Workshop on Automatic Summarization (2001)
22. Mikolov, T., Chen, K., Corrado, G., Dean, J.: Efficient estimation of word representations in vector space. CoRR abs/1301.3781 (2013)
23. Periwal, M.: Generating semantic sentences. In: Published in SSRN Electronics Journal (2017)
24. Rus, V., McCarthy, P., Lintean, M., McNamara, D., Graesser, A.: Paraphrase identification with lexico-syntactic graph subsumption. In: Proceedings of the 21th International Florida Artificial Intelligence Research Society Conference, FLAIRS-21, pp. 201–206 (2008)
25. Sahlgren, M.: An introduct ion to random indexing. In: Methods and Applications of Semantic Indexing Workshop at the 7th International Conference on Terminology and Knowledge Engineering, TKE 2005 (2005)

26. Sahlgren, M.: The word-space model: using distributional analysis to represent syntagmatic and paradigmatic relations between words in high-dimensional vector spaces (2006)
27. Salton, G., Wong, A., Yang, C.S.: A vector space model for automatic indexing. Commun. ACM **18**(11), 613–620 (1975)
28. Schütze, H.: Word space. Adv. Neur. Inf. Process. Syst. **5**, 895–902 (1993)
29. Sutskever, I., Vinyals, O., Le, Q.V.: Sequence to sequence learning with neural networks. Adv. Neural. Inf. Process. Syst. **27**, 3104–3112 (2014)
30. Tai, K.S., Socher, R., Manning, C.D.: Improved semantic representations from tree structured long short-term memory networks. In: Proceedings of the 53rd Annual Meeting of the Association for Computational Linguistics and the 7th International Joint Conference on Natural Language Processing (Volume 1: Long Papers), pp. 1556–1566. Association for Computational Linguistics, Beijing, China (2015)
31. Tellex, S., Katz, B., Lin, J., Fern, A., Marton, G.: Quantitative evaluation of passage retrieval algorithms for question answering. In: Proceedings of the 26th Annual International ACM SIGIR Conference, pp. 41–47 (2003)
32. White, L., Togneri, R., Liu, W., Bennamoun, M.: How well sentence embeddings capture meaning. In: ADCS, pp. 9:1–9:8. ACM (2015)
33. Yang, R., Zhang, J., Gao, X., Ji, F., Chen, H.: Simple and effective text matching with richer alignment features. In: Proceedings of the 57th Annual Meeting of the Association for Computational Linguistics, pp. 4699–4709. Association for Computational Linguistics, Florence, Italy (2019)
34. Tau Yih, W., Toutanova, K., Platt, J.C., Meek, C.: Learning discriminative projections for text similarity measures. In: CoNLL, pp. 247–256. ACL (2011).
35. Yin, W., Schütze, H., Xiang, B., Zhou, B.: Abcnn: Attention-based convolutional neural network for modelling sentence pairs. Trans. Assoc. Comput. Linguist. **4**, 259–272 (2016)
36. Zhang, C., Sah, S., Nguyen, T., Peri, D., Loui, A., Salvaggio, C., Ptucha, R.W.: Semantic sentence embeddings for paraphrasing and text summarization. In: 2017 IEEE Global Conference on Signal and Information Processing (GlobalSIP) abs/1809.10267, pp. 705–709 (2018)
37. Xu, S., Shen, X., Fukumoto, F., Li, J., Suzuki, Y., Nishizaki, H.: Paraphrase identification with Lexical, syntactic and sentential encodings. Appl. Sci. **10**, 4144 (2020)
38. Yinfei, Y., Yuan, Z., Chris, T., Jason, B.: PAWS-X: a Cross-lingual Adversarial Dataset for Paraphrase Identification. CoRR, Abs/ **1908**(11828), 1–6 (2019)
39. Mohamed, I., Hosam, W.: Exploring the recent trends of paraphrase detection. Int. J. Comput. Appl. **182**, 1–5 (2019)
40. Dhall, D., Kaur R., Juneja M.: Machine learning: a review of the algorithms and its applications. In: Singh, P., Kar, A.,Singh, Y., Kolekar, M., Tranwar, S. (eds) Proceedings of ICRIC 2019. Lecture Notes in Electrical Engineering, vol. 597, pp. 47–63. Springer, Cham

Design and Development of Software and Hardware Modules of Bioimpedance System Using LTSpice

K. M. Brajesh, Kirti Pal, and Munna Khan

Abstract One of the crucial components of each bioimpedance measuring device is a constant current source. Finite element numerical device simulation was used to analyze track to track capacitances and compared to measurements on developed PCB with soldering elements. Here, we present an analysis based on a modified circuit of a single constant current source. Finite element numerical system simulation was used to evaluate track to track capacitances and compared to measurements on built PCB without soldering elements. One of the most important components of the bioelectric impedance devices and Bio-Electrical Impedance Analyzer (BIA) is the current source circuit. There are many types of circuits, such as current source, enhanced current source, general impedance convertor, Wien bridge circuit, voltage pick-up amplifier. This paper presents the modules of bioimpedance creation of software.

Keywords Bioimpedance converter · Constant current source simulation LTspice · PCB design

1 Introduction

Bioelectrical impedance analysis (BIA) is a non-invasive, safe, inexpensive, and portable technique that requires minimal patient collaboration for rapid assessment of body composition and body water [1]. The human body has a variety of electrical features; it is known as bioimpedance as the category of properties related to the

K. M. Brajesh (✉) · K. Pal
Department of Electrical Engineering, Gautam Buddha University, Greater Noida, UP 201312, India
e-mail: nishujss16@gmail.com

K. Pal
e-mail: kirti.pal@gbu.ac.in

M. Khan
Department of Electrical Engineering Jamia, Millia Islamia University, New Delhi 110025, India
e-mail: mkhan4@jmi.ac.in

P. K. Singh et al. (eds.), *Recent Innovations in Computing*, Lecture Notes in Electrical Engineering 701, https://doi.org/10.1007/978-981-15-8297-4_16

opposition to an alternating current flow [2]. The way to measure is no different from the traditional method of impedance calculation, and it consists of applying a small alternating current signal and measuring voltage differences, and then determining the impedance by adjusting the amplitude and phase between the original and the reported signal. The unknown impedance can be determined by possible methods in any bioimpedance measurement system: either by applying a known voltage throughout the subject [3]. Instead measure the current through the subject or inject a known current into the subject and calculate the voltage across the subject [4]. There is an unknown impedance of contact between the electrode and the subject. This will degrade the operation of a low impedance voltage source [5]. Analog devices, pcb sheet and resistors were used to obtain the icop-07; normal saline was obtained from the local market. The project entitled—developing low-cost portable joint bioimpedance platform—has been successfully completed with two major parts measuring bioimpedance at constant frequency (50 kHz), and constant current source 4ma [6]. Analog devices, it is usually done in a series of four steps, cleaning, masking, etching and demasking. Sinusoidal signal wave generator and voltage-controlled constant current source are the main consideration of circuit creation for impedance analyzer, but here, we used function generator to generate sine wave signal. The function generator output was fed into the feedback voltage buffer for unit gain operating in the non-inverting mode [7]. This voltage buffer reduces the signal generator's loading effect. Impedance analysis turns on constant current injection, variable frequencies into the sample and the sample's subsequent V_{rms} calculation. In the current study, voltage-controlled constant current source was established. The output of the unit voltage gain buffer is connected to VCCS that works on the non-invert mode [8]. The controlled constant AC current was injected into the sample that had to be tested for impedance. As an impedance analyzer, the combination of unity gain voltage buffer and VCCS is given. The developed impedance analyzer was then used to analyze the biological sample's electrical behavior [9].

1.1 Importance of BIA Technique

The classical air- or fluid-displacement techniques are cumbersome, which greatly restrict their use outside the laboratory. On the other hand, the BIA technique is considerably more convenient to use in clinical routine settings [10]. The BIA using venous occlusion plethysmography showed wide acceptance as a routine screening test for major thrombosis used BIA to measure minimum venovenous bypass flow during liver transplantation to account for the inferior vena cava clamping [11]. A venovenous bypass relieves cardiac pressure and increases cardiac performance. The BIA thorax monitoring allows continuous determination of the volume of the stroke, contractility indices such as velocity and acceleration of the blood flow, systemic vascular resistance and index, cardiac output (CO), and cardiac index (CI), and fluid content of the thoracic [12]. Belott and Gilbert have further demonstrated that BIA provides other hemodynamic indices, such as systolic time internal, left cardiac work index, and pacemaker clinic diastolic index, and it takes only minutes to perform [13].

1.2 Development of BIA Technique

In the late nineteenth century, studies of bioelectric phenomena in human and animal tissue have begun. Bioelectric impedance (BI) measurements are based on Ohm's law: the current in a circuit is directly proportional to the voltage and inversely proportional to the resistance in a circuit of DC or impedance in a circuit of alternating current. AC is applied to the body or body segment by two electrodes [14]. The voltage signal from the body's surface is measured with the same or an additional two electrodes in impedance terms. In terms of physiological events, the resistance or impedance information of the body or body portion can be conveyed [15]. Between 1930 and 1950, a significant amount of research was done to develop the fundamental uses of impedance for measuring diverse forms of physiology in humans. The above research discussed the bioelectrical relationships impedance and physiological parameters such as thyroid function, basal metabolism, hormone levels, and blood flow. BI's ability to correctly represent shifts in blood volume was the target of criticism [16].

2 Simulation Using LT Spice Software

2.1 Wein Bridge with V-I Converter

A Wein bridge oscillator is a type of sine wave generating electronic oscillator. It can produce a wide range of frequencies. A Wien bridge oscillator is a type of sine waves producing electronic oscillator. It can contain a wide range of frequencies. The oscillator is based on a bridge circuit for impedance measurement originally developed by Max Wien In 1891. There are four resistors and two capacitors in the bridge [17].

2.2 Op-Amp-Based Constant Current Source

This section explains the circuit definition, function and simulations of the basic current source, modified current source and modified current source mirrored, while transient simulations display the behavior in the circuit; when a voltage input excites, AC simulation is conducted to find the range of frequencies for which the circuit operates as a current source. Current sources are critical to a bioimpedance system

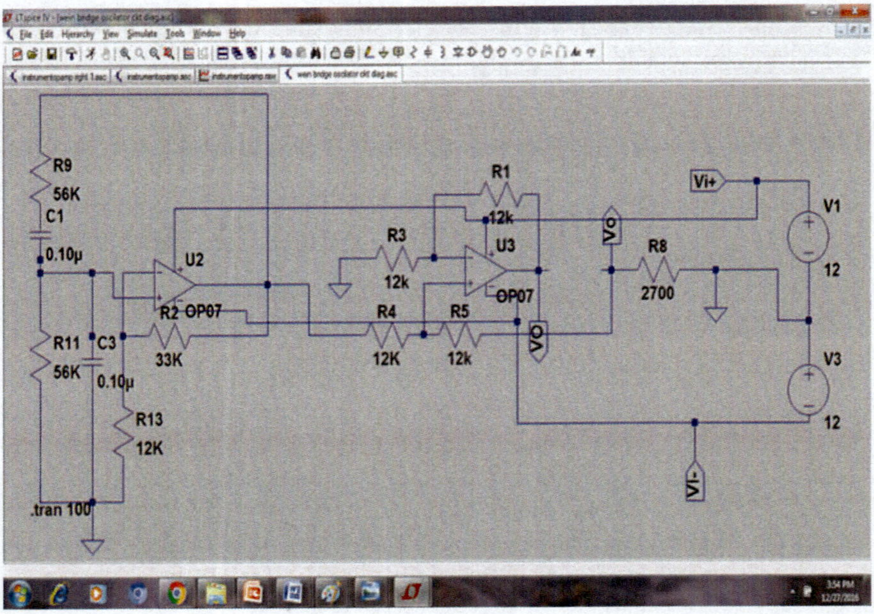

Fig. 1 Circuit diagram of VCCS

design [18]. It must drive a current of constant AC magnitude irrespective of the load connected to it, especially when the load is a human subject where the magnitude of the load impedance depends not only on the frequency of operation but also on other factors such as the subject's state and tissue whose impedance is calculated (the impedance of the whole body is greater than the impedance of the particular part of the body, electrode size, electrode form, etc., compared to the load impedance, it is important that the current source's output impedance is very small [19]. A load equal to the current source's output impedance can vary the circuit's output current. In that, the circuit will provide a constant current for the frequency 50 kHz range [20].

Software Simulation Result of Current Source: The current passed through the salt solution and the resulting voltage changes were reported at different frequencies with the aid of the oscilloscope. All measurements were conducted with 1 V input voltage and between 50 kHz frequency range.

Software: LTSPICE 401, Frequency: 50 kHz, Current: 4 mA.

2.3 Voltage Pick-Up (Instrumentation) Amplifier

The voltage amplifier is used to amplify the voltage of measurement across the current being applied. Band-pass filter finds the exact sine wave shape; we uses

Fig. 2 Circuit diagram of impedance analyzer using LTSpice software

a 50–200 kHz frequency band-pass filter [21]. An instrumentation amplifier is an effective voltage amplifier that amplifies varying voltages and suppresses typical mode signals. Depending on the topology, an instrumentation offers much higher typical mode rejection than operational amplifiers, making them a logical choice when working in noisy or very low voltage environment [22] (Figs. 3 and 4).

2.4 Post-signal Processing Output

Sinusoidal signal wave generator and voltage-controlled constant current source are the main consideration of circuit development for impedance analyzer, but here, we used function generator to generate sine wave signal [23]. The function generator output was fed into the feedback voltage buffer for unity gain operating in

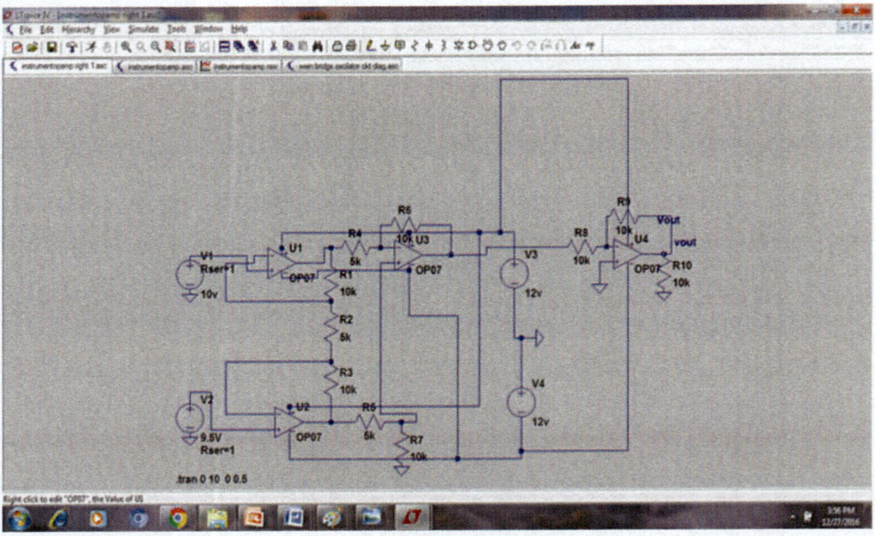

Fig. 3 Voltage pick-up(instrumentation) amplifier

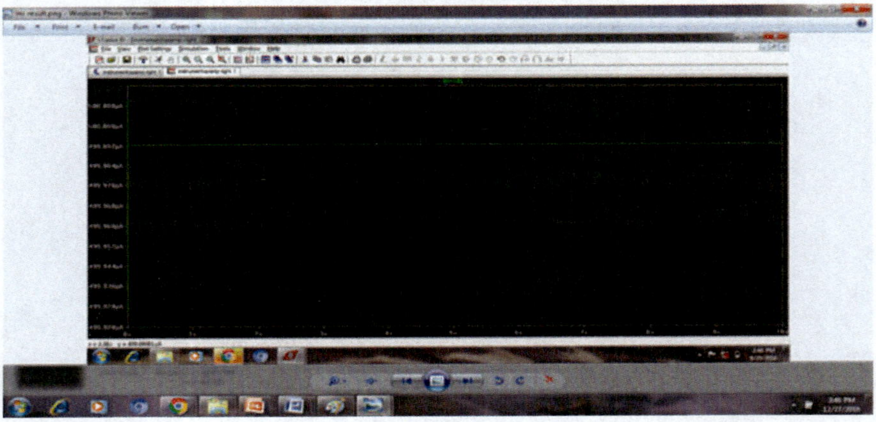

Fig. 4 Simulation result voltage output

the non-inverting mode shown in Figure. This stress buffer reduces the load signal generator effect. Impedance analysis turns on constant current injection, variable frequencies into the sample and the sample's subsequent V_{rms} calculation. Voltage-controlled constant current source has been developed in the current study. Unit output gains voltage buffer connected to VCCS that operates on non-inverting mode. The controlled constant AC current was injected into the sample that had to be tested for impedance. As an impedance analyzer, the combination of unity gain voltage buffer and VCCS is given. The developed impedance analyzer was then used to analyze the biological sample's electrical behavior [24].

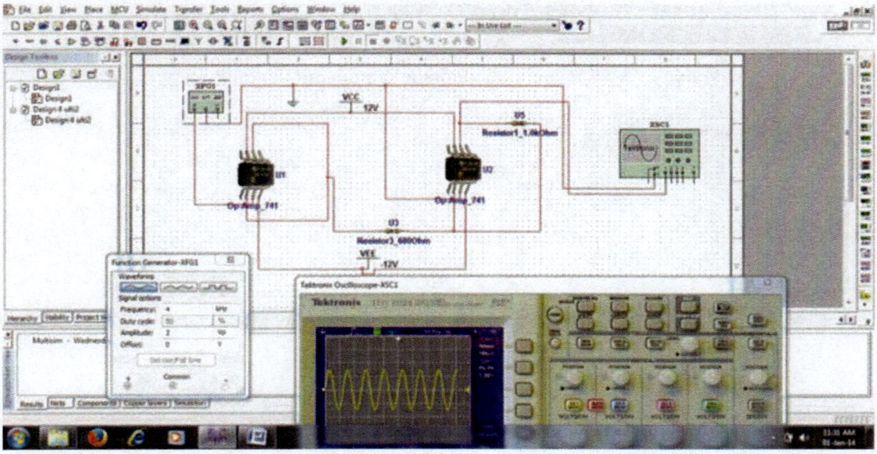

Fig. 5 Circuit diagram of impedance analyzer using LT SPICE 11.0 schematic software

3 Hardware Development of the System

3.1 Preparation of Electrodes

Electrode is prepared by two methods; the first is electrode prepared via printing using eagle and the second is electrode prepared via manually using marker.

3.2 Aluminum Ring Electrodes

Four-electrode BIA is a method of determining joint impedance that is comprehensive, reliable, easy to perform and inexpensive. Connected electrodes consist of material ag/agcl that is primarily used in bioimpedance applications. This procedure consists of applying a current (Io) through two electrodes and measuring the voltage between two other electrodes, and this approach excludes the impedance of the electrodes from the calculation of impedance [25]. This is true as long as the electrodes have relatively low impedance compared to the voltage detector circuit's input impedance in case of shallowness [26] (Fig. 6).

3.3 PCB Layout Design

See Fig. 7.

Fig. 6 Design of electrode shape-1

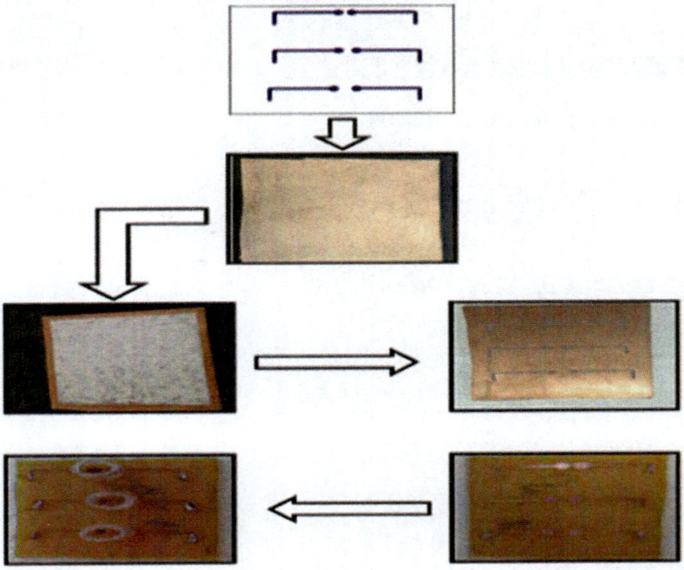

Fig. 7 Pictorial representation of different steps of electrode preparation. **a** Design of electrode using eagle. **b** Fresh PCB sheet. **c** Impregnated PCB. **d** Masking. **e** Demasking. **f** 3-well final electrode

3.4 PCB Design

Cleaning, masking, etching and demasking are usually done in a sequence of four stages.

A. **Cleaning:**
   It is the first step that is free of chemicals on the surface to be washed. A properly cleaned surface may result in the masked, poorly cleaned surface being properly attached, which may result in inaccurate final dimension. The material may additionally be immersed in basic cleaners or particular de-oxidizing results [27] (Fig. 8).

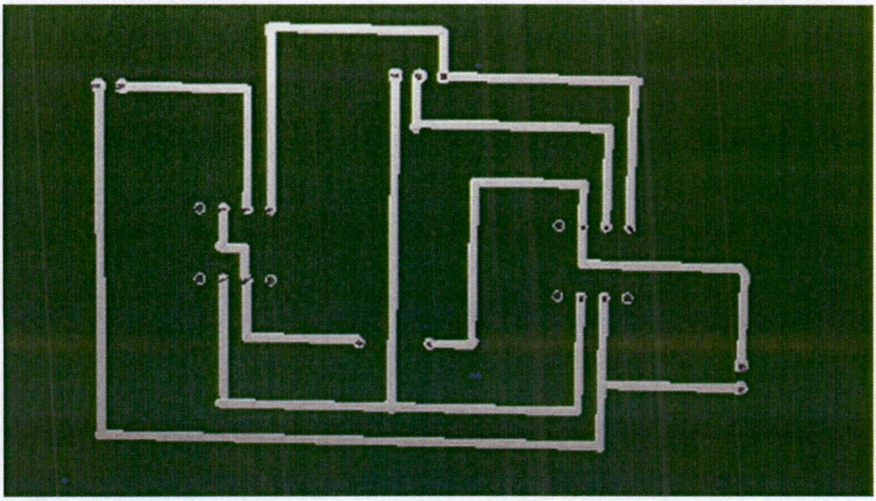

Fig. 8 Layout design of bioimpedance circuit (bottom view) using Ultiboard

B. **Masking:**
The masking material is added to the surface of the copper laminated PCB sheet in this process to ensure that only the area is needed for etching (Fig. 9).

C. **Etching:**
It is the process of applying the masking PCB layer immersed in the solution of ferric chloride (Etchant) to extract unmasked portion of copper by shaking the solution

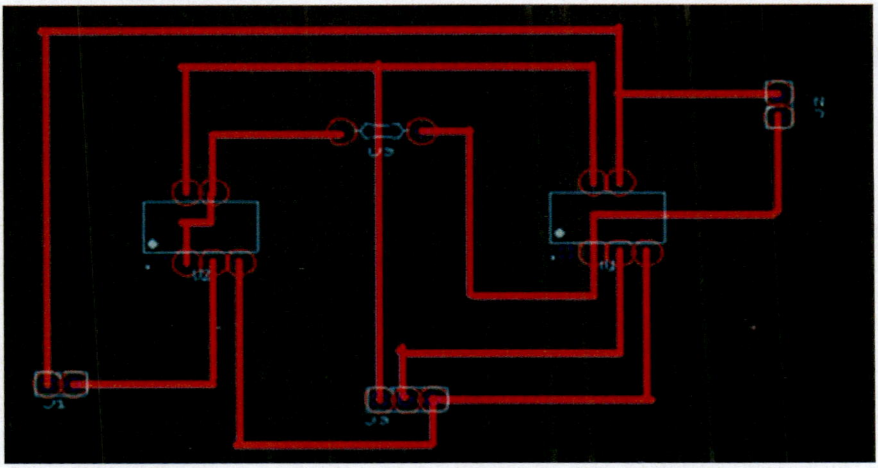

Fig. 9 Layout design of bioimpedance circuit (top view) using Ultiboard

$$FeCl_3 + Cu \rightarrow FeCl_2 + CuCl$$
$$FeCl_3 + CuCl \rightarrow eCl_2 + CuCl_2$$

In the above reaction, iron(III) chloride is first converted to copper(I) chloride and then to copper(II) chloride for PCB chip (Electrode) design [16].

D. **Demasking:**
Clearing the maskant substance is the combined technique. It is usually removed with a normal water wash [27].

4 Materıal and Methods

Materials: Analog devices and Noorwood, USA, acquired the IC OP-07. Local market procured the PCB board, resistors and regular saline. From Himedia, Mumbai, India, Glucose and Luria broth bertani were procured. Honyon International Inc., Hong Yang Chemical Corpn., China, received ethanol. During the entire study, purified water was used.

Methods: The project entitled—Developing Low Cost Portable Platform for Bioimpedance-based Diagnostics—has two major components; the first is electrode preparation, the second is impedance analyzer device circuit development, and the third is sample preparation and impedance analysis.

Required instruments: (1) Function generator. (2) Oscilloscope. (3) Cooling centrifuge-REMD (C-24BL). (4) Double beam UV spectrometer. (5) Magnetic stirrer. (6) Orbital shaker. (7) Digital multimeter [28].

4.1 Block Diagram of Bioimpedance System

The objective of the proposed research is to develop a system for measuring body segment electrical impedance using a combination of some low-cost, portable, reliable and easy design circuits [29]. The following diagram shows the machine block diagram and the configuration of the circuits. Bioimpedance system is designed by implementing the whole system to find bioimpedance, i.e. by injecting current and detecting voltage at different frequencies [30] (Fig. 10).

4.2 Experimental Setup for Impedance Analysis and Result

Monitor efficiency in comparison with other technologies is important and satisfactory. Measurement of bioimpedance is reliable at constant frequency (50 kHz).

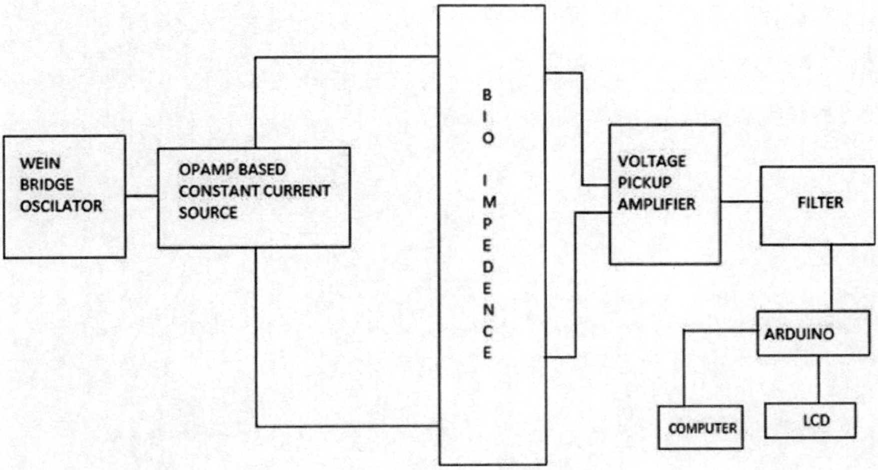

Fig. 10 Block diagram of low-cost bioimpedance measurement device

Monitor-measured impedance is highly accurate. For the measurement of fluid volume changes in knee (leg), a total of five experiments are carried out in this project. Joint contact with the elbow.

Impedance expressed in compression and decompression conditions at 50 kHz frequency varies linearly over time, tested impedance at 50 kHz.

Easily measured with aluminum electrode at 50 kHz frequency, transition in impedance from standing to sitting state.

Also, IPG technique is very useful in real-time measurement of changes in blood volume. In the subjects with lower limb knee joint and elbow joint edema due to hypoactivity, IPG technique is useful for calculating continuous mode changes in blood volume (Fig. 11).

5 Conclusion

In this paper various current sources that can be used for the calculation of bioimpedance are discussed. Bioimpedance measurement systems need a stable current source, have a high output impedance and can cover a wide range of frequencies, for the purpose of designing a 50 kHz frequency current source. The Wein bridge oscillator circuit for current source architecture has the easiest implementation among all the current sources. While a drawback is that its maximum frequency range is limited due to the amplifier's amplitude and phase shift characteristics, its many advantages outweigh its disadvantages. Therefore, the option of the most appropriate constant current source, 4 mA BIA, instrumentation amplifier can be made based on the necessary features. To develop the device that can assess bioimpedance,

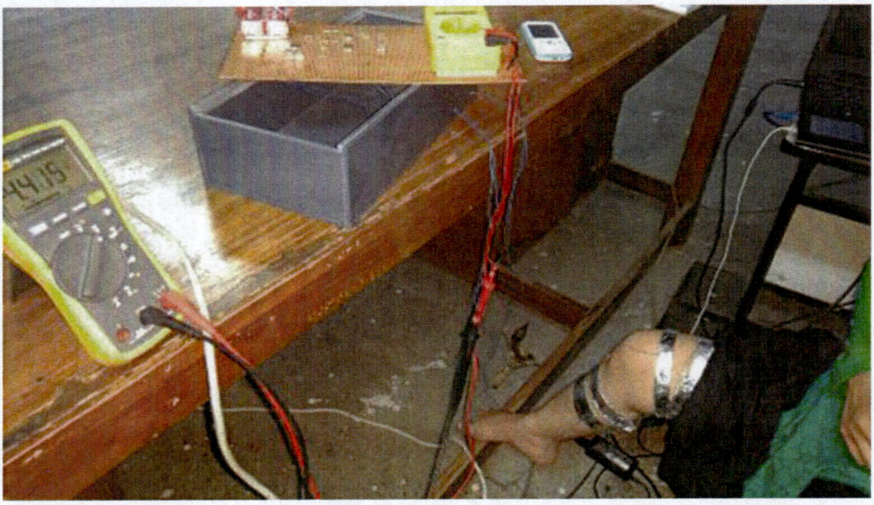

Fig. 11 Experimental setup for impedance analysis

the individual sub-circuits, i.e. the built current source, instrumentation amplifier and the rectifier circuits, are integrated.

References

1. Bera, T.K.: Bioelectrical impedance methods for non-invasive health monitoring: A review, hindawi publishing corporation, J. Med. Eng. 1–28 (2014)
2. Khan, M., Guha, S.K.: Prediction of electrical impedance parameters for the simulated leg segment of an aircraft pilot under G-stress. Aviat. space and environ. med. **73**(6), 558–564 (2002a)
3. Young, R.E., Sinha D.P.: Bioelectrical-impedance analysis as a measure of body composition in a west Indian population. The Am. J. Clin. Nutr. 1045–50 (1992)
4. Brajesh, K.m., Kirti, Pal., Khan, M.: Assessment of human joints using bioelectrical impedance technique. J. Stud. Indian Place Names, ISSN 2394-3114. **40**(10), 272–285, March-26- (2020)
5. Linda, B., Houtkooper, Lohman, T.G., Going, S.B., Hall, M.C.: Validity of bioelectric impedance for body composition assessment in children. Am. Physiol. Soc. 814–821 (1989)
6. Kushner, R.F., Kunigk, A., Alspaugh, M., Andronis, P.T.: Validation of bioelectric-impedance analysis as a measurement of change in body composition in obesity. Am. Soc. Clin. Nutr. 219–223 (1990)
7. Khan, M., Mahfooz, S., Khan, G.P.: Development of bioelectrical impedance analysis equations (BIA) equations to predict body composition of indian males. WAP J. Vol. **3**, 14–33, Jan (2013)
8. Gabriel, C.S., Gabriel, E., Corthout.: The dielectric properties of biological tissues: I.Literature survey. Phys. Med. Biol. **41**(11), 2231 (1996)
9. Miklavčič, D., Pavšelj, N., Hart, F.X.: Electric properties of tissues. Wiley Encyclopedia biomed, Eng (2006)
10. Montal, M., Mueller, P.: Formation of bimolecular membranes from lipid monolayers and a study of their electrical properties. Proc. Nat. Acad. Sci. **69**(12), 3561–3566 (1972)
11. Min, M., et al: An implantable analyzer of bio-impedance dynamics: mixed-signal approach [telemetric monitors]. Instrum. Meas., IEEE Trans. **51**(4), 674–678 (2002)

12. Gabriel, S., Lau, R., Gabriel, C.: The dielectric properties of biological tissues: II. Measurements in the frequency range 10 Hz to 20 GHz. Phys. Med. Biol. **41**(11), 22–51 (1996)
13. Segal, K.R., et al.: Estimation of extracellular and total body water by multiple-frequency bioelectrical-impedance measurement. Am. J. clin. Nutr (1991)
14. Kushner, R.F., Schoeller, D.A.: Estimation of total body water by bioelectrical impedance analysis. Am. J. Clin. Nutr. **44**(3), 417–424 (1986)
15. Nebuya, S., et al.: Estimation of the size of air emboli detectable by electrical impedance measurement. Med. Biol. Eng. Comput. **42**(1), 142–144 (2004)
16. Dudykevych, T., et al.: Impedance analyzer module for EIT and spectroscopy using undersampling. Physio. Meas. **22**(1), 19 (2001)
17. Ayliffe, H.E., Bruno Frazier, A., Rabbitt, R.: Electric impedance spectroscopy using microchannels with integrated metal electrodes. Microelectromech. Syst. J. **8**(1), 50–57 (1999)
18. Khan, M., Mehfuz, S., Khan, G.P.: Bioelectrical impedance analysis (BIA) for assessing TBW and FFM of indian females. Int. J. Comput. Eng. Res. **4**(3), 1–17 (2014)
19. Pawar, C., Khan, M., Saini, J.P.: Design and analysis of adjustable constant current source with multi-frequency for measurement of bioelectrical impedance. Int. J. Appl. Eng. Res. **13**(1), 262–267 (2018)
20. Khan, M., Vashisth, S., Vijay, R., Salhan, A.K.: Non-invasive measurement and subsequent analysis of human carotid pulse for ground-based simulation of G-stress. Int. J. Bioinf. Res. Appl. **12**(3), 227–237 (2016)
21. Munoz, D.R., Moreno, J.S., Escriva, C.R., Berga, S.C., Anton, A.E.N.: Constant current drive for resistive sensors based on generalized impedance converter. IEEE Trans. Instrum. Meas. **57**(10), 2290–2296
22. Brajesh, K.m., Pal, K., Khan, M.: Analysis of bioelectrical impedance measurements and Role of mathematical modeling in bone fracture healing. Nat. Conf. Rob. Mechatron. (NCORM2020), Jamia Millia Islamia, New Delhi March 3rd–4th, (2020)
23. Andrzejowski, P., Giannoudis, P.V.: The 'diamond concept' for long bone non-union management. J. Orthopaedics. Traumatology. **20**(1), (2019)
24. Ghiasi, M.S., Chen, J.E., Rodriguez, E.K., Vaziri, A., Nazarian, A.: Computational modeling of human bone fracture healing affected by different conditions of initial healing stage. BMC Musculoskeletal Disorders **20**(1), (2019)
25. Pawar, C., Khan, M., Saini, J.P.: Bioelectrical impedance measuring device based on principle of multifrequency and multi Segments. J. Stud. Indian Place Names, ISSN 2394-3114, **40**(10), 231–237, March-05- (2020)
26. Kashif, I.K., Sherwani, Kumar, N., Chemori, A., Khan, M., Mohammed, S.: RISE-based adaptive control for EICoSI exoskeleton to assist knee joint mobility. Rob. Auton. Syst. Vol. **124**:103354, 2020.
27. Pawar, C.: Assessment of human arm bioelectrical impedance using microcontroller based system. J. Int. J. Integr Eng. **12**(4), 172–181, Apr-30-(2020)
28. Khan, M., Sirdeshmukh, S.P.S.M.A., Javed, K.: Evaluation of bone fracture in an animal model using bio-electrical impedance analysis. Perspect. Sci. Vol. **8**, 567–569, (2016)
29. Khan, M., Guha, S.K.: Electrical impedance analysis for simulated arm blood pooling of an aircraft pilot under G-Stress. Aviat. Space Environ. Med. **74**(4), 406 (2002b)
30. Reggie, O.H., Khan, M., Pohlman, R.L., Schlub, J.: Leg resistance training: effects upon cardiovascular fitness (vo2 peak) and skeletal muscle myoplasticity. Int. J. Exercise Physiol. **7**(5), 27–43 (2004)

Investigations on Various Designs of Dielectric Resonator Antennas (DRA) for C Band Applications

Dishant Khosla and Kulwinder Singh Malhi

Abstract The overall growth of wireless communication is going on at such a rapid pace that a lot has been going on day to day basis leading to a number of changes in antenna designs to make their performance as per the required parameters. In this communication, various dielectric resonator antenna designs and their performance parameters like size, shape, feeding mechanism, bandwidth, substrate material and peak gain are discussed in the C band. DRA antenna has emerged as a good option in the microwave and millimeter frequency range due to the use of low loss dielectric material which reduces surface wave losses and metallic losses and in turn increases antenna efficiency. DRA due to its advantages like small size, wider bandwidth, much better efficiency, higher dielectric strength, low profile, higher power handling capacity, easy fabrication and low cost is widely used antenna.

Keywords DRA · 5G · Circularly polarized · VSWR · Dual band · Wireless devices · CP · DGS · UWB · MIMO

1 Introduction

Over the past few years, the technological growth of wireless communication has drastically increased across the globe [1]. The emphasis today is to develop multi-band compact antennas with low losses and higher efficiency. DRA antennas are a better option as it provides higher radiation efficiency and much wider impedance bandwidth [2]. The DRA geometries, material and feeding mechanism are important in deciding the performance of the antenna [3]. DRAs can be designed with different

D. Khosla (✉) · K. S. Malhi (✉)
Punjabi University, Patiala, Punjab, India
e-mail: dishant.coeece@cgc.edu.in

K. S. Malhi
e-mail: ksmalhip@gmail.com

D. Khosla
Chandigarh Group of Colleges, Mohali, Punjab, India

P. K. Singh et al. (eds.), *Recent Innovations in Computing*, Lecture Notes in Electrical Engineering 701, https://doi.org/10.1007/978-981-15-8297-4_17

structures to achieve desired characteristics [4]. While designing these antennas, first conventional structures such as rectangular, cylindrical are selected which are modified by proper optimization as per required performance characteristic parameters [5]. An antenna must operate on multiple bands, so that it could serve numerous applications [11]. A circularly polarized dual band rectangular DRA characteristics can be improved by changing the feeding mechanism and introducing a parasitic strip and using triangular ring-shaped aperture. Impedance bandwidth and axial ratio are varied by using parasitic strip [10]. Another cylindrical structure is cleaved into two halves and arranged to form a sigmoid shape, and metallic strip is used to get dual band CP response [6, 7]. By varying the size & location of metallic strip, the upper band 3 dB axial ratio is obtained for various 10 dB impedance passbands [8]. In another DRA structure, a compact dual band MIMO DRA is designed by stacking. Defected ground structure is being used in this to provide high isolation between the antenna ports [12]. Similarly, a single fed wide CP dual band antenna is designed by cutting two notches from cylindrical structure and employing arc-shaped slots to increase impedance bandwidth and axial ratio in the lower band [10]. Investigation of single DRA having multiple modes excited to support wideband or multiband behavior is done to achieve highly isolated multiple port antenna system [13–19]. Some of the high dielectric materials available are alumina, graphite, ceramic, etc. Using high dielectric materials and various gain enhancement techniques such as stacking of different dielectric materials, high gain (>7 dB) is achieved [20–22]. By using multiple ports to excite single DRA structure, we achieve wide band behavior (>4 GHz) [23]. For achieving high efficiency (>80%), DRA structure is designed with proper selection of dimensions, material, feeding techniques, etc. Gain is also an important factor in case of antenna design [24]. Further, a high gain cylindrical DRA is designed by using a dual annular patch, and metallic cylinder forms a cavity to achieve high gain and wide bandwidth. Stable radiation pattern is obtained with this design [25]. The antenna with large bandwidth is required. A U-shaped ultra-wideband dielectric resonator antenna is designed that is excited through microstrip feed line [26]. Two cylindrical dielectric resonators are fed through aperture coupling to obtain a compact DRA structure with good impedance bandwidth for wideband applications [9]. A hemispherical DRA is excited with conformal strips having multiple parasitic strips [28–31].

In this article, different DRA designs are discussed, and a comparison of their performance in terms of characteristics parameters is provided. Section 1 gives an introduction about recent literature of DRA. Section 2 reviews the different DRA designs. Section 3 compares their performance parameters of DRA design and the improvements that have been done to improve the antenna characteristics.

2 Various Dielectric Resonator Antenna Designs

A dual-stacked DRA is designed in which sapphire dielectric resonator is placed on the substrate and above that TMM13i dielectric resonator is placed. With the introduction of sapphire structure, the antenna size is reduced. The use of sapphire provides advantages like thermal insulation, light transmission and durability to the design [1].

Due to the designs physical properties like resistance to physical changes, chemical erosion and hardness, it is best suited for underground, rugged and long distance applications [32]. Through the use of stacking, the antenna gain is enhanced. The advantage of using rectangular DRA is that it provides better fabrication flexibility as compared to other geometries [33]. The FR4 is used for substrate with dielectric constant 4.4 & loss tangent 0.002. The substrate size is 50 × 50 mm^2. Sapphire used as lower dielectric has height 2.5 mm & dielectric constant 10. TMM13i used as upper dielectric has height 2.5 mm & dielectric constant 12.8. Figure 1 shows a dual-stacked rectangular DRA. The antenna designed works from 6.70 to 7.30 GHz and provides 5.2 dB gain and VSWR <2. The designed antenna provides a good radiation intensity of more than 16 dB in *E* plane & *H* plane, respectively. Antenna designed is a good candidate for mobile communication and future smartphones for 5G applications. The feed used is aperture coupled as it provides better isolation between antenna and feed network and avoids spurious modes. Higher gain, better radiation and impedance are achieved through this design [1].

An inverted umbrella-shaped dielectric resonator antenna is designed by notching a cylindrical dielectric from left and right side to form an inverted umbrella-shaped DRA [2]. By cutting the notches of a particular dimension, the permittivity and quality factor of the material are changed that results in wider bandwidth [34]. Notches reduce the quality factor and effective permittivity of the antenna and in turn result in wider bandwidth. An inverted umbrella-shaped cylindrical DRA is shown in Fig. 2. Impedance matching of −25.68 dB using single stub microstrip feed is achieved. Rogers RT/duroid 6010 dielectric material used has relative permittivity 10.2, height 5 mm with radius 20 mm and is mounted on the substrate of FR4 material. FR4 has thickness 1.6 mm, loss tangent 0.002 & dielectric constant 4.4. $HEM_{11\delta}$ mode is excited using single stub fed microstrip feed line. Feed type, feed location and antenna structure determine the mode. Both dielectric constant and physical dimension of the antenna decide its resonant frequency. The designed antenna operates in two bands (4.73–5.07 GHz and 5.27–6.22 GHz) and provides impedance bandwidth of 5.5%

Fig. 1 Dual-Stacked Rectangular DRA

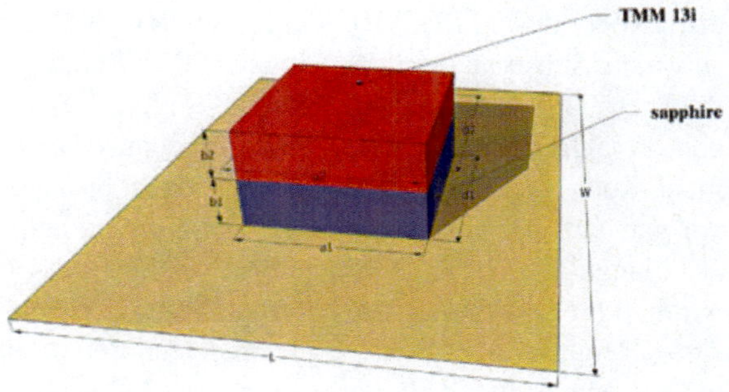

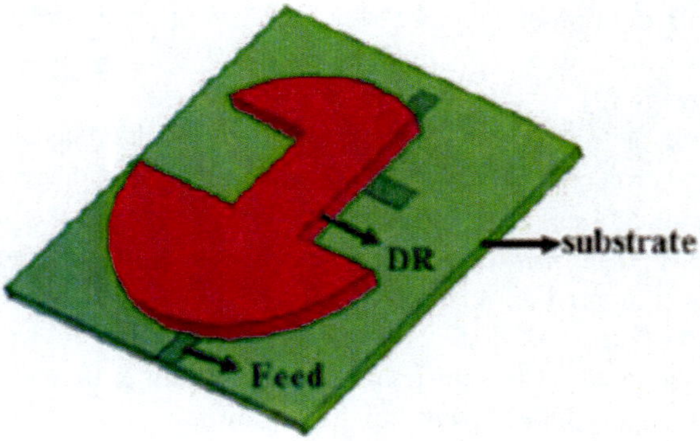

Fig. 2 Inverted umbrella-shaped cylindrical DRA

and 16.7%, respectively. Antenna provides impedance matching of −24 and −55 dB in the frequency band (4.73–5.07 GHz and 5.27–6.22 GHz).

Peak gains of 3.5 and 5.2 dBi are achieved at resonant frequency 4.88 and 5.6 GHz. Low profile, easy to fabricate antenna is designed with CDRA thickness of 5 mm [2].

A star-shaped dielectric resonator antenna is designed by inserting vacuum slots in the cylindrical DR, so as to increase resonant frequency by reducing the effective permittivity of the DRA. FR4 and Al_2O_3 are the materials used for substrate and DR having dielectric constant 4.4 & 9.8, respectively [3]. The feeding technique used is coaxial feed, and it is placed at any desired location to provide improve impedance matching and gain of the antenna [35]. The FR4 substrate has radius 60 mm and height 0.8 mm with loss tangent 0.02. The coaxial feed has radius inner conductor 0.6 mm, height 10.835 mm and air gap 11.67 mm, outer conductor radius 1.2 mm, height 6.3 mm. Inner conductor and outer conductor are made of PEC material and separated by Teflon. A star-shaped DRA is shown in Fig. 3. The designed antenna works from 3.516 GHz to 5.936 GHz. The simulated and measured results show peak gain of 7.8 dB and 6.4 dB, respectively, and 2.42 GHz bandwidth. The designed antenna is linearly polarized and can be used for military applications [3].

A surya yantra-shaped DRA is designed by using hexagonal-shaped conformal strip feed that is placed partially on the ground plane [4]. Inserting air gap reduces the DRAs effective permittivity and in turn increases the bandwidth of the antenna [28]. Roger duriod RO 3010 and Rogers RT 5880 are the material used for DR and substrate. Substrate has dimensions of 70 × 70 mm^2 with permittivity 2.2. Substrate thickness is 0.8 mm. In this antenna, a rectangular DR is fed by hexagon-shaped conformal feed that improves 29.6% impedance bandwidth. For bandwidth improvement, Koch fractal type geometry is used that provides 5% bandwidth improvement. Air gaps are introduced into the DRA with the help of cylindrical slot along with six triangular slots to further improve bandwidth of the designed antenna. Bandwidth of the DRA is decided by properly selecting the size and position of triangular slots. Height and shape of conformal feed decide the return loss performance of the antenna. Figure 4 shows a star-shaped DRA. The designed antenna provides an

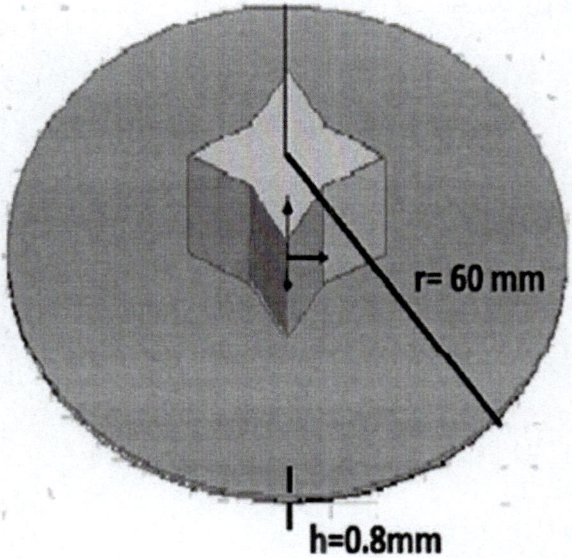

Fig. 3 Star-shaped DRA

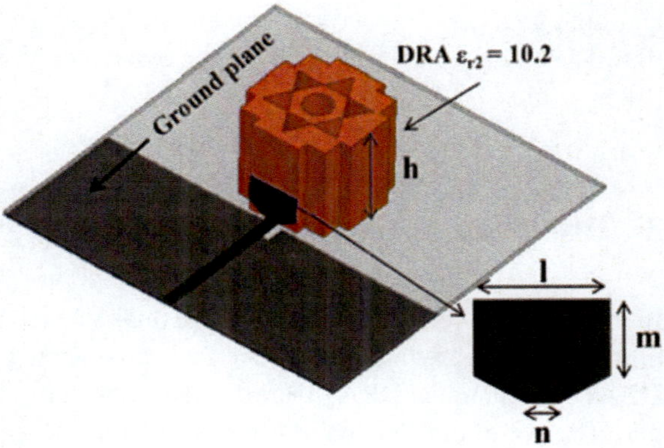

Fig. 4 Surya yantra-shaped DRA

impedance bandwidth of 113% from frequency 2.6 to 9.4 GHz. Gain of 4 dBi and radiation intensity of 98.6% is achieved with this design [4].

A rectangular DRA is designed in which input signal is fed through rectangular slots and excitation is provided using edge feed rectangular patch. A ceramic dielectric resonator is placed above ground plane & a rectangular feed is placed on bottom side. A cut in the shape of rectangle has been provided with 2 U-shaped cuts on ground plane. The substrate is of FR4 material with dimensions 40 × 40 × 1.6 mm^3 and loss tangent 0.02 & dielectric constant 4.4. The dielectric is made up of ceramic material with dimensions 22 × 22 × 6 mm^3 and has dielectric constant 9.8 & loss tangent 0.002. It provides a gain 4.5 dB at 5.7 GHz. Figure 5 shows a rectangular DRA. The designed antenna provides good radiation efficiency, enhanced bandwidth

Fig. 5 Rectangular DRA

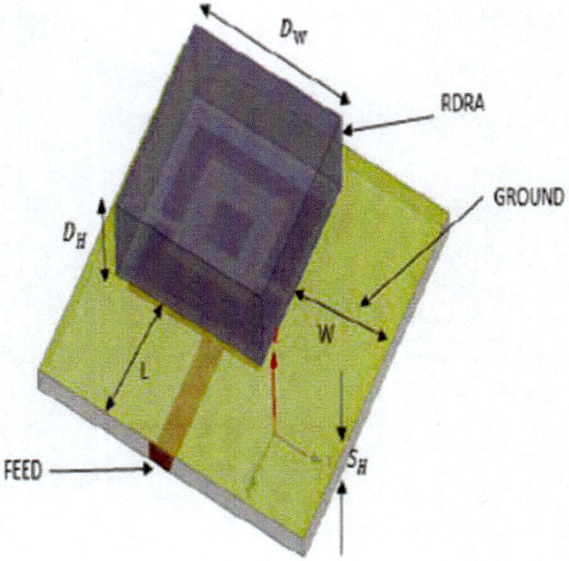

and high gain, so designed antenna can be used for radar and satellite communication. The antenna works perfectly in the two bands (3.8–4.5 GHz and 5.9–6.39 GHz) [5].

3 Discussions

In this paper, different DRA shapes like dual-stacked rectangular DRA [1], inverted umbrella-shaped cylindrical DRA [2], star-shaped DRA [3], surya yantra-shaped DRA [4], rectangular DRA [5] are discussed and how the changes in design impact the performance of the antenna in C band. DRA shape, feeding material, DRA material are the key parameters that decide the performance of the DRA. In [1], two rectangular dielectric materials sapphire and TMM13i are stacked and are fed through aperture coupling. The antenna designed provides gain 5.2 dB and VSWR <2 and operates from 6.70 to 7.30 GHz. The designed antenna provides a good radiation intensity of more than 16 dB in *E* plane & *H* plane. Better impedance matching, radiation pattern and gain are achieved through this design. In another design, by cutting notches in a cylindrical dielectric from left and right side to form an inverted umbrella-shaped DRA. $HEM_{11\delta}$ mode is excited through single stub fed microstrip feed line. Antenna operates in two frequency bands (4.73–5.07 GHz and 5.27–6.22 GHz) and provides impedance matching of −24 and −55 dB. Peak gain of 3.5 and 5.2 dBi is achieved at resonant frequency 4.88 and 5.6 GHz [2]. In one design, cylindrical DR is modified by inserting vacuum slots so as to form a star-shaped dielectric resonator antenna. As per simulated and measured results, peak gain of 7.8 and 6.4 dB is achieved. The

Table 1 Comparison of different DRA designs

Parameters	Dual-stacked rectangular DRA [1]	Inverted umbrella-shaped cylindrical DRA [2]	Star-shaped DRA [3]	Surya Yantra-shaped DRA [4]	Rectangular DRA [5]
DRA material	Sapphire & TMM13i	Rogers RT/duroid 6010	Al_2O_3	Rogers duroid RO 3010	Ceramic
Feed type	Aperture coupled	Microstrip feed	Coaxial feed	Conformal strip feed	Rectangular feed
Substrate Size (mm^3) & Material	50 × 50 × 0.8 FR4	50 × 50 × 5 FR4	Radius = 60 FR4	70 × 70 Rogers RT 5880	40 × 40 × 1.6 FR4
Band of operation	C Band	C Band	S Band & C Band	C Band & X Band	C Band
Bandwidth (MHz)	430	1670	2420	680	700 &490
Peak gain	5.2 dB	5.2 dBi	4.5 dB	4 dBi	4.2 & 4.6 dB

antenna works from 3.516 to 5.936 GHz and provides a bandwidth of 2.42 GHz [3]. By placing hexagonal-shaped conformal strip feed partially on the ground plane, a surya yantra-shaped DRA is obtained. Antenna provides radiation intensity of 98.6% and gain of 4 dBi from 2.6 to 9.4 GHz. An impedance bandwidth of 113% is achieved with this design [4]. In one design, the rectangular feed is placed below the ground plane, the dielectric resonator is placed above that, and two U-shaped cuts are provided on the ground plane. At 5.7 GHz, the antenna provides gain 4.5 dB and works perfectly from 3.8 to 4.5 GHz & 5.9 to 6.39 GHz [5]. By varying the basic DRA shapes like cylindrical or rectangular and introducing notches, parasitic strips, metallic strips, conformal strips in the basic antenna design its characteristics are improved to provide wider bandwidth, large gain and better impedance matching. The concept of using two dielectric provides improves radiation efficiency, impedance bandwidth and gain. The modes of antenna are determined with feed type and its position.

References

1. Bakshi, G., Vaish, A., Yaduvanshi, R.S.: Sapphire stacked rectangular dielectric resonator aperture coupled antenna for C-Band applications. Wirel. Pers. Commun. 895–905 (2019)
2. Vinodha, E., Raghavan, S.: A broadband inverted umbrella shaped cylindrical dielectric resonator antenna for "WLAN" and "C" band applications. Int. J. RF Microw. Aided Comput. Eng. **27**(6), 1–7 (2018)
3. Kumar, A., Kapoor, P.: Design and development of wideband dielectric resonator antenna for S and C band applications. In: IEEE Indian Conference on Antennas and Propagation, December 2018

4. Trivedi, K., Pujara, D.: Design and development of UWB dielectric resonator antenna. In: IEEE Indian Conference on Antennas and Propagation, December 2018
5. Ram, S.K., Roy S., Chakraborty, U.: A dual band microstrip antenna integrated with rectangular DRA for uplink and downlink C band communication. In: 4th International Conference on Recent Advances in Information Technology, March 2018
6. Gupta, A., Gangwar, R.K.: Dual-band circularly polarized aperture coupled rectangular dielectric resonator antenna for wireless application. IEEE Access, 2018
7. Varshney, G., Gotra, S., Pandey V.S., Yaduvanshi, R.S.: Inverted-sigmoid shaped multi-band dielectric resonator antenna with dual band circular polarization. IEEE Trans. Antennas Propag. 2018
8. Khan, A.A., Haizal, M., Aqeel S., Nasir, J.: Dual-band MIMO dielectric resonator antenna for WiMAX/WLAN applications. IET Microw., Antennas Propag. **11** (2017)
9. Zhou, Y.D., Jiao, Y.C., Weng, Z.B., Ni, T.: A novel single-fed wide dual-band circularly polarized dielectric resonator antenna. IEEE Antennas Wirel. Propag. Lett. **15** (2016)
10. Lu, L., Jiao, Y.C., Liang W., Zhang, H.: A novel low-profile dual circularly polarized dielectric resonator antenna. IEEE Antennas Wirel. Propag. Lett. (2016).
11. Pan, Y.M., Zheng, S.Y., Hu, B.J.: Design of dual-band omnidirectional cylindrical dielectric resonator antenna. IEEE Antennas Wirel. Propag. Lett. **13** (2014)
12. Abedian, M., Oraizi, H., Rahim, S.K.A. Danesh, S., Ramli, M.R. Mohammad, F., Jamaluddin, H.: Wideband rectangular dielectric resonator antenna for low-profile applications. IET Microw., Antennas Propag. **12**(1), 115–119 (2018)
13. Wong, K.L.: Planar Antennas for Wireless Communications. Wiley, New York (2003)
14. Dash, S.K.K., Khan, T., Borthakur, M.: Circularly polarized conical dielectric resonator antenna for X-band applications: an experimental study. In: Proceedings of the 14th European Radar Conference, 2017.
15. Wong, H., Luk, K.M., Chan, C.H., Xue, Q., So, K.K., Lai, H.W.: Small antennas in wireless communications. Proc. IEEE **100**(7), 2109–2121 (2012)
16. Lui, Y., Lui, H., Wei M., Gong, S.: A low-profile and high-permittivity dielectric resonator antenna with enhanced bandwidth. IEEE Antennas Wirel. Propag. Lett. (2015)
17. Feng K., Li N., Meng Q., Wang Y., Zhang J.: Study on dielectric resonator antenna with annular patch for high gain and large bandwidth. Chin. J. Electron. (2015)
18. Messaoudene, I., Denidni, T.A., Benghalia, A.: Ultra-wideband DRA integrated with narrow-band slot antenna. Electron. Lett. **50**(3), 139–141 (2014)
19. Majeed, A.H., Abdullah, A.S., Sayidmarie, K.H., Abd-Alhameed R.A., Noras, J.M.: Aperture-coupled asymmetric dielectric resonators antenna for wideband applications. IEEE Antennas Wirel. Propag. Lett. **13** (2014).
20. Lin, T.-Y., Chiu, T., et al.: A dual-band millimeter-wave high-gain dielectric resonator antenna using vertical assembly technology. In: IEEE CPMT Symposium, Japan (ICSJ), pp. 131–132, 2017.
21. Sharma, A., Sarkar, A., et al.: A polarization diversity substrate integrated waveguide fed rectangular dielectric resonator antenna. In: Proceedings of the Asia-Pacific Microwave Conference (APMC), IEEE, pp. 1–4, 2016.
22. Khosla D., Malhi, K.S.: Investigations on designs of dielectric resonator antennas for WiMax & WLAN applications. In: Proceedings of Fifth International Conference on Parallel, Distributed and Grid Computing (PDGC), pp. 646–651, December 2018
23. Palta, P., Sharma, M., Khosla, D., Goyal, S.: SIW-based leaky wave antenna: design and analysis for silicon. Int. J. Mobile Comput. Devices **4**(2), 20–25 (2018)
24. Sharma N., Khosla, D.: A compact two element U shaped MIMO planar inverted-F antenna (PIFA) for 4G LTE mobile devices. In: 5th IEEE International Conference on Parallel, Distributed and Grid Computing (PDGC-2018), pp. 838–841, December 2018.
25. Sharma, M., Singh, S., Khosla D., Goyal, S.: Waveguide diplexer: design and analysis for 5G communication. In: 5th IEEE International Conference on Parallel, Distributed and Grid Computing (PDGC-2018), pp. 586–590, December 2018.

26. Minhas, S., Khosla, D.: Compact size and slotted patch antenna for WiMAX and WLAN. Indian J. Sci. Technol. **10**(16), 1–5 (2017)
27. Kumar, N., Saini, G., Sahni, P., Khosla, D.: A compact multiband PIFA for personal communication handheld devices. Int. J. Conceptions Comput. Inf. Technol. **4**(4), 13–15 (2016)
28. Kaur, R., Khosla, D.: A wideband antenna for wireless communication devices. Int. J. Mod. Comput. Sci. **3**(3), 129–132 (2015a)
29. Kaur, R., Khosla, D.: Study of planar inverted-f antenna structures for various wireless devices. IJEEE **2**(3), 31–34 (2015b)
30. Sharma M., Singh, H.: SIW based Leaky wave antenna with semi C-shaped slots and its modeling, design and parametric considerations for different materials of dielectric. In: 2018 Fifth International Conference on Parallel, Distributed and Grid Computing (PDGC), pp. 252–258. IEEE, 2018.
31. Kaur, S.P., Sharma, M.: Radially optimized zone-divided energy-aware wireless sensor networks (WSN) protocol using BA (bat algorithm) IETE J. Res. **61**(2)170–179 (2015)
32. Sharma, M., Singh, S., Khosla, D., Goyal, S., Gupta. A.: Waveguide diplexer: design and analysis for 5G communication. In: 2018 Fifth International Conference on Parallel, Distributed and Grid Computing (PDGC), pp. 586–590. IEEE, 2018.
33. Kumar, A., Sharma, M.: Designing of ultra wide band microstrip antenna using triple slotted patch and DGS J. Telecommun., Switch. Syst. Netw. **5**(3), 12–19 (2019)
34. Sharma M., Singh, H.: Substrate integrated waveguide based Leaky wave antenna for high frequency applications and IoT. Int. J. Sens., Wirel. Commun. Control (2019)
35. Sharma, M., Singh, H.: A review on substarte intergrated waveguide for mmW. Circ. Comput. Sci., ICIC 2018, 137–138 (2018)

Planar Rectangular Micro-strip Patch Antenna Design for 25 GHz

Amandeep Kaur, Praveen Kumar Malik, and Ramendra Singh

Abstract The research article demonstrates the unconstrained optimization of planer rectangular micro-strip patch antenna for 5G communications. Proper dimensions of the antenna were adhering as per the accord and analogy of theoretical concept and adapted. The proposed antenna design is for the operation of the 5G band, especially at 25 GHz. The design of the antenna and its simulation are done on antenna simulation software and are optimized to have better return loss, transmission efficiency, directivity, and gain. For the entire band of operation of frequency as keen obtained results from the simulation are found close agreement with those obtained theoretically is unprecedented.

Keywords 5G · Bandwidth · Efficiency · Gain · Radiation power

1 Introduction

Factually, in this modern contemporary word, wireless transmitting devices archival now becomes compact in size and useful for different aspects of applications like IoT, artificial intelligence, mobile communication, and satellite communication and 5G networks. One of the significant challenges in wireless communication is the lack of frequency bandwidth as numbers of users are increasing day by day [1, 7]. To allay this problem, concept of millimeter wave technology and use description, which works on three different frequency bands 3300–3600 MHz, 24.25–27.5 GHz, and 27.5–29.5 GHz, is to fulfill the high data rate requirements of users that leads

A. Kaur · P. K. Malik (✉)
School of Electronics and Electrical Engineering (SEEE), Lovely Professional University, Phagwara, Punjab, India
e-mail: pkmalikmeerut@gmail.com

A. Kaur
e-mail: aman.dhaliwal18@gmail.com

R. Singh
Department of ECE, Inderprastha Engineering College, Ghaziabad, UP, India
e-mail: singh.ramendra23@gmail.com

P. K. Singh et al. (eds.), *Recent Innovations in Computing*, Lecture Notes in Electrical Engineering 701, https://doi.org/10.1007/978-981-15-8297-4_18

Table 1 Antenna design parameters

Parameters	Operating frequency	Dielectric constant	Substrate material	Loss tangent
Values	25 GHz	2.2	Rogers RT duroid	0.009
Parameters	Height of substrate	Length of Gnd (Lg)	Width of Gnd (Wg)	Length of Patch (Lp)
Values	1.6 mm	12.46 mm	14.34 mm	2.86 mm
Parameters	Width of patch (Wp)	Feed line length (Lfl)	Feed line width (Wfl)	Impedance of patch
Values	4.74 mm	2.156 mm	0.453 mm	144ohm

to challenging network design requirements. Apparently, in this type of structure, it is necessary to integrate high radio frequency design into a single antenna with a high degree of compactness [2, 9]. So, perusal based on network requirements needed for 5G, an antenna with less weight, small in size, and compatible with microwave circuits are highly regarded. Gravely planer rectangular micro-strip patch antenna plays a remarkable and veteran role in wireless communications [3]. The geometric shape of a micro-strip antenna consists of a radiating element on the dielectric substrate and a ground plane on the other side [13, 14]. There are several categories of the micro-strip patch antenna depending upon the shape of the patch, for example, the circular, square, triangular, and semicircular. However, the most common from all is rectangular patch antenna, as oratory [4, 8].

2 Micro-strip Antenna Design

A gambit rectangular micro-strip patch antenna for 5G applications is proposed in this article. The antenna has dimensions of 4.74 mm wide and 2.86 mm in length, which is very small in size. Substrate material used for antenna configuration is Rogers RT 5880 with a thickness of 1.6 mm, the dielectric constant of 2.2, and having a loss tangent of 0.0009. The antenna is simulated using antenna simulation software, and antenna performances are analyzed in terms of S-parameter, VSWR, gain, and radiation pattern and directivity. The proposed antenna is suitable to be operated in the frequency range of 5G communication, especially 25 GHz [5, 6].

In Fig. 1, proper validation and hypothesis of mathematical expression are taken care of cautiously and emphasized while designing the length and width of the antenna [10]. The impedance of the antenna is calculated certainly with the help of the transmission line concept. In Fig. 2, the width and length of the transformer are correctly matching the impedance of the patch by keeping cognizance with the port of 50 Ω [15–16].

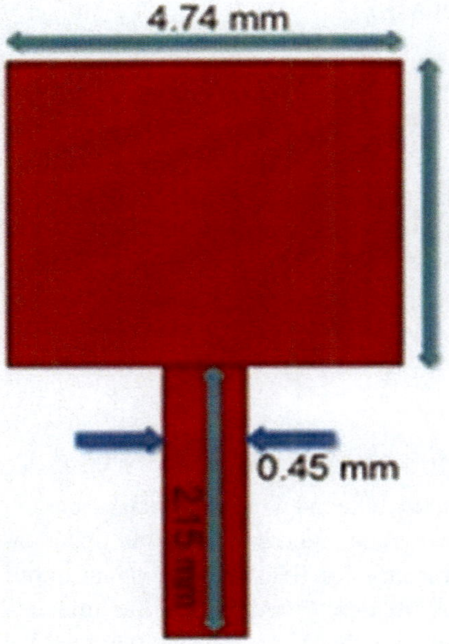

Fig. 1 Proposed antenna design top view

Fig. 2 Proposed antenna design patch, feed, and its excitation

Designing the antenna resonating frequency Fr, substrate dielectric constant and height of substrate h should be known utter. Synthesis of the wavelength of the antenna done by using the articles [10–12] (Table 1).

3 Result and Discussion

Consequently, the above antenna is designed in an antenna simulation software, and another performance parameter, including return loss, is verified with the help of a vector network analyzer (VNA). Performance parameters like gain voltage standing wave ratio and impedance of the antenna are simulated and reported using the software itself. One of the factual parameters which define the reflection coefficient or return loss of the antenna is shown in Fig. 3. Antenna design parameters are also listed in table [1–3]. Results obtained after the simulation are discussed as follows.

3.1 Return Loss

Reflection coefficient of any antenna as a performance parameter is defined by the equation of reflection coefficient. Where Zin is input impedance of the antenna and Zo is the impedance of transmission lines. Typical values of reflection coefficient are in between zero and one. Maiden, Fig. 3 depicts the efficacy efficiency return loss of rectangular patch antenna in dB. The scattering parameter analysis is quintessential for micro-strip antenna because it represents the loss of power reflected by the antenna. As per the theoretical analysis, this ratio value should be zero, and practically, it should be less than −10 dB as it complies in the design. It also reveals the operating bandwidth of the antenna, extensively from graphs, and it is evident that for operating frequency at 25 GHz, the value of significant return loss is less than −15 dB. From the graph, it can also be purportedly stated the antenna resonates for frequency 23.3–26.9 GHz with return loss less than −10 dB. The value of return

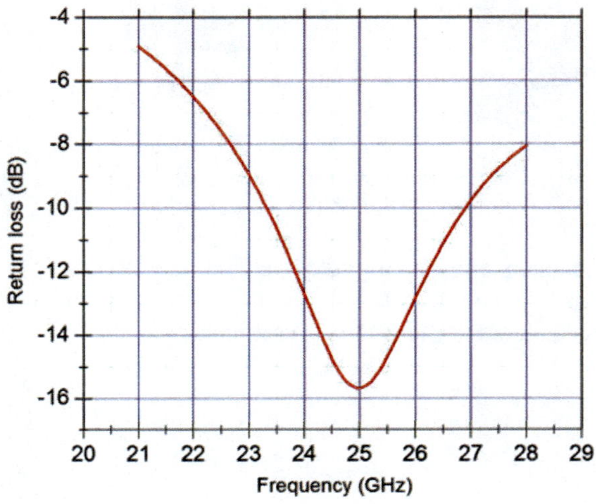

Fig. 3 Return loss of antenna in dB

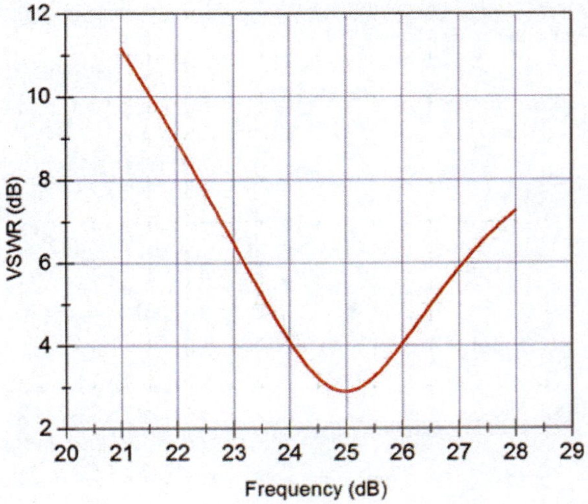

Fig. 4 Proposed antenna design patch, feed, and its excitation

loss for frequency 23.3 GHz is −10 dB, and till 26.9 GHz, it is −10 dB. With an important attribute, it is evident that the operating range of antenna would prevail at 3.6 GHz [26.9–23.3 GHz].

3.2 Voltage Standing Wave Ratio

Intuitively appealing voltage standing wave ratio (VSWR) inward is a function of reflection coefficient which is given in equation \ref{abc}, and it is defined in equation number \ref{abc1}.

In antenna, maximum power can transfer if antenna transmission line impedance matches the load impedance. The voltage standing wave ratio elicited and shows the impedance matching of the source with the load. The ideal value of VSWR should be unity. Figure 4 illuminates the VSWR vs. frequency response of the proposed antenna. It is found that VSWR for operating frequency 25 GHz is around 2 (2.88). Deficiency in VSWR is due to the disclaimer of the extended length of the patch.

3.3 Gain of Antenna

Figure 5 illustrates the solemn rectangular plot between the gain and directivity of proposed design and frequency in terms of GHz. It is perceived that gain should be more than 3 dB; graph cohesion shows that for the value of theta from −180 to +180, the highest value of directivity and gain is achieved for the proposed antenna which is around 4.7 dB at phi zero. Gain and directivity of the antenna, which come from

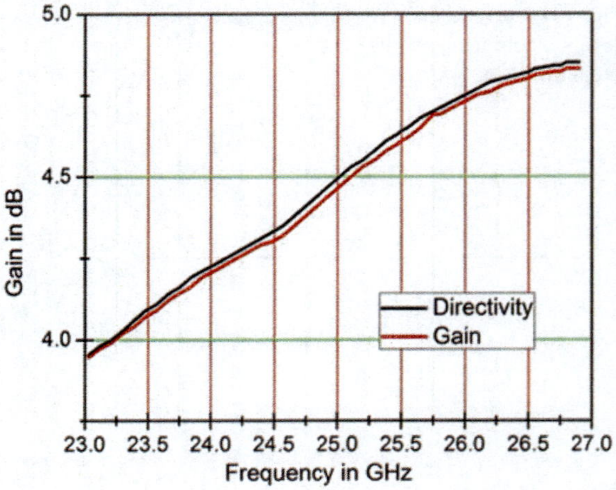

Fig. 5 Plot of directivity and gain versus frequency

the simulated result and measured gain at a different frequency, especially when the return loss is minimum, are vetted. It is clear from Fig. 5 that gain and directivity vary from 3.94 to 4.83 dB over the frequency range from 23.3 to 26.9 GHz.

3.4 Polar Plot Radiation of Proposed Design

A dominant gain in terms of the polar plot of the proposed design is shown in Fig. 6.

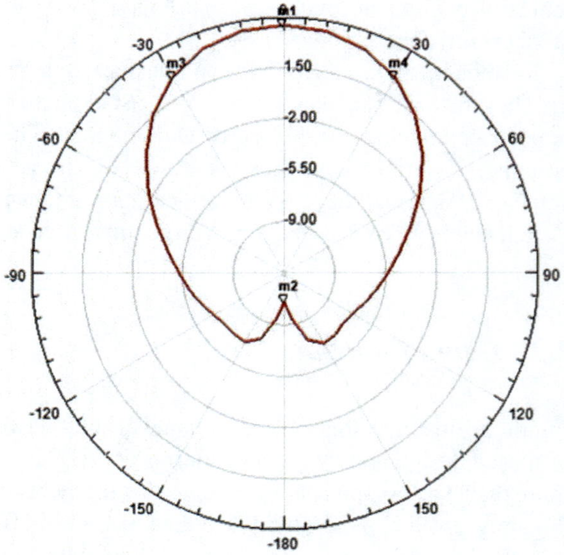

Fig. 6 Polar plot of proposed antenna

Fig. 7 Radiation pattern (3D)

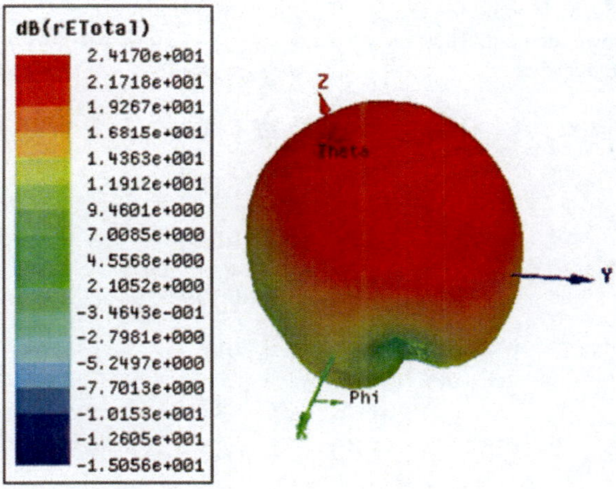

From Fig. 6, this can be quoted that the antenna is stunning directional in nature, and maximum radiation is in the direction of phi = 0°. The antenna also radiates its maximum in the direction of theta −30 to 30°. The value of the maximum gain elicited is around 4.47 dB readily. It is exciting, and up to the discretion of the reader to note that in the proposed antenna, there is the very bare value of power which is radiated in the back lobe, most of the power is transferred in the main lobe.

3.5 3D Radiation Pattern

Figure 7 shows the 3D radiation pattern of the proposed antenna. The discreet 3D plot shows the relation of strength of EM waves originated from the antenna is contiguous and to the source concerning values of theta and phi. As it is ought, antenna to radiate in one direction is buoyed.

Figure 8 displays the relation of radiated power and efficiency of the proposed design versus frequency of the operation. As stated earlier in the discussion and ascertain of return loss, the frequency of operation of the antenna is between 23.3 and 26.9 GHz. Insofar as the plot of radiated power and frequency is consider is taken between these ranges only. It is quoted that the radiated power of the antenna in between the ranges of 23.3–26.9 GHz is adequate and plenary. Notably, the radiated power is maximum at the frequency of resonance, i.e., around 25 GHz efficiently. An essential parameter of the transmitting antenna is its efficiency, and from Fig. 8, it is observed that in the operating range of 23.3 and 26.9 GHz, efficiency of the antenna is more than 90%.

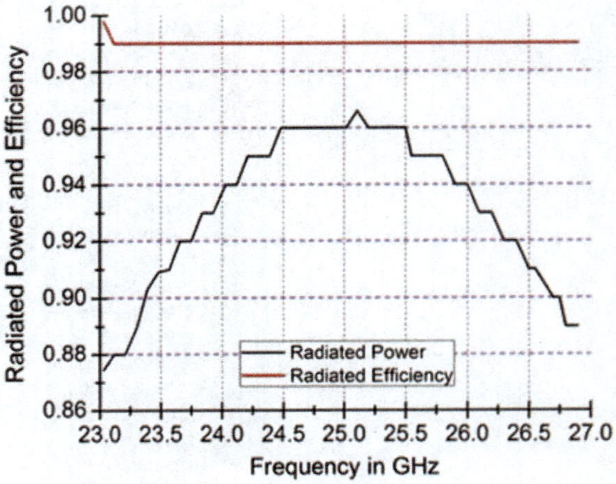

Fig. 8 Plot of radiated power and efficiency versus frequency

4 Conclusion

Consequently and unanimously, it is unveiled that a planer rectangular micro-strip antenna is designed for 5G wireless applications that efficiently operate from 23.3 to 26.9 GHz frequency band. All the performance parameters are also in accord and harmony with the theoretical concept. Radiation efficiency, gain, and the radiation pattern are the prominent performance of the proposed design. Eventually, it is speculated that the proposed antenna could be used efficiently for a higher frequency, especially for 5G in the range of the Ku band, as the antenna has suitable performance parameters for radiation purposes.

5 Conflict of Interest

The authors certify that they have no affiliations with or involvement in any organization or entity with any financial interest (such as honoraria; educational grants; participation in speakers' bureaus; membership, employment, consultancies, stock ownership, or other equity interest; and expert testimony or patent-licensing arrangements) or non-financial interest (such as personal or professional relationships, affiliations, knowledge, or beliefs) in the subject matter or materials discussed in this manuscript.

Acknowledgements This paper and the research behind it would not have been possible without the exceptional support of my supervisor, Dr. Praveen Kumar Malik. His enthusiasm, knowledge, and exacting attention to detail have been an inspiration and kept my work on track. This research was partially supported by Lovely Professional University, Phagwara, Punjab, India. We thank our colleagues from Lovely Professional University, Phagwara, Punjab, India, who provided insight and expertise that greatly assisted the research. Author wants to express their deep gratitude to Dr. Harish Parthasarthy, NSIT, New Delhi, for sharing their pearls of wisdom with us during the course of this research. Author would also express her deep gratitude to Inderprastha Engineering College, Ghaziabad, for sharing their pearls of wisdom with us during the course of this research. We are also immensely grateful to all our friends for their comments on the preparation of the manuscript, although any errors are our own and should not tarnish the reputations of these esteemed persons.

References

1. Balanis, C.A.: Antenna Theory: Analysis and Design, 3rd edn. WILEY Interscience (2009)
2. Harish, A.R., Sachidananda, M.: Antennas and Wave Propagation. Oxford Higher Education (2007)
3. Bahl, J., Bhartia, P.: Microstrip Antennas. Artech House, Inc. (1980)
4. Chen, T., Chen, Y., Jian, R.: A wideband differential-fed microstrip patch antenna based on radiation of three resonant modes. Int. J. Antennas Propag. **2019**, 7 p. Article ID 4656141 (2019)
5. Malik, P., Parthasarthy, H.: Synthesis of randomness in the radiated fields of antenna array. Int. J. Microw. Wirel. Technol. **3**(6), 701–705 (2011). https://doi.org/10.1017/S1759078711000791
6. Hossain, I., Noghanian, S., Pistorius, S.: A diamond shaped small planar ultra wide band (UWB) antenna for microwave imaging purpose. In: IEEE Antennas and Propagation Society International Symposium, pp. 5713–5716 (2007)
7. Malik, P.K., Parthasarthy, H., Tripathi, M.P.: Axisymmetric Excited Integral Equation Using Moment Method for Plane Circular disk. Int. J. Sci. Eng. Res., **3**(3), 1–3 (2012)
8. Cai, Y., Li, K., Yin, Y., Gao, S., Hu, W., Zhao, L.: A low-profile frequency reconfigurable grid-slotted patch antenna. IEEE Access **6**, 36305–36312 (2018)
9. Malik, P.K., Parthasarthy, H., Tripathi, M.P.: Alternative mathematical design of vector potential and radiated fields for parabolic reflector surface. In: Unnikrishnan, S., Surve, S., Bhoir, D. (eds.) Advances in Computing, Communication, and Control. ICAC3 2013. Communications in Computer and Information Science, vol 361. Springer, Berlin, Heidelberg (2013)
10. Malik, P.K., Wadhwa, D.S., Khinda, J.S.: A survey of device to device and cooperative communication for the future cellular networks. Int. J. Wirel. Inf. Netw. (2020). https://doi.org/10.1007/s10776-020-00482-8
11. Akram, S.V., Malik, P.K., Singh, R., Anita, G., Tanwar, S.: Adoption of blockchain technology in various realms: opportunities and challenges. Secur. Privacy. e109 (2020). https://doi.org/10.1002/spy2.109
12. Malik, P.K., Tripathi, M.P.: OFDM: a mathematical review. J. Today's Ideas—Tomorrow's Technol. **5**(2), 97–111 (2017). https://doi.org/10.15415/jotitt.2017.52006
13. Malik, P.K., Parthasarthy, H., Tripathi, M.P.: Analysis and design of Pocklingotn's equation for any arbitrary surface for radiation. Int. J. Sci. Eng. Res. **7**(9), 208–213 (2016). ISSN 2229-5518
14. Kaur, A., Malik, P.K.: Tri state, T shaped circular cut ground antenna for higher 'X' band frequencies. In: 2020 International Conference on Computation, Automation and Knowledge Management (ICCAKM), Dubai, United Arab Emirates, pp. 90–94 (2020)
15. Tiwari, P., Malik, P.K.: Design of UWB Antenna for the 5G mobile communication applications: a review. In: 2020 International Conference on Computation, Automation and Knowledge Management (ICCAKM), Dubai, United Arab Emirates, pp. 24–30 (2020)

Intelligent Opportunistic Routing Protocol in Wireless Sensor Networks: A Security Perspective

Deep Kumar Bangotra, Yashwant Singh, and Arvind Selwal

Abstract Wireless sensor networks (WSNs) applications have grown huge in the recent years. WSN is very popular due to its abilities to record, process, and transmit data related to various parameters of the environment. Due to WSN inherent nature, these networks are vulnerable to various types of attacks, especially at the network layer of the protocol stack in WSN. In this paper, an analysis about different security threats w.r.t different layers of the protocol stack using opportunistic routing (OR) is presented. The wireless nature and broadcast feature of WSN cause many attacks in the network layer and security of the transmitted data becomes an important issue. Different machine learning (ML) techniques are used for solving various security issues of the WSN. Also, in this paper, we focus on the analysis of different ML methods used for detection of security related attacks on the routing process in WSN.

Keywords Wireless sensor networks · Opportunistic routing · Machine learning · Trust-based management · Security attacks

1 Introduction

With the ever-increasing demand of data collection for analytical purposes, the WSNs have established themselves as ultimate solution for monitoring the physical environment to record the necessary parameters for data collection. This has become

D. K. Bangotra (✉)
Department of Higher Education, J&K Govt., Jammu, India
e-mail: deepbangotra.ap@gmail.com

Y. Singh · A. Selwal
Department of Computer Science and Information Technology, Central University of Jammu, Samba, Jammu and Kashmir 181143, India
e-mail: yash22222k1@gmail.com

A. Selwal
e-mail: arvind.cuj@gmail.com

P. K. Singh et al. (eds.), *Recent Innovations in Computing*, Lecture Notes in Electrical Engineering 701, https://doi.org/10.1007/978-981-15-8297-4_19

possible with the contemporary advancements in the domain of sensor network technologies. WSN is a collection of autonomous distributed motes scattered spatially in an environment to record the real conditions. WSNs' importance can be gauged from the fact that they are used in almost all applications ranging from civilian applications to military applications (battle field surveillance, pressure, humidity, moisture, wind speed, vibrations, heat, noise, etc.). The architecture of WSN is made of five layers where each layer is responsible for performing its intended function. The WSN layered architecture is presented in Fig. 1. The protocol stack is composed of five layers and network layer is one of the most important layers in the overall architecture of the WSN. This layer is responsible for the routing, topology management, and self-configuration in the WSN. It is responsible for link failures and delivers consistent updates to adjoining nodes. Though promising network operations at all times is a big task due to vigorously varying network topology, there is a considerable amount of difference between WSN routing protocols and traditional routing protocols. This leads to the need of optimizing network life by performing intelligent routing [1].

Routing is the method of finding best route for communication of the sensed data from source node to the destination node. One of the emerging and energy-efficient routing protocols for data communication is known as opportunistic routing (OR). OR is very capable of achieving faster and reliable delivery of data packets in WSN as compared to traditional routing algorithms. The OR protocol is principally based on the concept of using broadcast feature of the wireless medium. It focuses on choosing the best relay candidate from the list of forwarder nodes. The OR [1] is basically composed of three steps, i.e., (a) transmission using broadcast feature, (b) selection of the best relay candidate (prioritization among nodes using coordination protocol), and (c) selection of the node with highest ranking. The biggest strength of WSN lies

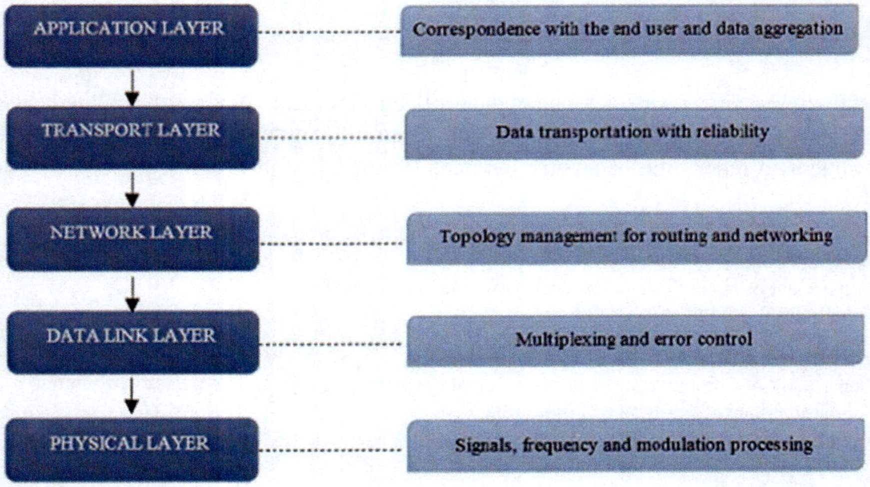

Fig. 1 WSN layered architecture [2]

in their abilities to monitor environment and communication of the recorded data. In contrast, the communication in the WSN is their biggest weakness and many attacks occur in network layer to disrupt routing and to affect the working of WSN. As WSNs are large networks composed of huge number of interconnected nodes but the safety and confidentiality issues avert its wide adoption. As these networks transmit unintelligible and vital data, the issues pertaining to confidentiality, reliability, and availability of the sensed data should be principal to their design. Therefore, a robust security mechanism should be required to mitigate the security issues related to the confidentiality, reliability, and availability of the sensed data [2]. Thus, security is very critical to routing in WSN.

In the recent past, many researchers have focused on developing secure routing by applying traditional methods based on cryptographic techniques. These methods are not efficient for resource constrained WSN, whereas in OR, a routing metric is used to decide the potential forwarders. The set of potential forwarders is called the candidate set. Next-hop forwarder will be a node from this candidate set based on next-hop selection metric [3]. These metrics are classified into two types, i.e., (a) end-to-end selection metrics and (b) local selection metrics. These metrics will not perform effectively, if a node has been affected by a malicious attack. For example, in a black hole attack, the affected node stops forwarding the packets toward other nodes. Such type of attacks needs special attention for secure routing algorithms.

The structure of the rest of the paper is as follows. The next section of the paper presents research related to the OR. The security aspect in WSN is presented in Sect. 3 including the security solutions for OR in WSN. Section 4 presents machine learning techniques for security issues in WSN. In Sect. 5, analysis of the presented work is done. In the end, Sect. 6 concludes the paper and discusses some future works.

2 Related Work

As far as OR in WSN is concerned, a significant amount of research has been done in the recent years. This division of the paper presents a brief outline of research work done by different researchers related to opportunistic routing protocol.

In 2003, Zorzi et al. [4] have developed a geographic random forwarding routing protocol for ad hoc and sensor networks that terms the forwarding approach based on the geographic location and the choice of the relay node by contention at the receiver end. The investigation of the performance in this multi-hop network situation is done by taking into account the count of steps to get to the destination. In 2005, Biswas and Morris [5] (ExOR) developed the primary protocol which essentially implements the opportunistic routing in wireless networks. This was based on the estimated transmission count (ETX) metric. The ETX was measured by hop count from source to the destination and data packet traveled through the minimum number of hops. In 2009, Li et al. [6] have worked on another routing protocol known as minimum transmission scheme–optimal forwarder list selection in opportunistic routing (MTS).

This protocol uses MTS instead of ETX as in ExOR. The MTS-based algorithm gives scarcer transmissions as compared to ETX-based ExOR, simple, practical, and effective opportunistic routing for short-haul multi-hop wireless networks [7]. In this protocol, the packet duplication rate is decreased. It is a simple algorithm and can be combined with other opportunistic routing algorithms. Spectrum-aware opportunistic routing (SAOR) [8] is another protocol that uses another cost metric known as the opportunistic link transmission (OLT) for positioning the nodes in the forwarder list. SAOR gives QoS like a reduced end-to-end delay and improved throughput than traditional routing protocols. In 2011, Xufei Mao et al. [6] have developed another energy-efficient opportunistic routing protocol that calculates the cost for individual node to onward data and then arranges the forwarding list. TMCOR [9] is another protocol that labels the balance between the cost metric and the security factor. TMCOR algorithm performs well by prohibiting the malevolent node to contribute in the network by adjudicating them based on the trust degree. This algorithm enhances security and reduces end-to-end packet delay. In 2015, Luo et al. [10] have presented ENSOR-Opportunistic Routing Algorithm for Relay node selection in WSN. This is another algorithm where the thought of energy-efficient node is implemented. The candidate list creation and arrangement of the nodes in that list is approved by energy saving via opportunistic routing algorithm. In 2016, Kumar and Singh [11] developed an energy-efficient opportunistic routing metric for WSN. This algorithm uses energy depletion factor for selection of the relay node. In 2017, Bapu and Gowd [12] proposed another algorithm based on quality of link. This link quality-based opportunistic routing algorithm for QoS provides improved network life span along with enhancement in security.

There are numerous concerns linked to security are present in WSN. Because of the absence of the chief authority and random deployment of motes in the network, the WSN is susceptible to safety fears. The sensor network will have to provide some security requirements in the form of network availability, data confidentiality, data integrity, data authentication, source localization, self-organization, and data freshness [13]. There are various traditional threats in WSN. These are tampering, exhausting, jamming, collision, routing control information, selective forwarding, sybil attack, sink hole, worm hole, hello flood, flooding, and malicious(replica) attack [14]. Out of many security attacks, malicious attack is one of the common attacks where a node is compromised and imitates as one of sensor nodes, misleading other nodes. Different protocols are offered by numerous authors to overcome security-related problems, i.e., SPINs, LEAP, TinySec, and ZigBee. In 2019, Alotaibi [15] presented an effective method termed as Hamming residue method (HRM) to lessen the malevolent occurrences of the security threat.

The security-related protocols for OR are broadly classified into three types, i.e., (a) game-based protocols, (b) trust-based protocols, and (c) other protocols [16]. In 2017, a safe and energy capable opportunistic routing protocol was proposed by Kumar [17]. This protocol is robust in identifying and isolating the malicious nodes. The trust management metric is used in this protocol. Another protocol, trust and location-aware routing (TLAR), was proposed [18] to communicate location apprehensive routing in WSN. SGOR [19] is one more routing protocol that offers both scalability and safety in WSN. There is another protocol known as trust-based security protocol (TSP) [20] which safeguards system from blackhole security threat. Cao et al. [21] presented a protocol (TDOR) in which the initial node estimates a secure route to be used to get to the target node plus unchanged path is chosen for sending of message. Darehshoorzadeh et al. [22] presented a unique way to recognize the performance of foul motes and furthermore support in gauging its impact on wireless mesh network. Another opportunistic routing protocol which is based on potential threat (PT) [23] works with a procedure that preferences the inward nodes on the basis of the features like prompt messages sending, status of motes, and their history.

OR protocols bank on the fact that all nodes in the network are secure and not affected by the attackers. However, all nodes in the network are not nonthreatening, the invaders may attack some nodes, and thus these nodes may go out of order. The pretentious nodes lead to unnecessary feasting of system assets and therefore affect the lifespan of the WSN. In order to safeguard the WSN from these malicious attacks, either nodes are to be secured from attacks or the routing is to be secured by using a new class of protocols, i.e., trust-based, reputation-based, or game-based protocols.

3 Security in WSN

A WSN is typically composed of several tiny-sized nodes which are infrastructure constrained, i.e., limited capacities in terms of perceiving, processing, storage, and communication. The sensor nodes are always operated through a small capacity battery, and they are always power constrained. Data acquired by sensor node is communicated to servers through the base station [2]. As shown in Fig. 1, the WSN uses layered architecture which typically comprises of five layers. There are certain features which make WSN unlike traditional wired network and more susceptible regarding security threats. These are (a) self-organization, (b) self-adaptive flow control, (c) resource restrictions, (d) centralized control, and (e) open environment. All these features require adoption of security mechanism so that WSN become efficient, secure, and reliable. The features like wireless medium of communication, the lack of physical protection, and the restrictions in the computing capabilities make WSN vulnerable to security threats. The majority of the attacks in WSN are classified as internal or external attacks. The way in which the nodes are controlled by the attacker decides whether it is an internal attack or external attack. As we are discussing the layered architecture of the WSN, the attacks with respect to each layer are presented in Table 1.

It is learned from Table 1 that each layer of the WSN architecture is vulnerable to security threats and there is an opportunity to develop secure communication protocols to mitigate the threats and to make the WSN reliable. Out of all five layers, the network layer is mainly affected by the security threats and requires attention. As routing is the function of the network layer, the majority of attacks will affect the routing process. The different attacks in the network layer will lead to

Table 1 Layerwise detail of attacks in WSN architecture

Sl. No.	Layer	Attacks
1	Physical layer	Eavesdropping, basic jammers, sensor tampering
2	Data link layer	intelligent jamming and collision
3	Network layer	Spoofed information, replay attack, choosy forwarding, blackhole, sinkhole, Sybil attack, node replication, wormhole
4	Transport layer	Desynchronization attack, data integrity, energy drain
5	Application layer	Attacks on reliability, malicious code attack
6	Multi-layer attack	Denial-of-service (DoS) attack and man-in-the-middle attack

creation of routing loops, increased network traffic, generating false error messages, and increase in data loss, compromising the transmission routes, eavesdropping on the falsely created links, undermining cryptographic protection, sending data to the false destination, and increasing the energy degradation. The special type of any path routing known as OR will also be affected by these attacks. These attacks are either internal attacks or external attacks. The attacker launches these attacks for diminishing the productivity of the network, dipping the routing precision, and poor usage of network resources. Therefore, to safeguard the WSN, numerous security procedures are to apply on opportunistic routing protocol for ensuring the security of nodes and the security of the data communicated.

3.1 Security in Opportunistic Routing (OR)

The broadcasting nature of the opportunistic routing protocol makes it highly vulnerable to security attacks. Due to this vulnerability, this promising protocol fails to achieve high reliability in terms of secure transmission of data packets from source node to the destination node in WSN. The presence of any malicious node in the network will affect the overall execution of the OR protocol. The OR protocol works in three steps, i.e., (a) broadcasting the data packets, (b) creation of the candidate list (prioritization among nodes, and (c) selection of the best node from candidate list to forward the data packet. All these three steps will not be executed effectively, if there is presence of malicious or compromised nodes in WSN. Thus, to take advantage of the basis on which the opportunistic protocol works, security mechanism in the network layer of the WSN protocol stack should be enhanced or made intelligent. In this section, different security protocols of OR is presented. Table 2 summarizes various security attacks on OR protocol that are presented by various researchers. From Table 2, it is clear that most of the security procedures used to make OR secure and efficient revolves around trust-based management method. Moreover, the potential of this method is recognized in the recent three to five years. Hence, there is not enough research work done on OR protocol security using trust- and reputation-based methods. From the literature, it is clear that energy efficiency and link state

Table 2 A brief background of secure opportunistic routing protocol

Sl. No.	Year	Authors	Protocol	Security attack/model	Security solution	Pros	Cons
1	2017	Kumar and Singh [17]	TPBOR	Blackhole, gray hole	Trust management	Increases the lifetime of the network by reducing security overhead	High in energy consumption
2	2016	Vamsi and Kant [18]	TLAR	Blackhole, packet modification attacks, badmouth attack	Trust management	Applicable even when there are resource constraint	Fails to work during multiple attacks on the network
3	2015	Lyu et al. [19]	SGOR	Grayhole, blackhole and location spoofing, attacks	RSS analysis, trust management	Scalability and network security	Sink node is vulnerable to attacks
4	2013	Woungang [20]	TSP	Blackhole attack	Trust management	Diminish the blackhole attack	Malicious nodes receive some messages
5	2018	Cao et al. [21]	TDOR	–	–	Messages are delivered with efficiency and reliability	Latency time is increased
6	2017	Salehi and Boukerche [25]	A novel packet salvaging model	Blackhole attack	Packet salvaging	Zero data packet overhead	Segregation of malicious nodes not possible

(continued)

Table 2 (continued)

Sl. No.	Year	Authors	Protocol	Security attack/model	Security solution	Pros	Cons
7	2016	Salehi et al. [22]	Modeling and performance evaluation of the security attacks for OR	Blackhole attack	–	Applicable on all types of OR protocol	Difficult to nullify malicious nodes in the network
8	2017	Chhabra et al. [23]	Security protocol based on potential threat (PT)	Blackhole attack	Game theory based	Reduced overhead in packet delivery and low packet dropping	Applicable to blackhole attack only
9	2017	Kumar et al. [26]	TEAROR	Blackhole attack and grayhole attack	Trust-based management	Security from blackhole and grayhole attacks	More parameters for trust calculation will lead to computational overhead

reliability have not been considered in case of security as far as OR is concerned. Therefore, there is a research opportunity to explore security features of OR using trust-based management metric.

4 Machine Learning-Based Techniques for Secure Routing in WSN

With the tremendous increase in the application of WSN and the ever-growing use of machine learning algorithms, the interdependence of both machine learning and WSN has solved significant issues present in the WSN, i.e., data processing and network optimization [24]. The machine leaning techniques are used to address various issues in the network layer especially related to security threats for safe communication of data in the network. These machine learning-based methods are used for network optimization and therefore work very well for security of the selected path for transmission of data in the network. As data transmission is very important and crucial task in the overall functioning of the WSN, there is a possibility of data loss or data tampering during transmission due to the possibility of security threats or attacks. Therefore, there is a need of security mechanism for protecting the sensed data during the transmission to the sink node in hop-to-hop communication. Blackhole attack, misdirection attack, wormhole attack, sinkhole attack, and hybrid anomaly are few attacks that may occur during the transmission of the data in the network layer.

According to the literature review, machine learning methods are available to detect different security threats in the WSN and therefore significantly improves security, reliability, and overall efficiency of the network. Table 3 summarizes various ML approaches for detecting security attacks in WSN.

From Table 3, it is observed that different machine learning techniques are used for detecting security attacks in the WSN. The use of these techniques has secured the data communication and therefore enhanced the reliability of the overall network.

5 Analysis

It is inferred from Tables 2 and 3, that the use of trust management methods for ensuring security of transmitted data using OR protocol and the use of different machine learning methods in securing routing process can be used together for effectively enhancing the security mechanism of OR protocol. As there is very less research work done for intelligent secure opportunistic routing in WSN, we may use Bayesian classifier, deep learning, random forest, etc., and machine learning technique for ensuring safe and secure transmission of data packets in WSN intelligently using OR protocol.

Table 3 Machine learning-based techniques for security issues in WSN

Sl. No.	Year	Authors	ML technique	Security attack/model	Remarks
1	2016	Wazid and Das [27]	*k*-Means	Blackhole and misdirection attack	Better accuracy is achieved
2	2013	Xie et al. [28]	*k*-NN	Random faults and online attacks	Low time complexity
3	2013	Garofalo et al. [29]	Decision tree	Sinkhole	High detection rate for environmental monitoring
4	2018	Gil et al. [30]	SVM + PCA (hybrid)	Outliers detection	Better accuracy
5	2017, 2015 & 2018	Feng et al. [31], Shahid et al. [32], Saeedi Emadi and Mazinani [33]	SVM	Anomaly detection, error detection and intrusion detection	Better accuracy with reduced time complexity
6	2014	Shamshirband et al. [34]	Q-learning	Denial of service (DoS)	Increase in lifetime of network
7	2015	Titouna et al. [35]	Bayesian	Outlier detection and trust management	Better accuracy
8	2015	Haque et al. [36]	Regression	Anomaly detection	Better accuracy
9	2016	Ma et al. [37]	Deep learning	Intrusion detection	Better accuracy

6 Conclusion and Future Scope

OR protocol provides better energy efficiency, reliability, and increase in network life time as compared to other routing protocols. This paper presents an analysis of diverse OR protocols by taking into consideration the security perspective and an analysis of different machine learning-based techniques for detecting various security threats at the network layer of WSN layered architecture. The OR protocol majorly uses trust management and reputation-based methods to provide security in data transmission. According to the present study, machine learning techniques are available for addressing security issues in routing but not much work has been done for securing the transmission of data using OR protocol. Therefore, intelligent secure opportunistic routing is an interesting research issue that requires attention.

Future scope of the work includes development of an intelligent and secures trust-based opportunistic routing algorithm for enhancing the security, energy efficiency, and reliability of the WSN.

References

1. Bangotra, D.K., Singh, Y., Selwal, A.K.: An intelligent opportunistic routing protocol for big data in WSNs. Int. J. Multimed. Data Eng. Manag. **11**, 15–29 (2020). https://doi.org/10.4018/IJMDEM.2020010102
2. Tomić, I., McCann, J.A.: A survey of potential security issues in existing wireless sensor network protocols. IEEE Internet Things J. **4**, 1910–1923 (2017). https://doi.org/10.1109/JIOT.2017.2749883
3. Kumar, N., Singh, Y., Singh, P.K.: Reputation-based energy efficient opportunistic routing for wireless sensor networks. J. Telecommun. Electron. Comput. Eng. Propos. **9**, 29–33
4. Zorzi, M., Rao, R.R.: Geographic random forwarding (GeRaF) for ad hoc and sensor networks: multihop performance. IEEE Trans. Mob. Comput. **2**, 1–11 (2003)
5. Biswas, S., Morris, R.: ExOR: opportunistic multi-hop routing for wireless networks. In: SIGCOMM'05 21–26 Aug 2005. ACM, Philadelphia, USA
6. Mao, X., Tang, S., Xu, X., et al.: Energy-efficient opportunistic routing in wireless sensor networks. IEEE Trans. Parall. Distrib. Syst. 22, 1934–1942. 1045-9219/11/$26.00
7. Lee, G.Y., Haas, Z.J.: Simple, practical, and effective opportunistic routing for short-Haul multi-hop wireless networks. IEEE Trans. Wirel. Commun. **10**, 3583–3588 (2011)
8. Lin, S., Chen, K.: Spectrum aware opportunistic routing in cognitive radio networks. In: 2010 IEEE Global Telecommunications Conference GLOBECOM 2010 IEEE Miami, FL, USA
9. Jie, Z., Huang, C., Xu, L., Wang, B., Chen, X, Fan, X.: A trusted opportunistic routing for VANET. In: Third International Conference on Networking and Distributed Computing. IEEE Computer Society, USA, pp. 86–90 (2012)
10. Luo, J., Hu, J., Wu, D., Li, R.: Opportunistic routing algorithm for relay node selection in wireless sensor networks. IEEE Trans. Ind. Inf. **11**, 112–121 (2015)
11. Kumar, N., Singh, Y.: An energy efficient opportunistic routing metric for wireless sensor networks. Indian J. Sci. Technol. **9** (2016). https://doi.org/10.17485/ijst/2016/v9i32/100197
12. Bapu, B.R.T., Gowd, L.C.S.: Link quality based opportunistic routing algorithm for QOS: aware wireless sensor networks security. Wirel. Pers. Commun. (2017). https://doi.org/10.1007/s11277-017-4586-4
13. Grover, J., Sharma, S.: Security issues in wireless sensor network—a review. In: 2016 5th International Conference on Reliability, Infocom Technologies and Optimization, (Trends Future Directions), pp. 397–404. https://doi.org/10.1109/ICRITO.2016.7784988
14. Core, C.: Security issues in wireless sensor networks. In: Columbia University Libraries. Columbia University Libraries, pp. 222–247 (2017)
15. Alotaibi, M.: Security to wireless sensor networks against malicious attacks using Hamming residue method EURASIP J. Wirel. Commun. Netw. **8** (2019). https://doi.org/10.1186/s13638-018-1337-5
16. Salehi, M., Boukerche, A.: Secure opportunistic routing protocols: methods, models, and classification. Wirel. Netw. **25**, 559–571 (2019). https://doi.org/10.1007/s11276-017-1575-1
17. Nagesh Kumar, Y.S.: Trust and Packet load balancing based secure opportunistic routing protocol for WSN. In: 4th IEEE Conference on Signal Processing, Computing and Control (ISPCC 2k17). IEEE, pp 463–467 (2017)
18. Vamsi, P.R., Kant, K.: Trust and location-aware routing protocol for wireless sensor networks. IETE J. Res. **62**, 634–644 (2016). https://doi.org/10.1080/03772063.2016.1147389

19. Lyu, C., Gu, D., Zhang, X., et al.: SGOR: Secure and scalable geographic opportunistic routing with received signal strength in WSNs. Comput. Commun. **59**, 37–51 (2015). https://doi.org/10.1016/j.comcom.2015.01.003
20. Gupta, S., Dhurandher, S.K., Woungang, I., et al.: Trust-based security protocol against black-hole attacks in opportunistic networks. In: International Conference on Wireless and Mobile Computing Network Communications, pp. 724–729https://doi.org/10.1109/WiMOB.2013.6673436
21. Cao, Y., Kaiwartya, O., Aslam, N., et al.: A trajectory-driven opportunistic routing protocol for VCPS. IEEE Trans. Aerosp. Electron Syst. **54**, 2628–2642 (2018). https://doi.org/10.1109/TAES.2018.2826201
22. Salehi, M., Boukerche, A., Darehshoorzadeh, A.: Modeling and performance evaluation of security attacks on opportunistic routing protocols for multihop wireless networks. Ad Hoc Netw. **50**, 88–101 (2016). https://doi.org/10.1016/j.adhoc.2016.07.004
23. Chhabra, A., Vashishth, V., Sharma, D.K.: A game theory based secure model against black hole attacks in opportunistic networks. In: 2017 51st Annual Conference on Information Sciences and Systems CISS 2017. https://doi.org/10.1109/CISS.2017.7926114
24. Bangotra, D.K., Singh, Y., Selwal, A.: Machine learning in wireless sensor networks: challenges and opportunities. In: PDGC 2018—2018 5th International Conference on Parallel, Distributed and Grid Computing
25. Salehi, M., Boukerche, A.: A novel packet salvaging model to improve the security of opportunistic routing protocols. Comput. Netw. **122**, 163–178 (2017). https://doi.org/10.1016/j.comnet.2017.04.019
26. Kumar, N., Singh, Y., Singh, P.K.: An Energy Efficient Trust Aware Opportunistic Routing Protocol for Wireless Sensor Network. **8**, 30–44 (2017). https://doi.org/10.4018/IJISMD.2017040102
27. Wazid, M., Das, A.K.: An efficient hybrid anomaly detection scheme using K-means clustering for wireless sensor networks. Wirel. Pers. Commun. **90**, 1971–2000 (2016). https://doi.org/10.1007/s11277-016-3433-3
28. Xie, M., Hu, J., Han, S., Chen, H.H.: Scalable hypergrid k-NN-based online anomaly detection in wireless sensor networks. IEEE Trans. Parall. Distrib. Syst. **24**, 1661–1670 (2013). https://doi.org/10.1109/TPDS.2012.261
29. Garofalo, A., Di Sarno, C., Formicola, V.: Enhancing intrusion detection in wireless sensor networks through decision trees. Lect Notes Comput Sci (including Subser Lect Notes Artif Intell Lect Notes Bioinformatics) 7869 LNCS: 1–15 (2013). https://doi.org/10.1007/978-3-642-38789-0_1
30. Gil, P., Martins, H., Januário, F.: Outliers detection methods in wireless sensor networks. Artif. Intell. Rev. **52**, 2411–2436 (2019). https://doi.org/10.1007/s10462-018-9618-2
31. Feng, Z., Fu, J., Du, D., et al.: A new approach of anomaly detection in wireless sensor networks using support vector data description Int. J. Distrib. Sens. Netw. 13 (2017). https://doi.org/10.1177/1550147716686161
32. Shahid, N., Naqvi, I.H., Bin, Q.S.: One-class support vector machines: analysis of outlier detection for wireless sensor networks in harsh environments. Artif. Intell. Rev. **43**, 515–563 (2015). https://doi.org/10.1007/s10462-013-9395-x
33. Saeedi Emadi, H., Mazinani, S.M.: A novel anomaly detection algorithm using DBSCAN and SVM in wireless sensor networks. Wirel. Pers. Commun. **98**, 2025–2035 (2018). https://doi.org/10.1007/s11277-017-4961-1
34. Shamshirband, S., Patel, A., Anuar, N.B., et al.: Cooperative game theoretic approach using fuzzy Q-learning for detecting and preventing intrusions in wireless sensor networks. Eng. Appl. Artif. Intell. **32**, 228–241 (2014). https://doi.org/10.1016/j.engappai.2014.02.001

35. Titouna, C., Aliouat, M., Gueroui, M.: Outlier detection approach using Bayes classifiers in wireless sensor networks. Wirel. Pers. Commun. **85**, 1009–1023 (2015). https://doi.org/10.1007/s11277-015-2822-3
36. Haque, S.A., Rahman, M., Aziz, S.M.: Sensor anomaly detection in wireless sensor networks for healthcare. Sens. (Switzerland) **15**, 8764–8786 (2015) https://doi.org/10.3390/s150408764
37. Ma, T., Wang, F., Cheng, J., et al.: A hybrid spectral clustering and deep neural network ensemble algorithm for intrusion detection in sensor networks. Sens. (Switzerland) **16** (2016). https://doi.org/10.3390/s16101701

OSEP: An Optimized Stable Election Protocol in Heterogeneous Wireless Sensor Networks

Samayveer Singh, Pradeep Kumar Singh, and Aruna Malik

Abstract The wireless sensor networks (WSNs) are attracting researchers because of their extensive variety of applications in different aspects. They perform monitoring responsibility in all kinds of the environment including harsh environments. Thus, a stable network is required for gathering data for a long duration from the monitoring areas. This constancy of the networks depends on the load balancing among the deployed nodes. The stability of networks can be enlarged by the clustering which executes a commanding role in the effective exploitation of energy degeneracy of their batteries and assistances in extending the network lifespan. The communication among the nodes and sink is consumed the highest energy which requires a procedure that can decrease the communication cost. In this paper, we deliberate an enhanced stable election protocol for improving the lifecycle of the heterogeneous WSNs. In the optimization of SEP protocol, three parameters are considered for effective cluster heads selection, namely the nodes remaining energy, node distance from the base station, and total network energy. In this methodology, a threshold-based formula is proposed which considers a different type of energies of the networks such as current energy of the networks, and preliminary and residual energy of the nodes which provides even load balancing. The proposed method contributes a dynamic clustering to lowering energy consumption and avoiding the load over the cluster heads. The number of active and dead, the sum of energy depletion, and the number of messages transferred to the control node matrices are considered to examine the enactment of the proposed scheme by using the MATLAB. After comprehensive

S. Singh · A. Malik
Department
of Computer Science & Engineering, Dr B R Ambedkar National Institute of Technology Jalandhar, Jalandhar, Punjab, India
e-mail: samayveersingh@gmail.com

A. Malik
e-mail: arunacsrke@gmail.com

P. K. Singh (✉)
Department of Computer Science & Engineering, ABES Engineering College, Ghaziabad, Uttar Pradesh, India
e-mail: pradeep_84cs@yahoo.com

P. K. Singh et al. (eds.), *Recent Innovations in Computing*, Lecture Notes in Electrical Engineering 701, https://doi.org/10.1007/978-981-15-8297-4_20

analysis, it has been evident that the projected scheme accomplishes superior to that of the existing methods.

Keywords Wireless sensor networks · WSNs · Network lifespan · Efficient clustering · Energy efficiency

1 Introduction

Nowadays, the collection and transferring information from one place to another place are automatic means that there is no need for human intervention. Thus, some lightweight devices are required which easily deployed in the harsh monitoring area without human intervention. Currently, the solution to this type of problem is wireless sensor networks (WSNs). It is one of the most significant challenging fields in networking. These networks are playing a significant role in industry and academia specifically in the field of communication, detecting, and calculation [1].

Nowadays, wireless sensor networks (WSNs) are one of the furthermost prevalent systems with their advances in electronic technology [1]. It is designed for the remote and harsh areas in our surroundings. The WSN can be expressed as an autonomous network of distinct devices that can communicate the actual information collected from a targeted area through wireless links. A WSN is a structure of dedicated transducers or sensor nodes with a well-controlled communication infrastructure for monitoring and collecting parameters from diverse locations where the sensors have been deployed. Normally, monitored parameters are environment temperature, a tier of humidity, sound intensity, pressure, illumination intensity, wind flow speed and direction, power-line drop in voltage, vibration intensity, pollutant tiers, patient body functions, etc. [2]. A WSN composed of multiple detection stations can be interpreted as sensor nodes, and each of sensor nodes is lightweight, small in size, and portable. Every sensor node is consisting of transducer, active or passive power source, transceiver, and microcomputer. The transducer senses the variations in a physical quantity from surroundings where they positioned and translate into electrical signs. The power source for each node can be regular supply or battery-based. The important function of the transceiver is to receive the commands from a base station or central computer and transmit the data to that location. The duty of the microcomputer just processes and stores the sensor output for analyzing. Generally, there are lot of applications, where we can deploy WSNs such as water quality monitoring, building health structure monitoring, surveillance of various types of buildings, gas leakage in the various gas and chemical plants, productivity, humidity, water requirement, fertilizer monitoring in the agriculture monitoring, health monitoring in medical application, wild animal monitoring the parks, fire detection in the multistory building, green park irrigation, and so on [3].

There are two broad categories of the WSNs such as homogeneous and heterogeneous. In the first one, all the deployed nodes have the same power capabilities whereas nodes have different capabilities in case of heterogeneous networks as energy, link capacity, memory, processing power, etc. Thus, based on the above discussion, we can categorize the heterogeneity into multiple categories, namely energy, link, and computational heterogeneity. In case of energy heterogeneity, heterogeneous networks consist of multiple types of batteries means they have different–different Volts or Jules batteries. Link heterogeneity means that sensor nodes have different capabilities of link for transferring data over the deployed networks like some of the networks have 10 Mbps and some of the links 50 Mbps bandwidth for transferring information. In the computational heterogeneity, sensor nodes have multiple types of microprocessor, memories, etc. Based on the above discussion, we can conclude that the nodes in the sensor networks have many restrictions like energy limitation, cost of the nodes, speed, processing power, etc. Thus, there is a dire need to develop energy effective and efficient algorithm which can overcome the above-discussed problems of the networks for various applications. Perhaps clustering plays a very effective role for enlightening the enactment of the WSNs by effective utilization of the heterogeneous resources like energy limitation, cost of the nodes, speed, processing power, etc.

In this paper, we deliberate an optimize stable election protocol for improving the lifecycle of the heterogeneous WSNs. In the improvement of SEP protocol, three parameters are considered for effective cluster heads selection, namely the nodes remaining energy, the distance among the base station and node, and total network energy. In this methodology, a threshold-based formula is proposed which considers a different type of energies of the networks such as current energy of the networks, and preliminary and residual energy of the nodes which provides even load balancing. The proposed method contributes a dynamic clustering to lowering energy consumption and avoiding the load over the cluster heads. The number of active and dead, the sum of energy consumption, and the number of messages transferred to the control node matrices are considered to examine the enactment of the proposed scheme by using the MATLAB. After a comprehensive analysis, it has been evident that the projected scheme accomplishes superior to that of the existing methods.

The association of the paper is defined as follows: In Sect. 2, the literature review of the existing clustering-based is discussed, and Sect. 3 discusses the various assumption for deploying networks, and radio and energy models. Section 4 discusses the proposed method. Section 5 discusses the results of the simulation of the proposed and the existing methods, and paper is concluded in Sect. 6.

2 Literature Review

Currently, WSNs are having many major issues such as limited size, low battery power, low connectivity, no replaceability of batteries, etc. We need some efficient schemes which can be implemented for solving the above-mentioned problems. In the literature, there are numerous clustering-based algorithms which try to balance the load among the clusters. The first clustering protocols are low energy adaptive clustering hierarchy (LEACH) protocol which tries to equilibrium the consignment among the clusters [3]. The working of LEACH divided into two phases, namely setup and steady phase. In the setup phase, nodes are deployed and possible cluster heads are selected for further processing, whereas the data collection is considered in the steady phase. LEACH is further improved in the form of power-efficient gathering in sensor information systems (PEGASIS) protocol [4]. It designed chains in the networks for gathering the information and data collection always stated from the farthest nodes to the near node from the base station. It created multiple chains in this process. The main disadvantage of this method is that these are fit only small networks, not for large networks. These protocols are defined for the homogeneous networks but after a time some of the heterogeneous protocols are discussed.

The first heterogeneous protocol is stable election protocol (SEP) which is discussed for the two-tier and multi-tier heterogeneity. The hetSEP [5] is the extension of the SEP protocol by integrating the two and three tiers of heterogeneity. It uses the probability formula for nominating the cluster heads. It increases the network overheads if the packet transmission is too long. After that, distributed energy-efficient clustering (DEEC) protocol is discussed by seeing the two and three tiers of heterogeneity. It also uses a probability formula for choosing the cluster heads by considering the residual and average energy. In DEEC, the energy of the networks is not used efficiently. In paper [6], the collection of the clusters is founded on the outstanding energy of the nodes and the networks. It only considered the three tiers of heterogeneity of the nodes in the deployed networks. These methods require the extra energy in the process of reclustering at the time of next rounds. To improving the further performance of the methods, Maheshwari et al. discussed a two-tier of grouping for collecting the data by considering the node degree [7]. The method increases the load of the sink and somehow creates the problem of the hotspot in the deployed networks. The methods discussed in papers [8–10] are considered heterogeneity of the multiple tiers. These papers are not considered the data aggregation process and chaining approach for efficient collection of data.

In paper [11], a fuzzy-based clustering approach is discussed for prolonging the lifetime of networks. It considered the numerous parameters for the distance between the nodes and sink, and residual energy of the clusters for cluster heads election. However, it agonizes the load balancing during the collection of the data. It is also not considered the data combination procedure. In paper [12], the authors discuss the hybrid routing technique based on the zone of the deployed fields. The clustering process is adopted by it is similar as defined in the LEACH protocol. This approach is not involving normal nodes in the clustering process. This method is further extended by the [13] by considering hybrid clustering concept and multi-hop data collection process. It does not use all the methods in the clustering process but only a few nodes have been used in the clustering process. This method grieves the consignment balancing the difficulty of the system. The paper [14] is also discussed for the heterogeneous protocols for prolonging the network lifetime. It considers the probability function for cluster heads election. It defines the two- and multi-tier heterogeneity in the deployed networks. The paper [15] discusses a routing-based technique which collects data on the bases of the chains. There are two phases of the process. In the first phases, nodes direct their data to the cluster heads, whereas in the second phase, clusters heads send their data to the sinks as they received data from the respective clusters by performing data aggregation. Wang et al. discuss a cluster routing protocol which elects the cluster head efficiently [16]. It also improved the node processing and inter-cluster routing problem. This method partially suffers from an effective cluster and data aggregation. Chen et al. discuss a chain-based hierarchical routing protocol in WSNs [17]. It divided monitoring areas into slighter fields to create restraints. It reduces excessive routes and also recovers energy. This method created many small chains in the networks. Linping et al. discuss an improved algorithm of PEGASIS protocol which is based on double cluster heads in WSN [18]. It uses low-tier clusters header for upgraded load balancing. This protocol is useful for large networks and executes better than the PEGASIS algorithm. The preliminary installations of this method create high overhead. Nadeem et al. discuss a gateway-based energy-efficient routing method for WSNs called M-GEAR by dividing the area into four different logical regions [19]. Saranraj et al. discussed the energy-efficient CH collection method for called OEECHS. The election of cluster heads (CHs) is based on consumed battery and distance [20]. This method undergoes from the load balancing difficult between the nodes. In the subsequent section, we considered the network conventions, proposed energy model, and energy overindulgence model which will assist in competent clustering.

3 Network System Assumptions, Energy Model, and Radio Dissipation Model

The fundamental presumptions of the proposed system model which will be used in the network design are given as follows:

- All the sensors have an ID and fixed location because nodes are stationary.
- The network consists of two types of the node as homogeneous and heterogeneous.
- The preliminary energies of the nodes are determined by the defined tier of heterogeneity.
- The connection between the sink and sensors are symmetric, and capacities, computational and memory powers are asymmetric.
- The sink is positioned in the center of the defined territory.

The proposed technique considered the three-tier heterogeneity networks with N number of nodes. The energies of tier-1, -2, and -3 nodes are designated as E_1, E_2, and E_3, respectively, with condition $E_1 < E_2 < E_2$, and their numbers are designated as N_1, N_2, N_3, respectively, with condition$N_1 > N_2 > N_3$. The energy of the system is calculated as follows:

$$E_{\text{Total}} = \gamma \times N \times E_1 + \gamma^2 \times N \times E_2 + \left(1 - \gamma - \gamma^2\right) \times N \times E_3 \quad (1)$$

E_{Total} can define tier-1, -2, and -3 heterogeneity using the value of γ which is a model parameter.

The tier-1 nodes are minimum in number, i.e., $\gamma \times N$, and having a minimum amount of energy denoted as E_1, where the range of model parameter γ is $0 \leq \gamma \geq 1$. The tier-2 nodes are less in number from tier-1 nodes, i.e., $\gamma^2 \times N$, and having more energy from tier-1 nodes denoted as E_2. The tier-3 nodes are less in number from tier-1 and tier-2 nodes, i.e., $\left(N - \left(\gamma * N + \gamma^2 * N\right)\right)$, and having more energy from tier-1 and tier-2 nodes denoted as E_3 that means they have a maximum amount of energy.

Tier-1 heterogeneity: For $\gamma = 0$, Eq. (1) defines the network has only one type of nodes. Thus, the energy of the networks is as follows:

$$E_{\text{Total}} = N * E_3 \tag{2}$$

where E_3 is the preliminary energy of the nodes. It is indicating the tier-3 nodes instead of tier-1 nodes. We impose a condition for converting the energy of tier-3 nodes into tier-1 nodes. It can be attained by using the given representation:

$$\gamma = \frac{E_3 - E_1}{\beta * f(E_2, E_3)} \tag{3}$$

where f and β are the functions of E_2 and E_3, and positive integer where $\beta > 1$. The function f can have either $(\mathrm{E_3} + \mathrm{E_2})$or $(\mathrm{E_3} - \mathrm{E_2})$.

Tier-2 heterogeneity: For $1 - \gamma - \gamma^2 = 0$, find the value of γ which defines two type of sensors, namely tier-1 and tier-2 nodes. The relation $1 - \gamma - \gamma^2 = 0$ is not selected arbitrarily which is diminished by the third term of Eq. (1). The relation $1 - \gamma - \gamma^2 = 0$ gives two solutions that are $\left(\left(\sqrt{5}\right) - 1\right)/2$ and $\left(\left(\sqrt{5}\right) + 1\right)/2$.

Where $\left(\left(\sqrt{5}\right) + 1\right)/2 > 1$ is not true by the bound value of the $0 \leq \gamma \geq 1$. Thus, the valid $\left(\left(\sqrt{5}\right) - 1\right)/2$ is the correct solution. This solution defines two types of sensor nodes with their preliminary energies E_1 and E_2.

Tier-3 heterogeneity: We have considered the upper and lower bound value of γ is $0 \leq \gamma \geq 1$ but in tier-2 heterogeneity, we have fixed the range value of the upper bond is $\left(\left(\sqrt{5}\right) - 1\right)/2$ denoted by γ_{ub}. Consider the lower bound of γ be γ_{lb}, i.e., which needs to be determined. The range of γ is $\gamma_{\text{lb}} < \gamma < \gamma_{\text{ub}}$ for three-tier heterogeneity, and consider function value f as $(E_3 - E_2)$ and γ from Eq. (3). Thus, the lower bound as

$$\gamma_{\text{lb}} < \gamma < \gamma_{\text{ub}}$$

Put the values of γ and γ_{ub} in the above equation and calculate the value of the γ_{lb} as follows:

$$\gamma_L < \frac{E_3 - E_1}{\beta * (E_3 - E_2)} < \left(\left(\sqrt{5}\right) - 1\right)/2 \tag{4}$$

Let $E_2 = \alpha_1 + E_1$ and $E_3 = \alpha_2 + E_2$ and using Eq. (4), we get the following relation.

$$\gamma_{\text{lb}} < \frac{\alpha_2 + \alpha_1}{\beta * \alpha_2}$$

$$\frac{\alpha_2}{\alpha_1} < \frac{1}{\beta * \gamma_{\text{lb}} - 1}$$

$$-\frac{\alpha_2}{\alpha_1} \geq \frac{1}{1 - \beta * \gamma_{\text{lb}}} \tag{5}$$

Subsequently, L.H.S. of Eq. (5) is negative; thus, just put $-\frac{\alpha_2}{\alpha_1} = 0$. We have the following relation:

$$1 - \beta * \gamma_{\text{lb}} < 0$$

$$\frac{1}{\beta} < \gamma_{\text{lb}} \tag{6}$$

Equation (4) can be written as

$$(E_3 - E_1) \leq \frac{\beta * \left(\left(\sqrt{5}\right) - 1\right)}{2} * (E_3 - E_2) \tag{7}$$

This inequality may be written as

$$\beta * \left(\left(\sqrt{5}\right) - 1\right) * E_2 - 2 * E_1 \leq \left(\beta * \left(\left(\sqrt{5}\right) - 1\right) - 2\right) * E_3 \tag{8}$$

The tier-1, -2, and -3 nodes energy is given as e_1, $e_2 = e_1 * (1 + \omega)$, and $e_3 = e_1 * (1 + \eta)$, respectively. And the value of coefficients ω and η are 0.06 and 0.11, respectively. Thus, the above-mentioned procedure defines the three-level heterogeneity model in WSNs.

Now, we deliberate a debauchery energy radio model to compute the energy depletion in the conveying and reception by the nodes during recognizing, transmission, and computational procedure. The energy collapse for conveying the L-bit message over the distance d is quantified as follows [3]:

$$E_{\text{TXS}} = L * \in_{\text{elec}} + L * \in_{\text{fs}} * d^2 \quad \text{if } d \leq d_0 \tag{9}$$

$$E_{\text{TXL}} = L * \in_{\text{elec}} + L * \in_{\text{mp}} * d^4 \quad \text{if } d > d_0 \tag{10}$$

where $\in_{elec}$, $\in_{\text{fs}}$ and $\in_{\text{mp}}$ are the vitality/energy decadent and d_0 is threshold distance which is given below:

$$d_0 = \sqrt{\frac{\in_{\text{fs}}}{\in_{\text{mp}}}} \tag{11}$$

The energies expended in receiving (E_{Rx}) and in sensing (E_{Sx}) are specified in (12) and (13) as follows:

$$E_{Rx} = L* \in_{\text{elec}} \tag{12}$$

$$E_{\text{Sx}} = L* \in_{\text{elec}} \tag{13}$$

In the next fragment, we deliberate the procedure for optimizing the electing of CHs which helps in load balancing.

4 OSEP: Optimized Stable Election Protocol

In this work, an enhanced stable election protocol (SEP) using a clustering algorithmic methodology for three-tier heterogeneity is proposed. This methodology elects appropriate cluster heads (CHs) using the three different parameters, namely the nodes remaining energy, node distance from the BS, and total network energy. The selection of the cluster heads method is used as a dynamic clustering procedure. The aim of this method is to elect cluster heads that consume less energy in inter- and intracluster communication. The complete process of the proposed method is separated into rounds, and CHs are designated for each round. Preliminarily, a defined percentage is used to electing the CHs, and then, several CHs are added according to the coverage of the number of deployed sensor nodes. It employes a condition of the vicinity of cluster heads. A cluster head may not be possible in the vicinity of other selected cluster heads in this scenario. The decision of selecting the CHs depends on the threshold value of the tier of the heterogeneity. This threshold value is associated with an arbitrary number which is generated between 0 and $E_{\text{average - nch}}/E_{i-\text{init}}$. If the threshold value > the generated number between 0 and $E_{\text{average - nch}}/E_{i-\text{init}}$, then the sensor node of a particular tier of heterogeneity turnsout to be a CHs for the present round. The proposed OSEP method threshold formula is given as follows:

$$E_i = \frac{E_{i-\text{curr}}}{E_{i-\text{init}}} \tag{14}$$

$$E_t = \left(E_{i-\text{res}} + r \text{ div } \frac{1}{p_i}\right) \times (1 - E_{i-\text{res}}) \tag{15}$$

$$T(n) = \begin{cases} \frac{p_i}{N - p_i\left[r \bmod \left(\frac{N}{p_i}\right)\right]} \times E_i \times E_t \text{ if } n \in G \\ 0 \qquad\qquad\qquad\qquad \text{otherwise} \end{cases} \tag{16}$$

where $E_{i-\text{curr}}$, $E_{i-\text{init}}$, and $E_{i-\text{res}}$ are the current energy of the networks, preliminary nodes energy, and node residual energy. The N and r are the numbers of sensor nodes and rounds. The E_i and E_t are the constant values which are defined in Eqs. (14) and (15).

In the proposed optimized SEP, a dynamic range is considered for the comparison with the threshold value for choosing the final cluster heads which are given as follows in Eq. (2).

$$\left[0,\ E_{\text{average - nch}}/E_{i-\text{init}}\right] \tag{17}$$

where $E_{\text{average}-\text{nch}}$ and $E_{i-\text{init}}$ are the node's average energy which is not CHs in the current round and nodes preliminary energy, respectively.

The total energy of the defined network which consists of three types of nodes, i.e., tier-1,-2, and -3, is represented by E_T as given below:

$$\gamma \times N \times E_1 + \gamma^2 \times N \times E_2 + \left(1 - \gamma - \gamma^2\right) \times N \times E_3 \tag{18}$$

$$N \times \left(\gamma \times E_1 + \gamma^2 \times E_2 + \left(1 - \gamma - \gamma^2\right) \times E_3\right) \tag{19}$$

$$E_1 \times N \times \left(\gamma + \gamma^2 \times E_2/E_1 + \left(1 - \gamma - \gamma^2\right) \times E_3/E_1\right) \tag{20}$$

Now, we will discuss the clustering process of the proposed optimized stable election protocol (OSEP) which contains three types of nodes. The E_1 is the preliminary energy of the tier-1 nodes. If all the nodes have preliminary energy E_1, then the total energy of the networks is $E_1 \times N$. Therefore, there is an increment factor of the energy according to Eq. (20) as $\left(\gamma + \gamma^2 \times E_2/E_1 + \left(1 - \gamma - \gamma^2\right) \times E_3/E_1\right)$. Means, heterogeneous nodes have $\left(\gamma + \gamma^2 \times E_2/E_1 + \left(1 - \gamma - \gamma^2\right) \times E_3/E_1\right)$ times more energy than the homogeneous nodes.

Generally, every sensor node becomes CH in case of homogeneous networks after $1/p_i$ rounds. Thus, the average cluster heads for homogeneous networks in a particular round will be $N \times 1/p_i$. But in the case of heterogeneous networks, every sensor node becomes CH after $1/p_i \times \left(\gamma + \gamma^2 \times E_2/E_1 + \left(1 - \gamma - \gamma^2\right) \times E_3/E_1\right)$ rounds. Thus, the average cluster heads for heterogeneous networks in a particular round will be $N \times 1/p_i \times \left(\gamma + \gamma^2 \times E_2/E_1 + \left(1 - \gamma - \gamma^2\right) \times E_3/E_1\right)$.

The threshold values of tier-1, -2 and -3 are fixed according to the preliminary energies of the respective tiers. Thus, each sensor of tier-1 converts into a CH once in every $1/p_{\text{opt}}$ rounds, and tier-2 and -3 sensor nodes turn into a cluster head $(1 + \alpha)$ and $(1 + \beta)$ times more than that of the tier-1 sensors in every $\left(\gamma + \gamma^2 \times E_2/E_1 + \left(1 - \gamma - \gamma^2\right) \times E_3/E_1\right)/p_i$ rounds, respectively. Thus, heterogeneity is despoiled the set constraints which is $N \times 1/\text{p}_i$. The

$E_0/\left(\gamma \times E_1 + \gamma^2 \times E_2 + \left(1 - \gamma - \gamma^2\right) \times E_3\right)$ is the weight of a node in the defined networks.

In the proposed OSEP protocol, the weighted probabilities of the tier-1, -2, and -3 nodes which are denoted by $p_{\text{level}-1}$,$p_{\text{level}-2}$, and $p_{\text{level}-3}$, respectively, and defined as follows:

$$p_{level-1} = \frac{p_i \times E_1}{\left(\gamma \times E_1 + \gamma^2 \times E_2 + \left(1 - \gamma - \gamma^2\right) \times E_3\right)} \tag{21}$$

It can be as follows:

$$p_{\text{level}-1} = \frac{p_i}{\left(\gamma + \gamma^2 \times E_2/E_1 + \left(1 - \gamma - \gamma^2\right) \times E_3/E_1\right)} \tag{22}$$

$$p_{\text{level}-2} = \frac{p_i \times E_1}{\left(\gamma \times E_1 + \gamma^2 \times E_2 + \left(1 - \gamma - \gamma^2\right) \times E_3\right)} \tag{23}$$

It can be as follows:

$$p_{\text{level}-2} = \frac{p_i}{\left(\gamma + \gamma^2 \times E_2/E_1 + \left(1 - \gamma - \gamma^2\right) \times E_3/E_1\right)} \tag{24}$$

$$p_{\text{level}-3} = \frac{p_i \times E_1}{\left(\gamma \times E_1 + \gamma^2 \times E_2 + \left(1 - \gamma - \gamma^2\right) \times E_3\right)} \tag{25}$$

It can be as follows:

$$p_{\text{level}-3} = \frac{p_i}{\left(\gamma + \gamma^2 \times E_2/E_1 + \left(1 - \gamma - \gamma^2\right) \times E_3/E_1\right)} \tag{26}$$

The $p_{\text{level}-1}$, $p_{\text{level}-2}$, and $p_{\text{level}-3}$ are the optimum weighted probabilities for the tier-1, -2, and -3 nodes. Thus, the threshold value of the tier-1 is given as follows:

$$T(n_{\text{level}-1}) = \begin{cases} \frac{p_{\text{level}-1}}{N - p_{\text{level}-1}\left[r \bmod \frac{N}{p_{\text{level}-1}}\right]} \times E_i \times E_t \text{ if } n_{\text{level}-1} \in G' \\ 0 \qquad \text{otherwise} \end{cases} \tag{27}$$

$$T(n_{\text{level}-2}) = \begin{cases} \frac{p_{\text{level}-1}}{N - p_{\text{level}-2}\left[r \bmod \frac{N}{p_{\text{level}-2}}\right]} \times E_i \times E_t \text{ if } n_{\text{level}-1} \in G'' \\ 0 \qquad \text{otherwise} \end{cases} \tag{28}$$

$$T(n_{\text{level}-3}) = \begin{cases} \frac{p_{\text{level}-3}}{N - p_{\text{level}-3}\left[r \bmod \frac{N}{p_{\text{level}-3}}\right]} \times E_i \times E_t \text{ if } n_{\text{level}-1} \in G''' \\ 0 \qquad \text{otherwise} \end{cases} \tag{29}$$

where G', G'', and G''' are the conventional of tier-1, -2, and -3 nodes which have not become CHs within last $\frac{1}{p_{\text{level}-1}}$, $\frac{1}{p_{\text{level}-2}}$, and $\frac{1}{p_{\text{level}-3}}$ rounds, respectively. The $T(n_{\text{level}-1})$, $T(n_{\text{level}-2})$, and $T(n_{\text{level}-3})$ are the threshold applied to type-1, -2, and

-3 nodes, respectively. Thus, deployed sensor nodes converted into dynamic clusters by using their probabilities and thresholds which helps in lifespan prolonging.

5 Experimental Results and Discussion

The simulated results between the proposed schemes with the existing SEP protocol [15], Mittal et al. [18], and hetSEP [5] are compared by considering the number of active and dead showing the sustainability and the lifetime of the proposed method in terms of first, half, and last node dead, the sum of energy depletion as the remaining energy, and the number of messages transferred to the control node as presenting the throughput of the proposed method matrices which are considered in this section. In the optimization of SEP protocol, three parameters are considered for effective cluster heads selection, namely the remaining energy of the nodes, node distance from the sink, and total network energy. In this methodology, a threshold-based formula is proposed which considers different types of energies of the networks such as current energy of the networks, and preliminary and nodes residual energy which helps in deciding the most eligible sensor nodes. The proposed network is considered 100 numbers of sensor nodes, sink at the center area, and preliminary energy of the normal 0.14 Jules, and results are simulated using MATLAB. The proposed network is considered three levels of heterogeneity by using three types of nodes, namely normal, advanced, and super as are 50, 30, and 20 in numbers, respectively, and their energies are 0.10 J, 0.16 J, and 0.21 J, respectively. The power depletion is 50 nJ/bits in running the circuit, 50 nJ/bit, 10 pJ/bit/m^2 to transmit the signal in the shorter distance, 0.0013 pJ/bit/m^4 to transmit signal in the longer distance. The energy consumption model also considered the packet length is 4000 bits, cluster range is 25 m, and the threshold distance is 75 m. We have commonly used 25 simulations and taken an average of all the simulations for calculating the final simulation results.

5.1 Performance Estimation for the Proposed Networks

A comparative analysis of the projected method and the SEP protocol [15, 18], hetSEP [5], existing methods in relationships of the number of active nodes, energy consumption, packets sent with reference of the number of rounds are discussed in this subsection. The number of active nodes concerning the number of rounds for SEP protocol [15], Mittal et al. [18], and hetSEP [5], and the proposed technique are illustrations as shown in Fig. 1.

The projected technique covers 2624 number of rounds before going to tie every node in the deployed networks, whereas SEP protocol [15], Mittal et al. [18], and hetSEP [5] covers 1742, 1819 and 2123 rounds, respectively. The network lifetime increment in Mittal et al. [18], hetSEP [5], and the proposed method is 04.42%, 21.87%, and 50.63%, respectively, in comparison with the SEP [15] protocol. Thus, it

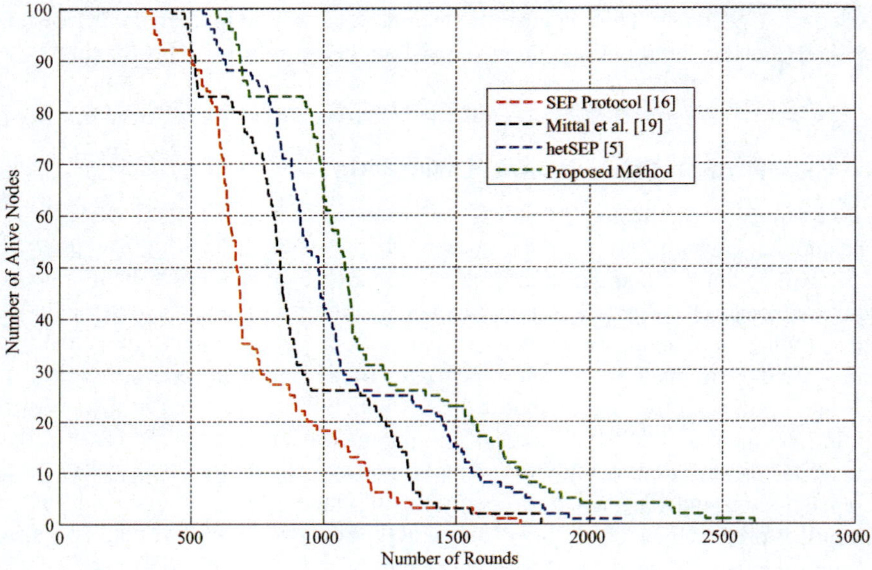

Fig. 1 Analysis of SEP protocol [15], Mittal et al. [18], hetSEP [5], and the proposed method in terms of alive sensor nodes and rounds

is evident from the results that the projected technique gives better results in respect of the existing techniques because the selection of CHs is efficiently due to the selected parameters. Furthermore, it decreases the communication cost in the data collection process and increases the lifetime of the networks. The proposed scenario of energy consumption helps in extending the lifetime by using different energy factors. The sustainability period of the Mittal et al. [18], hetSEP [5], and the proposed technique is 26.11%, 61.42%, and 77.45%, respectively, which is calculated by considering the first node dead information concerning SEP protocol [15]. Figure 1 also demonstrates the number half node dead for SEP [3], Mittal et al. [18], hetSEP [5], and the proposed method. Moreover, the half node dead (HND) of the Mittal et al. [18], hetSEP [5], and proposed method outperform by 24.55, 45.98, and 60.86%, as a comparison with SEP protocol [15], significantly, respectively.

Figure 2 shows the imitation consequences of the total energy dissipation regarding several rounds of the SEP protocol [15], Mittal et al. [18], hetSEP [5], and proposed method for three levels of heterogeneity. The total preliminary energy of the three levels of heterogeneity network is 14 J. The projected method is performing well than that of the SEP protocol [15], Mittal et al. [18], hetSEP [5], and existing methods because the proposed method sensing data for a longer period conserves very less energy during the data collection from the CHs and cluster members. It is also decreasing the cost of communication in a very effective manner.

The simulation results of the number of packets sent to the BS concerning the number of rounds are shown in Fig. 3 for SEP protocol [15], Mittal et al. [18], hetSEP [5] existing methods, and the proposed method using three levels of heterogeneity.

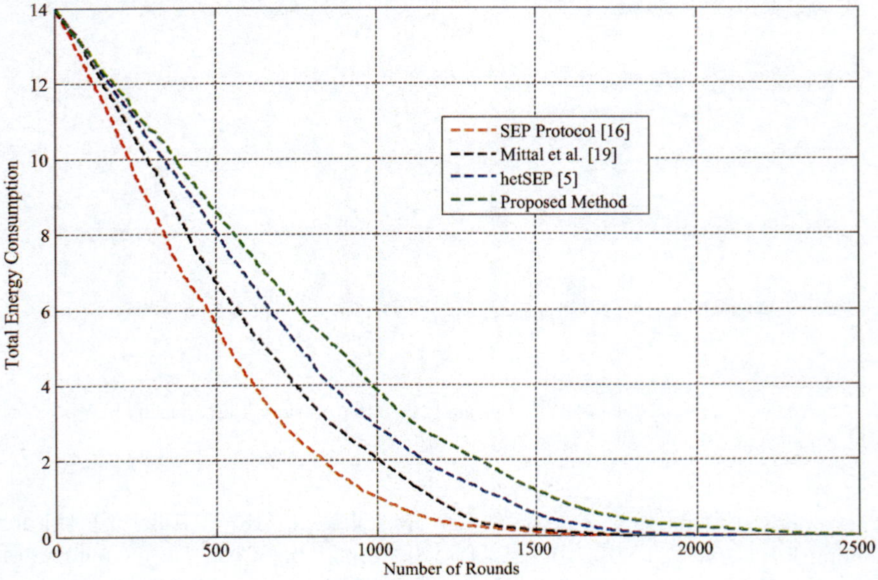

Fig. 2 Analysis of SEP protocol [15], Mittal et al. [18], hetSEP [5], and the proposed method in terms of energy consumption and rounds

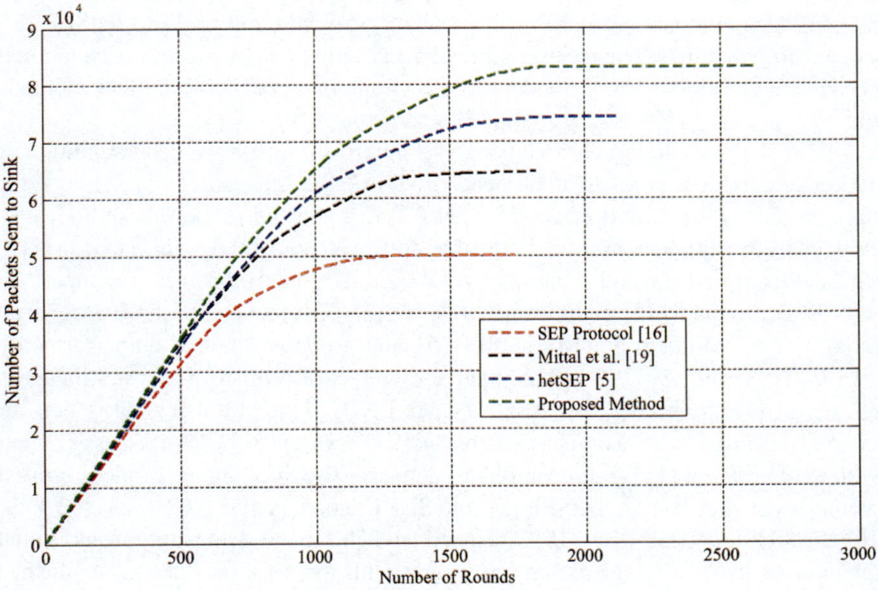

Fig. 3 Analysis of SEP protocol [15], Mittal et al. [18], hetSEP [5], and the proposed method in terms of packets sent to BS and rounds

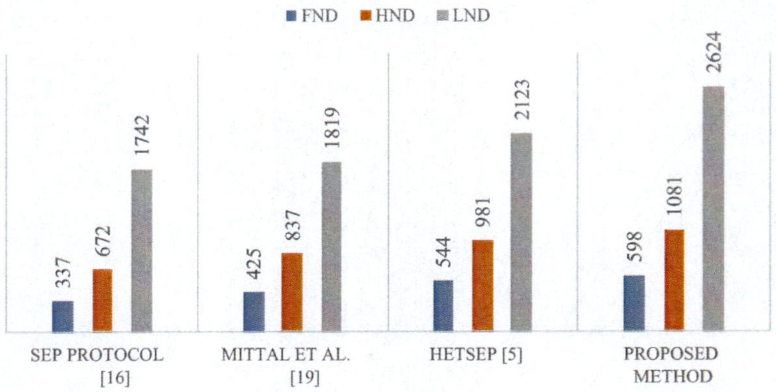

Fig. 4 Analysis in terms of the FND, HND, and LND of the network lifespan using SEP protocol [15], Mittal et al. [18], hetSEP [5], and the proposed method

The proposed method, SEP protocol [15], Mittal et al. [18], and hetSEP [5] are sending1.70×10^{-4}, 1.80×10^{-4}, 2.11×10^{-4}, and 2.62×10^{-4}, as the number of packets to the sink, respectively. It is concluded that the proposed method sent more number of packets to the sink as compared to SEP protocol [15], Mittal et al. [18], and hetSEP [5] because of the stability period increase by the proposed method. It is evident from the consequences that the projected method can transmit packets to the sink as a comparison to the SEP protocol [15], Mittal et al. [18], hetSEP [5], and existing methods. The more number of packets directed by the proposed method because of the alive time of the deployed nodes is more concerning other methods like SEP protocol [15], Mittal et al. [18], and hetSEP [5].

Figure 4 shows the proportional analysis in terms of last node dead (LND), half node dead (HND), and first node dead (FND) of the lifespan of the network for SEP protocol [15], Mittal et al. [18], hetSEP [5], and the proposed method. It is evident from Fig. 4 that the maintainable epoch of the Mittal et al. [18], hetSEP [5], and proposed method outstrips by 4.42, 21.87, and 50.63% as a comparison with SEP protocol [15] significantly, respectively. Moreover, the half node dead (HND) of the Mittal et al. [18], hetSEP [5], and proposed method outperforms by 29.17, 42.28, and 59.37%, as a comparison with SEP protocol [15], significantly, respectively. Furthermore, the last node dead (LND) of the Mittal et al. [18], hetSEP [5], and proposed method outperforms by 54.50, 67.66, and 86.77%, as an assessment with SEP protocol [15] significantly, respectively. The throughput for the proposed method, Mittal et al. [18], hetSEP [5], and SEP protocol [15] is 2.62×10^{-4}, 2.11×10^{-4}, 1.80×10^{-4}, and 1.70×10^{-4}, respectively. This method transmits an additional quantity of packets to the BS as associated with the existing ones. The lifespan prolonging of the Mittal et al. [18], hetSEP [5], and the proposed method is 26.11, 61.42, and 77.45%, as compared with the SEP protocol [15], deprived of addition energy of the network, i.e., 14 J, respectively.

It is evident from Table 1 that the maintainable epoch (also called the first node dead (FND) timing) of the Mittal et al. [18], hetSEP [5], and proposed method

Table 1 Analysis in terms of the network lifespan, energy dissipation, and throughput for SEP protocol [15], Mittal et al. [18], hetSEP [5], and proposed method

Protocols	Network lifetime			Energy consumption (J)	Throughput	% Increment in network lifetime
	FND	HND	LND			
SEP protocol [15]	337	672	1742	14	1.70×10^{-4}	–
Mittal et al. [18]	425	837	1819	14	1.80×10^{-4}	04.42%
hetSEP [5]	544	981	2123	14	2.11×10^{-4}	21.87%
Proposed method	598	1081	2624	14	2.62×10^{-4}	50.63%

outstrips by 4.42, 21.87, and 50.63% as a comparison with SEP protocol [15] significantly, respectively. Moreover, the half node dead (HND) of the Mittal et al. [18], hetSEP [5], and proposed method outperforms by 29.17, 42.28, and 59.37%, as a comparison with LEACH [3] significantly, respectively. Furthermore, the last node dead (LND) of the Mittal et al. [18], hetSEP [5], and proposed method outperforms by 54.50, 67.66, and 86.77%, as a comparison with SEP protocol [15] significantly, respectively. The throughput for the proposed method, Mittal et al. [18], hetSEP [5], and SEP protocol [15] is 2.62×10^{-4}, 2.11×10^{-4}, 1.80×10^{-4}, and 1.70×10^{-4}, respectively. This method transmits an additional quantity of packets to the BS as compared to the existing ones. The lifespan prolonging of the Mittal et al. [18], hetSEP [5], and the proposed method is 26.11, 61.42, and 77.45%, as compared with the SEP protocol [15], deprived of addition energy of the network, i.e., 14 J, respectively. The lifespan of the half node dead of the Mittal et al. [18], hetSEP [5], and the proposed method is 24.55, 45.98, and 60.86%, as compared with the SEP protocol [15], deprived of addition energy of the network, i.e., 14 J, respectively.

6 Conclusion

In this paper, an efficient clustering-based optimized stable election protocol for extending the lifetime of the WSNs is discussed. It has considered for heterogeneous networks and compared with the existing SEP protocol [15], Mittal et al. [18], and hetSEP [5]. The developed method used the remaining energy of nodes effectively and dynamic clustering technique which helps in increasing sustainable of the system accomplishment since the choice of nodes in the process of CH election is efficient which has higher residual energy, node distance from the base station, and total network energy. The lifespan prolonging of the Mittal et al. [18], hetSEP [5], and the proposed method is 26.11, 61.42, and 77.45%, as compared with the SEP protocol [15], deprived of addition energy of the network, i.e., 14 J, respectively. The simulation results of the proposed method demonstration lifetime are increased by 50.63% for 14 J network energy as compared with SEP protocol [15], respectively. The throughput for the proposed method, SEP protocol [15], Mittal et al. [18], and hetSEP [5] methods is 2.62×10^{-4}, 2.11×10^{-4}, 1.80×10^{-4}, and 1.70×10^{-4}, respectively. This method achieves better results than that of the SEP protocol [15], Mittal et al. [18], and hetSEP [5] methods.

References

1. Singh, S., Chand, S., Kumar, B.: Performance investigation of heterogeneous algorithms in WSNs. In: 3rd IEEE International Advance Computing Conference (IACC), 1051–1054, 2013
2. Singh, Y., Singh, S., Kumar, R.: A distributed energy-efficient target tracking protocol for three tier heterogeneous sensor networks. Int. J. Comput. Appl. **51**, 31–36 (2012)
3. Heinzelman, W.R., Chandrakasan, A.P., Balakrishnan, H.: An application-specific protocol architecture for wireless microsensor networks. IEEE Trans. Wirel. Commun. **1**, 660–670 (2002)
4. Lindsey, S., Raghavendra, C.S., Sivalingam, K.M.: Data gathering algorithms in sensor networks using energy metrics. IEEE Trans. Parall. Distrib. Syst. **13**, 924–935 (2002)
5. Singh, S., Malik, A.: hetSEP: Heterogeneous SEP protocol for increasing lifetime in WSNs. J. Inf. Optim. Sci. **38**, 721–743 (2017)
6. Singh, S., Malik, A., Kumar, R.: Energy efficient heterogeneous DEEC protocol for enhancing lifetime in WSNs. Eng. Sci. Technol., Int. J. **20**, 345–353 (2017)
7. Maheswari, D.U., Sudha, S.: Node degree based energy efficient two-tier clustering for wireless sensor networks. Wirel. Pers. Commun. **104**, 1209–1225 (2018)
8. Chand, S., Singh, S., Kumar, B.: Heterogeneous HEED protocol for wireless sensor networks. Wirel. Pers. Commun. **77**, 2117–2139 (2014)
9. Singh, S., Chand, S., Kumar, B.: Energy efficient clustering protocol using fuzzy logic for heterogeneous WSNs. Wirel. Pers. Commun. **86**, 451–475 (2016)
10. Singh, S., Chand, S., Kumar, B.: Multitier heterogeneous network model for wireless sensor networks. Telecommun. Syst. **64**, 259–277 (2017)
11. Singh, S., Chand, S., Kumar, B.: An energy efficient clustering protocol with fuzzy logic for WSNs. In: 5th International Conference-Confluence the Next Generation Information Technology Summit, 427–431, 2014
12. Mehra, P.S., Doja, M.N., Alam, B.: Stable period enhancement for zonal (SPEZ)-based clustering in heterogeneous WSN. Smart Innov., Syst. Technol. **79**, 887–896 (2018)
13. Mehra, P.S., Doja, M.N., Alam, B.: Stable period extension for heterogeneous model in wireless sensor network. Adv. Intell. Syst. Comput. **638**, 479–487 (2018)
14. Manju, Chand, S., Kumar, B.: Selective α-coverage based heuristic in wireless sensor networks. Wirel. Pers. Commun. **97**(2017) 1623–1636
15. Smaragdakis, G., Matta, I., Bestavros, A.: SEP: A Stable Election Protocol for Clustered Heterogeneous Wireless Sensor Networks, pp. 1–11. Boston University Computer Science Department (2004)
16. Singh, P., Paprzycki, M., Bhargava, B., Chhabra, J., Kaushal, N., Kumar, Y. (eds.) Futuristic Trends in Network and Communication Technologies. FTNCT 2018. Communications in Computer and Information Science, vol. 958. Springer, Singapore
17. Mittal, N., Singh, U., Sohi, B.S.: An energy-aware cluster-based stable protocol for wireless sensor networks. Neural Comput. Appl. **31**, 7269–7286 (2019)
18. Mittal, N., Singh, U.: Distance-based residual energy-efficient stable election protocol for WSNs. Arab. J. Sci. Eng. **40**, 1637–1646 (2015)
19. Nadeem, Q., Rasheed, M.B., Javaid, N.Z., Khan, A., Maqsood, Y., Din, A.: M-GEAR: gateway-based energy-aware multi-hop routing protocol for WSNs. In: Eighth International Conference on Broadband and Wireless Computing, Communication and Applications, 1–6 (2013)

20. Saranraj, G., Selvamani, K., Kanagachidambaresan, G.R.: Optimal energy-efficient cluster head selection (OEECHS) for wireless sensor network. J. Inst. Eng. (India) Series B, **100**(2), 349–356 (2019)
21. Qing, L., Zhu, Q., Wang, M.: Design of a distributed energy-efficient clustering algorithm for heterogeneous wireless sensor networks. Comput. Comms. **29**, 2230–2237 (2016)

Bot Detection in Social Networks Using Stacked Generalization Ensemble

Rahul Katarya, Raghav Mehta, Ryan Bansal, Pradyot Raina, and Mukul Mahaliyan

Abstract In recent times, the reach and influence of social media have grown tremendously across the entire globe. The ease of access, simplicity, publicity and reach offered by giant social networking sites have come to hold immense value nowadays. However, this has led to the widespread use of fake accounts or programmed bots in order to inflate one's social media popularity and further spread favourable content. Many recent studies have highlighted the impact of such bots in fields like advertising, commercial promotion and even elections. In this paper, we propose a method to detect bots on social networking sites and distinguish them from genuine user accounts by using a stacked learning approach whereby a convolutional neural network model is trained to feed forward to a machine learning model. This is achieved by using a supervised learning approach to build a layered classifier that makes predictions based on a user's profile information, tweets and activity information from a dataset of Twitter users. Our paper also analyses the comparative performance of many machine learning models applied to this problem.

Keywords Machine learning · Natural language processing · Deep learning · Convolutional neural network · Supervised learning

---

R. Katarya · R. Mehta (✉) · R. Bansal · P. Raina · M. Mahaliyan
Department of Computer Science and Engineering, Delhi Technological University, Shahbad Daulatpur, Main Bawana Road, New Delhi, Delhi 110042, India
e-mail: raghavmehta22@gmail.com

R. Katarya
e-mail: rahulkatarya@dtu.ac.in

R. Bansal
e-mail: ryan10bansal@gmail.com

P. Raina
e-mail: pradyotrainaout@gmail.com

M. Mahaliyan
e-mail: mukulmahaliyan@gmail.com

P. K. Singh et al. (eds.), *Recent Innovations in Computing*, Lecture Notes in Electrical Engineering 701, https://doi.org/10.1007/978-981-15-8297-4_21

1 Introduction

In today's world, social networking sites have acquired an integral position in the daily lives of millions of people globally. Social media has a massive hold on a significant part of the world's population and can thus be used, or misused, for influencing them. Celebrities, politicians and corporations are always looking to expand their social media footprint and achieve 'viral' publicity in order to attract people to their brands. This is because social media tends to broadcast content to millions of people all over the world in a matter of minutes. However, due to the massive overload of content on social media nowadays, people have started resorting to quick and fraudulent ways to boost their following and popularity. The most common way of falsely inflating one's online reach has been through the use of bots and fake accounts. 'Bots' are simply computer programmes that automatically produce or repost content and interact with humans on social networks.

In our system, we analysed the profile details, tweets and tweet metadata of numerous users on Twitter, including both bots and humans. To achieve this, we have proposed a layered machine learning approach consisting of stacked convolutional neural network (CNN) and machine learning models, i.e. the outputs of the neural network are fed forward to a machine learning algorithm to generate final predictions. This paper demonstrates the improvements in accuracy and efficiency achieved using this stacked learning model in addition to analysing and comparing the performance of various machine learning models applied to the problem of bot detection.

The novel method of bot detection in social networks proposed in this paper performs exceptionally well on unseen data, giving an accuracy of unto 99% in predicting whether a user is a bot or a human user. Thus, by deploying our proposed model, social networking sites can efficiently detect bots and fake accounts with exceptional accuracy and the nuisance caused by such fraudulent activities can be curbed.

1.1 Motivation

In recent years, people have realized just how important social networking sites like Twitter, Facebook etc. are as tools to manufacture and manage public opinion and narratives to suit themselves. And with the emergence of pre-programmed bots with which to flood such influential sites, ill-intentioned people and groups have found this task to be exponentially more comfortable and that much more effective. The motivation for this paper comes from the growing trend of news reports and studies highlighting the dire real-world consequences of bot activity on social media lately. There have been reports of bots spreading misleading information about the coronavirus [1], false claims about medicinal properties of cannabis [2], unsubstantiated conspiracy theories about the Australian bushfires [3] and being involved in many other menacing activities. In addition to these, various studies have highlighted and analysed the large-scale meddling by bots to influence opinions during elections around the globe in countries like Japan [4] and Finland [5].

1.2 Contributions of This Work

In this paper, we have proposed a system to tackle the problem of bot detection by identifying Twitter bots using their profile information, tweets and tweet metadata. The significant contributions of our proposed approach are:

- Cumulative analysis of neural networks and then their results are joined with user data and fed together into a machine learning model which generates the final prediction. This technique of feeding forward the predictions of the base model into another model is referred to as stacking.
- Applying the latest convolutional neural networks (CNN) to analyse and learn from the massive textual data in the dataset, i.e. millions of tweets posted by thousands of twitter users.
- Presenting a comparative analysis of various machine learning models based on their performance as observed in our proposed model measured based on various indicators such as precision, recall, F1 result and accuracy.

1.3 Research Challenges

The literature of twitter bot detection is vast, but the problem is still a very challenging task with bots becoming smarter with time. It is relatively simple to detect bots which show unusual behaviour such as retweeting very frequently. However, smarter bots curb to show unusual behaviour and act more humanly. It is harder for bots to show similar tweeting patterns as humans; thus, deep learning models are quite successful in detecting bot by textual analysis.

The state-of-the-art research approaches such as [6, 7] employ the latest deep learning techniques like Bi-LSTM-based approach for textual analyses for bot detection. A similar approach [8] models the user classification task differently at accounts level and tweet level, and comparisons based on the performance of each level are made. Our work combines the two levels and attempts to find unusual behaviour in the engagement of the user and also in tweets of the user. Bi-LSTMs are slightly better than LSTMs in the sense that each state has a memory of not just previously occurring series but also the memory of future occurring series. LSTMs perform well on time series data where the sequence of the sentence is essential. CNN has performed equally good or even better than LSTMs in text classification tasks [9] when feature detection is a significant and specific sequence to sequence learning tasks [10]. In the context of this problem, identification of features such as abuses, angry terms and hate speech is very useful in understanding the context of the tweet.

Table 1 Details of the dataset used for analysis

Class	Users	Tweets
Fake accounts #1	991	1,610,044
Fake accounts #2	3457	428,542
Fake accounts #3	464	1,418,557
Genuine accounts	3474	2,839,362

Thus, we employed CNN for textual analyses of tweets. Our proposed approach can achieve competitive performance compared to existing work [6, 7, 8, 11].

2 Dataset

The dataset was acquired from [12], which includes data of manually annotated genuine and fake twitter accounts along with tweet data corresponding to these accounts in CSV format. Three CSV files which consist of accounts information are Genuine accounts, Fake accounts #1, Fake accounts #2 and Fake accounts #3. Table 1 shows the distribution of users among the CSV files and the number of tweets corresponding to the users of each of these CSV files. The dataset consists of a total of 8386 accounts and 6,296,505 tweets of the accounts. The accounts' data initially consist of 40 attributes, while the tweet data consist of 25 attributes.

The feature selection step is a useful measure while addressing the following two significant concerns of data processing before training—namely null value cleaning and model training efficiency. In order to improve model training efficiency, it is necessary to reduce dimensionality by selecting an appropriate subset of the features in the dataset without compromising on model correctness and accuracy.

3 Proposed Approach

3.1 User-Level Modelling

Before performing the actual learning on our data, we comprehensively analyse our dataset in order to identify strongly correlated attributes better, highlight weak correlations and determine which features are most significant in the prediction of the target variable, i.e. 'bot' of the user metadata attributes, by elementary analysis tools and intuition, we are able to eliminate some like 'location', 'created_at', etc. as these attributes contain either majority. However, feature selection based purely on correlation is not always reliable. Thus, we apply the chi^2 test to determine the following five best or most reliable attributes out of all the attributes:

- 'Follower_count',
- 'Friend_count',
- 'Listed_count',
- 'Favorites_count',
- 'Statuses_count'.

3.2 Tweet-Level Modelling

LSTMs evaluate the relationship between consecutive time step data. LSTMs, as opposed to RNNs, keep a memory of previously occurring series (which here is the sequence of words) which it uses while feed forwarding from cell to cell. However, LSTM has a few demerits in the context of the problem. LSTMs are not the best model where feature detection is paramount. For example, searching for abuses, angry terms and named entities, ConvNet works better. LSTMs work better in tasks where the sequence of the sentence is important, tasks such as machine translation and question answering. Studies in [9] show a clear comparison. Thus, we select ConvNet as the text classification model. For using machine learning or deep learning algorithms on texts, the necessary step is to apply appropriate language modelling technique. We used Glove [13] word embedding for language modelling. Glove is a context-independent open-source embedding provided by Stanford University that captures semantic similarity between the large vocabulary of words. The input to the convolutional neural network model used during tweet-level analysis to generate the first level of predictions based on the textual data can be mathematically represented as:

$$\text{Input dimensions} = (\text{No of words in text}) \times (\text{Embedding length}) \tag{1}$$

After applying the NLP pipeline to all the tweets which accomplish the removal of punctuation and stop words(noise), the length analysis of all tweets individually revealed that the maximum length of a tweet as in our dataset is 30. Hence, we predefined the input shape of the embedding matrix as 30 × 300 as the glove embedding length is 300 and max tweet length is 30.

3.3 CNN Architecture

In the proposed model, the input layer receives a matrix of embedding sequences having dimension 30 × 300. Five convolution kernels or feature detectors or filters of heights [2-6] were initially selected. An array of five convolution operations is then applied to the embedding matrix corresponding to the five kernel sizes (heights) to extract the features. Each of the five Conv1D operations uses 200 filters to generate feature maps which are further activated using rectified linear unit (ReLU). A Conv1D operation with kernel height h on input having dimension $m \times n$ produces a feature map of dimensions $(m - h + 1) \times n$. Hence, the dimensions of the feature maps produced corresponding to the filter height 2, 3, 4, 5 and 6 are 29×300, 28×300, 27×300, 26×300 and 25×300, respectively. A GlobalMaxPool1D operation is applied to each of these five feature maps to reduce the parameters in the model. The pooled feature maps retrieved from these operations are concatenated to generate a single feature map which corresponds to an average of the five pooled feature maps. Dropout is an effective regularization technique that reduces overfitting and is used to decrease generalization error in deep neural networks. The feature map is thus subjected to a 10% dropout (D1) to avoid any possibility of overfitting the network. The first fully connected layer or dense layer (C1) along with activation function ReLU works on this feature vector resulting in a 1×128 vectors. This feature vector is subjected to another dropout of 20% to avoid overfitting of the network. The next fully connected layer or dense layer (C2) using softmax as activation function generates a probabilistic output determining whether the given input is a bot or not. An output greater than 0.5 predicts the input text is more likely a tweet tweeted by a spambot and an output less than 0.5 predicts that the tweet is more likely a tweet by a human. The architecture of the proposed convolutional neural network is depicted in Fig. 1.

The corresponding configurations of the various layers in the convolutional neural network are elucidated in the following table.

3.4 Methodology

The proposed overall design of our bot detection system consists of a two-layered stacked generalization ensemble with a convolutional neural network model feeding forward into a machine learning model. The following flowchart depicts the structure of our proposed model in detail (Fig. 2).

The composition of the stacked layers is as follows:

Level 0—A deep learning convolutional neural network has been used for model training, with 70% of the tweets with both spambots and non-spam bots sampled equally in the split. Predictions made with CNN on the remaining 30% tweets will form a feature for level 1. Feature engineering over tweet metadata (tweet information) revealed that the following attributes had a high correlation with our target 'bot' attribute:

- 'Retweet_count',
- 'Reply_count',
- 'Favorite_count',
- 'Num_hashtags',
- 'Num_urls',
- 'Num_mentions'.

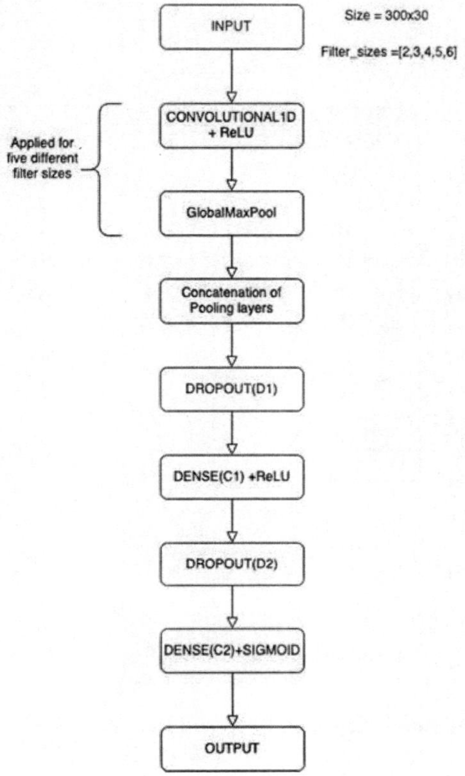

Fig. 1 Architecture of the proposed CNN used for tweet-level analysis

Since for each user, there are multiple tweets, we took the mean for each of the above attributes along with CNN predictions for the remaining 30 per cent tweets. The CNN tweet prediction will be named 'Tweet_prediction' as a feature in level 1.

Level 1—At this level, we have a combined dataset of users having user metadata attributes 'Follower_count', 'Friend_count', 'Listed_count', 'Favorites_count', 'Statuses_count' and 'Tweet_prediction'. Attributes corresponding to tweet metadata, namely 'Retweet_count', 'Reply_count', 'Favorite_count', 'Num_hashtags', 'Num_urls' and 'Num_mentions', are also taken as features at this level. This combined dataset of user metadata, tweet metadata and 'Tweet_prediction' is split into training and test sets with corresponding ratios of 70 and 30, respectively.

Finally, this 70% training dataset is used for training the following classifiers once using only user metadata and once combining user metadata with tweet metadata and the CNN prediction attribute 'Tweet_prediction'. Further, performances are compared.

- K Neighbors Classifier,
- Support Vector Classifier,
- Random Forest Classifier,

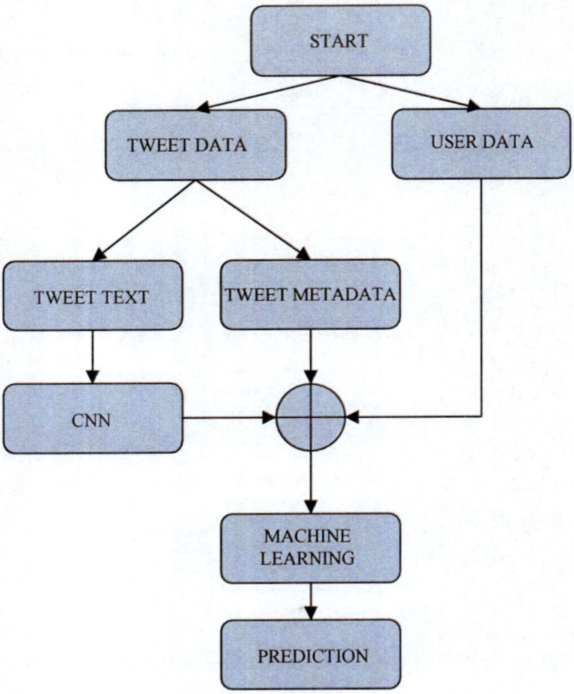

Fig. 2 Flowchart depicting the structure of the proposed model

- Ada Boost Classifier,
- Gradient Boosting Classifier,
- Gaussian NB,
- Linear Discriminant Analysis,
- Decision Tree Classifier,
- Multilayer Perceptron.

4 Results

In this section, we have compiled the results obtained for the proposed approaches as detailed in the previous sections. The performance of each method is measured based on various indicators including precision, recall, F1 result and accuracy. At first, we implemented and analysed the performance of a bot detection model based solely on the user information data without taking into account textual data in the form of tweets. This is the approach taken by a majority of existing bot detection solutions. The results achieved using this existing approach have been compiled and represented below (Table 2).

Table 2 Results obtained with user-level data only, i.e. follower count, tweet count, activity, etc.

Classifier	Accuracy	Precision	Recall	*F*-score
K neighbours	96.10	96.09	96.10	96.09
Support vector	94.78	94.76	94.78	94.60
Decision tree	97.47	97.48	97.47	97.48
Random forest	**97.93**	97.98	97.93	97.95
Ada boost	97.87	97.90	97.87	97.88
Gradient boosting	97.53	97.58	97.53	97.55
Gaussian NB	90.02	90.50	90.02	88.87
Linear discriminant	90.02	89.81	90.02	89.24
Multilayer perceptron	97.24	97.24	97.24	97.24

Bold signifies the best accuracy amongst the models

It can be observed that even though the results obtained are right, the underlying model is unreliable as it is solely based on user metadata which can easily be faked by buying followers and deploying bots. Thus, we can see that this method of classification is not robust since it does not take into account the actual tweets that the user has posted. The next set of results given below represents the performance of the novel method of bot detection proposed in this paper. These are the results obtained when both user-level and tweet-level data are taken into account in the manner detailed in this paper whereby tweet-level textual data is used in a convolutional neural network (CNN) classification model whose output is fed as a feature along with the user-level data features which are then fed together to a machine learning model that makes the final prediction on whether the user is a bot or a human (Table 3).

Table 3 Results obtained by combining user data with tweet metadata and CNN predictions using the proposed approach

Classifier	Accuracy	Precision	Recall	*F*-score
K neighbour	96.10	96.09	96.10	96.09
Support vector	94.78	94.76	94.78	94.60
Decision tree	99.48	99.48	99.48	99.48
Random forest	**99.54**	99.54	99.54	99.54
Ada boost	99.48	99.48	99.48	99.48
Gradient boosting	**99.54**	99.54	99.54	99.54
Gaussian NB	90.31	90.76	90.31	89.24
Linear discriminant	97.42	97.71	97.42	97.47
Multilayer perceptron	96.67	96.87	96.67	96.73

Bold signifies the best accuracy amongst the models

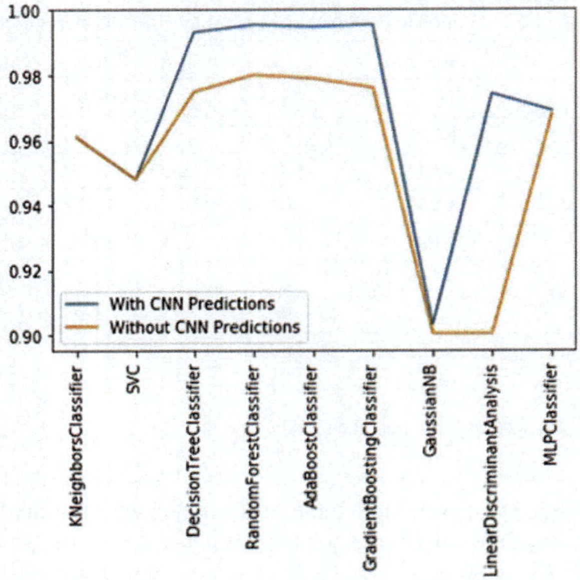

Fig. 3 Accuracy comparison of the proposed model with CNN-based stacking (blue line) and existing approaches without CNN-based text classification and stacking (orange line)

From the above data, it is evident that the approach proposed in this paper shows significantly improved results in terms of accuracy in generating the required predictions. By incorporating convolutional neural networks in the learning process for text classification, our model shows ~2% accuracy improvements across the different machine learning models with a maximum of more than 8% improvement in accuracy in the Linear Discriminant Analysis machine learning model. The improved performance of the proposed model over the existing approaches is represented in the graph below (Fig. 3).

Therefore, the observed results exhibit that the model proposed in this paper for detecting bots on social networking sites can help in significantly improving the existing mechanisms for bot detection. Thus, the technique proposed in this paper can be effectively used in real-world systems for fighting the menace of bots and spammers on social media.

5 Conclusions and Future Direction

Our proposed model gave state-of-the-art results with a relatively smaller dataset, around 3 million tweets of 4 thousand users for each class and several features, only 12. Some previous works in this also took leverage of oversampling to build a bigger dataset. Also, we combined the tweet-level and user-level classification by building a stacked ensemble model of CNN and various ML models, which can prove to be more scalable and robust on new unseen data. The methods of this research can be extrapolated to other social media platforms such as Instagram, Facebook and electronic mail services like Gmail for detecting social spambots, political bots, malicious propaganda bots, advertisement bots and medical bot. The classifier can be constructed using ensemble models combining the learning of two or more deep learning models for further improvement of the text classification model. A more generic bot detection model can be worked upon that would successfully identify different classes of bots.

References

1. State Department examination of Twitter found millions of coronavirus tweets pushed false information—The Washington Post, https://www.washingtonpost.com/technology/2020/02/29/twitter-coronavirus-misinformation-state-department/. Last accessed 2020/03/04
2. Allem, J.P., Escobedo, P., Dharmapuri, L.: Cannabis surveillance with Twitter data: emerging topics and social bots. Am. J. Public Health **110**, 357–362 (2020). https://doi.org/10.2105/AJPH.2019.305461
3. Bots and trolls spread false arson claims in Australian fires 'disinformation campaign' | Australia news | The Guardian, https://www.theguardian.com/australia-news/2020/jan/08/twitter-bots-trolls-australian-bushfires-social-media-disinformation-campaign-false-claims. Last accessed 2020/03/04
4. Mintal, J.M., Vancel, R.: (Un)Trendy Japan: Twitter bots and the 2017 Japanese general election. Polit. Cent. Eur. **15**, 497–514 (2020). https://doi.org/10.2478/pce-2019-0027
5. Rossi, S., Rossi, M., Upreti, B., Liu, Y.: Detecting political bots on Twitter during the 2019 Finnish Parliamentary Election. In: Proceedings of the 53rd Hawaii International Conference on System Sciences, vol. **3**, pp. 2430–2439 (2020). https://doi.org/10.24251/hicss.2020.298
6. Wei, F., Nguyen, U.T.: Twitter bot detection using bidirectional long short-term memory neural networks and word embeddings. In: 2019 First IEEE International Conference on Trust, Privacy and Security in Intelligent Systems, and Applications, pp. 101–109 (2020). https://doi.org/10.1109/TPS-ISA48467.2019.00021
7. Luo, L., Zhang, X., Yang, X., Yang, W.: Deepbot: a deep neural network based approach for detecting Twitter bots. IOP Conf. Ser. Mater. Sci. Eng. **719**, (2020). https://doi.org/10.1088/1757-899X/719/1/012063
8. Kudugunta, S., Ferrara, E.: Deep neural networks for bot detection. Inf. Sci. (Ny) **467**, 312–322 (2018). https://doi.org/10.1016/j.ins.2018.08.019
9. Yin, W., Kann, K., Yu, M., Schütze, H.: Comparative study of CNN and RNN for natural language processing. arXiv:1702.01923 (2017)
10. Gehring, J., Auli, M., Grangier, D., Yarats, D., Dauphin, Y.N.: Convolutional sequence to sequence learning. In: 34th International Conference on Machine Learning ICML 2017, vol. 3, pp. 2029–2042 (2017).
11. Zhao, C., Xin, Y., Li, X., Yang, Y., Chen, Y.: A heterogeneous ensemble learning framework for spam detection in social networks with imbalanced data. Appl. Sci. **10**, 936 (2020). https://doi.org/10.3390/app10030936
12. MIB Datasets: https://mib.projects.iit.cnr.it/dataset.html. Last accessed 2020/03/06
13. Pennington, J., Socher, R., Manning, C.D.: Glove: Global Vectors for Word Representation, pp. 1532–1543 (2014). https://doi.org/10.3115/V1/D14-1162

Internet of Things (IoT): Vulnerabilities and Remediation Strategies

Pooja Anand, Yashwant Singh, and Arvind Selwal

Abstract Iot being a transformative approach for imparting countless services raises consequential security flaws as well. These flaws germinate from the embedded vulnerabilities in IoT devices. The market is flooded with these vulnerable smart devices, which are easy to play with to remotely enter into an IoT system. This becomes more serious as communication protocols and Internet technologies were not devised to support IoT. In this paper, we mainly focus on the evolving vulnerabilities in IoT that can affect its sustenance in the long run. We also elaborated on the remediation strategies to be incorporated to lessen the fertility of the ground to launch numerous attacks. Finally, we conclude with the challenges and recommendations.

Keywords IoT · Threats · Vulnerabilities · Countermeasures

1 Introduction

The billions of smart devices have come under the IoT realm in 2020. The Internet of Things (IoT) enables remote controlling and monitoring of the environment by making sense of the data collected by ground devices like sensors, actuators that are connected to the Internet [1]. The rising spread of IoT services in people's lives is fabricated with numerous threats, specifically concerning security and privacy. The markets being flooded with insecure ill-designed IoT products raises an alarm to flesh and blood too. The heterogeneity and resource constraints prevailing in IoT devices have even turned the standard Internet security mechanisms insignificant for

P. Anand (✉) · Y. Singh · A. Selwal
Department of Computer Science and Information Technology, Central University of Jammu, Samba, Jammu and Kashmir 181143, India
e-mail: poojaanand892@gmail.com

Y. Singh
e-mail: yashwant.csit@cujammu.ac.in

A. Selwal
e-mail: arvind.csit@cujammu.ac.in

P. K. Singh et al. (eds.), *Recent Innovations in Computing*, Lecture Notes in Electrical Engineering 701, https://doi.org/10.1007/978-981-15-8297-4_22

them [2]. Furthermore, less-aware IoT manufacturers and users provide the fertile platform to the potential adversaries by offering their vulnerable day-to-day devices connected to a global network.

In general, it is found that we connect vulnerable things to the Internet for availing smart services [3]. The IoT devices with known vulnerabilities like obsolete OS versions, no update mechanisms, default passwords, and open ports come into the market [4]. The botnets like Mirai, Brickerbot, and Hijame exploit these loopholes to create the army of bots to shut down the various services for several hours [5]. Moreover, the hackers exploited baby monitors and toys [6] to get secret information like videos of baby monitoring [7], millions of voice recordings of parents and their children, accounts, passwords, etc. The adversaries are even easily reprogramming the firmware of IoT devices [8]. Consequently, in this article, we primarily focused on the major vulnerabilities in IoT devices that are being exploited by potential adversaries. Furthermore, we discussed the remediation strategies that can be adopted to safeguard from these loopholes.

The subsequent sections are organized as follows. In Sect. 2, we present the related work. In Sect. 3, we discussed the major vulnerabilities in an IoT system. Subsequently, we talked about the remediation strategies in Sect. 4. Following this, we concluded the paper in Sect. 5 with future scope.

2 Related Work

In [9], the evolving IoT vulnerabilities and possible remediation strategies are elaborated. Some attack scenarios are discussed as well. They have also given a data-driven empirical assessment focusing on IoT maliciousness. In context to the Cisco 7-layer reference model, Nia et al. [10] also discussed multiple attack scenarios and their mitigation ways. On similar lines, Makhdoom et al. [11] emphasized the threats at the layers of IoT and also discussed the attack approach of various malware. Furthermore, they included the security guidelines based on industry best procedures to apply the least security criteria in an IoT system.

Some authors, Chen et al. [12], recommended to embed intelligence and modularization in penetration testing tools for discovering vulnerabilities by using active strategy against IoT systems. In an alternate work, Visoottiviseth et al. [13] worked out a tool namely PENTOS for penetration testing in IoT devices. This tool has integrated various other tools like Metasploit, Kali, Nessus, etc., to provide a complete package for finding the vulnerabilities.

By using the three scans, Markowsky et al. [14] scanned the whole IoT system for vulnerabilities. Shodan was used to target the routers, Nmap to target connected printers, and Masscan for devices infected with Heart bleed bug. In an alternative work, Ko et al. [15] proposed a platform that allows sharing the vulnerabilities after assessing them.

Moreover, vulnerability assessment has become one of the modules in IoT testbeds. In [16], port scanning, fingerprinting, and vulnerability scanning are an integral part of the IoT testbed. Similarly, Siboni et al. [17] IoT testbed contain plugins like process enumeration, communication tampering, fingerprinting, port scanning, and vulnerability scan in security testing module.

Few machine learning-based vulnerability assessment solutions have also come across. The Dojo being an intelligent IoT vulnerability scanner launched by Bull-Guard is one among them. It is compatible with both Android and ios. It works by scanning all the devices in the Wi-Fi zone, examines the vulnerabilities, and grades them on multiple factors. IoTScanner, Censys, SeeSec–IoT Vulnerability Scanner, Bitdefender, and IotSploit also serve the same purpose.

3 IoT: Vulnerabilities

The security of IoT has become a matter of concern now. The security gaps namely vulnerabilities in IoT systems and their deployment configurations are exploited to launch deadly attacks. IoT is on the verge of losing its incredible potentials. In such a situation vulnerability assessment [18] must be the first thing to start with safeguarding IoT devices from these maturing threats. In the sequel, following Fig. 1, we particularize the vulnerabilities in IoT.

(a) *Physical Security of IoT devices*
Most of the IoT devices being in unattended surroundings could be exploited to get the side-channel information like power consumption and processing time [8]. The adversary can also make clones of IoT nodes which could be further used to launch several attacks. The credentials of those compromised nodes could also be easily accessed by them [11].

(b) *Open Debugging ports*
The open debug ports are exploited in every possible way to regulate the IoT system. From these vulnerable ports, the intruder can launch attacks like

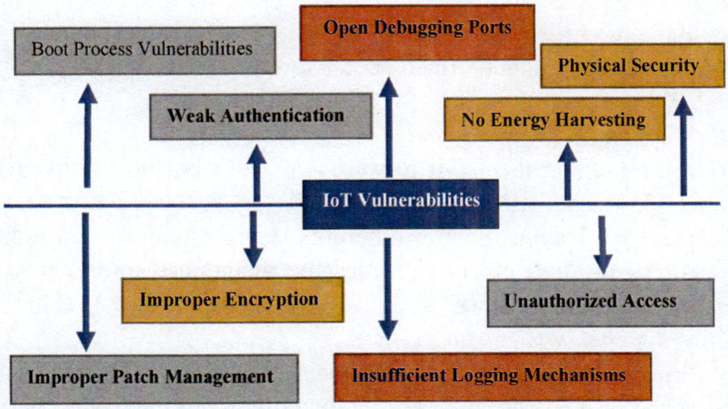

Fig. 1 Common vulnerabilities in IoT

injecting malicious code, falsely modifying the firmware, and bypassing their security [19]. The botnets like Mirai, Hajime, Brickerbot exploited the telnet port to launch variants of denial of service attacks.

(c) *No energy Harvesting*
The IoT nodes being resource-constrained are more prone to resource exhaustion attacks. As such, there is no way out to regenerate the energy of these power-constrained IoT nodes [20]. The advantage of this loophole is taken to deplete the IoT resources like CPU time, bandwidth, energy, and memory. As a result, legitimate users will be denied for several services as it may lead to battery drainage [21], node outage [22], and DoS attack.
(d) *Weak Authentication*
It has created many odds in implementing strong authentication mechanisms in resource-constrained IoT nodes. The smart actors easily enter the IoT system avoiding identity checks, thus exploiting the system in multiple ways. Some of the attacks framed out by ill-using this vulnerability are DDoS Attack, Dictionary Attack, Sybil Attack [23], Hello flood, and Homing Attacks [24].
(e) *Improper Encryption*
The wireless communication media creates more concern in terms of data leakage and privacy. It is found that almost all IoT devices use wireless media. Being an unreliable medium, personal information of the user is more prone to spoofing attacks. For instance, leakage of personal information could be life threatening in smart health services. The strong encryption mechanism can effectively prevent data leakage. However, implementing a robust crypto-algorithm becomes tough for resource-constrained IoT devices. This may result in attacks such as storage attacks, eavesdropping, and man-in-the-middle attack [9].
(f) *Unauthorized Access*
The IoT systems and data must be accessible only to authorized entities. For this, we need to implement a strong credential system. It is found that the IoT devices do not enforce sufficient complexity of passwords. Even, some devices continue with the default passwords and thus come in the trap of bad actors. The manufacturers provide an add-on by shipping devices with hardcoded passwords. Hence, easy unauthorized access threatens data and the security of the entire IoT system.
(g) *Insufficient Logging Mechanism*
With proper logging mechanism, we could make out intrusions and hacking attempts [25]. The administrators should log events related to successful and unsuccessful authentication attempts, login attempts, and authorization attempts. All the log information must be maintained in an encrypted form to further prevent the misuse.
(h) *Improper patch management*
For the smooth functioning of the device, there must be proper management of patches. These patches enhance the functioning of the device and prevent it from numerous threats. It is seen that IoT manufacturers do not put much effort

into providing security patches on time for growing vulnerabilities. Moreover, most of these devices are without any automated patch-update mechanism. The integrity of these patches and updates is also a matter of concern to further safeguard them. The false updates can badly ruin the functionality of the device and can affect the entire system in multiple ways [26].

(i) *Boot process vulnerabilities*
By exploiting the vulnerabilities of the boot process, an adversary can take control of the whole system. He can exploit the boot sequence, firmware, and the bootstrap loader to enter in. In an experimental setting, an attack was launched to exploit boot process vulnerabilities against the fitness tracker and Nest Thermostat [27].

4 IoT: Remediation Strategies

Some of the remediation strategies [9, 28–30] that could be adopted to deal with these growing vulnerabilities as shown in Fig. 2 are discussed in this section.

- *Side-Channel Analysis*
  We can detect malicious firmware and h/w trojans in a device using side-channel signal analysis. The signals like power consumption, timing signals, spatial requirements, and temperature are used for trojan detection. The same is compared with IC affected with trojan and the one with no trojans. The timing-based schemes detect trojans by using delay tests, the power-based schemes use active monitoring, and the social-temperature based use infra-red imaging mechanisms. On similar

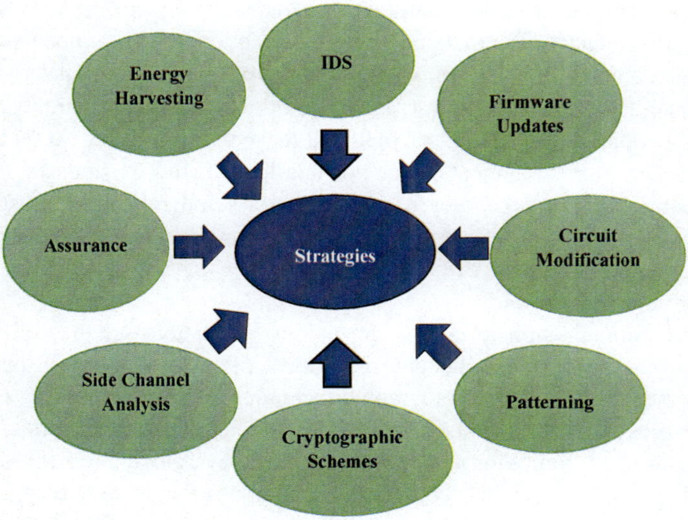

Fig. 2 Remediation strategies

lines, the side-channel information could be used to find the abnormal behavior of the machine that symbolizes the presence of malicious firmware in the system.

- *Policy-based mechanisms and intrusion detection systems*
  By incorporating the intrusion detection systems in an IoT environment, we can safeguard against security and privacy issues to a great extent. These detection systems can make out if there any essential policies being violated and thus can detect the intrusions. Furthermore, these systems are proved to be useful to guard against sleep deprivation and battery-drainage attacks by distinguishing unusual requests to the IoT node.
- *Circuit Modification*
  The physical, trojan, and side-channel attacks can also be defended by just changing the circuit. The integrated security mechanism within the physical hardware of IoT nodes enhances physical protection against those attacks. For example, in-home automation sensors, like, smoke detectors, many mechanical/electrical tamper-proofing mechanisms are imposed to protect them against tampering. On similar lines, the side-channel information could also be protected from adversaries by feeding them with tampered information by modifying the circuit. Some of the well-known approaches make use of randomized delay, infused noise, hamming weights, and modifying the cache architecture. Furthermore, by integrating PUF into the circuit, we can enable authentication, trojan detection, and device identification. A PUF is a physically unclonable function embedded into an IC, that is physically unclonable, tamper-proof, and unpredictable.

- *Securing Firmware Update*
  There are two ways to update the firmware either remotely or directly. In the former, the server broadcasts the signal (CMD) conveying the availability of a new firmware version. Then, the node with an updated firmware version broadcasts an advertisement (ADV) to its neighboring nodes. The willing nodes with ADV message compare the new firmware version with the existing one. If they need an update, they send the request (REQ) and thus will receive an update file. To securely update the firmware remotely, all these packets CMD, ADV, REQ, and data packet must be authenticated. For the direct firmware updates, e.g., via a USB cable, the firmware integrity, and the user's authentication, must be taken into consideration. The failure to this may enable an adversary to put the malicious one in place of legitimate device firmware.
- *Cryptographic Schemes*
  At the communication level, we can use cryptographic schemes, like, strong encryption, as a defense mechanism against attacks, like eavesdropping and routing attacks. The old cryptographic techniques could not be used to address these issues in resource-constrained IoT nodes. These techniques raise memory usage, power consumption, loss of packets, and delay. A lot many efforts have been put to develop lightweight cryptographic techniques for securing communication in IoT, e.g., PRESENT and CLEFIA. Still, there is a lack of public-key encryption techniques to meet lightweight IoT requirements while ensuring privacy.

- *De-patterning and decentralization*
  Another defense mechanism against anonymity and side-channel attacks is de-patterning and decentralization. By altering the traffic pattern of the data transmissions through fake packets, we can protect against side-channel attacks. Furthermore, decentralization ensures anonymity, in which the sensitive data is distributed through a spanning tree to not let any node a complete picture of the original data.
- *Software Assurance:*
  The software assurance lessens the vulnerabilities in both source and binary code of IoT software. Moreover, there is an alarming impact of ill-using IoT software. Thus, it must be an integral part of the software development life cycle. To handle these issues, Costin et al. have given an automated scalable framework for dynamic analysis. This framework aims at discovering the vulnerabilities in the embedded IoT firmware images. For the same, the authors used emulating firmware and Arachni the free penetration tool [28].
- *Security Protocols:*
  The resource-constrained nature of IoT devices makes them more and more vulnerable, as limited security mechanisms could be incorporated in them. To this end, efforts are put to develop energy-aware IoT ecosystems. One such architecture, i.e., 2EA architecture, is developed by Balasubramanian et al. In Energy-Aware-Edge-Aware (2EA) architecture, the IoT sensors can harvest their energy. For each sensor in an IoT network, the energy profile with power metrics is maintained. In the case of energy depletion, the victim node queries the energy profile and gets the most reliable node nearby. The optimal resource utilization is ensured by this scheme, based on the task arrival process.

5 Conclusion

The IoT based application has become more evident in this decade. With them, comes the growing cyber-attacks in terms of security and privacy of common people. In this paper, we discussed the major IoT vulnerabilities and their possible remediation strategies. We have found that there is a lack of remediation strategies for preventing physical access to the IoT devices, proper firmware updates, and managing the debug ports. Further, large-scale implementation and evaluation of these strategies in tangible IoT realms must be carried out soon.

References

1. Zanella, A., Bui, N., Castellani, A., Vangelista, L., Zorzi, M.: Internet of things for smart cities. IEEE Internet Things J. **1**(1), 22–32 (2014). https://doi.org/10.1109/JIOT.2014.2306328
2. Yosra Ben Saied: Collaborative security for the internet of things Sécurité Collaborative pour l ' Internet des Objets. Int. J. Comput. Appl. **135**(2), 23–29 (2013)

3. Hypponen, M., Nyman, L.: The Internet of (vulnerable) Things : on hypponen ' s law, security engineering, and IoT legislation. Technol. Innov. Manag. Rev. **7**(4), 5–11 (2017)
4. Corser, G., Fink, G.A., Aledhari, M.: IEEE Internet Technology Policy Community White Paper INTERNET OF THINGS (IOT) SECURITY. IEEE, no. February, pp. 1–13, 2017.
5. Kolias, C., Kambourakis, G., Stavrou, A., Voas, J.: DDoS in the IoT: Mirai and Other Botnets. Computer (Long. Beach. Calif). 79 (2017)
6. IoT connected teddy bear leaks millions of kids' conversations, exposed database to blame—TechRepublic. [Online]. Available: https://www.techrepublic.com/article/iot-connected-teddy-bear-leaks-millions-of-kids-conversations-exposed-database-to-blame/. Accessed: 03-Jan-2020.
7. Stanislav, M., Beardsley, T.: HACKING IoT: a case study on baby monitor exposures and vulnerabilities. September (2015)
8. Standaert, F.: Introduction to side-channel attacks. Secur. Integr. Circuits Syst. Springer Sci. Media, pp. 27–42 (2010). https://doi.org/10.1007/978-0-387-71829-3.
9. Neshenko, N., Bou-harb, E., Crichigno, J., Kaddoum, G., Ghani, N.: Demystifying IoT security: an exhaustive survey on IoT vulnerabilities and a first empirical look on Internet-scale IoT exploitations. IEEE Commun. Surv. Tutor. PP(c), 1 (2019). https://doi.org/10.1109/COMST.2019.2910750.
10. Nia, A.M., Member, S., Jha, N.K.: A comprehensive study of security of. IEEE Trans. Emerg. Top. Comput. **6750**(c), 1–19. (2016). https://doi.org/10.1109/TETC.2016.2606384.
11. Makhdoom, I., Abolhasan, M., Lipman, J., Liu, R.P., Ni, W.: Anatomy of threats to the Internet of Things. IEEE Commun. Surv. Tutor. PP(c), 1 (2018). https://doi.org/10.1109/COMST.2018.2874978.
12. Chen, C., Zhang, Z., Lee, S., Shieh, S.: In the IoT age. Computer (Long. Beach. Calif.) (2018)
13. Visoottiviseth, V., Akarasiriwong, P., Chaiyasart, S., Chotivatunyu, S.: PENTOS : penetration testing tool for Internet of Thing devices. In: Proceedings 2017 IEEE Region 10 Conference, pp. 2279–2284 (2017)
14. Linda, M.: Scanning for vulnerable devices in the Internet of Things. 463–467 (2015)
15. Ko, E., Kim, T., Kim, H.: Management platform of threats information in IoT environment. J. Ambient Intell. Humaniz. Comput. **9**, 1167–1176 (2017). https://doi.org/10.1007/s12652-017-0581-6
16. Sachidananda, V.: POSTER: towards exposing Internet of Things : a roadmap of the Negev. ACM **1**, 1820–1822 (2016)
17. S. Siboni, V. Sachidananda, Meidan, Y., Bohadana, M., Mathov, Y., Bhairav, S.: Security testbed for Internet-of-Things devices. IEEE Trans. Reliab., vol. PP, 1–22 (2018). https://doi.org/10.1109/TR.2018.2864536.
18. Anand, P., Singh, Y., Selwal, A., Alazab, M., Tanwar S., Kumar, N.: IoT vulnerability assessment for sustainable computing: threats, current solutions, and open challenges. In: IEEE Access **8**, 168825–168853 (2020) https://doi.org/10.1109/ACCESS.2020.3022842
19. Sachidananda, V., Toh, J., Siboni, S., Bhairav, S., Shabtai, A., Elovici, Y.: Let the cat out of the bag: a holistic approach towards security analysis of the internet of things. IoTPTS 2017—Proc. 3rd ACM Int. Work. IoT Privacy, Trust. Secur. co-located with ASIA CCS 2017, pp. 3–10 (2017). https://doi.org/10.1145/3055245.3055251.
20. Trappe, W., Howard, R., Moore, R.S.: Low-energy security: limits and opportunities in the internet of things. IEEE Secur. Priv. **13**(1), 14–21 (2015). https://doi.org/10.1109/MSP.2015.7
21. Ramesh, D., Rao, D.K.: Vampire attacks: draining life from wireless ad hoc sensor communication of networks. IEEE Trans. Mob. Comput. **3**(9), 1107–1110 (2014)
22. Matrosov, A., Rodionov, E., Harley, D., Malcho, J.: Stuxnet Under the Microscope pp. 1–72 (2010)
23. Rajan, A., Jithish, J., Sankaran, S.: Sybil attack in IoT: Modelling and defenses. In: International Conference on Advances in Computing, Communications and Informatics, ICACCI 2017, vol. 2017-Janua, pp. 2323–2327 (2017). https://doi.org/10.1109/ICACCI.2017.8126193.
24. Wallgren, L., Raza, S., Voigt, T.: Routing attacks and countermeasures in the RPL-based internet of things. Int. J. Distrib. Sens. Netw. **2013**(10), 1–11 (2013). https://doi.org/10.1155/2013/794326

25. Hernandez, G., Arias, O., Buentello, D., Jin, Y.: Smart Nest Thermostat : A Smart Spy in Your Home, pp. 1–8. Black Hat USA, Cisco (2014)
26. Basnight, Z., Butts, J., Lopez, J., Dube, T.: Firmware modification attacks on programmable logic controllers. Int. J. Crit. Infrastruct. Prot. **6**(2), 76–84 (2013). https://doi.org/10.1016/j.ijcip.2013.04.004
27. Arias, O., Wurm, J., Hoang, K., Jin, Y.: Privacy and security in Internet of Things and wearable devices. IEEE Trans. Multi-Scale Comput. Syst. **1**(2), 99–109 (2015). https://doi.org/10.1109/TMSCS.2015.2498605
28. Nia, A.M., Member, S., Jha, N.K.: A comprehensive study of security of. IEEE Trans. Emerg. Top. Comput. **6750**(c), 1–19 (2016).
29. Bertino, E., Islam, N.: Botnets and Internet. IEEE Comput. Soc. **18**, 76–79 (2017)
30. Alaba, F.A., Othman, M., Abaker, I., Hashem, T., Alotaibi, F.: Internet of Things security: a survey. J. Netw. Comput. Appl. **88**(April), 10–28 (2017)

DACHE: A Data Aggregation-Based Effective and Optimized Cluster Head Election Routing Protocol for HWSNs

Aruna Malik, Samayveer Singh, and Pradeep Kumar Singh

Abstract Most of the wireless sensor networks (WSNs) are having intrinsic resource limitations such as energy, link, and computational resources since energy is consumed by all the resources. Thus, there is a dare need to develop energy-efficient protocols that escalate the longevity of the network for collecting data over a long time. In this work, we propose a data aggregation-based effective and optimized CHs election routing protocol for heterogeneous WSNs. These networks consist of three levels of heterogeneity. In this method, a threshold-based formula is used which helps in optimized cluster heads (CHs) election. A new chain-based data aggregation process is also discussed in this paper which helps in the effective data gathering process. This threshold formula considered three criteria namely node and sink distance, node remaining energy, and total networks energy which decreases the energy depletion in both inter and intra communication between the sensor, CHs, and sink. This efficient data gathering process also removes the duplicate at the time of data collection from the deployed nodes and reduced the energy depletion and network overheads. The number of alive and dead, the sum of energy depletion, and a number of the message transferred to the control node matrices are considered to investigate the enactment of the proposed scheme by using the MATLAB. After a comprehensive analysis, it has been evident that the proposed method accomplishes healthier than that of the existing methods.

Keywords Energy efficiency · Network lifespan · Efficient clustering · Heterogeneous WSNs

A. Malik · S. Singh
Department of Computer Science & Engineering, Dr B R Ambedkar National Institute of Technology Jalandhar, Jalandhar, Punjab, India
e-mail: arunacsrke@gmail.com

S. Singh
e-mail: samayveersingh@gmail.com

P. K. Singh (✉)
Department of Computer Science & Engineering, ABES Engineering College, Ghaziabad, Uttar Pradesh, India
e-mail: pradeep_84cs@yahoo.com

P. K. Singh et al. (eds.), *Recent Innovations in Computing*, Lecture Notes in Electrical Engineering 701, https://doi.org/10.1007/978-981-15-8297-4_23

1 Introduction

Generally, the necessity of instant information is one of the most important needs of human beings in the current livelihood for taking instant decisions. Human gets instant information through some processing devices which can collect information and process it and then send that information to the other processing devices. In that way, humans can get information in a comprehensible manner and take instant decisions, accordingly. Nowadays, wireless sensor networks (WSNs) are one of the utmost popular systems with their advances in electro and mechanical technologies. These technologies help a momentous enlargement in WSNs in the arena of communication, sensing, and computation in industry and academia especially. Generally, the WSNs can survive in very different environmental conditions like in fire, flood, etc. [1]. Generally, there are lot of applications, where we can deploy WSNs such as water quality monitoring, building health structure monitoring, surveillance of various types of buildings, gas leakage in the various gas and chemical plants, productivity, humidity, water requirement, fertilizer monitoring in the agriculture monitoring, health monitoring in medical application, wild animal monitoring the parks, fire detection in the multistory building, and so on.

WSNs consist of low power, low size, and low complex devices which are deployed in the monitoring environment for collecting the information. The sensor nodes have small computing power, microprocessor, and bandwidth. Most of the sensors nodes have self-organizing capabilities means they have enough capabilities for self-automatic configuration. After deployment sensor nodes gathered data from the field, they forwarded that information to the base station. The sink or base station forwards that data to the server through the Internet or some other media where it is the further process. It is not essential that all the sensor nodes required to work at the same time. Some sensor can perform monitoring or some of the sensor nodes may go into the sleep state as required [1].

There are two broad categories of the WSNs such as homogeneous and heterogeneous. In the homogeneity, all the deployed nodes have the same power capabilities, whereas nodes have different capabilities in case of heterogeneous networks as energy, link capacity, memory, processing power, etc. Thus, based on the above discussion, we can categorize the heterogeneity into multiple categories namely energy, link, and computational heterogeneity [2]. In case of energy heterogeneity, heterogeneous networks consist of multiple types of batteries means they have different-different Volts or Jules batteries. Link heterogeneity means sensor nodes have different capabilities of link for transferring data over the deployed networks like some of the networks have 10 Mbps and some of the links 50 Mbps bandwidth for transferring information. In the computational heterogeneity, sensor nodes have multiple types of microprocessor, memories, etc. Based on the above discussion, we can conclude that the nodes in the sensor networks have many restrictions like energy limitation, cost of the nodes, speed, processing power, etc. Thus, there is a dire need to develop energy effective and efficient algorithm which can overcome the above-discussed problems of the networks for various applications. Perhaps clustering shows a very effective role for improving the enactment of the wireless sensor networks by effective utilization of the heterogeneous resources like energy limitation, cost of the nodes, speed, processing power, etc. [2].

Mostly, the sensor nodes are deployed in the environment using two different categories namely deterministic and non-deterministic which is based on the various applications. In the case of deterministic, all the nodes are deployed manually, whereas sensors are positioned randomly in the case of non-deterministic environment. Based on the above discussion, we can say sensor nodes have very limitations such as storing data capacity, batteries, bandwidth, and connections. Thus, an efficient and effective system is required for required which can remove such type of limitations. In this work, we propose a data aggregation-based effective and optimized CH election routing protocol for heterogeneous WSNs. These networks consist of three levels of heterogeneity. In this method, a threshold-based formula is used which helps in optimized cluster heads (CHs) election. A new chain-based data aggregation process is also discussed in this paper which helps in the effective data gathering process. This threshold formula considered three criteria namely node and sink distance, node remaining energy, and total networks energy which decreases the energy depletion in both inter and intra communication between the sensor, CHs, and sink. This efficient data gathering process also removes the duplicate at the time of data collection from the deployed nodes and reduced the energy depletion and network overheads. The number of alive and dead, the sum of energy depletion, and a number of the message transferred to the control node matrices are considered to investigate the enactment of the projected technique by using the MATLAB. After a comprehensive analysis, it has been evident that the projected technique achieves healthier than that of the existing methods.

The arrangement of the paper is as follows. Section 2 deliberates the literature review. In Sect. 3, the various assumption for deploying networks and models are deliberated. In Sect. 4, the projected optimized cluster head election routing protocol is discussed. Section 5 discusses the data collection with the data aggregation process at cluster heads. Section 6 deliberates the outcome and discussions. Lastly, Sect. 7 deliberates the achievements of the paper.

2 Literature Review

In WSNs, the extreme depletion of energy is in the transmission and reception of information instead of the sensing, computing, ideal, and others. There are many other limitations of the WSNs are as non-rechargeability, the limited size of nodes, non-replaceability of the nodes, batteries of nodes, etc., where everything depends on the batteries only. However, for solving such problems, there is a need for energy-efficient techniques which can solve such types of problems. There are a lot of energy balancing techniques among the sensor nodes which can equilibrium the load and helps in prolongation of the system lifetime. This load balancing can decrease the energy dissemination in the networks. The first clustering protocols are low energy adaptive clustering hierarchy (LEACH) protocol which tries to balance the load among the clusters [3]. The working of LEACH distributed into two phases namely setup and steady phase. In the setup phase, nodes are deployed, and possible CHs are selected for further processing, whereas the data collection is considered in the steady phase. LEACH is further improved in the form of power-efficient gathering in sensor information systems (PEGASIS) protocol [4]. It designed chains in the networks for accumulating the information and data collection always stated from the farthest nodes to the near node from the base station. It created multiple chains in this process. The main shortcoming of the method is that it is not appropriate for the huge size of networks. These protocols are defined for the homogeneous networks, but after a time, some of the heterogeneous protocols are discussed.

The first heterogeneous protocol is stable election protocol (SEP) which is discussed for the two-level and multi-level heterogeneity. The hetSEP [5] is the extension of the SEP protocol by incorporating the 2 and 3 levels of heterogeneity. It uses the probability formula for electing the cluster heads. It increases the network overheads if the packet transmission is too long. After that, distributed energy-efficient clustering (DEEC) protocol is discussed since the 2 and 3 levels of heterogeneity. It also uses a probability formula for electing the cluster heads by considering the residual and average energy. In DEEC, the energy of the networks is not used efficiently. In paper [7], the selection of the clusters is based on the nodes remaining energy and the networks. It only considered the three-level of heterogeneity of the nodes in the deployed networks. These methods require the extra energy in the process of reclustering at the time of next rounds. To improving the further performance of the methods, Maheshwari et al. discussed a two-level of grouping for collecting the data by considering the node degree [8]. This method increases the load of the sink and somehow creates the problem of the hot spot in the deployed networks. The methods discussed in papers [9–11] are considered heterogeneity of the multiple levels. These papers are not considered the data aggregation process and chaining approach for efficient collection of data. In paper [12], a fuzzy-based clustering approach is discussed for prolonging the lifetime of networks. It considered the many parameters for the distance between the nodes and sink, residual energy of the clusters for cluster heads election. However, this method undergoes load balancing during the data gathering and did not deliberate the information accumulation procedure.

In paper [13], the authors discuss the hybrid routing technique based on the zone of the deployed fields. The clustering process is adopted by it is similar as defined in the LEACH protocol. This approach is not involving normal nodes in the clustering process. This method is further extended by the [14] by considering hybrid clustering concept and multi-hop data collection process. It does not use all the methods in the clustering process, but only a few nodes have been used in the clustering process. It suffers the load balancing problem of the system. The paper [15] is also discussed for the heterogeneous protocols for increasing the network lifetime. It considers the probability function for selecting the cluster heads. It defines the two and multi-level heterogeneity in the deployed networks. The paper [16] discusses a routing-based technique which collects data on the bases of the chains. There are two phases of the process. In the first phases, the nodes direct their statistics to the master heads, whereas in the second phase, clusters heads send their data to the sinks as they received data from the respective clusters by performing data aggregation.

Wang et al. discuss a cluster routing protocol which elects the cluster head efficiently [17]. It also improved the node processing and inter-cluster routing problem. This method partially suffers from an effective cluster and data aggregation. Chen et al. discuss a chain-based hierarchical routing protocol in WSNs [18]. It divided monitoring areas into minor fields to create chains. It reduces excessive routes and also recovers energy. This method created many small chains in the networks. Linping et al. discuss an improved algorithm of PEGASIS protocol which is based on double cluster heads in WSN [19]. It usages low-level clusters header for upgraded load balancing. This protocol is useful for large networks and executes better than the PEGASIS algorithm. The initial installations of this method create high overhead. In [20], authors discussed many research issues related to energy consumption problems of the wireless sensor networks such as clustering, load balancing, multiple sink concepts, heterogeneity, and hot spot problem in the networks.

3 Network, Energy, and Radio Dissipation Model

In this section, the fundamental assumptions of the proposed system model are given as all the sensors have an ID and fixed location because the nodes are stationary, the sensor nodes can be heterogeneous its on the level of heterogeneity, the sensors have symmetric in terms of resource capabilities like connections, capacities, computational and memory powers, and the BS is situated in the middle of the defined territory.

A three-tier heterogeneity model which consists of N sensors is discussed in this section. The energies of level-1, -2, and -3 nodes are represented as e_1, e_2, and e_3, respectively, with condition$e_1 < e_2 < e_2$, and their numbers are denoted as N_1, N_2, N_3, respectively, by way of condition$N_1 > N_2 > N_3$. The network energy is as given as

$$e_{\mathbf{Total}} = \text{þ} \times N \times e_1 + \text{þ}^2 \times N \times e_2 + (1 - \text{þ} - \text{þ}^2) \times N \times e_3 \tag{1}$$

e_{Total} can define level-1, -2, and -3 heterogeneity using the value of þwhich is a parameter of the energy model.

The level-1 nodes are minimum in number, i.e. $\text{þ} \times N$and having a minimum amount of energy denoted as e_1 where the range of model parameter þis $0 \leq \text{þ} \geq 1$. The level-2 nodes are less in number from level-1 nodes, i.e. $\text{þ}^2 \times N$and having more energy from level-1 nodes denoted as e_2. The level-3 nodes are less in number from level-1 and -2 nodes, i.e. $(N - (\text{þ} * N + \text{þ}^2 * N))$and having more energy from level-1 and -2 nodes denoted as e_3 that means they have a maximum amount of energy.

Level-1 heterogeneity: For $þ = 0$, Eq. (1) defines the network has only one type of sensors and the total network energy as given below:

$$e_{\text{Total}} = N * e_3 \tag{2}$$

where e_3 is the nodes preliminary energy. It is indicating the level-3 nodes instead of level-1 nodes. We impose a condition for converting the energy of level-3 nodes into level-1 nodes. It can be attained by using the given representation:

$$þ = \frac{e_3 - e_1}{\beta * f(e_2, e_3)} \tag{3}$$

where f and β are the functions of e_2 and e_3, and positive integer where $\beta > 1$. The function f can have either $(e_3 + e_2)$ or $(e_3 - e_2)$.

Level-2 heterogeneity: For $1 - þ - þ^2 = 0$, find the value of þwhich defines two type of sensors namely level-1 and -2 nodes. The relation $1 - þ - þ^2 = 0$is not selected arbitrarily which is diminished by the third term of Eq. (1). The relation $1 - þ - þ^2 = 0$gives two solutions that are $\left(\left(\sqrt{5}\right) - 1\right)/2$ and $\left(\left(\sqrt{5}\right) + 1\right)/2$.

where $\left(\left(\sqrt{5}\right) + 1\right)/2 > 1$ is not true by the bound value of the $0 \leq \gamma \geq 1$. Thus, the valid $\left(\left(\sqrt{5}\right) - 1\right)/2$ is the correct solution. This solution defines two types of sensor nodes with their initial energies $e_1$and e_2.

Level-3 heterogeneity: We have considered the upper and lower bound value of þis $0 \leq þ \geq 1$, but in level-2 heterogeneity, we have fixed the range value of the upper bond is $\left(\left(\sqrt{5}\right) - 1\right)/2$ denoted by $þ_{\text{ub}}$. Consider the lower bound of þbe $þ_{\text{lb}}$, i.e. which needs to be determined. The range of þis $þ_{\text{lb}} < þ < þ_{\text{ub}}$for three-level heterogeneity and consider function value f as $(e_3 - e_2)$ and þfrom Eq. (3). Thus, the lower bound is

$$þ_{\text{lb}} < þ < þ_{\text{ub}}$$

Put the values of þand $þ_{\text{ub}}$in the above equation and calculate the value of the $þ_{\text{lb}}$as follows:

$$þ_{\text{L}} < \frac{e_3 - e_1}{\beta * (e_3 - e_2)} < ((\sqrt{5}) - 1)/2 \tag{4}$$

Let $e_2 = \alpha_1 + e_1$ and $e_3 = \alpha_2 + e_2$, and using Eq. (4), we get the following relation.

$$þ_{\text{lb}} < \frac{\alpha_2 + \alpha_1}{\beta * \alpha_2}$$

$$\frac{\alpha_2}{\alpha_1} < \frac{1}{\beta * þ_{\text{lb}} - 1}$$

$$-\frac{\alpha_2}{\alpha_1} \geq \frac{1}{1 - \beta * \mathfrak{p}_{\text{lb}}} \tag{5}$$

Subsequently, L.H.S. of Eq. (5) is negative, thus, just put $-\frac{\alpha_2}{\alpha_1} = 0$. We have the following relation:

$$1 - \beta * \mathfrak{p}_{\text{lb}} < 0$$

$$\frac{1}{\beta} < \mathfrak{p}_{\text{lb}} \tag{6}$$

Equation (4) can be as

$$(e_3 - e_1) \leq \frac{\beta * \left(\left(\sqrt{5}\right) - 1\right)}{2} * (e_3 - e_2) \tag{7}$$

This inequality may be written as

$$\beta * \left(\left(\sqrt{5}\right) - 1\right) * e_2 - 2 * e_1 \leq \left(\beta * \left(\left(\sqrt{5}\right) - 1\right) - 2\right) * e_3 \tag{8}$$

The energy of the level-1, -2, and -3 nodes is given as e_1, $e_2 = e_1 * (1 + \omega)$, and $e_3 = e_1 * (1 + \eta)$, respectively. Consider the value of coefficients ω, and η are 0.06 and 0.11, respectively. Thus, the above-mentioned procedure defines the three-level heterogeneity model in WSNs.

Now, we deliberate the energy dissipation model which compute and analyse the energy depletion in signal transmission and receiving for L-bit message over distance d. The energy collapse for transmitting is quantified as follows [3]:

$$E_{\text{TXS}} = L * \in_{\text{elec}} + L * \in_{\text{fs}} * d^2 \quad \text{if } d \leq d_0 \tag{9}$$

$$E_{\text{TXL}} = L * \in_{\text{elec}} + L * \in_{\text{mp}} * d^4 \quad \text{if } d > d_0 \tag{10}$$

where $\in_{\mathbf{elec}}$, $\in_{fs}$ and $\in_{\text{mp}}$ are the energy decadent and d_0 is the threshold, respectively. The d_0 is defined as follows:

$$d_0 = \sqrt{\frac{\in_{\text{fs}}}{\in_{\text{mp}}}} \tag{11}$$

The energies expended in sensing (E_{Sx})and receiving (E_{Rx})are quantified in (12) and (13) as follows:

$$E_{Rx} = L * \in_{\text{elec}} \tag{12}$$

$$E_{Sx} = L* \in_{\text{elec}} \tag{13}$$

In the next section, we deliberate the procedure for optimizing the electing the CHs which helps in load balancing.

4 Optimized Cluster Head Election Routing Protocol

In this work, we have optimized the Wang et al. protocol [17] using a clustering algorithmic methodology for three-level heterogeneity. This methodology elects suitable cluster heads (CHs) based on the three criteria namely the nodes remaining energy, node distance, and total network energy. The selection of the cluster heads method is used as a dynamic clustering procedure. The aim of this method is to elect cluster heads that consume less energy in inter and intracluster communication. In the proposed work, a data collection process with data aggregation is discussed that helps in lowering energy consumption during communication. The complete process of the proposed method is distributed into rounds, and CHs are selected for each round. Initially, a defined percentage is used to selecting the cluster heads, and then, several cluster heads are added permitting to the coverage of the number of deployed nodes. It employes a condition of the vicinity of cluster heads. A cluster head may not possible in the vicinity of other selected cluster heads in this scenario. The decision of selecting the CHs depends on the threshold value of the level of the heterogeneity. This threshold value is associated with an arbitrary number which is generated between 0 and $e_{\text{average - nch}}/e_{i-\text{init}}$. If the threshold value is > the generated number between 0 and $e_{\text{average - nch}}/e_{i-\text{init}}$, then the sensor node of a particular level of heterogeneity converts the CH for the present round. The proposed method threshold formula is given as follows:

$$e_i = \frac{e_{i-\text{curr}}}{e_{i-\text{init}}} \tag{14}$$

$$e_t = \left(e_{i-\text{res}} + r\,\text{div}\frac{1}{p_i}\right) \times (1 - e_{i-\text{res}}) \tag{15}$$

$$T(n) = \begin{cases} \frac{p_i}{N - p_i\left[r \bmod \left(\frac{N}{p_i}\right)\right]} \times e_i \times e_t & \text{if } n \in G \\ 0 & \text{otherwise} \end{cases} \tag{16}$$

where $e_{i-\text{curr}}$, $e_{i-\text{init}}$, and $e_{i-\text{res}}$ are the current energy of the networks, nodes preliminary energy, and node residual energy. The N and r are the numbers of sensor nodes and rounds. The e_i and e_t are the constant values which are defined in Eqs. (14) and (15).

In the proposed method, a dynamic range is considered for the comparison with the threshold value for choosing the final cluster heads which are given as follows in

Eq. (2).

$$\left[0,\ e_{\text{average - nch}}/e_{i-\text{init}}\right] \tag{17}$$

where $e_{\text{average}-\text{nch}}$ and $e_{i-\text{init}}$ are the node's average energy which is not CHs in the current round and preliminary energy of the nodes, respectively.

The total energy of the defined network which consists of three-types of nodes, i.e. level-1, -2, and -3, is represented by e_T as given as follows:

$$þ \times N \times e_1 + þ^2 \times N \times e_2 + (1 - þ - þ^2) \times N \times e_3 \tag{18}$$

$$N \times (þ \times e_1 + þ^2 \times e_2 + (1 - þ - þ^2) \times e_3) \tag{19}$$

$$e_1 \times N \times (þ + þ^2 \times e_2/e_1 + (1 - þ - þ^2) \times e_3/e_1) \tag{20}$$

Now, we deliberate the clustering process of the projected method which contains three types of nodes. The e_1 is the preliminary energy of the level-1 nodes. If all the nodes having an initial energy e_1, then the total network's energy is $e_1 \times N$. Therefore, there is an increment factor of the energy according to Eq. (20) as $(þ + þ^2 \times e_2/e_1 + (1 - þ - þ^2) \times e_3/e_1)$. Means, heterogeneous nodes have 2 $(þ + þ^2 \times e_2/e_1 + (1 - þ - þ^2) \times e_3/e_1)$times more energy than the homogeneous nodes.

Generally, every sensor node becomes CH in case of homogeneous networks after $1/p_i$ rounds. Thus, the average cluster heads for homogeneous networks in a particular round will be $N \times 1/p_i$. But in the case of heterogeneous networks, every sensor node becomes CH after $1/p_i \times (þ + þ^2 \times e_2/e_1 + (1 - þ - þ^2) \times e_3/e_1)$rounds. Thus, the average cluster heads for heterogeneous networks in a particular round will be $N \times 1/p_i \times (þ + þ^2 \times e_2/e_1 + (1 - þ - þ^2) \times e_3/e_1)$.

The threshold values of level-1, -2 and -3 are fixed according to the initial energies of the respective levels. Thus, each sensor of level-1 becomes a CH once in every $1/p_{\text{opt}}$ rounds, and level-2 and -3 sensor nodes turn into a cluster head $(1 + \alpha)$ and $(1 + \beta)$ times more than that of the level-1 sensors in every $(þ + þ^2 \times e_2/e_1 + (1 - þ - þ^2) \times e_3/e_1)\ /p_i$rounds, respectively. Thus, heterogeneity is despoiled the set constraints which is $N \times 1/p_i$. The $E_0/(þ \times e_1 + þ^2 \times e_2 + (1 - þ - þ^2) \times e_3)$is the weight of a node in the defined networks.

In the proposed method, the weighted probabilities of the level-1, -2, and -3 nodes are denoted by $p_{\text{level}-1}$,$p_{\text{level}-2}$, and $p_{\text{level}-3}$, respectively, and defined as follows:

$$p_{level-1} = \frac{p_i \times E_1}{(þ \times e_1 + þ^2 \times e_2 + (1 - þ - þ^2) \times e_3)} \tag{21}$$

It can be as follows:

$$p_{level-1} = \frac{\mathrm{p_i}}{(\text{þ}+\text{þ}^2\times e_2/e_1+(1-\text{þ}-\text{þ}^2)\times e_3/e_1)} \tag{22}$$

$$p_{level-2} = \frac{\mathrm{p_i}\times \mathrm{E_1}}{(\text{þ}\times e_1+\text{þ}^2\times e_2+(1-\text{þ}-\text{þ}^2)\times e_3)} \tag{23}$$

It can be as follows:

$$p_{level-2} = \frac{\mathrm{p_i}}{(\text{þ}+\text{þ}^2\times e_2/e_1+(1-\text{þ}-\text{þ}^2)\times e_3/e_1)} \tag{24}$$

$$p_{level-3} = \frac{\mathrm{p_i}\times \mathrm{E_1}}{(\text{þ}\times e_1+\text{þ}^2\times e_2+(1-\text{þ}-\text{þ}^2)\times e_3)} \tag{25}$$

It can be as follows:

$$p_{level-3} = \frac{\mathrm{p_i}}{(\text{þ}+\text{þ}^2\times e_2/e_1+(1-\text{þ}-\text{þ}^2)\times e_3/e_1)} \tag{26}$$

The $p_{\text{level}-1}$, $p_{\text{level}-2}$, and $p_{\text{level}-3}$ are the optimum weighted probabilities for the level-1, -2, and -3 nodes. Thus, the threshold value of the level-1 is given as follows:

$$T(n_{\text{level}-1}) = \begin{cases} \frac{p_{\text{level}-1}}{N - p_{\text{level}-1}\left[r \bmod \left(\frac{N}{p_{\text{level}-1}}\right)\right]} \times e_i \times e_t & \text{if } n_{\text{level}-1} \in G' \\ 0 & \text{otherwise} \end{cases} \tag{27}$$

$$T(n_{\text{level}-2}) = \begin{cases} \frac{p_{\text{level}-2}}{N - p_{\text{level}-2}\left[r \bmod \left(\frac{N}{p_{\text{level}-2}}\right)\right]} \times e_i \times e_t & \text{if } n_{\text{level}-1} \in G'' \\ 0 & \text{otherwise} \end{cases} \tag{28}$$

$$T(n_{\text{level}-3}) = \begin{cases} \frac{p_{\text{level}-3}}{N - p_{\text{level}-3}\left[r \bmod \left(\frac{N}{p_{\text{level}-3}}\right)\right]} \times e_i \times e_t & \text{if } n_{\text{level}-1} \in G''' \\ 0 & \text{otherwise} \end{cases} \tag{29}$$

where G', G'', and G''' are the conventional of level-1, -2 and -3 nodes which have not become CHs within last $\frac{1}{p_{\text{level}-1}}$, $\frac{1}{p_{\text{level}-2}}$, and $\frac{1}{p_{\text{level}-3}}$ rounds, respectively. The $T(n_{\text{level}-1})$, $T(n_{\text{level}-2})$, and $T(n_{\text{level}-3})$ are the threshold applied to type-1, -2 and -3 nodes, respectively. Thus, deployed sensor nodes converted into dynamic clusters by using their probabilities and thresholds which helps in lifespan prolonging.

5 Data Collection with Aggregation Process at Cluster Heads

In this section, a data aggregation method is discussed which can be applied during data collection from the monitoring filed. In this process, the duplicate or redundant data which is collected by the cluster heads will remove at the cluster heads by performing simple computations. This process only forwards the identical data to the next cluster heads, nodes, or sinks which will be the next nodes in the chaining approach. Let N be the number of sensor nodes in a specific cluster (α) that products the information packets $\alpha_{p1}, \alpha_{p2}, \alpha_{p3}, \ldots, \alpha_{p\delta}$. The D_{A1}, D_{A2}, and D_{A3} are signified the normal, sum, and sum of the accumulated data messages, respectively. The comprehensive impression of the information accumulation of the packets is given below:

Input: $Set\ of\ Clusters\ C = \{\alpha, \beta, \gamma \ldots.\}$ and BS
Output: $Aggregate\ data\ packets\ at\ CH\ or\ BS$
Begin
for every cluster $C = \{\alpha, \beta, \gamma \ldots.\}$
if $(\alpha_{p1} = \alpha_{p2} = \alpha_{p3} = \cdots = \alpha_{p\delta})$
// *all the sensors generated the exact same data packets*
$$D_{A1} = \left\{\left(\frac{\alpha_{p1}}{2^{q-1}}\right) + \left(\frac{\alpha_{p2}}{2^{q-1}}\right) + \left(\frac{\alpha_{p3}}{2^{q-2}}\right) + \left(\frac{\alpha_{p4}}{2^{q-3}}\right) + \cdots + \left(\frac{\alpha_{p\delta}}{2}\right)\right\}$$
// *q is number of nodes generated same data packets*
else if $(\beta_{p1} \neq \beta_{p2} \neq \beta_{p3} \neq \cdots \neq \beta_{p\delta})$
// *all the sensors generated the different data packets*
$D_{A2} = \beta_{p1} + \beta_{p2} + \beta_{p3} + \cdots + \beta_{p\delta}$ // *sum of the data packets*
else if $(\gamma_{p1} = \gamma_{p4} = \cdots = \gamma_{p\delta-1}) \neq \gamma_{p2} \neq \gamma_{p3} \neq \gamma_{p\delta}$
// *some sensors generated the exact same data packets and some sensor generated the different data packets*
$D_{A3} = D_{A1} + D_{A2}$
$$D_{A3} = \left\{\left(\frac{\gamma_{p1}}{2^{q-1}}\right) + \left(\frac{\gamma_{p4}}{2^{q-1}}\right) + \left(\frac{\gamma_{p5}}{2^{q-2}}\right) + \left(\frac{\gamma_{p6}}{2^{q-3}}\right) + \cdots + \left(\frac{\gamma_{p\delta-1}}{2}\right)\right\} + \{(\gamma_{p2} + \gamma_{p3} + \gamma_{p\delta})\}$$
end if
end for
End

6 Simulation Outcomes and Discussions

In this section, the proposed method outcomes are analysed with the existing Wang et al. [17] and heterogeneous Wang et al. methods by considering in terms of first, half, and last node dead, sum of energy consumption as the remaining energy, and several messages transferred to the control node as displaying the throughput of the proposed method matrices. Wang et al. is a new chaining methodology that diminishes power depletion for homogeneous networks [17]. First of all, we have implemented three levels of heterogeneity in the Wang et al. method called Hetero-Wang et al. method. The projected method has also applied for three-level of heterogeneity as well as chain-based data aggregation process is discussed. The proposed network is considered 100 numbers of sensor nodes, sink at the middle of the area, initial energy of the normal 0.5 Jules and outcomes are simulated using MATLAB. The proposed network is considered 3 levels of heterogeneity by using 3 types of nodes namely normal, advanced, and super as are 50, 30, and 20 in numbers, respectively, and their energies are 0.5 J, 1.0 J and 1.75 J, respectively. The power consumption is 50 nJ/bits in running the circuit, 10 pJ/bits/m^2 to transmit the signal in the shorter distance, 0.0013 pJ/bits/m^{4}to transmit signal in the longer distance. The energy consumption model also considered the packet length is 4000 bits, cluster radius is 25 m, and the threshold distance is 75 m. We have commonly used 25 simulations and taken an average of all the simulations for calculating the final simulation outcomes. The analysis of the outcome of the Wang et al. [17], Hetero-Wang et al., and the proposed method is deliberated by considering several active and dead, the sum of energy consumption, and a number of the message transferred to the control node matrices are considered to investigate the enactment of the proposed scheme.

Figure 1 shows the simulation outcomes of the projected technique and the Wang et al. [17], Hetero-Wang et al., existing methods in terms of alive nodes with reference of several rounds. The proposed technique covers 3530 number of rounds before going to die every node in the deployed networks, whereas the Wang et al. [17] and Hetero-Wang et al. existing methods cover 2748, and 2272, number of rounds, respectively. The network lifetime increment in Hetero-Wang et al. and the projected technique is 19.07%, and 24.37%, respectively, in comparison with the Wang et al. [17] method. Thus, it is manifested from the simulation outcomes that the projected technique gives better outcomes in respect of the existing techniques because the choice of CHs is efficiently due to the selected parameters. Furthermore, it decreases

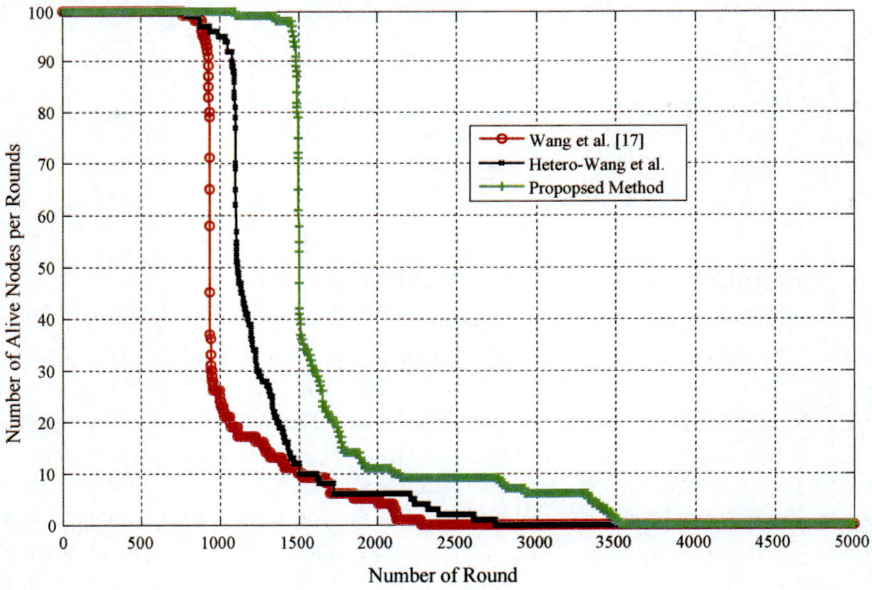

Fig. 1 Analysis of Wang et al. [17], Hetero-Wang et al., and proposed technique in the view of alive nodes versus rounds

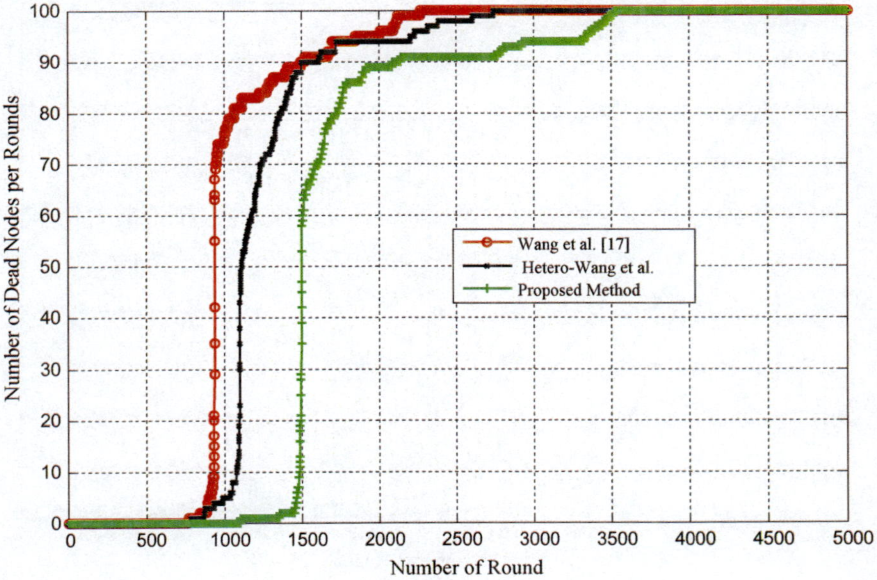

Fig. 2 Analysis of Wang et al. [17], Hetero-Wang et al., and proposed technique in the view of dead nodes versus rounds

the communication cost in the data collection process and increases the lifetime of the networks. The proposed threshold-based formula helps in extending the lifetime by using different energy factors. The sustainability period of the Hetero-Wang et al. and the projected technique is 20.95%, and 28.45%, respectively which is calculated by considering the first node dead information concerning Wang et al. [17]. Figure 2 shows the simulation outcomes of the proposed method and the Wang et al. [17], Hetero-Wang et al., existing methods in terms of dead nodes with reference of several rounds. The half node dead period of the Hetero-Wang et al. and the projected technique is 18.67% and 35.25%, respectively, which is computed by considering the three-level of heterogeneity concerning Wang et al. [17].

Figure 3 shows the imitation outcomes of the total energy dissipation concerning the number of rounds for three-level of heterogeneity. The initial total energy of the three-level of heterogeneity network is 50 J. The proposed method is performing well than that of the Wang et al. [17], Hetero-Wang et al. existing methods because the proposed method sensing data for a longer period and conserves very less energy during the data collection from the cluster heads and cluster members.

It is also decreasing the cost of communication in a very effective manner. The Wang et al. [17], Hetero-Wang et al. existing methods did not perform any data aggregation process at the sensor or cluster heads during data collection and communication using chaining approach.

Figure 4 indicating the normalized average energy dissipation regarding the number of rounds of the Wang et al. [17], Hetero-Wang et al., and the proposed

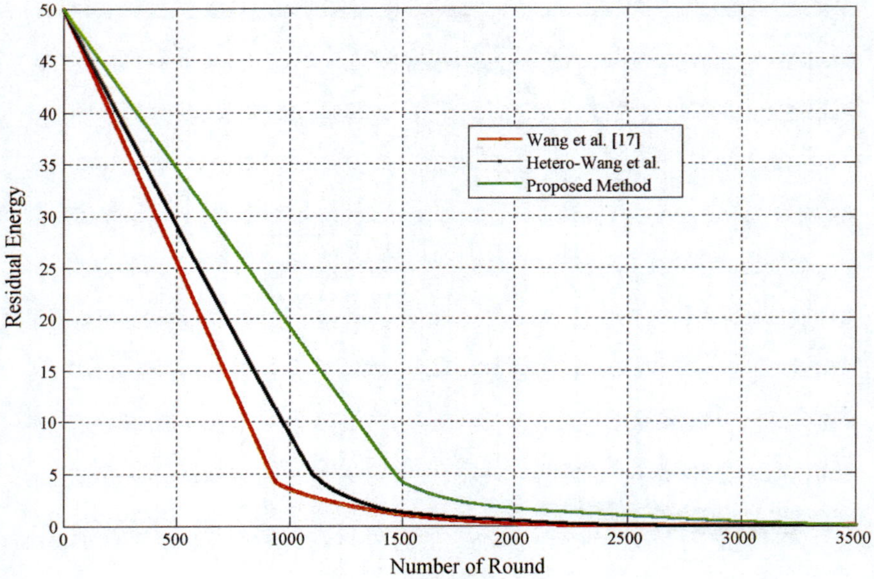

Fig. 3 Analysis of Wang et al. [17], Hetero-Wang et al., and proposed technique in the view of energy depletion versus rounds

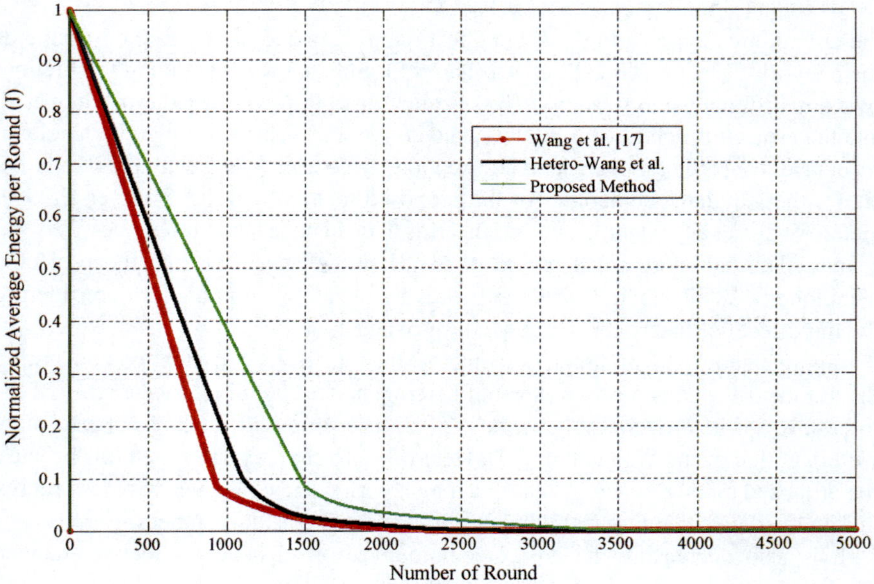

Fig. 4 Analysis of Wang et al. [17], Hetero-Wang et al., and proposed technique in the view of normalized average energy versus rounds

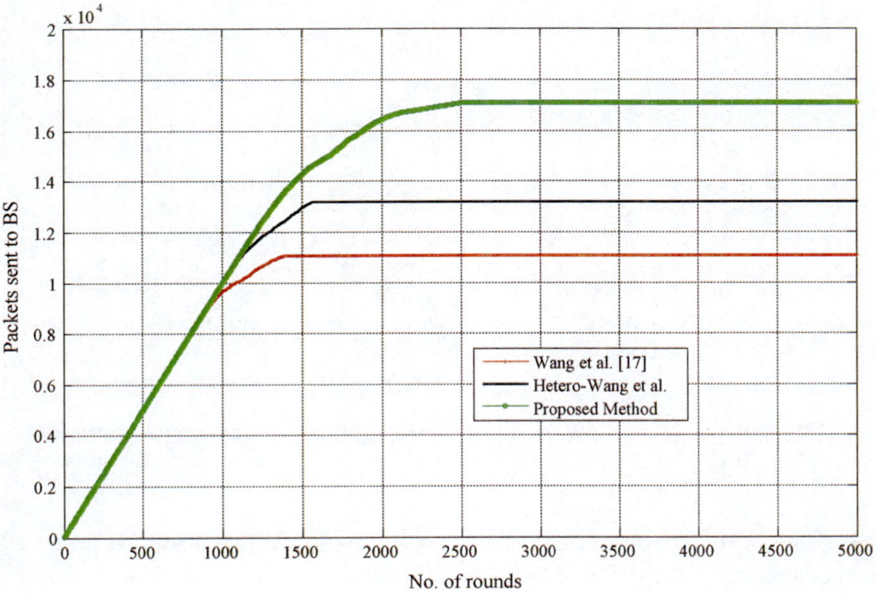

Fig. 5 Analysis of Wang et al. [17], Hetero-Wang et al., and proposed technique in the view of packets sent to BS versus rounds

method. Here, the proposed method has a more sustainable period than that of the Wang et al. [17], and Hetero-Wang et al. mean it collects data from more number of rounds. Thus, this data collection process saves energy while collecting the data from the environment and reducing communications cost networks.

The simulation outcomes of the quantity of packets sent to the BS regarding the number of rounds are shown in Fig. 5 for Wang et al. [17] and Hetero-Wang et al., existing methods and the proposed method using three-level of heterogeneity. The proposed method, Hetero-Wang et al., and Wang et al. [17] are 1.72×10^{-4}, 1.32×10^{-4}, 1.10×10^{-4}, the quantity of packets to the sink, respectively. It is evident from the outcomes that the proposed method can transmit more quantity of packets to the sink as a comparison to the Hetero-Wang et al., and Wang et al. [17] existing methods. The more quantity of packets sent by the proposed method because of the alive time of the deployed nodes are more concerning other methods like Hetero-Wang et al. and Wang et al. [17].

Figure 6 shows the qualified investigation in terms of last node dead (LND), half node dead (HND), and first node dead (FND) of the system lifespan for Wang et al. [17], Hetero-Wang et al., and proposed method for three-level heterogeneous networks. Figure 4 shows the ecological period (also called the FND timing) of the Hetero-Wang et al., and proposed method outstrips by 20.95%, and 28.45%, as a comparison with Wang et al. [17] significantly, respectively. Moreover, the half node dead (HND) of the Hetero-Wang et al. and proposed method outperforms by 18.67%, and 35.25%, as a comparison with Wang et al. [17] significantly, respectively.

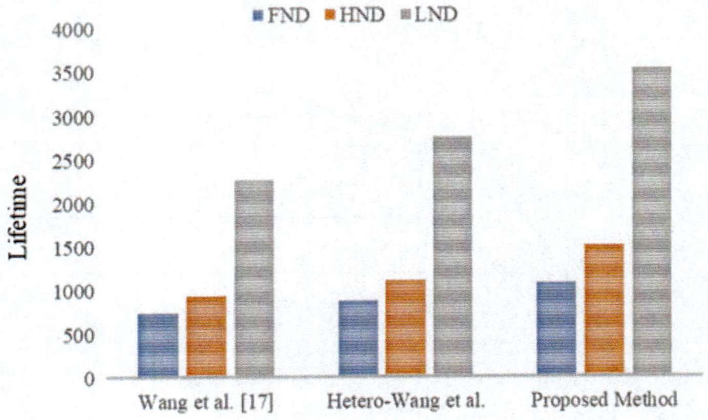

Fig. 6 Comparative analysis of Wang et al. [17], Hetero-Wang et al., and proposed technique in terms of FND, HND, and LND in terms of network lifespan for three-level heterogeneous networks

Furthermore, the last node dead (LND) of the Hetero-Wang et al. and proposed method outperforms by 19.07% and 24.37%, as a comparison with Wang et al. [17] significantly, respectively.

Table 1 shows the percentage an increase in the lifetime of the network, throughput, energy dissipation, the lifespan of the networks for Wang et al. [17], Hetero-Wang et al., and proposed a technique for three-level heterogeneous networks. The number of packets sent to BS for the projected technique, Hetero-Wang et al. and Wang et al. [17] is 1.72×10^{-4}, 1.32×10^{-4}, and 1.10×10^{-4}, respectively. This method transferred the more number of packets than that of the existing methods, and the increment in the stability of the proposed method and the Hetero-Wang et al. is 28.45% and 20.95% than that of the Wang et al. [17], respectively, for three-level heterogeneous networks.

Table 1 Comparative analysis of the proposed technique, Wang et al. [17], and Hetero-Wang et al. in terms of the network lifetime, throughput, energy dissipation, the lifespan of the networks for three-level HWSNs

Protocols	Network lifetime			Energy consumption (J)	Throughput	% Increment in network lifetime
	FND	HND	LND			
Wang et al. [17]	734	937	2272	50	1.10×10^{-4}	–
Hetero-Wang et al.	874	1112	2748	50	1.32×10^{-4}	20.95%
Proposed method	1087	1504	3530	50	1.72×10^{-4}	28.45

7 Conclusion

In this paper, a data aggregation-based effective and optimized CH election routing protocol for HWSNs is recommended. It considered homogeneous and heterogeneous networks for the proposed technique, Wang et al. [17], and Hetero-Wang et al. method. This method used a threshold-based dynamic clustering technique which efficiently elects the cluster heads. This method also proposed a new data collection method which aggregates the data at the cluster heads or the sensor nodes at the time of information gathering. The proposed scheme prolonged the sustainability period because of the collection of nodes for cluster heads in an efficient manner which has higher residual energy, node distance, and total network energy. As shown in the simulation outcomes, the lifetime is increased by 19.07% and 24.37% for 50 J energy of the network in the situation of Hetero-Wang et al. and projected technique concerning Wang et al. [17], respectively. The throughput for the projected method, Hetero-Wang et al. and Wang et al. [17] is 1.72×10^{-4}, 1.32×10^{-4}, and 1.10×10^{-4}, respectively. This projected technique achieves superlative among the Wang et al. [17] and Hetero-Wang et al. method.

References

1. Singh, S., Chand, S., Kumar, B.: Performance investigation of heterogeneous algorithms in WSNs. In: 3rd IEEE International Advance Computing Conference (IACC), pp. 1051–1054 (2013)
2. Singh, Y., Singh, S., Kumar, R.: A distributed energy-efficient target tracking protocol for three level heterogeneous sensor networks. Int. J. Comput. Appl. **51**, 31–36 (2012)
3. Heinzelman, W.R., Chandrakasan, A.P., Balakrishnan, H.: An application-specific protocol architecture for wireless microsensor networks. IEEE Trans. Wirel. Commun. **1**, 660–670 (2002)
4. Lindsey, S., Raghavendra, C.S., Sivalingam, K.M.: Data gathering algorithms in sensor networks using energy metrics. IEEE Trans. Parallel Distrib. Syst. **13**, 924–935 (2002)
5. Singh, S., Malik, A.: hetSEP: heterogeneous SEP protocol for increasing lifetime in WSNs. J. Inf. Optim. Sci. **38**, 721–743 (2017)
6. Qing, L., Zhu, Q., Wang, M.: Design of a distributed energy-efficient clustering algorithm for heterogeneous wireless sensor networks. Comput. Comms. **29**, 2230–2237 (2016)
7. Singh, S., Malik, A., Kumar, R.: Energy efficient heterogeneous DEEC protocol for enhancing lifetime in WSNs. Eng. Sci. Technol., Int. J. **20**, 345–353 (2017)
8. Maheswari, D.U., Sudha, S.: Node degree based energy efficient two-level clustering for wireless sensor networks. Wirel. Pers. Commun. **104**, 1209–1225 (2018)
9. Chand, S., Singh, S., Kumar, B.: Heterogeneous HEED protocol for wireless sensor networks. Wirel. Pers. Commun. **77**, 2117–2139 (2014)
10. Singh, S., Chand, S., Kumar, B.: Energy efficient clustering protocol using fuzzy logic for heterogeneous WSNs. Wirel. Pers. Commun. **86**, 451–475 (2016)
11. Singh, S., Chand, S., Kumar, B.: Multilevel heterogeneous network model for wireless sensor networks. Telecommun. Syst. **64**, 259–277 (2017)
12. Singh, S., Chand, S., Kumar, B.: An energy efficient clustering protocol with fuzzy logic for WSNs. In: 5th International Conference-Confluence the Next Generation Information Technology Summit, pp. 427–431 (2014)

13. Faisal, S., Javaid, N., Javaid, A., Khan, M.A., Bouk, S.H., Khan, Z.A.: Z-SEP: zonal-stable election protocol for wireless sensor networks. J. Basic Appl. Sci. Res. **3**(5), 132–139 (2013)
14. Khan, F.A., Khan, M., Asif, M., Khalid, A., Haq, I.U.: Hybrid and multi-hop advanced zonal-stable election protocol for wireless sensor networks. IEEE Access **7**, 25334–25346 (2019)
15. Smaragdakis, G., Matta, I., Bestavros, A.: SEP: A Stable Election Protocol for Clustered Heterogeneous Wireless Sensor Networks. Technical Report BUCS-TR-2004-022, Boston University Computer Science Department. pp. 1–11 (2004)
16. Tang, F., You, I., Guo, S., Guo. M., Ma., Y.: A chain-cluster based routing algorithm for wireless sensor networks. J. Intell, Manuf. **23**, 1305 (2012)
17. Wang, Z., Zhang, M., Gao, X., et al.: A clustering WSN routing protocol based on node energy and multipath. Cluster Comput. **22**, 5811–5823 (2019)
18. Chen, K.H., Huang, J.M., Hsiao, C.C.: Chiron: An energy efficient chain-based hierarchical routing protocol in wireless sensor networks. In: Proceeding of IEEE Symposium on Wireless Telecommunications (WTS-2009), Prague, pp. 1–5 (2009)
19. Linping, W., Wu, B., Zhen, C., Zufeng, W.: Improved algorithm of PEGASIS protocol introducing double cluster heads in wireless sensor network. In: IEEE International Conference on Computer, Mechatronics, Control and Electronic Engineering, pp. 148–151 (2010)
20. Singh, P., Paprzycki, M., Bhargava B., Chhabra J., Kaushal N., Kumar Y. (Eds.): Futuristic trends in network and communication technologies. In: FTNCT 2018. Communications in Computer and Information Science, Vol. 958. Springer, Singapore (2018)

Image Processing and Computer Vision

A Hybrid Approach with Intrinsic Feature-Based Android Malware Detection Using LDA and Machine Learning

Bilal Ahmad Mantoo

Abstract World is becoming small with the increase in the number of mobile phone users. The most influential and having huge market among mobile phones is android. Android is a software used in nowadays smart phones, which not only consists of operating system but also myriad number of key applications. These applications make large number of day to day tasks easy. There are millions of android applications in the market with over 3 billion or more downloads. The growing market of this platform not only invites smart phone users, but it also becomes a point of interest for black hat hackers. Hackers use this technology for large number of activities by spreading the android applications in this platform which are not actually android packages rather malicious codes or malware. Therefore, these malwares must be handled in a smart way; otherwise, they lead to huge loss. Different techniques have been used for detection of android malware which consists of network traffic analysis, static analysis, and dynamic analysis. In this paper, a combined approach of static, dynamic, and intrinsic features for android malware detection using k-nearest neighbor (k-NN), random forest, decision tree, SVM, and ensemble learning techniques. The calculation uses a publicly available dataset of Androtrack. The estimation results shows that both the decision tree and random forest classifiers produced accuracy of 99%. With the help of newly added feature and a different approach of preprocessing, i.e., linear discriminant analysis.

Keywords Dynamic analysis · Static analysis · Intrinsic features

1 Introduction

Malware or malicious software is defined as software intended to misrepresent and interrupt the mobile or computer applications, gather important information, and hence carry out malicious operations. These malicious operations consist of gaining access to secret information; stealthily steal the valuable information from the system,

B. A. Mantoo (✉)
Central University of Punjab, Bathinda 151001, India
e-mail: bilalbashir136@gmail.com

P. K. Singh et al. (eds.), *Recent Innovations in Computing*, Lecture Notes in Electrical Engineering 701, https://doi.org/10.1007/978-981-15-8297-4_24

show undesirable advertisement, and spy on the activities of the users [1]. Malware poses a huge threat to every technological field like cyber security of national infrastructure, service areas, commercial sectors, etc. Mobile devices like smart phones, tablets are performing almost all the tasks that a normal computer could perform. Android uses many open sources softwares, such as Torrents, play store, or directly downloads from third party markets, etc. [2]. Organizations like bank, intelligence, defense use applications that are implemented on this device. Confidential information is easily accessible via these devices that are fetched from the cloud.There is an increase in the number of malwares attacking these devices reported by AV-TEST.

2 Related Work

Plenty of studies have been voted for out on the detection of malwares. Some of the research paid responsiveness to the analysis of the evolution of malware ecosystem. Some paid a keen interest toward the behavioral, taint, and static analysis.In this section, we are going to brief about the approaches used in malware detection in android platform in various researches; first, we discuss about the static approaches, and dynamic approaches will be overviewed later.For detecting malware, various approaches have been proposed but the most important among these are static analysis-based approaches and dynamic-based analysis approaches.

2.1 Static Analysis-Based Approaches

There are various static-based approaches proposed so for detecting android malware using android manifest file. One of the research performs the static analysis on 3258 android applications. Extract permissions from all these applications and apply various machine learning algorithms. Naive Bayes classifier was found to be the best classification model among the classifiers applied [3]. Androtrack based on static analysis uses serial number from the certificates issued from certificate authority as a feature. This approach mainly checks the suspicious behavior of SMS hiding detects the malicious code commands in the code and analyzes the permission request [4]. Uses permissions and intents as features from different downloaded applications the feature set created was reduced using information gain algorithm. This dataset was given to different algorithms J48 and ID 3 classifier, and the result obtained shown that J48 algorithm gives the accuracy of 94% [5].

Machine learning technique uses permissions, intents, and API function calls as static features for malware detection [6].These features are combined, and then different algorithms have been applied including support vector machine, random forest, and neural networks. Results obtained from the experiments show the greater abilities for detecting android malware. Another simple low-cost approach for malware detection using manifest files. The results obtained show the method traces unknown behavior of malwares [7]. Randroid, in this approach permission, API's, and other key features using machine learning automatically detect malware and normal android applications. Various machine learning algorithms have been implemented like Support Vector Machine (SVM), Naïve Bayes (NB), Random Forest (RF), and Decision Tree (DT) to perform malware detection [8].

2.2 *Dynamic Analysis-Based Approaches*

Various dynamic analysis-based approaches for malware detection uses run time behavior of android applications. A novel-based approach for malware detection using the framework of client–server model for malware detection on server side. Client side watches on the applications, and the server side looks at the traces by means of suffix algorithm [9]. DroidScope provides the semantic view of detecting android malware by monitoring system calls in addition to process threads, monitoring kernel semantic changes [10].Another approach that creates a dataset from system call and applies three machine learning algorithms. Experiment shows the system call dataset and shows the better accuracy because of the implementation of Chi Square filtering [11].

Android malware detection using system call co-occurrence matrix and the results obtained from the experiments show that the co-occurrence matrix extracts more useful information than simple system call vector. Co-occurrence matrix shows the positive impact on TPR and negative impact on FPR [12].The system call frequency based on normalization which improves the accuracy of android malware detection. Results show that the system call frequency of some applications varies hugely. Some of them range between 10 K-20 K. However, some applications system calls frequency value reaches 200 K or even more [13].

3 Data Description

The dataset consists of 1552 applications which are obtained from Androtrack,[1] an open platform for malware and genuine applications are downloaded from Google Play store during the year 2018. For every application, we have extracted not only intrinsic features like size, versionCode, number of applications present in manifest file but their dynamic as well as static features. The picture of the suggested approach is given Fig. 1.

[1] www.malgenomeproject.org.

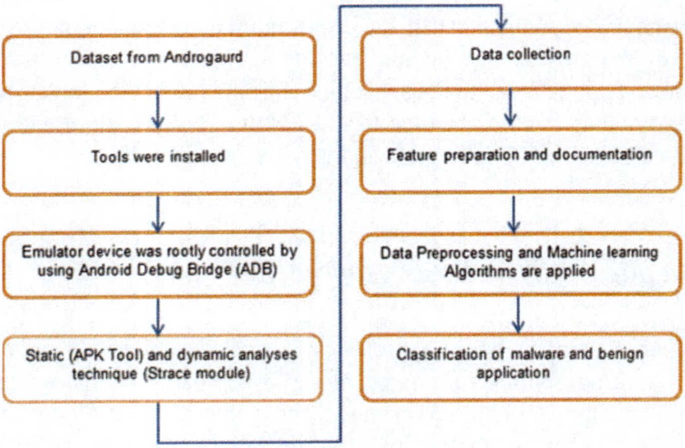

Fig. 1 Diagramatic view of malware detection

3.1 Static and Dynamic Application Features

The dataset consist of 20 dynamic features of different benign and malware Android applications. These attributes are extracted using variety of softwares that are present, like Gennymotion by installing each android apk file in the environment of Genny motion studio using *strace* command. These system calls are placed in an excel file for later processing as shown in Table 1. The dataset consists of 7 static features of 1552 applications containing equal sizes of individual classes. The static features are extracted using APK tool using command *apktool-d* [apk name] which decompiles the application and extracts the manifest file. The permissions present in manifest file are recorded in the same file Table 2.

3.2 Intrinsic Application Features

Intrinsic features consist of a brief information of android application, which includes size of an application, versionCode, and number of applications in manifest file shown in Table 3. These features are extracted using the APK tool.

4 Data Preprocessing

Preprocessing is a data mining approach that engulfs changing or converting raw data into desirable format, so that our machine learning algorithm works in better manner. Good results from machine learning algorithm can only be achieved by removing

Table 1 Dynamic features used in this work

S. No.	Feature	Type
01	Close	Int64
02	Read	Int64
03	Get time of day	Int64
04	Futex	Int64
05	Clock get time	Int64
06	Mprotect	Int64
07	Epoll_pwait	Int64
08	Receive from	Int64
09	Send to	Int64
10	Ioctl	Int64
11	Write	Int64
12	Pread64	Int64
13	Clone	Int64
14	Writev	Int64
15	Getsockopt	Int64
16	Getuid32	Int64
17	openat	Int64
18	socket	Int64
19	dup	Int64
20	prctl	Int64

Table 2 Static features used in this work

S. No.	Feature	Type
01	SMS	Int64
02	Phone	Int64
03	Storage	Int64
04	Contacts	Int64
05	Location	Int64
06	Camera	Int64
07	Microphone	Int64

Table 3 Intrinsic features used in this work

S. No.	Feature	Type
01	Size	Int64
02	versionCode	Int64
03	No. of files in manifest file	Int64

inconsistant, incomplete, null values from data, and this is done by using preprocessing approach. In data preprocessing, data goes via large number of steps like data cleaning which includes filling null values, data integration which includes that conflicting data should be combined, data transformation where data is normalized.

4.1 *Linear Discriminant Analysis*

Linear discriminant analysis is considered as an important approach of data preprocessing for pattern classification scenarios. The method works by plotting the higher dimensional data space into lower dimensional space with a greedy approach of finding best class of separable attributes that dips the computational cost. The aim of LDA is to classify objects into number of categories based on certain rules. With linear discriminant analysis, the approach looks for.

1. Features that find the best relationships between classes or objects.
2. Second, it looks for which is the best classification model that classifies the objects.

In some cases, classes might have number of features, using a single feature may overlap the result, so the requirement is to increase the number of features to avoid merging; hence, LDA best describes this principle.

5 Methodology and Data Analysis

5.1 *Initial Approach*

Feature selection is considered as the best and key aspect of increasing the performance of any machine learning algorithm. The most important features present in the dynamic analysis are the system calls which are present in every application shown in Fig. 2, and in case of static features, permissions which are considered as dangerous have been included.

5.2 *Feature Analysis*

From the analysis, the most important system calls that can classify into genuine and malware application are shown in Fig. 2. System calls are stored as per their intensity in an excel file to use them later in the process. As shown in Fig. 3, the frequency of features touches the level of 1.5 lakh in RCV from system call. The static features usually consist of permissions which are also stored in the same file. The permissions are stored as binary values "0" for the absence and "1" for the presence. The analysis shows that the frequency of malware applications touches to the highest levels.

The static features usually consist of permissions which are also stored in the same file. Permissions which are present in manifest file are stored either as "0" for the absence or "1" for the presence. The analysis shows that the frequency of applications touches to the highest levels. From the analysis, the most of the permissions that can classified into genuine permissions which are present in most of the android

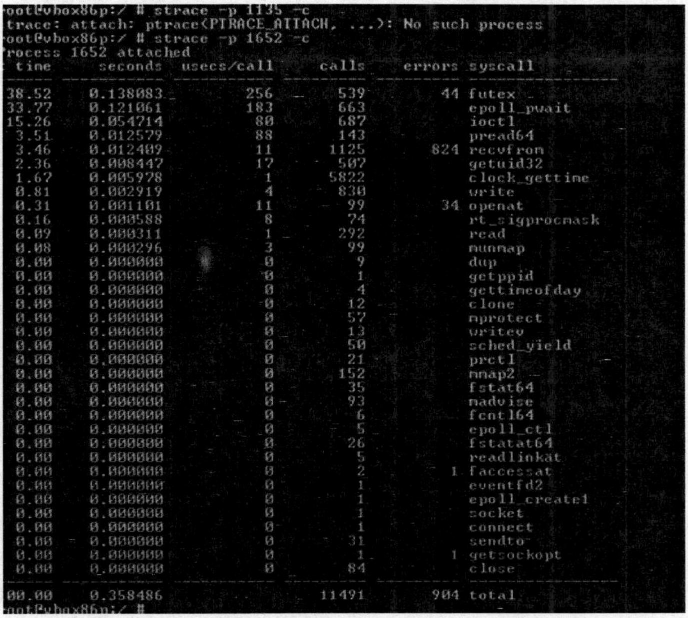

Fig. 2 Dynamic features of android benign application

application are shown in Fig. 4. The plot shows that malware applications request more number of permissions in comparison to benign applications.

Figure 5 shows the histogram of one of the system call Ioctl, in which most of the system calls lies in between 0 and 4 K. The analysis of a newly introduced feature size shows that the malware application size is very less as compared to the benign application as shown in Fig. 6, the size of benign applications goes till 90,000 KB, and most of them lies up to 30000 KB. Figure 7 shows the application of malware size that lies below 5000 KB.

6 Experiments and Results

Once the dataset is finalized, a variety of machine learning algorithms are used to evaluate the result. Initially, the algorithm was checked on logistic regression, and then, k-nearest neighbor where the value of "k" was taken as 5 by using a simple python script which emanates the value of ‘k’ graphically as shown in Fig. 8.

The dataset is also examined with decision tree irrespective of other machine learning algorithms which give the accuracy of 99%. In our problem, what decision tree is that it segregates the data which are malware and benign based on the significance of input variables among the features; the significance here is determined on the basis of frequencies of system calls, i.e., the features having more frequency of

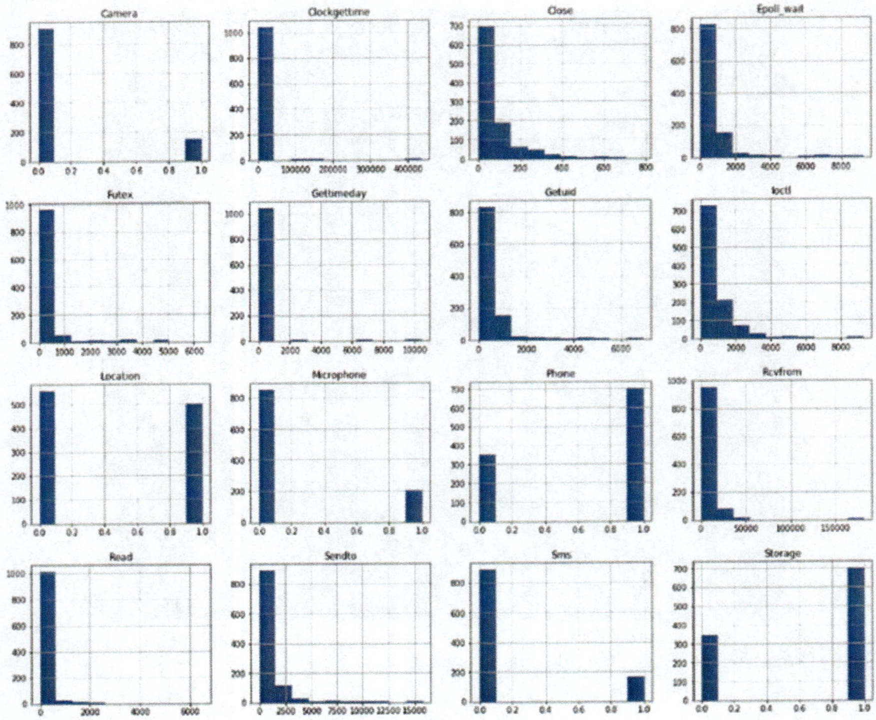

Fig. 3 Frequency of features

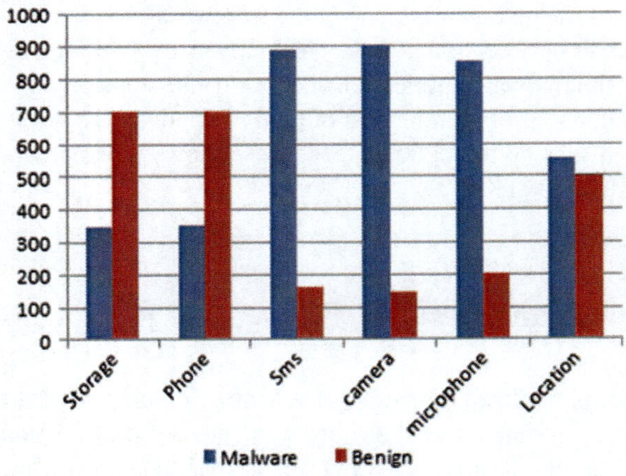

Fig. 4 Permissions present in malware and benign application

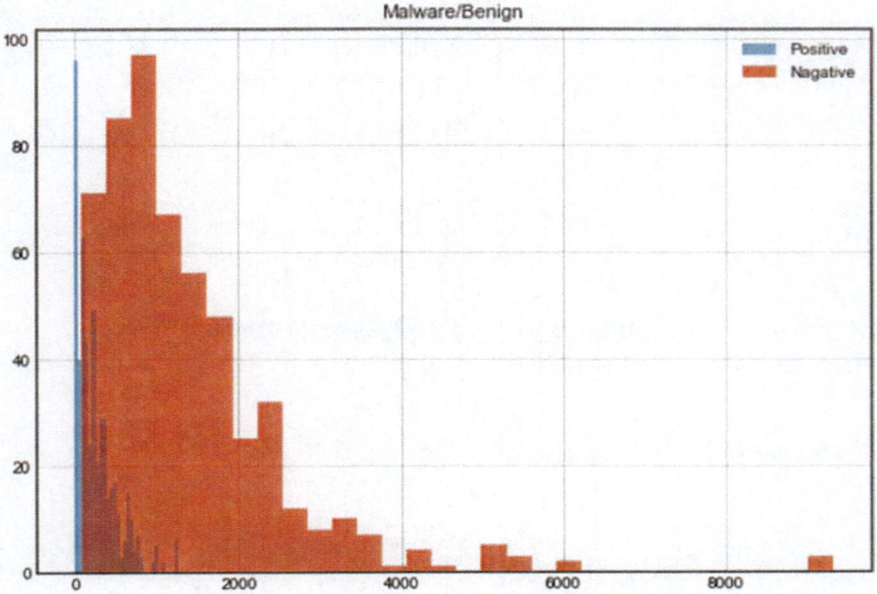

Fig. 5 Histogram of feature Ioctl present in malware and benign

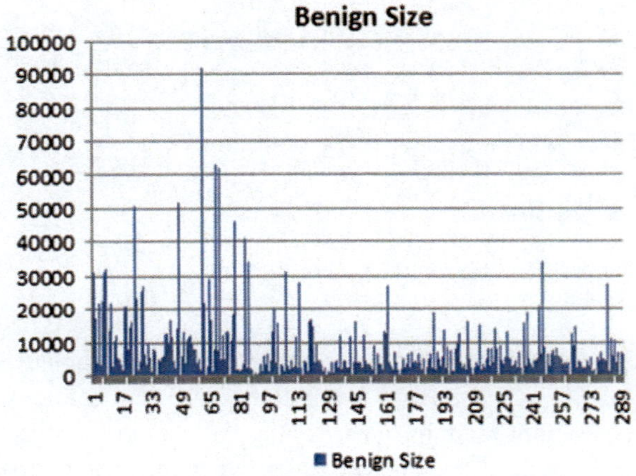

Fig. 6 Plot of the feature size used in benign applications

system calls is divide into one homogenous class which are heterogeneous to other sub tree and the features with lower frequencies into another sub tree.

As the decision trees are prone to over fitting as the tree gets deeper based on different outcomes, but in our case the pruning toward the over fitting is slightly less because of the only two cases. Coming toward random forest algorithm which

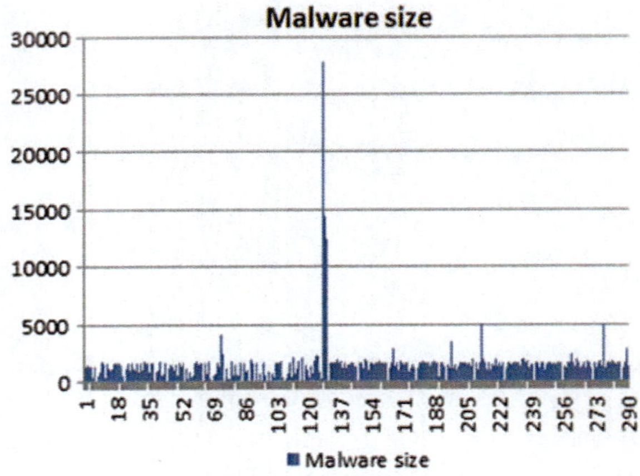

Fig. 7 Plot of the feature size used in malware applications

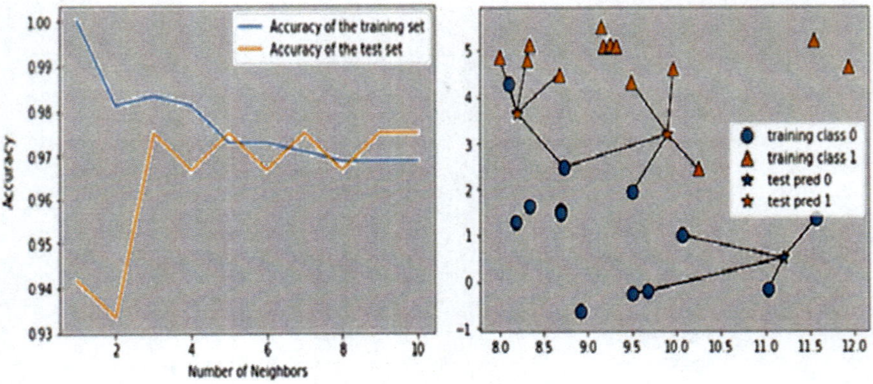

Fig. 8 Checking for the best value for k neighbors

overcomes the pruning drawback of decision tree and excels the decision tree algorithm by 0.4% in case, where LDA is not used but intrinsic feature was introduced; the increased accuracy of random forest is only the creation of different averaging multiple deep decision trees without actually going to the deep to avoid over fitting. Out of all, the algorithms applied SVM show the least values of accuracy 82.3% after k-NN and logistic regression, which is obvious as according to its behavior. SVM works fine with the data which is small and requires less to be trained. Due to the large dataset of 1552 applications, SVM shows the less accuracy, but its accuracy increases on applying LDA which reduces the dimensions of the dataset and hence gives the accuracy of 92.7%. The outcome of all other algorithms is also checked by varying the preprocessing approach, i.e., with LDA, but the results obtained where not as expected. Further, the effect of intrinsic feature size helps in increasing the accuracy of the algorithms shown in Table 4. As shown in Table 4, the accuracy of

Table 4 Static features used in this work

S. No.	Accuracy	LDA used	Feature size included
Logistic regression	97.0	Yes	Yes
Logistic regression	96.7	yes	No
Logistic regression	97.5	No	Yes
Logistic regression	95.3	No	No
k-NN	97.0	Yes	Yes
k-NN	96.8	Yes	No
k-NN	97.5	No	Yes
k-NN	95.3	No	No
Decision tree	91.5	Yes	Yes
Decision tree	91.0	Yes	No
Decision tree	99.1	No	Yes
Decision tree	97.1	No	No
Random forest	91.5	Yes	Yes
Random forest	91.7	Yes	No
Random forest	99.5	No	Yes
Random forest	98.4	No	No
SVM	92.7	Yes	Yes
SVM	89.9	Yes	No
SVM	88.5	No	Yes
SVM	82.3	No	No

the algorithms is improving when we are including the feature size and reduces a bit on its removal.

7 Conclusion and Future Scope

In this paper, different machine learning algorithms, like k-NN, logistic regression, decision tree, random forest, and SVM, etc., have been implemented over a dataset of 1552 android applications to identify malicious android application and assess the performance of each algorithm. Here, we implemented a simple approach for classifying android applications. The dataset consists of 40 features of static, dynamic, and intrinsic features. The dataset is then divided into training and testing sets. The training data is used to train the model. All the algorithms show the high accuracy, but out of all random forest outperforms and gives the accuracy of 99.5%. SVM gives the lowest accuracy but the introduction of LDA increases its accuracy by 10%. The addition of intrinsic feature also helps in increasing the accuracy. The improvement in this area is increasing the dataset and also using different machine learning classifiers. The same work could be evaluated using another dimensionality reduction algorithm (Principle Component Analysis) in future to check its impact on accuracy.

References

1. Xialoeiwang, Y.Z.: Accurate malware detection in cloud. springer plus, 123 (2015)
2. Handa, A.: Malware detection using data mining techniques. Int. J. Adv. Res. Comput. Commun. Eng. **5** (2015)
3. Ravi KiranVerma, K.P.: Ansroid Malware detection and security using machine learning. In: international Conference on I_SMAC(IoT Social Mobile,Analytics and Cloud), 618–623 (2017)
4. Hyun Jae Kang, J.W.J. Androtracker: Creator information based malware detection. In: International Conference on Technology on ioT, 7 (2017)
5. Muttoo, S.V.: An Android Malware detection framework based in intents and permissions. Defence Sci. J. **66**(6), 618–623 (2016).
6. Mengyu Qiao, A.H.: Merging permissions and API callsfor Android Malware detection. In: 5th IIAI International Congress on Advanced Informatics (2016).
7. Ryo Sato, D.C. Detecting Android Malware by Analyzing Manifest File. In Proceedings of the Asia-Pacific Advanced Network 2013 (Vol. 36, pp. 23–31). https://doi.org/10.7125/APAN.36.4 (2013). ISSN 2227-3026.
8. Koli, J.: Randroid: An ANdroid Malware detectionusing random machuine learning classifiers. In: IEEEE International Conference on Technologies for Smart-city Energy Security and power (2018)
9. TaeGuen Kim, B.K.: Runtime detection framework for Android Malware. Hindawi Mobile Information Systems, 2018. Article ID 8094314, 15. https://doi.org/10.1155/2018/8094314 (2018)
10. I.K.Yan, H.: Droidscope: seamlessly reconstructing the OS and dalvik semantic views for dynamic android. In Proceedings of the 21st USENIX Security Symposium (USENIX Security 12), pp. 569–584 (2012).
11. Sanya Chaba, R.K.: Malwarre detection approach for Android systems using system call logs (2016)
12. Xi Xiao, X.X.: identifying android malware with system call co-occurrence matrices. Trans. Emer. Telecommun. Technol. (2016), 27
13. Xiangli, C.D.: Detection of Android malware security on system calls. IEEE 978–1–4673–9613–4/16/$31.00 ©2016 (2016).

Towards Recognition of Normal Versus Pneumonia Infected Patients Using Deep Neural Network Technique

Deepak Kumar and Chaman Verma

Abstract Pneumonia disease treatment, mostly death records can be found in hospitals due to not early diagnosis or detection of disease in the world, which is of great concern. In this article, the radiological chest X-ray images dataset from the Kaggle website which includes various sample images was used here for classification purposes. Deep convolutional neural network approach was used for binary classification. In this technique, a pre-trained convolutional neural network model ResNet50 was used for extracting the features, then fine-tuned or using transfer learning via chest X-ray images for classification. In this research, three optimization algorithms named stochastic gradient descent (SGD), Adam, and Rmsprop were used. The uppermost prediction accuracy of the Adam with ResNet50 was achieved 99.34% and validated on the training ratio of 80:20. The Adam algorithm outperformed others in the prediction accuracy. The proposed deep learning model exhibited outstanding performance in predicting pneumonia from the X-ray image and could aid in better diagnosis of patients.

Keywords Pneumonia disease detection · Convolutional neural network · SGDM · Adam · Rmsprop

1 Introduction

Air pollution causes many diseases and according to the WHO, over 4 million premature deaths occur due to air pollution-related diseases including pneumonia [1]. As we know in the human body, lungs and respiratory systems are the main organs where the exchange of gases takes place and any irregularity can cause loss of life. In this case, prior identification of abnormality is a major apprehension to avoid

D. Kumar
Guru Kashi University, Talwandi Sabo, Punjab, India
e-mail: dr.d.k.mehta81@gmail.com

C. Verma (✉)
Eötvös Loránd University, Budapest, Hungary
e-mail: chaman@inf.elte.hu

P. K. Singh et al. (eds.), *Recent Innovations in Computing*, Lecture Notes in Electrical Engineering 701, https://doi.org/10.1007/978-981-15-8297-4_25

any threat to life. The most infectious disease of lung named pneumonia which can cause swelling to alveoli, leading to the death of most children worldwide [2], and especially, developing nations are facing these problems of air polluted diseases. Therefore, timely and accurate diagnosis is the need of the hour for those pneumonia infected people. In the last decade, more intense improvements can be witnessed in the field of image acquisition devices like X-ray, MRI scans, and CT scans are radiological machines and deep neural network models can take benefit from these medical images for disease classification propose. It uses convolution neural network algorithms for extracting the relevant features from given images and classifying them accordingly. In this article, we are going to use the chest X-ray dataset to extract significant features from chest images of pneumonia infected people and normal people for classification purposes. This will help in recognizing the unrecognized feature and deep learning will be going to analyze these. Our approach here to create a diagnostic tool that can infer through chest images (Fig. 1a, b) whether a patient is normal or pneumonia infected using deep neural network techniques—automated diagnostic tool approach. For this chest, the X-ray dataset is partitioned into training and testing datasets. The training data is passed through the model 'training stages' and validated against a tested dataset to analyze the performance.

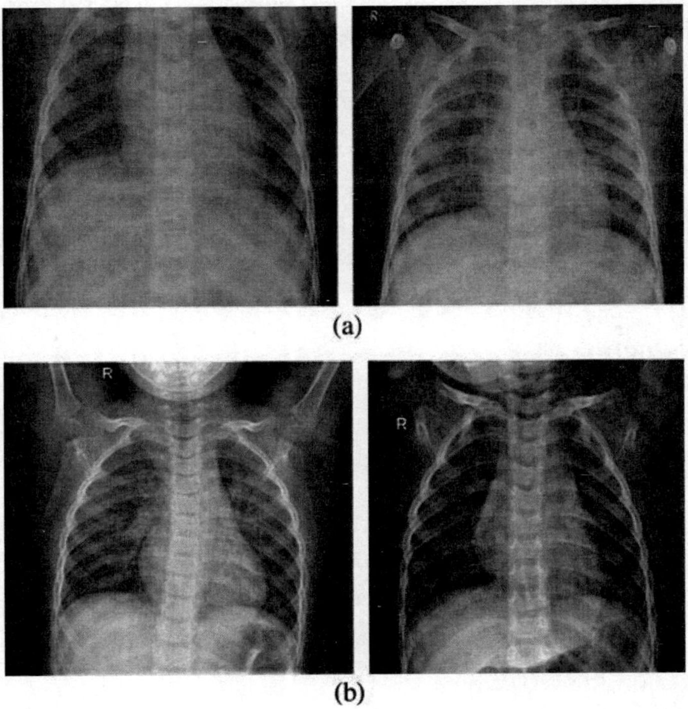

Fig. 1 **a** Pneumonia infected chest images, **b** normal person chest images

2 Material and Methods

2.1 System Configuration

The experiments are performed on the Intel Core I5 processor with NVIDIA 4 GB GPU 1050 with CUDA core support on 8 GB RAM. MATLAB 2019a support with deep neural network architecture packages is used for obtaining experiments results.

2.2 Dataset

X-ray scanned images dataset [3] is used for this article as depicted in Fig. 2. There are total of 5840 images and the authors used a total of 5216 images for training. The training dataset is categorized into two directories named normal and pneumonia consists of 1341 and 3875 images. For the testing purpose, 624 images are used validating the dataset. The resolution of images is 2090 × 1858.

Figure 3 reflects inherent medical images data is highly imbalanced which causes biasing to only pneumonia cases during prediction accuracy computation. For balancing the dataset, "data augmentation" techniques are used and this will also help in improving the robustness of the model.

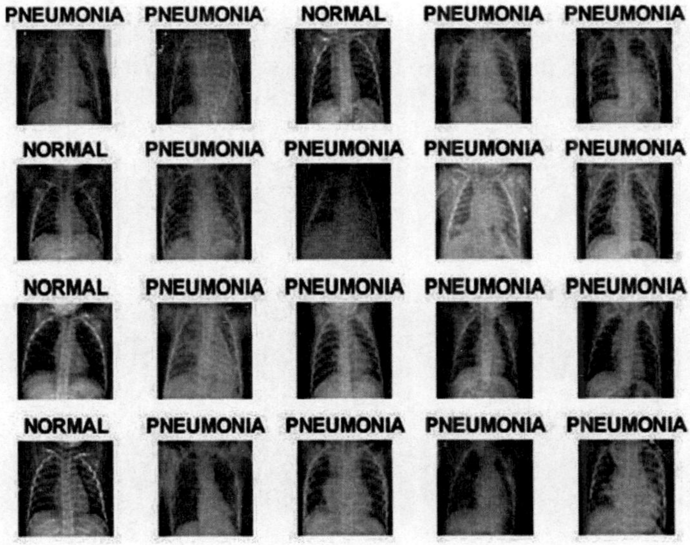

Fig. 2 Sampled chest X-ray images

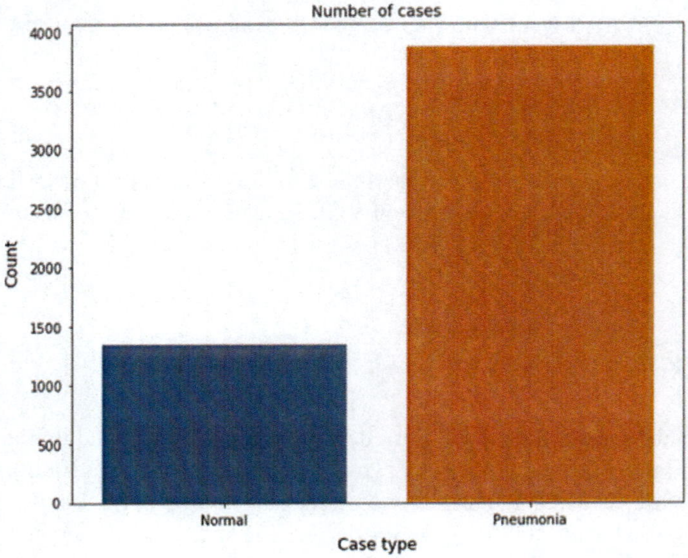

Fig. 3 Distribution of sample images

2.3 *A ResNet50—Convolutional Neural Network Architecture (CNN)*

The CNN structure can be corresponded with ordinary neural network structure. It can be considered as a stack of convolution layer, nonlinear layer (e.g. ReLu), pooling layer (e.g. max pooling) and a loss function (e.g. SVM/SoftMax) and at the end fully connected (FC) layer [4] as depicted in Fig. 4. In today's CNN scenarios, ReLu became a first choice to solve the vanishing gradient problem present in sigmoid

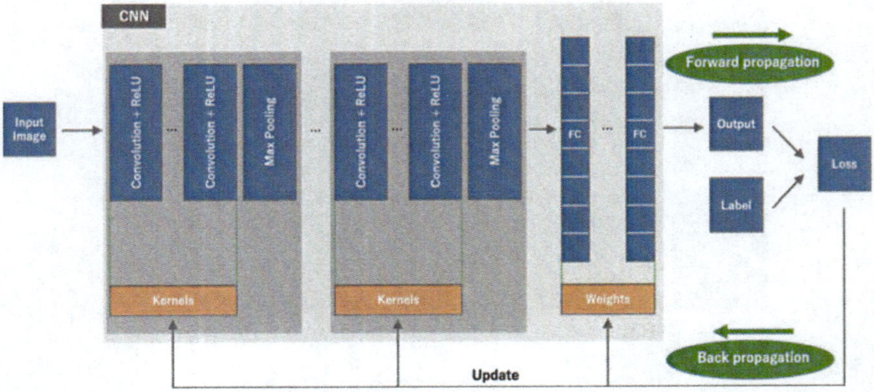

Fig. 4 CNN forward propagation technique

function and it solve the problem as positive inputs represented by gradient 1 and negative inputs represented by 0 gradient. A pooling operation is applied for introducing transition variance to a small shift by reducing in dimensionality—also called down-sampling of the feature map, after that distortion is performed to decrease the subsequent learnable parameters. The last layer fully connected layer, in this feature map of the pooling layer, transformed into a one-dimensional array and can be connected to subsequent fully connected layers. This repeated operation indicates the operation of dense layers. The final layer also called the final fully connected layer has the same number of output nodes as several classes as in figure last box named SoftMax layer with two display the two classes [5].

VGG, ResNet, GoogLeNet, and many more pre-trained models are exists [6]. ResNet is a familiar network model with auto-encoding, classification features, and won 2015 ImageNet Large Scale Visual Recognition Challenge (ILVRC) by less error rate.

The residual network 50 consists of a series of a block of three convolution layers by connecting the last fully connected layer for classification purpose as reflected in Fig. 5 [7]. The ResNet consists of four modules in which each module contains numerous residual parts with a fixed structure. It contains the convolutional layer, batch normalization, and activation function—ReLu function layer. There is always requirement of labelled training data and training a CNN from scratch is very tedious approach and we observe limited studies in research. Transfer learning is wider and easier approach for fine-tuned pretrained network or it can be used for feature extractor with the pretrained network [8, 9].

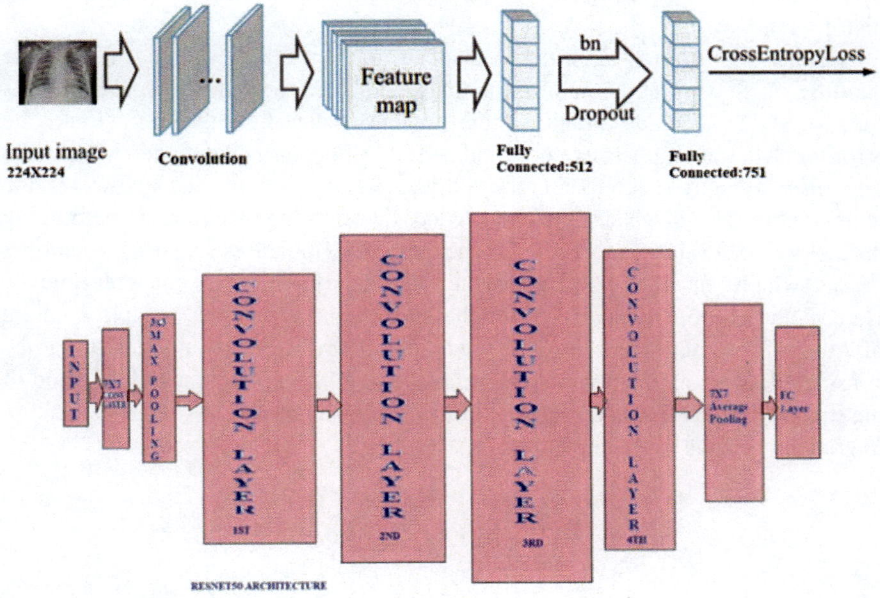

Fig. 5 ResNet50 pretrained model

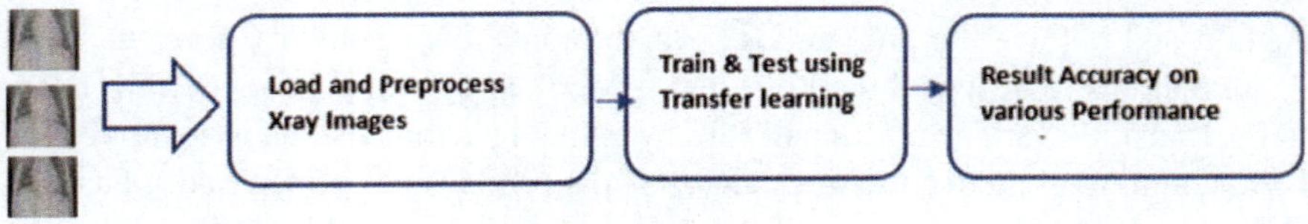

Fig. 6 Proposed method

2.4 Proposed Method

The workflow of experimental research as depicted in Fig. 6 radiological chest X-ray grayscale image is fed into the customized CNN-ResNet50 model to predict the X-ray image of pneumonia or a normal person. Images dataset are partitioned into training and testing images and dataset partitioned into 80% for training purpose and 20% for testing purpose. Pretrained ResNet50 model feature extraction layers are used for extracting the feature; only last classification layers are customized according to our requirement using transfer learning fine-tuning method. Fine-tuning is used here to learn the model by a stochastic gradient descent method [10–12], with an initial learning rate of 0.0001 and a moment of 0.9. When we want classification, accuracy remains the same then transfer learning with RMSprop with an initial learning rate of 0.001, moment—0.9 is used [13].

Some pre-processing has done on dataset images. In this process, all the images resolution is converted into 224 × 224 × 3 for input into CNN-ResNet50 as a prerequisite of ResNet50 neural network architecture then data augmentation is performed because the dataset is very highly unbalanced. This will balance the overall distribution.

Various optimization algorithms calculate gradient differently and in machine learning, SGD with momentum is a famous method for optimization problems. As for now, SGD is a unique variation of the gradient descent algorithm. It computes the error for each training data example and periodically updates the model. Mini-batch size plays a vital role in reducing the communication cost with SGD. However, the increment in mini-batch size typically reduces the convergence rate. In the training process with SGD, large value delivers accurate error gradient estimate to the learning process which converges slowly and small value at the cost of noise converge quickly. Here, defaults momentum is 0.9 for the experiment. Due to the adaptive nature of Adam, it calculates learning rates individually at various parameters. It can be considered as a combination of RMSprop and SGD because it uses the method of the square gradient for increasing the learning rate and method of SGD momentum to gradient moving average.

3 Experiments and Discussion

A recall by definition is a total number of true positives divided by sum of total true positive and false negatives or can be interpreted other way as following [14, 15]:

The model classified the data points true that are positives—true positive (TP).
The model classified the data points false that are positives—false negative (FN).
The model classified the data points false that are negatives—false positive (FP).

$$\text{Recall: TP}/(\text{TP} + \text{FN})$$
$$\text{Precision: TP}/(\text{TP} + \text{FP})$$

The recall is the model's ability to compute the data points of the dataset's interest and when a recall is high its effect also on precision will below or can be said vice versa and from Fig. 7 it becomes clear. On experiments when a recall is high precision is low. For the evaluation and effectiveness of projected work, we used an experimental approach here. Various hyperparameters are fine-tuned to achieve the well-performed model like augmentation of data, a variation of mini-batch sizes, and variation of optimization algorithms.

As it is evident in Fig. 8 that when the various optimization algorithms like SGD with 0.9 momentum, RMSprop, and Adam results are plotted with a variation of mini-batch sizes, Adam algorithm depicting better results with 99.36% on batch size 32.

Precision and recall can note down with mini-batch sizes variations. From Fig. 9, it is clear that on mini-batch size 32, both are depicting better results as higher accuracy 93.57%, and 94.21% with the SGDM optimization algorithm. By definition, $f1$ score is calculated with below formula:

$$F1\,\text{Score} = 2 * [(\text{Precision} * \text{Recall})/(\text{Precision} + \text{Recall})$$

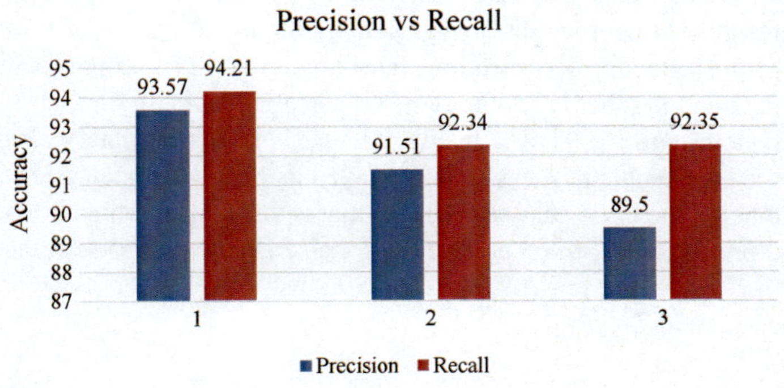

Fig. 7 Gradient descent algorithm prediction

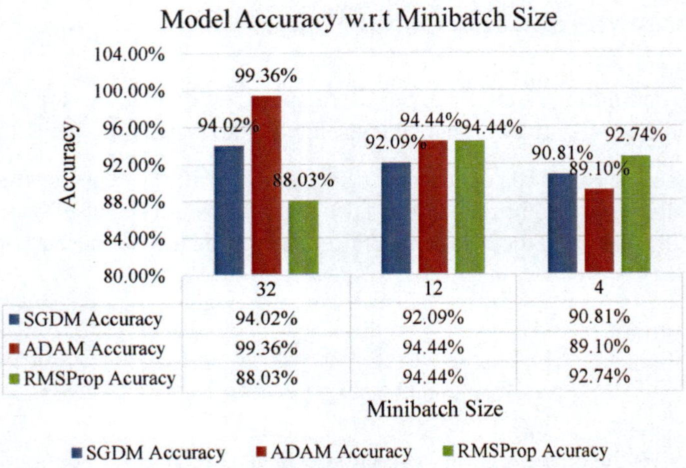

Fig. 8 Model accuracy with different gradient descent algorithm w.r.t minibatch sizes

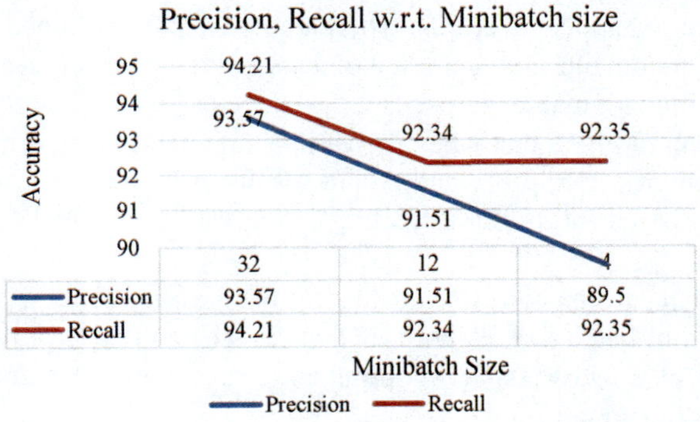

Fig. 9 Precision and recall computation result with various descent algorithm w.r.t batch sizes

To attain a balanced recall and precision, the $f1$ score should be significant as well. Therefore, our approach is always gaining the highest score of $f1$ and this is achieved by the following optimization algorithms. Adam's $f1$ score is better among all with 99.34 and in other cases the same scenario reflecting in Fig. 10.

A confusion matrix as shown in Fig. 11 that is depicting binary classification performance measurement. It has been displayed in the confusion matrix for various combinations for the binary classes [16–19] that is the total counting of true positive (TP), false positive (FP), true negative (TN), false negative (FN), e.g. the total no. of TP, FP, TN, and FN were 267, 1, 2, and 198, respectively, for X-ray pneumonia images classification.

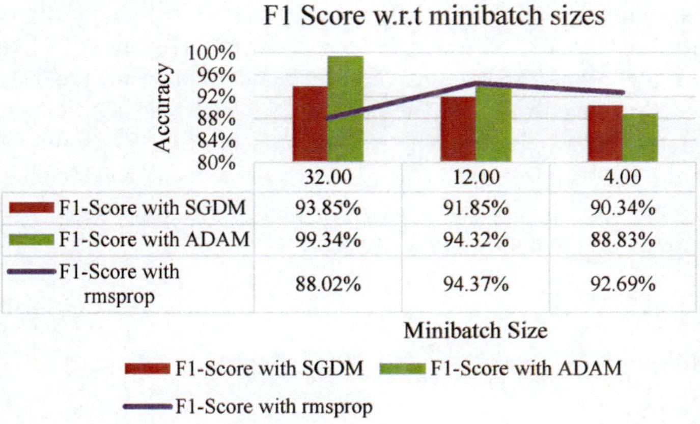

Fig. 10 $F1$-score w.r.t batch sizes

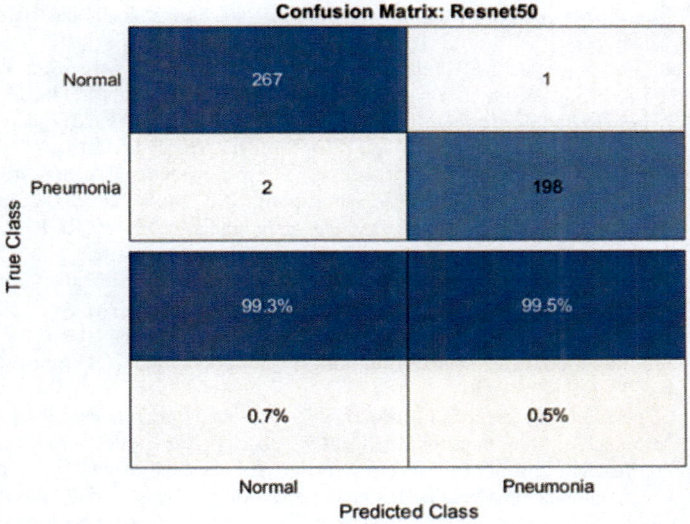

Fig. 11 Confusion matrix on mini batch size 32 and Adam optimization algorithm

4 Conclusion

The research work carried out here mainly focuses on the automatic diagnosis of pneumonia disease with the help of chest X-ray images using deep learning convolution neural network methods. This work will help in classifying the person whether he/she is infected with pneumonia or not. In this ResNet 50, the pre-trained CNN model was used here for feature extraction and classification purpose. The classification accuracy with Adam's optimization algorithm gave better accuracy among all gradient descent algorithms for chest X-ray pneumonia and normal person images. In this proposed method, transfer learning with ResNet-50 for classification purposes. The highest accuracy of 99.34% was achieved using training ratios of 80:20 using an optimization algorithm. The largest significant values of precision (93.57) and recall (94.21) were calculated with the batch size of 32 which also signifies the strength of the disease predictive model. The excellent f-score (93.85) computed with the SGDM algorithm that proved the balance between precision and recall. Therefore, the authors proposed this model be deployed on health clinics, care centres, and hospitals to support the nurses and doctors.

References

1. Koutina, M., Kermanidis, K.L.: Predicting postgraduate student's performance using machine learning techniques. In: IFIP Advances in Information and Communication Technology, p. 364 (2011)
2. World Health Organization: Household Air Pollution and Health [Fact Sheet]. WHO, Geneva, Switzerland (2018). https://www.who.int/newa-room/fact-sheets/detail/household-air-pollution-and-health
3. Chhikara, P., Singh, P., Gupta, P., Bhatia, T.: Deep convolutional neural network with transfer learning for detecting pneumonia on chest X-rays. In: Jain, L., Virvou, M., Piuri, V., Balas, V. (eds.) Advances in Bioinformatics, Multimedia, and Electronics Circuits and Signals. Advances in Intelligent Systems and Computing, vol. 1064. Springer, Singapore (2020)
4. Abdelhafiz, D., Yang, C., Ammar, R., et al.: Deep convolutional neural networks for mammography: advances, challenges and applications. BMC Bioinf. **20**, 281 (2019)
5. https://media.springernature.com/lw685/springer-static/image/art%3A10.1007%2Fs13244-018-0639-9/MediaObjects/13244_2018_639_Fig1_HTML.png?as=webp
6. Kaggle. https://www.kaggle.com/paultimothymooney/chest-xray-pneumonia. Accessed on 10 Feb 2020
7. Peng, J., Kang, S., Ning, Z., et al.: Residual convolutional neural network for predicting response of transarterial chemoembolization in hepatocellular carcinoma from CT imaging. Eur. Radiol. **30**, 413–424 (2020)
8. Mehta, D., Verma, C.: Prediction of cancer diagnosis patients from fine-needle aspirates using machine learning. In: Algorithms for Intelligent Systems, pp. 337–348, Springer, Berlin (2020)
9. Mehta, D., Verma, C.: Automatic leaf species recognition using deep neural network. In: International Conference on Evolving Technologies in Computing, Communications and Smart World. Lecture Notes in Electrical Engineering (LNEE), pp. 1–11, Springer, Berlin (2020)
10. Wang, Y., Wang, C., Zhang, H.: Ship classification in high-resolution SAR images using deep learning of small datasets. Sensors (Basel) **18**(9), 2929. Published online 2018 Sept 3 (2018). https://doi.org/10.3390/s18092929
11. Goodfellow, I., Bengio, Y., Courville, A.: Deep Learning. MIT Press, Cambridge (2016)
12. LeCun, Y., Bottou, L., Orr, G.B., Müller, K.R.: Efficient backprop. In: Orr, G.B., Müller, K.R. (eds.) Neural Networks: Tricks of the Trade, pp. 9–50. Springer, Berlin (1998)
13. Qian, N.: On the momentum term in gradient descent learning algorithms. Neural Netw. **12**, 145–151 (1999). https://doi.org/10.1016/S0893-6080(98)00116-6
14. Verma, C., Kumar, D., Illés, Z., Stoffova, V.: Automatic forecasting of student's province towards Information and communication technology awareness. IOP Conf. Ser. Mater. Sci. Eng. **872**, 012043 (2020). https://doi.org/10.1088/1757-899X/872/1/012043
15. Verma, C., Stoffová, V., Illés, Z., Kumar, D.: Towards prediction of student's guardian in the secondary schools for the real-time. In: Proceeding of ICRIC 2020. Lecture Notes in Electrical Engineering (LNEE), pp. 1–12. Springer, Berlin (2020)

16. Verma, C., Illés, Z., Stoffová, V.: Real-time prediction of development and availability of ICT and mobile technology in Indian and Hungarian university. In: Proceeding of ICRIC 2019. Lecture Notes in Electrical Engineering (LNEE), pp. 605–615, Springer, Berlin (2020)
17. Verma, C., Stoffová, V., Illés, Z.: Real-time prediction of student's locality towards information communication and mobile technology: preliminary results. Int. J. Recent Technol. Eng. **8**(1), 580–585 (2019)
18. Verma, C., Illés, Z., Stoffová, V.: Real-time classification of national and international students for ICT and mobile technology: an experimental study on Indian and Hungarian university. In: Journal of Physics, 14032, IOP Science, pp. 1–8 (2020)
19. Verma, C., Stoffová, V., Illés, Z.: Prediction of students' awareness level towards ICT and mobile technology in Indian and Hungarian university for the real-time: preliminary results. Heliyon **5**(6), 1–7 (2019)

Edge Detection Using Guided Image Filtering and Ant Colony Optimization

Akshi Kumar and Sahil Raheja

Abstract Edge detection is an important phenomenon in various classes of engineering problems. The classical methods which are based on the kernel designs are not very accurate and edges are falsely detected. Recent methods which are based on soft computing are adaptive in nature, and therefore using soft computing methods accuracy of detected edges can be improved. This paper presents an ant colony optimization-based edge detection process, and where edges are enhanced using guided filtering. The simulation results clearly show that the proposed scheme is much superior to recently proposed edge detection methods.

Keywords Edge detection · Guided filtering · ACO

1 Introduction

Image edge detection has been an important area of research, due to its applicability in various engineering and science problems. In wide sense, an edge can be defined as the boundary between various regions. Dominant edges can be detected using low level processing, however, for light edges, high level processing would be required. Traditional edge detection methods are based on kernels and they are independent of image characteristics [1]. Therefore, the performance is poor, and heavily depends on the threshold, thus for different image thresholds need to be set accordingly [2]. Moreover, the edge detection phenomenon is also dependent on lighting condition, object brightness, density of the pixels and noises. In general edge detection, comprises of three steps, in the first step, unwanted noise needs to be removed from the image, for the removal of noise, high pass filtering is performed, it is also customary to note that edges are also high frequency components, and therefore, during noise removal

A. Kumar
Department of Computer Engineering, Delhi Technological University, New Delhi, Delhi, India
e-mail: akshikumar@dce.ac.in

S. Raheja (✉)
Department of Information Technology, Delhi Technological University, New Delhi, Delhi, India
e-mail: sahilraheja78@gmail.com

P. K. Singh et al. (eds.), *Recent Innovations in Computing*, Lecture Notes in Electrical Engineering 701, https://doi.org/10.1007/978-981-15-8297-4_26

some edges may also get removed, therefore, trade-off exist between edge detection and noise removal mechanism. Hence, edge preservation mechanism is necessary while removal of noise. In the second stage, differential operator is applied for the detection of edges, and finally, in the last stage, edge localization is performed to find genuine edges. By keeping in the view of above, in this paper, we have considered guided image filtering for the enhancement of edges while reducing the noises.

For edge detection, soft computing techniques have recently gained much popularity, as due to controlling mechanism in edge detection [3, 4]. In recent past, ant colony optimization has been used for edge detection, and it has been found that ACO-based edge detection mechanism is a reasonably better technique with good accuracy [5–8]. However, ACO also fails to detect weak edges as in other techniques [9].

The objective of the paper is twofold, firstly, we will use guided filtering for edge enhancement, and secondly ACO will be used for edge detection.

2 Image Filtering

This section describes bilateral and guided filtering.

2.1 *Bilateral Filter*

Bilateral filter is an edge preserving image smoothening filter. In bilateral filter, the intensity of the particular pixels is replaced by the weightage average of the surrounding pixels. The chosen weights are based on Gaussian distribution. More importantly, the weights are dependent of pixels values, color, and depth, etc. The main advantage of the method is that it preserves the edges, however, on the downside, it produces staircase effect and image appears like carton, in addition to this, it also introduces false edges.

Considering and image to be filtered as '*p*' and guided image as '*I*', further considering pixels position as '*i, j*', then the joint bi-lateral filter kernel w^{bf} is given by [10]

$$w_{ij}^{bf}(I) = \frac{1}{N_K}\exp\left(-\frac{\left\|y_i - y_j\right\|^2}{\sigma_s^2}\right)\exp\left(-\frac{\left\|I_i - I_j\right\|^2}{\sigma_r^2}\right) \quad (1)$$

where x is the pixel coordinate and N_K is a normalizing factor such that $\sum_j w_{ij}^{bf}(I) = 1$. The parameters σ_s and σ_r control the sensitivity and color similarity, respectively.

2.2 Guided Image Filtering

This is an advance version of bilateral filter. But due to use of guided image, limitations of bilateral filtering can be suppressed significantly. Guided image sets basic guidelines for filtering, therefore, image artifacts are suppressed.

Guided Filter

In continuation of bilateral filters, and further considering output image as q. As discussed above, both I and p are given images beforehand, and they can be identical. The filtering output at a pixel 'i' can be written as [11]:

$$q_i = \sum_j w_{i.j}(I) p_j \tag{2}$$

where i and j are pixel indexes.

Now, we define the guided filter. We assume that q is a linear transform of I in a chosen window ω_k which is centered at the pixel k:

$$q_i = a_k I_i + b_k \quad \forall i \in \omega_k \tag{3}$$

In above equation, (a_k, b_k) are coefficients and are considered to be constant in chosen window ω_k.

Our aim is to define the linear coefficients (a_k, b_k), in term soft image parameters. For this, we model the output image q as the input image p and getting rid of the unwanted components like noise/textures:

$$q_i = p_i - n_i \tag{4}$$

As we are interested in minimizing the noise variance, i.e., $\sum_{i\in\omega_k} \sigma^2(n_i) = \sum_{i\in\omega_k} (q_i - p_i)^2 = \sum_{i\in\omega_k} (a_k I_i + b_k - p_i)^2$.

To counter the large values of a_k, we minimizes the following cost function, in the window ω_k:

$$E(a_k, b_k) = \sum_{i\in\omega_k} \left[(a_k I_i + b_k - p_i)^2 + \varepsilon a_k^2\right]. \tag{5}$$

Here, ε is a regularization parameter, the solution of Eq. 5 can be obtained using the ridge regression model [10, 11] and its solution is given by

$$a_k = \frac{\frac{1}{|\omega|}\sum_{i\in\omega_k} I_i p_i - \mu_k \frac{1}{|\omega|}\sum_{i\in\omega_k} p_i}{\sigma_k^2 + \varepsilon}, \text{ and}$$

$$b_k = \frac{1}{|\omega|}\sum_{i\in\omega_k} p_i - a_k \mu_k \tag{6}$$

Here, the mean and variance of I in window ω_k are μ_k and σ_k^2, respectively, and the number of pixels in ω_k is given by $|\omega|$. As in different windows, the values of (a_k, b_k) will be different, so after computing for all windows ω_k in the image, the output can be written as

$$q_i = \frac{1}{|\omega|} \sum_{k|i \in \omega_k} p_i(a_k I_i + b_k) \tag{7}$$

Noticing that $\sum_{k|i\in\omega_k} a_k = \sum_{k\in\omega_i} a_k$ due to the symmetry of the box window. Equations (3), (6) and (7) are the definition of the guided filter.

3 Image Edge Detection Using ACO

Edge detection can be accomplished using ant colony optimization. In any image, each pixel position is represented by (i, j) with intensity value $I_{i,j}$. Therefore, each image can be represented as 2D picture, on this picture, ants move randomly and build a pheromone matrix and based on matrix entries pixel is decided as edge and non-edge pixels. Initially, pheromone matrix $\tau^{(0)}$ has constant pre-defined values. The ant can move to its neighboring node (i, j) using the transition probability and mathematically expressed as [12]

$$p^n_{(l,m)(i,j)} = \frac{\left(\tau_{i,j}^{(n-1)}\right)^\alpha (\eta_{i,j})^\beta}{\sum_{i,j\in\Lambda_{(l,m)}} \left(\tau_{i,j}^{(n-1)}\right)^\alpha (\eta_{i,j})^\beta}, \; j \in \Lambda_i \tag{8}$$

$\tau_{i,j}^{(n-1)}$ is the pheromone value. Parameter $\Lambda_{(l,m)}$ denotes the all possible neighborhood nodes of the node (l,m), and $\eta_{i,j}$ is a heuristic parameter. The constants α and β control the associated parameters.

The heuristic information $(\eta_{i,j})$ is defined by

$$\eta_{i,j} = \frac{G_c(i, j)}{\sum_{i=1:M} \sum_{j=1:N} G_c(i, j)} \tag{9}$$

For the pixel $I_{i,j}$ under consideration (Fig. 1), the function $G_c(i, j)$ is

$$G_c(I_{i,j}) = f\left[\begin{array}{l} \left|I^P_{i-2,j-1} - I^P_{i+2,j+1}\right| + \left|I^P_{i-2,j+1} - I^P_{i+2,j-1}\right| + \left|I^P_{i-1,j-2} - I^P_{i+1,j+2}\right| \\ + \left|I^P_{i-1,j-1} - I^P_{i+1,j+1}\right| + \left|I^P_{i-1,j} - I^P_{i+1,j}\right| + \left|I^P_{i-1,j+1} - I^P_{i-1,j-1}\right| \\ + \left|I^P_{i-1,j+2} - I^P_{i-1,j-2}\right| + \left|I^P_{i,j-1} - I^P_{i,j+1}\right| \end{array}\right] \tag{10}$$

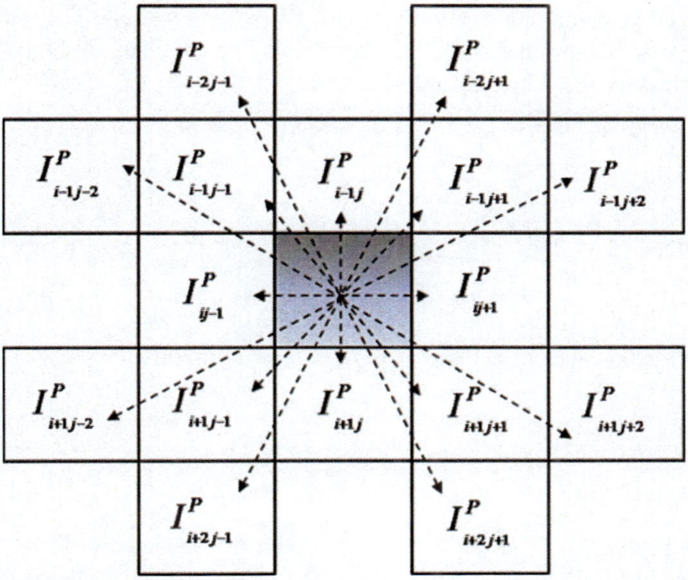

Fig. 1 Pictorial representation of clique $G_c(i, j)$

where $f(\cdot)$ in (10) is [9];

$$f(x) = 127x + \sin\left(\frac{\pi x}{2\lambda}\right) \text{ for } 0 \le x \le \lambda \tag{11}$$

The parameter λ control shapes.

The local update of pheromone matrix is given by

$$\tau_{i,j}^{(n-1)} = \begin{cases} (1-\rho)\tau_{i,j}^{(n-1)} + \rho\Delta_{i,j}^{k} & \text{if } (i,j) \in \text{vca} \\ \tau_{i,j}^{(n-1)} & \text{otherwise} \end{cases} \tag{12}$$

where 'vca' means 'visited current ant' and ρ is rate of evaporation of pheromone.

Second update which considers movement of all the ants in each construction step as in

$$\tau^{(n)} = (1-\psi)\tau^{(n-1)} + \psi\tau^{(0)} \tag{13}$$

Now, the parameter ψ represents the pheromone decay [7].

Decision Process

Step 1: Initialize $T^{(0)}$ as

$$T^{(0)} = \frac{\sum_{i=1:M}\sum_{j=1:N}\tau_{i,j}^{(n)}}{MN} \tag{14}$$

and fix the iteration index as $q = 0$.

Step 2: Separate the pheromone matrix $\tau^{(n)}$ into two class making use of threshold $T^{(q)}$ and obtain mean of two classes as

$$m_L^{(q)} = \frac{\sum_{i=1:M} \sum_{j=1:N} x\tau_{i,j}^{(n)}}{\sum_{i=1:M} \sum_{j=1:N} \tau_{i,j}^{(n)}} \text{ for } x \leq T^{(q)} \tag{15}$$

$$m_U^{(q)} = \frac{\sum_{i=1:M} \sum_{j=1:N} x\tau_{i,j}^{(n)}}{\sum_{i=1:M} \sum_{j=1:N} \tau_{i,j}^{(n)}} \text{ for } x \geq T^{(q)} \tag{16}$$

Step 3: Fix the index of iteration $q = q + 1$, and we modify the threshold as given below

$$T^{(q)} = \frac{m_L^{(q)} + m_U^{(q)}}{2} \tag{17}$$

Step 4: In the case of $\left|T^q - T^{(q-1)}\right| > \varepsilon$, after this move on to *Step 2*; else, the iteration method is discontinued and we make a binary decision on all pixel's location (i, j) in order to find out edge using:

$$E_{i,j} = \begin{cases} 1 \ \tau_{i,j}^{(n)} \geq T^{(q)} \\ 0 \ \text{ elsewhere} \end{cases} \tag{18}$$

4 Results

In most of the edge detection mechanism, the positions of the pixels do alters, therefore, measure like PSNR, SSIM fails, moreover, edge detection mechanism are not 100% accurate, therefore, both false positive and false negative edges are detected. Therefore, results are compared in terms of F-measure, and it is harmonic mean of precision and recall.

The precession (γ), recall (Γ) and F-measure are defined as:

$$\gamma = \frac{\text{TP}}{\text{TP} + \text{FP}}, \Gamma = \frac{\text{TP}}{\text{TP} + \text{FN}} \text{ and } F = \frac{2\gamma\Gamma}{(\gamma + \Gamma)}$$

where TP is true positive, P is total positive, FP negative and incorrectly identified, N is total negatives and so on (Fig. 2). F-measure is a test of accuracy in binary classification. It depends on both precession and recall to get test score. The maximum value of F is 1 with minimum as 0.

The simulation parameters are detailed in Table 1. The total numbers of ants which

Fig. 2 Characteristic matrix

	p	n
Y	**True Positives**	**False Positives**
N	**True Negatives**	**False Negatives**
Column Totals	**P**	**N**

Table 1 Simulations parameters

Parameters	Value
Pheromone matrix, τ_{init} (Initial values)	10^{-4}
Pheromone information, α (Weighting factor)	1
Heuristic information, β (Weighting factor)	0.1
Connectivity neighbourhood, Λ	8
Functions adjusting parameter, λ	10
Evaporation rate, ρ	0.1
Total number of ants	Vary
Total number of ant's movement-steps, S	40
Pheromone decay coefficient, ψ	0.05
Tolerance value, ε	0.1
Threshold, T	Adaptive

need to be taken depend on image size. Considering the size of image as $(M \times N)$ than, the numbers of ants are $(\sqrt{MN})$.

In Fig. 3, results for edge enhancement using guided filtering are shown. Here, four different images are shown in first column. Each row shows the results for an image under various mechanisms. Second column shows the gray scale images corresponding to its original image shown in first column; in third column, edge-enhanced image using the guided image filtering is shown. Finally, in the fourth column, gray scale version of the enhanced images is shown. It is clear from the results that the edge are more sharp and more clear in enhanced image as compared to original image. For accurate edge detection, it is desired that the intensity values in gray image should be either 0 or 255, but it is not possible to get exact binary image without using any threshold. Still, we expect that the histogram of the image should be more concentrate around 0 and 255. Therefore, in the next figure, four columns

Fig. 3 First column; original image, second column; gray scale original image, third column; enhanced image, fourth column; gray scale enhanced image

are shown, in the first column, original images is shown, and in second column, histogram of all the images are shown, in third column, enhanced images are shown, and finally, in the fourth column, histogram of enhanced images are shown (Fig. 4).

It is clear from the figure that the histogram of gray scale original image intensity is spread from 0 to 255, while in case of enhanced images, intensities values are more concentrated around 0 and 255, and the dominant part is more clearly visible as compared to background thus edges can be detected quite accurately.

To prove the applicability of the proposed edge detection methodology, the results are obtained on vast varieties of images and databases. In Fig. 5, first column shows that original considered images, and correspondingly edge detected images are shown in second column, in the third column, sharpen images are shown, while in fourth column, edge detection on sharpened images is shown. Each row in the figure shows the results for one type of original and sharpened image along with corresponding edge detected images. It is also observable that with proposed method more edges are detected. The notable areas are marked using circular and oval shapes, where clear difference in detected edges can be observed. However, more edges always not mean that correct edges are detected. Therefore, F-score is obtained for the four images, and minimum obtained F-score is shown in Table 2. From Table 2, it is clear that the proposed scheme is much better in comparison with recently/notable schemes.

It is also important to note that image enhancement is controlled using guided filter parameters.

Fig. 4 First column; original image, second column; histogram of original image, third column; enhanced image, fourth column; histogram of scale enhanced image

Fig. 5 First column; original image, second column; edge detected image, third column; enhanced image, fourth column; edge detected enhanced image

Table 2 Comparison with notable works

Method	Year	F-measure
Canny [13]	[1996]	0.49
Sobel [14]	[2009]	0.40
BEL [3]	[2006]	0.63
gPb [15]	[2011]	0.71
Sketch token [16]	[2013]	0.73
Structured forest [17]	[2013]	0.71
ACO [9]	[2017]	0.74
Proposed	[2019]	0.79

The sharpened image (I_s) is given by

$$S = I + \chi[I - \underbrace{I \otimes G}_{\text{Smoothening}}] \tag{19}$$

where

I: original image, G: guided filter and 'χ' is scaling factor.

In Fig. 6, 8 Lena images are shown with varying the sharpness levels 'χ', where the values of χ are 5, 10, 20, 30, 50, 75 and 100, respectively. Therefore, various levels of sharpened image can be used to detect desirable edges as required in various engineering and science applications.

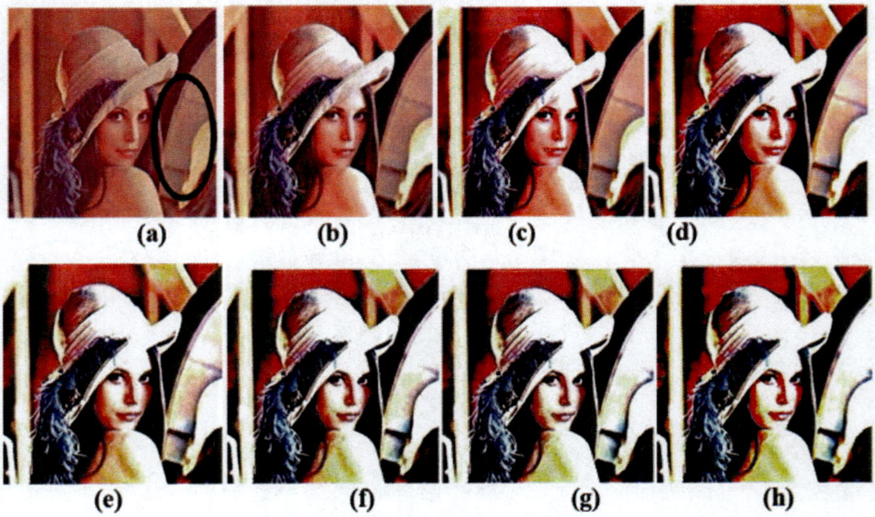

Fig. 6 Image sharpening using guided image filtering while varying the sharpness parameters

5 Conclusions

In this paper, an edge-enhanced ACO-based edge detection method is detailed and obtained results are compared with recently proposed methods in terms of F-score. ACO alone fail to detect some of the non-prominent edges, and to detect such edges, the edges of the image under consideration are enhanced using guided image filtering. It is found that the histogram of the enhanced image intensities more concentrated around 0 and 255. Thus, the dominant part is more clearly visible as compared to background. The edge detection of enhanced image is done using ACO, and results are compared by considering both original and enhanced images. It is found that the proposed method is much superior to recently proposed designs.

References

1. Singh, R.K., Shekhar, S., Singh, R.B., Chauhan, V.: A comparative study of edge detection techniques. Int. J. Comput. Appl. **100**, 19 (2014)
2. Maini, R., Aggarwal, H.: Study and comparison of various image edge detection techniques. Int. J. Image Process. (IJIP) **3**(1), 1–11 (2009)
3. Dollar, P., Tu, Z., Belongie, S.: Supervised learning of edges and object boundaries. In: 2006 IEEE Computer Society Conference on Computer Vision and Pattern Recognition, vol. 2, pp. 1964–1971 (2006)
4. Raheja, S., Kumar, A.: Edge detection based on type-1 fuzzy logic and guided smoothening. Evolving Syst. 1–16 (2019)
5. Dorigo, M., Thomas, S.: Ant Colony Optimization. MIT Press, Cambridge (2004)
6. Jing, T., Yu, W., Xie, S.: An ant colony optimization algorithm for image edge detection. In: Proceedings of the IEEE International, pp. 751–756 (2008)
7. Lu, D.S., Chen, C.C.: Edge detection improvement by ant colony optimization. Pattern Recogn. Lett. **9**, 416–425 (2008)
8. Gupta, C., Gupta, S.: Edge detection of an image based on ant colony optimization technique. Int. J. Sci. Res. (IJSR) **2**(6), 1256–1260 (2013)
9. Raheja, S., Kumar, A.: Edge detection using ant colony optimization under novel intensity mapping function and weighted adaptive threshold. Int. J. Integr. Eng. (to be appear in Feb 2020 issue)
10. Elad, M.: On the origin of the bilateral filter and ways to improve it. IEEE Trans. Image Process. **11**(10), 1141–1151 (2002)
11. He, K., Sun, J., Tang, X.: Guided image filtering. In: European Conference on Computer Vision, pp. 1–14. Springer, Berlin (2010)
12. Xiao, P., Li, J., Li, J.P.: An improved ant colony optimization algorithm for image extracting. In: Apperceiving Computing and Intelligence Analysis (ICACIA), 2010 International Conference, pp. 248–252 (2010)
13. Canny, J.: A computational approach to edge detection. IEEE Trans. Pattern Anal. Mach. Intell. **6**, 679–698 (1986)
14. Vincent, O.R., Folorunso, O.: A descriptive algorithm for sobel image edge detection. In: Proceedings of Informing Science & IT Education Conference, vol. 40, pp. 97–107 (2009)
15. Arbelaez, P., Maire, M., Fowlkes, C., Malik, J.: Contour detection and hierarchical image segmentation. IEEE Trans. Pattern Anal. Mach. Intell. **33**(5), 898–916 (2011)

16. Lim, J.J., Zitnick, C.L., Dollár, P.: Sketch tokens: a learned mid-level representation for contour and object detection. In: Proceedings of the IEEE Conference on Computer Vision and Pattern Recognition, pp. 3158–3165 (2013)
17. Dollár, P., Zitnick, C.L.: Structured forests for fast edge detection. In: Proceedings of the IEEE International Conference on Computer Vision, pp. 1841–1848 (2013)

A-VQA: Attention Based Visual Question Answering Technique for Handling Improper Count of Occluded Object

Shivangi Modi and Dhatri Pandya

Abstract Recently, visual question answering (VQA) system is being used in various research disciplines for extracting meaningful data from images and communicating this meaningful data to humans. Thus, to implement VQA system, computer vision and natural language processing fields are combined. Existing VQA system contains many open research issues such as improper count of occluded object problem, single word answer, time-specific answer problem and many more problems because of its wide assortment of utilizations and its more extensive region of research. In this paper, we present attention-based visual question answering (A-VQA) method for handling improper count of occluded object. A-VQA systems generate the textual answer by extracting image and textual features and applying multi-layer attention mechanism on these images and textual features. A-VQA system handle object recognition, counting, color and activity recognition types of visual questions. For training and evaluating A-VQA method, visual genome dataset is used.

Keywords Natural language processing · Computer vision · Visual question answering · Attention model · YOLOv3 · LSTM · Object detection · Occlusion

1 Introduction

Visual question answering (VQA) is framework through which appropriate response of a given question can be identified with image [1–6]. This framework gives bits of knowledge into the connections between two significant sources of data from computer vision and natural language processing fields [1–5]. In VQA, the framework

S. Modi (✉) · D. Pandya
Sarvajanik College of Engineering and Technology (SCET), Surat, Gujarat, India
e-mail: smodi0123@gmail.com

D. Pandya
e-mail: dhatri.pandya@scet.ac.in

P. K. Singh et al. (eds.), *Recent Innovations in Computing*, Lecture Notes in Electrical Engineering 701, https://doi.org/10.1007/978-981-15-8297-4_27

has picture and natural language question in textual format as an input and produces answer in textual format as an output [7].

VQA has pulled in a lot of thought because of its different uses in genuine everyday life such for visually disabled users, picture recovery, mechanical autonomy, art gallery and a lot more zones [1–5].

In visual question answering, there are two sorts of evaluation group [8, 9]. One is open-ended sort of evaluation group and another is multi-decision question answer type evaluation design [1, 10]. In open-ended sort of visual question answering, questions are composed into fundamentally twelve unique classifications, for example, object nearness, subordinate object acknowledgment, checking, shading characteristics, different traits, movement acknowledgment, sport acknowledgment, positional thinking, scene order, object utilities, conclusion understanding and foolish [7]. In multi-decision question answering type format, options are right, conceivable, famous or arbitrary.

We have distinguished certain issues in existing system of VQA such as improper count of occluded object problem, single word answer, time-specific answer problem and many more issues. A major issue of VQA system is that it was not handle occluded object counting type of problems. Thus, we propose an attention-based visual question answering (A-VQA) system that is using multi-layer attention mechanism to handle the improper count of occluded object problem. This system will take image and natural language question in a textual format and produce the answer. A-VQA system handles color, counting, object recognition and activity recognition types of visual questions. For training and evaluating A-VQA method, we are using visual genome [11] dataset.

In our proposed methodology, there are four phases for implementing visual question answering system. (A) In the initial phase of VQA, system is presented with a textual question and image. Textual question contains information about object to be queried from an image and which is asked by a person to get the correct answer from image. (B) The second phase is processing on both image and textual question and generates the image features and question encoding vector. Generating the semantics from image is part of computer vision and generating the question encoding vector is a part of NLP. Therefore, we can say that VQA is a mixture of both the fields. After getting both image feature vector and question encoding vector, third phase is applied. (C) In third phase, both these feature vectors are combined and generate the final combined feature matrix. (D) In fourth phase, we have applied dual attention mechanism to handle the improper count of occluded object problem. After applying attention mechanism, we are using softmax classifier to generate the final actual answer.

This paper aims to handle the improper count of occluded object problem through multi-layer attention mechanism. For that purpose, this paper is made as follows: In segment 2, different VQA methods are discussed. Area 3 talked about the proposed framework flow and implementation details. Final analysis of our A-VQA system is presented in Sect. 4.

2 Literature Review

This area gives a depiction of different visual question answering models and their methods.

2.1 VQA Methods

There are four different VQA methods are available which are (a) joint embedding method [8, 12], (b) attention method [9, 13–15], (c) compositional method [14, 16, 17] and (d) external knowledge-based method [15, 18, 19]. These methods are used to generate open ended type of answers. Joint embedding method extracting both textual and image features and then after these component vectors are together inserted into normal element vector. This regular component vector goes into classifier and it will generate answer. This technique centers around all the global features as opposed to concentrating on question explicit features. However, the attention method just focuses on question explicit features of a picture. Compositional model is used when questions are complex and questions require multi-step thinking to get answer appropriately. External knowledge-based method is useful where questions are based on common sense or they require some extra background knowledge to get answer properly. This extra background knowledge is provided in the form of supporting facts.

2.2 VQA Techniques

In this section, the summary of existing visual question answering techniques are discussed. These techniques are differing with each other through some parameters. Comparisons of various VQA techniques are discussed below with various parameters.

1. Talking with machine [8] technique uses joint embedding model as VQA model and FM-IQA as dataset. This technique handles action, object recognition, positions and commonsense types of questions. They are using CNN-GoogLeNet network for extracting image features and LSTM [20] network for extracting question features. However, this technique is not detecting small object.
2. Stacked attention networks (SAN) [9] technique uses DAQUAR-ALL, DAQUAR-REDUCED, COCO-QA and VQA as datasets with attention method. This technique handles object, color, count and location types of questions. They are using CNN-VGGNet [21] network for extracting image features and CNN/LSTM [20] network for extracting question features. However, this technique is not able to generate answer in sentence.

3. Graph structures [16] technique uses VQAv2 as datasets with attention method. This technique handles object, color and count types of questions. They are using object detector for extracting image features and RNN network for extracting question features. Scope of improvement of this technique is to improve performance of "Number" questions and to handle time-specific types of question answer.
4. FVQA: Fact-based visual question answering [18] technique uses FVQA as datasets and uses external knowledge base model. This technique handles object, scene and action types of questions. They are using fast R-CNN for extracting image features and RNN/LSTM network for extracting question features. Scope of improvement of this technique is to deal with more quantities of visual thinking kinds of inquiries.
5. Combined model network [14] technique uses CLEVR as datasets and uses both attention and compositional model. This technique handles object recognition, position and shape type of questions. They are using heuristic rule-based semantic parsing for extracting image features and Stanford dependency parser network for extracting question features. Scope of improvement of this technique is to predict answer in multiple words.
6. R-VQA [15] technique uses R-VQA as dataset and uses both attention and external knowledge-based model. This technique handles object, color, activity, position, scene, relation and commonsense type of questions. They are using CNN-ResNet-152 network for extracting image feature vector and LSTM network for extracting question feature vector. Scope of improvement of this technique is to predict answer in multiple words.

3 Attention-Based Visual Question Answering (A-VQA) Technique

Our proposed methodology as shown in Fig. 1, handles improper count of occluded object problem through multi-layer attention model. This system will handle activity recognition, color, counting and object recognition types of questions.

In our proposed work, there are four phases for implementing visual question answering system. In the first phase of VQA, system is presented with real picture and question in textual format. Textual question is related to an image and which is asked by a person to get the correct answer from image. At that point, the subsequent stage is preparing on both picture and textual question and produces the image features and question encoding vector. Creating the picture highlights is part of computer vision and generating the question encoding vector is a part of NLP. Therefore, we can say that VQA is a mixture of both the fields. After getting both image feature vector and question encoding vector, third phase is applied. In third phase, both these feature vectors are combined and generate the final combined feature matrix. In fourth phase, we have applied dual attention mechanism to handle the improper count of occluded object problem. After applying attention mechanism, we are using softmax classifier

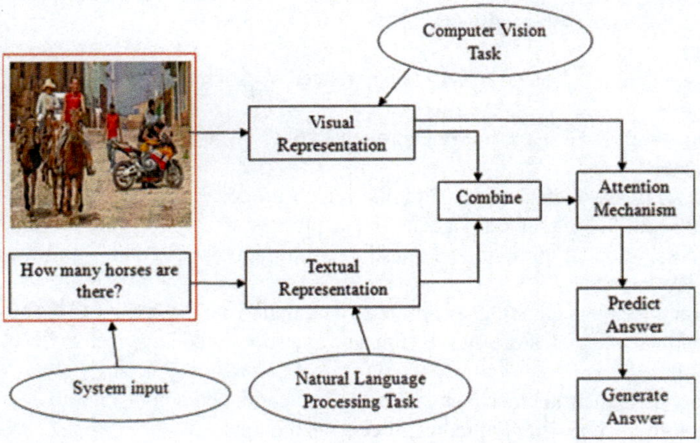

Fig. 1 Proposed work flow

to generate the final actual answer. These all phases are explained in pseudo code (Fig. 2).

where,

I	Input image
Q	Textual question
Fm	Image feature matrix
Ab	Anchor boxes
Bp	Bounding boxes
Rp	Coordinates of bounding boxes
CL	Class labels
Po	Potential objects
Vr	Visual representation matrix
Si	Three-different scale
Op	Object probability
Rj	Region
T	Tokenization
Ti	Set of tokens
We	Word embedding
Vi	Numerical vector indices
Qe	Question encoding vector
Cv	Combined output vector
Ov	All generated outputs
O	Final predicted output in textual format.

Pseudo code for proposed work
Begin
1. Take sample real picture I and image related question Q in textual format as an input
2. Processing on an input image I through RetinaNet-50 and YOLOv3 deep learning technique
3. Extract image feature matrix Fm and region proposal or set of bounding boxes Rp where p=1,2,3…n as an output from image processing
4. Processing on an input textual question Q through tokenization T and word embedding We
5. Extract question encoding vector Qe as an output from the series of LSTM cell.
6. Combine image feature matrix Fm and question encoding vector Qe through concatenation
7. Apply dual attention mechanism DA on final combined output vector Cv
8. Use softmax classifier to predict all generated outputs Ov
9. Generate final output O from all predicted outputs Ov
End

Pseudo code for RetinaNet-50
Begin
1. Select backbone network as ResNet-50 CNN
2. Pass an input image I to the ResNet-50
3. ResNet-50 extract image feature matrix Fm
4. Different sizes anchor boxes Ab are passes towards the feature maps Fm
5. Extract the bounding box coordinates Rp where p = 1,2,3…n and predict object classes CL where L= 1,2,3..n for all potential object Po
6. Get final output Vr which contains bounding box coordinates Rp where p = 1,2,3…n, Object class CL where L= 1,2,3..n and Object probability Op.
End

Pseudo code for YOLOv3
Begin
1. Pass an input image I to YOLOv3 Network
2. YOLOv3 network divides the input image into an R X R grid for each scale Si where i= 1,2,3
3. Pass each scale Si where i= 1,2,3 into 75 convolutional layers
4. Extract the image feature matrix Fm for each scale Si where i=1,2,3
5. Passes different sizes anchor boxes towards the feature map Fm
6. Predicts bounding boxes bp where p = 1, 2, 3...n and probabilities of object Op for each region Rj where j=1,2,3..n.
7. Get bounding box coordinate Rp where p = 1,2,3..n, object score Op and class confidence CL as an output Vr
End

Fig. 2 Pseudo code of proposed system

Pseudo code for Textual Question Processing
Begin
1. Input textual question Q is first converted into lowercase and remove punctuation mark
2. Generated proceed input textual question Q is then pass into tokenization T and generate set of tokens Ti where i=1,2…n
3. Each and every tokens Ti where i=1,2…n are then pass to the Glove word embedding We
4. Word embedding converts tokens Ti where i=1,2..n into numerical vector indices Vi where i=1,2..n and pass the sequence of indices into sequence of embedding
5. Finally each sequence of word embedding is fed sequentially to the LSTM network
6. LSTM network generate question encoding vector Qe which preserve the semantic similarity
End

Pseudo code for Attention Mechanism
Begin
1. Fed combined feature matrix (image feature matrix and question embedding vector) and bounding box coordinates to the single layer neural network
2. Apply softmax function to generate the attention distribution over the regions of image
3. Generate the attention weighted image feature map
4. Visual regions which have higher weights in attention map are more relevant to the question
End

Fig. 2 (continued)

4 Implementation and Analysis

In this section, we provide technical details of datasets and implementation that are used in our work for visual question answering. Moreover, we also compare the results of RetinaNet-50 and YOLOv3 object detection frameworks for the visual feature extraction task and show the results of visual feature representation and textual feature representation.

4.1 Dataset Details

We are using visual genome [11] dataset for training and evaluating A-VQA technique. Visual genome dataset handles six open ended types of questions. These six types of questions are ‘what,’ ‘who,’ ‘where,’ ‘how,’ ‘when’ and ‘why’. This dataset handles only real scene images. We are working with 80 number of object classes for training and testing A-VQA technique (Fig. 3; Table 1).

Q: how many horses are there?

Fig. 3 Testing image and question

Table 1 Name of object classes

Person	Bicycle	Car	Motorcycle	Airplane	Bus	Book
Train	Truck	Boat	Traffic light	Fire hydrant	Stop_sign	Clock
Frisbee	Bench	Bird	Baseball glove	Dog	Horse	Vase
Sheep	Cow	Elephant	Surfboard	Zebra	Giraffe	Scissors
Backpack	Umbrella	Handbag	Parking meter	Tie	Suitcase	Teddy bear
Skis	Bear	Bottle	Sports ball	Wine glass	Snowboard	Hair dryer
Bed	Toilet	Tv	Dining table	Laptop	Mouse	Toothbrush
Cat	Fork	Bowl	Skateboard	Sandwich	Carrot	Refrigerator
Kite	Knife	Banana	Tennis racket	Orange	Hot dog	Sink
Cup	Spoon	Apple	Baseball bat	Broccoli	Pizza	Microwave
Donot	Cake	Couch	Potted plant	Remote	Chair	Cell-phone
Keyboard	Oven	Toaster				

Computer vision task involves extracting the features of an input image and generate the region proposal. These both tasks are done by RetinaNet and YOLOv3 object detection techniques. These both networks are able to identify small objects as per literature survey. Therefore, we are implemented RetinaNet-50 and YOLOv3 network (Fig. 4).

Result analysis of RetinaNet-50 and YOLOv3 network

RetinaNet-50 network is able to identified too small objects, object class probability and generate bounding box coordinates. However, system is not able to identified occluded objects. Therefore, we are implementing YOLOv3 network.

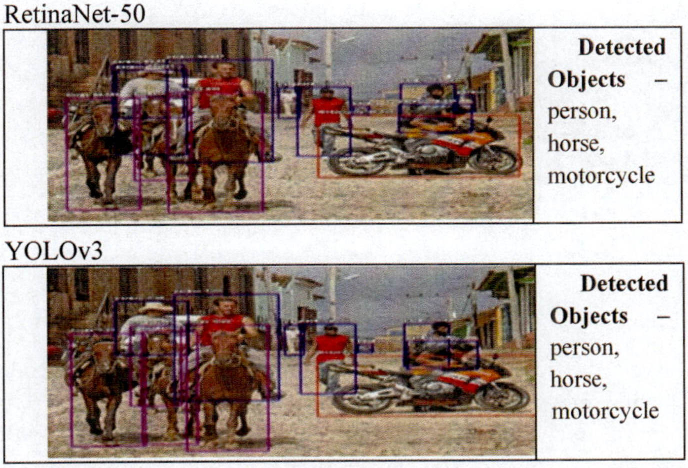

Fig. 4 RetinaNet-50 and YOLOv3 networks detected objects

4.2 Implementation Details

In this section, the implementation of visual question answering system using deep learning is mention. There are the following steps involved in this implementation.

Step 1 Take input image and textual question as an input.

Step 2 YOLOv3 network is able to identified too small object, faster than faster R-CNN and also able to identified occluded objects. Moreover, our motive is to correctly count occluded objects. Therefore, to generate correct answer, we are using YOLOv3 network for computer vision task.

Step 3 Question encoding

Visual question answering system takes textual question as an input. This textual question is in sequence of word format. However, machine does not understand this textual format. Therefore, there is a need to convert this textual information into vector format. To convert textual question into numerical vector form, first, we have implemented tokenization.

Q: how many horses are there?

Tokenization—Tokenization is the process of splitting the text into smaller parts called tokens. We are using NLTK 3.4 version for processing natural language processing task.

Answer of tokenization process is given below.

how, many, horses, are, there

Word embedding—Assign the numerical value to each token

how—29
many—30
horses—50
are—15

Fig.5 Attention mechanism result

there—16

GloVe Word embedding and LSTM—Generating question encoding vector which is semantic representation of a question.

Step 4 Combine image feature matrix and question encoding vector—after getting both image features matrix and question encoding matrix, we are performing element-wise multiplication operation to combine these both matrixes.

Step 5 Attention mechanism—we are predicting the answer through multi-step reasoning. In multi-step reasoning, we are applying multi-layer or duel attention mechanism on combined feature map and the output of the image processing. Dual attention mechanism is used to handle the improper count of occluded object problem. Testing result of dual attention mechanism is shown in Fig. 5.

Step 6 **Predict and Generate answer**—After applying attention mechanism, now we are predicting the output of visual textual question. To predict the output, we are using vqa_model to predict method which is taking attention weighted image feature map and question embedding as an input and generate 1000 top classified answers. After predicting the answers, we are fetching top three highest accurate answers from the all top 1000 predicted answers. Top-3 answers for this visual question are given below.

Answer generation
Answer—4, Accuracy—0.61.
Answer—2, Accuracy—0.22.
Answer—3, Accuracy—0.02.

4.3 Evaluation Metrics

We are evaluated our visual question answering system through directly comparing generated answer with database ground truth answer.

$$\text{Predicted answer} = \text{ground truth answer}$$

This evaluation metric is useful for knowing that whether the generated answer is correct or not.

4.4 Result Analysis

We are tested visual question answering system for answering the following types of questions with multiple images.

- Counting
- Color
- Object recognition
- Activity recognition

After testing with multiple images, we are analyzing their results on different aspects such as occluded object, small object detected or not while answering a counting type of question, accuracy of all types of questions, time taken for generating each type of visual question answer (Figs. 6 and 7).

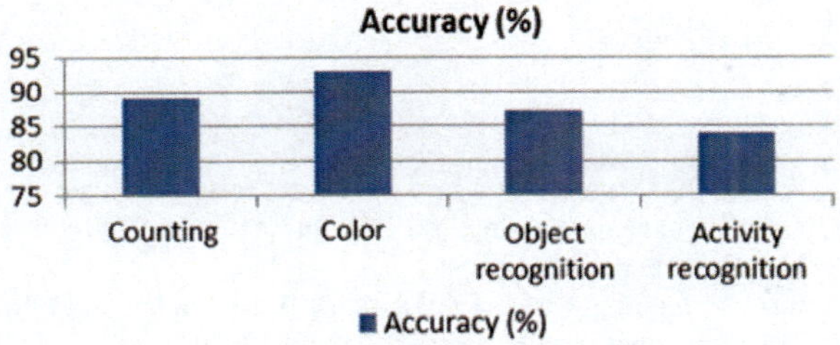

Fig. 6 Visual representation of accuracy for all types of visual question answer

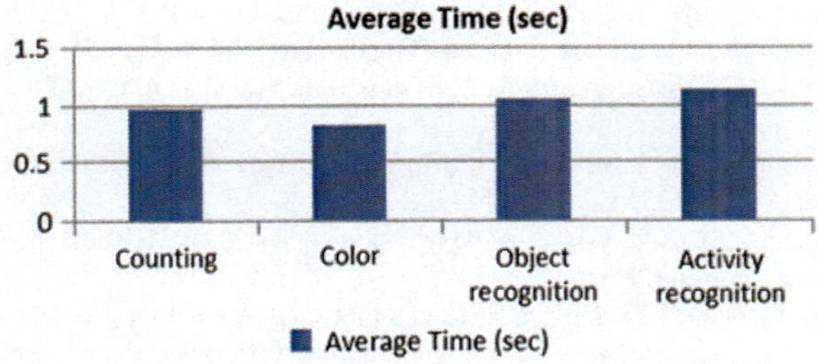

Fig. 7 Average time taken for answering all types of visual question answer

5 Conclusion

In this paper, we learned about VQA and various methods for implementing VQA. Then, we also discussed the need or applications of visual question answering system in real life. While concentrating all strategies of VQA, we have distinguished a few issues in existing VQA framework. The major issues include improper counting of occluded objects, small objects detection, single word answer problem and not able to handle time-specific visual question answer. Considering the issues that occur in visual question answering, we are motivated to handled occluded object counting type of visual question answering problems. To handle this counting object problem, we proposed attention-based visual question answering (A-VQA) method. This system will handle counting, object recognition, color, sport recognition and activity recognition types of questions. In our work, we are using visual genome dataset for training and evaluating VQA system. We have implemented visual feature extraction task, textual feature extraction task, element-wise multiplication operation on both image and question feature matrices. We are also implemented attention mechanism and finally we are predicting and generating answers based on various semantics such as counting, color, activity and object recognition types of the visual question.

References

1. Modi, S., Pandya, D.: VQAR: review on information retrieval techniques based on computer vision and natural language processing. In: 3rd International Conference on Computing Methodologies and Communication (2019)
2. Cambria, E., White, B.: Jumping NLP curves: a review of natural language processing research [review article]. IEEE Comput. Intell. Mag. **9**(2), 48–57 (2014)
3. Dudhabaware, R.S., Madankar, M.S.: Review on natural language processing tasks for text documents. In: Computational Intelligence and Computing Research (ICCIC), IEEE International Conference, pp. 1–5 (2014)
4. Wu, Q., Teney, D., Wang, P., Shen, C., Dick, A., van den Hengel, A.: Visual question answering: a survey of methods and datasets. Comput. Vis. Image Underst. **163**, 21–40 (2017)
5. Kafle, K., Kanan, C.: Visual question answering: datasets, algorithms, and future challenges. Comput. Vis. Image Underst. **163**, 3–20 (2017)
6. Agrawal, A., Lu, J., Antol, S., Mitchell, M., Zitnick, C.L.: Vqa: Visual question answering. Int. J. Comput. Vision **123**, 4–31 (2017)
7. Kafle, K., Kanan, C.: An analysis of visual question answering algorithms. In: Computer Vision and Pattern Recognition (2017)
8. Gao, H., Mao, J., Zhou, J., Huang, Z.: Are you talking to a machine? Dataset and methods for multilingual image question answering. In: Computer Vision and Pattern Recognition (2015)
9. Yang, Z., He, X., Gao, J., Deng, L., Smola, A.: Stacked attention networks for image question answering. In: Computer Vision and Pattern Recognition (2016)
10. Krizhevsky, A., Sutskever, I., Hinton, G.E.: ImageNet classification with deep convolutional neural networks. In: ACM Digital Library (2012)
11. Krishna, R., Zhu, Y., Groth, O., Johnson, J.: Visual genome: connecting language and vision using crowd sourced dense image annotations. In: Computer Vision and Pattern Recognition (2016)

12. Malinowski, M., Rohrbach, M., Fritz, M.: Ask your neurons: a neural-based approach to answering questions about images. In: Computer Vision and Pattern Recognition (2015)
13. Chen, K., Wang, J., Chen, L.C., Gao, H., Xu, W.: ABC-CNN: an attention based convolutional neural network for visual question answering. In: Computer Vision and Pattern Recognition (2016)
14. Aditya, S., Yang, Y., Baral, C.: Explicit reasoning over end-to-end neural architectures for visual question answering. In: Computer Vision and Pattern Recognition (2018)
15. Lu, P., Ji, L., Li, B., Zhang, W., Duan, N.: R-VQA: learning visual relation facts with semantic attention for visual question answering. In: Computer Vision and Pattern Recognition (2018)
16. Norcliffe-Brown, W., Vafeais, E., Parisot, S.: Learning conditioned graph structures for interpretable visual question answering. In: Computer Vision and Pattern Recognition (2018)
17. Hu, R., Andreas, J., Rohrbach, M.: Learning to reason: end-to-end module networks for visual question answering. In: Computer Vision Foundation (2017)
18. Wang, P., Wu, Q., Shen, C., Dick, A., Hengel, A.: FVQA: fact-based visual question answering. In: IEEE Transactions on Pattern Analysis and Machine Intelligence, p. 1 (2017)
19. Wu, Q., Wang, P., Shen, C., Dick, A.: Ask me anything: free-form visual question answering based on knowledge from external sources. In: Computer Vision and Pattern Recognition (2016)
20. Sundermeyer, M., Schlüter, R., Ney, H.: LSTM neural networks for language modeling. In: Thirteenth Annual Conference of the International Speech Communication Association (2012)
21. Simonyan, K., Zisserman, A.: Very deep convolutional networks for large-scale image recognition. In: Computer Vision and Pattern Recognition (2014)

Wireless Sensing with Radio Frequency Identification (RFID): Instrumental in Intelligent Tracking

Praveen Kumar Singh, Neeraj Kumar, and Bineet Kumar Gupta

Abstract Currently, all technologies and associated applications based on Internet of things (IoT) facilitate the user mobility and likely to contribute immensely in all future mobile applications. Radio frequency identifications (RFIDs) along with wireless sensor networks (WSNs) have potential to offer a great value addition on IoT platforms. Intelligent RFID tags powered through independent energy sources when attached to an object and networked through a wireless link allow seamless integration of some of the physical parameters like humidity and temperature in addition to their location data with the user information system on the IoT [1]. This paper provides a brief insight of the RFID technology with its sensing capabilities. It reviews design considerations of RFID-driven wireless sensors from their implementation perspective. It has primarily focused on intelligent operational logistics and its monitoring which have tremendous potential for both civil and defence logistic applications. A two-layer data networked architecture of RFID has been introduced over IoT platform in the proposal which is comprised of an asymmetric RFID tag-reader connectivity along with the interconnected RFID readers linked through the Wi-Fi or cellular networks [2]. This paper also discusses that how ultra wide band (UWB) RFID is considered today as a promising technique for cost-effective wireless sensing and ultra-low power mobile applications.

Keywords Chip-less · ICT · IoT · RFID tags · UHF · Tracking · UWB · Wireless sensors

---

P. K. Singh (✉) · B. K. Gupta
Department of Computer Application, Shri Ramswaroop Memorial University, Lucknow-Deva Road, Barabanki, India
e-mail: praveen.197505@yahoo.com

B. K. Gupta
e-mail: bkguptacs@gmail.com

N. Kumar
Department of CS and IT, Babasaheb Bhimrao Ambedkar University (A Central University), Satellite Campus, Tikarmafi, Amethi, India
e-mail: neerajmtech@gmail.com

P. K. Singh et al. (eds.), *Recent Innovations in Computing*, Lecture Notes in Electrical Engineering 701, https://doi.org/10.1007/978-981-15-8297-4_28

1 Introduction

The Internet of things (IoT) is an ongoing evolution of the ICT with an aim to connect the Internet cloud with the maximum feasible applications. Wireless sensing with radio frequency identification (RFID) has potential as one of the main technological driver to exploit the IoT benefits. All the upcoming RFID tags are likely to be equipped with sensors to interact with their carriers to monitor their adjoining surroundings apart from providing their location information [3]. IoT can facilitate these intelligent tags with a distinctive IP address or identification number associated with such objects, networked through the short-range wireless connectivity with accessible ICT infrastructures that permits the linkages between the physical worlds with the virtual objects over the Internet.

This constant technological evolution with numerous emerging applications can add tremendous value to both civil and operation logistics of defence. There are various IoT-related applications which are being used in logistics to track the associated objects. There are quite a few papers which highlight the significance of wireless sensor networks (WSNs) with RFID for intelligent tracking of both human and non-human objects with an aim to develop the processing of its operations. For instance, RFID-driven temperature loggers can be used to observe and track the distribution of temperature and transportation of objects. Such information is quite handy for prediction of shelf-life of many expendable objects to support the essential insights for intelligent logistics [4–5]. Even though, such applications have opened the yawning inroads to exploits the prospective of chip-less sensors, however, there are certain key issues which need to be addressed with due concerns like:

- To develop the associated peripherals for IoT platforms which can seamlessly integrate the wireless sensing technologies with the available ICT infrastructures;
- To support the integration of global information systems with RFID-driven wireless sensors; and
- Considerable expenditure of intelligent RFID tags and wireless sensing systems with enhanced functionalities to make it a cost-effective solution.

2 RFID as a Technology

A RFID system consists of RFID tags and readers also known as transponders and interrogators, respectively. There are two different types of RFID tags which are known as active and passive RFID tags based on the way they communicate with RFID readers through their respective power source. Active RFID tags hold an internal power source through its batteries. Consequently, a RFID tag's lifetime is restricted by its energy storage. On the other hand, passive RFID tags generally draw their energy through the signal of RFID readers themselves. In a classic passive RFID system, the tag is energized in an interrogating field which transmits the associated data to the RFID reader [6]. Frequency bands and operation principles of

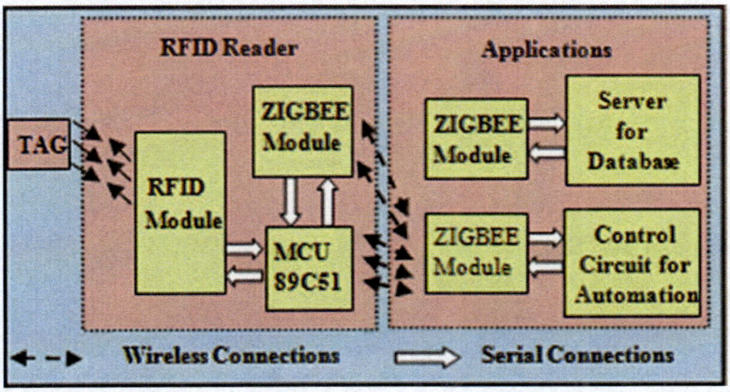

Fig. 1 Block diagram of a RFID system

RFID system fundamentally dominate their applications and performance against their data rates, operating ranges and their costs. Frequencies of RFID system vary from the low frequency (LF) up to microwave and ultra wide bands (UWB).

A block diagram of a basic RFID system has been illustrated in Fig. 1 which is comprised of two blocks, namely RFID reader and its application module. RFID reader consists of a RFID module and a microcontroller. Whenever an object having RFID tag comes into contact with the RFID reader and detected by it, microcontroller through its pre-stored data in ROM verifies and then compares it with the tag ID. In case, the TFID tag ID is legitimate, it is then further broadcasted. The RFID application unit is comprised of a Zig-Bee module having RS232 standard port which is used for a serial communication to a database server. Data transmission over the control circuit of the RFID reader and database server is further processed by a wireless communication [7]. At the same time, control circuit of the application unit receives the RFID tag ID which is mapped through the database server. In this system, all the objects tracked through the RFID tag are automated against their valid tag ID by the control circuit over a predefined profile.

Most of the prevalent passive RFID technologies function in either electromagnetic coupling (backscattering) or electrical/magnetic coupling. The far-field RFID backscattering tags operating in the UHF (860–960 MHz) frequency band are better suited for wireless monitoring and sensing applications. These RFID tags are much more powerful compare to the near-field tags which in turn ensures extended operating ranges, a superior data rate and a relatively smaller antenna dimension. The RFID tag captures continuous waves (CWs) energy from the reader, wherein a power converter can rectify the potential difference of electromagnetic energy over the antenna. The RFID tag transmits the data to RFID reader by means of mechanism of backscattering. The RF modulation is achieved by altering the impedance of antenna with respect to time, so that the RFID tag reflects back the incoming signal in approximation and via a pattern which can encode the tag's ID of a particular RFID [8]. Conversely, the active RFID can generally be linked with medium or short-range wireless applications in ISM band like Bluetooth and Zig-Bee. It typically employs conventional CW modulation by providing better performance over the passive RFIDs in respect to network throughput and operating distance though with higher complexities and power consumption.

3 Intelligent RFID Tag and Wireless Sensing

Now, let us focus on the key characteristics of RFID technology to monitor and track the mobile objects in the IoT context. After reviewing the basic principles of RFID, there is a need to look into sensing applications driven by the RFID and its associated technical challenges [9]. There are many technological features interface circuits, wireless transmissions, miniaturizations, etc., which hold merit to be discussed and thereby being discussed in succeeding paragraphs.

3.1 Sensors and Interface Circuits

The RFID sensing function becomes the basic parameter to draw the desired level of dividends from intelligent tag applications. Prevalent wireless sensor nodes consist of multiple components over the printed circuit board (PCB) offering a powered system in milli-watt with larger batteries for restricted lifetimes and such sensors mostly dominate the overall expenditure of these node. The advancements in RFID technology create the prospects to merge its sensing functionalities with existing ID tags which in turn considerably minimizes its size and the expenditure. Silicon-based sensors which are integrated over the IC chips become the most realistic solution to offer the finest trade-off in respect to performance and cost. For instance, over the complementary metal oxide semiconductor (CMOS) of RFID chip, temperature sensors can be implanted by employing bipolar junction transistors along with delta sigma readouts of conversion from analogue to digital [10–11]. CMOS temperature sensors are dependent over MOS operations which are serially connected having sub-threshold for all such passive RFID applications. However, it is still a challenge to integrate these sensors with their interface circuits over RFID tags powered wirelessly.

These sensors generally have higher consumption of power. For example, the of consumption of RFID applications' temperature sensors amounts to hundreds of μW, while gas and chemical sensors are known to be more complex and having consumption rates in terms of mW which may warrant additional heating provisions for the sensing film. Secondly, analogue sensors data have to be digitalized prior to its storage, processing and transmission. It necessitates low power ADCs to cope up with μW power level against such applications wirelessly powered [12]. In order to resolve this issue, ADCs with μW approximation register can be designed for RFID

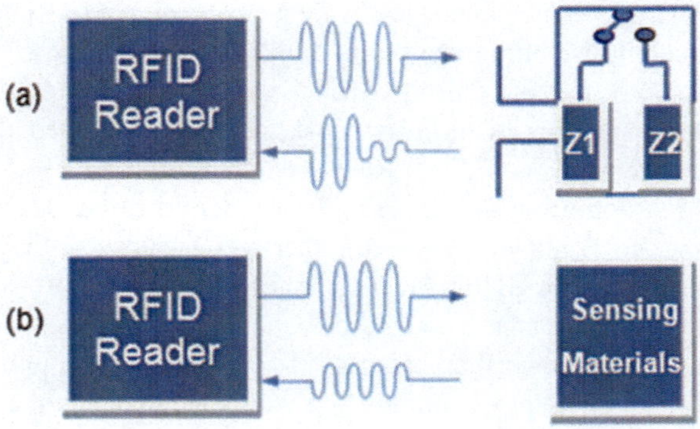

Fig. 2 **a** Traditional RFID backscattering and **b** RFID as a sensor

applications which can facilitate better scalability for performance and power as well as more promising energy efficiency.

In last few years, there have been substantial advancements in printing technologies and large area electronics which have facilitated in the realization of sensors and printed electronics over the diverse substrates mediums like organic substrates and papers. By integrating sensing fabrics like water absorbing fabrics against humidity sensors as well as carbon nanotubes against chemical sensors over the RFID tags, the variations in the electrical characters like impedance or capacitance can be seen through the RFID reader by a backscattered signal, as illustrated in Fig. 2a, b. The temperature and the humidity sensors are most appropriate wireless sensors with RFID against all objects preferably for expendable items in inventory tracking and monitoring applications [13].

3.2 *Wireless Transmission*

RFID technology future extensively lays on enhanced operating ranges as well as higher data rates. The applications in relation to wireless sensing offer asymmetric communications which are dominated by the transmitters. The sensor data which are digitalized not only create hurdles in RFID tag circuit designs, it also significantly increase the requirements for wireless transmission which may be more than 10 m as well as larger than 1 Mbps. The RFID UHF backscattering gives better performance amongst all passive RFID systems [14]. It may be even hundreds of kbps with less than 10 m ranges. In contrast, the active transceivers offer ideal performance which are either very complex or have higher amount of power consumptions in order to cope up with the passive RFID tags having μW power levels.

Amongst all potential RFID systems, the impulse UWB radio is one of the most promising solutions to prevail over most of the existing shortcomings of RFID systems with narrow bandwidth technology. Instead of using CW modulation, it employs ultra-short pulses over different time domains which is spread up to a few

GHz frequency band. The wideband signal demonstrates strong benefits of RFID to accomplish both precise tag localization and high speed identification. Recently, the most prominent chip-less RFID which incorporates the UWB technique for item-level tracking of ultra-low cost has been developed. These developments indicate the possibility of UWB RFID system to offer both localization of RFID tags having high-definition as well as a reliable identification, despite of various complexities in the RFID reader design. Another way to increase the read ranges is by employing semi-passive UHF RFID tags. These tags are comprised of a battery which is similar to the active tags. Though, the battery of these systems can only be used for communications rather than being used with fully active circuitry of associated transceiver [15]. These tags communicate with RFID reader like other passive tags; however, their batteries may be employed to enhance the sensitivity of such tags to RFID reader signals to increase the power of backscattered signal so that better communication ranges can be accomplished.

3.3 Miniaturization and Integration

Intelligent tags attached with any object are aimed to reduce the overall dimension of the system. The future sensor tags' integrations are likely to be beyond existing set up of chip on the system. It may be systems with multi-modules mounted over different substrates or materials. Therefore, it warrants for interconnecting and innovative solutions like system in package or with flexible electronics modules. In addition, the cost of existing RFID antennas and their integration comprised of almost 50% of the overall tag cost which demands for innovative mechanized processes that are required to be exploited. Upcoming technologies like on-chip antennas or printed antennas are estimated to facilitate the RFID system miniaturization and also to reduce its cost substantially. The adoptions of silicon-based RFID chips in the prevalent applications are likely to produce tremendous dividends. Heterogeneous integration may allow assembly of non-silicon materials and silicon circuits over a flexible substrate like sensors, IC, chips, displays and antennas [16]. It cannot only promote excellent exploitation of these upcoming technologies with additional functionalities, it also offer a cost-effective system solution with some other advantages like robustness, enhanced performances, better availability, compatibility, etc.

4 Passive RFID with Ultra Wide Band

4.1 Sensors and Interface Circuits

Unlike existing RFID tags which use ICs, chip-less RFID employs the radar principle in which the tag data of the RFID system is embedded into its electromagnetic

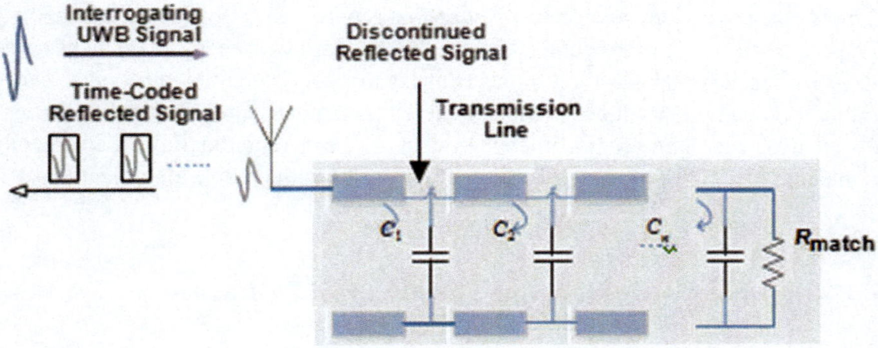

Fig. 3 The principle of chip-less ultra wide band (UWB) RFID tag

signature. It does not require any communication protocol as well as IC, wherein it has no connectivity between antenna and IC. It facilitates a high reliability RFID system with an ultra-low cost which performs an inkjet printing over various substrates on fully printable tags. It is done by encoding the particular ID number through varying impedances over the given transmission line which results in the alternate on and off keying data modulated in time domain by a UWB pulse reflections. An interrogation signal is received by the tag via a UWB antenna from a RFID reader. Thereafter, signal is processed thereafter over a transmission line, wherein it propagates further till it is reflected by a set of encrypted codes from the tag. The reader receives the signal and compares it with original interrogation signal by picking up sequential codes. The entire processing of the tag has been illustrated in Fig. 3 which mainly consists of a series of capacitors and a micro-strip line [17]. The UWB pulse received by the reader propagates over the micro-strip line till it is encountered by a shunt capacitor that may induce an impedance discontinuity by reflecting a part of the signal back. The reflected signal coding is denoted by the digit '1'. The signal keeps propagating forward till it arrives to a point where there will be no capacitor. This state of the particular signal will be represented through coding as '0' or else it will represent the coding by '1'.

4.2 *Passive UHF RFID Having Active UWB Ultra Transmitter*

The chip-less RFID tag is quite promising for low cost applications. Though, its read ranges and the encoding capacity of the system capacity will be limited. In order to address such limitations of this chip-less tag to accomplish an enhanced system capacity with better read ranges, an alternate approach by employing an active UWB transmitter which is powered through a UHF RF energy source may be employed. The power consumption of a UWB receiver is considerably high as well as very complex for its implementation in a system that is wirelessly powered such as RFID. It poses an aggressive duty cycling attribute that can cope with energy level in the tune of μW. Since the transmission of signal from tag to reader predominantly controls the traffic, it demands for a highly efficient energy transmitter with lower complexities [18]. It also necessitates for an efficient system to take on required traffic load as well as the hardware complexities in sensing applications with passive identifications.

5 Monitoring and Tracking Through IoT Platforms

5.1 System Considerations and Application Needs

A major concern in global transportation of expendable goods has been that almost half of them end up in a waste due to their inadequate tracking and monitoring while being transported. It also results in tremendous waste to both civil and military logistics in their global operations with serious financial implications. It necessitates having a desired level of sensing provisions to ensure appropriate measures to avoid such unwarranted wastage. The main causes of this wastage during transportation include biochemical changes, microbial infections, mishandling and physical damages. Introduction of RFID is gradually replacing the existing barcode system of auto-identification of objects aiming to reduce the labour costs and their processing time. Current advances of RFID systems with their networking and sensing capabilities permit constant real-time monitoring of goods in their logistic operations. Real-time monitoring of environmental conditions like humidity and temperature through RFID system during transportation of goods in logistic operations facilitate administrators to take appropriate measures to preserve such goods [19]. The RFID tags for this purpose can be either disposable like printed chip-less RFID tags or reusable like active nodes wireless sensors. In addition, the maintenance and system installation must be simple without requirement of any specialized knowledge or expertise. All such system considerations therefore demand a networked platform which can take on different sensors as well as diverse classes of RFID tags and interfaces.

5.2 System Architecture

An IoT platform which can take on the emerging ICT technologies on existing infrastructure with system compatibility is crucial for wireless sensing applications. Figure 4 illustrates the system architecture of an intelligent RFID tag system that may be deal for constant monitoring and tracking of mobile applications. A two-layer network hierarchy is illustrated in Fig. 5 that connects through an IP gateway to the Internet. Layer 1 in this network is basically a RFID system, wherein RFID

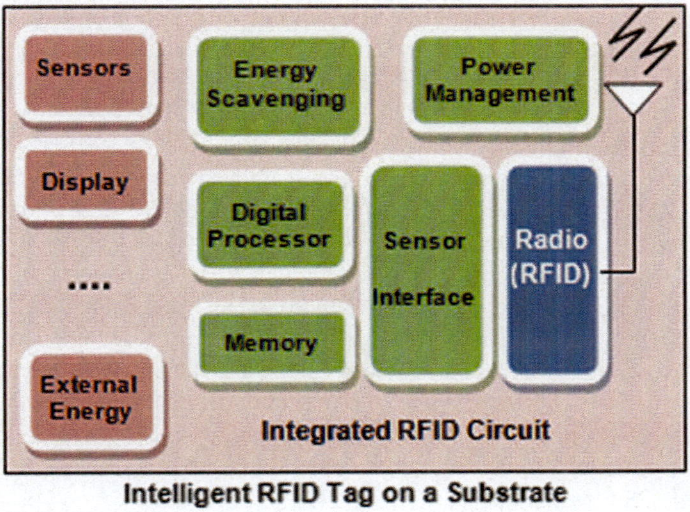

Fig. 4 An IoT platform with intelligent tag

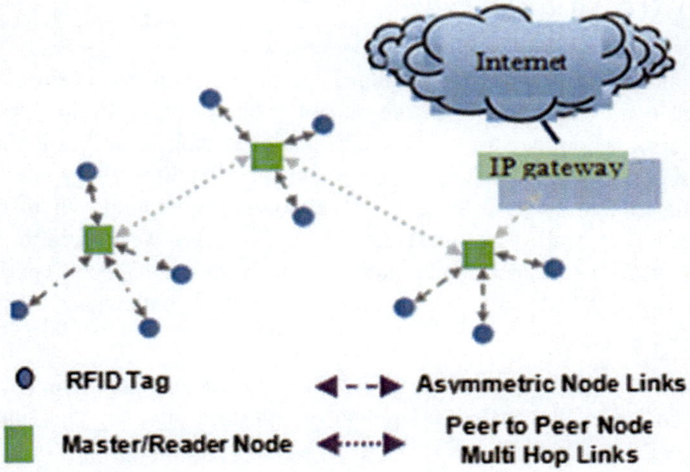

Fig. 5 Two layer hierarchical network of a RFID reader

tags are coordinated through a reader. In layer 2, there are self-organized ad hoc networks as a WSN between master nodes. Layer 1 commonly known as a RFID layer has asymmetrical links of heterogeneous networks which provides wide arrays of RFIDs available for their selection. These intelligent RFID tags are embedded in objects over different packaging like flexible substrates, a paper board or any other type of envelopments and communicate with a RFID reader through micro-powered wireless connectivity [20–21]. The intelligent RFID tags generally consist of a sensor interface, power harvesting block, memory block, a digital processor and also a radio transceiver. The radio interface in this chip-less intelligent RFID may be a typical integrated solution of far-field backscattering or near-field coupling over HF/UHF/any other existing industrial, scientific and medical (ISM) radio frequency bands.

The RFID layer in this system may include a non-standard RFID that typically presents certain unique benefits and lucrative features for sensing applications. For example, a RFID tag based on UWB technologies can offer precise positioning, rapid identification and a time domain sensing. Another pattern relates to printed chip-less RFID tags with intrinsic sensing features through RF measurements with respect to impedance variations over the unstable environmental parameters like changes in humidity and temperature. Sensors and tags in this layer are energy-efficient and very cost-effective solutions under autonomous energy stipulations. The master node in this system architecture acts like a reader to acquire the sensory data of the tag and establish Internet connectivity over diverse IoT platforms [21]. These linkages can be the cloud through different air interfaces like Wi-Fi or GSM/GPRS connectivity's along with GPS for the purpose.

5.3 *Key Technical Challenges*

The introduction of chip-less intelligent tags has potential to revolutionize both human and non-human objects in the pursuit of their constant real-time tracking and monitoring while they are on move. Unlike the simple identification feature in conventional RFIDs, the latest paradigm of chip-less intelligent tags not only offer multiple functionalities such as sensing, networking, computing, these tags have also reduced the overall system cost exponentially [22]. However, still there will be number of issues which need to be addressed to further enhance the performance of this promising RFID technology. Some of those key challenges which relate to several aspects include:

- Enhanced link performance in respect to read ranges and increased data rates.
- Improved integration and development of interface circuitry and sensors into chip-less RFID tags, and
- More sophisticated system integration in terms of dimensional optimization of sensors, integrated circuits (ICs) and antennas on these RFID tags.

6 Conclusion

This paper has reviewed the wireless sensing capabilities of chip-less state of the art intelligent RFIDs and its future trends in real-time object monitoring and tracking in mobile applications. There are tremendous prospects in wireless environmental monitoring supported by the proposed application which makes it very promising in implementation of the future RFID applications. Though, it has still many challenges to overcome with certain limitations of both active and passive RFID technologies. These tags also posses relatively lower data transmission rates as well as inadequate sensing functionalities. On the other hand, active tags may not have similar read ranges and speed restrictions compare to passive RFID tags [23–24]. These tags are relatively costly and also need batteries to operate. In this review, these specific constraints of existing active and passive RFIDs have been highlighted. It also particularly addresses such challenges and suggests the chip-less intelligent RFID as a viable solution in tracking and monitoring wireless sensing applications.

Besides, the proposed architecture can minimize the power needs of the RFID tags drastically and has potential to extend the estimated RFID's battery life almost beyond 8 years. A two-layer RFID wireless sensors data network architecture over an IoT platform suggested in this paper is comprised of an asymmetric connectivity amongst RFID readers connected through the cellular or Wi-Fi services via Internet which facilitate precise monitoring and tracking of mobile objects. Specific approaches to explain the wireless sensors with RFID have been used in this paper for active RFID operating on 2.4 GHz radio, and for a passive RFID, it employs UWB radio [25]. The hierarchical data network architecture as discussed has been though critical in faster adoption of fresh wireless sensing and RFID technologies, however, it facilitates the integrated chip-less intelligent RFID solutions over a networked information system which presents substantial promises in existing ICT infrastructures.

References

1. Sharma, D., Verma, S., Sharma, K.: Network topologies in wireless sensor networks: a review. Int. J. Electron. Commun. Technol. IJECT **4**(3), (2013)
2. Cheng, S.T., Chang, T.Y.: An adaptive learning scheme for load balancing with zone partition in multi-sink wireless sensor network. Expert Syst. Appl. **39**(10), 9427–9434 (2012)
3. Saabith, A.L.S., Fareez, M.M.M., Vinothraj, T.: Python current trend applications-an overview. Int. J. Adv. Eng. Res. Dev. **6**(10), 82–96 (2019)
4. Suryadevara, N.K., Mukhopadhyay, S.C., Kelly, S.D.T., Gill, S.P.S.: WSN-based smart sensors and actuator for power management in intelligent buildings. IEEE/ASME Trans. Mechatron. **20**(2), 564–571 (2015)
5. Shaikh, F.K., Zeadally, S.: Energy harvesting in wireless sensor networks: a comprehensive review. Renew. Sustain. Energy Rev. **55**(3), 1041–1054 (2016)
6. Rajasekaran, C., Jeyabharath, R., Veena, P.: Hardware-software reconfigurable techniques for wireless sensor network. Res. J. Appl. Sci. Eng. Technol. **8**(17), 1855–1862 (2014)
7. Omojokun, G.: A survey of Zigbee wireless sensor network technology: topology, applications and challenges. Int. J. Comput. Appl. **130**(9), 47–55 (2015)
8. Said, M.B., Kacem, Y.H., Kerboeuf, M., Amor, N.B., Abid, M.: Design patterns for self-adaptive RTE systems specification. Int. J. Reconfigurable Comput. 21 p. Article ID 536362 (2014)
9. Ciccozzi, F., Crnkovic, I., Di Ruscio, D., Malavolta, I., Pelliccione, P., Spalazzese, R.: Model-driven engineering for mission-critical IoT systems. IEEE Softw. **34**(1), 46–53 (2017)
10. Khalifeh, R., Yasri, M.S., Lescop, B., Gallee, F., Diler, E., Thierry, D., Rioual, S: Development of wireless and passivecorrosion sensors for material degradation monitoring in coastal zones and immersed environment. IEEE J. Ocean. Eng. **41**(15), 776–782 (2016)

11. Basagni, S., Naderi, M.Y., Petrioli, C., Spenza, D.: Wireless sensor networks with energy harvesting. In: Basagni, S., Conti, M., Giordano, S., Stojmenovic, I. (eds.) Mobile Ad Hoc Networking, Cutting Edge Directions, pp. 701–736. Wiley, Hoboken (2013)
12. Ciccozzi, F., Spalazzese, R.: MDE4Iot: supporting the internet of things with model-driven engineerin. In: Badica, C., El Fallah Seghrouchni, A., Beynier, A., et al. (eds.) Intelligent Distributed Computing X. IDC 2016, Studies in Computational Intelligence, vol. 678, no. 34, pp. 67–76. Springer, Cham (2016)
13. Shit, R.C., Sharma, S., Puthal, D., Zomaya, A.Y.: Location of things (LoT): a review and taxonomy of sensors localization in IoT infrastructure. IEEE Commun. Surv. Tutorials **20**(3), 2028–2061 (2018)
14. Portocarrero, J.M.T., Delicato, F.C., Pires, P.F., Batista, T.V.: Reference architecture for self-adaptive management in wireless sensor networks. In: Bouchachia, A. (ed.) Adaptive and Intelligent Systems, ICAIS 2014, Lecture Notes in Computer Science. Springer, Cham (2014)
15. Somov, A., Baranov, A., Savkin, A., Ivanov, M., Calliari, L., Passerone, R., Karpov, E., Suchkov, A.: Energy-aware gas sensing using wireless sensor networks. In: Picco, G.P., Heinzelman, W. (eds.) Wireless Sensor Networks, pp. 245–260. Springer, Berlin (2012)
16. Huang, H.Y.: Antenna sensors in passive wireless sensing systems. In: Chen, Z.N. (ed.) Handbook of Antenna Technologies. Springer, Singapore (2015)
17. Rodriguez-Zurrunero, R., Utrilla, R., Rozas, A., Araujo, A.: Process management in IoT operating systems: cross-influence between processing and communication tasks in end-devices. Sensors **19**(4), 805–814 (2019)
18. Carlos Mancilla, M., Olascuaga Cabrera, J.G., López Mellado, E., Mendez Vazquez, A.: Design and implementation of a robust wireless sensor network. In: Proceedings of the 23rd International Conference on Electronics, Communications and Computing (CONIELECOMP '13), Cholula, Mexico, pp. 230–235 (2013)
19. Durisic, M.P., Tafa, Z., Dimic, G., Milutinovic, V.: A survey of military applications of wireless sensor networks. In: 2012 Mediterranean Conference on Embedded Computing (MECO), Bar, Montenegro, pp. 196–199 (2012)
20. Boonma, P., Somchit, Y., Natwichai, J.: A model-driven engineering platform for wireless sensor networks. In: 2013 Eighth International Conference on P2P, Parallel, Grid, Cloud and Internet Computing, Compiegne, France, pp. 671–676 (2013)
21. Toor, A.S., Jain, A.K.: A survey on wireless network simulators. Bull. Electr. Eng. Inform. **6**(1), 62–69 (2017)
22. Paulon, A.R., Frohlich, A.A., Becker, L.B., Basso, F.P.: Wireless sensor network UML profile to support model-driven development. In: 2014 12th IEEE International Conference on Industrial Informatics (INDIN), Porto Alegre, Brazil, pp. 227–232 (2014)
23. Jacoub, J.K., Liscano, R., Bradbury, J.S., Fisher, J.: UML modeling of design patterns for wireless sensor networks. In: Proceedings of the 2nd International Conference on Sensor Networks, vol. 1: SENSORNETS, Barcelona, Spain, pp. 89–93 (2013)
24. Singh, P.K., Kar, A.K., Singh, Y., Kolekar, M.H., Tanwar, S.: Proceedings of ICRIC 2019, Recent Innovations in Computing, 2020. Lecture Notes in Electrical Engineering, vol. 597, pp. 3–920. Springer, Cham (2019)
25. Singh, P.K., Bhargava, B.K., Paprzycki, M., Kaushal, N.C., Hong, W.C.: Handbook of Wireless Sensor Networks: Issues and Challenges in Current Scenario's. Advances in Intelligent Systems and Computing, vol. 1132, pp. 155–437. Springer, Cham (2020)

Impact of Distortions on the Performance of Feature Extraction and Matching Techniques

Richha Sharma and Pawanesh Abrol

Abstract An image feature, such as edges and interest points, provides rich information on the image content and plays an important role in the area of image processing. These correspond to local regions in the image and are fundamental in many applications in image analysis. Raw data are complex and difficult to process without extracting or selecting appropriate features in advance. Feature extraction, a data reduction technique, is the transformation of large input data into a low-dimensional feature vector. It lowers the computational cost and also helps in controlling the issue of dimensionality. There are different methods of exacting features from an image and these techniques have different domains of applications. In this paper, four widely used feature detection algorithms, Harris, SURF, FAST, and BRISK feature detection algorithms are compared in terms of accuracy and time complexity for extraction and matching of feature points correctly. For this purpose, different types of transformations are added to the original images for computing the evaluating parameters like the number of features detected, matched features, and execution time required by each algorithm. Experimental results show that SURF performs better than other feature extractions and matching algorithms in terms of accuracy and run time.

Keywords Feature extraction · Feature point · SURF · BRISK · FAST · Harris

1 Introduction

In a recent couple of decades, real-time applications in feature extraction have been a challenging field for image processing specialists. It has wide use in the field of satellite imaging, 3D image reproduction, and computer vision fields [1]. The huge amount of data generated has grown multiple times with millions or trillions

R. Sharma (✉) · P. Abrol
Department of Computer Science and IT, University of Jammu, Jammu, Jammu and Kashmir, India
e-mail: richha09@gmail.com

P. Abrol
e-mail: pawanesh.abrol@gmail.com

P. K. Singh et al. (eds.), *Recent Innovations in Computing*, Lecture Notes in Electrical Engineering 701, https://doi.org/10.1007/978-981-15-8297-4_29

of data sets generated every day with Internet and social media. Feature extraction is a critical task that involves a huge amount of data as input and transforming it into an optimal feature set [2]. Steps in feature extraction consist of an initial set of data values that helps in selecting and extracting features intended to be informative and non-redundant. It facilitates the subsequent learning and generalization steps leading to better interpretations. In addition to lowering the computational cost, feature extraction also deals with the so-called curse of dimensionality [3]. In computer vision, feature detection and matching are two of the important problems and have a wide range of applications. An ideal feature detection technique should be robust to distortions, detecting the same features from a completely different image set. Also, it should be able to detect distinctive features matched with high probability. Distortion is a change or twist that makes an image appear different from the way it is. Distortion can be added using image transformation functions that map one set to another set after performing some operations. It can either change the size or rotate or change the position of the image. Scaling alters the size of an object either by compressing or expanding the dimension of the object. A scaling factor if less than 1 reduces the size of the object and if greater than 1 its size increases. Rotation turns a figure about a fixed point called the center of rotation without altering its shape and size. It turns the figure in a different direction, clockwise or anticlockwise. Translation moves every point of a figure by the same distance in a given direction along the straight line. The size, shape, and orientation of the figure are the same.

Feature matching, generally known as image matching is a part of many computer vision applications like image registration, object identification, and object recognition. It is done by establishing correspondences between two images of the same scene/object. Steps in image matching consist of detecting a set of interest points each associated with image descriptors from image data. After the extraction of features from the image set (original and distorted), corresponding feature points are matched. Matching shows that features are from the corresponding locations from completely different images.

In this paper, feature detection algorithms have been discussed and implemented for feature extraction and feature matching using different images. Features are detected using Harris corner detection, Speeded-Up Robust Features (SURF), Features from Accelerated Segment (FAST), and Binary Robust Invariant Scalable Key-points (BRISK) feature detection algorithms [4, 5]. Images with different types of background for feature extraction have been considered to calculate the comparison parameters for different feature detection techniques. Feature correspondences between the original images and the distorted images are shown. Based on this resultant feature values and total execution time for each algorithm, the comparison has been made.

The rest of this paper is organized as follows: Sect. 2 presents an overview of the feature detection and matching techniques. In Sect. 3, the methodology of the work done is explained. Section 4 presents the experimental results and performance analysis of selected techniques. The sensitivity of SURF, FAST, BRISK, and Harris against each of distortion parameters like rotation, scaling, and translation is shown and accuracy of matched points is evaluated. Section 5 concludes the paper.

2 Related Work

In this section, the feature detection techniques are briefly described. Different methods have been proposed for the extraction of features from images. The process of feature extraction is the conversion of large input data into a low-dimensional feature vector. It is a type of data reduction technique [6] and a necessary step to extract the informative feature from an image for object recognition and segmentation [7].

Scale invariant feature transform is a highly distinctive feature extraction technique, i.e., there is a high probability of matching a single feature from many images. This method is scale and rotation invariant (can tolerate up to about 60° out of plane rotation). It is partially invariant to changes in illumination [8, 9]. Features can be computed fast and efficiently.

Speeded-up robust feature detector emphases on blob-like structures in the image. It speed-up computations by fast approximation of (i) Hessian matrix, (ii) descriptor using integral images. SURF has three fundamental steps. Firstly, key feature points such as edges, corners, blobs, and T-intersection at distinctive regions are chosen in the image. The second step makes use of a feature vector to depict the surrounding neighborhood of each feature point [10–12]. A unique descriptor must be used. Finally, a fast Hessian detector is used for finding feature points. SURF descriptors are invariant to rotation and scaling with low computational time. SURF is fast in case of interest point localization and matching.

Harris corner detection algorithm detects feature points by the means of a local detecting window inside the image is shifted in different directions to determine the average variation in the pixel intensity [13–15]. The corner point is the center point of the window and a large variation in pixel intensity is seen on shifting the window in any of the directions.

Features from Accelerated Segment are a corner feature detection method that is known to be faster than many other well-known feature extraction methods. This method is not robust to high levels of noise [16, 17]. This method uses a circle of 16 pixels (a Brenham's circle of radius 3) to classify a corner.

Binary Robust Invariant Scalable Key-points (BRISK) estimates key-point scale in continuous scale-space neighborhood sampling (points are deterministic, equally spaced and in concentric circles) is used in this technique. Using a sampling pattern, short pairs, and long pairs are distinguished [18]. It is done based on a threshold value. Distance between short pairs are below a certain threshold d_max and long pairs in sampling pattern have distance above a certain threshold d_min. Long pairs are used in BRISK to determine the orientation and short pairs are used for the intensity comparisons that build the descriptor. This technique has a lower computational complexity.

Binary Robust Independent Elementary Features (BRIEF), a low bit rate descriptor, are introduced for image matching with random forest and random ferns classifiers [19]. It belongs to the family of binary descriptors such as LBP and BRISK, which only performs a simple binary comparison test using Hamming distance.

Oriented BRIEF is a binary descriptor [20, 21] based on BRIEF. It is fast, rotation invariant [22], and resistant to noise. It is appropriate for applications requiring high real-time and large-scale changes [23].

The reviewed literature showed the scope of exploring these methods for the analysis of feature matching accuracy and time is taken for different real-world input images [24] with added distortions. The work in the paper demonstrates the performance analysis of the accuracy of matched feature points in original and distorted images using four techniques namely SURF, FAST, BRISK, and Harris corner detector.

3 Methodology

The objective of this paper is to observe the impact of added distortions on the accuracy of matched features from original and distorted images. Data set as shown in Fig. 2 consists of real-world images with varying backgrounds. First, for each image in the input set, a set of interest points are selected and their local feature descriptors are constructed using all the four algorithms (SURF, FAST, BRISK, Harris). Then, by selecting the strongest feature point, representative points from the detected points are selected which deliver rich and distinguishing information about the image for recognition. For a descriptor, the representative point is an interest point having enough similar interest points. Features are extracted from each input image using all the selected algorithms with a metric threshold as 1000. Execution time and the number of detected features are noted for performance evaluation. After getting the extracted features, matching is done by using all techniques as shown in Fig. 1. Input image I_o (original image) is read and distortions are added to it. The resultant image is I_d. Features are extracted using different algorithms separately each for original and distorted images. Matching for the extracted features is done for the set of images (I_o, I_d) in each case.

Matching accuracy (%) is calculated as $(I_{mf}/I_{ef}) * 100$

where I_{mf} = Total number of matched features, I_{ef} = Total number of extracted features from the original image.

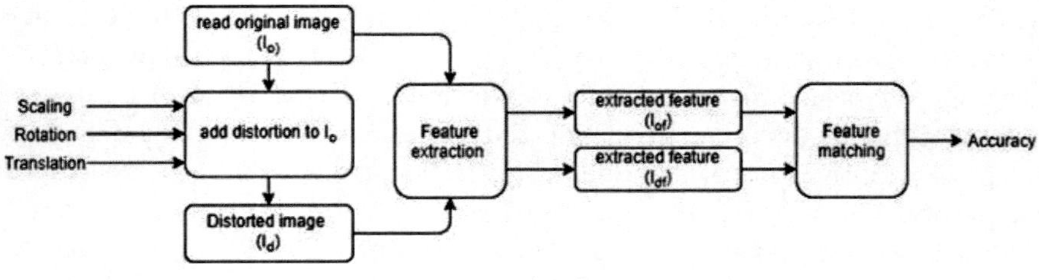

Fig. 1 Block diagram for finding accuracy of matched features

a b c d e

Fig. 2 Input image data set

4 Experimental Results and Analysis

The implementation is done in MATLAB R2018a. Image features are extracted from the input image using different feature extraction methods. The results obtained from comparing different feature detection algorithms based on the number of features extracted for each image and time complexity are presented in Table 1, where N_{fd} is the number of features detected.

As can be observed from Table 1, although BRISK detects the maximum number of features in most of the input images, the time efficiency is the lowest; SURF detects a good amount of features and time efficiency is better compared to other algorithms. Figure 3 shows the graphs plotted for comparison of different feature detection algorithms based on the number of detected features and based on time complexity, respectively.

Table 1 Time taken for feature detection using different input images

Algorithm	N_{fd}	Time (s)	N_{fd}	Time (s)	N_{fd}	Time (s)	N_{fd}	Time (s)	N_{fd}	Time (s)
	a		b		c		d		e	
SURF	296	0.043	36	0.039	244	0.070	669	0.042	464	0.042
FAST	325	0.044	17	0.038	317	0.095	524	0.042	475	0.039
BRISK	571	0.415	49	0.427	446	0.678	768	0.361	926	0.471
Harris	222	0.053	48	0.053	189	0.205	714	0.091	348	0.044

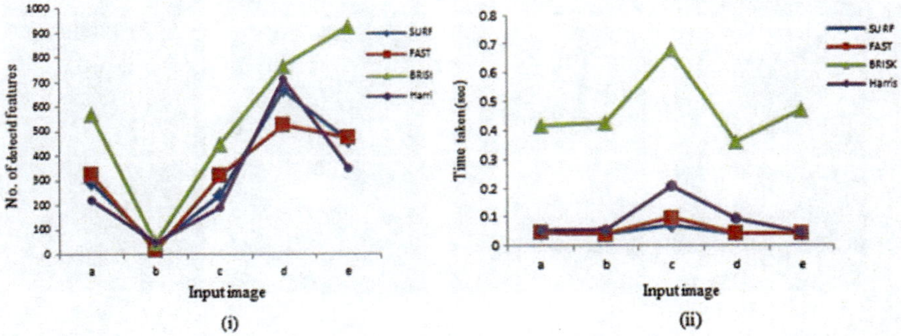

Fig. 3 Performance evaluation of (i) number of features detected, (ii) run time for image dataset using SURF, FAST, BRISK and Harris

Figure 4 shows the input image set along with resultant images with added distortions. Distortions are transformations applied to the images. Column (i) shows the original image in the set. (ii) Shows the result after scaling the input images by the factor of 1.2. Similarly (iii) shows the results of images when rotated to an angle of +30° and (iv) gives the resultant images when translated by the 15, 25 in x and y directions.

The resultant matched points obtained in the original and distorted image set after added distortions are shown in Fig. 5. Points from both the images are shown. Table 2 gives the number of extracted features in the original and distorted image sets with scaling of 1.2, rotation degree of +30, and translation of 15, 25 in x and y direction, respectively, for all algorithms. It also shows the accuracy of matched features using the number of features extracted from the original and distorted image set. Accuracy defines the percentage of correctly matched features.

Figure 6 shows the graph plotted for comparison of different feature detection algorithms based on accuracy for image transformation techniques. It is also observed that SURF has the best accuracy for feature matching among all the feature detectors for all transformations with a scaling factor of 1.2, rotation degree of 30, and translation factor of $x = 15$, $y = 25$. It should also be noted that the accuracy is a relative term that can vary from image to image. BRISK has poor accuracy in terms

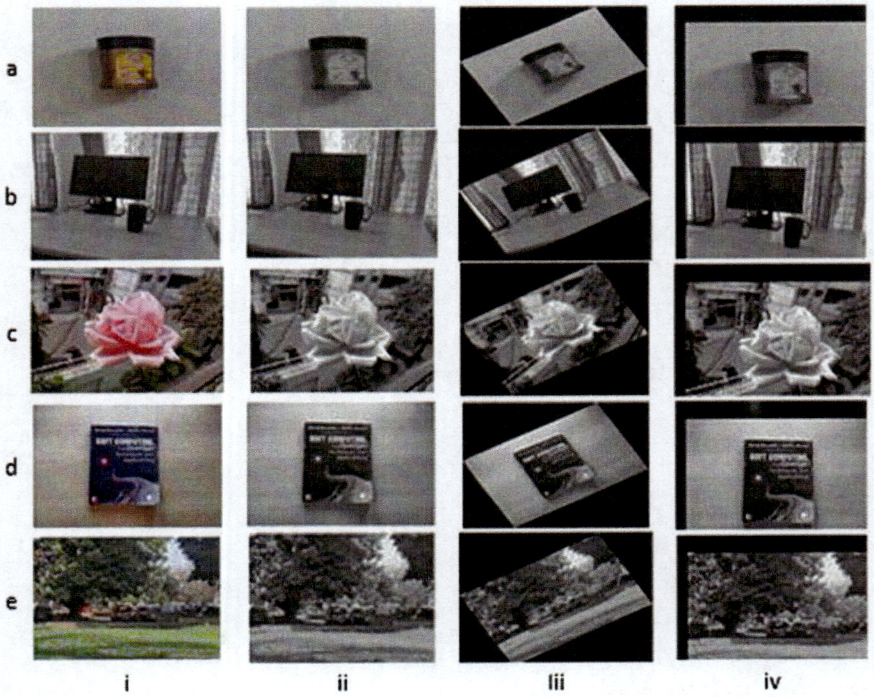

Fig. 4 Transformation on input image set [**a, b, c, d, e**]. (i) Original, (ii) scaled image with factor 1.2, (iii) rotated image with +30°, (iv) translated image with $x = 15$, $y = 25$

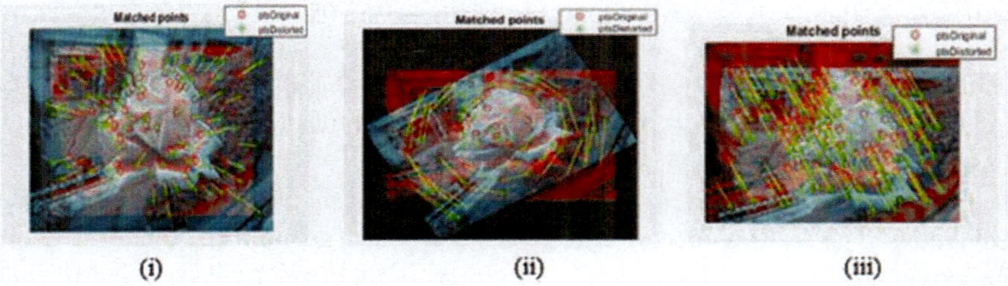

Fig. 5 Matching of feature points in the original and distorted image with (i) Scaling (1.2). (ii) Rotation (+30). (iii) Translation of $x = 15$, $y = 25$

Table 2 Performance of algorithms for original and distorted image set

Algorithm	Operation	Extracted features			
		Original image	Distorted image	Matched features	Accuracy (%)
SURF	Scaling	185	243	099	53.50
	Rotation	185	243	048	25.94
	Translation	185	189	157	84.86
FAST	Scaling	325	318	033	10.15
	Rotation	325	465	052	16.00
	Translation	325	303	237	72.92
BRISK	Scaling	571	607	056	09.80
	Rotation	571	775	023	04.02
	Translation	571	539	217	38.00
Harris	Scaling	222	261	023	10.36
	Rotation	222	265	018	08.10
	Translation	222	183	148	66.66

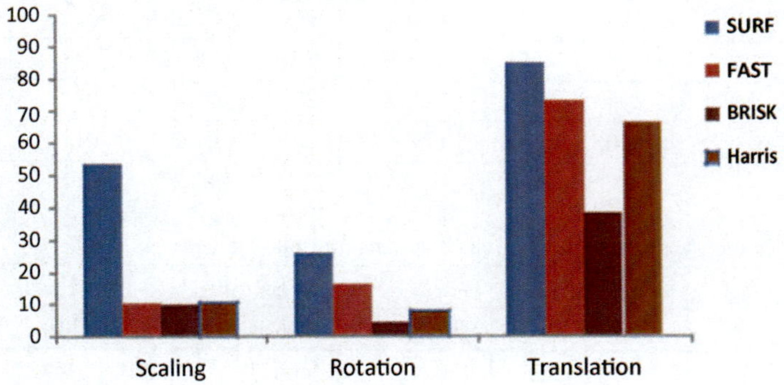

Fig. 6 Accuracy comparison of scaling, rotation, and translation

of feature matching between original and distorted image while FAST and Harris show the mediocre performance.

The results of transformations with varying factors of scaling, rotation, and translation are presented in Tables 3, 4, and 5, respectively, where MF is the number of matched features, and Acc % is the accuracy.

Table 3 shows the accuracy of matched features using all the four feature matching techniques. The scaling factor varies from 0.5 to 1.5. It is observed that SURF has the maximum accuracy in matching features while the performance of Harris and FAST is poor.

Similarly, it can also be observed from Tables 4 and 5 that SURF has maximum accuracy for rotation and translation with all values, i.e., SURF is more rotation and translation invariant as compared to other algorithms used in this implementation. It can also be observed that all algorithms performed better when rotation was 90 or

Table 3 Accuracy for comparing the image with its scaled image

Scaling factor	SURF		FAST		BRISK		Harris	
	MF	Acc (%)	MF	Acc (%)	MF	Acc (%)	MF	Acc (%)
0.5	027	14.595	02	0.615	13	02.277	1	0.45
0.8	066	35.676	19	05.846	31	05.429	8	03.604
1.2	099	53.514	33	10.154	56	09.807	23	10.36
1.5	107	57.838	03	00.923	417	73.03	3	01.351

Table 4 Accuracy for comparing the image with its rotated image

Rotation	SURF		FAST		BRISK		Harris	
	MF	Acc (%)	MF	Acc (%)	MF	Acc (%)	MF	Acc (%)
60	047	25.405	053	16.308	027	04.729	013	05.856
90	173	93.514	249	76.615	154	26.97	151	68.018
120	048	25.946	047	14.462	026	04.553	017	07.658
180	168	90.811	281	86.462	186	32.574	193	86.937
−60	048	25.946	048	14.769	037	06.48	014	06.306
−120	047	25.405	058	17.846	028	04.904	015	06.757

Table 5 Accuracy for comparing the image with its translated image

Translation factor	SURF		FAST		BRISK		Harris	
	MF	Acc (%)	MF	Acc (%)	MF	Acc (%)	MF	Acc (%)
[15, 25]	157	84.865	237	72.923	217	38.004	148	66.667
[−15, 25]	152	82.162	254	78.154	219	38.354	153	68.919
[15, −25]	131	70.811	188	57.846	177	31.349	113	50.901
[−15, −25]	130	70.270	195	60.000	179	31.349	114	51.351

180°. Translated images also showed better matching accuracy for all varying values. Image is translated in all four quadrants and resultant accuracy is calculated. It can be said that each algorithm performed better for translated images. The same results can be also observed from Fig. 7.

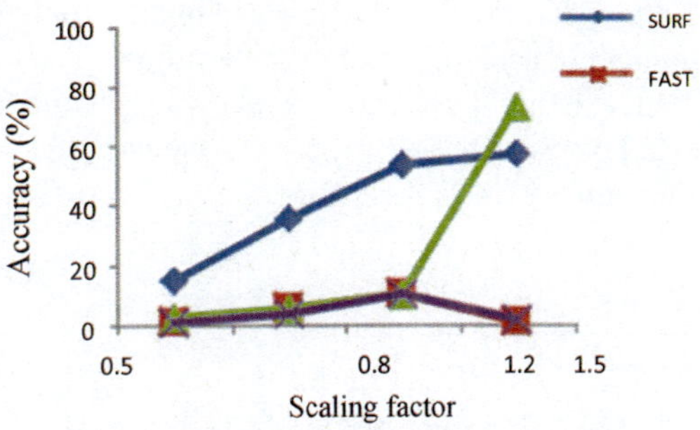

(i) Accuracy versus scaling factor plot

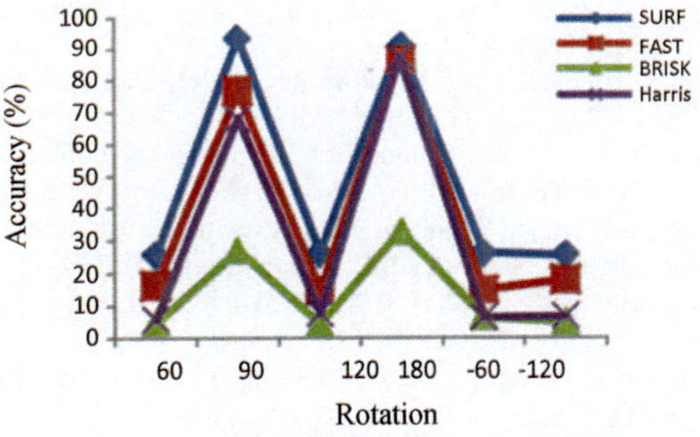

(ii) Accuracy versus rotation factor plot

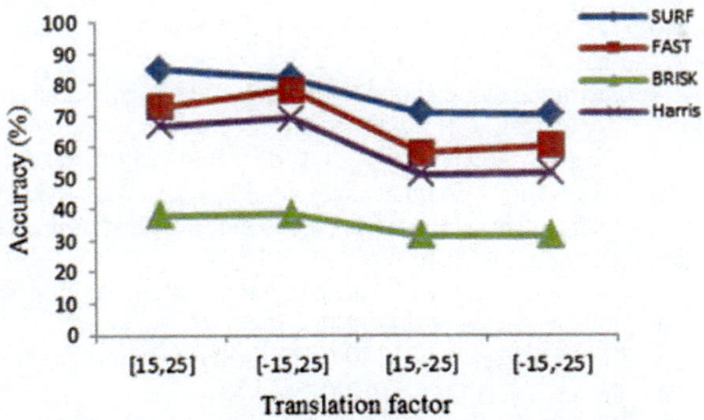

(iii) Accuracy versus translating factor plot

Fig. 7 Accuracy for added distortions. (i) scaling, (ii) rotation, (iii) translation

5 Conclusion

The work presented in this paper focused on evaluating and comparing the performance of four different feature detection and matching techniques (SURF, FAST, BRISK, and Harris). These algorithms were evaluated for different transformations, scaling, rotation, and translation applied to original images for finding the number of detected features, matching accuracy, and execution time. The results revealed that SURF performed better than other algorithms showing invariance to added distortions. Harris and FAST showed the average performance. BRISK performed worst in case of feature matching.

References

1. Bhosale Swapnali, B., Kayastha Vijay, S., Harpale Varsha, K.: Feature extraction using SURF algorithm for object recognition. Int. J. Tech. Res. Appl. **2**(4), 197–199 (2014)
2. Banerjee, A., Mistry, D.: Comparison of feature detection and matching approaches: SIFT and SURF. Glob. Res. Dev. J. Eng. **2**(4) (2017)
3. Sharma, A., Abrol, P., Lehana, P.K.: Accuracy of point cloud estimation for tiny objects. Int. J. Mod. Comput. Sci. **4**(3), 142–147 (2016)
4. Ghosh, P., Pandey, A., Pati, U.C.: Comparison of different feature detection techniques for image mosaicing. ACCENTS Trans. Image Process. Comput. Vision **1**(1), 1–7 (2015)
5. Goel, R., Kumar, V., Srivastava, S., Sinha, A.K.: A review of feature extraction techniques for image analysis. Int. J. Adv. Res. Comput. Commun. Eng. **6**(2), 153–155 (2017)
6. Juan, L.: A comparison of SIFT, PCA-SIFT, and SURF. Int. J. Image Process. **3**(4), 143–152 (2009)
7. Tian, D.P.: A review on image feature extraction and representation techniques. Int. J. Multimedia Ubiquitous Eng. **8**(4), 385–396 (2013)
8. Kumar, G., Bhatia, P.K.: A detailed review of feature extraction in image processing systems. In: International Conference on Advanced Computing and Communication Technology, pp. 5–12 (2014)
9. Kumar, P., Biswas, A., Chandra, M.: Feature extraction methods. J. Telecommun. **1**(2), 11–15 (2010)
10. Kumar, R.M.: A survey on image feature descriptors. Int. J. Comput. Sci. Inf. Technol. **5**(6), 7668–7673 (2014)
11. Bheda, D., Joshi, M., Prof, A., Agrawal, V.: A study on features extraction techniques for image mosaicing. Int. J. Innovative Res. Comput. Commun. Eng. **2**(3), 3432–3437 (2014)
12. Kumbhar, P.: A survey on feature selection techniques and classification algorithms for efficient text classification. Int. J. Sci. Res. **5**(5), 1267–1275 (2016)
13. Medjahed, S.A.: A comparative study of feature extraction methods in images classification. Int. J. Image Graph. Sig. Process. **3**, 16–23 (2015)
14. Pachouri, K.K.: A comparative analysis & survey of various feature extraction techniques. Int. J. Comput. Sci. Inf. Technol. **6**(1), 377–379 (2015)
15. Panchal, P.M., Panchal, S.R., Shah, S.K.: A comparison of SIFT and SURF. Int. J. Innovative Res. Comput. Commun. Eng. **1**(2), 323–327 (2013)
16. Salahat, E., Qasaimeh, M.: Recent Advances in Features Extraction and Description Algorithms: A Comprehensive Survey. arXiv preprint arXiv:1703.06376v1 (2017)
17. Saleem, Z.: A comparative analysis of SIFT, SURF, KAZE, AKAZE, ORB, and BRISK. In: International Conference on Computing, Mathematics and Engineering. IEEE (2018)

18. Calonder, M., Lepetit, V., Ozuysal, M., Trzcinski, Strecha, T.C., Fua, P.: BRIEF: computing a local binary descriptor very fast. IEEE Trans. Pattern Anal. Mach. Intell. **34**, 1281–1298 (2012)
19. Kharabe, S., Nalini, C.: A review of feature extraction methods in image processing. Int. J. Innovative Res. Manage. Eng. Technol. **3**(4), 131–135 (2018)
20. Sharma, A., Abrol, P., Lehana, P.K.: AgroVaid: a user-friendly agriculture system for enhanced farmer interaction. Int. J. Sci. Tech. Advancements **2**(2), 47–51 (2016)
21. Karami, E., Prasad, S., Shehata, M.: Image matching using SIFT, SURF, BRIEF, and ORB: performance comparison for distorted images. arXiv preprint arXiv:1710.02726 (2017)
22. Wang, R., Shi, Y., Zhang, W., Cao, W., Wang, X.: GA-ORB: a new efficient feature extraction algorithm for multispectral images based on geometric algebra. IEEE Access **7**, 71235–71244 (2019)
23. Chen, Z., Hu, Y., Zhang, Y.: Effects of compression on remote sensing image classification based on fractal analysis. IEEE Trans. Geosci. Remote Sens. **57**(7), 1–14 (2019)
24. Nixon, M., Aguado, A.: Feature Extraction and Image Processing for Computer Vision, 4th edn. Academic Press Elsevier, London (2020)

Realization of a Robust Watermarking System in Spatial Domain

Ishrat Qureshi, Shabir A. Parah, Nazir A. Lone, Nasir Hurrah, and G. J. Qureshi

Abstract In this paper, we have proposed a watermarking system that is based on the spatial domain and is blind and robust in nature. This scheme is developed to withstand most of the image processing attacks. The watermark has been embedded in the cover image by modification of the DC coefficients calculated in the spatial domain. This method directly calculates DC coefficients in the spatial domain. The values of pixels can be changed/modified in the spatial domain in accordance with available watermark information. Since we have avoided the time-consuming transform operation, i.e., Discrete Cosine Transform, the computational efficiency is very high. For embedding a watermark bit, a particular image is disintegrated into 8 × 8 blocks, followed by further division of each block into two 4 × 4 blocks. After calculation of the DC coefficient of each 4 × 4 block, the watermark bit is embedded by modifying the DC values such that the DC coefficient of one block becomes greater than the other. The output results prove our proposed technique is highly robust to commonly occurring signal processing attacks. Experimental results obtained against numerous signal processing attacks are represented in terms of quality measuring parameters like PSNR, SSIM, and BER to check the efficiency and execution of our scheme.

Keywords Robustness · Watermarking · Imperceptibility · Computational complexity

I. Qureshi
Electronics and Communication Department, Affiliated College of University of Kashmir, Srinagar, Jammu and Kashmir, India

S. A. Parah (✉) · N. A. Lone · N. Hurrah
Department of Electronics and IT, University of Kashmir, Hazratbal, Srinagar, Jammu and Kashmir, India
e-mail: shabireltr@gmail.com

G. J. Qureshi
Department of Higher Education, Government of Jammu and Kashmir, Srinagar, Jammu and Kashmir, India

P. K. Singh et al. (eds.), *Recent Innovations in Computing*, Lecture Notes in Electrical Engineering 701, https://doi.org/10.1007/978-981-15-8297-4_30

1 Introduction

Millennials and people of present times are considered digital natives, they are more tech-savvy than their previous generations. Yet, they are easy targets for cybercriminals. A large percentage of millennials across the world use public wi-fi connections, they are always ready to answer survey questions, almost all of them install third-party apps, and almost everyone once in a while provides access to files while we are online. Technology has made its way in almost every sector and institution of life and the results have been fascinating. The smart incorporation of technology is an essential part of success and survival in this era of the internet. We see technology influencing almost every sector be it commerce, finance, research and development, entertainment industry, healthcare, legacy industries like hospitality, construction, and retail, etc. These days we incorporate technologies to improve our operations like AI (Artificial Intelligence), ML (Machine Learning), etc. to make a better impact. Industries and economies have grown and are growing at a tremendous rate because of the changes in technological availability. Healthcare is in crisis particularly in developing nations and rural areas but despite problems health care is improving due to the sub-sectors like biotech, health data management, etc. Technology has drifted the financial system towards financial tech, e-banking, and cryptocurrencies. Similarly, real-estate has drifted towards online brokerages and other high-end online techniques. Similarly, legacy industries have been here for generations now but they are also overcoming their obstacles by using online techniques and resources [1–6]. The use of the internet in these sectors involves data handling which comprises sending, receiving, and transferring information day in and day out, from billions of individuals and gadgets. In several complex networks and systems, huge data is shared in real-time to carry out different automatic or non-automatic tasks. This kind of data may include patient records, banking/financial records, and secret or classified documents in various forms like images, text files, audios, and videos, gained in numerous ways and processed and stored by corresponding systems. With the exponential growth of the Internet and the multimedia systems in distributed environments, access to multimedia data has become much easier. User-generated content like digital images, audios or videos, and different kinds of data generated through all online interactions, is the lifeblood of social media. However, current technology does not protect their copyrights effectively [7–12]. In this world of internet, smartphones, social media, and the availability of cost-free state-of-the-art multimedia editing tools almost everyone is sharing data (images, videos, audios, movies, etc.) online. This opens doors for people who want to misuse it by copying it, editing it, and finally reproducing it without legal authority and permission. The fact is, whenever we share/put anything on the internet space, it is prone to theft. So this is also the time of massive data out breaches caused by various cyber-attacks. In the times of unlimited data, there are massive data out breaches and heightened vulnerability. A single data breach alone compromises the data of billions of people. A massive breach can cost billions of dollars in losses and can have a devastating impact on those people whose personal details are stolen and compromised. In a single attack billions of encrypted credit card records, logins, passwords, and other classified information can be compromised. So, multimedia security becomes an important issue in nowadays Internet world. Everyone is doing more to keep personal information safe online and major IT firms are investing more on online security as a precaution in this age of increasing vulnerability mainly the concentration of their efforts for developing security strategies has been in sectors like digital banking, entertainment industry, healthcare, research, army, and security department, social media, ongoing cryptocurrency projects like LIBRA. Various technologies are used to take care of the copyright and security issues and amongst the most effective technologies that contribute to developing an effective copyright protection system is digital watermarking [13–19]. Digital watermarking is a technique of embedding copyright right information called watermark into the multimedia, which is to be facilitated with copyright protection, without leaving any trail to the human audio-visual system. Robustness of a watermarking system is a very important parameter; however, it is usually traded off with computational complexity.

In our paper, we have proposed a novel blind and robust watermarking method in which watermark information is embedded directly by changing/modifying the pixel values of an image. Our proposed technique/algorithm calculates the DC coefficient in the spatial domain directly. According to the watermark information and the modification factor, the values of pixels can be properly modified in the spatial domain. Our proposed algorithm keeps the distribution feature of the frequency coefficients and avoids the errors which may result from the frequency transformation. Our proposed system for watermarking is not difficult and has extended efficiency.

Main contributions of the proposed work are:

1. High robustness against signal processing attacks.
2. Because no transform is used so the computational complexity is very less.
3. Fair degree of imperceptivity.

2 Proposed Embedding System

Figure 1 shows the block diagram of the proposed watermarking system Conversion from the RGB color space to the YCbCr color space and then a watermark is finally embedded into the luminance component Y.

The watermark embedding in the luminance component is carried out in the steps as follows:

Step 1: First, we divide the luminance component into 8×8 non-overlapping blocks and let an arbitrary block be denoted by 'P'.

Step 2: Divide the pixel block 'P' further into two 4×8 sub-blocks by separating the even location and odd location pixels as depicted in Fig. 2 The even location and odd location pixels sub-blocks are respectively denoted by P_{odd} and P_{even}.

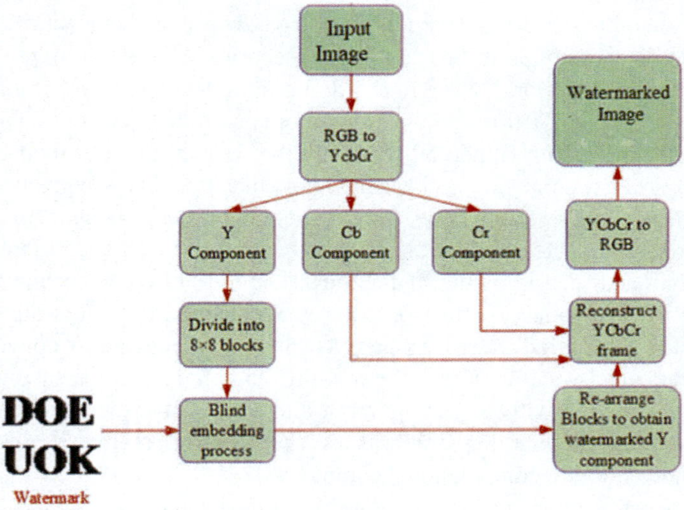

Fig. 1 Block diagram of the proposed watermarking system

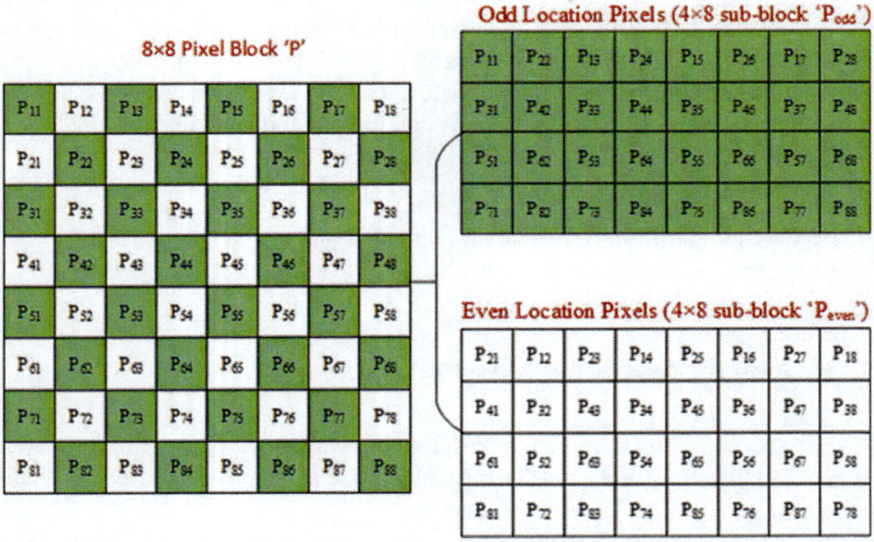

Fig. 2 Division of an 8 × 8 block into two sub-blocks

Step 3: Compute the DC coefficient of the sub-blocks P_{odd} and P_{even}. using Eqs. 1 and 2 respectively. Let the DC coefficient of P_{odd} and P_{even} sub-blocks be denoted by DC_{odd} and DC_{even} respectively

$$DC_{odd} = \frac{1}{\sqrt{4*8}} \sum_{x=0}^{3} \sum_{y=0}^{7} P_{odd}(x, y) \quad (1)$$

$$DC_{even} = \frac{1}{\sqrt{4*8}} \sum_{x=0}^{3} \sum_{y=0}^{7} P_{even}(x, y) \quad (2)$$

It is shown from Eqs. 1 and 2 that a simple averaged sum of all pixel values of that block DC coefficient of the block can be obtained directly instead of computing the DCT of Block.

The watermark is embedding by modifying the DC coefficients of the two sub-blocks such that DC_{odd} becomes greater than DC_{even} in case the watermark bit is '1' and the reverse is the case for watermark bit '0'.

Step 4: Embedded the watermark bit '1' or '0'

For *Watermark bit = 1* do the following

If $DC_{odd} < DC_{even}$ then
  $DC'_{odd} = DC_{even}$
  $DC'_{even} = DC_{odd}$
End

To raise the robustness of the system, the difference between the two DC coefficients is forced to be greater than a predefined embedding factor '*E*' as:

If $DC'_{odd} - DC'_{even} < E$ then
  $DC'_{odd} = DC'_{odd} + E/2$
  $DC'_{even} = DC'_{even} - E/2$
End

End of embedding bit '1'

For *Watermark bit = 0* do the following

If $DC_{odd} > DC_{even}$ then
  $DC'_{odd} = DC_{even}$
  $DC'_{even} = DC_{odd}$
End

To raise the robustness of the system the difference between the two DC coefficients is forced to be more than a predefined embedding factor '*E*' as:

If $DC'_{even} - DC'_{odd} < E$ then
  $DC'_{even} = DC'_{even} + E/2$
  $DC'_{odd} = DC'_{odd} - E/2$
End

End of embedding bit '0'

Step 5: Compute the required modifications needed to be brought to the pixels of the two sub-blocks as:

$$\Delta_{odd} = \frac{DC'_{odd} - DC_{odd}}{\sqrt{4 * 8}}$$

$$\Delta_{even} = \frac{DC'_{even} - DC_{even}}{\sqrt{4 * 8}}$$

Step 6: Compute the modified sub-blocks as:

$$P'_{odd} = P_{odd} + \Delta_{odd}$$

$$P'_{even} = P_{odd} + \Delta_{even}$$

Step 6: Combine the modified sub-blocks P'_{odd} and P'_{even} to get the watermarked pixel block P'.

Step 7: We repeat Step 2 to Step 6 until watermark bits are embedded in all the blocks of the luminance component and result in the watermarked luminance component.
The watermarked image is obtained by changing the watermarked luminance component from YCbCr color space to the RGB color space.

3 Extraction Process of the Watermark

The digital watermark extraction process is similar to the watermark embedding process and is summarized in the following steps:

Step 1: Firstly we divide the luminance component into 8 × 8 non-overlapping blocks and let an arbitrary block be denoted by '*P*'.

Step 2: Divide the pixel block '*P*' further into two 4 × 8 sub-blocks by separating the even location and odd location pixels. The even location and odd location pixels sub-blocks are respectively denoted by P_{odd} and $P_{even.}$

Step 3: Compute the DC coefficient of the sub-blocks P_{odd} and P_{even}. using Eqs. 1 and 2 respectively. Let the DC coefficient of P_{odd} and P_{even} sub-blocks be denoted by DC_{odd} and DC_{even} respectively.

Step 4: Extract a watermark bit as:

If $DC_{odd} > DC_{even}$ then
Watermark bit = 1
Else
Watermark bit = 0
End

Step 5: We repeat Step 2 to Step 6 until watermark bits are extracted from all the blocks of the luminance component and result in the extracted watermark.

4 Experimental Results

The output results of the proposed technique demonstrate that our algorithm performance and execution is efficient against common signal processing attacks. For carrying out our work we used a system with core i7 configuration and MATLAB 2016a. we have used various types of images (grayscale, colored, and medical images) of dimensions 512 × 512 and watermarks of dimensions 64 × 64 for verifying/testing our proposed technique. We have tested our algorithm against some common signal processing attacks like Median Filtering (MF), Salt and Pepper noise (S&P), Histogram Equalization (HE), Gaussian Noise (GN), JPEG Compression and Low pass Filtering (LPF). Figure 3 depicts the original images and corresponding watermarked images in which the watermark logo was embedded. The quality of the watermarked image is said to be good if its PSNR is above 36 dB. We have recorded the PSNR of our proposed technique and it is above 40 dB and it is greater than the thresh hold value of 36 dB.

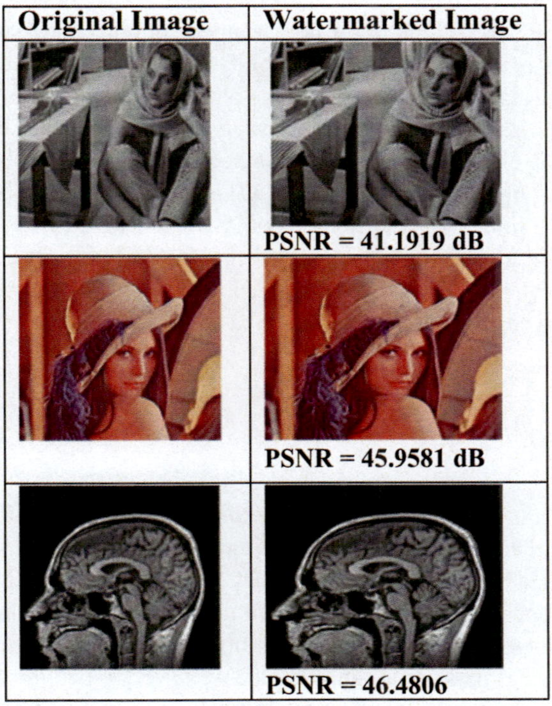

Fig. 3 Original images and corresponding watermarked images at $E = 25$

Table 1 Various objective parameters

Images	Embedding strength (E) = 25		
	PSNR	SSIM	BER
Barbara (grayscale)	41.1919	0.974977	0
Plane	46.1809	0.988315	0
Baboon	43.9273	0.998162	0
Lena	45.9581	0.999264	0
Medical image	46.4806	0.987011	0

Table 2 Comparison with techniques [16] and [17]

Images		Baboon	Lena	Airplane	Peppers
Proposed method	PSNR	43.9273	45.9581	46.1809	43.3266
	SSIM	0.998162	0.999264	0.988315	0.998544
Method of Das et al. (2014)	PSNR	40.2446	41.7801	40.7932	41.0122
	SSIM	0.9893	0.9704	0.9872	0.9733
Method of Kalra et al. (2015)	PSNR	36.1103	42.0109	39.7644	42.6843
	SSIM	0.9754	0.9787	0.9851	0.9816

While analyzing the technique we did the imperceptibility and robustness analysis of our proposed method. The imperceptibility of the original images against the watermarked images was carried out and the output was achieved in terms of PSNR and SSIM. In the same way, Robustness analysis was done and the outputs were achieved in terms of BER. The results obtained for no attacks are shown below in Table 1 and the comparison with [16] and [17] in Table 2. The comparison has been performed for the imperceptivity in terms of PSNR and SSIM.

It is clear from the two tables that our technique performs better. We have also tested our technique against various noise attacks and it has been observed that we are able to observe the watermark in all the cases (Table 3).

5 Conclusion

In this scheme, a highly efficient watermarking system capable of producing high quality watermarked images has been presented. Watermark has been embedded in the cover image by modification of the DC coefficients calculated in the spatial domain. The proposed method directly computes the DC coefficient in the spatial domain. The values of pixels are efficiently modified in the spatial domain in accordance with the watermark information. Since we have avoided the time-consuming transform operation, i.e., Discrete Cosine Transform, The output results prove that our technique is highly robust to commonly occurring attacks. The results obtained

Table 3 Attack analysis results for lena image

Attack analysis for LENA		
Attack	Percentage BER	Logo detected
Median filtering	11.0107	
Salt and pepper noise	1.09863	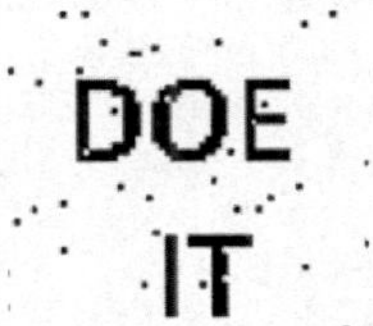
Histogram equalization	0	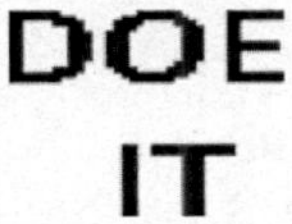
Gaussian noise	1.68457	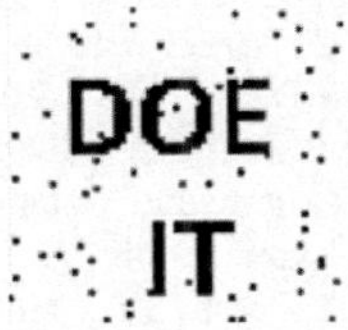
JPEG compression	19.458	
Low pass filtering	3.56445	

for numerous attacks are presented in the form of parameters like PSNR, SSIM, and BER to test the execution and efficiency of the scheme. we have compared our simulation results with different existing procedures and the comparison results have depicted that the proposed scheme outshines the schemes under comparison.

References

1. Hurrah, N.N., Parah, S.A., Loan, N.A., Sheikh, J.A., Elhoseny, M., Muhammad, K.: Dual watermarking framework for privacy protection and content authentication of multimedia. Future Gener. Comput. Syst. (2019). https://doi.org/10.1016/j.future.2018.12.036
2. Hurrah, N.N., Loan, N.A., Parah, S.A., Sheikh, J.A.: A transform domain based robust color image watermarking scheme for single and dual attacks. In: Fourth International Conference on Image Information Processing (ICIIP) (2017). https://doi.org/10.1109/ICIIP.2017.8313677
3. Akhoon, J.A., Parah, S.A., Sheikh, J.A., Loan, N.A., Bhat, G.M.: Information hiding in edges: a high capacity information hiding technique using hybrid edge detection. Multimed. Tools Appl. https://doi.org/10.1007/s11042-016-4253-x
4. Ahad, F., Parah, S.A., Sheikh, J.A., Loan, N.A., Bhat, G.M.: Information hiding in medical images: a robust medical image watermarking system for E-healthcare. Multimed. Tools Appl. (2015). https://doi.org/10.1007/s11042-015-3127-y
5. Singh, P.K., Kar, A.K., Singh, Y., Kolekar, M.H., Tanwar, S.: Proceedings of ICRIC 2019. Recent Innovations in Computing. Lecture Notes in Electrical Engineering, vol. 597. Springer Nature, Switzerland (2020). https://doi.org/10.1007/978-3-030-29407-6
6. Parah, S.A., Sheikh, J.A., Bhat, G.M.: A secure and efficient spatial domain data hiding technique based on pixel adjustment. Am. J. Eng. Technol. Res. **14**, 33–39 (2014)
7. Loan, N.A., Parah, S.A., Sheikh, J.A., Bhat, G.M.: Robust and blind watermarking technique in DCT domain using inter-block coefficient differencing. J. Digit. Signal Process. (2016). https://doi.org/10.1016/1.dsp2016.02.005
8. Ahad, F., Parah, S.A., Sheikh, J.A., Bhat, G.M.: On the realization of robust watermarking system for medical images. In: 12th IEEE India International Conference (INDICON) on Electronics, Energy, Environment, Communication, Computers, Control (E3-C3), Jamia Millia Islamia, New Delhi, pp. 17–20 (2015)
9. Loan, N.A., Parah, S.A., Sheikh, J.A., Akhoon, J.A., Bhat, G.M.: Hiding electronic patient record (EPR) in medical images: a high capacity and computationally efficient technique for e-health care applications. J. Biomed. Inform. **73**, 125–136 (2017)
10. Hurrah, N.N., Parah, S.A., Sheikh, J.A.: A secure medical image watermarking technique for e-healthcare applications. In: Handbook of Multimedia Information Security: Techniques and Applications (HMIS), Chapter. Springer, Berlin (2019)
11. Parah, S.A., Sheikh, J.A., Bhat, G.M.: On the realization of a secure, high capacity data embedding technique using joint top-down and down-top embedding approach. Elixir Comp. Sci. Eng. **49**, 10141–10146 (2012)
12. Akhoon, J.A., Parah, S.A., Sheikh, J.A., Loan, N.A., Bhat, G.M.: A high capacity data hiding scheme based on edge detection and even odd plane separation. In: Annual IEEE India Conference (1NDICON). IEEE. 978.11-4673-6540-6115. https://doi.org/10.1109/INDICON.2015.7443595
13. Afzal, I., Parah, S.A., Hurrah, N.N., Song, O.Y.: Secure patient data transmission on resource constrained platform. Multimed. Tools Appl. (MTAP) **79**(18) (2020). https://doi.org/10.1007/s11042-020-09139-3
14. Singh, P.K., Pawłowski, W., Tanwar, S., Kumar, N., Rodrigues, J.J.P.C.: Proceedings of First International Conference on Computing, Communications, and Cyber-Security (IC4S 2019). Lecture Notes in Networks and Systems, vol. 121, pp. 3–917. Springer Nature, Singapore (2020). https://doi.org/10.1007/978-981-15-3369-3
15. Agarwal, N., Singh, A.K., Singh, P.K.: Survey of robust and imperceptible watermarking. Multimed. Tools Appl. **78**, 8603–8633 (2019). https://doi.org/10.1007/s11042-018-7128-5
16. Das, C., Panigrahi, S., Sharma, V.K., Mahapatra, K.K.: A novel blind robust image watermarking in DCT domain using inter-block coefficient correlation. AEU-Int. J. Electron. Commun. **68**(3), 244–325 (2016)
17. Talwar, K., et al.: Adaptive digital image watermarking for color images in frequency domain. Multimed. Tools Appl. **74**(17), 6849–6869 (2015)

18. Hurrah, N.N., Parah, S.A., Sheikh, J.A.: Embedding in medical images: an efficient scheme for authentication and tamper localization. Multimed. Tools Appl. (MTAP) **79**(16) (2020). https://doi.org/10.1007/s11042-020-08988-2
19. Hurrah, N.N., Parah, S.A., Sheikh, J.A., Turjamaan, F.A., Mohammad, K.: Secure data transmission framework for confidentiality in IoTs. In: Ad Hoc Networks, vol. 95, p. 101989 (2019). https://doi.org/10.1016/j.adhoc.2019.101989

Investigations on Mathematical Modeling of Imaging Infrared (IIR) Missile

Rahul Kakkar, Sohni Singh, Joginder Singh, Sumeet Goyal, Dishant Khosla, and Manvinder Sharma

Abstract The IIR is the most advanced technology and the design is applicable to many missiles which are projected to exist in the next decade. Infrared technology has replaced the radar-guided missiles. The missile guidance system based on infrared technology or the IIR seekers does not provide any indication that they are tracking the missile and therefore called as passive devices. It has become very necessary to study the defense strategy related to Infrared technology. The effectiveness of the overall missile system is determined by the individual components and the key parameters that describe them. The design equations corresponding to the efficiency of different components used in the substems are discussed. The IIR seeker is a very critical element used in the guidance system of the missiles. Many disciplines related to the IIR missile such as signal processing, optics, microelectronics, manufacturing, and stabilization are integrated together and are discussed in the paper.

Keywords IIR missile · IR seeker · Missile subsystem · Signal processing

R. Kakkar (✉) · S. Singh · J. Singh · S. Goyal · D. Khosla · M. Sharma
Chandigarh Group of Colleges, Landran, Mohali, Punjab, India
e-mail: cgccoe.appsc.rk@gmail.com

S. Singh
e-mail: sohni.3841@cgc.edu.in

J. Singh
e-mail: joginder.appsci@cgc.edu.in

S. Goyal
e-mail: sumeet.coeapplied@cgc.edu.in

D. Khosla
e-mail: dishant.coeece@cgc.edu.in

M. Sharma
e-mail: manvinder.sharma@gmail.com

P. K. Singh et al. (eds.), *Recent Innovations in Computing*, Lecture Notes in Electrical Engineering 701, https://doi.org/10.1007/978-981-15-8297-4_31

1 Introduction

In recent times, Infrared technology has replaced the radar-guided missiles and therefore Infrared guided missiles have become the most important weapon in the warfare. The missiles guided by IR technology damaged the maximum number of aircrafts in almost 20 years of war area [1]. So it has become very necessary to study the defense strategy related to Infrared technology. In this type of guidance system, infrared light emission is used to track the missile and then follow it [2, 3]. Radar seekers used in the missile act as active devices as they clearly indicate that they are tracking or following the enemy missile. But the missile guidance system based on infrared technology or the IIR seekers does not provide any indication that they are tracking the missile and therefore called passive devices unlike the radar missiles [4]. IIR missile seekers are used at longer ranges and for sneak attacks since their tracking is not visible. IR devices were used first time in the era of World War II. The spatial properties of the sources detected in imaging missile are used in discriminating between the target and clutter. The countermeasures are deployed to overcome the limitation of temporal filtering used in infrared missiles due to which the spatial properties of the source becomes difficult to extract. The infrared seeker employs discriminants in large numbers which leads to the difficult development of countermeasure to overcome the threats of IIR seekers [5, 6]. There are eight subsystems in the missile system which is used to target the aircraft [7]. These subsystems are seeker, fuse, warhead, propulsion, and flight control and data link. This is shown in Fig. 1.

The location of the IIR seeker used in the missile is in the forward. The function of the seeker is to provide commands to the flight control so that the interception of the aircraft target can take place in the missile. The infrared seeker consists of several components [8]. The signal processor is used to work on the intensity of signals and for generating the commands. The seeker needs to be protected from the weather and forces and this job is done by infrared dome. The target energy needs to be focused on the detector which is done by the optical system of IIR seeker used in the missile [9–11]. The target jitter is minimized by the stabilizing system of the IIR missile and to set the optical path between the point and the target [12]. The intensity

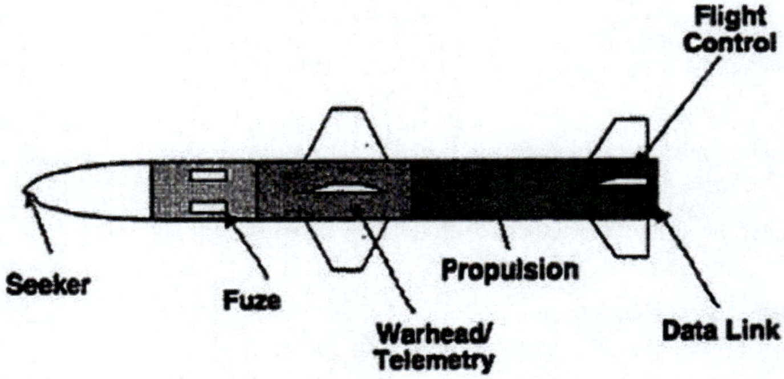

Fig. 1 Missile subsystem

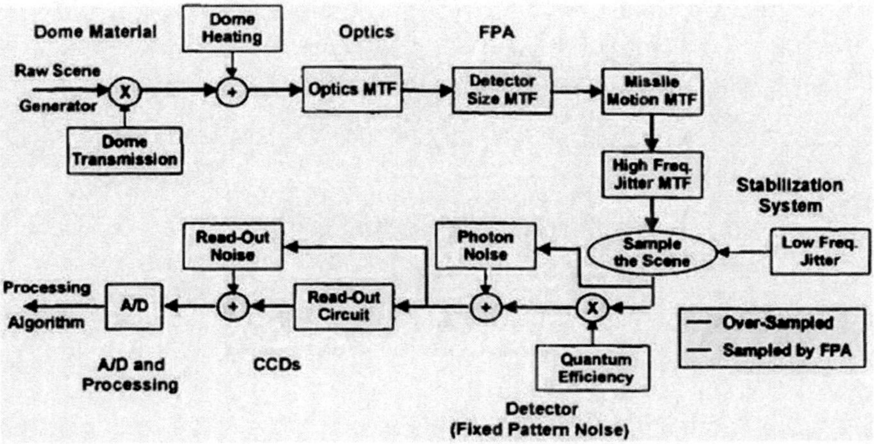

Fig. 2 Missile seeker flow diagram

signals are generated by the infrared detector. It also does the collection of target photons. The IR guided missiles track and detect the target passively either hitting it or burning it out. The missile head is always towards the target [11, 13]. The missile does not shift instantaneously if the target moves and therefore the guidance system of the missile must be robust in order to direct and intercept the target [14]. The missile guidance laws determine the precision and effectiveness of the interception of missiles. In guidance related to the target, the sensor collects the data related to the target and make it real-time available. The missile is navigated towards the target direction according to the data provided by IR sensor [15–17]. The flow diagram of the seeker which determines the whole function of missile seeker is shown in Fig. 2.

2 Design Equation of IIR Missile

The supersonics speeds in the missiles of the enemy are so high that the domes used in the missiles must be able to withstand it. The dome must be hard enough so that it does not degrade with the damages provided by the environment such as dust, rain, etc. The aerodynamic heating creates the thermal noise at supersonic speeds which because noise source for infrared detectors [18]. Therefore, the materials having low emissivity are used [19]. The optics of the IIR seeker are shielded by the dome. The overall transmission of the missile is improved by using the dome equation. The dome structure is given in Fig. 3.

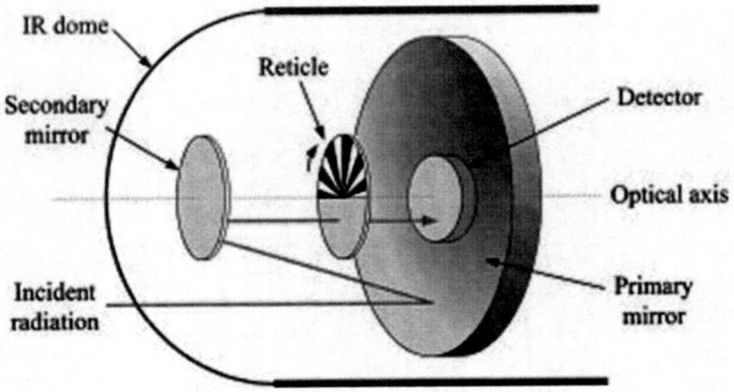

Fig. 3 Dome structure in IR missile

The thermal pressure and stress in the dome determine the thickness of the infrared dome [20]. The dome equation is given as

$$I = I_0 \tau_{\text{dome}} = I_0 e^{-(\alpha_{\text{dome}})(t_{\text{dome}})} \tag{1}$$

τ_{dome} is denoted as dome transmission
α_{dome} is absorption coefficient and
t_{dome} is denoted as the thickness of dome.

The properties of the material used in the dome and the geometry of the dome defines the heat flux $Q_{\text{lim.}}$ The heat flux is given as

$$Q_{\text{sTlim}} \equiv \frac{2R_H^{'}}{t_{\text{dome}}} \tag{2}$$

$$R_H^{'} \equiv \frac{\sigma(1-\nu)k}{\alpha_T E} \tag{3}$$

ν is denoted as Poisson's ratio,
k is thermal conductivity,
E is denoted as Young's modulus and
σ_f is denoted by failure stress.
The actual heat flux is given by

$$Q_{\text{actual}} = h_{\text{ST}}(T_{\text{ST}} - T_{\text{iw}}) \tag{4}$$

Here,
h_{ST} is the coefficient of heat transfer,
T_{iw} denotes the temperature of the interior wall of the dome,
T_{ST} denoted the temperature of the outer wall and is given as

$$T_{ST} = T_a(1 + 0.17M^2) \tag{5}$$

T_a is the ambient temperature and M is the match number.

The pressure differential between the outer and inner walls of the dome determines the thickness of the dome. The expression is given as

$$\frac{t_{dome}}{R} = 0.7 \times DSF^{\frac{2}{3}} \times \left(\frac{\Delta P}{\sigma_f}\right)^{\frac{2}{3}} \tag{6}$$

The design safety factor is given by *DSF* which is assumed to be 4. The dome thickness can be minimized by using the equations given above [21, 22]. The reflectivity of the dome material is given by *R*. The emissivity of the dome ε_{dome} is given as

$$\varepsilon_{dome} = \frac{(1-r)(1-e^{-\alpha_{dome} t_{dome}})}{1 - re^{-\alpha_{dome} t_{dome}}} \tag{7}$$

3 IR Detectors and Signal Processing

The detectors used in the IIR missile seekers are quite different from others as they use more than one detector for the reconstruction of image. If single detector is used, then it faces the problem of low sensitivity. Therefore the scanning of the images should be done in line rather than using single detector for a better Field of View (FOV). Focal plane arrays are used to provide higher sensitivity and need not to be scanned [23]. The design of the seeker using focal plane arrays is shown in Fig. 4. There is a beam splitter inserted in the optical path to split the light to next FPA. The support lenses are placed to collect the light [24].

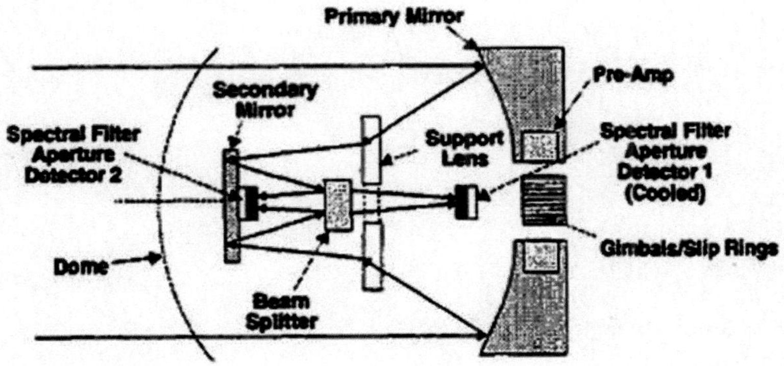

Fig. 4 Design of a FPA seeker used in IIR missile

The multiplexing readouts are used in FPA for high sensitivity which adjusts the current gain. The readout gives the current gain the signal and is dependent on the transconductance of MOSFET given by g_m, the dynamic resistance of IR detector given by R_{det} and C_{tot} which is denoted by total capacitance. The readout A_1 is given as

$$A_1 = \frac{g_m}{g_{m,load}} \eta_{inj}$$

$$\eta_{inj} = \frac{g_{m,load} R_{det}}{1 + g_{m,load} R_{det}} \frac{1}{1 + \frac{j\omega C_{tot} R_{det}}{1 + g_m R_{det}}} \tag{8}$$

The technology of the infrared detectors used in IIR missiles is growing very rapidly in the overall development of the seeker. The detector chips used in the seeker should be integrated, powerful, and much smaller. The IR detectors are integrated on the electronic readouts placed on the focal plane since the charged couple devices have been invented. The image resolution is one of the important factors of the IR detector to have a perfect scanned image of the target aircraft. The number of detectors used and the optical aperture of the infrared seeker are the functions of image resolution. Since the invention of the silicon integrated circuits, the size and complexity of the IR detectors used in IIR missiles have been significantly reduced [25]. The signal processing in the subsystem of infrared seeker used in IIR missiles will be digital in nature. The digital systems whose quality is determined by the speed of the signal processing. The material properties define the quantum efficiencies of the IIR missile. The quantum efficiency is increased by using more than one reflective layer. An active layer is included in the IR detector which absorbs all the radiation. 80–90% efficiency is achieved if the material used in the IR detector is HgCdTe for dual bands. The expression used for computing the quantum efficiency is given as

$$A = (1 - R)\left(1 - e^{-ad}\right)/\left(1 - Re^{-ad}\right)$$

$$T = \left(1 - e^{-d/L}\right)L/d$$

$$\eta_i = \frac{\int_{E_L}^{E_H} \frac{dN}{dE} P(E) dE}{\int_{h\nu}^{E_{F+h\nu}} \frac{dN}{dE} dE}; \; P(E) = \left(1 - \sqrt{\frac{E_F + \phi_B}{E}}\right), \; \frac{dN}{dE} = N_\nu E^{\frac{1}{2}} \tag{9}$$

where,
A is absorptance,
T denotes transport efficiency,
d is the thickness of layer and
L denotes the mean free path.

4 Missile Guidance Subsystem

Missile guidance means how the missile navigates their path accurately to the target. The basic idea of the missile guidance is shown in Fig. 5.

The operations performed by the missile guidance include tracking and detecting the target, guide the missile, and suppress the noise. The fixed pattern noise is suppressed by electronic sensors. While processing in space, the registration of the scene is a very important step. The sensors used in IIR missiles should be carefully analyzed. The CNR (contrast to noise ratio) and the revealing and hiding of the features must be carefully addressed [26]. The efficiency of the scene registration is determined by the following equation

$$\mathrm{MFO}(\vec{x_0}, \vec{v}) = \frac{1}{N_x N_y} \sum_{k\omega} \frac{S^*\left(\vec{k}, \omega\right) D\left(\vec{k}, \omega\right) e^{i \vec{k} . \vec{x_0}}}{P\left(\vec{k}, \omega\right)}$$

$$S^*\left(\vec{k}, \omega\right) = \mathrm{MTF}\left(\vec{k}\right) f^*\left(\omega - \vec{k} . \vec{v}\right) \tag{10}$$

Here,

The matched filter output is given by MFO,

S^* is the expected signal FFT,

k is wave number,

v is velocity,

$\mathrm{MTF}\left(\vec{k}\right)$ is the transfer function of the system.

Detection of the target is the next step in scene registration. This involves two operations: first, the filtering operation is applied and then the threshold operation. The convolution of original image $I(x, y)$ and transfer function $h(x, y)$ is done in the filter to obtain the processed image $I_o(x, y)$. The target signal is enhanced using the filter and the unwanted signals are suppressed. The filter equation is given as

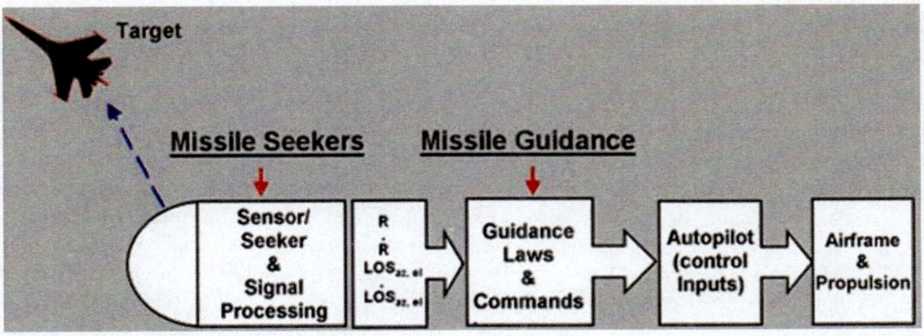

Fig. 5 Missile guidance concept

$$I(x, y) = \int I_0(x_0, y_0)h(x - x_0, y - y_0)\mathrm{d}x\mathrm{d}y \quad (11)$$

The interception of the target is done using the guidance laws in the missile guidance. The acceleration command of the missile is given for the proportional navigation of the missile and for the target to properly intercept. There are many factors on which the acceleration command of the missile, A_{m_β} depends such as β, the line of sight angle, velocity V_c, navigation gain N, and the acceleration of the target A_{T_β}. The expression is given as

$$A_{m_\beta} = N V_\mathrm{c}\dot{\beta} + \frac{N}{2} A_{T_\beta} \quad (12)$$

5 Conclusion

Infrared technology has replaced the radar-guided missiles and therefore Infrared guided missiles have become the most important weapon in warfare. The missile guidance system based on infrared technology or the IIR seekers do not provide any indication that they are tracking the missile and therefore called as passive devices unlike the radar missiles. Since these missiles have always their head towards the target, they either catch the target or burn it out. Therefore, the guidance system of the missile must be robust so that the target is detected easily. The equations related to the missile guidance system are discussed in the paper. Many missile guidance laws govern the missile intercept which further determines the effectiveness of the missile.

References

1. Bae, T.W., Kim, B.I., Kim, Y.C., Ahn, S.H.: Jamming effect analysis of infrared reticle seeker for directed infrared countermeasures. Infrared Phys. Technol. **55**(5), 431–441 (2012)
2. Virtanen, K., Raivio, T., Raimo, P.: Modeling pilot's sequential maneuvering decisions by a multistage influence diagram. In: Proceedings of the AIAA Guidance, Navigation, and Control Conference, Montreal, Canada, pp. 1–11 (2001)
3. Armon, T., Janjoris, R., Pieter, S., et al.: Dynamic scripting with team coordination in air combat simulation. In: Modern Advances in Applied Intelligence: 27th International Conference on Industrial Engineering and Other Applications of Applied Intelligent Systems, IEA/AIE 2014, Kaohsiung, Taiwan, June 3–6, 2014, Proceedings, Part I, vol. 8481 of Lecture Notes in Computer Science. Springer, Berlin, pp. 440–449 (2014)
4. DiMarco, J.S., Kemper Jr, P.J., Pringle, L.N.: Closed-loop guidance of imaging infrared missile seekers. In: Infrared Imaging Systems: Design, Analysis, Modeling, and Testing X, vol. 3701. International Society for Optics and Photonics, pp. 254–265 (1999)

5. Glasgow, B., Bell, W.: The future of anti-aircraft imaging infrared seeker missile threats. In: 1999 IEEE Aerospace Conference. Proceedings (Cat. No. 99TH8403), vol. 4. IEEE, pp. 457–465 (1999)
6. Gray, G.J., Aouf, N., Richardson, M.A., Butters, B., Walmsley, R.: An intelligent tracking algorithm for an imaging infrared anti-ship missile. In: Technologies for Optical Countermeasures IX, vol. 8543. International Society for Optics and Photonics, p. 85430L (2012)
7. Cox, L.J., Batten, M.A., Carpenter, S.R., Saddleton, P.A.B.: Modeling countermeasures to imaging infrared seekers. In: Technologies for Optical Countermeasures, vol. 5615. International Society for Optics and Photonics, pp. 112–119 (2004).
8. Gray, G.J., Aouf, N., Richardson, M.A., Butters, B., Walmsley, R., Nicholls, E. (2011). Feature-based tracking algorithms for imaging infrared anti-ship missiles. In: Technologies for Optical Countermeasures VIII, vol. 8187. International Society for Optics and Photonics, p. 81870T (2011)
9. Hall, C.S., Alongi, A.J., Fortner, R.L., Fraser, L.K.: Development of a fire and forget imaging infrared seeker missile simulation. In: Signal and Image Processing Systems Performance Evaluation, Simulation, and Modeling, vol. 1483. International Society for Optics and Photonics, pp. 29–38 (1991)
10. Willers, C.J., Wheeler, M.S.: The validation of models in an imaging infrared simulation. In: 2007 SBMO/IEEE MTT-S International Microwave and Optoelectronics Conference. IEEE, pp. 250–254 (2007)
11. Kemper Jr, P.J., DiMarco, J.S.: Imaging infrared seeker signal processing overview: image processing, adaptive thresholding, and track processing. In: Infrared Imaging Systems: Design, Analysis, Modeling, and Testing X, vol. 3701. International Society for Optics and Photonics, pp. 266–273 (1999)
12. Vashisht, S., Jain, S.: SOOCS: SDN-enabled opportunistic offloading and charging scheme in multi-UAV ecosystem. Int. J. Commun. Syst. **32**(8), e3939 (2019a)
13. Vashisht, S., Jain, S.: Location aware network of drones for consumer applications: challenges and solutions. IEEE Consum. Electron. Mag. **8**(3), 68–73 (2019b)
14. Geng, Y.G., Zhang, M.Q.: Miniaturization technique of dual field optical system in imaging infrared seeker. Infrared Laser Eng. **36**(6), 887 (2007)
15. Sharma, M., Singh, H.: Substrate integrated waveguide based leaky wave antenna for high frequency applications and IoT. Int. J. Sens. Wireless Commun. Control **9**, 1 (2019). https://doi.org/10.2174/2210327909666190401210659
16. Barth, J., Fendt, A., Florian, R., Kieslich, W.: Dual-mode seeker with imaging sensor and semi-active laser detector. In: Infrared Technology and Applications XXXIII, vol. 6542. International Society for Optics and Photonics, p. 65423B (2007)
17. Mohindru, V., Bhatt, R., Singh, Y.: Reauthentication scheme for mobile wireless sensor networks. Sustain. Comput. Inf. Syst. **23**, 158–166 (2019)
18. Flournoy, J.T., Towry, E.R., Deep, N.S.: Automated imaging infrared seeker performance evaluation system. In: Characterization, propagation, and simulation of infrared scenes, vol. 1311. International Society for Optics and Photonics, pp. 212–218 (1990)
19. Sakarya, D.U., Bayram, A.: Optical design of dual mode seeker for short-wave infrared and four quadrant detectors in missile application. In: Physics and Simulation of Optoelectronic Devices XXVII, vol. 10912. International Society for Optics and Photonics, p. 109121K (2019)
20. Yi, C.A.I., Xu, H.U.: State of the art and future trend of detectors for infrared imaging seekers. Infrared Laser Eng **1**(001)
21. Kaur, S.P., Sharma, M.: Radially optimized zone-divided energy-aware wireless sensor networks (WSN) protocol using BA (bat algorithm). IETE J. Res. **61**(2), 170–179
22. Sharma, M., Singh, S., Khosla, D., Goyal, S., Gupta, A.: Waveguide diplexer: design and analysis for 5G communication. In: 2018 Fifth International Conference on Parallel, Distributed and Grid Computing (PDGC). IEEE, pp. 586–590 (2018)
23. Kim, H.Y., Kang, S.J., Jhee, H.J.: The cost optimization solution for developing the image infrared (IIR) missile seeker operated under various environments. J. Korea Inst. Inf. Commun. Eng. **23**(4), 365–373 (2019)

24. Jeong, D.G., Park, J.S., Lee, J.H., Jun, D.S., Son, S.H.: Study on effects of roll in flight of a precision guided missile for subsytem requirements analysis. J. Korea Soc. Simul. **28**(2), 131–137 (2019)
25. Sharma, M., Singh, H.: SIW based leaky wave antenna with semi C-shaped slots and its modeling, design and parametric considerations for different materials of dielectric. In: 2018 Fifth International Conference on Parallel, Distributed and Grid Computing (PDGC). IEEE, pp. 252–258 (2018)
26. Kim, T.H., Kim, J.H., Kim, P.: New guidance filter structure for homing missiles with strapdown IIR seeker. Int. J. Aeronaut. Space Sci. **18**(4), 757–766 (2017)

Recommendation Systems Based on Collaborative Filtering Using Autoencoders: Issues and Opportunities

Ria Banerjee, Preeti Kathiria, and Deepika Shukla

Abstract With the advancement of deep-learning-based approaches, complex problems under Artificial Intelligence can now be addressed in a comparatively easier way. One such application domain is Recommendation Systems. Recommendation systems are powerful tools in this age of data explosion for providing meaningful insights from data. Collaborative Filtering is one of the popular approaches for building recommendation systems and extensive literary works suggest that it is very effective. In recent years deep-learning-based models have been bounteously applied for the development of recommendation systems using collaborative filtering. Autoencoders are a deep-learning based neural architecture which can be used for implementing collaborative filtering. This paper presents a survey of different autoencoder based models which employ collaborative filtering methodology for making recommendation systems. The paper initially provides an understanding of models and thereafter summarizes various works reported in the literature in the light of the methodology used, taxonomy, datasets used for experimentation, limitations and results reported.

Keywords Deep learning · Collaborative filtering · Autoencoders

1 Introduction

In the environment of Big Data and E-commerce, machine learning has widened its scope of applications. Table 1 shows the relationship between E-commerce, Big Data Analytics (BDA) and Neural Networks (NN) as three entities intermixed and

R. Banerjee · P. Kathiria (✉) · D. Shukla (✉)
Nirma University, Ahmedabad, Gujarat, India
e-mail: preeti.kathiria@nirmauni.ac.in

D. Shukla
e-mail: deepika.shukla@nirmauni.ac.in

R. Banerjee
e-mail: 19mca168@nirmauni.ac.in

P. K. Singh et al. (eds.), *Recent Innovations in Computing*, Lecture Notes in Electrical Engineering 701, https://doi.org/10.1007/978-981-15-8297-4_32

Table 1 Big data analytics in E-commerce

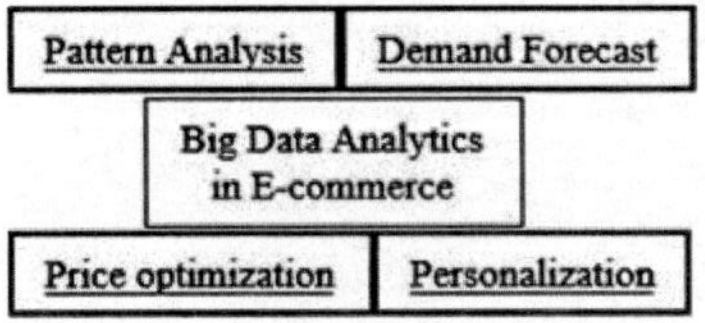

Fig. 1 Companies use BDA to track user patterns to provide price optimization and personalization, thus forecasting demand

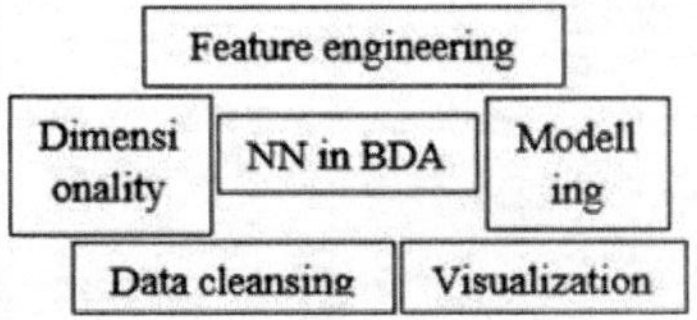

Fig. 2 NNs are employed to handle manipulations on enormous amount of data

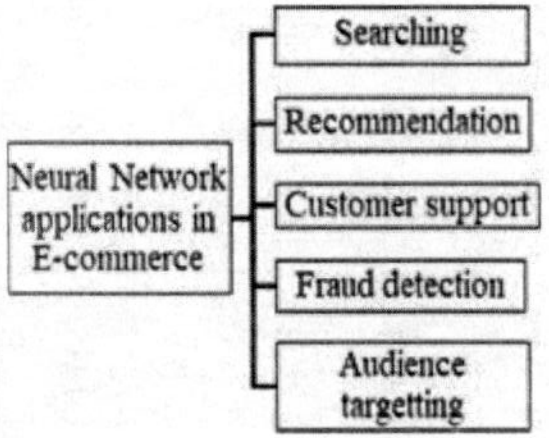

Fig. 3 NN applications in E-commerce activities

interdependent, with the goal of creating a profitable domain for the E-commerce industry (Figs. 1, 2 and 3).

As the pool of online content widens and deepens, users get a plethora of choices. But this often poses 'paradox of choice'. Recommendation systems are expert systems that harness the power of data to enhance the decision-making process for the user.

Recommendation systems gather information in three ways: *Implicitly, Explicitly* [1] and by *Hybrid* methods. In *Implicit* feedback, user behaviour is captured from the actions made by the user while accessing the application. It does not require user intervention. In *Explicit* feedback, information is gathered by explicit user intervention. Hybridization of feedback is achieved in two ways: (i) User is allowed to participate in explicit feedback gathering, and (ii) implicit data as a check on explicit feedback.

Recommendation systems can be broadly classified into Content-based [2] (based on the features extracted from the items that the user has rated positively).

Collaborative filtering based [3, 4] (recommends items to the target users by mapping similar users) which can be further categorized into item-based and user-based

- Item-based: Find the correlation between items based on user's previous rating.
- User-based: Finds a correlation between target user's and other users' profile.

The third major category of recommendation systems is Hybrid [5], which is a combination of both content-based and collaborative filtering (CF).

Major issues in recommender systems are cold start, shilling attacks, synonymy, grey sheep, Limited Content Analysis and Overspecialization, privacy, latency, Evaluation and Availability of Online Datasets, context awareness, sparsity and scalability [6, 7]. However, due to the inherent nature of CF, cold start, sparsity, and shilling attacks are the most prominent problems. The 1990s saw information overload problems [3] as a major challenge in context-based recommender systems. Amazon was one of the earliest commercial applications of CF. Over the years CF algorithms have advanced in complexity and performance. The algorithms under CF are grouped as:

1. Memory-based [8, 9]: they employ statistical techniques in the user-item matrix to find the nearest neighbours of the users and make recommendations.
2. Model-based [10]: Huge datasets are compressed and fed into a model that generate recommendations. Matrix Factorization (MF) and clustering are examples of traditional methods. In the recent past deep-learning-based methods have also gained popularity [11, 12]. Algorithms under deep learning can be of the following type:

   (a) Multi-layered Perceptron (MLP): it is a feed-forward NN architecture with hidden layers, where back-propagation is used for learning.
   (b) Convolutional NN (CNN) [12, 13]: It incorporates convolution (a mathematically represented linear operation) on data projected in a grid.
   (c) Recurrent Neural Networks (RNN) [12, 14]: These NN are good for modelling sequential data. They can remember former computations.

(d) Restricted Boltzmann Machines (RBM) [12, 15]: These generative stochastic NN with 2 layers produce a random probability distribution of the input.
(e) Autoencoders (AE) [16]: AE is an unsupervised NN model attempting to reconstruct its input data in the output layer.

Lots of content has been published where the surveys are provided for deep-learning based recommendation systems that use CNN, RNN, or RBM as the architecture, whereas this paper focuses on providing a survey of literature and approaches that use CF for recommender systems using autoencoders as the architecture.

1.1 Research Challenges and Future Direction

It is established by this study that the benchmark datasets are available for research in the area of recommender systems, however, it is strongly recommended that more datasets pertaining to other fields of recommendation should be generated so that the recommender systems for other fields can also be developed. One striking observation that has come up during this survey is that the literature available for this area lacks in comparative analysis wherein all the models are compared experimentally. The comparative analysis of all models can be carried out experimentally so that better insights can be projected and application-specific choice of model can be suggested.

The rest of the paper is organized as Sect. 2 discusses AEs and their types. Section 3 presents the literature available for the subject of the paper, various modes for evaluating the approaches and datasets used for comparing the approaches in the area. Section 4 throws the light on a comparison between various works reported in the literature for recommendation systems using CF and AEs. Section 5 concludes the study with major insights and throwing open-ended research issues and challenges.

2 Autoencoders

In general, Autoencoders consists mainly of four main parts [16]: Encoder, bottleneck, Decoder, and reconstruction Loss method. The encoder learns the way to input dimensions, compresses the input into an encoded representation. The bottleneck layer is used as a salient feature representation of the input data. The bottleneck layer represents the data in the lowest possible dimensions. The decoder reconstructs the encoded data into a representation that is as close to the input as possible. The reconstruction loss method is responsible for making out the losses incurred during this encode-decode process. Training an AE means minimizing the network's loss using back propagation (Fig. 4).

A vanilla AE is a simple two-layered NN with one hidden layer, where the length of the encoder and decoder is the same, and that of the hidden layer is smaller than

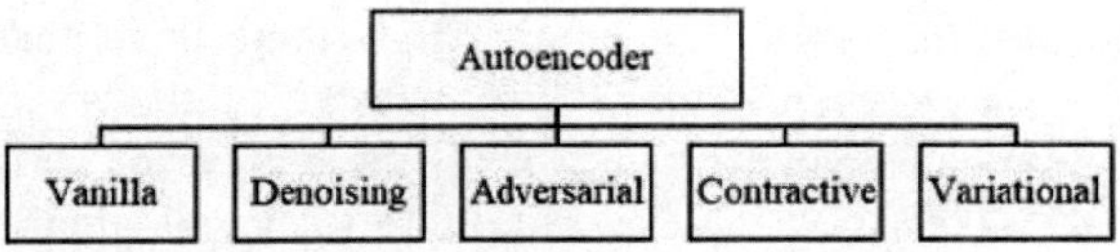

Fig. 4 Types of autoencoders

them. A Denoising AE [17] introduces noise in the input, and then it reconstructs this corrupted data to produce cleaned output. Adversarial AEs consists of a discriminator and a generator. The discriminator generates the probability of a point x belonging to the data distribution, while a generator generates data that is fed to the discriminator intending to fool the discriminator [18]. Contractive AE [19] adds a regularizer to the objective function of the AE. This regularizer belongs to the Frobenius norm of the Jacobian mix. Variational AE maps the input to a distribution, where the bottleneck consists of a mean vector and a standard deviation vector.

3 Literature Survey

The recent past has seen a shift from matrix factorization and towards deep learning approaches with CF for recommendation systems. Of the numerous architectures available in the literature, autoencoders exactly fit the CF problem. This section discusses and critically reviews various models proposed in the literature that use AE as underlying architecture, and then presents factors of evaluation for these models. This section also discusses prominent datasets used in research in recommender systems.

One of the integrations of AEs in CF was introduced as AutoRec [20], a state-of-the-art model for recommendation. It has two variations: *I-AutoRec* (item based), and *U-AutoRec* (user based). It takes as input partial vectors of either users or items and aims to project it to a lower dimensional latent space in the hidden layer. It projects the reconstructions to the output layer, which are the recommendation results. It uses identity and sigmoid activation functions. By comparing RMSE values, authors conclude that I-AutoRec outperforms U-AutoRec on MovieLens 1M and 10M dataset.

Denoising Autoencoders (DAE) [17] introduce the concept of adding noise to the input. Strub and Mary [21] introduced an AE model that computes non-linear matrix factorization from sparse rating inputs. It has two versions: *Uencoder* and *Vencoder*. It converts the missing values to zero in the input and backpropagated layers. This is known as masking noise, which helps make recommendations. Input is sparse representation of users and items, and gives a dense output, hence addressing sparsity issues. The authors report V-AutoRec outperforms *Uencoder* and *Vencoder.*

Marginalized Stacked DAE (mSDA) [22] uses linear denoisers to marginalize out random feature corruption. It uses Stochastic Gradient Descent (SGD). This architecture addresses the issues of scalability, high-dimensional features, and high computation costs. *Deep CF* (DCF) model intermixes MF methods with marginalized DAEs, where the time complexity is $O(tN)$ (t is the number of iterations, N is the number of ratings). *mSDA-CF*, a variant of mSDA, stacks marginalized DAEs, and updates latent features on the layers given by the average of total number of layers and 1. The authors claim that this significantly improves time complexity.

Collaborative Denoising AE (*CDAE*) [23] is a generalized model of several other state-of-the-art. It generalizes Latent Factor Model [24, 25] and Factorized Similarity Model. CDAE has $I + 1$ node in the input layers, where I is the number of items, and the last node is a user-specific node. The hidden layer has a bias node, and the output has I nodes. Layers are fully-connected. CDAE uses SGD to learn parameters.

Based on Stacked DAE (SDAE), a CF model was developed [26]. It is user-based in nature and uses Gaussian noise to corrupt the data. It uses output of hidden layers for training process. Built upon it is CF-based Stacked Denoising AE model *CF-SDA*. It calculates the difference between the latent similarity obtained from the aforementioned model and the surface similarity. It employs Sigmoid and Identity mapping activation functions, Mean Squared Error as loss function and Adam optimizer.

Recommendation via *Dual AEs* (*ReDa*) [27] is a representational learning framework that learns latent features from user-item data and aims to minimize deviations in the training data. It uses stacked AEs to reach an optimal global solution. *ReDa* uses Sigmoid function during encoding and decoding. The algorithm requires calculation of partial derivatives of the variables used in the equations of optimizations. The model uses gradient descent for optimization, until convergence, which is not guaranteed every time as the optimization problem proposed is not convex in nature.

Collaborative Adversarial AE (*CAAE*) [28] is a framework that is based on Generative Adversarial Networks (GAN) [18]. A GAN has a discriminator NN and a generator NN. The discriminator generates the probability of a point belonging to a data distribution, and the generator generates data to feed in the discriminator, to fool it. CAAE consists of a discriminator that uses L2 regularization and is implemented with Bayesian Personalized Ranking (BPR) for learning relative preferences on items. It has a positive and negative item generator AE. SGD is used for the learning process.

Fine-Grained Collaborative AE (*FG-ACAE*) [29] is an adversarial framework for non-linear model optimization, where the noises iteratively minimize and maximize the loss function. This framework is applied to CDAE [23], where the noise mixing layer replaces denoising layer. Identity activation function is used in the mixing layer, loss function is cross-entropy, and mini-batch gradient descent for optimization.

Mult-VAE^{PR} [30] extends Variational AE (VAE) [31, 32] for CF for implicit feedback incorporating non-linear probabilistic model. The model samples K-dimensional latent representations from the Gaussian prior, and non-linear function f_θ are applied to it to produce probability distribution over the items. Softmax function is used for output normalization. The model uses variational inference to minimize the Kullback-Leiber (KL) divergence. Using multinomial likelihood for the data distribution and adjusting the over-regularized VAE objective gives significant improvement over state-of-the-art baselines. It uses Bayesian inference for parameter estimation.

Sequential Variational AE (SVAE) [33] conditions each event to the previous event, thus modelling temporal dependencies. The authors argue that there is a recurrent relationship between the current datapoint and the previous one.

Queryable Variational AE (*Q-VAE*) [34] models the joint probability of the user's preferences and the conditional probability of other similar user's preferences. It models the log joint probability of the subset of partial preferences of the user, where the partition of this subset is arbitrary. Authors claim this is different from the existing VAE based models. It uses KL divergence for regularizing posterior distributions.

RecVAE [35] is built upon *Mult-VAE*^{PR}. It changes the encoder of *Mult-VAE*^{PR} to a denoising encoder. The decoder uses Softmax activation. It makes a convex combination of standard Gaussian prior and a regularization term in the form of KL divergence. The training takes alternatively on the encoder and the decoder. Monte-Carlo sampling is used for log-likelihood and KL divergence. It uses cross-entropy as the main component of the loss function. The authors claim it outperforms other methods.

AutoSVD++ [36] is a hybrid CF model which uses Contractive Auto-Encoders (CAE) [19] and SVD++ [37]. Atop this is, a new hybrid model that takes implicit feedback as input. The model takes high-level feature representations from CAE, which is integrated with AutoSVD++ model. The model tries to minimize the regularized squared error loss. It uses SGD to learn parameters. Authors claim that their model outperforms *mSDA-CF* and *U-AutoRec*. They claim that this model is scalable.

Semi Auto-Encoder (*SA-HCF*) [38] proposes an architecture where the output layer is shorter than the input layer which makes incorporating side information easier. Based on it, the authors have designed a hybrid CF model. The authors claim it can solve the problem of rating prediction and ranking prediction. For ranking prediction, the user's partial observed vector and profile vector is used, and for rating prediction item's partially observed vectors and explicit ratings are used.

Hybrid collaborative filtering (HCF) model with semi-stacked denoising AE (Semi-SDAE), referred to as *HCF-SS* [39] is a hybrid model. The Semi-SDAE takes three inputs: the original, the noise-corrupted, and the side information data. Semi-SDAE can incorporate extended side information as input, without changing the dimension of the output layer. The model uses two layers of Semi-SDAE which is incorporated into MF. The parameters are learnt using SGD. The authors demonstrate that *HCF-SS* model outperforms *AutoSVD++*, *ReDa* and *SA-HCF*.

AutoCOT [40] is based on cooperative training (COT). It implements AutoRec and combines COT model with it. The mediator model generates a mixture distribution of real and generated user browsed data, and aims to optimize the KL divergence, and the generator model optimizes the JS divergence between them. The mediator and generator are one hidden layered architecture. The training process occurs iteratively on them. Authors claim AutoCOT can alleviate sparsity issues in large datasets.

Deep Heterogeneous AEs (*DHA*) [41] utilize information from multiple domains for accuracy improvement. It uses SDAE and RNN to extract latent features from non-sequential and sequential data respectively. It uses two Long Short Term Memory networks for the sequential data. The model incorporates two DHAs. It uses coordinate descent for optimizations and SGD for learning weight matrix and bias vectors. The author's experimentations demonstrate competitive model performance.

Built upon U-AutoRec, is an AE based CF (*ACF*) framework that provides top-N recommendation. It employs two AEs: *PR-Net and IL-Net* [42]. *PR-Net* employs pairwise regularization and samples negative samples, while *IL-Net* captures the items which are more likely to be true negatives. *PR-Net* is fed explicit feedback and *IL-Net* is fed the implicit feedback. SGD is used to learn models. The authors claim that this framework can also be extended for one-class CF by altering the inputs.

EASE^{R} [43] is a linear model with zero hidden layers. It takes a user-item interaction matrix X (generally binary) as the input. The diagonal of the item-item weight matrix must be constrained to zero. This model uses square loss between X and predicted scores. EASE^{R} states that the constrained convex optimization problem has a closed solution. The authors state the computational complexity of the algorithm as $O(|I|^3)$ where I is the number of items. Authors claim the model's superior accuracy.

AE algorithm for CF (AE-CF) [44] presents a combination of AE with clustering. The scoring data is combined with the user data to construct a scoring matrix, for extracting user features. This is used to plot the clusters, and using k-means recommendations are made to the user. This algorithm narrows the search space. The AE is used here for dimensionality reduction. It uses Pearson's correlation coefficient.

3.1 Matrices Used for Evaluation of Approaches

Models surveyed in this paper use Root Mean Squared Error (RMSE), Mean Absolute Error (MAE), Precision and Recall, Mean Average Precision (MAP), Normalized Discounted Cumulative Gain (NDCG), Mean Reciprocal Rank (MRR) as evaluation metrics. These are very well-known metrics and have been thoroughly studied in the literature.

3.2 Datasets

Table 2 mentions the highlights of the major datasets used in research related to recommendation systems and existing in the literature.

Table 2 Datasets used in the models

Name	Domain	#users	#item	#ratings
Movielens (ML)100K	Movies	943	1682	100,000
Movielens (ML)1M	Movies	6039	3883	1,000,209
Movielens (ML)10M	Movies	71,567	10,681	10,000,054
Movielens (ML)20M	Movies	136,677	20,108	20,000,000
Netflix	Movies	463,435	17,769	56,900,000
Book-crossing (BC)	Books	278,858	271,379	1,149,780
Advertising	Services	448,158	737	N.A
Yelp	Business	9600	7000	243,000
MillionSong data (MSD)	Songs	571,355	41,140	33,600,000
Watcha	Movies	1391	1927	101,037

4 Comparison of Various Approaches Surveyed in the Study

This section provides a comparison of various approaches reported in the last decade and existing in the literature for making the recommendation based on CF and AE as architecture. The approaches are reviewed based on the model of AEs used in the approach, the category of information taken for applying CF, the datasets used. Table 3 also summarizes the methodology adapted, results reported, and their comparison with other contemporary approaches in the light of evaluation metrics (Table 4).

5 Contribution of the Survey

In this paper, multiple models based on AE architecture for collaborative filtering are surveyed. It forms an informative basis for decision making when trying to select a model for implementation. This survey also highlights the limitations in these models, which can inspire innovation of new models that will overcome those limitations. Similarly, the plus points of the models surveyed can inspire new models to build upon one or more of them, therefore combining the positives of multiple models.

6 Conclusion and Future Work

In this paper, a thorough survey has been conducted on AE based models that employ CF methods for recommender systems. Through the survey, it is well evident that the research in the field of recommender systems is well flourishing and has a great scope of applications. In the literature, it is proposed that, as the number of layers in the AEs

Table 3 Comparison of various approaches surveyed in the study

Refs.	Model	Basis	Dataset	Description	Limitation	Metrics	Results
[20]	AutoRec	User and Item	Ml1M Ml10M, Netflix	Vanilla AE	Does not guarantee performance [23]	RMSE	No. of hidden units determine error
[45]	mSDA-CF	Item–Item	Ml100K ML1M, BC, Advertising	Uses MF method with mSDA	Includes only ratings, not review text. High computational cost	RMSE	Reduced RMSE, better stability
[23]	CDAE	User –User	Movielens, Netflix, Yelp	Generalizes other state-of-the-art	Poor performance with sparse inputs [34]	Precision and Recall, MAP	Outperforms other compared models
[26]	CF-SDA	User –User	ML1M	SDAE based, computes latent and surface similarity	High RMSE values which indicate dissimilarity between recommendations	RMSE, Jaccard similarity	Standardization improves accuracy
[27]	ReDa	Item–Item	ML100K, ML1M, DB, DM	Based on representational learning	Organizes user's rating behaviour in a sparse matrix [46], convergence not guaranteed always	MAE, RMSE	Denser data gives better results
[28]	CAAE	Item–Item	Ciao, Watcha, Ml100K	Employs GAN, implements BPR	The training time observed is comparatively very high	Precision and Recall, nDCG, MRR	Accuracy improves with BPR
[29]	FG-ACAE	Item–Item	Ciao, ML1M	Designs minimax game for CDAE	Less efficient on sparse dataset	Hit ratio and NDCG	Almost converges on 500 epochs

(continued)

Table 3 (continued)

Refs.	Model	Basis	Dataset	Description	Limitation	Metrics	Results
[30]	Mult-VAE^{PR}	Item–Item	ML20M, Netflix, MSD	Implicit feedback used in non-linear probabilistic model	Input and output vectors are sparse, thus may lead to instability during training [35]	Recall and truncated nDCG	Robust, less sensitive to hyperparameter choice
[34]	Q-VAE	Both user, item based	ML1M, Netflix prize (2M users)	Combines joint and conditional likelihood	Does not perform well on Recall	Precision, Recall, MAP, NDCG	Fast and smooth convergence
[35]	RecVAE	User –User	ML20M, Netflix, MSD	Extends Mult-VAE^{PR}, adds DAE	Might not be pragmatically better-performing	Recall and NDCG	Improves Mult-VAE^{PR}
[36]	AutoSVD + +	User –User	ML100K, ML1M	Hybrid of CAE and SVD++	No incorporation of any content information [38]	RMSE	Time complexity reduced, scalable
[38]	SA-HCF	Item–Item	ML100K. ML1M	Asymmetrical architecture	Only considers explicit feedback	RMSE	Improved rating prediction
[39]	HCF-SS	Item–Item	ML1M, MovieLens-HetRec	Uses semi-SDAE, on asymmetrical architecture	Requires input from three sources, might be computationally expensive	MAE and RMSE	Low error as training set increases
[40]	AutoCOT	User–item	ML100K, ML1M	Uses cooperative learning	Requires more iteration for a smaller dataset	Precision and Recall	Alleviates sparsity

(continued)

Table 3 (continued)

Refs.	Model	Basis	Dataset	Description	Limitation	Metrics	Results
[41]	DHA model	User–item	ML100K, ML10M	Uses SDAE and RNN	Rich text as input, some versions may require higher time steps	MAP, Recall	Better Recall and Precision
[42]	ACF	Item–Item	Ciao, Watcha, Ml100K, ML1M	Extends U-AutoRec, employs PR-Net, IL-Net	Larger amount of training time than almost all compared methods	Precision, Recall, NDCG, reciprocal rank	Outperforms compared methods
[43]	$EASE^R$	Item–Item	ML20M, Netflix, MSD	No hidden layer has L2 normalization	Author's implementation may require too much memory	Recall, NDCG	Shorter training time
[44]	AE-CF	User–User	ML1M	Combines AE with clustering	Dependent on specific iterations	MAE	Alleviates cold start and sparsity

Table 4 Acronyms used in the table and their full forms

MF	Matrix factorization
LFM	Latent factor model
FSM	Factorized similarity model
GAN	Generative adversarial networks
BPR	Bayesian personalized ranking
CAE	Contractive autoencoders
SVD	Singular value decomposition

is increased from two to four, they perform slightly better. Also, the resultant setup behaves more robustly when the number and type of hyperparameters are changed. Through the survey, it has also been found that HCF-SS, $EASE^R$. HCF-SS and AE-CF perform great, in terms of evaluation metrics such as RMSE, MAE, Precision and Recall and NDGC when used with MovieLens as a dataset. A future work might be that a combination of two or more models can be implemented, for example, $EASE^R$ with AE-CF, to see their synergetic effect on recommendations made.

References

1. Núñez-Valdéz, E.R., et al.: Implicit feedback techniques on recommender systems applied to electronic books. Comput. Hum. Behav. **28**(4), 1186–1193 (2012)
2. Isinkaye, F.O., Folajimi, Y.O., Ojokoh, B.A.: Recommendation systems: principles, methods and evaluation. Egypt. Inf. J. **16**(3), 261–273 (2015)
3. Ekstrand, M.D., Riedl, J.T., Konstan, J.A.: Collaborative filtering recommender systems. Found. Trends Hum. Comput. Interact. **4**(2), 81–173 (2011)
4. Terveen, L., Hill, W.: Beyond recommender systems: helping people help each other. In: HCI in the New Millennium, vol. 1, pp. 487–509 (2001)
5. Burke, R.: Hybrid recommender systems: survey and experiments. User model. User Adap. Inter. **12**(4), 331–370 (2002)
6. Khusro, S., Ali, Z., Ullah, I.: Recommender systems: issues, challenges, and research opportunities. In: Information Science and Applications (ICISA) 2016. Springer, Singapore, pp. 1179–1189 (2016)
7. Xin, Y.: Challenges in recommender systems: scalability, privacy, and structured recommendations. Diss. Massachusetts Institute of Technology (2015)
8. Adomavicius, G., Tuzhilin, A.: Toward the next generation of recommender systems: a survey of the state-of-the-art and possible extensions. IEEE Trans. Knowl. Data Eng. **17**(6), 734–749 (2005)
9. Bobadilla, J., et al:. Recommender systems survey. Knowl. Based Syst. **46**, 109–132 (2013)
10. Do, M.-P.T., Van Nguyen, D., Nguyen, L.: Model-based approach for collaborative filtering. In: Proceedings of the 6th International Conference on Information Technology for Education (2010)
11. Yang, Z., et al.: A survey of collaborative filtering-based recommender systems for mobile internet applications. IEEE Access **4**, 3273–3287 (2016)
12. Zhang, S., et al.: Deep learning based recommender system: a survey and new perspectives. ACM Comput. Surv. (CSUR) **52**(1), 1–38 (2019)

13. Yan, A., et al.: CosRec: 2D convolutional neural networks for sequential recommendation. In: Proceedings of the 28th ACM International Conference on Information and Knowledge Management (2019)
14. Devooght, R., Bersini, H.: Collaborative filtering with recurrent neural networks. arXiv preprint arXiv:1608.07400 (2016)
15. Salakhutdinov, R., Mnih, A., Hinton, G.: Restricted Boltzmann machines for collaborative filtering. In: Proceedings of the 24th International Conference on Machine Learning (2007)
16. Goodfellow, I., Bengio, Y., Courville, A.: Deep Learning. MIT Press. https://www.deeplearningbook.org (2016)
17. Vincent, P., et al.: Extracting and composing robust features with denoising autoencoders. In: Proceedings of the 25th International Conference on Machine Learning (2008)
18. Goodfellow, I., et al.: Generative adversarial nets. In: Advances in Neural Information Processing Systems (2014)
19. Rifai, S., et al.: Contractive auto-encoders: explicit invariance during feature extraction (2011)
20. Sedhain, S., et al.: Autorec: autoencoders meet collaborative filtering. In: Proceedings of the 24th International Conference on World Wide Web (2015)
21. Strub, F., Mary, J.: Collaborative filtering with stacked denoising autoencoders and sparse inputs (2015)
22. Chen, M., et al.: Marginalized denoising autoencoders for domain adaptation. arXiv preprint arXiv:1206.4683 (2012)
23. Wu, Y., et al.: Collaborative denoising auto-encoders for top-n recommender systems. In: Proceedings of the Ninth ACM International Conference on Web Search and Data Mining (2016)
24. Salakhutdinov, R., Mnih, A.: Probabilistic matrix factorization. In: Advances in Neural Information Processing Systems, pp. 1257–1264 (2007)
25. Koren, Y., Bell, R., Volinsky, C.: Matrix factorization techniques for recommender systems. Computer **42**(8), 30–37 (2009)
26. Suzuki, Y., Ozaki, T.: Stacked denoising autoencoder-based deep collaborative filtering using the change of similarity. In: 2017 31st International Conference on Advanced Information Networking and Applications Workshops (WAINA). IEEE (2017)
27. Zhuang, F., et al.: Representation learning via dual-autoencoder for recommendation. Neural Netw. **90**, 83–89 (2017)
28. Chae, D.-K., Shin, J.A., Kim, S.-W.: Collaborative adversarial autoencoders: an effective collaborative filtering model under the GAN framework. IEEE Access **7**, 37650–37663 (2019)
29. Yuan, F., Yao, L., Benatallah., B.: Adversarial collaborative neural network for robust recommendation. In: Proceedings of the 42nd International ACM SIGIR Conference on Research and Development in Information Retrieval (2019)
30. Liang, D., et al.: Variational autoencoders for collaborative filtering. In: Proceedings of the 2018 World Wide Web Conference (2018)
31. Kingma, D.P., Welling, M.: Auto-encoding variational bayes. arXiv preprint arXiv:1312.6114 (2013)
32. Rezende, D.J., Mohamed, S., Wierstra, D.: Stochastic backpropagation and approximate inference in deep generative models. arXiv preprint arXiv:1401.4082 (2014)
33. Sachdeva, N., et al.: Sequential variational autoencoders for collaborative filtering. In: Proceedings of the Twelfth ACM International Conference on Web Search and Data Mining (2019)
34. Wu, G., Bouadjenek, M.R., Sanner, S.: One-class collaborative filtering with the queryable variational autoencoder. In: Proceedings of the 42nd International ACM SIGIR Conference on Research and Development in Information Retrieval (2019)
35. Shenbin, I., et al.: RecVAE: a new variational autoencoder for top-N recommendations with implicit feedback. In: Proceedings of the 13th International Conference on Web Search and Data Mining (2020)
36. Zhang, S., Yao, L., Xu, X.: AutoSVD++: an efficient hybrid collaborative filtering model via contractive auto-encoders. In: Proceedings of the 40th International ACM SIGIR Conference on Research and Development in Information Retrieval (2017)

37. Koren, Y.: Factorization meets the neighborhood: a multifaceted collaborative filtering model. In: Proceedings of the 14th ACM SIGKDD International Conference on Knowledge Discovery and Data Mining (2008)
38. Zhang, S., et al.: Hybrid collaborative recommendation via semi-autoencoder. In: International Conference on Neural Information Processing. Springer, Cham (2017)
39. Zou, H., et al.: Hybrid collaborative filtering with semi-stacked denoising autoencoders for recommendation. In: 2019 IEEE International Conference on Dependable, Autonomic and Secure Computing, International Conference on Pervasive Intelligence and Computing, International Conference on Cloud and Big Data Computing, International Conference on Cyber Science and Technology Congress (DASC/PiCom/CBDCom/CyberSciTech). IEEE (2019)
40. Bai, R., et al.: AutoCOT-autoencoder based cooperative training for sparse recommendation. In: 2018 IEEE 15th International Conference on e-Business Engineering (ICEBE). IEEE (2018)
41. Li, T., et al.: Deep heterogeneous autoencoders for collaborative filtering. In: 2018 IEEE International Conference on Data Mining (ICDM). IEEE (2018)
42. Chae, D.K., Kim, S.W., Lee, J.T.: Autoencoder-based personalized ranking framework unifying explicit and implicit feedback for accurate top-N recommendation. Knowl.-Based Syst. **176**, 110–121 (2019)
43. Steck, H.: Embarrassingly shallow autoencoders for sparse data. In: The World Wide Web Conference (2019)
44. Chu, H., et al.: Towards a deep learning autoencoder algorithm for collaborative filtering recommendation. In: 2019 34rd Youth Academic Annual Conference of Chinese Association of Automation (YAC). IEEE (2019)
45. Li, S., Kawale, J., Fu, Y.: Deep collaborative filtering via marginalized denoising auto-encoder. In: Proceedings of the 24th ACM International on Conference on Information and Knowledge Management (2015)
46. Yi, B., et al.: Deep matrix factorization with implicit feedback embedding for recommendation system. IEEE Trans. Ind. Inf. **15**(8), 4591–4601 (2019)

An Algorithm to Recognize and Classify Circular Objects from Image on Basis of Their Radius

Bhim Sain Singla, Manvinder Sharma, Anuj Kumar Gupta, Vandana Mohindru, and Sunil Kumar Chawla

Abstract For the computer vision, fast and accurate detection of an object is challenging. Detecting a circular object in a cluttered image has always been a problem. Circular object detections has wide applications in the field of biometrics, automobile and other mechanical production industries. The traditional existing circular object detection are maximum likelihood estimation (MLE) and voting-based methods. The voting based methods have high memory requirements and more computational complexity while these are less sensitive to noise. MLE approach consumes less memory and are efficient in terms of computational complexity but these approaches are more prone to noise. This paper proposes modified Hough transform based algorithm for detection of circular objects within other shaped objects also it can identify circular objects on basis of diameter. The proposed algorithm worked efficiently and detected the circular objects on basis of diameters with very less computational time and less memory consumption.

Keywords Circular detection · Modified Hough transform · CHT · Circular object

---

B. S. Singla (✉)
College of Engineering and Management, Punjabi University, Patiala, Punjab, India
e-mail: bhim.pup@gmail.com

M. Sharma · A. K. Gupta · V. Mohindru · S. K. Chawla
Chandigarh Group of Colleges, Landran, Mohali, Punjab, India
e-mail: manvinder.sharma@gmail.com

A. K. Gupta
e-mail: anuj.coecse@cgc.edu.in

V. Mohindru
e-mail: vandana.coecse@cgc.edu.in

S. K. Chawla
e-mail: sunil.3550@cgc.edu.in

P. K. Singh et al. (eds.), *Recent Innovations in Computing*, Lecture Notes in Electrical Engineering 701, https://doi.org/10.1007/978-981-15-8297-4_33

1 Introduction

The task of object detection is encountered in many practical applications and situations. Rolling elements have great applications in the area of technical reality and while considering the natural sciences, biological shapes are mostly circular and that is because they are very easy to optimize because of their ratio of area to perimeter. There are many problems that can arise while detecting the circular objects which include difficulty to distinguish them and their detection in real-time [1–3]. The problem of distinguishing the objects is due to many factors such as low contrast, partial object visibility, illumination, etc. the real-time detection of the circular objects becomes difficult due to technological issues such as sorting and classification. Circular and elliptical shape detection is quite a common task in computer vision and in the recognition of images [4]. There are many methods for detection which depend upon the conversion of greyscale image to the binary image and then using the technique of edge detection. There are many shape descriptors which are very useful in detecting the ellipses and circles. The ellipticity of the shape can also be measured with the help of these shape descriptors [5, 6]. These are certain edge detection methods of low level that do not guarantee the uninterrupted boundaries of objects. The analysis of image becomes difficult for such cases especially when the image is noisy. Then came the contour grouping algorithms that aimed to connect all the edges of the parts of the object which is detected. But only the significant curves were detected using the contour grouping algorithm [7–9]. Then there were many methods designed based on active contours for the detection of objects whose boundaries are not defined by gradients. The moving objects could also be tracked using active contours. The difference between the features of object and background is calculated by the energy function [10]. Another system efficiently used in deep learning is license plate recognition (LPR) [11]. This technique is widely used in the digital world due to rapid growth in vehicles. The surveillance system is also a major concern in the digital era. D-CAD technology is efficiently used to address the issue based on the concept of deep learning [12]. A very powerful and effective tool for detecting the parametric curves in images is the Hough Transform (HT). This algorithm is able to recognize different shapes in the image. It was first introduced by Paul Hough in 1962 and then IBM patented it. In 1972, the Hough Transform was modified and named as Generalized Hough Transform (GHT). The Circular Hough Transform (CHT) is the extended version of General Hough Transform. The voting process is implemented in this algorithm where the mapping of the edge points in the image is done with the parameter space that is appropriately defined [13–15]. The Circular Hough Transform (Fig. 1) is used to identify circles in an image whose center pint and radius are defined. The accumulation of votes in the three-dimensional parameter space is very necessary for determining a circle. Few improvements in the CHT algorithm were proposed in the literature of image processing taking in view the memory and computational requirements [16]. The parameter space could also be used for different radii as shown in Fig. 2.

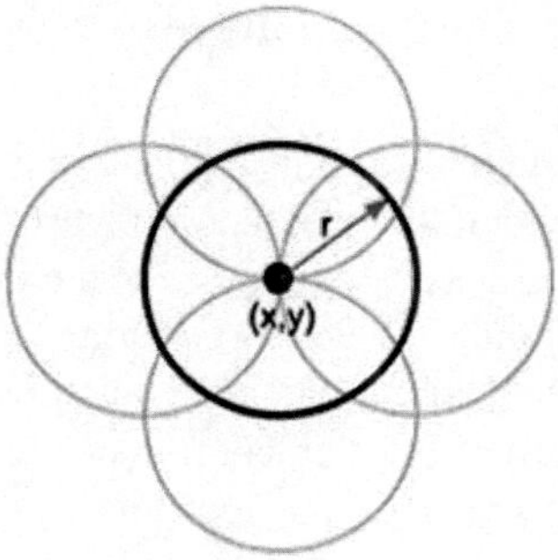

Fig. 1 Basic CHT

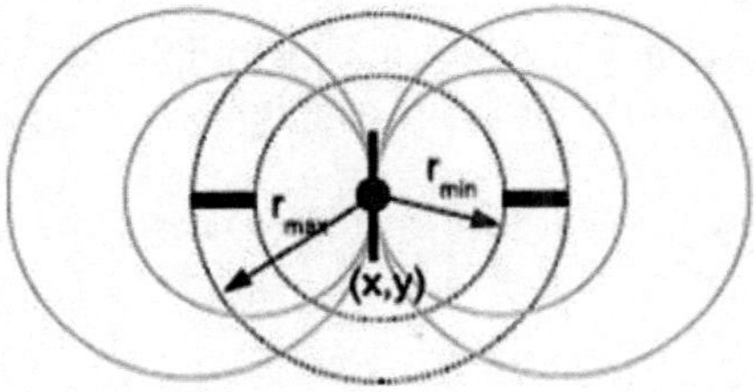

Fig. 2 Multiple radii in single parametric space

The location of the circle is detected with three basic steps. First is the convolution and non-maximal suppression [17, 18]. Second one is the contour tracing and the third one is finding the radius and center of the circle. The block diagram is as shown in Fig. 3.

Firstly, the grey-scale image is given the input to the block. Here, each pixel obtains its vertical and horizontal directions with the help of convolution kernels. The strength at the edges of the images can also be detected using the horizontal and vertical convolution kernels. The non-maximal suppression is done to thin the edges obtained from the direction of the pixels and edge strength. The contour tracing of the thin edges is done in the second stage of the algorithm. This is done by finding the arcs through pixel directions in each of the pixels which proves them to be a part of the circle. The edge pixels and spurious points which do not satisfy the part of the circle are discarded. Only the arcs which are the part of the circle are retained [19]. In the last stage, the center and radius of the circle is determined. In this way, the circle are located and detected.

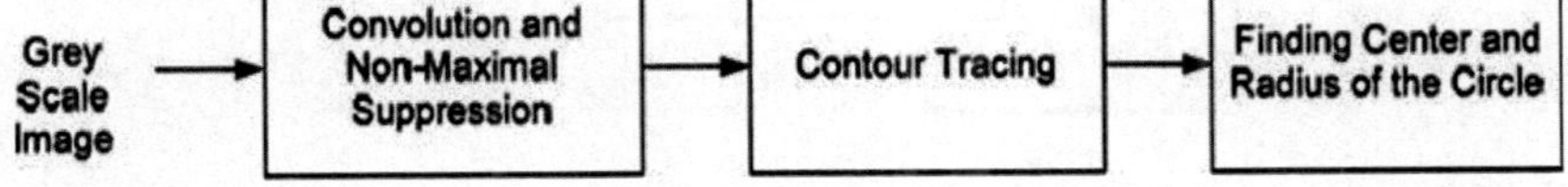

Fig. 3 Algorithm stages

2 Circular Hough Transform: Literature Review

The Circular Hough Transform has its main application in extracting circular objects from the image. It doesn't matter if the circle is complete or not. The elongated ellipses are ignored and the transform is selective only for the circles. Radial symmetry is required for searching the circular objects. The radius and the centric of the circle can also be measured with the Circular Hough Transform. The transform works on the basis of vote. Whenever the Cartesian coordinate encounters pixel with greater than zero intensity, one vote is given to the coordinate. As the image is transformed into the circular of specific radius, the votes get accumulated at the center of the circle [20]. The maxima in the transform can be calculated by counting the highest number of votes in the coordinate and in this way the center of the circle within the image can be found out.

The parameters of the circle are determined with the help of Hough Transform. The parametric equations of the circle are as follows

$$x = a + R\cos(\theta) \tag{1}$$

$$y = b + R\sin(\theta) \tag{2}$$

Here,

R is the radius of the circle;

Θ sweeps through full 360° in the coordinates (x, y);

Center of the circle is defined by (a, b).

The perimeter of the circle is traced if most of the points on the perimeter are known. Three parameters are needed to find out to describe the circle which are (a, b, R).

If the radius R of the circle is known, there will be a common center point of all the parameter circles. The locus points of the circle are (a, b) which fall on the circle of radius R with centers at (x, y). The Hough accumulation array is used to find out the true center point which is common to all the parameters circle. This is as shown in Fig. 4.

Each point in the left geometric space generates a circle in the right geometric space. All these circles intersect at the center of the circle (a, b).

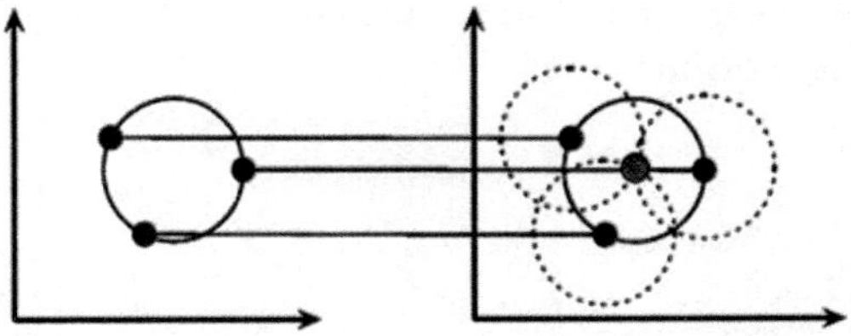

Fig. 4 Geometric-scale representation to search a circle with fixed R

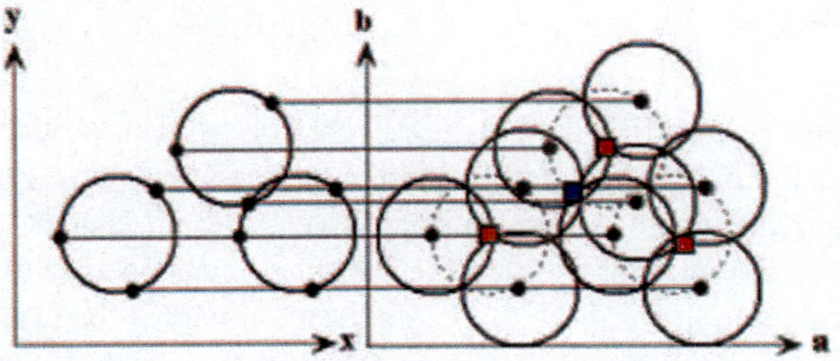

Fig. 5 Geometric-scale representation to search a circle with multiple *R*

The same technique can be used to find out multiple circles with the same radius. Figure 5 depicts the geometric scale representation to search a circle with multiple *R*.

The red cells represent the center points in the parameter space. The blue cells represent the overlapping of circles with the help of which the spurious centers can also be found out. When the circles in the original image are matched, then the spurious circles can be removed. Each point in the left geometric space generates a circle in the right geometric space [21, 22]. All these circles intersect at the center of the circle (a, b).

The locus points present in the parameter scale falls on the cone surface if the radius of the circles is known. A cone surface is produced in the parameter space by each (x, y) point. All the maximum cone surfaces intersect at the accumulation cell by corresponding triplets (x, y, R). The conical surface is generated in the parameter space for (x, y) point. At each level 'r' a circle is constructed having different radius. A three-dimensional accumulation matrix can be used to search the circle having unknown radius. Figure 6 shows the geometric scale representation to search multiple circles with fixed *R*.

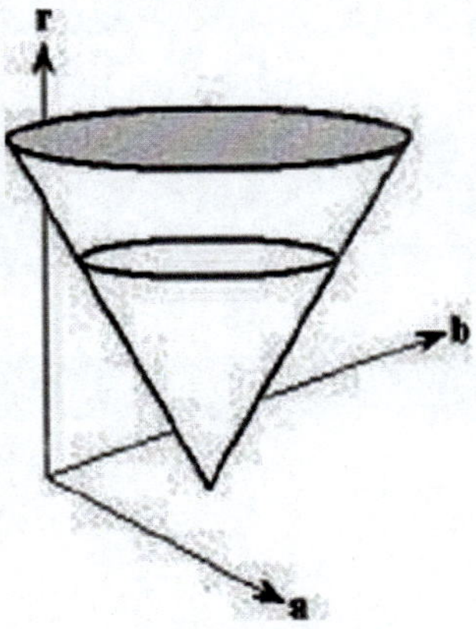

Fig. 6 Geometric-scale representation to search multiple circles with fixed *R*

3 Design Flow

The flowchart describes different steps that were followed in order to identify the circular object in the image. Figure 7 shows the flowchart of steps. Firstly the reference image was read which contained objects of different shapes. This cluttered image is used for the identification of the circular object in the image. Then the image is converted into the grayscale. The grayscale image is further converted into the binary image. This conversion of image from grayscale to binary is the most important step and which greatly defines the detection of the circular image. Then, the radius of the circles are defined to be detected [23, 24]. Then, the circular objects are being detected using the modified algorithm through Circular Hough Transform (CHT). This method is quite efficient as it decreases the computational time and storage required for its implementation. And then in the last step finally, the circular objects are detected. All the steps given in the algorithm clearly defines the flow or process followed while identifying the circular objects.

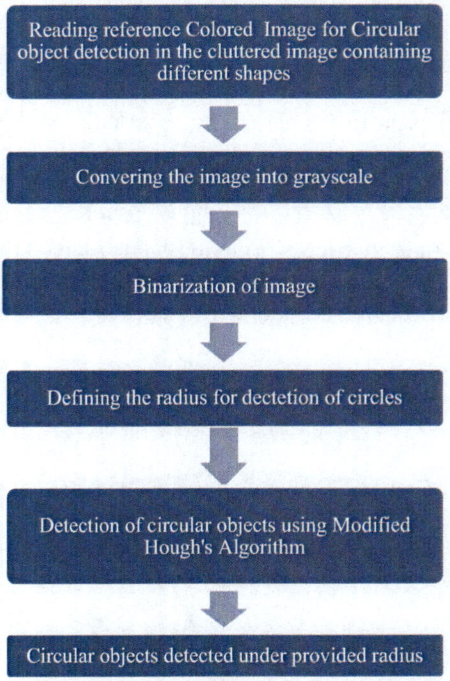

Fig. 7 Flowchart of method of detection

4 Results

In a work, the algorithm uses a modified approach for the detection of circular objects. The following results show the different design steps that are followed. Figure 8 shows the input image from which the circular objects are to be detected. The image has different shapes in it. Figure 9 shows the result after converting the image to grayscale. Using the design flow the modified algorithm is compiled and run. In Fig. 10 only those circular objects are detected which have the radius of 25. In Fig. 11 the circular objects with radius 40 are detected and similarly in Fig. 12 represents the circular object detection with radius 80. Figure 13 shows the results as detection of circular objects from the different shaped objects in image with very less time. The circular objects identified in Fig. 7 are of different radius. All the circular objects present in the image are detected with very less computational and storage time. The objects which are marked are identified as circular objects while the others are not identified as they are not circular. The part of the circular object is also identified in the non-circular objects as they contain the circular images. In this way, all the circular images are identified and detected whether fully as a complete circle or partially on the non-circular objects. This proves the accuracy and efficiency of the algorithm being used.

Fig. 8 Input cluttered image with different shaped objects

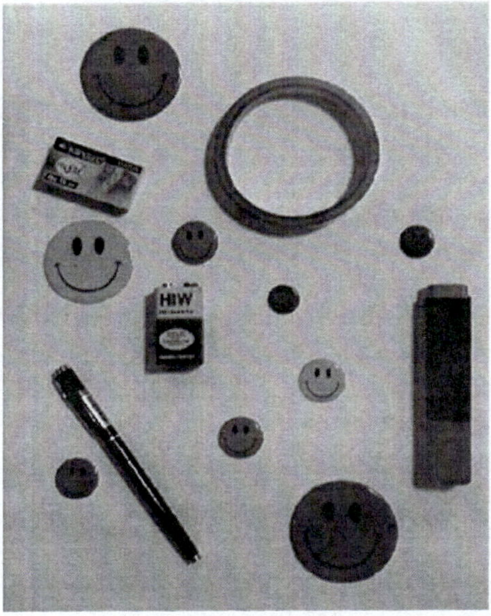

Fig. 9 Image after conversion to grayscale

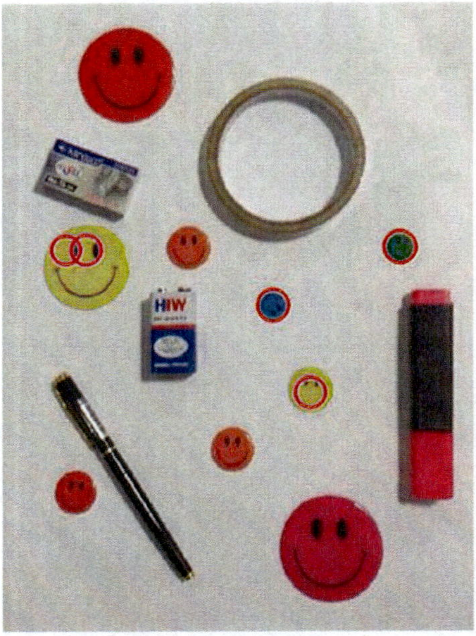

Fig. 10 Detection of circular objects with radius = 25

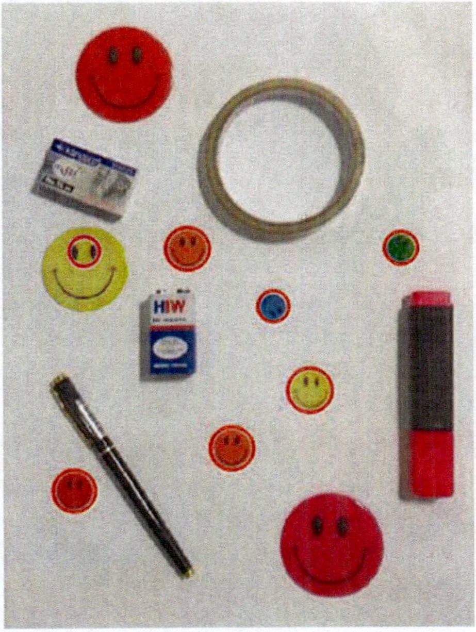

Fig. 11 Detection of circular objects with radius = 40

Fig. 12 Detection of circular objects with radius = 80

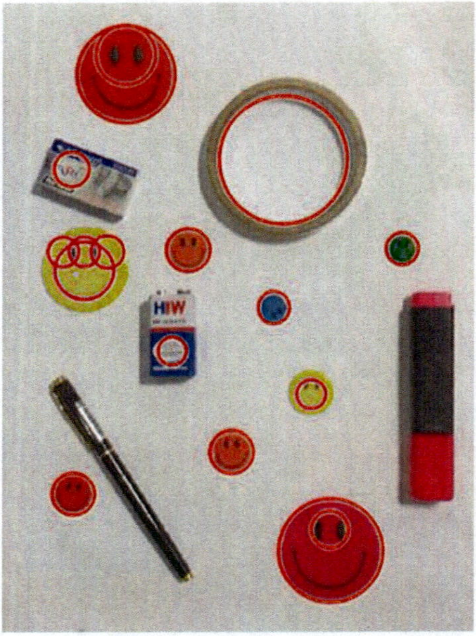

Fig. 13 Detection of all circular objects

5 Conclusion

In this paper, a modified and novel method for detection of circular objects under specific radius in different shaped objects is proposed and discussed which produces results in less computational time, unlike the other methods which come different sorts of limitations. The discussed approach is able to produce results without requiring any additional hardware or any additional input image. As the hardware comes with its own complexities and therefore do not provide efficient solution. As the computing time of algorithm is less, it can be connected with a hardware that can respond and perform required operation in real-time after detection. The software version overcomes all the limitations. The algorithm can be used for various applications including rejecting non-circular shaped objects as well as oversized circular objects in an industry.

References

1. Landau, U.M.: Esimation of a circular arc center and its radius. Comput. Vision Graph. Image Process. **38**, 317–326 (1986)
2. Crawford, J.F.: A non-iterative method for fitting circular arcs to measured points. Nucl. Instrum. Methods Phys. Res. **211**, 223–225 (1983)

3. Karimäki, V.: Effective circle fitting for particle trajectories. Nucl. Instrum. Methods Phys. Res. **A305**, 187–191 (1991)
4. Thom, A.: A statistical examination of the megalithic sites in Britain. J. Roy. Statist. Soc. Ser. A General **118**, 275–295 (1955)
5. Kasa, I.: A circle fitting procedure and its error analysis. IEEE Trans. Instrum. Meas. **25**, 8–14 (1976)
6. Coath, G., Musumeci, P.: Adaptive arc fitting for ball detection in robocup. In: APRS Workshop on Digital Image Computing, Brisbane, Australia, Feb 2003, pp. 63–68
7. Atherton, T.J., Kerbyson, D.J.: Size invariant circle detection. Image Vis. Comput. **17**, 795–803 (1999)
8. Kerbyson, D.J., Atherton, T.J.: Circle detection using Hough transform filters. Image Process. Appl. 370–374 (1995)
9. Kaur, S.P., Sharma, M.: Radially optimized zone-divided energy-aware wireless sensor networks (WSN) protocol using BA (bat algorithm). IETE J. Res. **61**(2), 170–179 (2015)
10. Duda, R., Hart, P.: Use of the Hough transform to detect lines and curves in pictures. Commun. ACM **15**(1), 11–15 (1972)
11. Kumar, K., Sinha, S., Manupriya, P.: D-PNR: deep license plate number recognition. In: Proceedings of 2nd International Conference on Computer Vision & Image Processing. Springer, Singapore, pp. 37–46 (2018)
12. Vashisht, S., Jain, S.: An energy-efficient and location-aware medium access control for quality of service enhancement in unmanned aerial vehicular networks. Comput. Electr. Eng. **4**(75), 202–217 (2019)
13. Sharma, M., Singh, S., Khosla, D., Goyal, S., Gupta, A.: Waveguide diplexer: design and analysis for 5G communication. In: 2018 Fifth International Conference on Parallel, Distributed and Grid Computing (PDGC). IEEE, pp. 586–590
14. Gupta, A.K., Sharma, M., Khosla, D., Singh, V.: Object detection of colored images using improved point feature matching algorithm. Cent. Asian J. Math. Theory Comput. Sci. **1**(1), 13–16 (2019)
15. Yip, R., Tam, P., Leung, D.: Modification of Hough transform for circles and ellipses detection using a 2-dimensional array. Pattern Recogn. **25**, 1007–1022 (1992)
16. Sharma, M., Singh, H.: SIW based leaky wave antenna with semi C-shaped slots and its modeling, design and parametric considerations for different materials of dielectric. In: 2018 Fifth International Conference on Parallel, Distributed and Grid Computing (PDGC). IEEE, pp. 252–258 (2018)
17. Pao, D.C.W., Li, H.F., Jayakumar, R.: Shapesrecognition using the straight line Hough transform: theory and generalizaion. IEEE Trans. Pattern Anal. Mach. Intell. **14**, 1076–1089 (1992)
18. Sharma, M., Singh, S., Khosla, D., Goyal, S., Gupta, A.: Waveguide diplexer: design and analysis for 5G communication. In: 2018 Fifth International Conference on Parallel, Distributed and Grid Computing (PDGC). IEEE, pp. 586–590 (2018)
19. Chernov, N., Lesort, C.: Least squares fitting of circles and lines. J. Math. Imaging Vision (to appear)
20. Duda, R.O., Hart, P.E.: Use of the Hough transform to detect lines and curves in pictures. Commun. ACM **15**, 11–15 (1972)
21. Sharma, M., Singh, H.: Substrate integrated waveguide based leaky wave antenna for high frequency applications and IoT. Int. J. Sens. Wirel. Commun. Control **9**, 1 (2019). https://doi.org/10.2174/2210327909666190401210659
22. Berman, M., Culpin, D.: The statistical behaviour of some least squares estimators of the centre and radius of a circle. J. Roy. Stat. Soc. Ser. B Stat. Methodol. **48**, 183–196 (1986)
23. Gupta, A.K., Sharma, M., Khosla, D., Singh, V.: Object detection of colored images using improved point feature matching algorithm. Cent. Asian J. Math. Theory Comput. Sci. **1**(1), 13–16
24. Sharma, M., Singh, H., Singh, S., Gupta, A., Goyal, S., Kakkar, R.: A novel approach of object detection using point feature matching technique for colored images. In: Proceedings of ICRIC 2019. Springer, Cham, pp. 561–576 (2020)

A Systematic Review on Various Techniques on Image Segmentation

Diksha Thakur, Nitin Mittal, and Saumya Srivastava

Abstract In the present situation, image processing is one of the huge developing fields. It is a strategy which is ordinarily used to improve raw image which is collected from different assets. It is a sort of sign processing. The image segmentation is considered one of the most essential processes in the production of images. Some of these methods use only the histogram of the gray scale, some use spatial information, while others use conceptual approaches of the fuzzy set. In noisy environments, these methods are not appropriate. Some work has been done using the method of MRF, which is robust to noise but involves computing. This paper gives an outline of image processing techniques. The fundamental concern of this paper is to characterize different methods utilized in various periods of image processing.

Keywords Image processing · Segmentation · Threshold · Acquisition · Enhancement

1 Introduction

Image segmentation is a very essential and inspiring tool for image processing. Divide an object into relevant sections with similar properties and characteristics using the image segmentation technique. The main goal of segmentation is to simplify a picture's representation in a clear and easy way to understand. Segmentation is a mechanism by which an object is grouped into homogeneous units with respect to one or more features; it is an important task in the analysis of images [1–3]. The purpose of image segmentation is to divide an image into numerous segments with similar

D. Thakur (✉) · N. Mittal · S. Srivastava
Electronics and Communication Engineering Department, Chandigarh University, Mohali, India
e-mail: Dikshathakur.phd@gmail.com

N. Mittal
e-mail: mittal.nitin84@gmail.com

S. Srivastava
e-mail: Saumyasrivastava65@gmail.com

P. K. Singh et al. (eds.), *Recent Innovations in Computing*, Lecture Notes in Electrical Engineering 701, https://doi.org/10.1007/978-981-15-8297-4_34

attributes or features. The basic uses of image segmentation are: content-based object recovery, medical imaging, detection and recognition of objects, video surveillance, automated traffic control systems, etc. Two main forms of local segmentation and global segmentation can be classified.

Segmentation subdivides a picture into its constituent region or element. Methods of image segmentation are classified based on the discontinuity and similarity of two properties [4]. Based on this, picture segmentation property is classified as segmentation based on edged [5–7] and segmentation based on area [8–11]. The methods of segmentation based on pixels are considered to be techniques based on boundaries or edges. The edge-based segmentation approach is designed to overcome image segmentation by detecting edges or pixels between different regions with rapid intensity transfer and extracting and connecting them to the boundaries of closed objects. The effect is an object that is binary. The theory, gradient-based approaches and gray histogram-based approaches are two key edge-based segmentation approaches [12].

The thresholding method was discovered to be the most common method out of all the current methods used to segment different image kinds. Thresholding is the simplest and most capable methods of image segmentation that can discriminate against objects from the background by setting pixel-level thresholds. It can not only display well-defined areas with minimal overlap and aggregation efficiency, but also provide original prediction or preprocessing for more complex segmentation approaches [13]. It is possible to break down thresholding approaches into parametric and nonparametric classifications [14–16].Usually, the parametric approach is time-consuming, whereas the nonparametric is more accurate and, as a consequence, more attention is paid to determining the appropriate limit by optimizing certain norms such as inter-class variance, entropy and error rate [16–18].

2 Image Processing

Image processing is spreading in different fields. Image processing is a strategy which is ordinarily used to improve crude image which is gotten from different assets [19]. It is a method to change an image into computerized structure also and execute certain activities on it, so as to make an improved image or to theoretical profitable data from it. It is a sort of sign administration where image is information and yield is likewise an image or highlights related to image. The reason for image processing is conveyed into a few gatherings which are given underneath.

Visualization: Image preparing is utilized to distinguish those items which are not perceivable.

Image sharpening and restoration: In image processing, different procedures are connected on the image to create a superior image.

Image retrieval: By image preparing, client can identify just that segment of the image which is applicable to the client.

Pattern measurement: Numerous components in an image are estimated.

Image Recognition: Substances in an image are perceived. Image processing utilizes numerical systems for preparing of images. Two strategies utilized for processing of images are simple image preparing and digital image processing.

2.1 Analog Image Processing

This processing strategy utilizes electrical sign for any change required in the image. Analog processing incorporates two-dimensional simple sign. In this methodology, images are adjusted by changing the electrical sign. It is primarily utilized for printed versions like concerning printing reason and for photography.

2.2 Digital Image Processing

In this system, preparing of image is finished by advanced PCs. Right off the bat by means of scanner-digitizer image are changed over into computerized structure and after that further preparing is done on the image. Computerized image processing utilizes numerous methods like amendment, arranging of the information, enhancement methodology to make image with better quality [20]. Essentially, there are primarily four activities utilized in advanced image processing like image preprocessing, division of image, highlight extraction, order of image (Fig. 1).

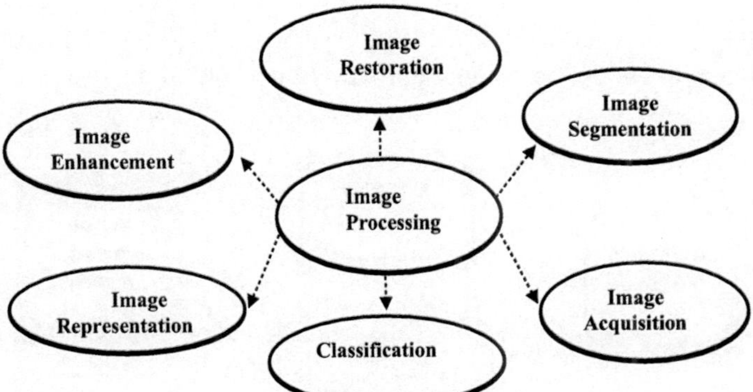

Fig. 1 Different techniques of image processing [20]

3 Image Segmentation

Segmentation means dividing of image into different districts or parts. In image segmentation, an image is partitioned into subparts as indicated by the pixels of the client or the issue being comprehended [21]. It isolates the image into pixels. Image segmentation separates the image in such a way along these lines that it turns out to be very exact. Essentially, this methodology is utilized for examination of substances, outskirts what's more, extra records which are applicable for processing [22]. The result of image segmentation is a lot of segments that together spread the all-out image or group of forms expelled from the image. The goal of segmentation is to streamline or to change the show of image in such a way that is progressively huge and simple to assess. It delivers the better appearance of image. Division of images is accomplished for compression of image, acknowledgment of articles and altering reason. For image division, image edge techniques are connected. Segmentation designates mark to every pixel in the image, such that pixel having comparative name can share unmistakable highlights [23] (Fig. 2).

Usually, the present segmentation technique can be summarized as more than four primary classifications: region, edge, cluster, threshold, ANN and watershed-based methods [10, 24, 25].

3.1 *Various Methods for Image Segmentation*

3.1.1 Thresholding Techniques

Thresholding is a crucial method in object segmentation uses. The basic idea of thresholding is to choose an optimal gray-level threshold to distinguish objects of

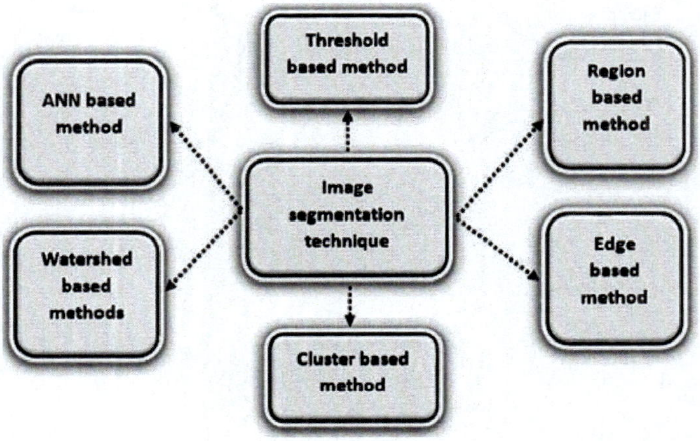

Fig. 2 Different techniques of image segmentation [22]

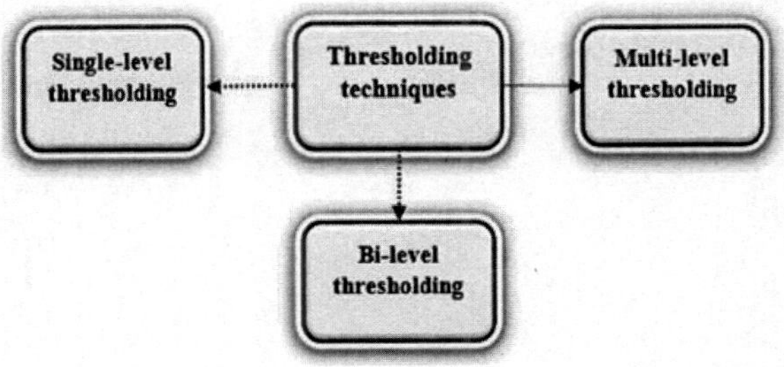

Fig. 3 Different thresholding techniques [26]

interest from the context based on their distribution at gray level. While humans can easily distinguish an object from a complex context, distinguishing it from image thresholds is a difficult task. The most straightforward strategy for division is thresholding (Fig. 3).

This methodology changes a dim-scale image into paired image any place the two are allotted to pixels. These focuses are beneath and on upper side of the clear limit esteem. In this technique, an edge worth is utilized that limits are acquired from histogram of the first image. The estimation of the histogram is determined by identification of edges. So, limit worth is exact just if the identification of edges is precise. Division performed by means of thresholding has lesser computations identified with different techniques. This method not gives fitting outcomes in complex condition [26].

3.1.2 Region-Based

This strategy bunches together certain articles utilized for division [27, 28]. Locale-based division strategy is utilized with this technique. That district must be as one with one another on which division needs to perform. It is otherwise called comparability-based division (Fig. 4).

The fringes are perceived to perform division. Each progression takes at any rate one pixel for processing reason. Subsequent to applying the procedure shading and

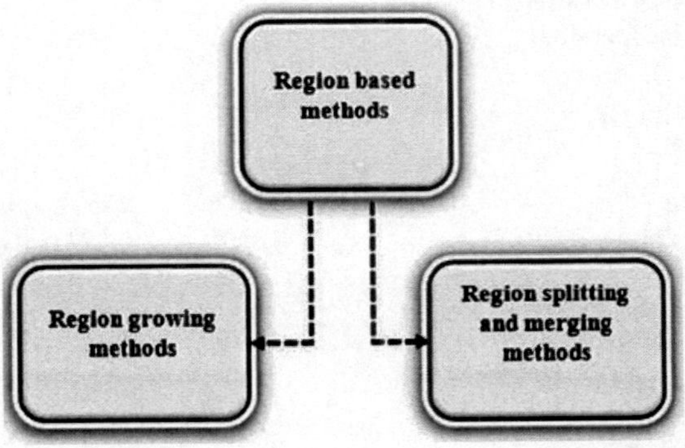

Fig. 4 Different region-based methods [28]

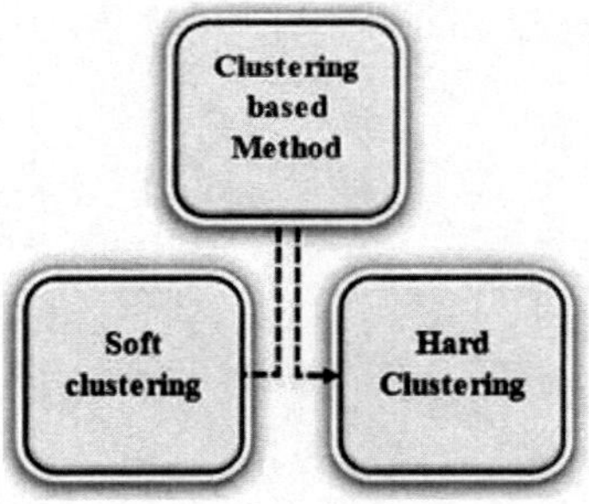

Fig. 5 A different feature-based clustering [30]

surface of the image is adjusted and after that a vector is made from the edge stream. At that point, further processing is connected on these edges [29].

3.1.3 Feature-Based Clustering

Another approach to perform division is grouping. In this plan, an image is changed into histogram. After that, bunching is performed on it [30] (Fig. 5).

Pixels of the shading image are bunched for division utilizing an unaided method fluffy *C*. This is connected for customary images. On the off chance that it is an uproarious image, it results in fracture.

3.1.4 Edge-Based

Another system for division is edge recognition strategy. To recognize dissimilarities from the image, edges are distinguished. To perceive pixel esteems, edges are drawn and after that these edges are contrasted and another pixel. In the edge finder strategy, it is not obligatory that identified edge ought to be close with one another. In this technique, initially the data about edges are removed and after that marking is accomplished for pixels (Fig. 6).

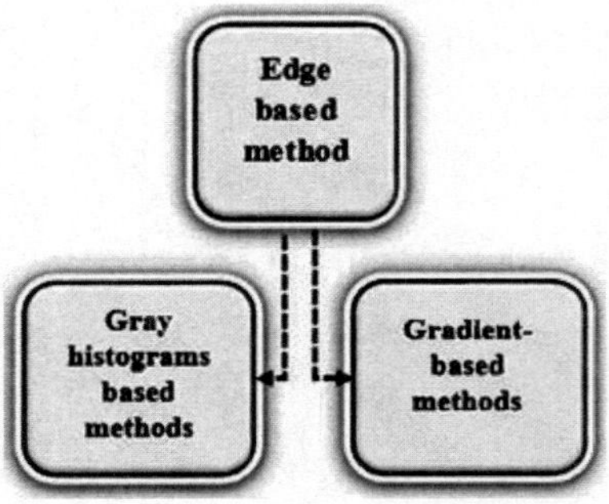

Fig. 6 Different edge-based methods [32]

This technique additionally gets the data from the frail limit [31]. The procedure of division may likewise be performed by edges. As the edges are not shut with one another, there are a few holes among the edges. So, connecting is performed to fill the hole between the edges [32].

3.1.5 Model-Based

This technique is based on Markov random field. For color segmentation, inbuilt region constraint is used. To define accuracy of edges, MRF is joined with edge detection [33]. This method contains the relations among color components.

4 Comparison of Various Methods for Image Segmentation

Table 1 provides a contrast between different techniques of segmentation by presenting a brief explanation of each approach and its merits and demerits.

Table 1 Comparison of several segmentation approaches

Segmentation technique	Explanation	Merits	Demerits
Thresholding technique	Based on image's histogram peaks to find specific threshold value	No need for prior information, easiest way	Spatial specifics are not considered, highly dependent on peaks
Region-based technique	It is based on separating image into standardized regions	More protected to noise, it is beneficial when it is stress-free to describe resemblance standards	Costly technique in terms of time and memory
Watershed technique	It is based on topological clarification	Results are constant; detected boundaries are uninterrupted	Difficult calculation of pitches
Edge-based technique	It is based on discontinuity detection	Good for superior contrast images between objects	Not suitable for errors
PDE-based technique	Based on differential equations	System is quickest and suitable for time-critical uses	More difficulty in computing
Clustering technique	Divided into homogeneous bunches	Therefore, fuzzy uses partial membership to make actual problems more valuable	It is not easy to define the membership function
ANN-based technique	It is based on the decision-making system simulation	No need to write complicated programs	More consumption of time in exercise

5 Image Compression

Image compression connotes compression of the records among the computerized images [34]. Image compression wipes out duplication of the information with the goal that it will be put away and transmitted in a powerful manner. Image compression may be lossy and lossless. In lossless compression when compression the nature of information stays reliable. In lossy compression, the nature of information diminishes in the wake of applying the compression methods. Lossless compression is for the most part utilized for therapeutic imaging, specialized drawing substance, authentic purposes and so on. Lossy methodologies are utilized in those conditions in which minor loss of value is adequate to achieve an impressive decrease in bit rate. The most far-reaching system for compression is JPEG which packs full shading or dark-scale images. This strategy isolates the image into eight by eight squares. These squares are partitioned in such a way that no covering is framed among them. JPEG utilizes discrete cosine change procedure for compression [35]. There is another method for compression known as wavelet change. Through wavelet, information is partitioned into various recurrence segments and after that further investigation is accomplished for every segment. Wavelets have points of interest over customary Fourier methodologies in analyzing physical conditions.

6 Classification of Image Processing

Classification of images is utilized to remove the data from the images, mark and pixels from the images. So as to perform grouping, numerous images of similar items are required. A proper order plan and satisfactory measure of preparing tests are rudiments for a compelling grouping. Fundamentally, order framework intentionally relies upon client's necessities [36]. There are various arrangement methodologies open like fake neural systems, master frameworks, fluffy rationale and so forth. Different sorts of grouping calculations like according to pixel, sub pixel, per field. Per-pixel order is generally utilized technique. Subpixel calculation strategies reduced with the fluctuated pixel issue. These give larger amount of precision. For fine three-dimensional goals, information per field's characterization is the best alternative. The characterization systems are either directed or unaided. In regulated arrangement, ghostly marks which are acquired from preparing tests are utilized to order an image. Mark record is effectively made from the given preparing tests; further with the assistance of multivariate grouping instruments, image is arranged. In solo arrangement, the yield relies upon machine with no cooperation with the client. In this procedure, pixels have a place with same classification assembled into one class. The accompanying outline portrays the working of directed and solo arrangement methods. In the managed characterization as a matter of first importance, tests are gathered; at that point, these examples are assessed. After this, a mark record is made (Fig. 7).

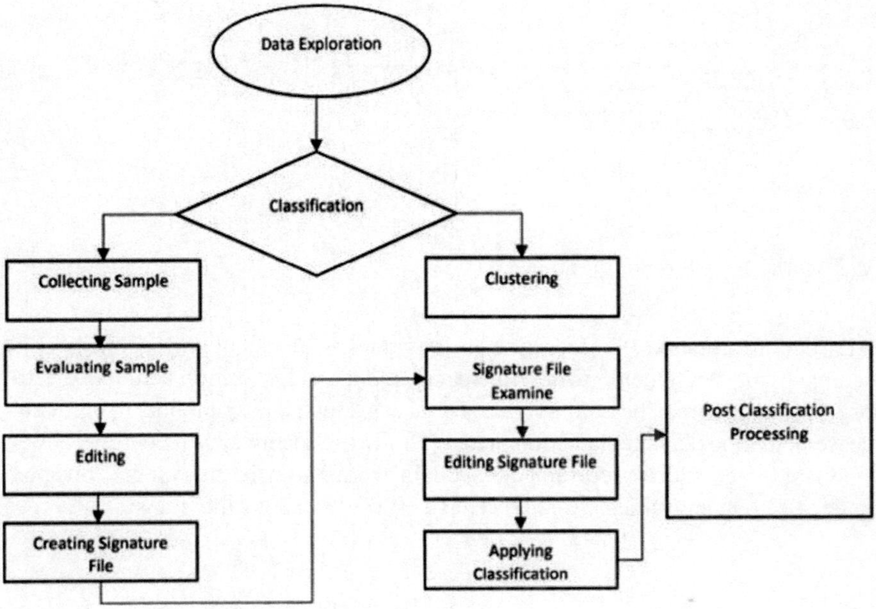

Fig. 7 Different classification techniques [36]

After making of mark record, different characterization strategies are connected on the mark document to arrange an image. The solo characterization manages grouping. In this arrangement, no examples are gathered for further processing. All work is finished by PC with the assistance of different calculations. There are different components identified with arrangement which are critical to get victories. These components are high caliber detected information, substantial system for grouping, aptitude and experience of the examiner. A fitting arrangement plot and a sufficient measure of preparing tests are essentials for a successful characterization.

7 Image Restoration

Image restoration is a strategy through which an undermined and loud image is handled so that an ideal image is built [37]. Accordingly, reclamation reconstructs those images whose quality is pillaged because of clamor or framework blunder. There are different reasons for corruption, for example, clamor from the sensor, camera misfocus and barometrical aggravation. There are two kinds of strategies utilized to re-establish the image. One strategy is to demonstrate the image whose quality is corrupted by means of certain reasons. Another strategy is known as image improvement; it expands the nature of image by applying different channels (Fig. 8).

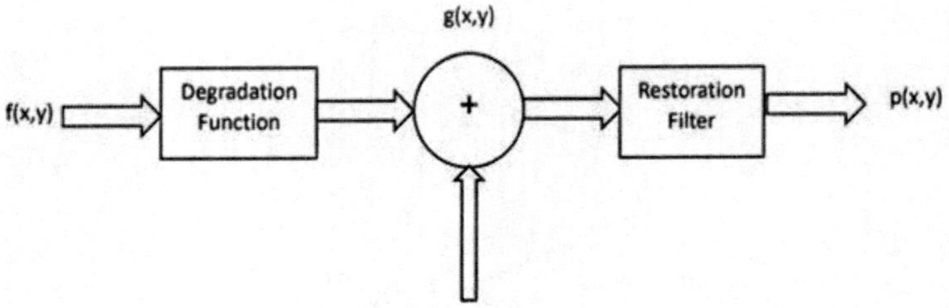

Fig. 8 Image restoration [37]

Earlier information of corruption is important to re-establish the image. The accompanying figure demonstrates the debasement and restoration action. Restoration of the images may be accomplished by means of two kinds of model in particular debasement model and reclamation model [38]. In the accompanying outline, $f(x, y)$ is the first image which is corrupted by certain exercises. After this on the corrupted image, different capacities are connected so as to re-establish the image.

8 Image Acquisition

The main period of each perception plan is the image obtaining stage. At the point when the image is acquired, then different procedures are connected on the image. Fundamentally, an image securing is a procedure through which images are recovered from different assets. The most widely recognized technique for image securing is constant procurement strategy. This technique makes a pool of records which are prepared naturally. An image securing technique makes 3D geometric information [39].

9 Image Representation

Image representation means changing over the crude information in such a way that PC processing can apply on it. Essentially, two kinds of procedures are utilized to speak to the photographs: limit representation and district representation. Limit representation shows the interior state of the image. Area representation is utilized when the fundamental concern is about the inside properties. Relying on level of preparing of images by means of machine, there are four techniques for image representation, for example, pixel-based, block-based, region-based and hierarchical-based. Image representation is suitable for the development of elements, learning-based models which must be removed from image databases that are made utilizing predefined choice standards [40].

10 Image Enhancement

Image enhancement improves the image showing quality. Now and again when image is caught from different assets, then the nature of images is not generally excellent because of hindrances. Image improvement changes segments of the photographs with the goal that clearness of images can be expanded. The data substance of the images will likewise be expanded by changing the visual effect. This strategy is utilized for dissecting the image, highlighting extraction and showing the images. Calculations which are utilized for this procedure are dependent on applications and intelligent. There are some improvement strategies, specifically differentiation extending, clamor sifting and histogram alteration. Spatial space systems work with pixels. In this system, the estimations of pixels are changed to accomplish the ideal improvement. It contains different strategies who's working straightforward subject to the pixels of the images. Recurrence space strategies are fitting for the images which depend on recurrence systems, and it takes a shot at the symmetrical transformation of the image as opposed to the image itself [41].

11 Conclusion

Image processing is utilized to improve the nature of the image that is taken from different assets. This paper talks about different image processing strategies like image representation, division, pressure, securing, image enhancement and so forth. These strategies are utilized in various zones. The strategy that we are picking relies on the application zone. Each system has its own upsides and downsides. The various image processing implementations of each process were also evaluated and analyzed, the dissimilar image processing methods are used, the limitations are checked, and the system is used.

12 Future Scope

There are various image related to image segmentation have been made till now a day for enhancement yet at the same time there is further necessity for improvement, which may be cultivated by means of computerized reasoning frameworks for streamlining that can create agreeable outcome. Later on, compelling image improvement procedures utilizing man-made brainpower will be grown with the goal that enhancement results may be accomplished in adjusted way which would give better outcomes to streamlining.

References

1. Pratt, W.K.: Digital Image Processing. Wiley, New York (1978)
2. Pavlidis, T.: Structural Pattern Recognition. Springer, Heidelberg/New York (1977)
3. Rosenfeld, A., Kak, A.C.: Digifal Picture Processing, 2nd edn. Academic Press, New York/London (1982). (Vol. 2)
4. Gonzalez, R.C., Woods, R.E.: Digital Image Processing, 2nd edn. Publishing House of Electronics Industry, Beijing (2007)
5. Hertz, L., Schafer, R.W.: Multilevel thresholding using edge matching. Comput. Vis. Graph. Image Process. **44**, 279–295 (1988)
6. Kohler, R.: A segmentation system based on thresholding. Comput. Graphics Image Process. **15**, 319–338 (1981)
7. Wang, S., Haralick, R.M.: Automatic multithreshold selection. Comput. Vis. Graph. Image Process. **25**, 46–67 (1984)
8. Baukharouba, S., Rebordao, J.M., Wendel, P.L.: An amplitude segmentation method based on the distribution function of an image. Comput. Vis. Graph. Image Process. **29**, 47–59 (1985)
9. Carlotto, M.J.: Histogram analysis using scale-space approach. IEEE Trans. Pattern Anal. Machine Intell. PAMI-**9**, 121–129 (1987)
10. Kapur, J.N., Sahoo, P.K., Wong, A.K.: A new method for grey level picture thresholding using the entropy of the histogram. Comput. Vis. Graph. Image Process. **29**, 273–285 (1985)
11. Kittler, J., Illingworth, J.: Minimum error thresholding. Pattern Recognit. **19**, 41–47 (1986)
12. Kang, W. X., Yang, Q. Q., Liang, R. R.: The comparative research on image segmentation algorithms. In: IEEE Conference on ETCS, pp. 703–707 (2009)
13. ShiandJ, N., Pan, J.: An improved active contours model for image segmentation by level set method. Optik **127**(3), 1037–1042 (2016)
14. Oliva, D., Cuevas, E., Pajares, G., Zaldivar, D., Osuna, V.: A multilevel thresholding algorithm using electromagnetism optimization. Neuro Comput. **139**, 357–381 (2014)
15. Akay, B.: A study on particles warm optimization and artificial bee colony algorithms for multilevel thresholding. Appl. Soft Comput. J. **13**(6), 3066–3091 (2013)
16. Osuna-Enciso, V., Cuevas, E., Sossa, H.: A comparison of nature inspired algorithms for multi-threshold image segmentation. Expert Syst. Appl. **40**(4), 1213–1219 (2013)
17. Kurban, T., Civicioglu, P., Kurban, R., Besdok, E.: (2014). Comparison of evolutionary and swarm based computational techniques for multilevel color image thresholding. Appl. Soft Comput. J. **23**, 128–143 (2014)
18. Sarkar, S., Das, S., Chaudhuri, S.S.: A multilevel color image thresholding scheme based on minimum cross entropy and differential evolution. Pattern Recogn. Lett. **54**, 27–35 (2015)
19. Kulkarni, P.M., Naik, A.N., Bhadvankar, A.P.: Review paper on image processing techniques. Int. J. Sci. Res. Dev. **3**(10) (2015). http://doi.org/10.1109/ICSensT.2012.6461695
20. Mahmud, S.A., Mohammed, J.B., Hasan, M.S., Alzghool, M.: A survey of digital image processing techniques in character recognition. IJCSNS Int. J. Comput. Sci. Netw. Security **14**(3) (2014)
21. Rao, K.M.: Overview of image processing. Reading Images (2006)
22. Kaur, A.: A review paper on image segmentation and its various techniques in image processing. Int. J. Sci. Res. **3**(12) (2014)
23. Aly, A.A., Deris, S.B., Zaki, N.: Research review for digital image segmentation techniques. Int. J. Comput. Sci. Inf. Technol. **3**(5) (2011)
24. Chiranjeevi, K., Jena, U.: Fast vector quantization using a Bat algorithm for image compression. Eng. Sci. Technol. Int. J. **19**(2), 769–781 (2016)
25. Karri, C., Jena, U.: Image compression based on vector quantization using cuckoo search optimization technique. Ain Shams Eng. J. (2016) (in press)
26. Baradez, M.O., McGuckin, C.P., Forraz, N., Pettengell, R., Hoppe: A robust and automated unimodal histogram thresholding and potential applications. Pattern Recogn. **37**(6), 1131–1148 (2004)

27. Yogamangalam, R.: Segmentation techniques comparison in image processing. Int. J. Eng. Technol. (IJET) **5**(1) (2013)
28. Kaganami, H., Beiji, Z.: Region based segmentation versus edge detection. Intell. Inf. Hiding Multimedia Signal Process. 1217–1221 (2009)
29. Ma, M.T., Manjunath, B.S.: Edge flow: a framework of boundary detection and image segmentation. IEEE Trans. Image Process. **9**(8), 1375–1388 (2000). https://doi.org/10.1109/cvpr.1997.609409
30. Li, D., Zhang, G., Wu, Z., Yi, L.: An edge embedded marker-based watershed algorithm for high spatial resolution remote sensing image segmentation. IEEE Trans. **19**, 2781–2787 (2010)
31. Shih, F.Y., Cheng, S.: Adaptive mathematical morphology for edge linking. Inf. Sci. **167**(4), 9–21 (2004)
32. Comaniciu, D., Meer, P., Robust analysis of feature spaces color image segmentation. In: Proceedings of the IEEE CVPR Conference, pp. 750–755 (1997)
33. Luo, J., Cray, R.T., Lee, H.C.: Incorporation of derivative priors in adaptive Bayesian color image segmentation. In: Proceedings of the ICIP'97, Vol. 3, pp. 58–61, Oct 26–29, 1997 Santa Barbara, CA. http://doi.org/10.1109/ICIP.1998.727372
34. Dhawan, S.: A review of image compression and comparison of its algorithms. Int. J. Electronic. Commun. Technol. **2**(1), (2011)
35. Wallace, G.K.: The JPEG still picture compression. Standard Comm ACM **34**(4) (1991)
36. Lu, D., Weng, Q.: A survey of image classification methods and techniques for improving classification performance. Int. J. Remote Sens. **28**(5), 823–870. http://doi.org/10.1080/01431160600746456
37. Li, P., Li, H.O.: Fuzzy techniques in image restoration research—a survey. Int. J. Comput. Cogn. **2**(2), 131–149 (2004)
38. Maru, M.: Image restoration techniques: a survey. Int. J. Comput. Trends Technol. **3**(12) (2014)
39. Moustakides, G., Briassoulis, D., Psarakis, E., Dimas, E.: 3D image acquisition and NURBS based geometry modelling of natural objects. In: Advances in Engineering Software, pp. 955–969 (2000)
40. Kuriakose, B., Preena, K.P.: A Review on 2D image representation methods. Int. J. Eng. Res. Technol. (IJERT) **4**(4) (2015)
41. Kaur, G.: Image enhancement and its techniques, a review. Int. J. Computer Trends Technol. (IJCTT) **3**(12) (2014)

On Performance Analysis of Biometric Methods for Secure Human Recognition

Annu Sharma, Shwetank Arya, and Praveena Chaturvedi

Abstract In recent years, the necessity of secure and reliable human identification has led to increasingly fast growth in development and demand of biometric systems. Human recognition is the technique for identifying the person using their biological, chemical, and behavioral characteristics. Biometric system is a computer-based automatic system to establish identity of the users by using their biological and physiological traits. The most popular traits in modern applications are biological aspects of the prospective user for identification. Although using chemical traits of the human for identification is more accurate and reliable, but these are very difficult to achieve. In this paper, performance of automatic human recognition system is presented based on various parameters like users psychology, easiness of use, security, reliability, and market share. Furthermore, various analysis and comparison of different notable biometric techniques are discussed in tabular format. It has been observed that these systems provide authentication and recognition but security of these systems at template level is also one of the challenges for designers.

Keywords Biometric · FAR · FRR · Gait · DNA · Face recognition · Verification · Identification · Hand geometry

1 Introduction

Biometric system is a computer-based automatic system to establish identity of the users by using their biological traits and biometry is combination of two terms "Bio" and "metric," where bio means biological, and metric means measurement based on physiological and behavioral characteristics. Biometric refers to a measurable

A. Sharma (✉) · S. Arya
Gurukul Kangri Vishwavidalaya, Haridwar, India
e-mail: annumca01@gmail.com

P. Chaturvedi
Gurukul Kangri Vishwavidalaya Campus, Dehradun, India

P. K. Singh et al. (eds.), *Recent Innovations in Computing*, Lecture Notes in Electrical Engineering 701, https://doi.org/10.1007/978-981-15-8297-4_35

characteristic that is unique to an individual such as a fingerprint, palm recognition, facial structure, the iris or speech [1]. The need for identification with a high degree of security has motivated the use of biometric systems in various applications such as managing access to devices like mobile phones, laptop, to confirm a claimed identity against an existing credential, such as at border control, managing accessible materials such as financial and medical care services. The main objective of biometric technology is recognizing people based on their physical and behavioral traits. Biometric identification offers user convenience, better accountability, and higher efficiency.

Biometric system is an automatic system for human recognition, and it works by capturing the required characteristics from human body by using a sensor. After capturing the desired trait, unique feature points are extracted by using a suitable technique. These feature points are denoted with the help of numerical values, and a collection of feature points represents a feature template [2]. The feature templates of an enrolled user are stored in a database for further usage. The enrolled users are identified or recognized by using the same procedure as discussed earlier. The process of recognition is done by using matching algorithm which matches proposed template with stored template of the user in the database. Basic components of biometric system are presented in Fig. 1. There are two modes identification or verification on the basis of which a biometric system works. In the identification mode proposed template is matched with all the stored templates in the database to identify the given user. Searching in this mode which is carried out in 1: N manner. On the other hand, when a given template of a user is matched against stored template of the same user then system is said to be working in verification mode. Search is carried out in verification mode in a 1:1 manner. The majority of the identification applications are in

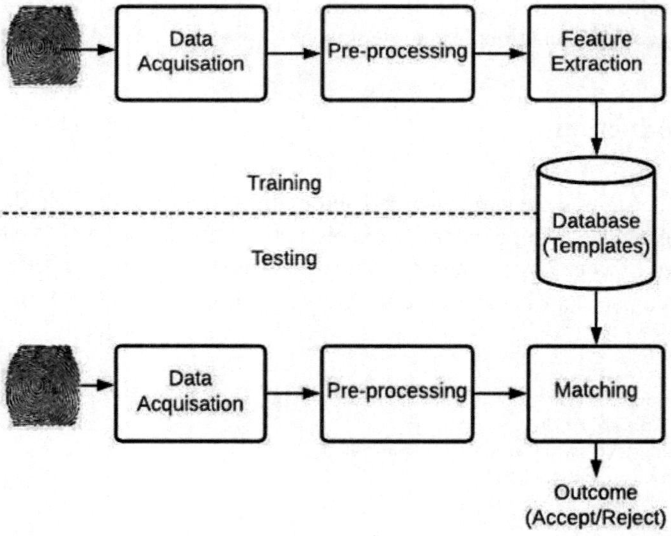

Fig. 1 Modules of a biometric system

law enforcement, forensics, and intelligence. The applications include identifying faces from mug shots, surveillance images, newspapers, photographs, and images of deceased people [2, 3]. Verification applications include control access to apartment or secure buildings, verifying identities during point of sale transactions and continuous verification of identity at computer terminals or in secure facilities. In India, biometric devices are being deployed in various sectors such as Aadhar cards, local security, law enforcement, banking system, and as per studies, the Indian biometric market value is expected to grow by huge number. Biometric devices has also find its way into defense, consumer electronics, attendance system in school and offices, and fingerprint biometric is mainly used in India in these systems for identification and verification. As far as development in biometric in India, recently, RBI has made it mandatory that all banking related activities network has to be linked with Aadhar card [4] in order to improve the security of system and also to promote cashless transaction.

2 State of the Art

Biometric system is a powerful recognition tool that involves conventional imaging system for automatic human recognition. For many years, the history of fingerprint recognition has been told in many ways. It was believed that on the palm side of each person's hands and on the soles of each person's feet are prominent skin features that single him or her out from everyone else in the world [5]. These feature when they come in contact with some object they leave behind impressions of its shapes which are known as prints; accordingly, we have fingerprint, palmprint, footprint, etc. These impressions are used to identify the individuals since thousands of years in several cultures. It has been reported that friction ridge skin impression was used for identity of an individual in China during 300 B.C and in A. D. 702 in Japan. In India, there are references to the nobility using friction ridge skin as signatures in A.D. 1637 to demonstrate authenticity of authorship when writing an important document; it was mainly reserved for royalty. The use of friction ridge skin as a signature in China, Japan, India, and possibly other nations prior to European discovery is thus well documented. Water marking can also be used for security of the fingerprint templates [6]. Many automatic human recognition techniques like fingerprint, face recognition, iris scan, retina scan exist in the literature among all of these fingerprint is the most widely used technique which covers almost 50% of market share. According to [7], fingerprint is used for identification since 7000 to 6000 BC by ancient Assyrians and Chinese in the form of print on pottery and clay. It is reported that Babylon (1792–1750 BC) used fingerprint for identification of criminals. The quantitative identification through fingerprint and facial measurements was first proposed by Henry Faulds, William Herschel, and Sir Francis Galton in 1880s. During mid-1800, the fact was established that no two fingerprints have same ridge pattern. In the year 1888, minutiae features of fingerprint for matching were introduced by sir Francis Galton. Fingerprint identification method was accepted as a personal identification method in early twentieth century which included fingerprint acquisition; fingerprint classification and matching were developed. In 1920, the biometric was introduced in forensic identification by Edmond Locard [8, 9]. In 1960s, fingerprint recognition system was used for access control and financial transaction. During 1970s, hand geometry systems came into existence; the hand geometry scanner was introduced for identification and was deployed in large scale projects, especially in government sectors [10]. In the year 1885, Alphonse Bertillon recommended the use of Iris for human recognition [11]. James Doggart, a British ophthalmologist in the year 1949, suggested that uniqueness of iris patterns can be used in the same way as the fingerprints for biometric identification but no algorithm was proposed to support the idea [12]. During 1991, John Daugman came with first algorithms for iris recognition and patented it. With the demonstration of this algorithm, the iris recognition system gained its popularity in the late 1990s and is being used till date [13]. In late 1980s, automated face recognition system were developed. In 1991, Turk and Pentland developed the eigenfaces techniques, which lead to real-time automated face recognition systems [14].

3 Performance of Automatic Recognition System

The success of any biometric system depends on number of criteria. For biometric recognition, feature set of individual is taken and compared by the process of verification or identification; it is very rarely that the two feature set of same biometric traits of user are exactly same. This may be due to sensor imperfection [15], alteration in user biometric characteristics, changes in ambience condition, etc. The variability observed in biometric feature set of an individual is referred to as intra-class variation and that between feature set of two different individuals is known as inter-class variation. There must be small intra-class and large inter-class variation. In most of the biometric systems, similarity scores are used to measure the performance of biometric system. The similarity score is used to indicate the degree of similarity between two biometric feature sets. When two biometric samples are matched, a similarity score is returned. Performance of a biometric system deals with the quantifiable assessment of the accuracy and other characteristics of the system. Template matching is a key to the system and affects the precision and efficiency of the whole system directly. Performance of a biometrics system can be measured for three tasks: ROI detection, verification, and identification [16]. Two standard matching error rates are concerned with AFIS; these are the false acceptance rate (FAR) and fall rejection rate (FRR). For verification, if we have a population of N different people, the system can be assessed using the False Acceptance Rate (FAR; those situations where an impostor is accepted) and the False Rejection Rate (FRR; those situations where a genuine user is incorrectly rejected). Both errors have a bad effect and need to be weighted carefully to make sure the optimal mix this necessary trade-off between them is usually established by adjusting a decision threshold. Depending on the threshold value, the score can either be interpreted as authentic score or an impostor score, authentic if the two samples matches and impostor score if there is mismatch between the two templates based on value of threshold. When an impostor score exceeds the threshold value η, it results into a false accept of biometric trait of person on the basis of which False Accept Rate (FAR) is calculated as the fraction of impostor scores that exceeds the threshold value η, whereas when a the score is genuine but it falls below the threshold value η; it results in a false reject of biometric trait of person, and the False Reject Rate (FRR) of a system is calculated as the fraction of genuine scores falling below the threshold value η. Accuracy of biometric systems is measured by the comparison of frequency of false accepts versus true accepts, called the False Accept Rate (FAR) and True Accept Rate (TAR) [3, 16]. Biometric systems are generally optimized toward minimizing the false accept rate.

4 Analysis of Various Biometric System

Various biometric systems are available for human recognition which has their own merits and demerits. Table 1 shows a brief qualitative analysis of different biometrics, and Table 2 briefly describes approximation of various performance parameters of each biometric. An introduction of each biometric is being presented in the following paragraphs.

4.1 Fingerprint as Biometric

Fingerprint biometric is most widely, and efficient technique being used in modern era of human identification. It is being used for attendance system, providing security in devices like personal computers, online authentication in banking activities. It is also being used for making smart cards like Aadhar cards [4] and passports. System works by capturing image of index finger of a person and then storing the extracted template into database. Feature point of the finger is represented by a triplet (x, y, θ), where x, y are positional parameters, and θ is angular position of the feature [17, 18]. Due to ease of use and less price, fingerprint covers around 50% of the market

Table 1 Performance-based comparative analysis of various biometric systems

Sr. No.	FAR	FRR
Fingerprint	1%	0.01%
Face	1%	10%
Iris	0.94%	0.99%
Hand geometry	0.1%	3.9%
Palmprint	1.45	1.16
Voice	2–5%	5–10%

Table 2 Comparative analysis of various biometric Systems

Modality	User's participation	Spoof vulnerability	Accuracy	Cost	Ease of use
Fingerprint	Required	Medium	High	Very economical	Simple
Face	Not necessary	High	High	Costlier	Complex
Iris	Required	Low	Very high	Costlier	Very complex
Hand geometry	Required	High	Low	Economical	Simple
Gait	Not necessary	Low	Low	Costlier	Complex
Voice	Required	High	Moderate	Costlier	Complex
Signature	Required	High	Moderate	Costlier	Complex
DNA	Required	Very low	Very high	Very expensive	Very complex

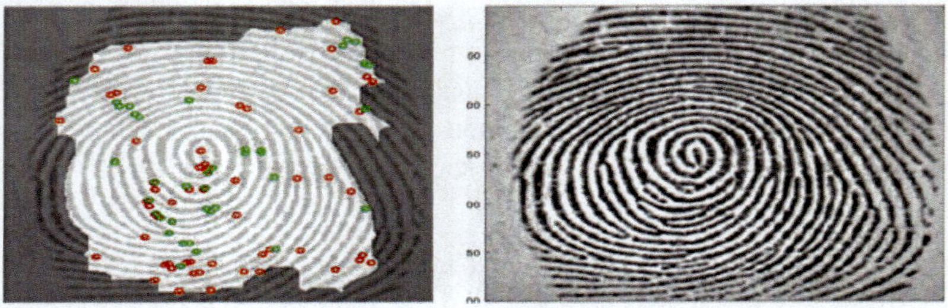

Fig. 2 ROI of fingerprint image depicting minutiae points

share as compared to other biometrics which has 10–20% share in the global market (Fig. 2).

4.2 Recognition via Face

Face Recognition (FR) biometric system is widely used in criminal identification, law enforcement, security systems, image processing applications, human computer interaction, smart card, e-passport. The system works by detecting the patterns, shapes, and shadows in the face; the process involves tracking, detection, analysis, and synthesis. The various face recognition algorithms proposed till date can be broadly categorized viz—Feature based and Appearance based [14]. Properties and geometric relations such as that the areas, distances, angles between the facial feature points are used as descriptor for face recognition. In geometry-based face recognition various methods were proposed, which include filtering and morphological operations, Hough transform methods, and deformable templates. Appearance-based methods consider the global properties of face image intensity pattern, and

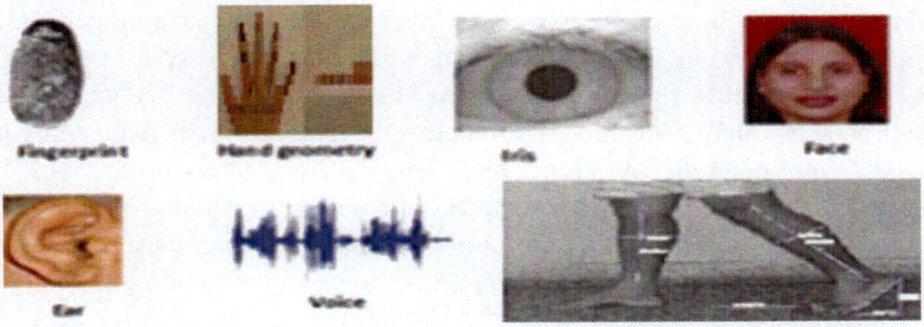

Fig. 3 A view of various biometric modalities

appearance-based algorithm proceeds by computing basis vectors to represent the face data efficiently. Most of the popular algorithm is based on appearance of the face, which include PCA [14], LDA [19], ICA, and LFA [20], PCA being widely used one as it shows the most encouraging results for face recognition [21]. The principle component analysis using higher-order statistics is the underlying mathematics for this facial pattern recognition. In PCA, the system function by projecting the face images onto a feature space that spans the significant variations among known face images. The significant faces are known as "eigenfaces" because they are eigenvectors (principal components) of set of faces [22]. This is the most popular approach and is based on eigenfaces that represent the differences between the face under recognition and the enrolled ones in the database.

$$\text{Covariance }(W, Z) = \sum \frac{(wi - W_{\text{avg}})(zi - Z_{\text{avg}})}{(q - 1)} \tag{1}$$

The Attribute Vectors (AVs) taking out method is the significant phase in FR. Because the vector space is much high, it is not easy process to find the Covariance Matrix (COVM) for these appearance. PCA technique shows a great significance to achieve this which is based on image matrices. The image matrices and its transformation have vital significance in image AVs extraction as well as in the reduction of dimensions [23]. PCA is used to take out AVs as well as decrease the sizes of image dataset. Image processing and analysis are generally seen as operations on two dimensional (Fig. 3).

4.3 Recognition via Iris

Wherever the biometrics is required for large accuracy and search speed iris recognition finds its place as the best performing technique. It has been used extensively used in many commercial projects like ID cards, passports, border surveillance [4]. Structurally distinct character and uniqueness of individual iris allows them to be used for human recognition purposes [5]. The recognition is done on the basis of analysis of random pattern of iris image; these pattern are different for each individual. Iris recognition works by performing a test of statistical independence between two iris codes, in order to decide whether they arise from the same or from different irises [24]. Image processing algorithms are used to detect the area of the iris that is obscured by eyelids, eyelashes, reflection from eyeglasses. The algorithms which are based on iris identification use the methods such as Gabor wavelet encoding, Exclusive–OR bit vector, binarization based on zero crossings, and hamming distance similarity metrics. An iris image is defined by its iris code and a corresponding mask and is ready for matching. In matching two iris codes, Daugman's approach computes a fractional hamming distance between iris codes [13, 5, 25].

5 Other Biometric Trait Used for Recognition

Traits like hand, ear [26], speech, way of walking (gait), signature [27], DNA of a person all these are used as biometric for human recognition depending upon their applications and the security and time requirement of the system.

5.1 Hand Geometry Recognition

Hand geometry is being used for identification since 1970s. Hand geometry biometric captures the shape of the hand and fingers for identification purpose [28]. The device uses an optical camera or flatbed scanner to capture digital image of hand and fingers. The captured sample is processed into a biometric template and compared to a reference template in the enrollment database. These biometric systems binarize the captured image into black and white so these biometric systems are insensitive to changes in surface of the hand such as cuts, burns, tattoos, scrapes, etc. The performance of the hand geometry system varies depending upon the algorithm used which is based on contours or on geometric features. The system is successfully deployed in various applications such as physical access control, attendance data collection but is not suitable for national Id and surveillance applications [29].

5.2 Signature as Biometric

Signature biometric is socially and legally accepted worldwide for human recognition; online signature biometric is mostly used in banking sector, and for legal document verification, it can be easily taken by pointer-based devices such as digitizing tablets which provide coordinate information (horizontal x and vertical y position) and also considers the pen pressure and pen angle. Euclidean distance and Mahalanobis distance are used as distance measures for signature verification [30, 31].The process of online signature considers the global feature approach in which the time functions of the different signatures are directly matched by using elastic distance measures such as statistical modeling Hidden Markov Models (HMM) [32] or with Dynamic Time Warping(DTW). The deployment of this biometric modality remains a challenge due to low universality, low permanence, inter-class variations, and also its vulnerability to attacks.

5.3 Voice as Biometric

Voice biometric is considered as most economic biometric as it does not require any special device or transmission system. It is widely used in case of remote biometric systems with the deployment of VoIP [33]. It finds its application in remote access control by phone; speaker spotting and Forensic speaker recognition. Voice generation is a complex process which includes personal characteristics as well as environmental and sociolinguistic variables [23]. The main source of voice biometric is the linguistic content based on the use of linguistic content; there are two different types of voice recognition technologies—Text-dependent technology and text-independent technology. In text dependent technology, user is required to speak specific key phrase and is the main subject of biometric access control and voice authentication applications. In text independent, the user is recognized without text-based approach and is mainly used in speaker detection and forensic voice recognition.

5.4 Gait as Biometric

Human recognition is based on physiological or behavioral characteristics. Gait biometric comes under behavioral biometrics; it is based on the pattern of shape and motion of a walking person being captured by the camera. Niyogi and Adelson in the early 1990s developed human recognition system using gait technique [34, 35]. Gait means that a person can be identified by the manner of his walking. The walk of a person is a periodic activity with each gait cycle covering the left foot forward and right foot forward strides. The gait biometric considers the shape and the dynamics. The configuration of the persons as they carry out different gait phases refers to shape, and the rate of transition between these phases is the dynamics which is human motion recognition.

5.5 Commercial Aspects of Biometric

During the past decade, there has been a significant growth in biometric systems both in commercial and government sector but still the problem of recognition remains a

Table 3 Comparative analysis of various biometric as per market share

Biometric modalities	Market share (%)
Fingerprint	48.8
Face	19.0
Hand based	10.4
Iris scan	6.2
Voice	2.7
Signature	6.2
Keystroke	0.4
Middleware	11.9

challenge as far as security of the system is concerned. The speedy growth in human recognition system deployment during the past decade has led to the development of novel capturing devices (sensors), modern feature extraction techniques, enhanced matching algorithms and its innovative applications [36]. With the use of these techniques, the biometric system market is growing tremendously with the widespread use of biometric modalities both in government as well as commercial sector for identification and verification. Use of biometric in military and defense, healthcare, banking and finance, passport, e-commerce, and Aadhar card are the key areas of the market. Based on the technology being used in recognition which includes fingerprint recognition, face recognition, hand recognition, voice recognition, Signature recognition, DNA recognition, and others, the biometric system market can be segmented and analyzed for its market share. The greatest market share of biometric technologies is covered by fingerprint biometric; in the year 2010, it was estimated at $2.7 billion, and by the end of 2020, it will be around $12 billion. After fingerprint recognition, the face, retina an iris, voice patterns, vein, and signature pattern together captures the rest of the market share (second largest) [37]. During 2010, it was estimated of $1.4 billion and is expected to reach around $10 billion by the end of 2020, and the overall biometric system market is expected to reach around $35 billion by 2021 (Table 3).

6 Conclusion

Authentication is one of the essential activities in modern day computing for which various tools and techniques are available. In this paper, an analysis of various biometric systems for authentication has been presented. A qualitative and quantities comparative analysis of different biometric systems has been presented with respect to various parameters. It is found that there is a trade-off between cost and accuracy of these systems. It is noticed that fingerprint biometric is widely used due to its compactness and cost; on the other side, DNA is most reliable and costlier biometric. Fingerprint systems are beneficial to use in simple applications like attendance while DNA is used in sophisticated applications like forensic sciences. Other biometric systems are also equally good but have their own limitations like interclass or intra-class variations. It is observed that this biometrics may have better accuracy if combined with one another to build multibiometric system. Most of the biometric has been found to be vulnerable to spoof attacks. Security of these systems from attackers is a major concern for researchers. The deployment of multispectral biometric system for proper spoof detection by using other properties of the biological trait like spectral analysis will improve the performance and security of the system. This biometrics may be analyzed with respect to various types of attack.

References

1. Jain, A.K., Ross, A., Prabhakar, S.: An introduction to biometric recognition. IEEE Trans. Circuits Syst. Video Technol. **14**(1), 4–20 (2004) (Special Issue on Image- and Video-Based Biometrics)
2. Jain, A.K., Ross, A.: Introduction to Biometrics. Springer, New York (2011)
3. Egan, J.: Signal Detection Theory and ROC Analysis. Academic Press, New York (1975)
4. Government of IndiaUnique Identification Authority of India (2011). https://uidai.gov.in/
5. Daugman, J.: High confidence visual recognition of persons by a test of statistical independence. IEEE Trans. Pattern Anal. Mach. Intell. **15**(11), 1148–1161 (1993)
6. Agarwal, N., Singh, A.K., Singh, P.K.: Survey of robust and imperceptible watermarking. Multimed. Tools Appl. **78**, 8603–8633 (2019). https://doi.org/10.1007/s11042-018-7128-5
7. Jain, A.K., Maltoni, D., Maio, D., Prabhakar, S.: Handbook of Fingerprint Recognition. Springer, New York (2003)
8. Lee, H. C., Gaensslen, R.E.: Advances in Fingerprint Technology, 2nd edn. Elsevier Publishing, New York (2001)
9. Locard, E.: Numerical Standards and Probable Identifications. J. Forensic Ident. **45**(2), 136–163 (1995)
10. Miller, B.: Vital Signs of Identity, p 22. IEEE Spectrum (1994)
11. Polski, J, Ron S, Robert G.: The report of the international association for identification, standardization II committee., National Institute of Justice 233980 (2011)
12. Doggar, J.H.: Ocular Signs in Slit-lamp Microscopy. Kimpton, London (1949)
13. Daugman, J.G.: Biometric Personal Identification System Based on Iris Analysis. US Patent 5, 291, 560 (1994)
14. Turk, M., Pentland, A.: Eigenfaces for recognition. J. Cogn. Neurosci. **3**(1), 71–86 (1991)
15. Singh, P.K., Bhargava, B.K., Paprzycki, M., Kaushal, N.C., Hong, W.C.: Handbook of wireless sensor networks: issues and challenges in current scenario's. Adv. Intelli. Syst. Comput. **1132**, 155–437 (2020) (Springer: Cham, Switzerland)
16. Wayman, J.: Fundamentals of Biometric Authentication Technologies. National Biometric Test Center Collected Works 1997–2000. University Press, San Jose (2000)
17. Cappelli, R.: Handbook of Fingerprint Recognition. Springer, New York (2003)
18. Sharma, A, Shwetank, A, Praveena, C.: Multispectral image fusion system based on wavelet transformation for secure human recognition. J. Int. Adv. Sci. Technol **28**(19), 811–820 (2019)
19. Lu, J., Plataniotis, K.N., Venetsanopoulos, A.N.: Face recognition using LDA-basedalgorithms. IEEE Trans. Neural Netw. **14**(1), 195–200 (2003)
20. Xie, C., Savvides, M., Vijaya Kumar, B.V.K.: Kernel Correlation Filter Based Redundant Class-Dependence Feature Analysis (KCFA) on FRGC2.0 Data. In IEEE Workshop on Analysis and Modeling of Faces and Gestures (AMFG), pp. 32–43 (2005)

21. Zhao, W., Chellappa, R., Phillips, P.J., Rosenfeld, A. (2003). Face recognition: a literature survey. ACM Comput. Surv. **35**, 399–458.
22. Pratap, N., Shwetank: Development of spectral signatures and classification using hyperspectral face recognition. J. Interdisc. Math. **23**(2), 453–462 (2020)
23. Przybocki, M.A., Martin, A.F., Le, A.N.: Nist speaker recognition evaluation chronicles, Part 2. In: Proceedings of IEEE Odyssey (2006)
24. Daugman, J. (2003).The importance of being random: statistical principles of iris recognition. IEEE Trans. Pattern Anal. Mach. Intell. **36**(2), 279–291
25. Daugman, J.: How iris recognition works. IEEE Trans. Circuits Syst. Video Technol. **14**(1), 21–30 (2004)
26. Hurley, D.J., Nixon, M.S., Carter, J.N.: Force field feature extraction for ear biometrics. Comput. Vis. Image Underst. **98**, 491–512 (2005)
27. Jain, A.K., Griess, F.D., Connell, S.D.: On-line signature verification. Pattern Recogn. **35**(12), 2963–2972 (2002)
28. Gonzalez, S., Travieso, C.M., Alonso, J.B., Ferrer, M.A.: Automatic biometric identification system by hand geometry. In: Proceedings of the 37th Annual International Carnahan Conference on Security Technology, pp. 281–284 (2003)
29. Akkermans, A.H.M., Kevenaar, T.A.M., Schobbenx, D.W.E.: Acoustic ear recognition for person identification. In: Fourth IEEE Workshop on Automatic Identification Advanced Technologies (AutoID'05), pp. 219–223 (2004)
30. Kholmatov, A., Yanikoglu, B.: Identity authentication using improved online signature verification method. Pattern Recogn. Lett. **26**(15), 2400–2408 (2005)
31. Yang, L., Widjaja, B.K., Prasad, R.: Application of Hidden Markov models for signature verification. Pattern Recogn. **28**(2), 161–170 (1995)
32. Fierrez-Aguilar, J., Krawczyk, S., Ortega-Garcia, J., Jain, A.K.: Fusion of local and regional approaches for on-line signature verification. In: Proceedings of IWBRS. Springer LNCS-3781, pp. 188–196 (2005)
33. Ramos-Castro, D., Gonzalez-Rodriguez, J., Ortega-Garcia, J.: Likelihood ratio calibration in a transparent and testable forensic speaker recognition framework. In: Proceedings of IEEE Odyssey (2006)
34. Niyogi, S.A., Adelson, E.H.: Analyzing gait with spatiotemporal surfaces. In: Proceedings of IEEE Workshop on Non-Rigid Motion, pp. 24–29 (1994)
35. Liu, Z., Sarkar, S.: Improved Gait recognition by Gait dynamics normalization. IEEE Trans. Pattern Anal. Mach. Intell. **28**(6), 863–876 (2006)
36. Sharma A, Shwetank A, Praveena C.: A novel image compression based method for multispectral fingerprint biometric system. Procedia Comput Sci **171**, 1698–1707 (2020) (Elsevier)
37. Biometric System Market Report, Report Code: SE 3449 [available at] https://www.marketsandmarkets.com/Market-Reports/fingerprint-sensors-market-169519533.html (2019)

Automatic System for Text Detection and Localization Using Cellular Automata

Sukhdev Singh and Monika Pathak

Abstract The cellular automata are a discrete and abstract way of representing dynamic model which change state based on some relationship with other members in the system. These patterns are used to map an object within the image. The present research used the cellular automata-based system to detect the text region in the natural seen image. Automatic text detection is gaining attention day by day due to its versatile range of applications. The present research used cellular automata to develop a system to detect Gurmukhi text in the natural scene images. The natural scene may contain signboard images, text on banners, text written on walls and text viable on any vehicle. The Gurmukhi script has its own set of unique features which helps to classifier to recognize. The efficiency of the text recognition heavily depends on the text extracted from the images. An algorithm was developed to detect and localize the text (in the present study, only Gurmukhi script is considered for study) in the natural scene images. In absence of benchmark dataset of natural scene images, we have developed our own dataset to test the efficiency of the system. The system was tested with well-known matrices named as recall metric, precision metric and P-value.

Keywords Cellular automata · Text detection · Natural scene images · Text localization

1 Introduction

The automatic text extraction system is capable to detect and localize the position of the text in the image. It is a complicated task because the image may contain text like objects or patterns. Moreover, orientation and variation in size and color itself make

S. Singh (✉) · M. Pathak
Department of Computer Science, Multani Mal Modi College, Patiala, India
e-mail: tomrdev@gmail.com

M. Pathak
e-mail: monika_mca@yahoo.co.in

P. K. Singh et al. (eds.), *Recent Innovations in Computing*, Lecture Notes in Electrical Engineering 701, https://doi.org/10.1007/978-981-15-8297-4_36

text complicate to detect in the image. The text detection system generally performs a series of actions in phase manner to recognize text which usually includes pre-processing recognition (feature extraction and classification) and post-processing (methods to improve result outputs). The figure shows three different steps of the text extraction system. The text extraction from natural scene images has a wide area of applications. Some of the applications are navigation system to assist blind persons, automatic vehicle number plate detection system, traffic law enforcement, automatic system to draw the attention of a driver toward traffic signals, automatic system to draw the attention of a driver toward traffic signals, preservation of historical document, system to overcome the language barrier, etc.

All three phases consist of various operations which are varying in nature and choice of operation depending on the nature of the problem undertaken. The output of one phase becomes input to the next phase and so on. It makes the system dependent on each other. The performance of one process affects the performance of the next process. Keeping the dependency nature, we have developed and tested these phases separately and integrated at a later stage (Fig. 1).

Pre-processing: The system used pre-processing to overcome computation overhead and performed gray scaling, image binarization, noise suppression, detection and localization of text in the image, word skew detection and correction and character segmentation. The binarization is most significant which actually used to suppress background [3]. The algorithms based on discontinuity segment the text on the basis of sudden variation in gray values, whereas algorithms based on similarity of gray values are based on computation of threshold, region splitting and growing. In the

Fig. 1 Logical design of text extraction system

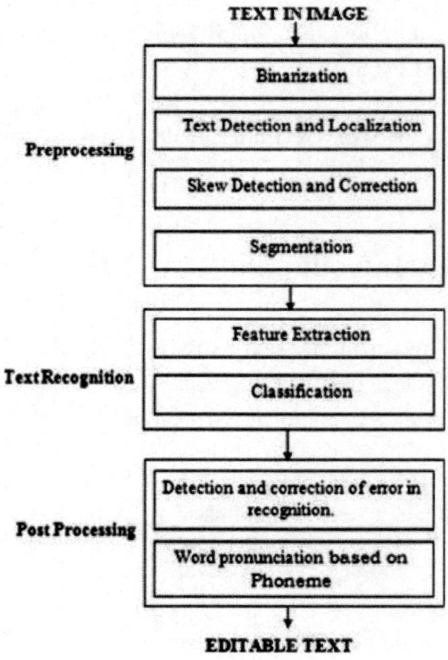

case of threshold algorithms, the pixel intensity value is compared with some fixed value which is called threshold value. If the intensity value of the pixel is above a threshold value, then the pixel value is set to be the value of background or foreground or vice versa (Table 1).

Text Detection and Localization: Text detection is a complex task as there are a lot of text like patterns (such as window, fences, tree and tree leaves) in the image which may mislead detection of text. There is need to develop a system which can detect text even in the presence of text like patterns. There are various methods available in literature such as edge-based methods, connected components-based methods and cellular automata-based methods [2, 3].

Text Recognition: Text recognition is the outcome of feature extraction and classification. It helps to identify and understand the text detected by the system. Most of the natural language processing system used feature extraction and classification as an integral part of the recognition. The feature extraction refers to the process of gathering the geometrical or the statistical characteristics of the individual character. Every language has its own features which can help to identify the characters. A set of good features contain discriminating information, which can distinguish one object from other objects [4].

Classification: It is a decision-making process that used features as a yardstick to classify the characters into different categories. Particularly in the case of Gurmukhi script, the characters have unique features in upper, middle and lower zones which help the classifier to distinguish the character for identification. The well-known classification methods are statistical, neural networks based, binary classification tree and nearest neighborhood classifier. The classifier used in the present study is k-nearest neighbor (KNN). The features extracted from the characters are tested by the KNN classifier for recognition.

Post-Processing: The system has used two impotent post-processing methods. The first method deals with error detection and correction, and the second is dedicated to the word pronunciation.

2 Literature Review

The purpose of the literature review is to understand and analyze previously exiting methods and techniques in the domain of the problem considered for the study. It is observed in the literature review that the most of the work is found on machine-printed document images and handwritten document images of Indian, and very few attempts have noticed on text extraction from natural scene images for the Indic script. Discussion about few literature studies is listed below.

Murthy et al. [5] developed a text extraction system using 2D Haar discrete wavelet transformation and k-means clustering. Morphological operators were used to distinguishing the text and non-text regions. The system was tested on 150 images that contained text in Hindi and English.

Table 1 Results of text detection and localization algorithm

Input Image	Output Image

	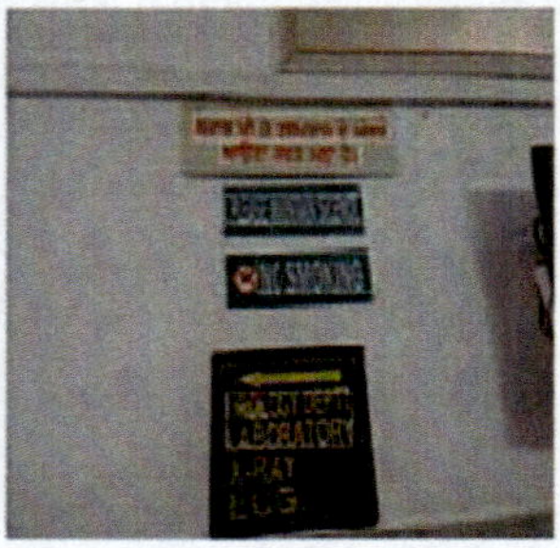

Agnihotri et al. [6] worked on the Devanagari handwritten text recognition system. They used zone directional feature and special fitness functions for the recognition of Devanagari character recognition. The system was tested and showed precisions 85.78% (match) and 13.35% (mismatch).

Ghosh et al. [7] introduced a feature extraction system for an online handwritten character recognition system for two scripts, namely Bengali and Devanagari scripts. The features of handwritten characters were extracted which were based on unique basic strokes. The stroke information such as writing direction, slope, curvature, curliness and the standard deviation of x and y coordinate features was considered for feature extraction. The recognition accuracy of the proposed system is 78% for the Bengali script and 77.23% for the Devanagari script.

Hasan et al. [8] proposed a system which can pronounce Bengali text automatically detected from on road signs. The system has three modules: text detection and extraction, optical character recognition and speech synthesis. The first module worked on data acquisition, pre-processing, text detection and localization. The OCR module involves character segmentation, features extraction and character classification using artificial neural networks. The speech synthesis module converts the recognized text into voice streams. The experimental results gave 96% accuracy with 820 total numbers of characters of Sulekha font. It gave 98% accuracy with 820 total numbers of characters of Sulekha.

Sethi et al. [8] used a structural approach to recognize the number of Devanagari script. The segmentation techniques were based on horizontal and vertical segmentation where skew correction was carried for right and left slants. The decision tree was used as analysis tool to access the presence/absence of these primitives and their interconnection. Bansal et al. [9] proposed OCR for the Devanagari script. Various character features such as vertical bar feature, horizontal zero-crossings and number of positions of the vertex points were considered for recognition. The character classification was carried by a tree classifier. The testing results showed 93% accuracy. Sharma et al. [10] proposed the Gurmukhi recognized system for isolated handwritten Gurmukhi using neocognitron. The neocognitron defined a hierarchical multilayered neural network used for classification. Total 15,000 character images were considered for testing. The learned images have shown 91.77% accuracy whereas unlearned images showed 93.79% accuracy. Gatos et al. [11] proposed a technique to detect text from indoor and outdoor images. The technique was used to isolate background using binarization methods. The binarization method segments the text region from the background, and then commercial OCR was used for testing purposes. Tran et al. [12] proposed an approach for finding text in images by using ridges at several scales. The proposed method did not depend on a particular alphabet. The experimental results showed good detection.

Jindal et al. [13] proposed a method to segment the Gurmukhi touching characters from the middle. For segmentation, various classifications were carried out which were based on the structural properties of the characters.

Saini et al. [14] described a corpus-based transliteration system for the Punjabi language. They developed a new system for the first time of its kind for Shahmukhi script of Punjabi language. The proposed system for Shahmukhi to Gurmukhi transliteration was implemented with various research techniques based on language corpus. The corpus analysis program was run on both Shahmukhi and Gurmukhi corpora for generating statistical data for different types such as character, word and n-gram frequencies. This statistical analysis was used in different phases of transliteration.

3 Cellular Automata Method

The cellular automata make use of regular lattice to represent an object. It consists of regular cells which change value in discrete-time span. The text detection system proposed by Konstantin et al. [15] mapped Roman characters in cellular pattern. A general cellular expression can be written as:

$$CA = (\text{Cells, Cell space, Cells state, Neighborhoods, Rules})$$

where the cell has finite state and change value using transmission function. It is the function which determines the state of each cell based on the state and present neighborhood. The arrangement of cells around central cell defines neighborhood. Most of the literature defines two neighborhoods named Von Neumann and Moor neighborhood. The present study used Moor neighborhood to map character set. In Moore neighborhood, the central cell is surrounded by eight neighborhood cells which cover all the direction positions of pixel around the central pixels. The neighborhood consists of total 9 which may change their state by transmission function. The figure shows cell arrangement in the Moor neighborhood (Fig. 2).

$$N(x_0, y_0) = \{(x, y) : |x - x_0 \leq r, |y - y_0|| \leq r\}$$

where r represents neighborhood cells, and (x_0, y_0) represents central cell.

Algorithm for text detection and localization: Algorithm is based on 2D lattice of cells, where every cell represents a pixel. The size of the cell depends on the object to be mapped in the problem. The input images considered for mapping are

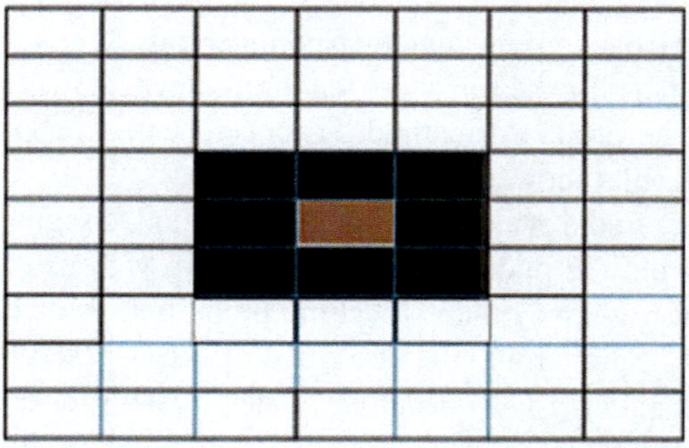

Fig. 2 Moore neighborhood with $r = 1$

binary images where cell may be black or white. These two colors represented as two different states. Following is the algorithm used to detect and localize text region in the image.

Algorithm to detect text region in natural scene images

1. Morphological Dilation: It is applied to map character edges of every text line. The variable $I(x, y)$ represents intensity at central pixel. Following equation represents intensities of other neighbor pixels in Moor neighborhood. The morphological dilation is represented as this mask neighborhood.

$$I_{(x,y)} \rightarrow I_{(x-5y)}, I_{(x-4,y)}, I_{(x-3,y)}, I_{(x-2,y)}, I_{(x-1,y)}, \\ I_{(x,y)}, I_{(x+1,y)}, I_{(x+2,y)} I_{(x+3,y)}, I_{(x+4,y)}, I_{(x+5,y)}$$

Carry out morphological operation for 12 connected neighborhood of 7 × 7 neighborhood mask.

$$I_{(x,y)} \rightarrow I_{(x-3y)}, I_{(x-2,y)}, I_{(x-1,y)}, I_{(x,y)}, I_{(x+1,y)}, I_{(x+21,y)}, I_{(x+3,y)}, \\ I_{x,y-3}, I_{x,y-2}, I_{(x,y-1)}, I_{(x,y+1)} I_{(x,y+2)}, I_{(x,y+3)}$$

2. Carry out morphological dilation in neighborhood for mask size 7 × 7.

$$I_{(x,y)} \rightarrow I_{(x-3y)}, I_{(x-2,y)}, I_{(x-1,y)}, I_{(x,y)}, I_{(x+1,y)}, I_{(x+2,y)}, I_{(x+3,y)}, \\ I_{x,y-3}, I_{x,y-2}, I_{(x,y-1)}, I_{(x,y+1)} I_{(x,y+2)}, I_{(x,y+3)}$$

2. Detecting text region and non-text regions: In this process, we use various sizes of Moor neighborhood which are capable to map expected text region in the image. It helps to suppress small objects and helpful to estimate word size. After various iterations, we come to close to the size of the Moor neighborhood with size = 16 for detecting small character and size = 3 detecting large characters size words (Fig. 3).

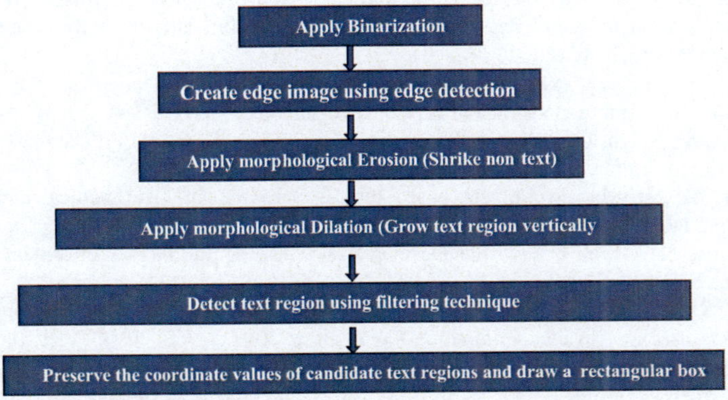

Fig. 3 Steps to detection of text in natural scene images

$$N^{M}_{(x_0 y_0)} = \{(x, y) :: |x - x_0| \leq r1, |y - y_0| \leq r2\}, \text{Where } r1 = 16,\ r2 = 3$$

Steps to detection of text in natural scene images are implemented into following phases:

The dataset of approximately 4500 images was created and used for training and tested on the system. Most of the images contain Gurmukhi text. Below are a few sample images along with visual results:

Recall metric: Recall metric represents correctness of the system and defines how correctly system works. It is computed using following

$$R = \frac{\text{Number of Rectangles (Correctly Detected)}}{\text{Number of Rectangles in Actually exsit}}$$

Precision metric: The precision metric specifies about number of false alarms as shown as below:

$$p = \frac{\text{Number of Rectangles (Correctly Detected)}}{\text{Total Number of Rectangles Detected}}$$

Ground truth images were also created to find the efficiency and accuracy of the system. The performance metrics recall and precision are used to evaluate the performance of the proposed algorithm. The proposed text detection and localization algorithm are evaluated for performance. The average value for recall $= 0.835$, and the average value for $p = 0.880$.

References

1. Ntirogiannis, B.G., Konstantinos, Pratikakis, I.: Binarization of textual content in video frames. In: International Conference on Document Analysis and Recognition (ICDAR). IEEE, pp. 673–677 (2011)
2. Jiang, X., Mojon, D.: Adaptive local thresholding by verification-based multi-threshold probing with application to vessel detection in retinal images. In: Pattern Analysis and Machine Intelligence, IEEE, pp. 131–137 (2003)
3. Yi, C., Tian, Y.: Text extraction from scene images by character appearance and structure modeling. In: Computer Vision and Image Understanding, pp. 182–194 (2013)
4. Lehal, G.S., Singh, C.: A Gurmukhi script recognition system. In: Proceedings of th 15th International Conference on Pattern Recognition, vol. 2, IEEE (2000)
5. Murthy, V.S.V.S., et al.: Content based image retrieval using Hierarchical and K-means clustering techniques. Int. J. Eng. Sci. Technol. **2**(3), 209–212 (2010)
6. Agnihotri, V.P.: Offline handwritten Devanagari script recognition. Int. J. Information Technol. Comput. Sci. **4**(8), 37–42 (2012)
7. Ghosh, R., Bhattacharyya, D., Bandyopadhyay, S.K.: Segmentation of online Bangla handwritten word. In: Advance Computing Conference, 2009. IACC 2009. IEEE International, IEEE, pp. 658–663 (2009)
8. Hasan, M.A.M., Alim, M.A., Islam, M.W.: A new approach to Bangla text extraction and recognition from textual image. In: 8th International Conference on Computer and Information Technology, ICCIT, pp. 202–209 (2005)

9. Sethi, I.K., Chatterjee, B.: Machine recognition of constrained hand printed Devanagari. Pattern Recognition **9**, 69–75 (1977)
10. Bansal, V., Sinha, R.M.K.: Integrating knowledge sources in Devanagari text recognition system. IEEE Trans. Syst. Man Cybern. Part A: Syst. Humans **30**(4), 500–505 (2000)
11. Gatos, B., Pratikakis, I., Kepene, K., Perantonis, S.J.: Text detection in indoor/outdoor scene images. In: Proceedings of the First Workshop of Camera-based Document Analysis and Recognition, pp. 127–132 (2005)
12. Tran, H., Lux, A., Nguyen, H.L.T., Boucher, A.: A novel approach for text detection in images using structural features. Pattern Recogn. Data Mining **3686**, 627–635 (2005)
13. Jindal, M.K., Lehal, G.S., Sharma, R.K.: A study of touching characters in degraded Gurmukhi text. In: World Academy of Science, Engineering and Technology, pp. 121–124 (2005)
14. Angadi, S.A., Kodabagi, M.M.: A texture based methodology for text region extraction from low resolution natural scene images. Int. J. Image Process. **3**(5), 229–245 (2009)
15. Zagoris, K., Pratikakis, I.: Scene text detection on images using cellular automata. In: 10th International Conference on Cellular Automata for Research and Industry. Santorini Island, Greece, 24–27, 2012

Obstacle Avoidance in Robot Navigation Using Two-Sample T-Test-Based Obstacle Recognition

Neerendra Kumar and Zoltán Vámossy

Abstract In this paper, a solution to the obstacle avoidance problem in autonomous robot navigation is presented using two-sample t-test. A procedure to perform t-test on two-independent LASER scan distance-range vectors is provided. The proposed strategy compares the LASER scan readings of the obstacles in the navigation path. Similarity of the obstacles is recognized on the basis of the acceptance of the null-hypothesis. Robot navigation results are simulated using Turtlebot-Gazebo simulator and MATLAB software. A discussion is made on the results produced using the proposed strategy with the previously existing results in the literature.

Keywords LASER scan · Obstacle avoidance · Obstacle recognition · Robot navigation

1 Introduction

In robotic operations, the navigation of the robot is among the most required operations in case of mobile robots. Due to this importance, the robot navigation is among the key research area for the researchers in this field (e.g. [1, 2]). The view-based robot navigation and localization play a significant role for autonomous robots [3]. Using random and heuristic search, an algorithm to optimize the route plan for robot navigation is described in [4]. However, the algorithm by [4] may face challenges in unknown environments. Vision systems can provide solutions for mobile robot navigation [5]. Yet, the complexity of the robot system is raised by using vision systems. To overcome the complexity, a LASER-based solution is considered by [6] for obstacle recognition during robot navigation. The obstacle recognition is used

N. Kumar (✉)
Department of Computer Science & IT, Central University of Jammu, Jammu, India
e-mail: neerendra.kumar@phd.uni-obuda.hu

N. Kumar · Z. Vámossy
John Von Neumann Faculty of Informatics, Óbuda University, Bécsi út 96/B, 1034 Budapest, Hungary
e-mail: vamossy.zoltan@nik.uni-obuda.hu

P. K. Singh et al. (eds.), *Recent Innovations in Computing*, Lecture Notes in Electrical Engineering 701, https://doi.org/10.1007/978-981-15-8297-4_37

for obstacle avoidance in [7]. However, in [6, 7], the obstacle recognition is based on the comparison of the difference of LASER scan ranges with an arbitrary value.

The remaining of the paper is structured in the following manner: The considered research problem is introduced in Sect. 2. Section 3 presents the scientific background of the t-test. The solution to the research problem of Sect. 2 is provided in Sect. 4. Section 5 consists of experimental results acquired by applying the solution given in Sect. 4. Finally, the conclusion of the paper is given in the Sect. 6.

2 Problem Definition

In [7], the Algorithm 2 of [7] has presented a solution for obstacle recognition and avoidance using the difference of the standard deviations of LASER scan distance-range vectors. In Algorithm 2 of [7], the statements 18, 23, and 25 have been stated as given in (1), (2), and (3), respectively.

$$S(c) \leftarrow \sigma_R; \tag{1}$$

$$\sigma_r \leftarrow |(S(c) - S(k))|; \tag{2}$$

$$\text{if } (D_{c,k} < P_t \& \sigma_r < \sigma_t \& T_d > T_t) \text{ then} \tag{3}$$

In (3), the if statement includes logical "AND" of the three conditions. In the if statement, the second condition (i.e., $\sigma_r < \sigma_t$) is to compare the difference of standard deviations (σ_r) of two "LASER scan distance-ranges" with σ_t (an arbitrary constant value). Although, to set σ_t value is problem specific. Hence, standardization of the condition $\sigma_r < \sigma_t$ is a research problem.

Applied statistical techniques are available in literature (e.g., in [8]). Using "t-test", two independent samples may be compared to test their similarity on several aspects [9]. In this paper, LASER scan distance-ranges are compared using t-test.

3 Scientific Background of the T-Test

3.1 The Student's T

Let a sample of data is given in (4) as follows:

$$x = \{x_1, x_2, x_3, \ldots x_m\}. \tag{4}$$

For the sample data as taken in (4), William Sealy Gosset (who used the pseudonym as "Student" [7]) defined the statistics "t" as follows:

$$t = \frac{(\overline{x} - \mu)\sqrt{m}}{\sigma_x} \tag{5}$$

where m, $\overline{x}$, μ, and represent the sample-size, sample-mean, and mean of the population, respectively. In (5), the statistics t is known as "Student's t" having the degree of freedom (d_f) defined by (6).

$$d_f = m - 1 \tag{6}$$

σ_x is the standard deviation (as defined in (7)) of the sample.

$$\sigma_x = \sqrt{\frac{\sum_{k=1}^{m} (x_k - \overline{x})^2}{m - 1}} \tag{7}$$

3.2 Difference of the Sample-Mean and Mean of the Population Using T-Test

For a sample, the t-test has the following steps:

(i) Set H_0 (the null-hypothesis) as follows:

H_0: "There is no significant difference between the sample-mean and mean of the population."

(ii) Compute "t" using (5).

(iii) For the df obtained by (6), compare the value of "t" with the tabulated value of "t" (say t_c) at the desired level of significance.

If ($|t| \leq t_c$) then "H_0" is accepted Else "H_0" is rejected.

For different d_f values, the corresponding values of t_c are presented in Table 1 for 1% and 5% significance levels.

3.3 Difference of Two Means Using T-Test

Consider two independent samples x (as taken in (4)) and y (as given in (8)).

$$y = \{y_1, y_2, y_3, \ldots y_n\}. \tag{8}$$

Table 1 Critical values of t at 1% and 5% levels of significance

d_f	$t_{.01}$	$t_{.05}$	d_f	$t_{.01}$	$t_{.05}$	d_f	$t_{.01}$	$t_{.05}$	d_f	$t_{.01}$	$t_{.05}$
1	63.66	12.71	10	3.17	2.23	19	2.86	2.09	28	2.76	2.05
2	9.92	4.30	11	3.11	2.20	20	2.84	2.09	29	2.76	2.05
3	5.84	3.18	12	3.06	2.18	21	2.83	2.08	30	2.75	2.04
4	4.60	2.78	13	3.01	2.16	22	2.82	2.07	40	2.70	2.02
5	4.03	2.57	14	2.98	2.14	23	2.81	2.07	60	2.66	2.00
6	3.71	2.45	15	2.95	2.13	24	2.80	2.06	80	2.64	1.99
7	3.50	2.36	16	2.92	2.12	25	2.79	2.06	100	2.63	1.98
8	3.36	2.31	17	2.89	2.11	26	2.78	2.06	120	2.62	1.98
9	3.25	2.26	18	2.88	2.10	27	2.77	2.05	>120	2.58	1.96

Following [8], the statistics "t" can be defined by (9).

$$t = \frac{(\overline{x} - \overline{y})}{\sigma_s \sqrt{\frac{1}{m} + \frac{1}{n}}} \tag{9}$$

where m and n are the sizes of the x and y samples considered in (8), respectively. x and y are the means of the samples x and y, respectively. The value of σ_s is given by (10).

$$\sigma_S = \sqrt{\frac{1}{m+n-2}\left[\sum_{i=1}^{m}(x_i - \overline{x})^2 + \sum_{j=1}^{n}(y_j - \overline{y})^2\right]} \tag{10}$$

Here,

$$d_f = m + n - 2 \tag{11}$$

To compare the two samples, the "t-test" has the following steps:

(i) Set H_0 (the null-hypothesis) as follows:

H_0: "$\overline{x}$ and $\overline{y}$ (the means of the two samples) do not differ significantly."

(ii) Compute "t" by using (10).

(iii) For the d_f obtained using (11), accept or reject H_0 following the step (iii) of the Sect. 3.2.

3.4 Tabulated (Critical) Values of "T" to Apply T-Test

For different d_f values, the critical values of "t" (t_c) are given in Table 1 (as presented in [8, 10, 11]). In Table 1, $t_{.01}$ and $t_{.05}$ represent t_c at 1% and 5% levels of significance, respectively.

4 Obstacle Recognition and Avoidance Using T-Test

For the application of "t-test", the statements 18, 23, and 25 in the Algorithm 2 of [7] are replaced with the statements presented in (12), (13), and (14), respectively.

$$S(c) \leftarrow R; \tag{12}$$

$$h \leftarrow NZtTest(S(c), S(k)), \tag{13}$$

$$\text{if } (D_{c,k} < P_t \& h = 0 \& T_d > T_t) \text{ then} \tag{14}$$

where

"NZtTest" = The procedure defined in Algorithm 1.

S(c) and *S(k)* = Two distance-range vectors as input arguments to the procedure "NZtTest".

In Algorithm 1, x and y are the two input parameters corresponding to the two input arguments, $S(c)$ and $S(k)$ of NZtTest, in (13).

Algorithm 1 Procedure for the t-test to compare two distance-range-vectors.

procedure `NZtTest`(x; y)
$m \leftarrow$ size of x
$n \leftarrow$ size of y
$d_f \leftarrow$ m + n -2
$t_c \leftarrow$ tabulated value (Table 1) at desired level of significance and d_f.
$\overline{d_x} \leftarrow \frac{1}{m}\sum_{i=1}^{m} x_i$
$\overline{d_y} \leftarrow \frac{1}{n}\sum_{j=1}^{n} y_j$
Compute σ_s by equation (10).
Compute t using equation (9).
if ($|t| \leq t_c$) **then**
    Return 0
else
    Return 1
end if
end procedure

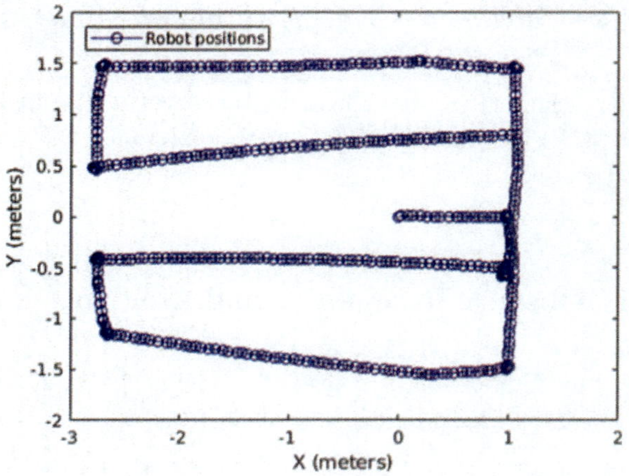

Fig. 1 Robot path obtained using the *t*-test

5 Experimental Results and Discussion

To get the results of implementation of the "t-test" in Algorithm 2 of [7], the Gazebo-world presented in [7] is used. Further, the start location of the robot is taken approximately as (0, 0). The "t-test" has been applied to obtain the difference of the two "LASER scan distance-range vectors" having 5% significance level.

Using the results of the t-test, Fig. 1 depicts path travelled by the robot. The following are the two key observations:

i. Similar to the Algorithm 2 of [7] (without modifications for "t-test"), the modified Algorithm (applying "*t*-test") in Sect. 4 identifies the similarity of the obstacle wall, successfully. Consequently, the robot does not enter into the repetitive paths.
ii. The modified Algorithm 2 of [7] (using "t-test") can reverse the robot's angular velocity nearly at the coordinates (1, −0.5) (Fig. 1), for the first time. Whereas, the Algorithm 2 of [7] (without modifications for the "t-test") had reversed the angular velocity nearly at the coordinates (1, −1:5).

The modification (i.e., using "*t*-test") in Algorithm 2 of [7] can early identify the re-occurrence of similar obstacle. The "*t*-test" uses standard statistical technique to check the similarity of two samples. Therefore, the difference of two LASER scan distance-range vectors needs not to compare with any arbitrary constant like σ_t used in the statement 3.

6 Conclusion

The LASER scan distance-range vectors have been compared using the two-sample *t*-test. Two distance-range vectors have been found as similar if

"Computed *t*" ≤ "Critical value (tabulated) of *t*".

For the obtained degree of freedom, the comparison between the computed "t" and critical value (tabulated) of "t" is obtained at the desired significance level.

By including the "t-test", the Algorithm 2 (presented in [7]) is modified. Subsequently, the necessity to compare "the difference of the two LASER scan distance-ranges" with "an arbitrary constant value" is removed.

References

1. Matveev, A.S., Magerkin, V.V., Savkin, A.V.: A method of reactive control for 3D navigation of a nonholonomic robot in tunnel-like environments. Automatica **114**(108831), 1–11. (2020) http://www.sciencedirect.com/science/article/pii/S0005109820300297
2. Wang, D., Hu, Y., Matitle, T.: Mobile robot navigation with the combination of supervised learning in cerebellum and reward-based learning in basal ganglia. Cogn. Syst. Res. **59**, 1–14 http://www.sciencedirect.com/science/article/pii/S1389041719304723. (2020)
3. Oishi, S., Inoue, Y., Miura, J., Tanakatitle, S.: SeqSLAM ++: view-based robot localization and navigation. Robot. Autonomous Syst. **112**, 13–21 (2019). http://www.sciencedirect.com/science/article/pii/S092188901730684X
4. Xu, Y., Guan, G., Song, Q., Jiang, C., Wangtitle, L.: Heuristic and random search algorithm in optimization of route planning for Robot's geomagnetic navigation. Comput. Commun. **154**, 12–17 http://www.sciencedirect.com/science/article/pii/S0140366420300803. (2020)
5. Łaganowska, M.: Application of vision systems to the navigation of mobile robots using markers. Transp. Res. Proc. **40**, 1449–1452 (2019). http://www.sciencedirect.com/science/article/pii/S2352146519303709
6. Kumar, N., Vámossy, Z.: Robot navigation in unknown environment with obstacle recognition using laser sensor. Int. J. Electrical Comput. Eng. **9**(3), 1773–1779 (2019). http://ijece.iaescore.com/index.php/IJECE/article/view/13723/12876
7. Kumar, N., Vámossy, Z.: Obstacle recognition and avoidance during robot navigation in unknown environment. Int. J. Eng. Technol. **7**(3), 1400–1404 (2018). https://www.sciencepubco.com/index.php/ijet/article/ view/13926/6363
8. Snedecor, G.W., Cochran, W.G.: Statistical methods. Iowa State University Press, Ames (1989)
9. Fritz, M., Berger, P.D.: Chapter 2—Comparing two designs (or anything else!) using independent sample T-tests. In: Improving the user experience through practical data analytics. Morgan Kaufmann, Boston, pp 47–69 (2015). http://www.sciencedirect.com/science/article/pii/B9780128006351000021
10. Student: Probable error of a correlation coefficient. Biometrika **6**(2–3), 302–310 (1908). https://doi.org/10.1093/biomet/6.2-3.302
11. Fisher, L., Mcdonald, J.: 3—Two-sample t-test. In: Fixed effects analysis of variance. Probability and mathematical statistics: a series of monographs and textbooks, pp 21–35. Academic Press

On Data-Driven Approaches for Presentation Attack Detection in Iris Recognition Systems

Deepika Sharma and Arvind Selwal

Abstract With the development of modern machine learning-based techniques for accurate and efficient classification, the paradigm has shifted to automatic intelligent-based methods. The iris recognition systems constitute one of the most reliable human authentication infrastructures in contemporary computing applications. However, the vulnerability of these systems is a major challenge due to a variety of presentation attacks which degrades their reliability when adopted in real-life applications. Hence, to combat the iris presentation attacks, an additional process called as presentation attack detection mechanism is integrated within the iris recognition systems. In this paper, a review of the modern intelligent approaches for iris presentation attack detection (PAD) mechanisms is presented with a special focus on the data-driven approaches. The presented study shows that the machine learning-based approaches provides better classification accuracy as compared to conventional iris PAD techniques. However, one of the open research challenge is to design the robust intelligent iris PAD frameworks with cross-sensor and cross-database testing capabilities.

Keywords Iris recognition · Presentation attack · Presentation attack detection · Deep learning · Machine learning

1 Introduction

With the growth of secured authentication in the contemporary world, there arises a major concern of security threat. These threat issues need to be addressed by developing robust and highly efficient pattern recognition systems. Therefore, biometric-based authentication is preferred over traditional approaches of human recognition. Biometrics is a science of recognizing human on the basis of their distinctive physiological, chemical, and behavioral traits [6]. A variety of biometrical traits have

D. Sharma (✉) · A. Selwal
Department of Computer Science & Information Technology, Central University of Jammu, J&K 181143, India
e-mail: sharmadeepika749@gmail.com

P. K. Singh et al. (eds.), *Recent Innovations in Computing*, Lecture Notes in Electrical Engineering 701, https://doi.org/10.1007/978-981-15-8297-4_38

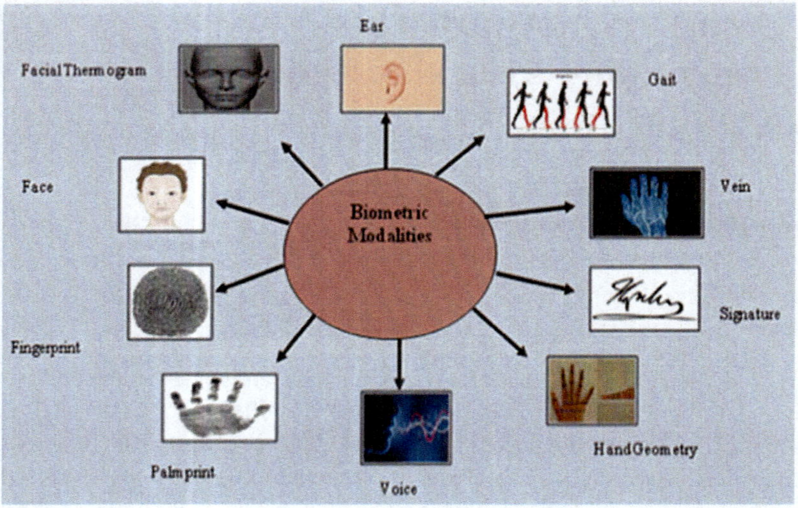

Fig. 1 A variety of biometrical traits in human body

been identified in human being, which include physical, behavioral, and chemical characteristics as shown in Fig. 1.

The biometric systems provide better security and are more suitable than classical approaches of human recognition [11]. As classical approaches are knowledge-based (what you remember like passwords, PIN code) or token-based (what you possess like ID cards, Smart cards), where these representations of identity are vulnerable to a variety of attacks. By using biometric technology, it is possible to establish the identity and integrity on the basis of who we are rather than what we remember or what we possess. In this article, we focus our study on the iris biometric modality. Iris is the region that is bounded by the pupil and the white portion of the eye. The iris texture carries the distinctive information that is useful for recognition of a person. The iris-based recognition systems are widely deployed in many user applications as because of following advantages, i.e., highly scalable, no physical contact with the system, minimal false acceptance rate, etc. [7]. With the high rate of usability, these systems are more prone to attacks and most commonly the presentation attack. In presentation attack, an adversary can present the fake iris data to the system in order to bypass the iris sensor module. Hence, in order to combat these attacks, a process called PAD mechanism or liveness detection is used (Galbally, Julian, & Cappelli, n.d.). In this article, we present the recent intelligent-based approaches which have been widely employed in order to mitigate iris presentation attacks. The rest of the paper is organized as follows: Sect. 2 presents a brief introduction of various iris presentation attacks. The modern intelligent approach-based countermeasures are reviewed in Sect. 3. The publically available iris datasets are introduced in Sect. 4. Sect. 5 focuses on the major research challenges in the iris PAD mechanisms. Finally, the conclusions is presented in the Sect. 6.

2 Iris Presentation Attacks

In iris presentation attack, the sensor module is spoofed by presenting either the artifacts or an image of original eyes with the cooperation of the authorized user. A variety of techniques are used to spoof the sensor of iris recognition system by presenting fake iris traits [12]. The taxonomy of iris presentation attacks is depicted in Fig. 2.

i Artifacts Attacks

(a) Paper Printouts: Paper printout is the simplest method of spoofing the iris biometric system. The iris printouts may be created in a variety of ways. There is no pre-established fact about the accuracy of colored or black and white printouts. However, the commercial sensors may be used to generate the iris artifacts by creating a whole in the place of pupil to mimic the exact iris traits. An illustration of real and corresponding print of an image taken from LivDet Iris Warsaw 2017 dataset is shown in Fig. 3.

Thalheim et al. [17] used an inkjet printer and matte paper to attack the iris biometric system where they printed an image of resolution 2400 × 1200 dpi and also cut in the center to provide a hole to represent the pupil of eyes. They were successful to spoof the Panasonic Authenticam BM-ET100 sensor by presenting this

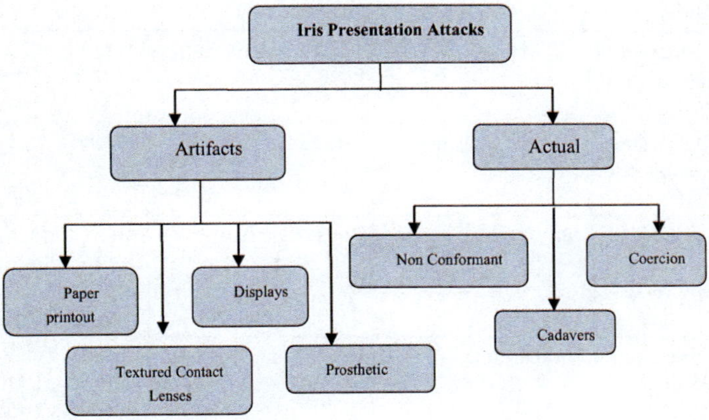

Fig. 2 Classification of Iris presentation attacks

Fig. 3 An example of iris printout-based attack

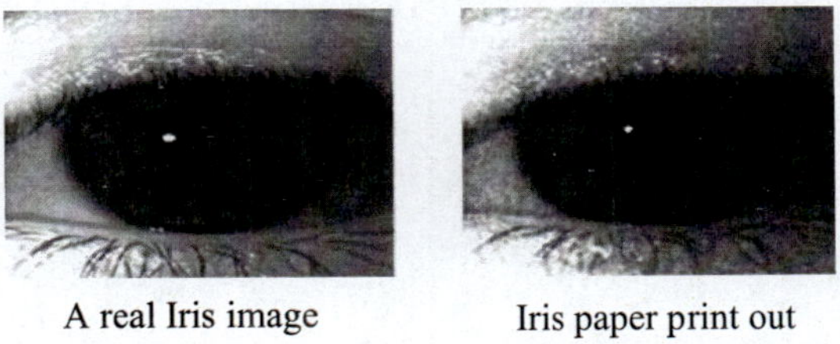

fake iris trait. Similarly, Ruiz-Albacete et al. [15] used two types of attacks where they spoofed the iris recognition system by using fake and original images both for enrollment and testing phase with FMR 0.1% and FNMR of 17%. They achieved a success rate of 34% and 37%, respectively, for type 1 and type 2 attacks 15 (Table 1).

(b) Textured Contact Lenses: In this method, the contact lenses are constructed such that carry a visual texture within them. The basic difficulty with these instruments is that the texture in these lenses sometimes overlays with the natural texture of iris; therefore, it creates a problem of mixed textures of both artificial and natural sources of iris trait. Another problem is that these lenses move on the surface of eye where the mixture of these images varies from one case to another case. An example of textured contact lense and real image is shown in Fig. 4. Moreover, this attack looks reasonable in standard terms, but it is difficult and expensive to achieve the desired success rate [10].

(c) Displays: Video display of an electronic screen has been used by attackers who spoof an recognition system by using a still image extracted from the video source. An example of an iris image and its display in an electronic device as used for presentation attack is shown in Fig. 5. Display attack works well when both the electronic screen and sensor module functions in the same wavelength range (Galbally et al., n.d.).

Table 1 A summary of various intelligent-based approaches for iris PAD detection

Year	Author	Technique	Dataset	Performance
2015	Raghavendra and Busch [13]	M-BSIF	VSIA and LivDet Iris 2013	ACER of 0.29% for VSIA and 1.27% for LivDet Iris 2013
2015	Rigas and Komogortsev [14]	Eye movement	Private Database	EER of 3.4%
2017	Czajka et al. (Czajka & Bowyer, 2018)	CNN	Private database	An ACER above 95% (89%) for recognition of left/right (upright/upside-down)
2018	Bineet Kaur (Kaur et al., 2019)	Discrete orthogonal moment-based invariant feature set	IIITD Contact Lens Iris (IITD-CLI), IIITD Iris Spoofing (IIS)), Clarkson LivDet 2015, and Warsaw LivDet 2015	The results obtained are of 100% for IIITD-CLI and 99.48% for Clarkson datasets, respectively
2019	Bineet Kaur et al. 2019 (Kaur, 2020)	Discrete orthogonal moment-based invariant feature-set	IIITD Contact Lens Iris (IITD-CLI), IIITD Iris Spoofing (IIS)), Clarkson LivDet 2015, and Warsaw LivDet 2015	The results obtained are of 100% for IIITD-CLI and 99.48% for Clarkson datasets, respectively
2019	Waleed S.-A et al. [4]	LBP	CASIA-Iris-Syn and ATVS-FIr- DB	100%ACA

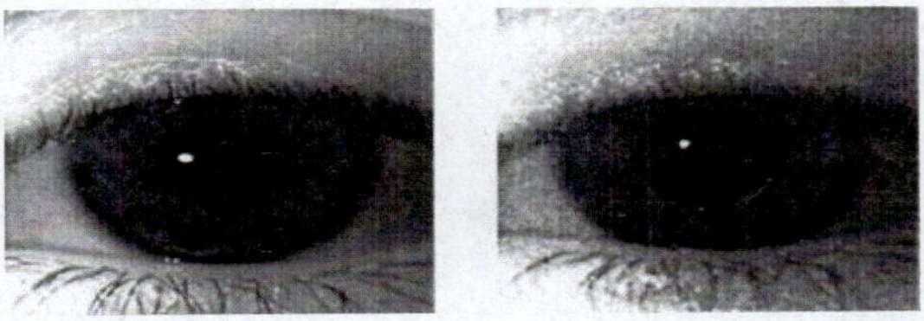

Fig. 4 An example of iris printout-based attack

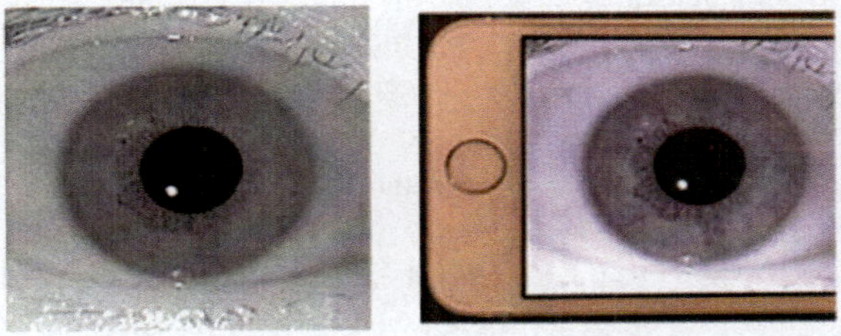

Fig. 5 Display iris presentation attack

(d) Prosthetic Eyes: The ocularists create handcrafted eyes which look like the eyes of the living human being. However, it is a very time consuming and requires rich knowledge to create prosthetic eyes which may the iris recognition system. The resultant iris looks very similar to the original eyes which were captured by commercially available sensors in the near-infrared imaging. The literature shows that the prosthetic eyes may be created in compliance of ISO/IEC 19794-6:2011 which may be further used for the iris recognition system. However, very less number of such types presentation attacks have been attempted; therefore, success rates of these attacks is minimal.

ii Real Eyes Attacks

(a) Non-Conformant Use: However, it is necessary for a user to cooperate during both the enrollment and the testing phases, but it is easiest way to exploit the iris recognition system. The intentional iris spoofing may involve excessive eyelid closure which may results in capturing the lesser number of features thus making incorrect match [2].

(b) Cadavers: The concept of using organs of a non-living human being for presentation attacks has been noticed from many movies. It is possible that iris cadaver may be obtained from the non-living organisms up to month of death by using commercially available sensors at a temperature of approximately 6 degree celcius. Sansola [16] was able to use an IriTech IriShield MK 2120U system to report that the post-mortem iris recognition is possible for 11 days after death of a person.

(c) Coercion. It is a spoofing attack on iris sensor by using the act of threat or force. This type of attack is a possible vulnerability for presentation attack; however, there is no reported article which shows the success of these types of attacks [2].

3 Iris Presentation Attack Detection Mechanisms

In iris biometric systems, PA attacks are the attacks which include the presentation of fake iris data as input to the system in order to interfere the normal working of the system. The term liveness detection is also used in the literature for classifying the bonafide and the artifacts used for making spoof attack successful [1]. The PAD techniques are divided into three main classes as shown in Fig. 6.

In this subsection, we review the intelligent-based techniques which uses machine or deep learning approaches. Raghavendra and Busch [13] proposed an iris liveness detection technique which is based on M-BSIF and linear SVM classifiers. The proposed scheme consists of four major components such as iris segmentation, where optimal path of contours is traced by using the vertbi algorithm. In periocular region extraction step, the image is resized to a dimension of 120×120 pixels. Then, in PAD algorithm, two important components are used namely; M-BSIF filters and SVM classifiers. The M-BSIF kernels are defined using a set of natural image segments. The experiments were performed on VSIA and LivDet Iris2013 database where the results provide 0.29% and 1.27% ACER for both the datasets, respectively. Rigas and Komogortsev [14] proposed a technique which utilizes the movement of eye cues for the iris PAD. The fundamental distortion which arises due to movement of eye signal and because of this the functional and structural discrepancies was detected between the paper-printed iris and a iris of a natural eye. The experiments involved the execution of print-attacks against an eye-tracking device, which also results in the collection of the eye movement signal. The experiments were performed on large dataset collected from 200 different subjects, which contains the signals for both the real print-attack eye movement. The proposed PAD method results in a higher spoof detection performance with a ACR of 96.5% and EER of 3.4%. Czajka et al. [3] present a novel method for iris PAD which recognizes the correct right/left and upside/upright down direction of iris trait images. It is used to counteract the presentation attacks which are directed for creating spoof identities

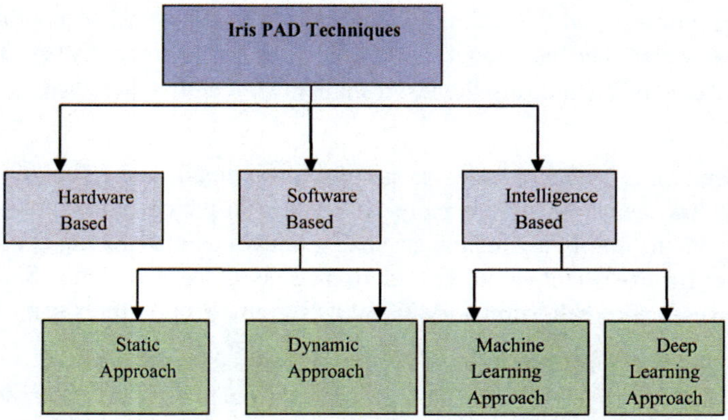

Fig. 6 A taxonomy of iris PAD techniques

by moving both either an iris image or sensor during the process of image acquisition. Both the approaches were evaluated on the same data, by making use of same evaluation protocol. The handcrafted features were classified by using SVM, and the CNN model was used for feature learning which is based on data-driven features. A dataset which contains 20,750 iris images was used for training. In addition, a dataset containing 1939 images acquired from 32 different subjects was used for testing the purposed iris PAD technique. The proposed technique was evaluated in the scenario of same-sensor and cross-sensor so as to generalize the unknown hardware. The proposed approach results in an ACER above 95% in the case of left/right orientation and 89% for upright/upside-down orientation, whereas ACER 99% in the case when images were captured by the same sensor. The proposed approach, however, performed better in the case of same-sensor scenario but failed to generalize in the case of unknown sensor. Kaur [8] presented a iris PAD method which uses "discrete orthogonal moment" based invariant feature vector consisting of Tchebichef, Dual-Hahn, and Krawtchouk moments. The feature was made invariant to rotation, scale, and translation. to overcome the problem of variations during sample acquisition. The normalized images were divided into 32 non-overlapping sub-regions of each of equal size of 32 × 32. A feature set of 8160 features was extracted from 32 blocks till 15th order. But, in the case of Dual-Hahn moments, the feature vector was constructed by using a fusion of 8160 local and 8160 global features. Then, for the classification purpose, the method used KNN classifier to predict whether it is real or fake iris. The performance of the proposed approach was evaluated on four iris PAD databases: Clarkson LivDet 2015, IIITD Contact Lens Iris (IITD-CLI), Warsaw LivDet 2015, and IIITD Iris Spoofing (IIS). The results obtained were of 99.48% for Clarkson datasets and 100% for IIITD-CLI, respectively. As a future research, the authors suggest that the biometric templates may be used to create synthetic spoofing datasets. Kaur et al. [9] proposed a novel PAD technique in the uncontrolled environment acquired by using multiple commercially available sensors. The method extracts feature set which are rotation invariant and are based on continuous moment. These features were extracted by using "Zernike moments" and "polar harmonic transforms" which give variation in local intensity in given image patches. The approach used "circular Hough transform (CHT)" for iris segmentation and pupil localization. Then, Daugman's homogeneous rubber sheet model was used for normalization of iris or converting the samples into fixed size. Finally, for the classification, KNN classifier was used in the proposed technique. The method is evaluated on four iris PAD databases: Clarkson LivDet Iris 2015, IIITD-CLI, IIITD Iris Spoofing, and Warsaw LivDet Iris 2015. The proposed technique was evaluated by using performance metrics like APCER, NPCER, and ACER as per ISO/IEC 30,107-3 standard for PAD methods. The proposed method results in comparable accuracy with respect to other state-of-the-art iris PAD methods. The proposed algorithm lacks the testing against the new material scenario, and accuracy may be further improved by using deep level features. Waleed et al. [4] presented an efficient method for iris livness detection in which there is no need of segmentation and normalization step that are employed in traditional iris livness detection system. Wavelet packets (WPs) were used for segmenting the iris image into wavelet approximation and detail channels.

Then, the entropy values were extracted from both the wavelet channels LBP images of the channels. The statistical features were extracted from the segmented wavelet channels. In particular, the global features were extracted by using the entropy. The entropy of wavelet channels and LBP features is fused together, and then, a feature selection operation is applied. The SVM is used as a pattern classifier for discriminating the real and fake iris image. The CASIA-Iris-Syn and ATVS-FIr-DB databases were used for experimentation of the proposed approach. The authors claimed 100% average classification accuracy (ACA) which is better than other state-of–the-art methods but high computational load. However, the authors have not evaluated the proposed method in terms of other metrics like ACER, EER, etc. The generalization of this method in other types of iris attack and cross-sensor testing is missing in this work.

4 Iris Databases

In this section, we provide a brief introduction of the various openly available databases that are used for iris PAD mechanisms evaluation. An iris database consists of structured collection of iris images mainly used for the development and the evaluation of the iris PAD algorithms. The most widely deployed databases are from openly available liveness detection competition series: LivDet Iris 2013 [19], LivDet Iris 2015 [18], LivDet Iris 2017 [18], as well as CASIA and ATVS- Fir DB. A brief summary of these commonly used iris PAD datasets is explained as follows:

(i) LivDet Iris 2013 DB: It was the first iris liveness detection competition which was held in 2013. This database consists of 4000 sample images which are captured from around 500 different iris traits, and it is further divided into three datasets acquired at three universities, namely Clarkson University, University of Notre Dame, and University of Warsaw.

(ii) LivDet Iris 2015 DB: The 2015 dataset was the second dataset of iris liveness detection competition series [20]. Similar to LivDet Iris2013 DB, spoof attacks are represented with printed iris images and patterned contact lenses. The images were collected by exploiting the sensor by presenting a fake iris trait and Warsaw University compiled a dataset by using printed artifacts. The Dalsa and LG datasets consist of 20 different images with patterned contact lenses. Out of 20 images, 15 were available for training, and remaining five images were used for testing purpose only.

(iii) LivDet Iris 2017 DB: This dataset was collected at Clarkson University. The iris images were captured by an LG IrisAccess EOU2200 camera. The dataset was created for extending the LivDet Iris 2015 DB. The Clarkson dataset is divided into three modules: The first module has iris images from genuine users. The second part consists of patterned contacts with various types. The last one has printed iris images. The dataset is divided into two set: one is training, and other is testing. The training set comprises of 2469 live iris images from 25 subjects, whereas testing set uses additional unknown fake image types.

(iv) ATVS-Fir DB: The iris ATVS-Fir DB was collected by the Biometric Recognition Group—ATVS. It consists of real and fake iris images with BMP format and size of 640 × 480 pixels. The database has images of 50 subjects in which both the eyes of an individual were considered. There are total 1600 images in the dataset in which 800 are real and remaining are of fake iris trait.

(v) CASIA: The Casia iris database is provided by the Chinese Academy of Sciences (CAS). It consists of three subsets, namely CASIA-Iris-Twins, CASIA-Iris-Lamp, and CASIA-Iris-Interval. The dataset contains total 22,035 iris images collecting from at least 700 subjects. The iris images are in jpeg format with gray level of 8 bit.

(vi) IIT Delhi: This iris dataset was collected at IIT Delhi by the Biometrics Research Laboratory. The images were captured by using JPC1000 digital CMOS camera19. The dataset consists of 1120 iris images of 224 users. Among these 224 users, 176 are male and remaining are female users. The resolution of iris images is 320 × 240 pixels.

5 Open Research Issues

The study reveals that the iris recognition systems are usually attacked by presenting artifacts while in other cases with real eyes. To countermeasure these iris presentation attacks, a variety of solutions are available in the literature with their merits and demerits. However, with the development of machine learning models, data-driven PAD approaches are becoming popular which provide better classification accuracy for iris liveness detection, but one of the open issue is availability of the sufficient amount of iris anti-spoofing datasets. Moreover, majority of the intelligent PAD approaches uses automatic feature extraction based on convolutional neural network (CNN) and image feature descriptor which are evaluated on openly available iris anti-spoofing datasets like datasets like LivDet Iris 2013DB, LivDet Iris 2015 DB, ATVS-Fir, etc. This leads to the problem of validation of data-driven iris PAD techniques in cross-sensor and cross-database scenario. As pointed out by Kaur [8] in her research work that the biometric templates can be exploited to create synthetic spoofing databases for iris liveness detection system. Similarly, Waleed et al. [4] in their research raised a concern that the generalization of the LBP based iris PAD technique in other types of iris attack and cross-sensor testing may be further undertaken as a research work.

6 Conclusions

We have presented a review of typical iris presentation attacks based on artifacts and real eyes. The main focus in this article was to countermeasure iris presentation attacks by using intelligent data-driven PAD approaches based on image feature descriptors or automatic feature extraction using CNN. These approaches provides better classification accuracy to discriminate live and fake iris traits, but the major concern is the availability of sufficient amount of training datasets to built the intelligent PAD models. Furthermore, the validation of intelligent data-driven iris PAD models in cross-sensor and cross-database scenario is another open research issue.

References

1. Bori, A., Galbally, J.: Anti-spoofing : Iris Risks of Biometric Spoofing, pp. 1–13. https://doi.org/10.1007/978-3-642-27733-7
2. Czajka, A., Bowyer, K.W.: Presentation attack detection for iris recognition: an assessment of the state-of-the-art. ACM Comput. Surv. **51**(4). https://doi.org/10.1145/3232849
3. Czajka, A., Bowyer, K.W., Krumdick, M., Vidalmata, R.G.: Recognition of image-orientation-based Iris spoofing. IEEE Trans. Inf. Forensics Security **12**(9), 2184–2196 (2017). https://doi.org/10.1109/TIFS.2017.2701332
4. Fathy, W.S.A., Ali, H.S.: Entropy with local binary patterns for efficient iris liveness detection. Wireless Personal Commun. **102**(3), 2331–2344 (2018). https://doi.org/10.1007/s11277-017-5089-z
5. Galbally, J., Julian, F., Cappelli, R.: Handbook of Biometric Anti-Spoofing (Second; S. Marcel, M. S. Nixon, J. Fierrez, & N. Evans, eds.). Springer (n.d.)
6. Jain, A.K., Flynn, P., Ross, A.A.: Handbook of Biometrics (A. K. Jain, P. Flynn, & A. A. Ross, eds.). Springer, London (2008)
7. Jain, A.K., Ross, A., Prabhakar, S.: An introduction to biometric recognition. IEEE Trans. Circuits Syst. Video Technol. **14**(1), 4–20 (2004). https://doi.org/10.1109/TCSVT.2003.818349
8. Kaur, B.: Iris spoofing detection using discrete orthogonal moments. Multimedia Tools Appl. **79**(9–10), 6623–6647 (2020). https://doi.org/10.1007/s11042-019-08281-x
9. Kaur, B., Singh, S., Kumar, J.: Cross-sensor iris spoofing detection using orthogonal features. Comput. Electrical Eng. **73**, 279–288 (2019). https://doi.org/10.1016/j.compeleceng.2018.12.002
10. Kohli, N., Yadav, D., Vatsa, M., Singh, R., Noore, A.: Detecting Medley of Iris Spoofing Attacks using DESIST Naman Kohli (1994)
11. Maltoni, D., Maio, D., Jain, A., Salil, P.: Handbook of Fingerprint Recognition (Second). Springer, London (2009)
12. Menotti, D., Chiachia, G., Pinto, A., Schwartz, W.R., Pedrini, H., Falc˜ao, A.X., Rocha, A.: Deep representations for Iris, Face, and Fingerprint. IEEE Trans. Information Forensics Security **10**(4), 1–16 (2015). https://doi.org/10.1109/TIFS.2015.2398817
13. Raghavendra, R., Busch, C.: Robust scheme for iris presentation attack detection using multi-scale binarized statistical image features. IEEE Trans. Information Forensics Security **10**(4), 703–715 (2015). https://doi.org/10.1109/TIFS.2015.2400393
14. Rigas, I., Komogortsev, O.V.: Eye movement-driven defense against iris print-attacks. Pattern Recogn. Lett. **68**, 316–326 (2015). https://doi.org/10.1016/j.patrec.2015.06.011
15. Ruiz-Albacete, V., Tome-Gonzalez, P., Alonso-Fernandez, F., Galbally, J., Fierrez, J., Ortega-Garcia, J.: Direct attacks using fake images in iris verification. Lecture Notes in Computer Science (Including Subseries Lecture Notes in Artificial Intelligence and Lecture Notes in

Bioinformatics), 5372 LNCS, pp. 181–190 (2008). https://doi.org/10.1007/978-3-540-89991-4_19
16. Sansola, A.K.H.: Postmortem iris recognition and its application in human identification. ProQuest Dissertations and Theses, 70 (2015)
17. Thalheim, L., Krissler, J., Ziegler, P.: Biometric Access Protection Devices and their Programs Put to the Test, pp. 1–9 (2009)
18. Yambay, D., Becker, B., Kohli, N., Yadav, D., Czajka, A., Bowyer, K.W., Federico, N., et al.: LivDet Iris 2017 - Iris Liveness Detection Competition 2017, pp. 733–741 (2017)
19. Yambay, D., Doyle, J.S., Bowyer, K.W., Czajka, A., Schuckers, S.: LivDet-Iris 2013 – Iris Liveness Detection Competition 2013 (2013)
20. Yambay, D., Walczak, B., Schuckers, S., Czajka, A.: LivDet-Iris 2015 – Iris Liveness Detection Competition 2015, pp. 1–6 (2017). https://doi.org/10.1109/ISBA.2017.7947701

Anomaly Detection and Qualitative Analysis of Diseases in Tomato

Meenakshi Sood, Anjna, and Pradeep Kumar Singh

Abstract Disease detection in the crops is a difficult work as crops get affected due to various attacks from different bacteria, fungi and viruses. The disease symptoms on the infected crop plant can be seen as usually color change, circular spots, specks and hollow areas having concentric rings. This paper proposes a solution for identification of crop diseases (i.e., bacterial and fungal diseases) in tomato cash crop of Himachal Pradesh. Detecting the disease at an early stage enables the farmers to act and treat the plants at the appropriate time and effectively. Accurate and timely detection of plant diseases can help mitigate the agriculture loss experienced by the local farmers. An initial evaluation system and statistical analysis proposed in this work show a positive impact. The dataset has been created by the authors by collecting real-time pictures from various fields of Himachal Pradesh state which contains images with different diseases for tomato plant. The proposed approach provides efficient result that can lead to connection between farmers and agriculturists.

Keywords Bacterial diseases · Texture features · Statistical analysis

1 Introduction

The production of vegetables and fruits faces significant losses due to infection that may be bacterial, virus and fungi [1]. Diseases in plants have become a considerable issue as they can cause serious loss of agricultural products, both quality and quantity, and negatively influencing the country's economy [2, 3]. The most important reason

M. Sood
NITTTR Chandigarh, Chandigarh, India

Anjna
Department of Electronics & Communication Engineering, Jaypee University of Information Technology, Waknaghat, Solan, HP, India

P. K. Singh (✉)
ABES Engineering College, Ghaziabad, UP, India
e-mail: pradeep_84cs@yahoo.com

P. K. Singh et al. (eds.), *Recent Innovations in Computing*, Lecture Notes in Electrical Engineering 701, https://doi.org/10.1007/978-981-15-8297-4_39

which leads to destruction of crops is plant diseases; by taking care of the diseases in earlier stages enables us to overcome the crop yield loss [1]. Previously, plant disease identification and detection are done by the farmers naked eye observation which requires continuous monitoring of plant, and this method is costly, tedious, less accurate and tough task [3].

Managing diseases is a challenging task for farmers as plant diseases cause major production and economic losses of agricultural products. So, a technique is required that is less time-consuming and provides accurate and cost-effective solution for farmers to monitor their crops.

To make this task more accurate and efficient, various studies are carried out to detect diseases automatically by applying image processing methods. For production, it is used to develop automatic decision support system for examining the quality of yield.

While monitoring large crop fields for detecting the symptoms of diseases as soon as they appear on the plant, automatic plant disease detection may prove beneficial. Hence, classification and identification of crop diseases by using techniques of image processing are essential so as to identify the diseases correctly in initial stage which helps in taking action to prevent the losses and hence improve the agricultural yield [15].

2 Related Work

Prior many approaches have been adopted by distinct researchers for plant disease detection; some of the plant disease detection and classification-related work are elaborated in this section.

Work related to disease detection and classification of plant leaf:

Abu-Naser S.S. et al. (2008) describe expert system developed for plant disease analysis using two different methods namely step-by-step detailed and graphical authentic methods, and it brings distinct approach of plant disease analysis to the user.

Al-Bashish et al. (2011) proposed a solution for plant leaf diseases detection and classification. In this methodology, for the input RGB leaf image, a structure of color transformation is created, for segmentation k-means clustering technique is used, and texture features are evaluated and passed through a pre-trained neural network. The developed neural network classifier executes well and can strongly detect/classify the diseases with 93% accuracy. Arivazhagan et al. (2013) refined an image processing design for color transformation structure creation, masking of green pixels followed by segmentation mechanism, and texture features are calculated for the diseased segmented area and cross through the classifier that can profitably detect and classify the diseases with 94% accuracy.

Chouhan et al. (2018) introduced a scheme for automatic recognition and classification of plant leaf diseases named as bacterial foraging optimization-based radial basis function neural network (BRBFNN).

They use bacterial foraging optimization for accrediting optimal weight to RBFNN that boost the network's speed and accuracy to identify the plant leaf-infected areas by distinct fungal diseases like early blight, late blight, leaf spot, leaf curl, cedar apple rust and common rust.

The suggested mechanism obtains greater accuracy in disease identification. Kaur et al. (2016) presented an image processing approach to detect and classify the leave diseases by adopting k-means segmentation and classification performed by SVM. The fungal and bacterial diseases considered are Alternaria alternata, leaf spot, bacterial blight and anthracnose. The suggested approach accuracy result varies from 96.77% to 98.42%. Mainkar et al. (2015) describe a solution to find, diagnosis and segregate plant leaf diseases automatically using image processing techniques. They used GLCM, k-means and BPNN techniques, these accurate leaf disease detection is performed with a little computational effort. Mokhtar et al. (2015) proposed an efficient approach based on SVM technique to recognize two types of tomato leaf diseases, namely early blight and powdery mildew. The experimental result shows that the proposed scheme is able to correctly classify the diseased images with 99.5% accuracy.

Muthukannan et al. (2015) suggested a way for plant leave disease classification by adopting techniques, such as radial basis function networks (RBFs), feedforward neural network (FFNN) and learning vector quantization (LVQ). The proposed algorithms were tested for classification of diseased leaf images of bitter gourd and bean. The system performance is analyzed by adopting parameters like accuracy, recall ratio, precision and F_measure, so this approach provides better result. Naik et al. (2015) focus on image processing approach to detect and quantify the cabbage diseases. This system is for grading the cabbage and classification of cabbage diseases based on the numerical results obtained after the analysis of the cabbage diseased area by extracting color features from the image. Patil et al. (2011) discussed tomato leaf diseases color feature extraction; they used color as feature to extract the features of the images by calculating color moments CM1, CM2 and CM3, which are further used to classify tomato diseases. P. Priya et al. (2015) present a study to detect agricultural product diseases by using different feature extraction techniques.

Sabrol et al. (2016) suggested image processing and soft computing approaches for automatic recognition and classification of plant diseases. Five types of tomato diseases, namely bacterial leaf spot, bacterial canker, late blight, septoria leaf spot and tomato leaf curl, are considered in this work. Three classifiers, namely adaptive neuro-fuzzy inference system, fuzzy inference system and multi-layer feedforward backpropagation neural network, are adopted for categorizing healthy and diseased tomato. With multi-layer feedforward backpropagation neural network, the accuracy of 87.2% is achieved. Singh Malti et al. (2017) describe the study for detection of diseases in plant leaf of beans and tea by adopting image processing methods. The suggested approach consists of steps: image acquisition, pre-processing for image enhancement, segmentation for finding diseased area, feature extraction and classification. Singh Vijai et al. (2017) present an algorithm for image segmentation technique and the review on distinct disease classification approaches that can be used for automatic disease detection and classification of plant leaf. The suggested algorithm is tested on banana, jackfruit, beans, lemon, potato, mango, tomato and sapota. It is also used for the identification of diseases belongs to these categories. By this method, optimum results were obtained with less computational efforts. Sagar et al. (2007) proposed an algorithm for tomato plant disease detection by applying image segmentation and multi-class SVM for classification. They took four diseases, namely bacterial spot, early blight, Septoria leaf spot and iron chlorosis for testing their algorithm. Various parameters such as energy, entropy, correlation and homogeneity are used for classification purpose. This approach gives better classification accuracy which is 93.75%.

Related work to plant (fruit/vegetable) disease detection and classification:

Dhakate et al. (2015) propose neural network method for diagnosis of pomegranate fruit/leaves affected by different fungal and bacterial diseases. They focused on diseases like bacterial blight, leaf spot, fruit rot and fruit spot. The pomegranate images affected with these diseases are pre-processed for image enhancement, to figure out the diseased area segmented by applying k-means algorithm; texture features are extracted and fed to neural network for diagnosis of the disease. Pujari et al. (2014) considered powdery mildew fungal disease symptom affected on different products like grape, mango, chili, beans, wheat and sunflower, for disease recognition and classification purpose. The color and texture features are extracted thereafter fed to knowledge-based and ANN classifier. The average classification accuracy by applying ANN classifier for color features is 70.48%, for texture features is 70.07%, and for combined features is 76.61%, and using knowledge-based classifier has increased to 71.92%, 80.60% and 87.80%.

From the literature study, it is revealed that most prominently used features are contrast, correlation, energy, entropy and homogeneity, so authors have chosen these parameters to identify various plant diseases and excluded other features. To identify the various diseases of plants, the analysis of various parameters is mentioned below.

This work presents tomato fruit and leave disease identification using image processing techniques to monitor tomato diseases based on color space segmentation and feature analysis. In this work, various bacterial-fungal diseases of tomato are discussed and identified on the basis of the parameters evaluated by the feature extraction method. Six distinct types of diseases are focused in this work such as anthracnose, bacterial spot, bacterial canker, bacterial speck, early blight and late blight.

Fig. 1 Dataset images **a** Healthy tomato **b** Healthy tomato leaf **c** Bacterial canker diseased tomato **d** Bacterial speck diseased tomato

3 Materials and Methods

3.1 Dataset

For the dataset preparation, real images of both healthy/diseased tomatoes along with the leaves are collected from various farm fields of Solan district. By using a digital camera with 16 megapixel resolution, images are captured so as to have better image quality. The dataset consists of 100 images of fruit and leaves of tomato, and one of each kind is depicted in Fig. 1.

3.2 Methodology

The proposed methodology for disease identification and classification consists of following steps:

In image acquisition, real-time data of healthy/diseased tomato plants is taken by a digital camera from the various villagers of Shimla and Solan districts, engaged in the farming of these crops. As the real-time collected images may contain noise, so enhancing of these images is done. Pre-processing includes image resizing and contrast enhancement, all images are resized to 256*256 pixels, and by applying histogram equalization, the contrast of these images is enhanced. The segmentation approaches based on pixel locality are applied to segment tomato and its infected part. The resized and enhanced image is further segmented by implementing k-means clustering algorithm so as to get the diseased portion of the plant. The various features as obtained from literature review are calculated for healthy, diseased image and segmented portion of the image. The features extracted for various diseases were statistically analyzed, and variation in the features based on the type of diseases is studied. The type of diseases is depicted on the basis of these feature values. The distinct features like contrast, energy, correlation, entropy and homogeneity are computed. The description of these features is explained below [17, 19] (Fig. 2).

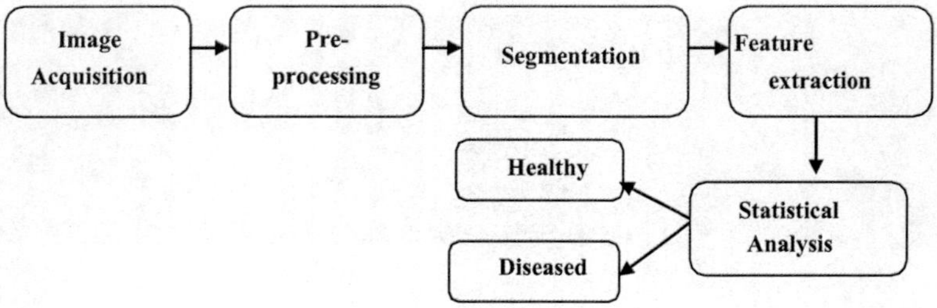

Fig. 2 Block diagram of proposed methodology [1–3, 7]

Contrast: It defined as the term which represents a measure of the intensity contrast between a pixel and its neighbor over the whole image. For a constant image, the value of contrast is 0.

Correlation: How a pixel is correlated to its neighbor pixels over the entire image is what correlation tells. For positively correlated image, its value is 1, and for negatively correlated image, its value is −1. The range of is correlation = [− 1 1]

Energy: Uniformity is measured by energy; more homogeneous the image is that means its energy value is larger. The image is supposed to be a constant image if the energy value nearer to 1. Range = [0 1] 1 for a constant image.

Entropy: Randomness of intensity image or disorder within a region is measured by the entropy.

Homogeneity: The similarity of pixels is measured by the term homogeneity; a diagonal gray-level co-occurrence matrix gives homogeneity of 1. For a diagonal GLCM, Range = [0 1] (Fig. 3).

3.3 Common Tomato Diseases

Tomato plants are susceptible to certain diseases, and cause of these diseases depends on many factors like plant variety, soil property, maintenance, the current season and the local climate (Table 1).

Table 2 explains the different diseases of tomato fruits with the description about the disease. Diseases that commonly affect tomato plants are listed in Table 2.

Table 2 shows various bacterial and fungal diseases of tomato with their symptoms and images. Having knowledge of symptoms related to specific disease, we can identify the diseases accurately. Bacterial canker disease symptoms can be seen as yellow or brown spots, slightly raised, surrounded by a persistent white halo ("bird's eye spot"). Bacterial speck disease symptoms appear as dark brown to black lesions on leaves and fruit. Bacterial spot disease indicates slightly sunken spots enlarge and turn brown. Anthracnose symptoms appear on ripe fruit as circular spots having center tan in color. The symptoms of early blight disease can be seen as small dark

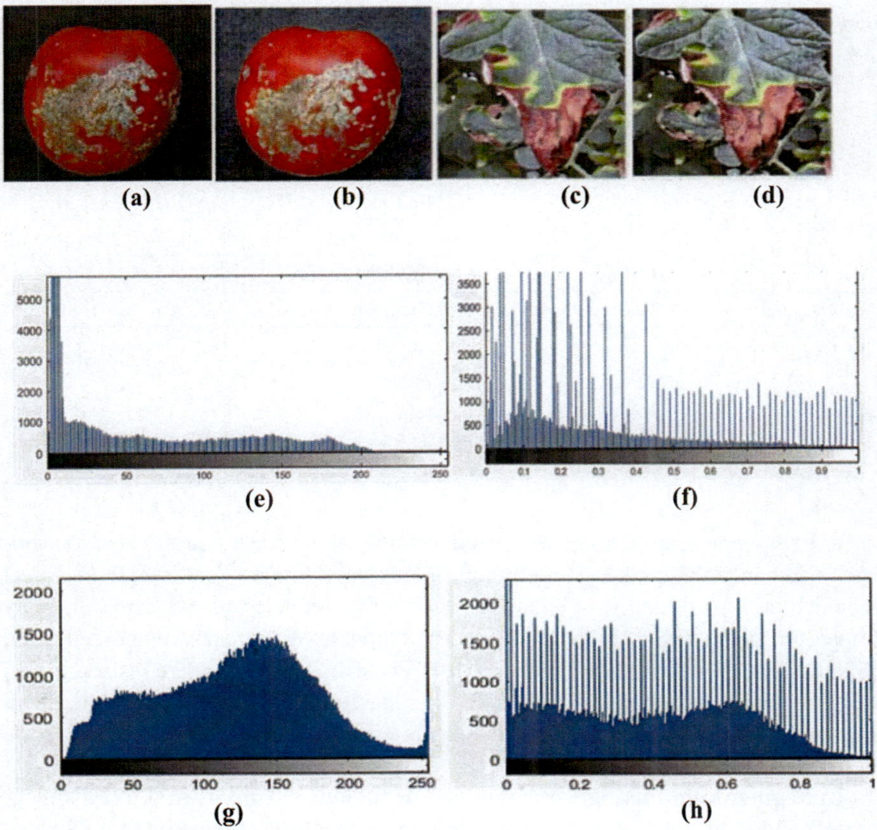

Fig. 3 Pre-processing results **a** Original tomato image **b** Enhanced image of original tomato **c** Original bacterial canker tomato leaf **d** Enhanced image of bacterial canker tomato leaf **e** Histogram of image-a **f** Histogram of equalized image-b **g** Histogram of image-c **h** Histogram of equalized image-d

spots consisting of concentric rings. Late blight symptoms on leaves have brown lesions on the upper leaf surface and on fruit green to dark brown spots appear [16] (Fig. 4).

4 Results and Discussion

MATLAB platform is adopted for conducting experiments on different images of healthy/diseased tomato by implementing various algorithms. Detection and categorization of healthy as well as diseased plants of tomato are achieved for selected bacterial/fungal diseases (Fig. 5).

Table 1 Various parameter analyses for tomato disease identification

Affected images	Features	Parameters	References
Leaves	Texture Features	Contrast	[6, 7, 10, 19]
		Correlation	[6, 10, 20]
		Energy	[6, 7, 10, 19, 20]
		Entropy	[6, 19, 20]
		Homogeneity	[5, 6, 9, 15, 16]
	Color Features	Color Moments	[12]
Fruit/Vegetable	Texture Features	Contrast	[5, 13]
		Correlation	[13]
		Energy	[5, 13]
		Entropy	[5]
		Homogeneity	[13]

(a) Pre-processing (Histogram equalization): A pre-processing phase is considered to enhance the quality of the input tomato leave images as the main objective of image pre-processing is to enhance, smooth and remove noise that is caused by defects of camera flash or high lights. Since the input images are raw images collected directly from the farms, they may be of different size, so images were resized to 256 $\times$ 256 resolution, so as to utilize the storage capacity or to reduce the computational time in the later processing [20]. In the pre-processing step, histogram equalization method has been applied to improve the contrast of the original image.

(b) Segmentation using k-means: Segmentation is done using k-means technique, and by this technique, enhanced original image is segmented into number of clusters (can be 3 or 5). The best cluster which depicts the disease clearly is taken for feature extraction process [21].

(c) Qualitative analysis

The purpose of extracting the features is to reduce the original dataset of each image including texture and color. The major five texture features extracted for normal as well as diseased tomato for the diseases, namely bacterial canker, bacterial spot, bacterial speck, anthracnose, early blight and late blight (Fig. 6).

Table 3 shows the average values of five extracted features for all the collected images infected by the five various diseases. For a normal/healthy tomato, the value of contrast is 0.344, correlation is 0.958, energy is 0.096, entropy is 7.466, and homogeneity is 0.882, and these average values are compared with the feature values of the diseased tomato and observed that if the feature values of contrast, energy and homogeneity are above and correlation and entropy are lower than the normal tomato feature values, then tomato is considered as diseased.

Healthy images are having high contrast value, which is up to 0.35 for our dataset. On the other side, different categories of diseased tomato are having lesser contrast value. Minimum value is identified for bacterial spot with an amount of nearly 0.2.

Table 2 Various Tomato Diseases affecting the fruit and leaf

Disease Category	S. No.	DiseaseName	Images: Tomato	Images: Leave	Symptoms
Bacterial Diseases	1	Bacterial canker			Fruit spots are like bird's-eye.
	2	Bacterial spot			On fruit small, dark slightly increased dots.
	3	Bacterial speck			Tiny, dark brown to black spots.
	4	Anthracnose			Circular, small slightly raised spots
Fungal Diseases	5	Early Blight			Black dark circular spots consisting of concentric rings.
	6	Late Blight			Green to dark brown lesions, cover large part of fruit/leaf.

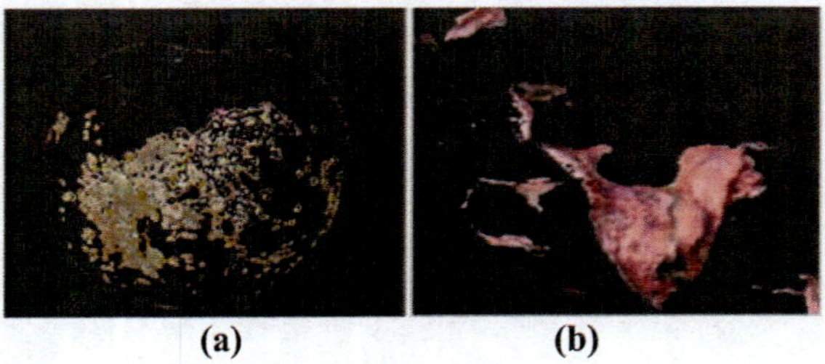

Fig. 4 Segmentation results **a** Segmented image of tomato affected by bacterial canker **b** Segmented image of tomato leaf affected by bacterial canker

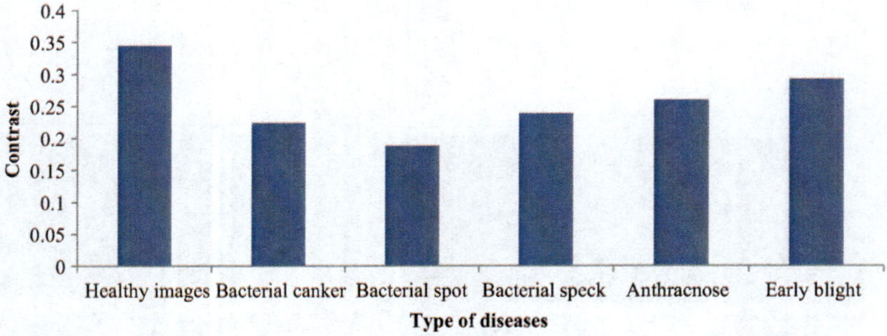

Fig. 5 Contrast feature for healthy and diseased tomato (fruit)

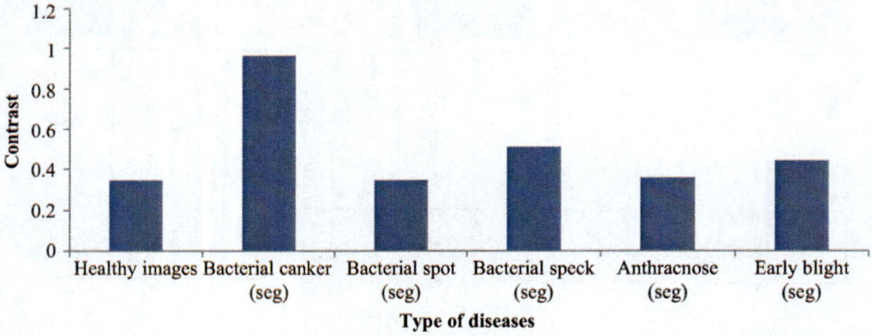

Fig. 6 Contrast for segmented area of diseased tomato (fruit)

The average values of contrast for the segmented infected area by various diseases are 0.958 for bacterial canker, 0.345 for bacterial spot, 0.509 for bacterial speck, 0.356 for anthracnose and 0.439 for early blight. It is observed that threshold value for contrast is 0.345, and above this value, tomato is said to be affected by the disease (Fig. 7).

Table 3 Features values (Avg.) for normal/diseased tomato

Affected Plant part	Normal/Diseased	Disease Name	Contrast	Texture Features (Avg. values)			Homogeneity
				Correlation	Energy	Entropy	
Tomato	Normal		0.344	0.958	0.096	7.466	0.882
	Diseased	Bacterial canker	0.958	0.962	0.882	7.466	0.949
		Bacterial spot	0.345	0.928	0.444	3.436	0.960
		Bacterial speck	0.509	0.929	0.400	3.676	0.951
		Anthracnose	0.356	0.920	0.539	2.723	0.958
		Early blight	0.439	0.937	0.458	3.250	0.951
Leaves	Normal		0.388	0.959	0.077	7.663	0.864
	Diseased	Bacterial canker	0.671	0.915	0.374	4.131	0.913
		Bacterial spot	0.979	0.833	0.442	3.216	1.274
		Bacterial speck	1.154	0.797	0.478	3.153	0.901
		Early blight	0.685	0.904	0.475	3.440	0.926

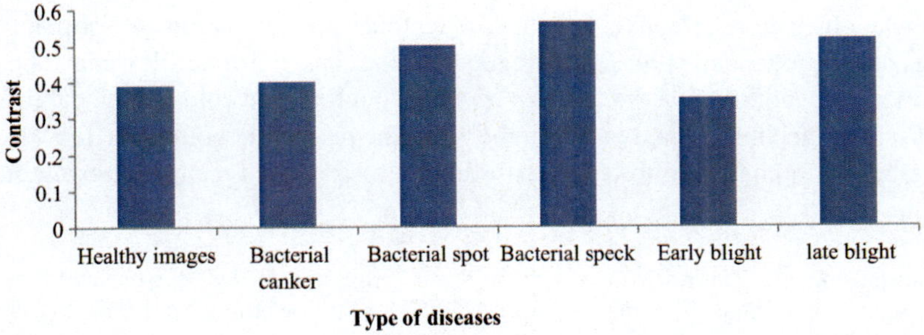

Fig. 7 Contrast for healthy and diseased tomato (leaves)

For a healthy tomato leaf, the contrast value is 0.388 for our dataset, and for various diseases, contrast value is higher than the normal/healthy tomato leaf.

From Fig. 8, it is concluded that healthy images are having contrast value up to 0.388 for our dataset and the correlation of diseased images is more than the normal images, and hence, these images are defected.

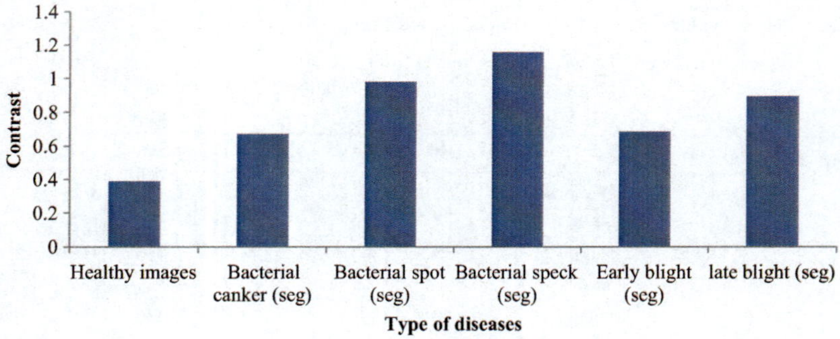

Fig. 8 Contrast for segmented area of diseased tomato (leaves)

5 Conclusion

This research aims at developing a vigorous framework for tomato crop, detecting their diseases and then classifying them. In this work, we are targeting tomato plants as a primary crop and detecting their main diseases. This approach is convenient and can significantly detect crop plant diseases with a little computational effort. This will help farmers to identify the disease and provide efficient solution for specific disease control. Our proposed framework can extract useful features from image and perform classification. By applying image processing approaches, tomato crop disease recognition becomes more effective. In future, techniques applied on tomato for plant and leaves may be extended other sets of vegetables like capsicum, cauliflower and beans. However, depending upon the size of vegetable, contrast and color's applicability of the discussed techniques may vary slightly. So, the researcher could take the various vegetables as a challenge to test the usefulness of proposed technique available in literature.

Acknowledgements This research work is a part of the project funded by Department of Environment, Science & Technology, Himachal Pradesh (DEST), under the project No. Env.S&T(F)/R&D project/2017/4657.

References

1. Abu-Naser, S.S., Kashkash, K.A., Fayyad, M.: Developing an expert system for plant disease diagnosis. J. Artif. Intell. **1**(2), 78–85 (2008)
2. Al-Bashish, D., Braik, M., Bani-Ahmad, S.: Detection and classification of leaf diseases using k-means-based segmentation and neural-networks-based classification. Information Technol. J. **10**(2), 267–275 (2011)
3. Arivazhagan, S., Shebiah, R.N., Ananthi, S., Varthini, S.V.: Detection of unhealthy region of plant leaves and classification of plant leaf diseases using texture features. Agric. Eng. Int. CIGR J. **15**(1), 211–217 (2013)

4. Chouhan, S.S., Kaul, A., Singh U.P., Jain, S.: Bacterial foraging optimization Based Radial Basis Function Neural Network (BRBFNN) for identification and classification of plant leaf diseases: an automatic approach towards plant pathology. IEEE Access **6**, 8852–8863 (2018)
5. Dhakate, M., Ingole, A.B.: Diagnosis of Pomegranate Plant Diseases using Neural Network, Computer Vision, Pattern Recognition, Image Processing and Graphics (NCVPRIPG). In: Fifth International Conference on Computer Vision, Pattern Recognition, Image Processing and Graphics (NCVPRIPG) IEEE (2015)
6. Kaur, I., Aggarwal, G., Verma, A.: Detection and classification of disease affected region of plant leaves using image processing technique. Indian J. Sci. Technol. **9**, 1–13 (2016)
7. Mainkar, P.M., Ghorpade, S., Adawadkar, M.: Plant leaf disease detection and classification using image processing techniques. Int. J. Innov. Emerg. Res. Eng. **2**(4), 139–144 (2015)
8. Mokhtar, U., Ali Mona, A.S., Hassenian, A.E., Hefny, H.: Tomato leaves diseases detection approach based on support vector machines. In: International Computer Engineering Conference (ICENCO) IEEE, pp. 246–250 (2015)
9. Muthukannan, K., Latha, P., Pon Selvi, R., Nisha, P.: Classifcation of diseased plant leaves using neural network algorithms. ARPN J. Eng. Appl. Sci. **10**(4), 1913–1919 (2015)
10. Naik, D., Shaikh, R., Shetti, S., Praveen, K., Kanakaraddi, S.G., Jahagirdhar, S.: Detection and quantification of disease in cabbage using clustering and RGB colour features. Int. J. Emerg. Technol. Comput. Sci. Electronics **14**(2), 194–199 (2015)
11. Patil, J.K., Kumar, R.: Color feature extraction of tomato leaf diseases. Int. J. Eng. Trends Technol. **2**(2), 72–74 (2011)
12. Priya, P., D'souza, D.A.: Study of feature extraction techniques for the detection of diseases of agricultural products. Int. J. Innov. Res. Electrical Electronic. Instrumentation Control Eng. **3** (2015)
13. Pujari, J.D., Yakkundimath, R., Byadgi, A.S.: Recognition and classification of produce affected by identical looking powdery mildew disease. Acta Technologica Agriculturae **2**, 29–34 (2014)
14. Sabrol, H., Kumar, S.: Fuzzy and neural network based Tomato plant disease classification using natural outdoor images. Indian J. Sci. Technol. **9**(44), 1–8 (2016)
15. Shankar, R., Harsha, S., Bhandary, R.: A Practical Guide to Identification and Control of Tomato Diseases (2014)
16. Shankar, R., Harsha, S., Bhandary, R.: A Practical Guide to Identification and Control of Pepper Diseases (2014)
17. Singh, M.K., Chetia, S.: Detection and classification of plant leaf diseases in image processing using MATLAB. Int. J. Life Sci. Res. **5**(4), 120–124 (2017)
18. Singh, V., Mishra, A.K.: Detection of plant leaf diseases using image segmentation and soft computing techniques. Information Process. Agriculture **4**(1), 41–49 (2017)
19. Vetal, S., Khule, R.S.: Tomato plant disease detection using image processing. Int. J. Adv. Res. Comput. Commun. Eng. **6**(6), 293–297 (2017)
20. Xie, C., He, Y.: Spectrum and image texture features analysis for early blight disease detection on eggplant leaves. Sensors **16**, 676 (2016)
21. Zaka-Ud-Din, M., et al.: Classification of disease in Tomato plants' leaf using image segmentation and SVM. Tech. J. Univ. Eng. Technol. (UET) Taxila, Pakistan **23**(2) (2018)

To Study the Effect of Varying Camera Parameters in the Process of Matching and Reconstruction

Anuj Kumar, Naveen Kumar, and Rakesh Kumar Saini

Abstract In computer vision, there are many methods of reconstructing the image point. 3D reconstruction from digital image sequence of scene or object is a difficult and important task in computer vision. However, such a reconstruction requires a large computational effort for finding the point of correspondence between different views. Furthermore, the accuracy should not be reduced in case of noisy data. There is one important technique to digitization of physical object which is binocular stereo vision. From two subsequent digital images of the physical object taken from different viewpoints, we can make a 3D virtual model for the physical object by using this approach. Basically, the common processes for binocular stereo vision comprise digital image acquisition, camera's calibration, feature point extractions, feature points matching and 3D reconstructions. In this paper, we have discussed the problem of varying camera parameter in the process of matching and reconstruction. we have the studied the problem and how it affects when the camera parameter varies in the process of matching; two types of parameter are as follows: intrinsic parameter and extrinsic parameter; image setup and some equation help to set the parameter, and a robust algorithm is proposed for reconstructing free-form space.

Keywords Binocular stereo · Epipolar constraint · Intrinsic parameter · Extrinsic parameter · Disparity map · Calibration

A. Kumar
Department of Computer Application, SRCM, Muzaffarnagar, India
e-mail: dixit.anuj12008@gmail.com

N. Kumar (✉) · R. K. Saini
School of Computing, DIT University, Dehradun, Uttarakhand, India
e-mail: naveen.it23@gmail.com

R. K. Saini
e-mail: rakeshcool2008@gmail.com

P. K. Singh et al. (eds.), *Recent Innovations in Computing*, Lecture Notes in Electrical Engineering 701, https://doi.org/10.1007/978-981-15-8297-4_40

1 Introduction

Most recently working areas are camera calibration and stereo matching systems based on the binocular stereo vision camera. In camera calibration, we have to estimate the parameter of camera that is very useful to acquire good 3D positioning reconstructions accuracy. This paper investigates camera parameters that inference the system reconstruction accuracy. The camera parameters associated with a stereo imaging setup can be classified into two main categories: (i) intrinsic camera parameters and (ii) extrinsic camera parameters. The effect of extrinsic camera parameters in a stereo imaging system can be tackled by the epipolar geometry constraint. However, intrinsic camera parameters, like focal length, lens distortion, image resolution, etc., interfere directly in the 3D reconstruction. In this context, it will be interesting to study the effects of these parameters in the 3D reconstruction.

In set of 2D digital image that considered as an input (arbitrary perspective views) and parameters of the viewing geometry, the output represents 3D object that defines by structural parameter of the model. This type of reconstruction is performed by input in the form of individual point (pixels), and groups of pixels represent some geometric feature in the image like as line, curve, etc. However, it is very difficult to process because the relative depth is missing during the projection of 3D scene on to the projection plane. In traditional method, 3D real-time depth information was extracted from image using sensor, whereas passive stereo vision model represents a significant error in processing un-texture region which is frequent in indoor world.

The 3D points lie on the line which starts from the center of projection, and it moves in the direction of its corresponding projected points in an image plane. We have only one image or view; the corresponding unique 3D points cannot be reconstructed as different 3D points lie on the same projected line. So if we want unique reconstruction, then we use at least two images. It provides the intersection of two projection lines through several projection planes that yield the unique 3D point.

There are following basics steps for stereo view 3D reconstruction process:

- Check the corresponding 2D features of the projected 3D scene or object onto the view planes.
- Find out the parameters of these projected 2D geometrical features in both view planes.
- Establish the correspondence between these projected 2D features.
- Use inverse perspective equations for computing the 3D structural parameters.
- Show the 3D object using line drawings or shading model.

The 2D projections or input image generate hurdles in the problem of reconstruction of a 3D object. These projections mainly depend on the relative position of the object with effect to the cameras, line of sight and other parameters of the cameras. The 3D reconstruction problem is divided into two subproblems: correspondence and triangulation. In the problem of correspondence includes the corresponding features in different images. It is formulated as: Given different images or views of a 3D object, find different points or features in one view which may be referred as the same points or features in other views. While the triangulation problem yields the location of 3D points from its two or more projections. The main difficulties in the field of multiview reconstruction involve: establishing the correspondence between pair of images and obtaining inverse mathematical functions based on camera models (perspective, weak perspective, affine, etc.)

Each object in real world is 3D, and this is one of the main issues in computer vision to acquisition of 3D information of real-world scene from 2D image. The acquisition is divided into two methods: active method and passive method. The passive method calculates the inverse problem of process of projecting a 3D scene onto a 2D image plane, and here is an advantage to find the three-dimensional information with no effect to the scene. And the active method emits the radio or light energy from source. The passive technique is referred to as motion, stereo, shading, texture, contours and so on. Basically, there are two categories of inverse problem: photometric (optical) and geometrical. And all these types of inverse problems are used for image formation process.

1.1 Binocular Stereo

Binocular stereo or two-view stereo is based on the person stereo vision, and it is a non-contact passive sensing method. A pair of images of a 3D object (scene) is obtained from two different viewpoints under perspective projection as illustrated in Fig. 1. To obtain a 3D object from these perspective images, a distance measurement (depth) from a known reference coordinate system is computed based on triangulation. The main problem in binocular stereo is to find the correspondences in stereo pair called the stereo correspondence problem. In this section, some concepts

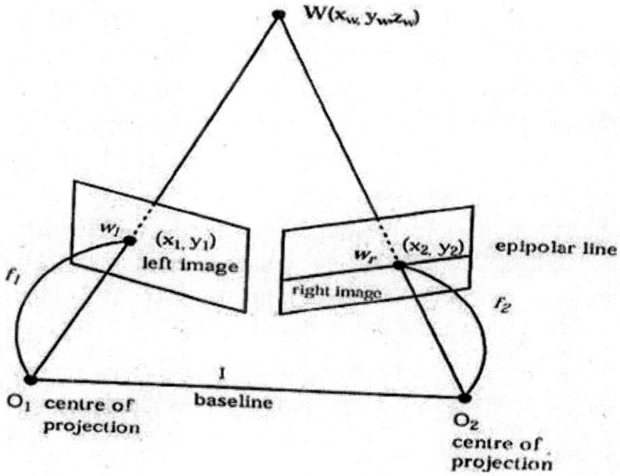

Fig. 1 Binocular stereo image formations

(camera model, epipolar geometry, rectification and disparity map) related to stereo correspondence problem have been introduced.

2 Literature Review

Several studies [1, 13, 16, 20] were carried out for a surface reconstruction from the scanned data. However, the study split into two methods: Voronoi based and mesh free. Voronoi methods are used with Delaunay triangle [5] to reconstruct the surfaces, and according to mesh-free algorithm, the surfaces are reconstructed by using B-splines [22], radial basis functions [18], and PDE and MLS [9] techniques. Along with these techniques, data point acquisition is proving to be a very useful step. In most of the hardware system [7, 21], a light wave technique was utilized in [7] for 3D surface point reproduction. A MS Xbox kinect sensor is used to get the volumetric depiction, and a similar kind of MS kinect sensor camera was used in [21] 3D inside scene recreation. Computational geometry devices were created in [1] for the surface demonstration from the valuable information. To again construct surfaces and to get correspondences all the while, some notable finite set of instruction [6, 11, 12] which depends on the volumetric reproduction is displayed from the 2D pictures existed in writing. A strategy for spaces carving and voxel's coloring dependent on the perceive ability and the stability of the voxels in the picture was created [12]. In the direct discrete minimization (DDM), details of graph cut methodologies were likewise proposed in [8, 11, 17] for building up the corresponding within the images. The entirety of the proposed strategies acquires disparity maps with exact shapes [2, 19] but limited depth precision. Be that as it may, this limitation was removed in [4]. Other calculation for the reconstruction is of 3D scenes from just two perspectives and depends on minimum line correspondences introduced in [15]. Another method for a precise 3D shape estimation of different separate items was exhibited in [10] utilizing stereo vision methodology. Actually, the excellence of the system is capacity of the system of playing out full-field three-dimensional shape estimation with high precision even within the sight of discontinuities and different separate districts. Numerous applications like acknowledgment, robot vision and activity require a substitute portrayal of three-dimensional questions in a minimized structure. Curve skeleton is an alternate representation from average surface representation [3]. Curve skeleton incorporates a few models: virtual route, enrollment, liveliness, transforming, logical investigation, shape acknowledgment and shape recovery. On account of 2D, skeleton is named as medial axis, but in 3D, these are called mean surface. Curve skeleton is an uncommon 1D representation of a 3D object whose recreation is additionally a troublesome subject because of a not-well-presented issue. Aside from the above representation techniques, endeavors were additionally made for reproducing the 3D surface utilizing skeletons of the object [7, 14].

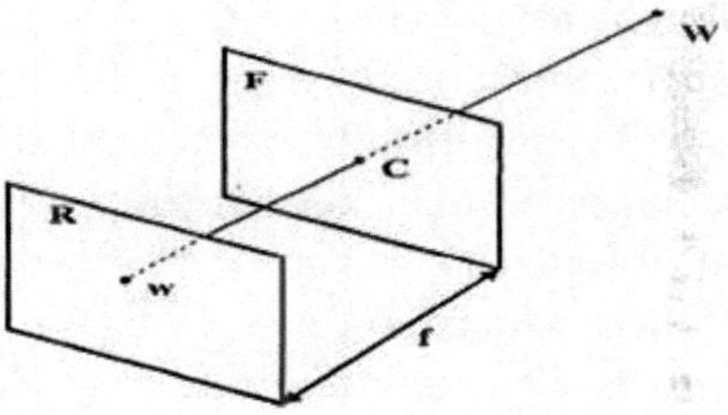

Fig. 2 Pinhole camera model

The Pinhole Camera Model.
The pinhole camera model describes the mathematical relationship between the coordinates of a 3D point and its projection onto the image plane.

As shown in Fig. 2, basically it consists of two screens: retinal plane and focal plane. The retinal plane (Re) is where the 2D image is formed; the focal plane (F) placed in center is called optical center C. Here f is the focal length of the camera based on both planes which are parallel from certain distance. A straight line connects the point W. It is world point. This point W mapped 2D image by using perspective projection. Since retinal plane and focal plane are parallel to each other, the points lie on focal plane and it has no image on retinal plane.

Epipolar Constraints:
The epipolar constraint makes sure that the epipolar lines are corresponded to the horizontal scan lines and here shifted horizontally all the corresponding point in both the images and reduced the two-dimensional search space into one-dimensional image.

If there are two cameras, every points of these cameras are $W = [x_w\ y_w\ z_w, 1]^T$ of the real-world reference frame can be projected to the suitable image frame (w_l and w_r) using the transformation matrix $P = \mathrm{TiT_l}$ as known form above equation.

$$W_r = P_r W,$$
$$W_l = PW$$

In order to correct the image according to the epipolar constraints, these left projection matrix P_l & right projection matrix P_r have to be used in the following particular conditions:

1. Focal length must be equal for both cameras which is necessary.
2. Focal plane must be same for both cameras which is also necessary
3. And the optical centers of camera need to be constant.
4. Each point in left image and right image needs to be same for correspondence of the vertical coordinate.

P_l and P_r can be further calculated using these, and obtained image can be transformed according to the epipolar constraint.

Rectification Process:
The rectification process is used to determine distance of an object in triangular-based stereo vision. However, binocular disparity is the process of corresponding depth of an object to change its position, especially in case of different camera view, when relative position is known for the camera.

It is a transform process that is used onto same plane image for two or more images. Generally, ratification is used for correspondence analysis. Correspondence processes the used rectification in very simple way. It defines two new perspective matrices that store the optical centers with the baseline contained in the focal planes. This ensures that the epipoles are at infinity, and hence, these epipolar lines are parallel.

3 Disparity for Three-dimensional Reconstructions

The disparity is the distance between two points in the rectified object that is image. It is actually applied in three-dimensional reconstruction, because it is proportional to the 3D world coordinate and distance between the cameras.

3.1 Disparity Map

Once both the stereo images are rectified, the next step is to find the correspondence between them. For this, between left image and the right image, it shows the similarity or different measures. **Sum of Square Differences** (SSD) used measure for are based, the **Sum of Absolute Differences** (SAD), the Normalized Cross Correlation (NCC) and the census. Here coordinates are defined as

$$\text{SSD (d)} = \sum_i \sum_j [I_l(x + d + i, y + j) - I_r(x + i, y + j)]$$

$$\text{SSD (d)} = \sum_i \sum_j [I_l(x + d + i, y + j) - I_r(x + i, y + j)]$$

$$\text{NCC }(d) = \frac{C(I_l I_r) - \sum_i \sum_j \mu_l \mu_r}{\sum_i \sum_j \sigma_l \sigma_r}$$

Here d represents the disparity between left and right images, and I_l and I_r represent the stereo pair in left and right images at a point (x, y) in the right image. μ_l and μ_r represent the mean intensities of left image as well as right image in the corresponding windows, and σ_l and σ_r are standard deviations in the windows, respectively. $C(I_l, I_r)$ is a cross-correlation between the corresponding windows:

Fig. 3 Disparity

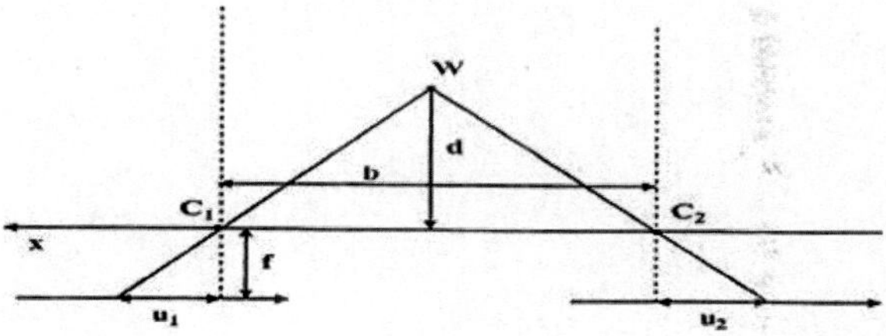

$$C(I_l, I_r) = \sum_i \sum_j I_l(x + d + i, y + j) I_r(x + i, y + j)$$

In Fig. 3, the disparity d can be found by minimize difference measures or by maximize similarity measures, but due to ill-posed nature of the problem, one cannot find the unique correspondence.

4 Camera's Parameters

As to change a point of the 3D world coordinate into a 2D purpose of the image planet, the information of quality camera is needed. Basically, two types of camera parameter, that is intrinsic and extrinsic, are also called internal and external, respectively. Intrinsic parameter defines the interior geometric and optical attributes of the camera, and extrinsic defines the position and direction of the camera in the world framework. The consequences for the framework reconstruction exactness are analyzed using results of simulation. At the end, we analyzed three parameter errors and system reconstruction errors, and the affected accuracy of the system is given. However, intrinsic camera parameters, like focal length, lens distortion, image resolution, etc., interfere directly in the 3D reconstruction accuracy which is somewhat bigger, and relationship of them is direct. The mistake in outer parameter straightforwardly impacts remaking exactness by influencing pattern separation and the point between cameras.

As observed in Fig. 4, the system for displaying at least two cameras comprises three separate coordinate scheme, (x_w, y_w, z_w) is the world reference frame and (x_c, y_c, z_c) is the camera frame with the focuses on origin and the imaging frame (X, Y). A 3D point, given in same world directions, can be changed over into the camera frame, by a revolution r_{ij} and an interpretation t_j which is communicated by the outward parameter as

$$\begin{pmatrix} x_c \\ y_c \\ z_c \end{pmatrix} = T_l \begin{pmatrix} \mathrm{x_w} \\ \mathrm{y_w} \\ \mathrm{z_w} \end{pmatrix} \text{Where } T_l = \begin{pmatrix} \mathrm{r_{11}\ r_{12}\ r_{13}\ t_1} \\ \mathrm{r_{21}\ r_{22}\ r_{23}\ t_2} \\ \mathrm{r_{31} r_{32}\ r_{33}\ t_3} \end{pmatrix}$$

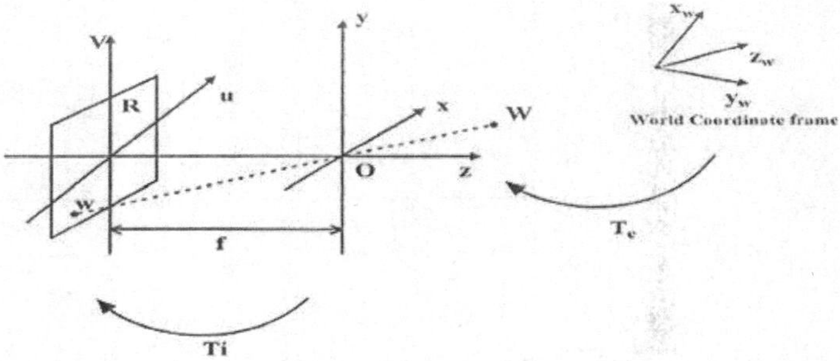

Fig. 4 Intrinsic and extrinsic parameters of camera

(I) Intrinsic Parameters:

It is the most important parameter of the camera that for each camera, it characterizes the change from picture plane coordinate to pixel coordinate. If the intrinsic camera parameters are known, it is possible to obtain a metric reconstruction. If you obtain matrix reconstruction, it is necessary know intrinsic parameter. This metric has five parameters; this calibration can be obtained through off-line calibration with a calibration object.

At that point, it is changed over to the two-dimensional picture planes utilizing the intrinsic coordinates. These are specifically the central length f, the rule point (u_0, v_0), which is the focal point of the picture plane, and (k_0, k_1) the size of pixels in millimeter (mm) or $\alpha = f/k_0$ and $\beta = f/k_1$. The change utilizing inherent parameter is as per the following:

$$\begin{pmatrix} X \\ Y \\ Z \end{pmatrix} = T_i \begin{pmatrix} \mathrm{x_c} \\ \mathrm{y_c} \\ \mathrm{z_c} \end{pmatrix} \text{ Where } T_i = \begin{pmatrix} \alpha 0 u_0 \\ 0 \beta v_0 \\ 001 \end{pmatrix}$$

Since (X, Y, Z) is uniform, and Z variables are divided X, Y, Z variable in order to find out coordinate of pixels X' & Y' and on the focal plane the points are $Z = 0$ and $\mathrm{z_c} = 0$. These pixels cannot be changed into image plane coordinate because it is divided by zero that is not define and this point connected by in straight; by this reason does not intersect by optical center to image plane, whereas image plane is parallel to it. There are following equations that given a point in world coordinate to convert into 2D image plane.

$$\begin{pmatrix} X \\ Y \\ Z \end{pmatrix} = T_l T_i \begin{pmatrix} x_w \\ y_w \\ z_{\;w} \\ 1 \end{pmatrix}$$

The information given by internal and external parameter of camera permits for rectifying of images and ensures the epipolar constraint.

(II) Extrinsic Parameters (R, T)

The extrinsic parameter describes the relative position and orientation of the two cameras. It defines the 3D world coordinate to 3D camera coordinates. It also defines the position of the camera's heading in world coordinates and camera center. R represents the rotation matrix, and T represents the position of the origin of the world coordinate.

5 Calibration

Stereo vision is used to find out the depth from camera image when we compare one or more views of same scene. When we compute the result, each 3D point corresponds to a pixel in one of the images. Binocular stereo process applies on only two images, which taken with cameras that are differentiated by a horizontal distance called "baseline." Calibrating the stereo camera defines the actual unit of 3D world points such as millimeters, relative to the cameras. So many applications of this process include measurement, aerial photogrammetric and robot navigation. The calibrated stereo camera combination is of following steps.

1. To the calibrated stereo camera system
2. To correct a pair of stereo images
3. To compute disparity
4. And reconstructing the three-dimensional point cloud.

5.1 Calibration of a Stereo Camera

The calibration of stereo camera provides the intrinsic estimation of every camera. It involves the translation and rotation of the second camera corresponding to the first camera. All the parameters of camera are used to rectify a stereo pair of image and to make the two images parallel. The rectified image calculates the disparity map required to reconstruct 3D view. When you calibrate the stereo method, you drew multiple pairs of a calibration design from different views. The whole design of calibration can be shown in every image. You must save the image in a format that not has use lossy compression such as PNG format. JPEG image format is not in lossy format. The lossy format reduces the calibration accuracy. When the camera is set in parallel position, then epipolar line turns into parallel. This means that, the left image point is exactly same in the right image on the line (Fig. 5).

When you want to calibrate your stereo camera, then the camera lenses make some distortion. This distortion in the image makes straight lines in the real world appear curved (Fig. 6).

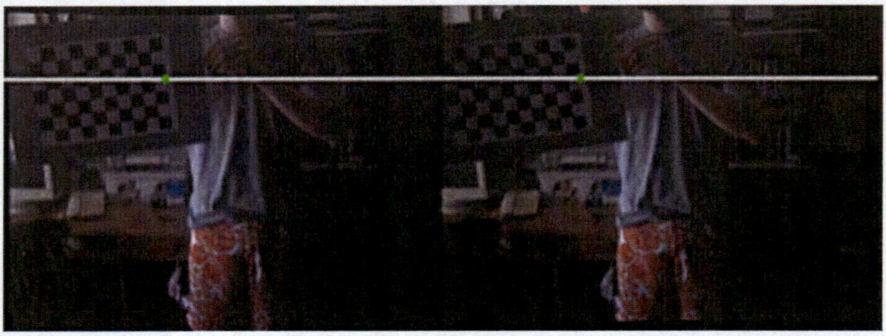

Fig. 5 Calibrated image

Fig. 6 Uncalibrated stereo images

The above two images are taken from uncalibrated stereo. In this picture, the cameras are not perfectly parallel that mean more or less parallel.

6 Experimental Results

A stereo imaging system was simulated in order to observe the effect of stereo rig parameters in the reconstruction error. We have given our emphasis on the two parameters mainly:

1. Baseline distance
2. Focal length.

In order to see the effect of these two parameters, we have considered the reconstruction of a quadratic curve, in which these parameters have been perturbed to see the effect on the 3D shape of the curve in stereo reconstruction.

The original 3D plot of the synthetic shapes has been shown in Fig. 7. Generally, the projected images in both views correspond to the different samplings of 3D object. Thus, we have evaluated the reconstruction algorithm with those synthetic objects whose data points correspond to different samplings of the 3D free-form objects. We

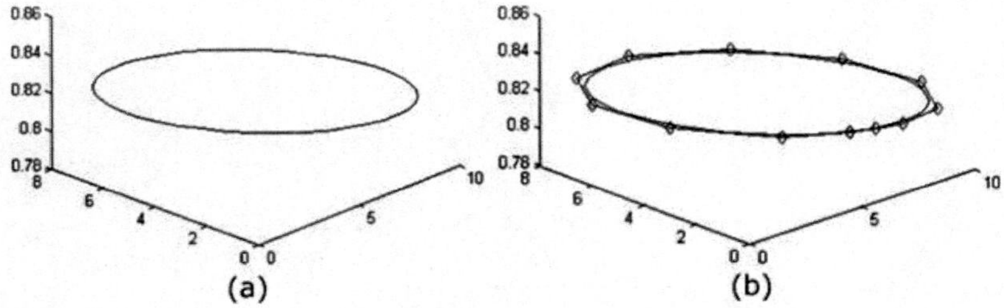

Fig. 7 Original and reconstructed elliptical curve in space with perturbed stereo parameters

Table 1 Effect of focal length

Change in focal length (in percentage)	Absolute error in reconstruction of an ellipse (voxels)	Absolute error in reconstruction of a helix (voxels)
1%	0.25	1.40
2%	0.98	3.58
5%	3.5	7.88
10%	17.25	32.40

have added the white Gaussian noise with varying standard deviations to the original data and perturbed the focal length and baseline distance a little bit in order to access the effect on reconstructed result.

In this experiment, the original and reconstructed objects are found to have a similar shape. The proposed algorithm is equally effective for the reconstruction of curve in case of a small variation in the stereo rig parameters. Moreover, from a qualitative point of view, we consider the following two tables to evaluate the effect of stereo parameters in the reconstruction of an ellipse and a helix using stereo vision technique.

Table 1 represents absolute error in reconstruction of an ellipse and helix when changing focal length parameter of camera (Fig. 8).

In Table 2, we have shown the mean error in reconstruction of an ellipse and helix when changing the baseline distance parameter of camera (Fig. 9).

7 Conclusion

In this paper, we studied the effect of varying camera parameters in the process of matching and reconstruction and gave some results. This paper shows if the recent methods are adopted using inverse geometric reconstruction, then the problem can be modeled as a stereo vision where the effect of stereo rig parameters is negligible if the deviation in the parameters is quite small. However, the error increases significantly

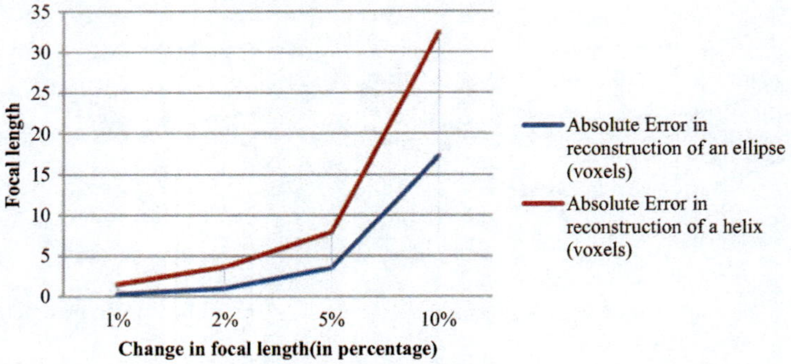

Fig. 8 Focal length versus change in focal length percentage

Table 2 Effect of baseline distance

Change in baseline (in percentage)	Mean error in reconstruction of an ellipse (voxels)	Mean error in reconstruction of a helix (voxels)
1%	1.08	1.58
2%	3.42	3.14
5%	6.22	11.00
10%	33.44	38.96

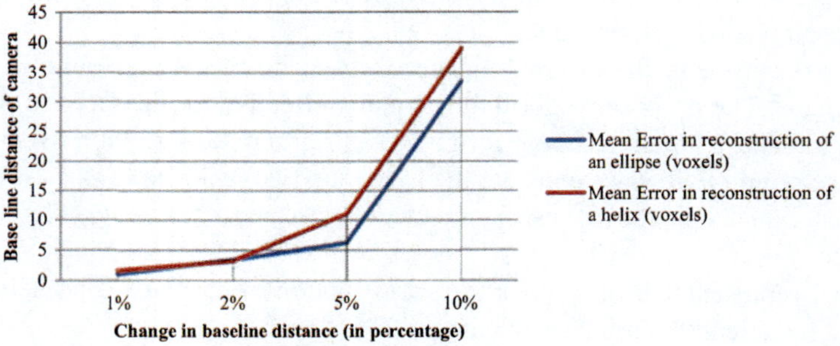

Fig. 9 Baseline distance of camera versus baseline distance of camera in percentage

if there is a large perturbation in the stereo rig parameters. In future, this problem can be studied in real-time stereo. Such an implementation will be useful in many applications such as ball trajectory tracking in sports event, missile and robot path planning.

References

1. Amenta, N., Choi, S., Kolluri, R.: The power crust, unions of balls and the medial axis transform. Computational Geometry: Theory and Applications **19**, 127–153 (2001)
2. P. K. Atrey, A. De, and N. Rajpal. Polynomial representation of 2-D image boundary contours. In IEEE Region 10 Conference on Computers, Communications,Control and Power Engineering (TENCON '02), volume 1, pages257–260, Beijing, China, 2002.
3. Bhatt, D., Mohammad, S.: Validation on medial axis transforms objects. Computer-Aided Design & Applications **9**(4), 517–529 (2012)
4. Y. Boykov and V. Kolmogorov. Computing geodesics and minimal surfaces via graph cuts. In International Conference on Computer Vision, 2003.
5. F. Cazals and J. Giesen. Delaunay triangulation based surface reconstruction: ideas and algorithms. Springer, 2006.
6. O. Faugeras and R. Keriven. Complete dense stereovision using level set methods. In European Conference on Computer Vision (ECCV), pages 379–393,
7. E. Frigerio, M. Marcon, and S. Tubaro. Surface reconstruction using 3D morphological operators for objects acquired with a multi-kinect system. In MIRAGE, volume 9, 2013.
8. H. Ishikawa and D. Geiger. Occlusions, discontinuities, and epipolar lines instereo.In European Conference on Computer Vision (ECCV), 1998.[63] I. Isuppli. The teardown: The kinect for xbox 360. IET Engineering Technology, 6(3):94–95,
9. M. Kazhdan, M. Bolitho, and H. Hoppe. Poisson surface reconstruction. In Proceedings of the Fourth Euro graphics Symposium on Geometry Processing, 2006.
10. Kieu, H., Pan, T., Wang, Z., Le, M., Nguyen, H., Vo, M.: Accurate 3D shape measurement of multiple separate objects with stereo vision. Measurement Science Technology **25**, 1–7 (2014)
11. V. Kolmogorov and R. Zabih. Multi-camera scene reconstruction via graph cuts. In European Conference on Computer Vision (ECCV), 2002.176
12. Kutulakos, K.N., Seitz, S.M.: A theory of shape by space carving. International Conference on Computer Vision, Kerkyra **1**, 307–314 (1999)
13. M. S. Lee, G. Medioni, and P. Mordohai. Inference of segmented overl aping surfaces from binocular stereo. IEEE Transactions on Pattern Analysis and Machine Intelligence (PAMI), 2002.
14. J. Liu, X. Zhang, and F. Blaise. Distance contained skeleton for volume reconstruction. In Proceedings of Asian Conference on Computer Vision (ACCV2004), pages 581–586, 2004.
15. Mosaddegh, S., Fofi, D., Vasseur, P.: Two view line-based motion and structure estimation for planar scenes. Electronic Letters on Computer Visionand Image Analysis **11**(1), 28–40 (2012)
16. P. J. Narayanan, P. W. Rander, and T. Kanade.Constructing virtual worlds using dense stereo. In European Conference on Computer Vision (ECCV), Bombay, 1988.
17. S. Roy. Stereo without epipolar lines: A maximum-flow formulation.International Journal of Computer Vision, 34:147–161, 1999.
18. M. Samozino, M. Alexa, P. Alliez, and M. Yvinec. Reconstruction with voronoi centered radial basis functions. In Proceedings of the Fourth Euro graphics Symposium on Geometry Processing, pages 51–60, 2006.
19. Singh, V.K., Atrey, P.K., Kankanhalli, M.S.: Coopetitive multicamera surveillance using model predictive control. Springer Journal of Machine Vision and Applications **19**(5–6), 375–393 (2008)
20. C. K. Tang and G. Medioni.Curvature-augmented tensor voting for shape inference from noisy 3D data. IEEE Transactions on Pattern Analysis and Machine Intelligence (PAMI), 2002.
21. Tiangang, S., Zhou, L., Xinyang, D., Yi, W.: 3D surface reconstruction based on kinect sensor. International Journal of Computer Theory and Engineering **5**(3), 567–573 (2013)

22. H. Wendland.Scattered data approximation, volume 17.Cambridge University Press, 2005.
23. G. Zeng, P. Sylvain, L. Quan L., and L. Maxime. Surface reconstruction by propagating 3D stereo data in multiple 2D images. In European Conference on Computer Vision (ECCV), Berlin, pages 163–174. Springer-Verlag, 2004

E-Learning Cloud and Big Data

Loop Holes in Cookies and Their Technical Solutions for Web Developers

Talwinder Singh and Bilal Ahmad Mantoo

Abstract Session hijacking is the term used to describe the theft of session cookies, i.e., sniff the cookies and use those to impersonate the end user. A cookie is a small-sized text file sent by the Web server to the user's browser and is store at the client side. When a user visits a Web site first time, the Web server generates a fresh cookie. The Web site uses that cookie to track the movements of an authorized user. Main threats of cookies are session fixation attack, cross-site scripting (XXS) attack, session sniffing attack, cookies cloning attack, and cookies controlling malware. The hacker sniffs the network traffic for cookies and uses same to impersonate the user. With performing session hijacking attack, the attacker acts as actual user on Web. In this paper, we are going to discuss some of the technique that helps in optimizing the cookie attacks in Web applications.

Keywords Cookies · Session id · Session hijacking · XSS · Session sniffing · Session fixation

1 Introduction

A Web application is a user-interactive program that runs on the Web server [18], the user interacts with the Web server using online forms, usually JavaScript requests, shopping cart, etc [3]. The Web server provides credentials (a combination of unique id and password or key) to each user for their identification on the Web server. When a user authenticates on the Web server with his credential (i.e., user id and password), the Web server creates a unique identifier other hand session identifier. Web server stores a cookie at the client side which contains the session identifier only; the status of the user (i.e., login or logged out) is also stored at the server side, in their session variable by "setPerimeter" method. Using the "setPerimeter" method, key and value are stored at the server (in server's log). After establishing the session identifier, users do not require to enter their credentials, i.e., user id and password, every time. The

T. Singh (✉) · B. A. Mantoo
Central University of Punjab, Bathinda 151001, Punjab, India
e-mail: talwindermann41@gmail.com

P. K. Singh et al. (eds.), *Recent Innovations in Computing*, Lecture Notes in Electrical Engineering 701, https://doi.org/10.1007/978-981-15-8297-4_41

user sends each request to the server with his/her unique identifier (session id), and server identifies the user with their session id. The session identifier is valid until the user closes the Web app or logout from the Web application.

Cookies: A cookie is a small piece of file stored at client side in the Web browser. Cookies contain user's session id or personal information of the user (e.g., user preference, age, language preferences) [16]. Typically, cookies are used by the Web server to personalize the dynamic Web pages for the particular client. For example, when a user first visits on any Web app or site, the Web server of the same generates a unique id for a particular user (known as session id). Session id also placed in a cookies store at the user's browser [12]. When the user visits again on the same Web site, then his/her browser sends the request with their unique session id, the Web server identifies the user with the user's session id only. There is no requirement of user id and password to identify the user after session establishment.

According to their functionalities, the cookies are following types:

1.1 Session Cookies

Session cookies are used to store the session identifier at the client side. These are used by Web servers to know the user's status (log in or logged out) on the Web site. When the user closes the browser, it automatically deletes. Sometimes, session cookies are used to count the user's clicks, like online shopping, etc. In an online shopping site, when the user selects an item then it will be shown in their cart. When the user lost their session or closes the browser and then revisits the same site, then the user checks that the cart is empty.

1.2 Persistent cookies

Persistent cookies are stored for a long period of time [6]; it is not deleted when the user closes the browser. Persistent cookies are deleted after a specific time (i.e., 6 months or after year) as shown in Fig. 1. When the user visits a Web site, the first time it has a default value, whenever he/she personalizes the site to fit his/her preferences (like language, menu preference, bookmarks, or favorites of visiting Web sites, etc.), then persistent cookies will be updated accordingly. Persistent cookies are used to remember the user's preferences next time to visit the same site [8, 20].

1.3 Third-party cookies

Third-party cookies are those cookies, which are written by third-party Web sites, which may not visit by the user. These types of cookies are used to collect and store

Name	Value	Domain	Path	Flags	Expiry date	
JSESSIONID	19mv60u564xq31wz0ve2dbpvpo	172.16.16.16	/corporate		At end of session	Edit
PHPSESSID	b4qc2k5hkuq9h2o3anefhjb111	14.139.60.229	/		At end of session	Edit
OGP	-5061451:	.google.com	/		13/Sep/2018 14:30:26	Edit
1P_JAR	2018-08-14-09	.google.com	/		13/Sep/2018 14:33:56	Edit
NID	136=oPOcYmVZOfUFERuRGaVBfYIHNVSejMOz_M-Fd-T HoDqn3cJThQ7RI0McFcAe93kQoy2S1r5god7YcJMPu4TRF	.google.com	/	httpOnly	13/Feb/2019 14:34:00	Edit
GOOGLE_ABUSE_EXEMPTION	gmail=CoMBAAIriVdGxCMuAREnbOIAl2WLIoCnAYVVL m-CilhFjNsXzqgjGP7XTXLPpRJ9rcwIKgqCA2Haw9mlegD	.google.com	/		At end of session	Edit
AID	AJHaeXIXyYmkIReKwKGiiSyOh1cKxy4p0qR77nhQ84Y	.google.com	/ads	httpOnly	5/Feb/2020 13:30:00	Edit
COIC	CgxmaXJlZm94LWItYWIiP3RleHQvaHRtbCxhcHBsaWNhd Glvbi94aHRtbCt4bWwsYXBwbGljYXRpb24veG1sO3E9MC	.google.com	/complete/search	httpOnly	13/Feb/2019 04:14:05	Edit

Fig. 1 Various parameters of cookies shown by "cookie-manager" browser's extensions

information about the user's online searching or his/her interests. When the user visits on a Web site, the third-party cookies collect various user's searching/interest information and sold collected information to the advertisers [4]. These types of cookies typically collect marketing-relevant information like age of user's, gender, his/her location, user's occupation, interests, what he/she searched on the Internet, etc., so the marketers provide his/her custom or relevant advertisements. For example, when we search on the Internet, i.e., jobs, products, or loan then we receive emails from advertisers containing information.

1.4 Super cookies

Super cookies are browser cookies, which are stored permanently on the user's computer. These cookies are used to track the technologies; these do not rely on HTTP cookies. The super cookies cannot be deleted the same as regular cookie [2, 6]. These types of cookies are used to store the browsing history, ads related data [3, 14], and authentication data.

1.5 Zombie cookies

Zombie cookies/ Ever cookies are HTTP cookies, that are automatically regenerated after deletion [6]. It may be stored online into the visitor's browser. Web analytical companies use these cookies to trace Internet usage or searching (page visited) for marketing research [3].

2 Structure of Cookie

There are six parameters of cookies. These parameters are as follow [6]:

i. Name of cookie, i.e., PHPSID, JSESSIONID, AID, etc. The cookie's name is shown in Fig. 1; name is used to identify the cookie at client side.
ii. Value of cookie, i.e., 16 or 32-bit alphanumeric string for session id. But in the case of persistent cookies value is an encrypted text (may more than 16/32 bits).
iii. Domain of the server from where the client's browser sends the cookie, i.e., 14.139.60.229 and .google.com as shown in Fig. 1.
iv. The server's path, where the browser sends the cookies. The default value (path) of this parameter is "/". Both path and domain parameters are used to determine the scope of cookies.
v. The demand for flags to activate in the cookie, there are two flags available in cookies, i.e., "httpOnly" and "Secure".

   a. If "httpOnly" flag is active, then JavaScript methods at the client side cannot access the particular cookie. The main advantage of this flag is that the cookie is not affected by XSS attack.
   b. If the flag "secure" is active, then cookies are being encrypted. If any attacker grabs or sniffs the cookie, then he is not able to read the cookie, thus the "secure" flag insures that cookie data is safe [6]. On the other hand, if "secure" flag is active, then the man-in-the-middle (MITM) attack may not be possible on the same cookie [9]. If between the Web server and browser HTTPS (Secure connection) is established, then secure flag is active.

vi. Expiration time of cookies. Two types of deadline, one is "At end of session" (for session cookies) [11] and the other is "specific date" (for persistent cookie) [20]. Persistent cookies show specific date to expire (shown in Fig. 1). A persistent cookie automatically expires on this specific date, and session cookies are automatically invalidated when user close the browser.

3 Web Session

Web session is temporary information exchanged between the client's browser and the Web server, involving with each request and responses (exchange between client and server) [11]. When a user visits a particular Web site the first time, the Web server generates a unique identifier (session id) for each user to differentiate their all activities. When the user successfully authenticates on the same Web site by their credentials (i.e., username and password) [12], the status of user, user id, user name, etc., are added in their session variable. At the client side, cookie only contains reference no. of the session variable (i.e., session id), other data of the user like his status on Web site, password, username, etc., all are stored at the server side. The client's browser will get the session id to identify its session with the server. However,

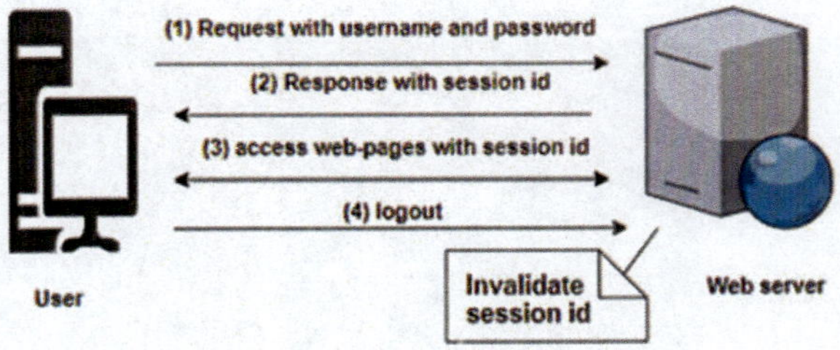

Fig. 2 Life cycle of session creation to session invalidation

the session information is temporary, whenever the user clicks on the logout button of the Web site, the Web server kills the Web session by invalidating the session identifier.

Figure 2 shows the full life cycle of session creation to session invalidation. In first step, the user sends the request (with login credential) to the Web server, if the user is authenticated successfully, then the server creates session id and responds this session id to the user (step 2).

The session id is also stored in cookies; user accesses the Web site with their established session id. Web server identifies the user with their session id and responds accordingly, i.e., server responds only to the actual user (because user name is obtained from session). When the user sends the request for logout, the Web server kills the session or invalidates the session id (step 4).

4 Main threats of cookies

4.1 *Sniffing for cookies*

There are so many Web sites which use HTTP protocol to transfer the packets between client and server. Most Web sites use HTTPS only for login, after this they continue with HTTP protocol. In HTTP protocol, cookies are visible and are prone to attacks such as session hijacking attack. The attackers capture the packets that pass between client and server, for getting cookie which contains session id. To capture the packets, following tools are used.

a. Wire shark
b. Fiddler
c. Ferret
d. Cain and able
e. kismet.

Figure 3 shows the sniffed cookie by ferret. The ferret is tool of Linux used to grab the cookies passing between client and the Web server.

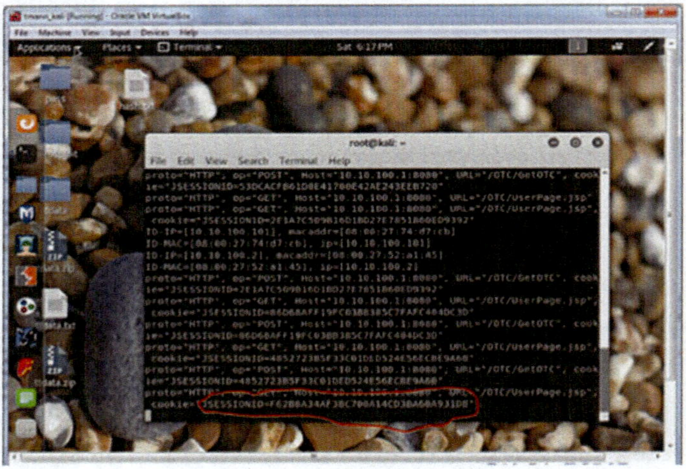

Fig. 3 Sniffing cookies by ferret

4.2 Cross-Site Scripting (XSS) attack

This type of attack steals the cookies at client side. In this, attacker makes malicious JavaScript code which runs at victim's machine to collect the cookies and send these cookies to attacker [4]. The XSS attacks are of three main types:

Stored XSS: The attacker permanently stores his/her JavaScript code at target serer, i.e., log file, database, or comment field [6]. When user accesses the affected Web page, then malicious code will execute at client side [13].

Reflected XSS: The malicious code (attacker's payload) is also a part of victim's request which is sent to the Web server. The Web server reflects back the response including malicious code from the HTTP request [6, 14]. The reflected malicious code is executed at victim's side. The attackers use phishing emails, social engineering attempts, and malicious links to include the XSS payload in victim's request [13].

DOM- based XSS: In this attack, the malicious code is not sent to the Web server [6], the code is at the client side. This attack is possible if the client-side script of the application is responsible to write the data by Document Object Model (DOM) [15]. The attacker injects his/her payload in DOM, when the data is read from the DOM then payload is being executed [1].

4.3 Session fixation attack

This attack exploits the vulnerability of the session id management system of the Web application. At the Web server, when the user authenticating successfully, the

Fig. 4 Steps of session fixation type attack

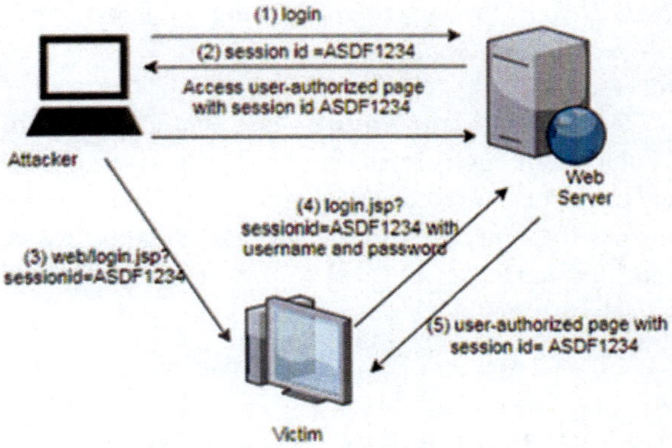

Web server set values in the old session variable (i.e., same session id), it does not create a new session variable for the user [17]. In this type of attack, the attacker establishes the reference no. of session variable (i.e., session id) in the session cookie, before the login attempt of the user. Typically before the user's login, the session fixation attack may start [15, 17]. The basic execution of session fixations attack is explained in the following steps with example:

i. First of all, the attacker visits the login page
ii. The Web server places the reference no. of a session variable (session id) in the cookie at the attacker's machine. As shown in Fig. 4, the Web server replies to the attacker with a session id = ASDF1234.
iii. The attacker makes a link for login containing same session id (i.e., ASDF1234) and sends the same link to the victim.
iv. The victim enters their credentials on the attacker's link and sends the request to the server including attacker's established session id without changing it (as shown in Fig. 4).
v. After authenticating the victim's, the Web server does not create a new session variable; it assigns the values (i.e., status logged in, user id, etc.) in the same session variable. As shown in Fig. 4, the Web server responds the authorized home page to the actual user with the same session id.
vi. After successfully authenticating the victim on Web server, the attacker may able to access the victim's authorized Web page with the same session identifier which he/she already knows.

4.4 *Data theft attack*

In these types of attacks, the attacker may directly gain the cookies (which contain session identifier) from victim's hard drive [19]. User's cookies are also stored in "user" directory in their computer. In case of Google Chrome browser, the cookies are stored in "Cookies.file" residing in path: *AppData\Local\Google\Chrome\User*

Data\Default. And in case of Mozilla Firefox cookies are stored in "Cookies.sqlite" file residing in: *AppData\Roaming\Mozilla\Firefox\Profiles\nrcwpjdj.default* directory.

There are various ways to gain the cookies from victim's hard drive, i.e., by PenDrive, through email, remote sharing over network or Internet, malware attack, etc [2]. For example, an attacker may sit behind the user, when the user not accessing his/her machine, then the attacker directly copies cookies of the user from their machine and make clones of cookies using grabbed information at attacker machine to impersonate the victim.

4.5 Predictable Session Identifier

The attacker understands the session identifier's creation process and structure of session id; an attacker can generate or predict a valid session identifier's [11] and access the Web application services using the same predicted session identifier. Some Web servers generate session identifiers using predictable information, i.e., username, IP address, timestamp, etc. [7] which is vulnerable.

Also, the attacker can generate the session identifier using brute-force techniques and test the same, until he/she successfully gains authorized access to the Web server.

After getting the valid session identifier it is time to make a clone of cookie. The cookies are cloned by hamster or browser extensions. Hamster is a tool of Linux, which replaces the session cookies with grabbed cookies. On the other hand, the attacker is able to make the clone of cookies using browser's extensions, i.e., cookie editor, cookie manager, etc.

If an attacker makes a clone of the cookie and sends a fake request to the Web server, then Web server responds user authorized page to the attacker. So, the attacker successfully impersonates the actual user.

5 Remedy for session hijacking

If a session id is stored in cookies, then lot of security issues arise. If an attacker successfully gains the active session id of the victim's, then attacker is able to perform the session hijacking attack. Therefore, we require securing the session identifier or session cookie with the following suggestions.

5.1 Use HTTPS protocol

Using full HTTPS/SSL support, the cookies are protected from session sniffing attacks. Some Web sites like online banking or e-commerce Web sites offer their services with an encrypted connection, because they do not want to lose any financial information to hackers [11]. In this, the hackers cannot sniff the session identifier to perform session hijacking attacks, because the data is transferred over the encrypted network [9]. This approach reduces the chances of sniffing on the network level, but the approach does not guarantee to protect session data at the client side, attackers get session data by XSS attacks, data theft attack (attacker directly copies the cookie from victim's side), etc.

5.2 *Use multiple cookies to store the session id*

The session id is stored in a particular cookie, for example, the session identifier is stored in "JSESSIONID" named cookie in case of java server pages (JSP), and in case of PHP session identifier is stored in "PHPSESSID" named cookie by default [10]. For performing the session hijacking attack, the attacker requires default cookie (used to store session id), i.e., JSESSIONID, PHPSESSID, ASPSESSIONID, etc.

If we store the session information in multiple parts at the client side, i.e., session id stored in multiple cookies, then the chances of session hijacking may be less. But, the hackers grab the entire cookies shared between server and client. Using the entire cookies information, the attacker makes the cookies clone. And the attacker sends the request to the Web server along with the entire clone of cookies, then the Web server will allow to access the user's account [9].

5.3 *One-Time Cookie approach*

The author [16] and [5] proposed one-time cookie (OTC) approach. OTC is used to authenticate the user on Web server. The value of one-time cookie is matched on every request and updated on every response from the Web server (Fig. 5).

If an attacker captures current cookie (OTC) and makes the clone of cookie using that information and sends the request to the server before the victim, then Web server updates a new cookie (OTC) at attacker's machine. So, the actual user may not able to access the Web server, because he/she has earlier or old OTC.

5.4 *Session Identifier should be strong*

Session id should be created as strong, randomness, and long session id created with a strong algorithm is very important. This idea may reduce the probability of session prediction by brute-force attacks.

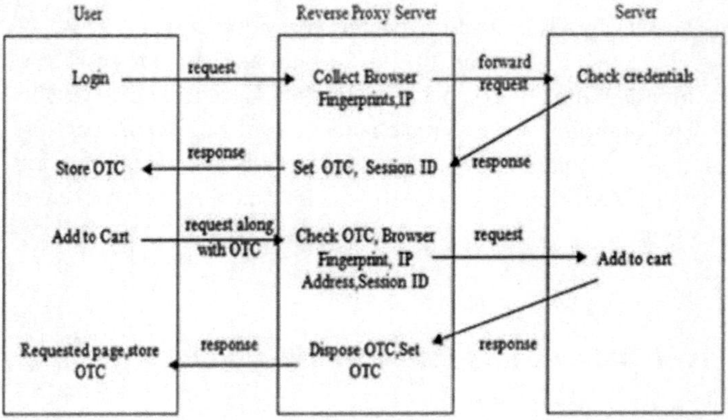

Fig. 5 Process of one-time cookie approach

5.5 Use short expiration time

Over session identifier, short expiration time may decrease session hijacking chances. In each limited period of time (i.e., 10 sec to 60 sec), session variable should be changed. If an attacker grabs the session id, then he is bound to make a clone of the cookie within a limited period of time. After passing the limited period of time, session id should be automatically invalidated.

5.6 Store Client's information at the server side

Client on sending the request passes his information, i.e., browser's finger print, IP address, MAC address, stored key, or signature, etc. This information can be used as session credential. In this approach, session credential is not bound on the session cookie only. The Web server will establish key or signature at client side in local storage.

6 Conclusion

In this paper, all types of cookies and parameters of cookies (how they effect on cookie) are discussed in detail. Threats of cookies and how the attacker get valid or active cookie are discussed. Possible prevention or safeguarding is suggested.

References

1. Acunetix: (n.d.). Types of XSS: Stored XSS, Reflected XSS and DOM-based XSS. (Acunetix). Retrieved January 18, 2019, from https://www.acunetix.com/websitesecurity/xss/
2. Bisson, D.: Social Engineering Attacks to Watch Out For. (Tripwire) Retrieved from tripwire.com, 2015, March 23. https://www.tripwire.com/state-of-security/security-awareness/5-social-engineering-attacks-to-watch-out-for/
3. Braun, A.: What Are Supercookies, Zombie Cookies, and Evercookies, and are they a Threat (make tech easier), 2018, October 2. Retrieved from https://www.maketecheasier.com/supercookies-zombie-cookies-evercookies/
4. Cross-site_Scripting_(XSS), 2018, June 5. Retrieved 1 7, 2019, from www.owasp.org: https://www.owasp.org/index.php/Cross-site_Scripting_(XSS)
5. Dacosta, I., Chakradeo, S., Ahamad, M., Traynor, P.: One-Time Cookies: Preventing Session Hijacking Attacks with Stateless Authentication Tokens. Georgia Institute of Technology, School of Computer Science. Georgia (2012)
6. Dutko, J.: Types of computer cookies. Retrieved January 15, 2019, from CRU Solutions, 2018, August 16. https://crusolutions.com/blog/how-types-of-computer-cookies-affect-your-online-privacy/
7. Endler, D.: Brute-Force Exploitation of Web Application Session IDs. iDefence The power of Intelligence, 40, Chantilly. iDEFENSE Inc, Virginia, United States of America, 2001, November 1.
8. Juels, A., Jakobsson, M., Jagatic, T.N.: Cache Cookies for Browser Authentication. In: 2006 IEEE Symposium on Security and Privacy (S&P'06), p. 5. IEEE, Berkeley/Oakland, CA, USA (2006).
9. Kumar, V.: Three Tier Verification Technique to Foil Session Sidejacking Attempts. Second Asian Himalayas International Conference on Internet (AH-ICI). IEEE , Kathmandu, Nepal (2011)
10. Nathani, B.C., Adi, E.: Website vulnerability to session fixation attacks. J. Information Eng. App. II **7**, 32–36 (2012)
11. Palmer, C.: Secure Session Management with Cookies for Web. Retrieved October 14, 2018, from crypto.stanford.edu, 2008, September 10. https://crypto.stanford.edu/cs142/papers/web-session-management.pdf
12. Park, J.S., Sandhu, R.: Secure Cookies on the Web. IEEE **4**(4), 36–44 (2000)
13. Reflected cross-site scripting. (n.d.). (PortsWigger web security). Retrieved January 18, 2019, from https://portswigger.net/web-security/cross-site-scripting/reflected
14. Rouse, M.: supercookie. (TechTarget), 2017, February 28. Retrieved from https://searchsecurity.techtarget.com/definition/supercookie
15. Ruzicka, V.: Session Fixation Attack, 2017, February 20. Retrieved August 9, 2018, from www.vojtechruzicka.com. https://www.vojtechruzicka.com/session-fixation-attack/
16. Sathiyaseelan, A.M., Joseph, V.: A Proposed System for Preventing Session Hijacking with Modified One Time Cookie. IEEE, pp. 451–454, 2017, March
17. Singh, R., Kumar, D.S.: A study of cookies and threats to cookies. Int. J. Adv. Res. Comput. Sci. Softw. Eng. VI **3**, 339–343 (2016)
18. Takahashi, H., Yasunaga, K., Mambo, M., Kim, K., Youm, H.Y.: Preventing Abuse of Cookies Stolen by XSS, pp. 85–89. CPS (Confrene Pulisher Services), Seoul, South Korea (2013)
19. Techopedia: Data Theft, 2016, April 27. Retrieved from Techopedia.com: https://www.techopedia.com/definition/26274/data-theft
20. What are persistent cookies used for? (n.d.). Retrieved January 2019, 19, from allaboutcookies.org. https://www.allaboutcookies.org/cookies/persistent-cookies-used-for.html

Fog Computing Enabled Healthcare 4.0

Shaheen Parveen, Pawan Singh, and Deepak Arora

Abstract Industry 4.0 has revolutionized the utilization of information technology (IT) in the contemporary scenario. Different disruptive technologies have imparted their contribution in the success of Industry 4.0. In the transition from Industry 1.0 to 4.0, the ideas from mechanical to electrical engineering, from electrical to telecommunication and information technology and then artificial intelligence have switched abruptly. Artificial intelligence (AI), big data analytics (BDA) and machine learning (ML) contributed its leading role. The competence of artificial intelligence brings machine at the capability to handle the human health issues. In Healthcare 4.0, the whole efforts are to resolve the multi-dimensional problems of treatments faced in daily life. Healthcare 4.0 is the rebellion paradigm where hospital-centric environment, artificial intelligence and computing resources are integrated to provide the real-time treatment to patients. The fog computing can play the major role for real-time treatment. This paper illustrates the applicability of fog computing in Healthcare 4.0 and the transition of hospital-centric healthcare (HCH) system to patient-centric care (PCC). The solution for tracking health record is also proposed in this paper by analysing the case study of diabetic patient in India.

Keywords Industry 4.0 · Disruptive technologies · Healthcare 4.0 · Hospital-centric healthcare system · Patient-centric care

S. Parveen (✉) · P. Singh · D. Arora
Department of Computer Science & Engineering, Amity School of Engineering and Technology, Amity University Uttar Pradesh, Lucknow Campus, Lucknow, India
e-mail: shaheenparveen1407@gmail.com

P. Singh
e-mail: pawansingh51279@gmail.com

D. Arora
e-mail: darora@lko.amity.edu

P. K. Singh et al. (eds.), *Recent Innovations in Computing*, Lecture Notes in Electrical Engineering 701, https://doi.org/10.1007/978-981-15-8297-4_42

1 Introduction

Earlier, we had the hospital-centric healthcare (HCH) system where people used to suffer basic issues and problem of health. With the introduction and development of cloud computing, we are in the era of patient-centric care (PCC) where technology is integrated with the hospital-centric environment to avail the benefit of it. As like Industry 4.0, there is a revolution of Indian healthcare system from hospital-centric to patient-centric care shown in Fig. 1. Initially, there are no such devices which are used to assist in patient treatment. Healthcare 1.0 can be seen as resource lacking phase of the hospitals which occurred from 1970 to 1990. From 1991, digital tracking and imaging system are used by doctors for treatment with the advancement of information technology and computing devices which remained till 2005. From 2006 to 2015, electronic health record (EHR) was used in place of patient data chart, and then from 2016, the Healthcare 4.0 made the impact on Indian healthcare system with the trend of cloud, BDA, AI and ML which is expected to achieve its target of high tech and high touch facility with the estimated budget of $6000 million by 2020.

In the USA, it is supposed that 90% of health care is planned and turn its value system towards the Healthcare 4.0. Figure 2 shows the per capita government health expenditure in India as compared to other countries. A number of IoT devices such as wearable medical sensors and services are being developed to facilitate the Healthcare 4.0. Internet of Things (IoT) has evolved into Internet of Everything (IoE) where process encapsulates people, things and data into a single package. In this process, cloud computing has come into existence to store, process, compute, visualize and analyse the huge data set of patients, but it is failed to address all the problems which are being discussed as follow:

1. The time taken by the cloud server for processing and sending the result to its client's machines and devices is not acceptable for real-time sensitive data.
2. In case of unavailability, power or machine failure of cloud system may be life-threatening for patient.

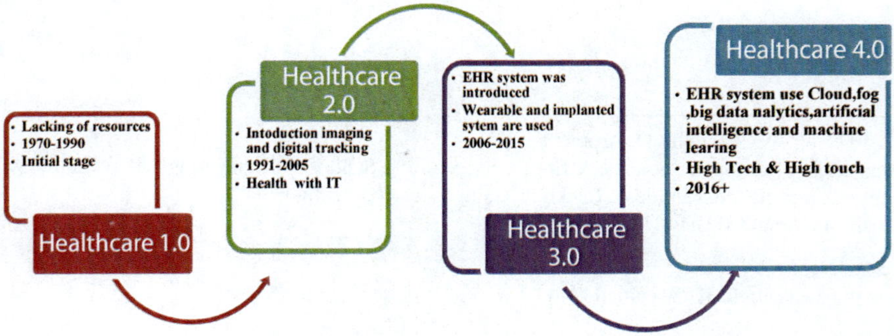

Fig. 1 Industrial revolution of Indian Healthcare system

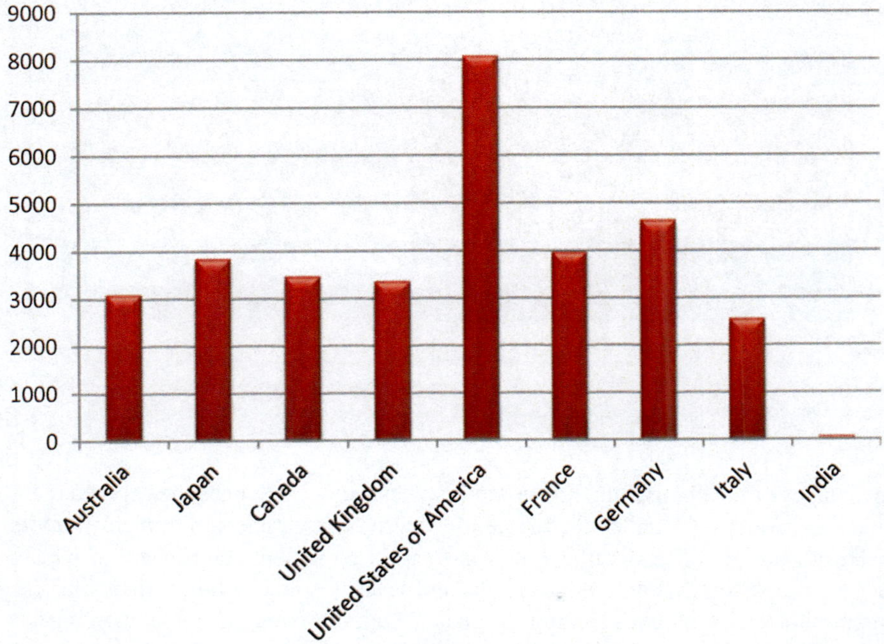

Fig. 2 Per capita government health expenditure

3. The scalability of data may lead to the problem of delay in processing and storage capacity.
4. Privacy of patient's data can be invaded.

The aforesaid issues may be resolved with the help of new paradigm of cloud computing called fog computing. Fog computing is the latest extension of cloud computing which was first coined by CISCO in 2012 and founded the OpenFog Consortium in 2015 by Cisco Systems, Princeton, Intel, ARM Holdings, Microsoft and Dell for promoting the interest and implementation of fog computing. Fog nodes are the distribution of the network processing devices such as routers, switches, gateways and hubs in the proximity of end users to attain the high throughput with low latency of the network which is basically the decentralization of cloud computing or data centres. Figure 3 shows the architecture of fog computing proposed by CISCO. Fog nodes are acting as a middleware and extensible services of cloud computing or data centres. Where cloud services include the Infrastructure as a Service (IaaS), Platform as a Service (PaaS), Software as a Service (SaaS) and new added service named Container as a Service (CaaS), fog computing is adding Quality of Services (QoS) such as low latency, business agility, data privacy and security and low energy consumption which are leading towards the smart cities including smart health care, smart homes, smart traffic system and smart manufacturing and production houses. The real-time application such as flight, oil production and heath centres can take the

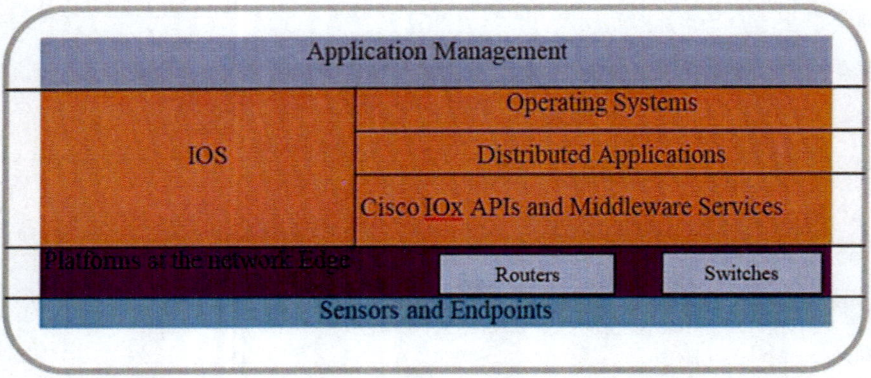

Fig. 3 Fog computing proposed by CISCO

advantages of fog computing. Fog computing has many challenges to deal with real-time sensitive data which could be major concern of researchers in implementation of Healthcare 4.0. The Healthcare 4.0 is focused on patient-centric care in which all healthcare centres such as large and mid-level hospitals, clinics, dispensaries, nursing homes, rural areas lacking of medical support, patient's home, workplaces and remote healthcare system are integrated. Fog computing which plays a major role in providing the service to these integrated networks is well known as telehealth. The organization of this paper is as follow: Section 2 illustrates the role of fog computing in Healthcare 4.0; in Sect. 3, the literature review has been summarized; Sect. 4 emphasizes the applications and services; in Sect. 5, research challenges of Healthcare 4.0 are discussed; in Sect. 6, we discussed the case study of diabetic patient and proposed a solution for tracking their health record; and finally, in Sect. 7, we concluded the paper.

2 Fog Computing in Healthcare 4.0

The cloud computing has some drawbacks due to which the fog computing model or paradigm was proposed. Basically, it is the distribution of middleware devices in such a way that we could minimize the network latency, optimally utilize the bandwidth without having network congestion and optimize the energy consumption in whole network. Resource management is the key concept of fog computing which tackles the basic operations of cloud computing such as storage, computation and resource sharing in proximity or close to the end users. The routing devices such as routers and gateways could be used for this purpose so that end users get the high throughput with low latency. The major applications of fog nodes can be perceived in real-time scenarios such as health care and augmented reality. Fogging is enabling the Internet of Things (IoT) and cloud security systems. From wearable gadgets to home

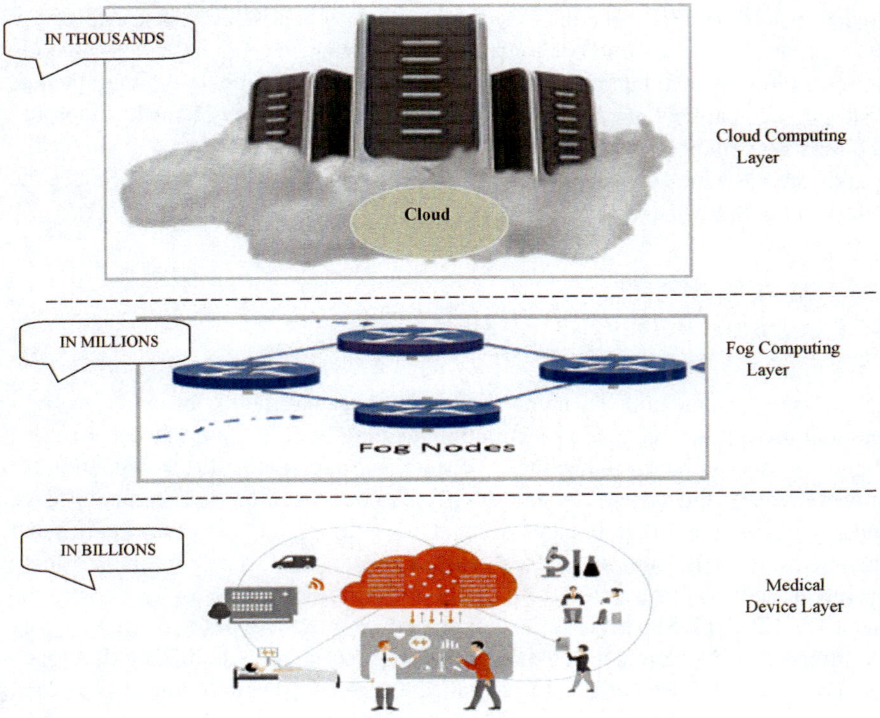

Fig. 4 Three-tier architecture of fog in Healthcare 4.0

appliances, all could be connected to Internet which may be the biggest achievement of fog computing. Centralization of cloud may lead to the security concern to data loss which could be improved by the help of fogging or distribution of those data in an efficient manner of virtualization of the network.

The three-tier architecture as in Fig. 4 of fog computing summarizes the concept of this paradigm in Healthcare 4.0 industry. There are thousands of cloud servers which are connected with millions of fog nodes which in turn connected with end users or IoT. Fog computing is also giving the flexibility to cloud data centres without extension of cloud servers itself. There are billions of end users or IoT devices which are directly connected with its proximate fog nodes, so it is also balancing the load of cloud services. The concept of fog in the Healthcare industry is enabling the IoT and distributing the workload of cloud computing. Fog devices reduce the network traffic by taking the load of end users at their edge of network which in turns provides low latency. By distribution of fog devices, one need not to provide energy all the time to keep cloud working for better response, so in this way, we can also optimize the energy consumption by keeping work at the requested time by end users. Thus, the energy consumption by the aggregated fog devices must be less than to centralized cloud server. The major role of fog layer could be seen in real-time application such as seeking the live report of a critical patient at the health centre by the distant located doctors where only the data need to store at cloud which is send back, and rest of the computation is performed on fog layer itself. In this way, fog layer does not only mitigate the workload but also save times to process and compute the intermediate result. In medical field, it could be a milestone to keep us aware of any health issues and for its treatment. A number of sensors and services are developed to enable the Healthcare 4.0 which can avail the benefit of fog paradigm as well and produce the best result with low latency.

3 Literature Review

Table 1 shows some author's name along with year and their contributed work in the field of health care and fog computing. Mutlag et al. [1] presented the issues, challenges and problems in healthcare system and implemented fog computing for improvement of the system. Dash et al. [2] also discussed the importance of edge and fog computing in health care. Kumari et al. [3] explain the opportunities and challenges of fog computing for Healthcare 4.0 in details. Some contributions help the researchers to get the architecture, applications and advantages of fog computing. Mouradian et al. [4–8] presented the comprehensive literature of fog by categorizing its different fields. Gupta et al.'s [9] contribution is remarkable as they developed the iFogSim toolkit to model, analyse and simulate the fog environment [10]. Paul et al. [11] proposed the health care which is based on fog and IoT monitoring system in his work. Nandyala et al. [12] define how smart home and hospitals can make the way from cloud to fog and IoT-based real-time U-healthcare monitoring system. Stantchev et al. [13] described the role of fog in enabling of servitization in healthcare business trend. Elmisery et al. proposed that fog nodes are capable to maintain privacy of patient's data for cloud-based healthcare system [14]. Some authors also discuss the IoT-based solution in healthcare system. Kumar et al. [15, 16] proposed a decision-based model to support personal health record through mobile to monitor

Table 1 Summary of important literatures

Authors	Years	Contribution of work
Mutlag et al. [1]	2019	Enabling technologies for fog computing in healthcare IoT systems
Kumari et al. [3]	2018	Fog computing for Healthcare 4.0 environment: opportunities and challenges
Paul et al. [11]	2018	Fog computing-based IoT for health monitoring system
Nandyala et al. [12]	2016	From cloud to fog and IoT-based real-time U-healthcare monitoring for smart homes and hospitals
Stantchev et al. [13]	2015	Smart items, fog and cloud computing as enablers of servitization in health care
Kumar et al. [17]	2019	Decision model for PHR-based diabetes monitoring system

the diabetes [17]. Al-Joboury introduced the MQTT protocol in implementation of fog-cloud-based healthcare system [18].

4 Applications and Services

The applications of fog computing in Healthcare 4.0 are oriented towards user, and services of the same are oriented towards developer.

The application is developed to fetch the data from individual patient on timely basis to keep the record and check the health issues of that patient. These applications include ECG monitoring, glucose level sensing, blood pressure and body temperature monitoring. After collecting the data from the patient, data are analysed by fog computing layer to provide the eHealth system. There are some community established which are concentrated to provide real-time assistance to this system. There are many diagnostic apps and medical calculators available which aggregate the data from the user of these apps and devices. It can efficiently help the patient in critical situation such as alerting for heart attack and stroke as well.

Services available for the smart healthcare system include eHealth, mHealth, medical implant, assistance to special need patient such as aged, incapable and the knowledge of adverse drug reaction. The services of smart wheelchair and smart gloves are provided to those patients who are incapable to move and hear, respectively. The fog-based medical implant can function better to stimulate pacemaker efficiently. We have knowledge-based system and cloud-based EHR which can be used to aware of the wrong drug consumption. The mobile health or mHealth and eHealth system are developed for those patients who have busy schedule to visit the hospital regularly. All these services incorporating with fog computing are providing the best result to patient.

5 Research Challenges of Healthcare 4.0

There are several issues and challenges in research which need to be overcome before the implementation of fog layer in healthcare environment. Real-time data need to be managed properly for handling dissimilar data format such as text and images. There can be seen the fluctuation of data between fog and cloud computing which require administrative monitoring and proper setup of the environment. Regular data collection can be accommodated through scaling of such community centre which provides easy and time-saving assistance to patients. Patients must not require visiting the hospital regularly, so such facilities and services must be developed which can help to collect, process and after analysing the data produce the result for treatment without any significant delay. The integration of fog, cloud and devices or sensors at user end must be secure such that the privacy of patient data must not be breached through unauthorized access. The interoperability of data requires such standards, protocols and regulation which should be followed by each organization in Healthcare 4.0. There must be user-friendly and feedback-based devices which collect the suggestion for designing the product and services. It would help developer to serve the patient requirement better.

6 Future Directions for Diabetic Patients Diagnosis Consultation and Data Analysis

Although fog devices can be used almost every sphere of medication, we take the example of diabetes for the expansion of medical centre or installation of ATM-like machines in those regions where this disease is getting prevalent. Figure 5 shows the death rate per 100,000 populations due to diabetes in 2016 of different states of India. The data was made public by the Indian Council of Medical Research, Public Health Foundation of India and Institute for Health Metrics and Evaluation in 2017. Fog computing may be used as the recording of the quick health report by giving the sample of blood on these medical centre or ATM-like machine in future. Sensors on these medical centre or ATM-like machine may be used to collect or aggregate data without taking much time from its users, and fog layers would provide the immediate result of them. The final result may be sent to cloud or connected mobile devices of users for details, and further actions such as diet plan and exercises may be suggested with the help of machine learning and artificial intelligence.

Fog devices would play the major role by saving the computation time and energy of cloud server. Figure 6 illustrates the role of fog layer in healthcare system which processes the end patient request within minutes. Most of the diseases are diagnosed by taking the sample of blood, so we can develop those mobile devices, medical centres or ATM-like machines for collecting the sample of blood. Patient or end

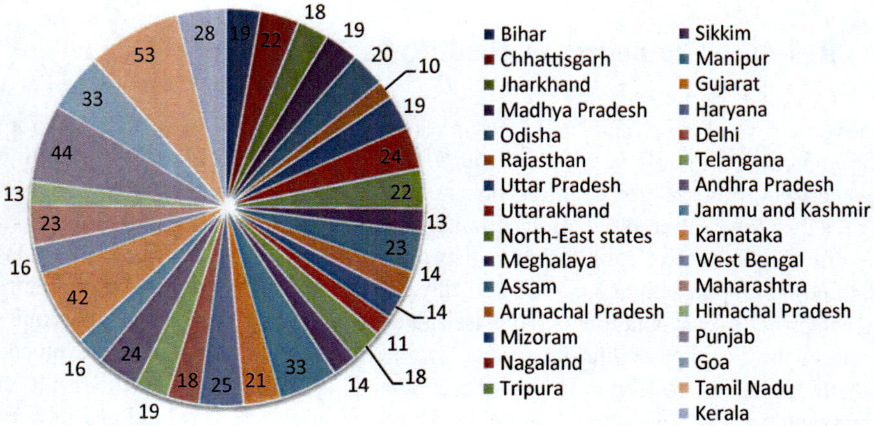

Fig. 5 Death rate due to diabetes in different states of India in 2016

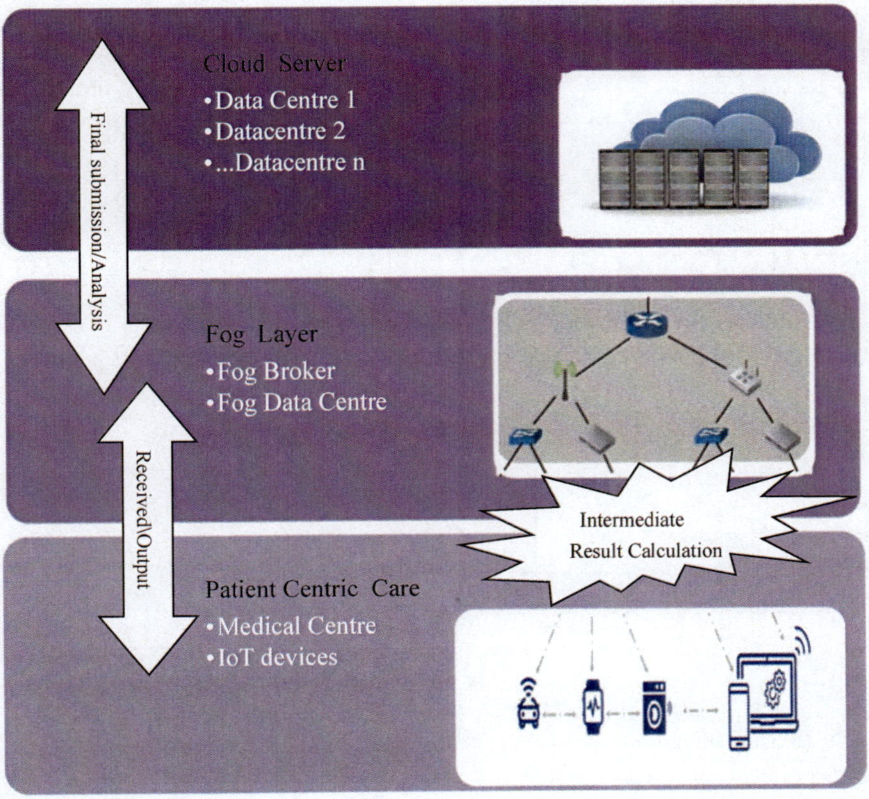

Fig. 6 Working of fog computing in Healthcare 4.0

user can logon through devices for giving their blood sample by their biometric authentication where they need not to tell the detail of him/her. These devices will collect the blood sample, and after calculating the intermediate and quick report on fog computing devices, the result may be sent according to the requirement. User can customize their requirement after login and share the same with his/her doctor. The flow diagram of this proposed solution is shown in Fig. 7.

7 Conclusion and Future Work

There are certain challenges and research issues in implementation of fog in Healthcare 4.0 such as data management, scaling of the services, security, privacy, interoperability and standardization of data flowed in network and efficient design of product and service by taking patient feedback. All these concerns are discussed in this paper.

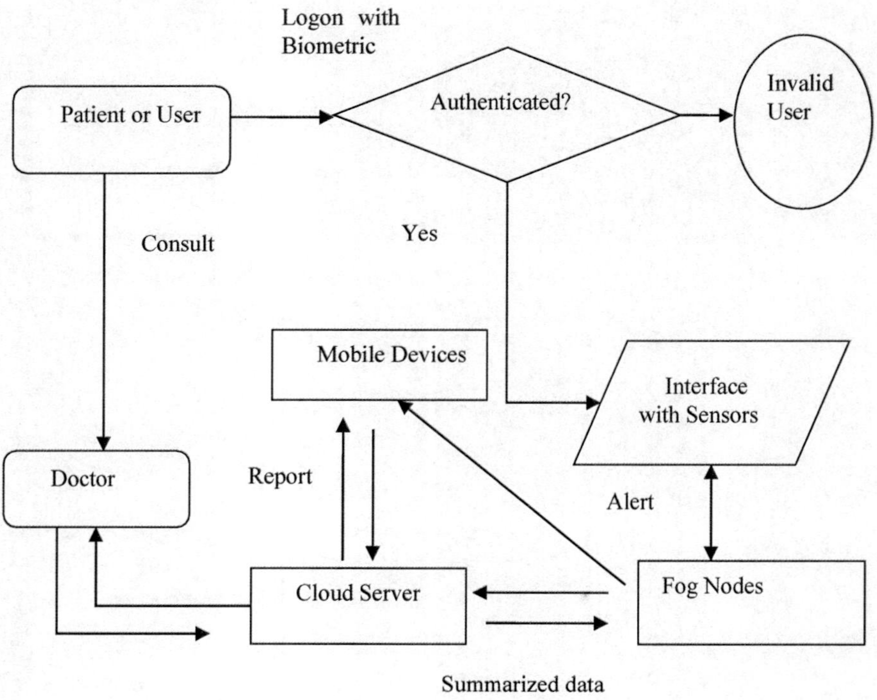

Fig. 7 Data flow diagram of fog in health care

We have also described the literature review, role, applications and services of fog computing in Healthcare 4.0. The architecture of fog layer in Healthcare 4.0 has been presented which can explain how the fog layer works between cloud and medical device layer. This paper summarizes all the basic requirement of fog computing in implementation of Healthcare 4.0. The system is also proposed which can be taken into consideration in the deployment of better healthcare features in future.

References

1. Mutlag, A.A., Ghani, M.K.A., Arunkumar, N., Mohammed, M.A., Mohd, O.: Enabling technologies for fog computing in healthcare IoT systems. Futur. Gen. Comput. Syst. **90**, 62–78 (2019)
2. Dash, S., Biswas, S., Banerjee, D., Rahman, A.R.: Edge and fog computing in healthcare—a review. Scalable Comput.: Pract. Exp. **20**(2), 191–205 (2019)
3. Kumari, A., Tanwar, S., Tyagi, S., Kumar, N.: Fog computing for healthcare 4.0 environment: opportunities and challenges. Comput. Electr. Eng. **72**, 1–13 (2018)
4. Bonomi, F., Milito, R., Zhu, J., et al.: Fog computing and its role in the Internet of Things. In: The Proceedings of the First Edition of the MCC Workshop on Mobile Cloud Computing, pp. 13–16. ACM, Helsinki, Finland (2012)

5. Vaquero, L.M., Rodero-Merino, L.: Finding your way in the fog: towards a comprehensive definition of fog computing. ACM SIGCOMM Comput. Commun. Rev. **44**(5), 27–32 (2014)
6. Mahmud, R., Kotagiri, R., Buyya, R.: Internet of Things—Technologies, Communications and Computing, 16th edn. Springer, Singapore (2018)
7. Parveen, S., Singh, P., Arora, D.: Fog computing research opportunities and challenges: a comprehensive survey. In: Singh, P.K., Pawłowski, W., Tanwar, S., Kumar, N., Rodrigues, J.J.P.C., Obaidat, M.S. (eds.) Proceedings of First International Conference on Computing, Communications, and Cyber-Security (IC4S 2019), LNNS, vol. 121, pp. 171–181. Springer, Singapore (2020)
8. Sunyaev, A.: Fog and edge computing. In: Internet Computing, Principles of Distributed Systems and Emerging Internet-Based Technologies, pp. 237–264. Springer Cham (2020)
9. Mouradian, C., Naboulsi, D., Yangui, S., Glitho, H., Morrow, M.J., Polakos, A.P.: A comprehensive survey on fog computing: state-of-the-art and research challenges. IEEE Commun. Surv. Tutor. **20**(1), 416–464 (2018)
10. Gupta, H., Dastjerdi, A.V., Ghosh, S.K., Buyya, R.: iFogSim: a toolkit for modeling and simulation of resource management techniques in the Internet of Things, edge and fog computing environments. Softw. Pract. Exp. **47**(9), 1275–1296 (2016)
11. Paul, A., Pinjari, H., Hong, W.H., Seo, H.C., Rho, S.: Fog computing-based IoT for health monitoring system. J. Sens. **2018**, 1–7 (2018)
12. Nandyala, C.S., Kim, H.K.: From cloud to fog and IoT-based real-time U-healthcare monitoring for smart homes and hospitals. Int. J. Smart Home **10**(2), 187–196 (2016)
13. Stantchev, V., Barnawi, A., Ghulam, S., Schubert, J., Tamm, G.: Smart items, fog and cloud computing as enablers of servitization in healthcare. Sens. Transducers **185**(2), 121–128 (2015)
14. Elmisery, A.M., Rho, S., Botvich, D.: A fog based middleware for automated compliance with OECD privacy principles in internet of healthcare things. IEEE Access **4**, 8418–8441 (2016)
15. Mahmud, R., Koch, F.L., Buyya, R.: Cloud-fog interoperability in IoT-enabled healthcare solutions. In: ICDCN '18: Proceedings of the 19th International Conference on Distributed Computing and Networking, pp. 1–10 (2018)
16. Kertesz, A., Pflanzner, T., Gyimothy, T.: A mobile IoT device simulator for IoT-fog-cloud systems. J. Grid Comput. **17**(3), 529–551 (2018)
17. Kumar, Y., Yadav, G., Singh, P.K., Arora, P.: A PHR-based system for monitoring diabetes in mobile environment. In: Paiva, S. (ed.) Mobile Solutions and Their Usefulness in Everyday Life. EAI/Springer Innovations in Communication and Computing. Springer, Cham (2019)
18. Al-Joboury, I.M., Al-Hemiary, E.H. F2CDM: Internet of things for healthcare network based fog-to-cloud and data-in-motion using MQTT protocol. In: Sabir, E., García Armada, A., Ghogho, M., Debbah, M. (eds.) Ubiquitous networking. UNet 2017. Lecture Notes in Computer Science, vol. 10542. Springer, Cham (2017)

DAMS: Dynamic Association for View Materialization Based on Rule Mining Scheme

Ashwin Verma, Pronaya Bhattacharya, Umesh Bodkhe, Akhilesh Ladha, and Sudeep Tanwar

Abstract In data warehousing, view selection (VS) is an important aspect. Optimal VS needs to be materialized in order to minimize the overall data retrieval time. To support the same, performance metrics like memory constraints to save materialized views, query execution time, and query workloads needs to be addressed to reduce the overall retrieval time. As far as static view materialization (VM) is concerned, pre-computing strategies are required to execute the query workload prior to VM, but the approach is not scalable for small disk sizes. In the current era, the memory requirement is humongous to store pre-computed views in the materialized query table (MQT) that adds an overhead to view maintenance cost and disk sizes. To address the aforementioned issues, the authors propose a novel VM scheme *DAMS*. *DAMS* operates in three phases. In the first phase, the scheme chooses a materialized view in a dynamic and on-demand basis to reduce the query processing time. Then, in the second phase, a novel attribute selection algorithm is proposed based on association rule mining (ARM) technique in VS to address historical queries. It selects a candidate view from a pool of such views. As the number of queries is large, the proposed algorithm reduces the computational latency in fetching the view result. Finally, selected views are prioritized by grouping items as clusters set based on support and confidence metrics to speed up VM operations.

A. Verma · P. Bhattacharya (✉) · U. Bodkhe · A. Ladha · S. Tanwar
Institute of Technology, Nirma University, Ahmadabad, Gujarat, India
e-mail: pronoya.bhattacharya@nirmauni.ac.in

A. Verma
e-mail: ashwin.verma@nirmauni.ac.in

U. Bodkhe
e-mail: umesh.bodkhe@nirmauni.ac.in

A. Ladha
e-mail: akhilesh.ladha@nirmauni.ac.in

S. Tanwar
e-mail: sudeep.tanwar@nirmauni.ac.in

P. K. Singh et al. (eds.), *Recent Innovations in Computing*, Lecture Notes in Electrical Engineering 701, https://doi.org/10.1007/978-981-15-8297-4_43

Keywords View materialization · View selection · Candidate views · Materialized query table · Least recently used

1 Introduction

The rise in data generation by devices has raised challenges to explore optimal ways to visualize data. Earlier, visualization is conducted through static tools like tableau, Microsoft Excel and other related applications [12]. Today, the collected datasets are humongous with multiple attributes [10]; thus, VS is a critical issue. Manual approaches to determine the most appropriate view through a list of offered represented views are tedious and time-consuming. Thus, there is a pressing need to dynamically allocate views based on on-demand query fired by users. The problems are that there are numerous factors on which a particular query is raised like similarities in view generations, informational semantics, grouping of data and aggregation. This requires a scanning of the entire MQT every time a query is fired which is cost-ineffective.

A static or dynamic approach can be used for materialization of view. Static VM materialize views before the execution of workload and remains as it is till the last statement of workload, in between if database object update the relation or tables that will not immediately propagated for the remaining workload, which create inconsistency in the database, in contrast to dynamic view will be updated automatically when the database object modifies the database, we can understand the concept of VM process in Fig. 1. Incremental approaches were proposed by authors that define multiple performance metrics like accuracy in view updates, misses in base tables and latency in fetching queries. The performance parameter includes optimizes a query sequence and involves complexity of view, memory constraints and selectivity of view [8]. A materialized view is an object of database which contains a result of the query, which may be local copy of data located remotely or may be a join result or a summary of an aggregate function. The collection of views is effectively chosen in such a way that most of the query can be answered, through which we can reduce the overall execution time of the query. The drawback of VM is that they need to be refreshed timely whenever an update happens in the base tables. Thus, the overall system is not scalable as the number of users increases. Table 1 presents a comparative analysis of different existing approaches for VM.

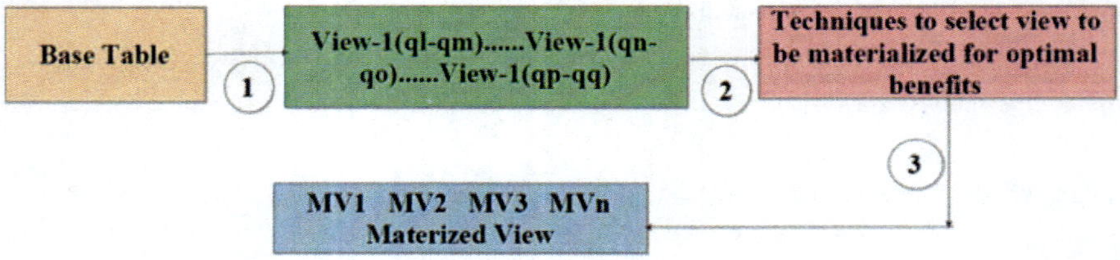

Fig. 1 Sequence flow of view materialization

Table 1 Comparative table for existing approaches for VM

Authors	Years	Objective	Pros.	Cons.
Rashid et al. [21]	2010	Reducing the maintenance cost of materialized view in object relational DBMS	Incremental model to improve query performance cost	Results limited to static view models, hence not scalable with increasing users
Anter et al. [3]	2012	A hybrid integrated system to materialized view based on past queries	Distribution of queries based on user choices	More parameters need to be considered like query execution time and query fetching time in conjunction with user choices
Zlamaniec et al. [28]	2015	Increase the SPARQL (SPARQL protocol and resource description framework) endpoint availability	Selection of view is based on the access patterns and frequency of access to support real-time experience	Optimization is limited toward data retrieval methods for SPARQL database only
Yoshifumi et al. [19]	2017	A PROforma-based approach on view updates, computation-based calculations of temporal updates	Cartesian product and join views became updatable	Not applicable on relational DBMS as view adaptability is managed by INSTEAD OF triggers
Kumar et al. [18]	2017	Materialized view based on discrete genetic operators	Particle swarm optimization (PSO) to select top k views from the multidimensional lattice for materialization	The approach is compared with heuristic and local optimum is achieved with low convergence rate during subsequent iterations
Jindal et al. [16]	2018	Expression of queries on cluster datasets to compute common sub-expressions for view materialization	Reusability of common sub-expressions to expedite the execution time of view fetch	Redundancy in different sub-expressions during clustering of datasets

(continued)

Table 1 (continued)

Authors	Years	Objective	Pros.	Cons.
Azgomi et al. [6]	2018	Game theory with players' view maintenance costs and query processing costs to decide play-off function to reach Nash equilibrium	Convergence toward equilibrium state is fast and efficient resulting in reduced query processing costs	Whenever payoff does not increase with subsequent iterations, the game reaches a local optimal equilibrium point which is far from ash convergence
Ye et al. [27]	2018	A multi-view clustering method to reduce the effects of noisy views during clustering process	Allows assignment of small weights to improve clustering performance	Focuses on clustering the relevant views, still we need to find out different attributes which will be considered for the view materialization
Gosain et al. [13]	2018	Priority-based VM scheme for data warehouse	PSO for selection of prioritized set of queries for VM on data cube	Approach is not suitable for dynamic view selections
María et al. [11]	2019	Assertions are designed with the help of materialization to ensure tuples in view follow cross-row constraints	Suited for high complex view models and large-sized queries	The model is not scalable in terms of access cost as the number of queries increases in the system

In the literature, many authors proposed solutions to address challenges in VM like scalability, information extraction and retrieving patterns from raw data. The various conditions where a view is needed to be materialized [21], materialization in hybrid integration [3] and the optimization of materialization in SPARQL are discussed [28]. PROforma-based approach for view updation [19] and use of discrete genetic operator and optimization technique is used to materialize the view [18]. Overlapping sub-expression [16] to identify the common expression for materialization and game theory [6]-based framework for selection of view is discussed. Priority-based [13] materialization method, multi-clustering [27] to remove noisy view from materialization process and cross-row constraints [11] in materialization are addressed. Data mining techniques [22], frequent pattern analysis [24], are proposed to support ARM applications like word embedding in textual data [15], extract labels in intrusion detection system [5] and decentralized storage patterns in blockchain [17]. The authors in [26] addressed the problems of selecting view dynamically and dropping of MV. Researchers in [4, 7] use a clustering method to find the cluster of closely related queries. Phan et al. [20] describe automated, dynamic materialized query table management scheme that materializes views.

Materialized view can be managed by least recently used (LRU) policy. Hossein et al. [6] proposed a game theory approach where one player is greedy for high query processing time and the other layer is greedy for the high view maintenance cost. The main advantage of the approach is flexibility. However, if the number of parameters increases the system becomes complex to find materialized set.

1.1 Motivation

As the amount of data on the server is huge, therefore, for better performance views need to be processed and stored in an efficient manner in a systematic storage system where it can be frequently accessed based on business logics. So, to minimize the query execution time we need, we cannot bring the whole relation or database in memory, and we need to focus only on the selected portion which is sufficient to answer the maximum query workload. Many authors propose data mining [4], pattern analysis [23, 25], label extraction [5] and game theory approaches [6] to address VM issues. However, reducing the size of relevant data in memory is important. Hence, motivated by the same, *DAMS* proposes a dynamic VM approach which has twofold benefits. First, to reduce the query execution time by materializing the selected candidate views through proposed attribute selection algorithm based on association rules, and then, view prioritization is achieved based on clustering views on confidence and support metric results.

1.2 Research Contributions

The proposed research contributions are now as follows:

- A novel attribute selection algorithm is presented based on association rule mining in VS to address historical queries.
- Selected views are clustered-based support and confidence matrix and then prioritize for materialization.

1.3 Organization of the Paper

The contributed paper work is organized into five sections. Section 1 gives introduction and motivation behind the paper, and Sect. 2 presents the key terminologies in view materialization process. Section 3 confers the proposed approach and architecture. Section 4 presents the research challenges and future scope in the materialization process and solving associated attribute sets. Finally, Sect. 5 presents the conclusion at the end.

2 View Materialization: Key Terminologies

For view materializing, we need to consider both space and processing time for a query as a constraint in the system. As the database object updates any information in the base table of the database, we need to propagate those changes in the view. Hence, the key challenge is to maintain the consistency in the materialized view. Some systems re-compute the materialized view from the scratch, which is not desirable as re-computations in base tables consume more time. The re-computations of the views need to be done in an incremental fashion.

2.1 Performance Evaluation Parameters for View Materialization

Selectivity of View: Selectivity of view can be defined as follows [21]

$$\text{View\_selectivity} = \frac{N_{\text{rowquery}}}{N_{\text{rowview}}}$$

where N_{rowquery} denotes the number of qualified rows and N_{rowview} denotes the number of rows existing in the view. The views are calculated with the help of a different parameter and constraint which results in a different view selectivity for the same database.

To understand this better consider the following example:

```
CREATE VIEW sales-view-1 AS SELECT att-1, att-2 FROM sales
WHERE att-3=k1 HAVING SUM(att-4)>k2.
```

Suppose the total number of rows in the base table is 5000 and the number of rows in the sales-view-1 is 1500 and we change the constraints to WHERE att-3 > $k1$ and att-3 < $k2$ HAVING AVG(att-4) > $k3$ GROUP BY att-5. And let us say the resulted rows in sales-view-1 become 1000, and now you can understand the difference in the view selectivity due to different constraints on the same database table.

View selectivity (sales-view-1) = 1500/5000 = 0.3

View selectivity (sales-view-1) = 1000/5000 = 0.2

View Complexity: It can be defined as a result of join between two or more relations and combination of WHERE and GROUP BY clause, because view with JOIN another combination of clause is complex and takes time to compute. Understand the complexity of both sales-view-1 and sales-view-2.

```
CREATE VIEW sales_view-1 AS SELECT att-1, att-2 FROM sales
WHERE att-3=k1.
CREATE VIEW sales_view-2 AS SELECT att-1, att-2 FROM sales
WHERE att-3=k1 HAVING AVG(att-4)>k2 GROUP BY att-5.
```

Database Size: It depends on the organization and their business requirement, how frequently data is accessed from multiple resources for analysis and fact finding. A small organization relatively stores small size database as compared to the organization with huge amount of data stored in different geographical locations. So, the size of materialized view also depends on organization and its requirement.

Query Optimization: Based on the view calculated and stored in the materialized query table, few conditions are applied to check whether a view is capable of providing the solution or not; if it is not, we need to fetch the result from the original base table. So generally we prefer one of the two approaches, In the first approach, we match the attribute of the query with the view, or alternatively in second approach we can compare the join attribute, selection condition and aggregate function between the query from the workload and materialized view. The following is depicted through an example on defining a view.

```
CREATE VIEW Employee_Department_V1 as SELECT E.
E.employee_name, D.department_name,
D.department_location, M.employee_name as
Contractor FROM Employee E JOIN Department D us
ing(department_no) join Employee M on (M.employee_no=
D.manager);
Query: SELECT E.employee_name, D.department_name FROM Em
ployee E, Department D WHERE D.department_no=101 and
M.employee_no=D.manager;
```

A VS problem is based on parameter size of the view, frequency of view update [9], query prioritization [14] and cost function. The approaches of dynamic view materialization are depicted in Fig. 2.

3 *DAMS:* The Proposed Approach

In the proposed approach, a heuristic framework is presented. Workload is considered as a sequence of statements $\{s_1, s_2, \ldots, s_n\}$ to be executed in a specific order. Figure 3 shows the VM process. For view selection, mainly four steps are involved:

1. Preprocessing of queries to generate attribute matrix
2. Identify the association rules
3. Identify clusters
4. View selection.

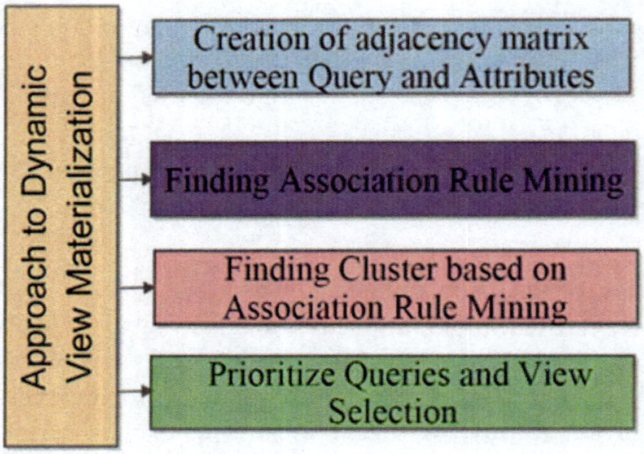

Fig. 2 Approaches to dynamic view materialization

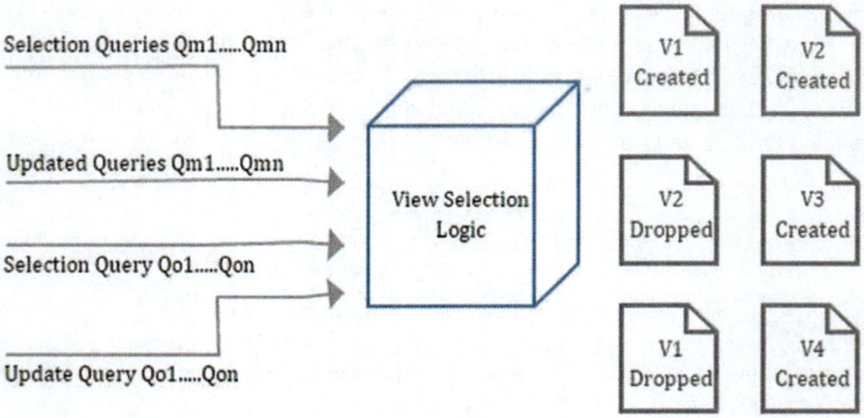

Fig. 3 View creation

3.1 Attribute Matrix Generation Table

Attribute matrix generation algorithm generates attribute matrix (M_{kj}) where k is the number of queries in the workload (Q), j is the number of attributes present in the relation/table represented by (A), and Att_{set} is the attribute set which is created from the attribute; initially, it contains φ and after the completion of this algorithm Att_{set} ends up with collection of sets of single attribute which belongs to Queries Q_j.

Table 2 *DAMS*: generated attribute matrix table

Q/A	At_1	At_2	At_3	At_4	At_5	At_6	At_7	At_8	At_9	At_{10}	At_{11}
Q_1	1	1	1	0	0	0	0	1	0	1	1
Q_2	0	1	1	1	1	1	1	0	0	0	0
Q_3	1	0	1	0	0	0	0	1	0	1	1
Q_4	0	0	1	1	1	1	0	0	1	0	0
Q_5	1	0	0	1	0	0	1	1	1	1	1
Q_6	0	1	1	0	0	0	1	0	1	0	0

Algorithm 1 Algorithm for Attribute Matrix

Input: Queries(Q) {1.......k}
Attribute(A) {1.......j}
$Att_{set} = \phi$
Output: Attribute Matrix M_{kj}

```
for Q 1 to k do
    for A 1 to j do
        if A[j] ∈ Q_k then
            M_kj ← 1
            Att_set = Att_set ∪ A[j]
        else
            M_kj ← 0
        end if
    end for
end for
```

That is, $\text{Att}_{\text{set}} = \{\{\text{At}_1\}\{\text{At}_2\}\{\text{At}_3\}\{\text{At}_4\}\ldots\{\text{At}_j\}\}$. Based on values of M_{kj} generated by algorithm 1, a dimension table is presented in Table 2. All possible attributes present in the query workload $\{Q_1, Q_2, Q_3, Q_4, Q_5, Q_6\}$ are depicted in the column, and individual row represents the queries. In cell, a value '1' indicates the presence and '0' indicates the absence of attribute in the specific query. Using this attribute matrix, we can co-relate the set of queries in the workload and identify the clusters.

3.2 Mapping Association Rules to Find Clusters

The main issue is to find the view which is associated with maximum number of queries, i.e., can provide result to major portion of the workload.

Consider the queries:

```
SELECT SUM(salary) from EmployeeDB where Branch_office='Mumbai'
and Joining_year=2019 GROUP BY department.
```

In the above query, we first identify the rows with office at 'Mumbai' and year 2020 and then take average of salary using aggregate function and then arranging them on the basis of department, the same query can also be considered if we first group the data on the basis of department with branch Mumbai and year 2019 and then apply aggregate function. So, we need to find the result faster. This is depicted in Fig. 4.

Consider the set of items ITEM(I) = Bread, Milk, Butter which are frequent part of your transaction, and same is shown in Table 3, where presence and absence of attribute are shown with numeric values 1 and 0, respectively, and from a collection of transaction we can identify the closely related attribute which occurs in transaction more frequently. In online transaction, this rule will help the customers to take decision on the current trends irrespective of the requirement sometimes, and in this

Fig. 4 Proposed architectural diagram for view materialization

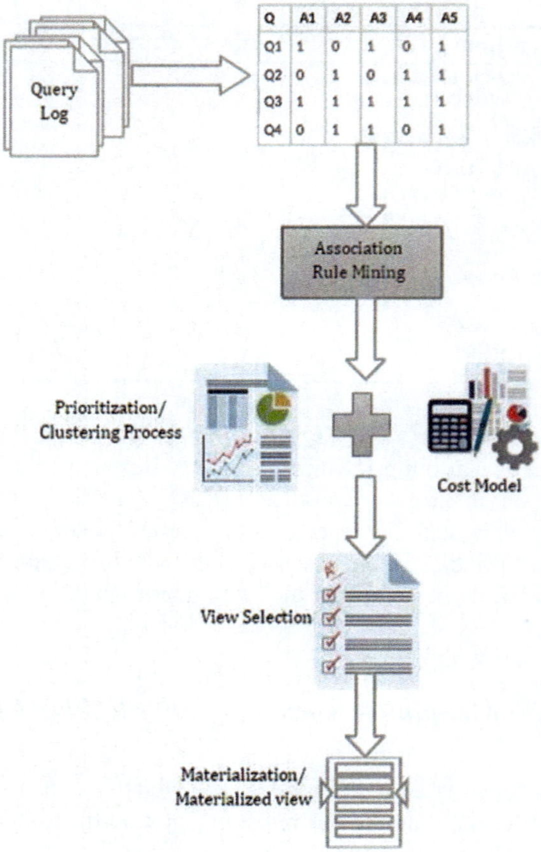

Table 3 Mapping of generated associated values to rule sets

Transaction ID	Bread	Milk	Butter
1	0	1	0
2	1	1	1
3	1	1	0
4	0	0	1

way, you get suggestion on e-commerce and social media platform [4]. The row in the attribute matrix represents in the following manner.

$$\mathrm{Q1} = 10100, \mathrm{Q2} = 00011$$

We can understand the presence and absence of attribute in the bit vector. Now to find the co-relation between multiple queries, we can make use of bit-wise logical operator.

3.3 Clustering and Prioritizing Queries

Based on bit vector supporting a query set, we identify the attribute, create a cluster of it, then prioritize them based on the frequency in the transaction and create a cluster based on the decision support system for view selection. Consider Table 4 storing the bit value in the matrix for each query and attribute.

According to Agrawal et al. [1], ARM problem is defined as: Let $A = \{\mathrm{At}_1, \mathrm{At}_2, \mathrm{At}_3, \mathrm{At}_4, \ldots\}$ be a set of attributes from the query workload which belongs to a transaction we call it items. Let $Q = \{Q_1, Q_2, Q_3, Q_4, Q_5, Q_6, \ldots\}$ be a set of queries in the workload. To understand this, consider each query with a unique identification which contains subset from the item set I, and A rule is defined as $A \rightarrow B$ where A, $B \subseteq S$ and $A \cap B = \mathrm{NULL}$ [2].

To identify the interesting co-relation from the set of transactions, constraint can be applied. The key constraint to draw any conclusion in this system is use of minimum support and confidence as a threshold.

Support: The support supp($A \Rightarrow B$) of an items $A \cup B$ is defined as the frequency of data A and B comes together in a transaction data set.

$$\mathrm{Support}(A \Rightarrow B) = \frac{n(A \cup B)}{N}$$

Table 4 To generate clusters and prioritize query sets

Query	Freq	At_1	At_2	At_3	At_4	At_5
Q_1	1	1	0	1	0	0
Q_2	1	0	1	1	0	0
Q_3	1	1	0	0	1	0
Q_4	1	0	1	0	1	1
Q_5	1	1	0	1	0	0
Q_6	1	0	1	1	1	1
Q_7	1	1	0	0	1	1
Q_8	1	0	1	1	1	0

where N is count of total transaction. If in a total of 1000 transaction the frequency of A and B comes together 150, then Support $(A \cup B) = 150/1000 = 0.15$ or 15%.

Confidence: The confidence is defined as association between item A and B, i.e., how likely B is selected if A is considered.

$$\text{Confidence}(A \Rightarrow B) = \frac{\text{Support}(A \Rightarrow B)}{\text{Support}(A)}$$

For example, if the value of confidence is $0.15/0.15 = 1$, i.e., (100%) it means that for 100% of the transactions, if attribute B is present, then A is also present, so we consider both the attributes in the same cluster. The reason behind the association rule is to create cluster based on the priority of the view. To prioritize, we have chosen A-priori algorithms, because of its efficient approach to categorize dataset items [2]. To understand the concept, consider hypothetical data along with the frequency in the query workload.

Consider the item set of one and two attributes initially.

1. Attribute set: $\{\{At_1\}, \{At_2\}, \{At_3\}, \{At_4\}, \{At_5\}\}$
2. Attribute set $\{\{At_1, At_2\}, \{At_1, At_3\}, \{At_3, At_4\}, \ldots$ so on$\}$
3. Attribute set: contains combination of 3 attributes together and so on, so this attribute set will depend on the number of attributes present in the table.

So, we first find the frequent set with 1 attribute alone in the database which we call L_1 and likewise we calculate till L_k.

L_k = attribute set of K frequent attribute.

Step 1: find the frequent set with 1 attribute in the database.

$L_k = \{\{At_1\}, \{At_2\}, \{At_3\}, \{At_4\}, \{At_5\}\}$

Next step to identify the frequent set with 1 attribute by observing L_1.

$C_2 = \{\{At_1, At_2\}, \{At_1, At_3\}, \{At_1, At_4\}, \{At_1, At_5\}, \{At_2, At_3\}, \{At_2, At_4\}, \{At_2, At_5\}, \{At_3, At_4\}, \{At_3, At_5\}, \{At_4, At_5\}\}$

Now scan the attribute table for $\{At_i, At_j\}$ existing in a Q_i in the attribute table, where $i, j \in \{At_1, At_2, At_3, At_4, At_5\}$ and $Q_i \in$ query workload.

$L_2 = \{\{At_1, At_2\}, \{At_1, At_3\}, \{At_1, At_4\}, \{At_1, At_5\}, \{At_2, At_3\}, \{At_2, At_4\}, \{At_2, At_5\}, \{At_3, At_4\}, \{At_3, At_5\}, \{At_4, At_5\}\}$

On passing minimum support and confidence on L_2, we generate C_3.

C_3 is generated by combining all possible sets of L_2.

$C_3 = \{\{At_1, At_2, At_3\}, \{At_1, At_2, At_4\}, \{At_1, At_2, At_5\}, \ldots$ and so on$\}$

Using C_3, we will get L_3 and let us assume only two sets finally qualify the minimum threshold.

$L_3 = \{\{At_2, At_3, At_4\}\ \{At_2, At_4, At_5\}\}$

$C_4 = \varphi$ and so we stop here for clustering of attribute because no further possible attribute set is capable of qualifying threshold value of set L_k, and we also calculate the support and confidence of all attribute set.

For example, attribute set $\{At_2, At_4, At_5\}$

$$\text{Confidence}(\{At_2, At_4\} \Rightarrow \{At_5\})$$

$$\text{Confidence}(\{At_2, At_5\} \Rightarrow \{At_4\})$$

$$\text{Confidence}(\{At_4, At_5\} \Rightarrow \{At_2\})$$

$$\text{Confidence}(\{At_2, At_4\} \Rightarrow \{At_5\})$$

$$\text{Confidence}(\{At_5\} \Rightarrow \{At_2, At_4\})$$

$$\text{Confidence}(\{At_2\} \Rightarrow \{At_4, At_5\})$$

$$\text{Confidence}(\{At_4\} \Rightarrow \{At_2, At_5\})$$

$$Support(\{At_2, At_4\} \Rightarrow \{At_5\})$$

and so on, and we can calculate the other support rule for this above case.

This result can help us find more relevant and co-related attribute set which helps prioritize the strong associations and how strongly they are connected together; this helps us to keep them in a same cluster and can be resolved in a same view.

Algorithm 2 Algorithm for Attribute selection using candidate set

Input: Queries(Q) {1.......k}
Attribute(A) {1.......j}
Att_{set}
C_m(candidate set)={}
L_m(minimal set)=ϕ

Output: Candidate set C of attribute

```
while Q_m ≠ φ do
    for Att_set ← 1 to j do
        for Q ← 1 to k do
            if ∀ Att_set ∈ A, Att_set ∈ Q_k then
                n(Att_set) ← n(Att_set) + 1
            end if
        end for
        C_m ← map(n(Att_set),j)
    end for
    for ∀Att_set ← 1 to j do
        if C_m[j] ≥ min_count then
            L_m[j] ← C_m[j]
        end if
    end for
    C_m+1 ← L_m X L_m
    j ← length(C_m+1)
end while
```

3.4 View Selection by Attribute Selection Using Candidate Sets

Attribute matrix generation algorithm is presented that generates possible candidate views for materialization. The following Algorithm 2 is used for finding the candidate set of attributes which are frequent and need to be materialized for efficient access and storage, where k is the number of queries represented by (Q), j is the number of attributes present in the relation/table represented by (A), and $\mathrm{Att_{set}}$ is the attribute set which was calculated in above Algorithm 1 and will be updated in each iteration using Cartesian product of previous sets which is combination pair of all the previous set element, i.e., state after 1st iteration $\mathrm{Att_{set}} = \{\{\mathrm{At_1}, \mathrm{At_2}\}, \{\mathrm{At_1}, \mathrm{At_3}\}, \{\mathrm{At_1}, \mathrm{At_4}\}, \{\mathrm{At_1}, \mathrm{At_5}\}, \{\mathrm{At_2}, \mathrm{At_3}\}, \{\mathrm{At_2}, \mathrm{At_4}\}, \{\mathrm{At_2}, \mathrm{At_5}\}, \{\mathrm{At_3}, \mathrm{At_4}\}, \{\mathrm{At_3}, \mathrm{At_5}\}, \{\mathrm{At_4}, \mathrm{At_5}\}\}$ state after 2nd iteration $\mathrm{Att_{set}} = \{\{\mathrm{At_1}, \mathrm{At_2}, \mathrm{At_3}\}, \{\mathrm{At_1}, \mathrm{At_2}, \mathrm{At_4}\}, \{\mathrm{At_1}, \mathrm{At_2}, \mathrm{At_5}\}, \ldots$ and so on$\}$. C_m is the candidate set which counts the occurrence of the attribute in the query, and those which qualify minimum support count in the occurrence of queries are shifted in least count set L_m where m denotes the number of set required to find out the minimum attribute which is considered for materialization process.

Now we materialized only those view which are capable of providing answers to multiple queries, and will reduce the candidate set. Query order permutation and frequency of past queries are also used to reduce the overall query processing.

4 Research Challenges and Future

In VM, refresh needs to be performed constantly on the database object whenever an update occurs. These updates need to be propagated at real time to simultaneous users. The challenge lies in identifying the updates that are consistent among all users. The research challenges are now presented as follows

- **Partial versus Complete Refresh of View**: The refreshes are sometimes partial, where the updates are applied as they happen and the view is presented. This increases the efficiency of query updates but may introduce inconsistency among parallel users operating simultaneously. The refreshes are sometimes complete, where all updates first occur on the VM and then the view is presented to users. This allows consistency among all users for a common view but increases the latency of the query. Fine-tuning an appropriate balance is a difficult task.
- **Cost of Materialization**: Another challenge is establishing a trade-off among limited memory in constrained environments against the query execution time. Increasing memory usage reduces the execution time of fetching queries, which is not feasible in low-powered environments.
- **Attribute Selection**: Another challenge is identifying the proper sets of frequent attributes during VM. Frequent attributes may be pre-fetched so that latency reduces. The challenge is to determine the estimated range value of attributes that needs to be pre-fetched.

As rule-based queries have casual dependencies, the cost of querying view increases. To address the same, the authors as part of future work would propose machine-learning-based approaches on graph-based models to classify a set of candidate views to be materialized. This would reduce the overall query and maintenance cost of updating views. To address the same, queries are clustered as graph datasets. A priority algorithm for the selection of materialized views would be proposed. This reduces the overall storage and query processing cost.

5 Conclusion and Future Work

Optimal VS requires materialization in order to minimize the overall data retrieval time. In the proposed approach *DAMS*, based on assigned workloads, similarity and dissimilarity of features are grouped into cluster sets based on confidence values. To define the candidate view, we use clusters of related queries which further solve multiple queries. If candidate views are in number, it will take more time to identify which candidate is merged and at the same time ensure the capability to answer multiple queries. The concept rule mining provides much stronger associated attribute set. It also helps in the clustering of related attribute which is frequent in queries in a database for a given amount of time period.

References

1. Agrawal, R., Imieliundefinedski, T., Swami, A.: Mining association rules between sets of items in large databases. SIGMOD Rec. **22**(2), 207–216 (1993)
2. Agrawal, R., Srikant, R.: Fast algorithms for mining association rules. In: Proceedings of 20th International Conference on Very Large Data Bases VLDB1215 (08 2000)
3. Anter, S., Zellou, A., Idri, A.: personalization of a hybrid integration system: creation of views to materialize based on the distribution of user queries. In: 2012 IEEE International Conference on Complex Systems (ICCS), pp. 1–7. IEEE (2012)
4. Aouiche, K., Jouve, P.E., Darmont, J.: Clustering-based materialized view selection in data warehouses. In: Manolopoulos, Y., Pokorný, J., Sellis, T.K. (eds.) Advances in Databases and Information Systems, pp. 81–95. Springer, Berlin (2006)
5. Arab, M., Sohrabi, M.K.: Proposing a new clustering method to detect phishing websites. Turkish J. Electr. Eng. Comput. Sci. **25**(6), 4757–4767 (2017)
6. Azgomi, H., Sohrabi, M.K.: A game theory-based framework for materialized view selection in data warehouses. Eng. Appl. Artif. Intell. **71**, 125–137 (2018)
7. Bhattacharya, P., Tanwar, S., Bodke, U., Tyagi, S., Kumar, N.: Bindaas: blockchain-based deep-learning as-a-service in healthcare 4.0 applications. IEEE Trans. Netw. Sci. Eng. 1–1 (2019). https://doi.org/10.1109/TNSE.2019.2961932
8. Bhattacharya, P., Tanwar, S., Shah, R., Ladha, A.: Mobile edge computing-enabled blockchain framework—a survey. In: Singh, P.K., Kar, A.K., Singh, Y., Kolekar, M.H., Tan-war, S. (eds.) Proceedings of ICRIC 2019. pp. 797–809. Springer International Publishing, Cham (2020)
9. Bhattacharya, P., Tiwari, A.K., Srivastava, R.: Dual buffers optical based packet switch incorporating arrayed waveguide gratings. J. Eng. Res. **7**, 1–15 (2019)

10. Bodkhe, U., Bhattacharya, P., Tanwar, S., Tyagi, S., Kumar, N., Obaidat, M.S.: Blohost: blockchain enabled smart tourism and hospitality management. In: 2019 International Conference on Computer, Information and Telecommunication Systems (CITS). pp. 1–5 (Aug2019)
11. Cavero Barca, J.M., Sánchez, B.V., García de Marina, P.C.: Evaluation of an implementation of cross-row constraints using materialized views. ACM SIGMOD Record **48**(3), 23–28 (2019)
12. Ehsan, H., Sharaf, M.A.: Materialized view selection for aggregate view recommendation. In: Chang, L., Gan, J., Cao, X. (eds.) Databases Theory and Applications, pp. 104–118. Springer International Publishing, Cham (2019)
13. Gosain, A., Madaan, H.: Efficient approach for view materialization in a data warehouse by prioritizing data cubes. IET Softw. **12**(6), 498–506 (2018)
14. Gosain, A., Madaan, H.: Query Prioritization for View Selection, pp. 403–410 (07 2018). https://doi.org/10.1007/978-981-10-3373-540
15. Hemmatian, F., Sohrabi, M.K.: A survey on classification techniques for opinion mining and sentiment analysis. Artif. Intell. Rev. 1–51 (2017)
16. Jindal, A., Karanasos, K., Rao, S., Patel, H.: Selecting subexpressions to materialize at data-center scale. Proc. VLDB Endow. **11**(7), 800–812 (2018)
17. Kabra, N., Bhattacharya, P., Tanwar, S., Tyagi, S.: Mudrachain: blockchain-based frame-work for automated cheque clearance in financial institutions. Futur. Gen. Comput. Syst. **102**, 574–587 (2020)
18. Kumar, A., Kumar, T.V.: Materialized view selection using discrete genetic operators-based particle swarm optimization. In: 2017 International Conference on Inventive Systems and Control (ICISC), pp. 1–5. IEEE (2017)
19. Masunaga, Y.: An intention-based approach to the updatability of views in relational databases. In: Proceedings of the 11th International Conference on Ubiquitous Information Management and Communication, p. 13. ACM (2017)
20. Phan, T., Li, W.: Dynamic materialization of query views for data warehouse workloads. In: 2008 IEEE 24th International Conference on Data Engineering, pp. 436–445 (April 2008)
21. Rashid, A.N.M.B., Islam, M.S.: An incremental view materialization approach in ordbms. In: 2010 International Conference on Recent Trends in Information, Telecommunication and Computing, pp. 105–109 (March 2010)
22. Sohrabi, M.K., Akbari, S.: A comprehensive study on the effects of using data mining techniques to predict tie strength. Comput. Hum. Behav. **60**, 534–541 (2016)
23. Sohrabi, M.K., Azgomi, H.: Evolutionary game theory approach to materialized view selection in data warehouses. Knowl.-Based Syst. **163**, 558–571 (2019)
24. Sohrabi, M.K., Hasannejad, M.H.: Association rule mining using new fp-linked list algorithm (2016)
25. Srivastava, A., Bhattacharya, Singh, A., Mathur, A., Prakash, O., Pradhan, R.: A distributed credit transfer educational framework based on blockchain. In: 2018 Second International Conference on Advances in Computing, Control and Communication Technology (IAC3T), Allahabad, India, pp. 54–59. IEEE (2018)
26. Xu, W., Theodoratos, D., Zuzarte, C., Wu, X., Oria, V.: A dynamic view materialization scheme for sequences of query and update statements. In: Song, I.Y., Eder, J., Nguyen, T.M. (eds.) Data Warehousing and Knowledge Discovery, pp. 55–65. Springer, Berlin (2007)
27. Ye, Y., Liu, X., Yin, J.: Multi-view clustering with noisy views. In: Proceedings of the 20182nd International Conference on Computer Science and Artificial Intelligence, pp. 339–344 (2018)
28. Zlamaniec, T., Chao, K.M., Godwin, N., Shah, N., Farmer, R.: A framework for workload-aware views materialisation of semantic databases. In: 2015 IEEE 12th International Conference on e-Business Engineering, pp. 15–22. IEEE (2015)

Multimodal Sentiment Analysis of Social Media Data: A Review

Priyavrat, Nonita Sharma, and Geeta Sikka

Abstract With the exploration of the Internet, social media platforms provide users to share their views toward various products, people, and topics. Nowadays, social media platforms have not limited their users to post or share only text but also images and videos to express their opinion or other social media activity. Multimodal sentiment analysis is an extension of sentiment analysis that is used to mine the heterogeneous type of unstructured data together. This paper gives a review of sentiment analysis and various studies contributed in the field of multimodal sentiment analysis and also discusses some important research challenges in the sentiment analysis of the social media data.

Keywords Sentiment analysis · Multimodal sentiment analysis · Social media

1 Introduction

With the exploration of the Internet and social media applications, the amount of online data is increasing exponentially from the last two decades. When we talk about the type of online data, it has a heterogeneous nature exhibiting different types of data like text, video, images, and GIFs. The online data can be used to mine human behavior and represent their emotional state because the data is generated by the people. Also, online data has given birth to some new research areas like data storage, mining, analysis, predictions, and a lot more. Likewise, this paper reviews

Priyavrat (✉) · N. Sharma · G. Sikka
Dr. B. R. Ambedkar National Institute of Technology, Jalandhar, Punjab, India
e-mail: priyavrat.cs.18@nitj.ac.in

N. Sharma
e-mail: nonita@nitj.ac.in

G. Sikka
e-mail: sikkag@nitj.ac.in

P. K. Singh et al. (eds.), *Recent Innovations in Computing*, Lecture Notes in Electrical Engineering 701, https://doi.org/10.1007/978-981-15-8297-4_44

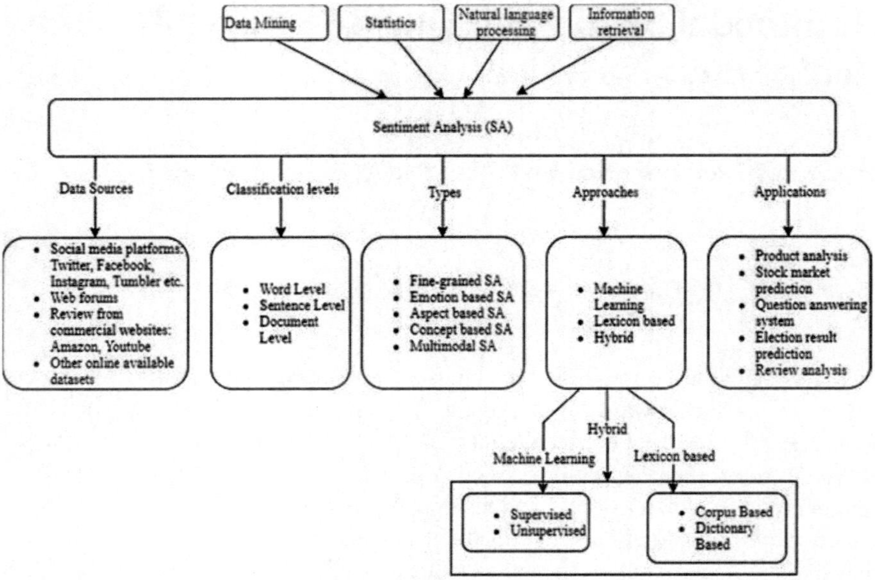

Fig. 1 Sentiment analysis at a glance

the work done on mining human behavior from social media data and explores the potential and research capabilities in the field of multimodal sentiment analysis.

Figure 1 illustrates various data sources, approaches, types, classification levels, and applications of sentiment analysis. Sentiment analysis is an interdisciplinary use of many other fields such as statistics, natural language processing, data mining, Web mining, and information retrieval as shown in Fig. 1. There are three main approaches to perform sentiment analysis named as machine learning, a lexicon-based, and hybrid approach [1]. All these approaches contain different techniques for the sentiment analysis process. Figure 1 gives a quick review of sentiment analysis and represents some important topics discussed in this paper.

Presently, there are so many applications of sentiment analysis, almost in every domain like product analysis, text to speech synthesis, tracking sentiment timelines in online forums and news, stock market prediction, answering questions, and summarizing conversation [2, 3]. Along with these applications, sentiment analysis is also highly involved in politics where it is used for election result forecasting, mining results of online campaigns, finding popularity of an electoral and gauging mood of public for a new bill or a new scheme.

Before the year 2000, little research had been done on people's sentiments and their opinion. The main reasons for the popularity of sentiment analysis and opinion mining are its wide range of applications, new challenging problems that were never discussed before, and the availability of the huge amount of opinionated data on the Web [4].

2 Sentiment Analysis

Sentiment analysis is the computational study of people's opinions, feelings, sentiments, attitudes, and moods. The main task of sentiment analysis is to find polarity (positivity, negativity, and neutrality) in a given text. Sentiment analysis is a very general term, and many other terms like opinion mining, emotion analysis, opinion extraction, sentiment extraction, sentiment classification, sentiment mining, subjectivity analysis, affect analysis, and review analysis are under its umbrella [4]. This section discusses sentiment analysis and its various types or variants as illustrated in Fig. 1.

2.1 Sentiment Analysis or Opinion Mining?

Some studies have reported that both sentiment analysis and opinion mining are the same terms used for sentiment classification. Further, the term sentiment analysis is commonly used in industry and both sentiment analysis and opinion mining are frequently employed in academia [4]. However, the dictionary meaning of sentiment and opinion is quite different.[1,2] The sentiment is defined as "an attitude, thought, or judgment prompted by feeling," and opinion is defined as "a view, judgment, or appraisal formed in the mind about a particular matter." The definitions indicate that sentiment is more of a feeling whereas an opinion is more of a person's view about something. For example, the sentence "I am concerned about the current state of the economy" expresses a sentiment, whereas the sentence "I think the economy is not doing well" expresses an opinion.

2.2 Fine-Grained Sentiment Analysis

In performing sentiment analysis, most of the studies use binary classification categories (positive or negative) or three classification categories (positive, negative, and neutral). Fine-grained sentiment analysis is a typical breakdown of sentiments in more discrete classes (e.g., highly negative, negative, neutral, positive, highly positive). Fine-grained sentiment analysis is often used in the comparative analysis of sentences and systems where there is a need to prioritize customer complaints.

[1] https://www.merriam-webster.com/dictionary/sentiment.

[2] https://www.merriam-webster.com/dictionary/opinion.

2.3 *Emotion-Based Sentiment Analysis*

Emotion-based sentiment analysis tries to extract specific emotions from the text rather than positive and negative classes. Paul Ekman gave six basic emotions that are used in various studies to mine emotions from data [5–7].

2.4 *Aspect-Based Sentiment Analysis*

Aspect-based sentiment analysis is also called a feature-based sentiment analysis in which a text is first broken down into some aspects, attributes, or features and then some sentiments are assigned to these aspects. For example, the camera quality of the Nokia phone is good but the sound quality is very bad. In the above sentence, camera and sound are two features for the Nokia phones, in which sentiment for camera feature is positive and sentiment for sound is negative.

2.5 *Concept-Based Sentiment Analysis*

In concept-based sentiment analysis, words in opinionated texts having similar meanings are considered as the same concept. Concept-based sentiment analysis focuses on semantic analysis of text through the use of semantic networks or Web ontologies, which allow the recognition of conceptual and affective information associated with textual opinions, e.g., "Flat rate was the best thing that happened this year, @UBER bring it back!!!". In this sentence, first of all, the algorithm will detect a category of a concept like "price" from the similar words network and then sentiment classification of the above sentence will be done by considering "price" as a keyword. So, the above sentence gives a positive review for the "price" of Uber.

2.6 *Multimodal Sentiment Analysis*

Multimodal sentiment analysis is different from traditional sentiment analysis where the focus was only on the mining sentiment of the text. As the name suggests, multimodal means multiple types of data (includes text, images, videos, audio, GIFs). So, multimodal sentiment analysis means finding sentiments of the data having multiple modalities as shown in Fig. 2.

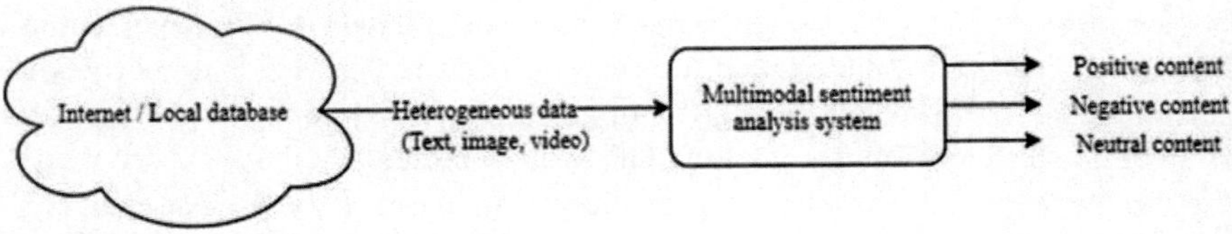

Fig. 2 Multimodal sentiment analysis (MSA)

3 Literature Review

3.1 Multimodal Sentiment Analysis

This subsection discusses some studies which make use of two modalities to mine sentiments of data from the various sources. Kumar and Garg [8] found sentiment polarity and the score of text and images (typographic and infographic) using SentiBank and R-CNN. Huang et al. [9] proposed a new image and text sentiment analysis model. In this model, this work used the visual and semantic attention mechanism with a mixed fusion scheme for image and textual data. Poria et al. [10] proposed a model for the sentiment analysis of YouTube videos and textual content. In this study, decision-level and feature-level fusion techniques were used to combine emotion-oriented information of different modalities like audio, visual, and text. Kasturai and Satoh [11] proposed a novel approach for the Instagram and Flickr images that exploit latent correlations among multiple views: sentiment view constructed using SentiWordNet and visual as well as textual views. Baecchi et al. [12] used two different models CBOW-LR and CBOW-DA-LR to find sentiments of text and image information, respectively, from the Twitter dataset. The novel model they proposed outperforms the FSLM and SentiBank models. Poria et al. [13] discussed the feature extraction process from different modalities and also explained that how these features can be used in multimodal sentiment analysis. This work used two different video datasets for the affective analysis of visual, textual, and audio data. Chen et al. [14] proposed a novel multimodal analysis architecture for multimodal sentiment analysis of visual, text, and audio data. This study demonstrated the effectiveness of the proposed approach to publicly available CMU-MOSI dataset. Perez et al. [2] integrated linguistic, audio, and visual features to identify sentiment in online Spanish videos from YouTube and English review videos from ExpoTV. In the paper, various linguistic, audio, and visual features of online videos were discussed and also found that the results obtained by using the three modalities (text, audio, and visual) at a time have greater accuracy than one or two modalities. Fang et al. [15] proposed a multimodal probabilistic graphical model to face the problems faced in multimodal sentiment mining for entities in social media. The model tried to find a correlation between textual and image modalities and also the association between different aspects and related opinions. You et al. [16] gave a new model called the competitive vector autoregression (CVAR) model. The model was a forecasting system for the United States House race and the presidential elections. CVAR combines visual and textual information from Flickr data and analyzes it. Jianqiang et al. [17] used a deep convolution neural network for sentiment classification of five different Twitter datasets. This work reported that the proposed model performs better than the baseline and currently available models. Asghar et al. [18] introduced a rule-based framework to mine sentiments of tweets related to phone products such as Samsung, Huawei, Nokia, and iPhone. Ji et al. [19] used Twitter sentiment analysis for measuring public health concerns related to epidemics and communicable diseases. Khodabaksh et al. [20] used previous Twitter data of a user timeline to predict upcoming personal life events that he can post on Twitter. Ruz and Mascareno [21] used Twitter data related to the 2010 Chilean earthquake (dataset 1) and the 2017 Catalan independence referendum (dataset 2) to assess the performance of five supervised machine learning classifiers. The work reported that SVM and RF (random forest) as best performing classifiers for dataset 1 and dataset 2, respectively. Singh et al. [22] advocated that emotional analysis of tweets can be used to improve planning and rescue operations and services during some unusual circumstances. This work used tweets related to Las Vegas Shooting incident 2017 in the study. Azar and Lo [23] made use of tweets which were referencing to the Federal Reserve Bank, and the study advocated that the content of tweets can be used to predict future returns, even after controlling market factors. Xie et al. [24] found that Twitter is an influential tool in political communication, and the study showed that emotional appraisals on tweets by the candidate are correlated with a large number of retweets shared by users. Furthermore, some studies [25–34] also used Twitter sentiment analysis to predict election results.

3.2 Use of Emoji and Emoticon

Many researchers claimed an improved accuracy of their classification model by using the extra emotional content like emoji and emoticons in the text. Therefore, this subsection highlights some of these types of studies.

Asghar et al. [5] integrated cognitive-based emotion theory (given by Psychologist Paul Ekman) with sentiment analysis techniques to detect and classify different types of six emotions from text. In this study, emoticons, slang words, and emotion words from the text play a major role in improving the accuracy of the previous text classification results by 4.59%. Gomes and Casais [35] contributed the research in analyzing the threat-oriented feelings from Facebook and YouTube comments. This work analyzed the comments regarding anorexia nervosa awareness using text analysis as well as emojis. The study showed the greater representativeness of positive feelings expressed by users through the quantification of emojis. Bahri et al. [36] gave a novel algorithm based on "emotion score learning" for identifying sentiments of a given text-oriented sentence. The proposed algorithm EmScore computes sentiment score of emoticons, while SentE is used for sentiment classification of 1000 tweets. This work reported 91.1% of accuracy for sentiment classification of the selected number of tweets. Hogenboom et al. [37] proposed a novel method to exploit sentiments of Dutch tweets and forum messages, which contain emoticons. This study found an improved polarity classification performance for paragraph- and sentence-level sentences. Spina [38] gave detailed research work on the use of emoticons on Twitter data, i.e., how they are used in the interactions. The paper also discussed some variables that influence as well as affect their use in online interactions.

3.3 Fusing Results of Multimodal Sentiment Content

Fusing multiple modalities to output a single polarity score is a challenging task in the multimodal sentiment analysis. Generally, there are three types of fusing techniques used in multimodal sentiment analysis named feature-level fusion, decision-level fusion, and hybrid-level fusion. Feature-level fusion also called early fusion combines all the features from different modalities during the feature extraction step and puts all those features in a single feature vector. This feature vector is then used to train the model for the classification process. On the other hand, decision-level fusion also called late fusion treats data from each modality into its classification algorithm, and final classification result is obtained through fusing each result into a single decision vector. Hybrid-level fusion is made of both feature-level and decision-level fusion techniques. In hybrid-level fusion technique, first feature-level fusion is done on two modalities and then decision-level fusion is applied to combine the initial results with the new modalities. This subsection discusses some of the studies related to modality fusion in multimodal sentiment analysis.

Huddar et al. [39] proposed an attention-based inter-modality fusion scheme for the fusion of audio, video, and text data from two different datasets CMU-MOSI and IEMOCAP. Corchs et al. [40], in this paper, fused the text and visual data by late and early fusion schemes. The study found that an ensemble method based on late fusion scheme and deep learning-based feature representation gives highly improved results. Tran and Cambria [41] used both feature-level and decision-level fusion methods to fuse the visual, audio, and textual modalities from the YouTube video dataset. This work found that feature-level fusion scheme was better than other fusion schemes based on recall and precision. Cerezo et al. [42] proposed a novel fusion methodology that can fuse any number of modalities from different affective information sources like text, emoticons, and facial expression. Yu et al. [43] proposed an entity-sensitive attention and fusion network (ESAFN). This network models the inter-modality interactions, i.e., text–image alignments and the intra-modality interactions, i.e., entity–text and entity–image alignments. Song et al. [44] proposed a method to fuse audio and facial expressions using the decision-level fusion scheme. This work recognized only 43.40% accuracy using ANN and k-NN classifiers at the fusing stage. Williams et al. [45] showed that proposed intermediate-level feature fusion outperforms other fusion techniques and achieves 74% accuracy for the binary classification of video and text modalities.

Table 1 gives a summarized comparative view of the above literature study on the

Table 1 Comparative analysis of various studies

Paper detail	Data source	Modality	Technique	Accuracy achieved
[17]	Twitter STSTd, SE2014, STSGd, SED, and SSTd dataset	Text	Deep CNN	85.63% (avg.)
[18]	Tweets related to mobile products	Text	Rule-based technique	92.57% (avg.)
[19]	Twitter	Text	NB, multinomial Naïve Bayes, SVM	70% (avg.)
[20]	Twitter	Text	Recurrent neural networks	Not mentioned
[21]	Twitter	Text	NB, SVM, RF	81% for dataset 1, 85% for dataset 2
[22]	Twitter	Text	NRC emotion lexicon, Bayesian change point detection algorithm	Not mentioned
[23]	Twitter	Text	SentiWordNet dictionary	Not mentioned
[24]	Google, Twitter, and Facebook	Text	Kalman filter	Not mentioned
[25]	Twitter	Text	No. of tweets shared, SentiStrength, NRC, LIWC lexicons, and Naïve Bayes classifier	Not mentioned
[26]	Twitter	Text	SentiStrength and NRC	Not mentioned
[27]	Twitter (Obama–McCain dataset)	Text	NB	80%
[28]	Twitter	Text	No. of followers, tweets shared, and KLOUT score	Not mentioned
[29]	Twitter	Text	LIWC dictionary	1.65% (MAE)
[31]	Twitter	Text	Bayesian optimization	83%
[32]	Twitter	Text	CNN	Not mentioned

(continued)

Table 1 (continued)

Paper detail	Data source	Modality	Technique	Accuracy achieved
[33]	Twitter	Text	Opinion and AFINN lexicon	4.50% (MAE)
[34]	Twitter	Text	Trivial, multinomial Naïve Bayes (MNB), AdaBoost MNB (ADA-MNB), SVM, ADA-SVM	5.85% (MAE)
[8]	Twitter	Text + Image	SentiStrength and SentiBank lexicon, and regions with convolutional neural networks (R-CNNs)	91.32%
[9]	Getty, Twitter, and Flickr	Text + Image	Deep multimodal attentive fusion (DMAF)	86.9% for Getty, 76.3% for Twitter, and 88% for Flickr datasets
[10]	YouTube	Text + Audio + Video	ELM, SVM, ANN, and SenticNet and EmoSenticNet lexicons	80%
[11]	Flickr and Instagram	Text + Image	SVM, and SentiWordNet and WordNet lexicons	74.77% for Flickr and 73.66% for the Instagram datasets
[12]	Twitter	Text + Image	Continuous bag of words and logistic regression	Multiple accuracy achieved for different datasets with different methods. Maximum reported: 83.01%
[13]	YouTube and IEMOCAP	Text + Audio + Video	CNN, SVM, ELM classifiers	88.60% for YouTube and 73.37% for IEMOCAP datasets

(continued)

Table 1 (continued)

Paper detail	Data source	Modality	Technique	Accuracy achieved
[14]	CMU-MOSI dataset	Text + Audio + Video	Gated multimodal embedding—LSTM	75.7%
[2]	YouTube and ExpoTV	Text + Audio + Video	SVM	64.86% for YouTube and 75% for ExpoTV datasets
[15]	Flickr	Text + Image	Affinity propagation algorithm, SentiStrength, SentiWordNet, iFeel, VADER	Not mentioned
[16]	Flickr	Text + Image	Autoregression and Sentiment140	Not mentioned
[35]	Facebook and YouTube	Text + Emoji/Emoticons	Dictionary method	Not mentioned
[36]	Twitter	Text + Emoji/Emoticons	AFINN-111 dictionary	91.1%
[37]	Twitter	Text + Emoji/Emoticons	SentiWordNet	94% for Dutch tweets and 90% for English tweets
[38]	Twitter	Text + Emoji/Emoticons	Ime4 and ImerTest	Not mentioned
[39]	CMU-MOSI and IEMOCAP datasets	Text + Audio + Video	Bidirectional LSTM (biLSTM), CNN, gated recurrent unit	80.33% for CMU-MOSI and 80.87% for IEMOCAP datasets
[40]	DATA345 and DATA5 datasets	Text + Image	NB, SVM, BN, k-NN, decision tree classifiers	76% for DATA345 and 81% for DATA5
[41]	YouTube	Text + Audio + Video	ELM, SVM, ANN, SenticNet lexicon	77.10%

(continued)

Table 1 (continued)

Paper detail	Data source	Modality	Technique	Accuracy achieved
[42]	Instant messages	Text + Emoji/Emoticons + Video	RIPPER, SVM, C4.5, NB, multilayer perceptron	Different success rates for different evaluation criteria. Max. 94.12% reported (on axis criteria)
[43]	Twitter	Text + Image	LSTM, CNN	>70%
[44]	eNTERFACE'05 and SAVEE datasets	Audio + Video	ANN, CNN, k-NN	43.40%
[45]	CMU-MOSI	Text + Audio + Video	CNN, LSTM, biLSTM	74%

basis of data source, modality, technique used, and accuracy achieved in different studies.

4 Results and Discussion

This section gives some insights based on the literature reviewed in this paper. All the insights analyzed are based on the data source, modality, technique used, and accuracy achieved in different studies present in Table 1. This paper comprises a comparative analysis of thirty-eight (38) papers related to sentiment analysis as well as multimodal sentiment analysis.

Figure 3 illustrates data sources used in different studies. This figure shows five different data sources from where data was extracted. Six (6) studies [2, 9, 11, 24, 35, 39] show a combined data source which means authors extracted data from more than one source and one study [42] used its own tool's instant messaging data to analyze multimodal sentiments in the conversation. Results show that most of the studies used Twitter as a data resource.

Figure 4 illustrates various modalities used in the papers. Results show that maximum papers used single modality to mine sentiments of users from the collected data. Multimodal sentiment analysis (Text + Image or Text + Audio + Video or Text + Emoticons) is done only half times of the single modal sentiment analysis. So, it is clear that multimodal sentiment analysis needs more attention among practitioners, academicians, and researchers.

Figure 5 shows the accuracy level achieved in the papers that has been reviewed. Three studies [29, 33, 34] measured the accuracy of their work based on mean average error (MAE). So, these studies are not mentioned in Fig. 5. Accuracy is the main concern for sentiment analysis, but most of the studies have reported their classifier's

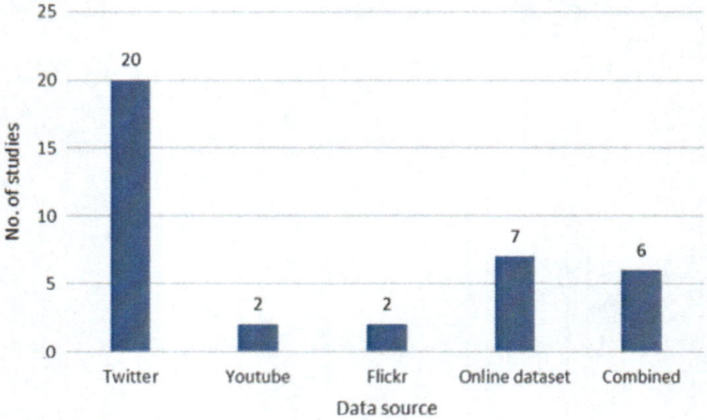

Fig. 3 Different data sources used in studies

Fig. 4 Different modalities used in studies

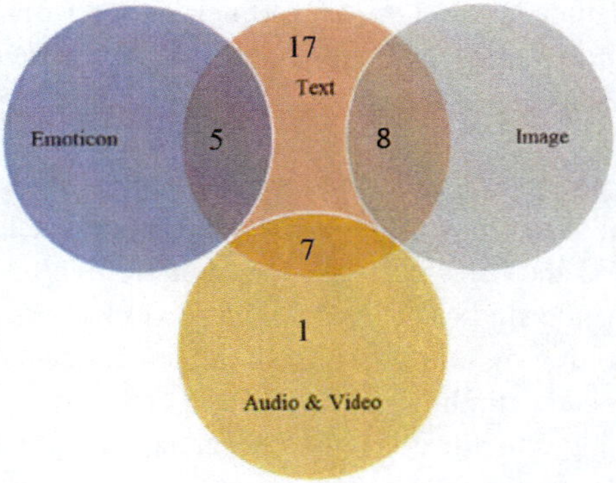

Fig. 5 Accuracy level achieved in different studies

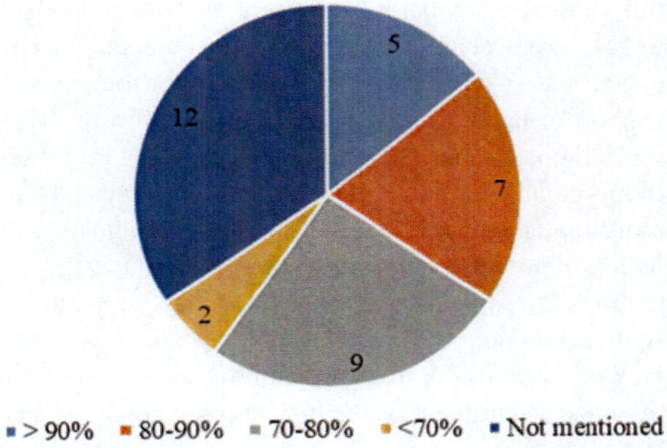

accuracy between 70 and 90%, which cannot be considered a very high accuracy level achieved in this digitized world. Therefore, more work is needed in this field.

5 Research Challenges in Multimodal Sentiment Analysis of Social Media Data

Multimodal sentiment analysis and classification of social media data are an emerging research area, and a lot more work has to be done in this research field. Social media is a platform that provides an ongoing event's information and discussions over various trending topics worldwide. Social media data is present in various forms such as messages, comments, replies, tweets, and timeline posts on the Web. But unfortunately, mining the online data is still a challenging task because it contains text mostly embedded with images or video. This heterogeneous type of data is not easy to fuse because it is not easy to determine which modality should contribute most in the fusion process. Another problem in mining online data is a complex sentence structure. Sometimes, both positive and negative sentiments toward one or more targets are expressed in some sentences where the system fails to distinguish that what is being said about whom, e.g., "I love the actor that you hate." This sentence contains both positive and negative sentiments toward the actor. Also, one more problem in multimodal sentiment analysis is the accuracy of the sentiment analysis model. Presently, the accuracy achieved by any sentiment analysis model (on the basis of Table 1) is not so high and it is limited to only 70–90% in most of the studies. Therefore, more work is needed to achieve a high level of accuracy in mining sentiments from the heterogonous type of data. Further, another problem faced in finding sentiments of social media data is noisy data such as slang words and ungrammatical text. Moreover, some text messages also contain Uniform Resource Locator (URL) links to some Web pages which contain image and video information which can explore some more hidden sentiments of the users, but generally, these links are removed before sentiment analysis process during text preprocessing step. So, Tumasjan et al. [30] suggested that embedded URL links should also be included during the analysis process. In addition, different social media apps provide huge support of language selection for the comfort of their users and there is hardly any multilingual model available that can analyze sentiments written in multi-languages. In this context, some studies [25, 46, 47] suggested that more work has to be done on multilingual tweets. Moreover, another problem in sentiment analysis of social media data is detecting sarcasm. Sarcasm is a statement where somebody hurts other's feelings by saying something that seems to be positive. Moreover, in a sarcastic sentence, people try to express their negative feelings using positive words, which makes it difficult for machines or programs to understand the context of the situation in which a sentiment or feeling was expressed.

6 Future Directions

Multimodal sentiment analysis is a new and unexplored area of research. Multimodal sentiment analysis of social media data includes analysis of the various parts of a message or post over social media platforms that may be a single sentence or a sentence embedded with emoticon or image or a video. Text is not enough to find the sentiment of a user because people are not limited to share text on their timeline or a reply. They are highly involved in sharing emoticons, pictures, and videos also. Therefore, if someone wants to perform a sentiment analysis of social media data then he or she can go one step forward and could consider all the modalities present in such data during the sentiment analysis process. Further, most of the social media user's posts are in their native languages; therefore, sentiment analysis of this specific language should also be considered by introducing new language-specific lexicons. Also, for proper utilization of social media posts, slang words and misspelled words should be replaced by their root words. Therefore, strong lexicons should be developed and considered for maintaining the substantiality of data. Moreover, nowadays social media content contains a large number of troll and sarcastic messages. Dealing with sarcastic posts and trolls should also be considered because they can affect the accuracy of a classification model.

7 Conclusion

This paper emphasizes the significance of sentiment analysis and its heavy use in mining human behavior from social media data. This paper gave a detailed review of multimodal sentiment analysis and the use of emoticons and result fusion. After a quick discussion on comparative analysis of the studies, the paper highlighted the research challenges in multimodal sentiment analysis of social media data and gave some future directions which could be beneficial for new upcoming practitioners, researchers, and academicians. In most of the studies, the accuracy level achieved is not so high. Therefore, the paper concludes that more work is required to be done in the field of sentiment analysis as well as in multimodal sentiment analysis for better interpretation of sentiments and emotions from social media platforms.

References

1. Hussein, D.M.E.D.M.: A survey on sentiment analysis challenges. J. King Saud Univ. Eng. Sci. **30**, 330–338 (2018). https://doi.org/10.1016/j.jksues.2016.04.002.
2. Perez Rosas, V., Mihalcea, R., Morency, L.P.: Multimodal sentiment analysis of Spanish online videos. IEEE Intell. Syst. **28**, 38–45 (2013). https://doi.org/10.1109/MIS.2013.9
3. Pang, B., Lee, L.: Opinion mining and sentiment analysis. Found. Trends® in Inf. Retr. **2**, 1–135 (2008). https://doi.org/10.1561/1500000011.
4. Liu, B.: Sentiment analysis and opinion mining. Synth. Lect. Hum. Lang. Technol. **5**, 1–167 (2012). https://doi.org/10.2200/S00416ED1V01Y201204HLT016
5. Asghar, M.Z., Khan, A., Bibi, A., Kundi, F.M., Ahmad, H.: Sentence-level emotion detection framework using rule-based classification. Cognit. Comput. **9**, 868–894 (2017). https://doi.org/10.1007/s12559-017-9503-3
6. Hallsmar, F., Palm, J.: Multi-class sentiment classification on Twitter using an Emoji Training Heuristic, pp. 1–27 (2016)
7. Wood, I.D., Ruder, S.: Emoji as emotion tags for Tweets. Proc. Lr. 2016 Work. Emot. Sentim. Anal. 76–79 (2016)
8. Kumar, A., Garg, G.: Sentiment analysis of multimodal twitter data. Multimed. Tools Appl. (2019). https://doi.org/10.1007/s11042-019-7390-1
9. Huang, F., Zhang, X., Zhao, Z., Xu, J., Li, Z.: Image–text sentiment analysis via deep multimodal attentive fusion. Knowl.-Based Syst. **167**, 26–37 (2019). https://doi.org/10.1016/j.knosys.2019.01.019
10. Poria, S., Cambria, E., Howard, N., Huang, G.B., Hussain, A.: Fusing audio, visual and textual clues for sentiment analysis from multimodal content. Neurocomputing. **174**, 50–59 (2016). https://doi.org/10.1016/J.NEUCOM.2015.01.095
11. Katsurai, M., Satoh, S.: Image sentiment analysis using latent correlations among visual, textual, and sentiment views. In: ICASSP, IEEE International Conference on Acoust. Speech Signal Process. Proc. 2016-May, pp. 2837–2841 (2016). https://doi.org/10.1109/ICASSP.2016.7472195

12. Baecchi, C., Uricchio, T., Bertini, M., Del Bimbo, A.: A multimodal feature learning approach for sentiment analysis of social network multimedia. Multimed. Tools Appl. **75**, 2507–2525 (2016). https://doi.org/10.1007/s11042-015-2646-x
13. Poria, S., Peng, H., Hussain, A., Howard, N., Cambria, E.: Ensemble application of convolutional neural networks and multiple kernel learning for multimodal sentiment analysis. Neurocomputing **261**, 217–230 (2017). https://doi.org/10.1016/j.neucom.2016.09.117
14. Chen, M., Wang, S., Liang, P.P., Baltrušaitis, T., Zadeh, A., Morency, L.P.: Multimodal sentiment analysis with word-level fusion and reinforcement learning. In: ICMI 2017 Proceedings of 19th ACM International Conference on Multimodal Interact. 2017-January, pp. 163–171 (2017). https://doi.org/10.1145/3136755.3136801
15. Fang, Q., Xu, C., Sang, J., Hossain, M.S., Muhammad, G.: Word-of-mouth understanding: entity-centric multimodal aspect-opinion mining in social media. IEEE Trans. Multimed. **17**, 2281–2296 (2015). https://doi.org/10.1109/TMM.2015.2491019
16. You, Q., Cao, L., Cong, Y., Zhang, X., Luo, J.: A multifaceted approach to social multimedia-based prediction of elections. IEEE Trans. Multimed. **17**, 2271–2280 (2015). https://doi.org/10.1109/TMM.2015.2487863
17. Jianqiang, Z., Xiaolin, G.U.I., Xuejun, Z.: Deep convolution neural networks for Twitter sentiment analysis. IEEE Access. **6**, 23253–23260 (2018). https://doi.org/10.1109/ACCESS.2017.2776930
18. Asghar, M.Z., Khan, A., Khan, F., Kundi, F.M.: RIFT: a rule induction framework for Twitter sentiment analysis. Arab. J. Sci. Eng. **43**, 857–877 (2018). https://doi.org/10.1007/s13369-017-2770-1
19. Ji, X., Chun, S.A., Wei, Z., Geller, J.: Twitter sentiment classification for measuring public health concerns. Soc. Netw. Anal. Min. **5**, 1–25 (2015). https://doi.org/10.1007/s13278-015-0253-5
20. Khodabakhsh, M., Kahani, M., Bagheri, E.: Predicting future personal life events on Twitter via recurrent neural networks. J. Intell. Inf. Syst. (2018). https://doi.org/10.1007/s10844-018-0519-2
21. Ruz, G.A., Henríquez, P.A., Mascareño, A.: Sentiment analysis of Twitter data during critical events through Bayesian networks classifiers. Futur. Gener. Comput. Syst. **106**, 92–104 (2020). https://doi.org/10.1016/j.future.2020.01.005
22. Singh, N., Roy, N., Gangopadhyay, A.: Analyzing the emotions of crowd for improving the emergency response services. Pervasive Mob. Comput. J. **58**, 101018 (2019)
23. Azar, P.D., Lo, A.W.: The wisdom of twitter crowds: predicting stock market reactions to FOMC meetings via twitter feeds. J. Portf. Manage. **42**, 123–134 (2016). https://doi.org/10.3905/jpm.2016.42.5.123
24. Xie, Z., Liu, G., Wu, J., Tan, Y.: Big data would not lie: prediction of the 2016 Taiwan election via online heterogeneous information. EPJ Data Sci. **7** (2018). https://doi.org/10.1140/epjds/s13688-018-0163-7
25. Jaidka, K., Ahmed, S., Skoric, M., Hilbert, M.: Predicting elections from social media: a three-country, three-method comparative study. Asian J. Commun. **29**, 252–273 (2019). https://doi.org/10.1080/01292986.2018.1453849
26. Kušen, E., Strembeck, M.: Politics, sentiments, and misinformation: an analysis of the Twitter discussion on the 2016 Austrian Presidential Elections. Online Soc. Netw. Media **5**, 37–50 (2018). https://doi.org/10.1016/j.osnem.2017.12.002
27. Awwalu, J., Bakar, A.A., Yaakub, M.R.: Hybrid N-gram model using Naïve Bayes for classification of political sentiments on Twitter. Neural Comput. Appl. **31**, 9207–9220 (2019). https://doi.org/10.1007/s00521-019-04248-z
28. Ahmed, S.: My name is Khan: the use of Twitter in the campaign for 2013 Pakistan General Election. In: 2014 47th Hawaii International Conference on System Sciences, pp. 2242–2251 (2014). https://doi.org/10.1109/HICSS.2014.282
29. Tumasjan, A., Sprenger, T.O., Sandner, P.G., Welpe, I.M.: Predicting Elections with Twitter: What 140 Characters Reveal About Political Sentiment, pp. 178–185 (2010)

30. Tumasjan, A., Sprenger, T.O., Sandner, P.G., Welpe, I.M.: Election forecasts with Twitter: how 140 characters reflect the political landscape. Soc. Sci. Comput. Rev. **29**, 402–418 (2011). https://doi.org/10.1177/0894439310386557
31. Awais, M., Hassan, S.U., Ahmed, A.: Leveraging big data for politics: predicting general election of Pakistan using a novel rigged model. J. Ambient Intell. Humaniz. Comput. (2019). https://doi.org/10.1007/s12652-019-01378-z
32. Heredia, B., Prusa, J.D., Khoshgoftaar, T.M.: Social media for polling and predicting United States election outcome. Soc. Netw. Anal. Min. **8**, 1–16 (2018). https://doi.org/10.1007/s13278-018-0525-y
33. Khatua, A., Khatua, A., Ghosh, K., Chaki, N.: Can #Twitter-Trends predict election results? Evidence from 2014 Indian general election. In: Proceedings of Annual Hawaii International Conference on System Sciences, 2015-March, pp. 1676–1685 (2015). https://doi.org/10.1109/HICSS.2015.202
34. Bermingham, A., Smeaton, A.F.: On using Twitter to monitor political sentiment and predict election results. Sentiment Analysis where AI meets Psychol. Work. Int. Jt. Conf. Nat. Lang. Process., pp. 2–10 (2011)
35. Gomes, R.F., Casais, B.: Feelings generated by threat appeals in social marketing: text and emoji analysis of user reactions to anorexia nervosa campaigns in social media. Int. Rev. Public Nonprofit Mark. **15**, 591–607 (2018). https://doi.org/10.1007/s12208-018-0215-5
36. Bahri, S., Bahri, P., Lal, S.: A Novel approach of sentiment classification using emoticons. Procedia Comput. Sci. **132**, 669–678 (2018). https://doi.org/10.1016/j.procs.2018.05.067
37. Hogenboom, A., Bal, D., Frasincar, F., Bal, M., De Jong, F., Kaymak, U.: Exploiting emoticons in polarity classification of text. J. Web Eng. **14**, 022–040 (2015)
38. Spina, S.: Role of Emoticons as Structural Markers in Twitter interactions. Discourse Process. **56**, 345–362 (2019). https://doi.org/10.1080/0163853X.2018.1510654
39. Huddar, M.G., Sannakki, S.S., Rajpurohit, V.S.: Multi-level context extraction and attention-based contextual inter-modal fusion for multimodal sentiment analysis and emotion classification. Int. J. Multimed. Inf. Retr. (2019). https://doi.org/10.1007/s13735-019-00185-8
40. Corchs, S., Fersini, E., Gasparini, F.: Ensemble learning on visual and textual data for social image emotion classification. Int. J. Mach. Learn. Cybern. **10**, 2057–2070 (2019). https://doi.org/10.1007/s13042-017-0734-0
41. Tran, H.N., Cambria, E.: Ensemble application of ELM and GPU for real-time multimodal sentiment analysis. Memetic Comput. **10**, 3–13 (2018). https://doi.org/10.1007/s12293-017-0228-3
42. Cerezo, E., Hupont, I., Baldassarri, S., Ballano, S.: Emotional facial sensing and multimodal fusion in a continuous 2D affective space. J. Ambient Intell. Humaniz. Comput. **3**, 31–46 (2012). https://doi.org/10.1007/s12652-011-0087-6
43. Yu, J., Jiang, J., Xia, R.: Entity-sensitive attention and fusion network for entity-level multimodal sentiment classification. IEEE/ACM Trans. Audio Speech Lang. Process. **28**, 429–439 (2020). https://doi.org/10.1109/TASLP.2019.2957872
44. Song, K.S., Nho, Y.H., Seo, J.H., Kwon, D.S.: Decision-level fusion method for emotion recognition using multimodal emotion recognition information. In: 2018 15th International Conference on Ubiquitous Robotic UR 2018, pp. 472–476 (2018). https://doi.org/10.1109/URAI.2018.8441795
45. Williams, J., Comanescu, R., Radu, O., Tian, L.: DNN multimodal fusion techniques for predicting video sentiment. In: Proceedings of Grand Challenge and Workshop on Human Multimodal Language, pp. 64–72 (2018). https://doi.org/10.18653/v1/w18-3309
46. Bose, R., Dey, R.K., Roy, S., Sarddar, D.: Analyzing political sentiment using Twitter data. Smart Innov. Syst. Technol. **107**, 427–436 (2019). https://doi.org/10.1007/978-981-13-1747-7_41
47. Burnap, P., Gibson, R., Sloan, L., Southern, R., Williams, M.: 140 characters to victory? Using Twitter to predict the UK 2015 General Election. Elect. Stud. **41**, 230–233 (2016). https://doi.org/10.1016/J.ELECTSTUD.2015.11.017

CrO_2 Half Metal-Based Magnetic Tunnel Junction and Its Application for Digital Computing

Muzafar Gani, Khurshed A. Shah, Shabir A. Parah, and Altaf A. Balki

Abstract The half metal (CrO_2) ferromagnet has attracted immense research interest because of its large Curie temperature of 390 K and 100% spin polarization. Silicene a two-dimensional monolayer material shows outstanding magnetic and electronic properties. We have calculated the spin dependent transport of electron in our modeled device consisting of two CrO_2 electrodes and out of plane silicene as scattering region. We have simulated the device in ATK software which uses non-equilibrium greens function and density function theory for calculation of the transport characteristics. The device is simulated to obtain IV curve and transmission spectrum. Furthermore, spin injection efficiency and tunneling magnetoresistance are calculated from the obtained transport characteristics. The transmission spectrum is found to be in agreement with the IV curve. The spin degree of freedom is expected to increase the speed, decrease power dissipation and add nonvolatility to devices. Also some basic logic functions have been realized from the modeled device.

Keywords Spintronic devices · Logic gates · Spin transport · CrO_2 · Half metal

1 Introduction

Spintronics also termed as spin electronics is an evolving technology, which utilizes spin of an electron and the associated magnetic moment in combination with the charge of an electron. Astonishing results have been achieved in the electronic industry especially in data storage. Since the conventional CMOS technology is expected to reach the scaling limit posed by short channel and electrostatic effects, the scaling in CMOS devices is in agreement with Moor's law [1] from past three

M. Gani · S. A. Parah (✉) · A. A. Balki
Department of Electronics and Instrumentation Technology, University of Kashmir, Srinagar 190006, Jammu and Kashmir, India
e-mail: shabireltr@gmail.com

K. A. Shah
Department of Physics, S P College, Cluster University Kashmir, Srinagar 190001, India

P. K. Singh et al. (eds.), *Recent Innovations in Computing*, Lecture Notes in Electrical Engineering 701, https://doi.org/10.1007/978-981-15-8297-4_45

decades. The scaling limits has compelled the researchers for alternative to CMOS [2]. Spintronic devices because of their low power, high speed and nonvolatility have become a prominent choice for replacement of CMOS [3].

Magnetic tunnel junction (MTJ) constitutes of three layers, a barrier layer which is sandwiched in between the two ferromagnetic layers. The MTJ device shows small resistance, when the ferromagnetic layers are set to same magnetization configuration and large resistance when the ferromagnetic layers are set to opposite magnetization configuration. The change in resistance of the device by shifting the associated magnetic arrangements in ferromagnetic layers from antiparallel to parallel magnetic arrangements is due to diverse fermi functions that are associated with the spin up/down electrons that traverses through the device. The fermi function is a consequence of electronic configuration of the magnetic material, which results in magnetic configuration dependence of electrical resistance. The effect is termed as tunneling magnetoresistance. Spintronic devices like MTJ, magnetic domain walls, ferroelectric tunnel junction, spin FET, spin valve, etc., are expected to revolutionize the electronic industry by increased processing speed, decreasing power consumption and instant switching of devices [4].

The rutile structures of transition metal dioxides display interesting magnetic and electric properties. The band gap in transition metal dioxides varies from large value in TiO_2 to a good metallic conductor like RuO_2. CrO_2 is one of the important members of the family due to its unique magnetic properties. It has metallic and magnetic properties and corresponds to itinerant ferromagnets. CrO_2 shows highest Curie temperature of 390 K as compared to other transition metal oxides, it also presents half metallicity having 2 μB the magnitude of magnetic moment per chromium atom. One of the spin channels in case of half metals is metallic, while the other is semiconducting or insulating [5–7]. This results in unique transport properties of half metals.

In conventional electronics, the charge property of an electron is responsible for the representation of digital information (bit 0/1). But this paradigm undergoes fundamental limitation. Since charge of an electron is a scalar quantity consists of magnitude only. Thus, the logic can only be differentiated by magnitude of the charge, e.g., the existence of charge represents logic 0 and the absence of charge represents logic 1. This scheme is adopted in MOSFET transistor, when the charge is present in the channel, the transistor belongs to ONN state which represent logic bit 0. The charge depleted channel results in OFF state of the transistor, which represents logic 1. In order to switch the device between the logic levels, the charge in the channel needs either to be removed or accumulated, which results in the current flow and the associated power dissipation equal to I^2R, that is unavoidable in conventional electronics.

Instead spin, a pseudo vector with fixed magnitude but variable direction, the direction of electrons spin can be restricted to two states when placing the electron in magnetic field. These two states can be used to encode digital logic bit 1/0. Switching of the logic states merely requires flipping the spin of electron. Hence, the power losses are reduced by eliminating I^2R loss that belong to conventional devices. Energy dissipation associated with flipping of spin state of electron (logic

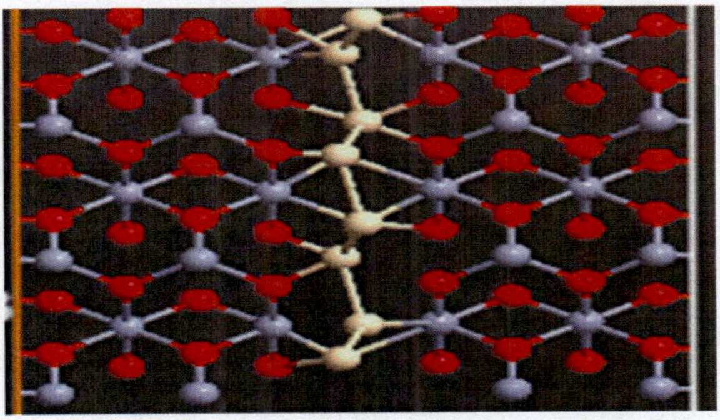

Fig. 1 Modeled device consisting of CrO_2 electrodes and out of plane silicene scattering region

switching) belongs to the difference in the energy between the polarization states which are distinguished by Zeeman energy ($g\mu_B B$), which may be reduced to fewer than $k_B T$ (thermal energy).

MTJs with out of plane scattering region are promising devices for the high density information storage as well as higher thermal stability [8]. They may also be helpful in dipping the required critical current density for reversal of magnetic configuration by spin transfer torque (STT) without altering thermal stability [9, 10]. There are many theoretical and experimental reports on out of plane MTJs [11–14]. Silicene due to its high mobility and other spin properties has attracted the research attention for use in spin devices [15]. In this work, we have carried out modeling and simulation of device consisting of CrO_2 electrodes and out of plane silicene as shown in Fig. 1. Spin dependent transport characteristics have been obtained. TMR and spin injection efficiency have been calculated from the obtained transport characteristics. Also, some basic logic functions are realized from the modeled device.

2 Models and Methods

The modeling and simulation of the device was carried out by ATK software which employs density function theory (DFT) [16] and non-equilibrium greens function (NEGF) in combination for electron transport calculation in materials at molecular level [14, 17]. The spin-polarized generalized gradient-approximation (SGGA) approximation is adopted since the flow of current is due to electrons through the device which is spin polarized. The transmission spectrum and I-V characteristics were calculated from the ATK software. In order to get the accurate results, the parameters used for simulation were set as follows: density-mesh cut-off adjusted to 155 Ry, single-zeta basis set and double-zeta basis set is used for electrodes and silicene scattering region, respectively. The k-point sampling was set to $3 \times 3 \times 100$ for accurate transport calculations. The transport calculations were carried out for parallel magnetization configuration (PC) and antiparallel magnetization configuration (APC). In PC, both the electrodes were set to spin up magnetization configuration, and in APC, one of the electrodes is set to spin up magnetization configuration and the other one to spin down magnetization configuration.

3 Result and Discussions

The current through the device is calculated in terms of transmission spectrum ($T(E, V_B)$). Transmission spectrum gives the probability of electron to cross through the device at energy E and applied bias voltage of V_b. The below equation represents the current flowing through the device [18].

$$I \uparrow (\downarrow) = \frac{e}{h} \int T \uparrow (\downarrow)(E, V_B) * [F(E - \mu_r) * F(E - \mu_l)] * dE \qquad (1)$$

where $I\uparrow(\downarrow)$ denotes spin up (spin down) current, $T\uparrow(\downarrow)$ (E, V_B) denotes the transmission spectrum for spin up (spin down) electrons, F is Fermi–Dirac distribution and $\mu_{r(l)}$ denotes chemical potential of left(right) electrode.

The variation of the current through the device with applied bias is shown in Fig. 2. In parallel magnetization configuration case, there is increase in the spin up current through the device with the applied bias, but the spin down current is small and remains almost same with the variation in applied bias. This results in high spin injection efficiency. In antiparallel configuration case, the spin up spin (down) current through the device remains almost same, which results in low spin injection efficiency.

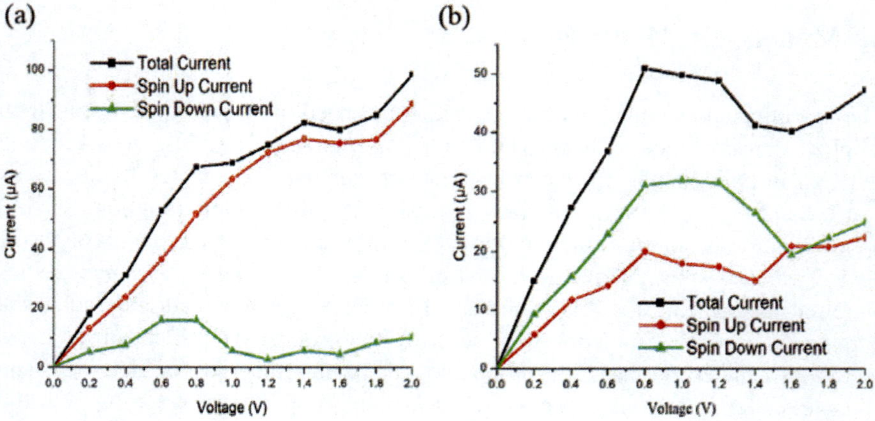

Fig. 2 IV plot for **a** parallel magnetization configuration and **b** antiparallel magnetization configuration for modeled device

We examine transmission spectrum to validate the achieved IV characteristics. The transmission spectrum for both parallel/antiparallel configuration is depicted in Fig. 3. Figure 3a represents the transmission spectrum for parallel configuration, and the green and blue curves indicate the spin up and spin down channel's transmission coefficient, respectively. In case of PC, for spin up channel near fermi level, the transmission peaks increases with applied bias voltage, while for spin down channel, the transmission is very small. Similarly for APC, for both spin up spin (down) channel, the transmission is small and almost same, and hence, these results justify the obtained IV characteristics.

The TMR given by $((I_p - I_{ap}) * 100)/I_p$ (where I_{ap} and I_p represent the net current in antiparallel and parallel configuration) versus voltage is represented in Fig. 4. The maximum of 120% is obtained at 2 V. Also, the spin injection efficiency given by $(I\uparrow - I\downarrow)/(I\uparrow + I\downarrow)$ (where $I\uparrow/I\downarrow$ represent spin up/spin down current) versus applied bias is presented in Fig. 5, for parallel/antiparallel configuration. The peak value of 100% efficiency is obtained at 1.20 V for PC that is due to negligible spin down current at 1.20 V.

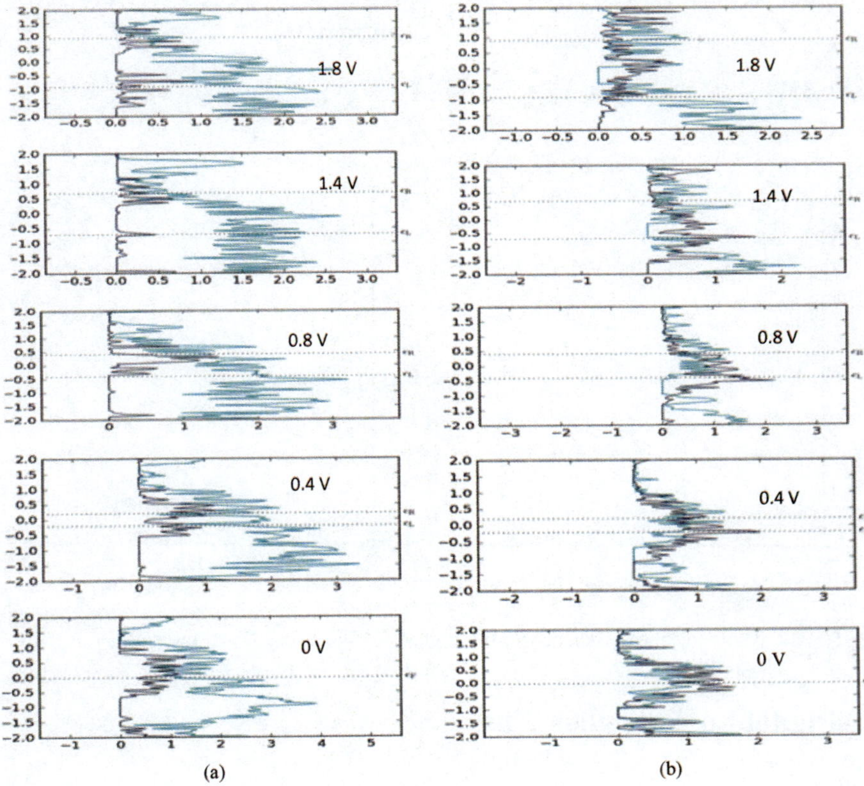

(a) (b)

Fig. 3 Transmission spectrum of the modeled device with **a** parallel magnetization configuration and **b** antiparallel magnetization configuration

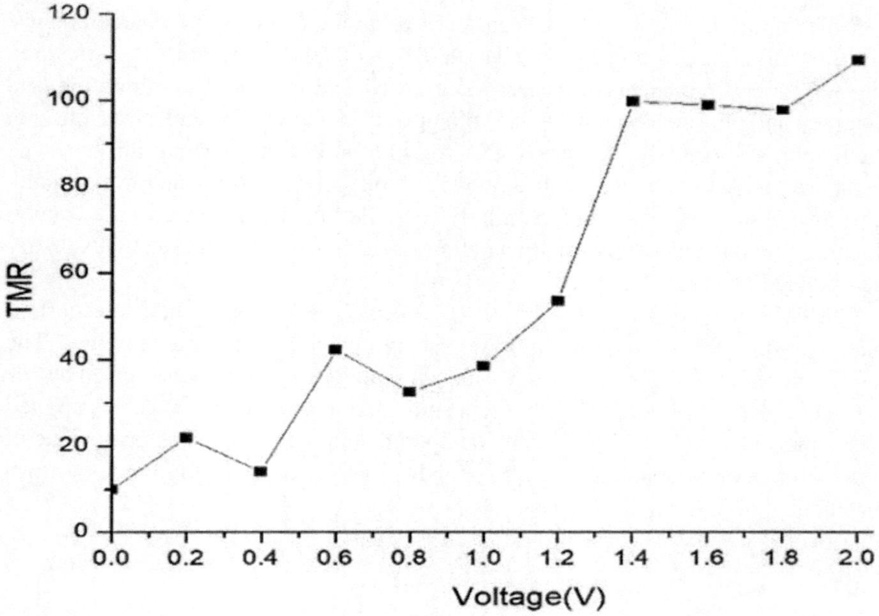

Fig. 4 TMR versus voltage plot

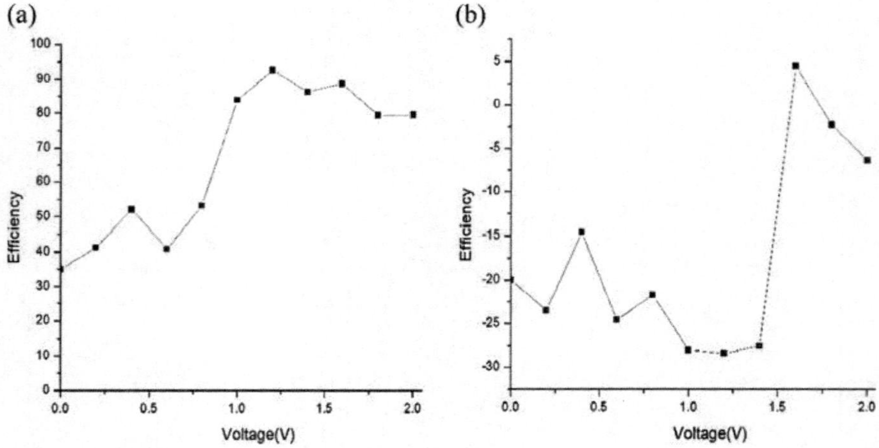

Fig. 5 Plot of spin injection efficiency versus applied voltage **a** for PC and **b** for APC

4 Digital Logic Applications

We have implemented some basic logic functions using the simulated device. The input of the device is represented by magnetization of the electrode, spin up configuration belongs to logic 1, and spin down configuration belongs to logic 0. The output

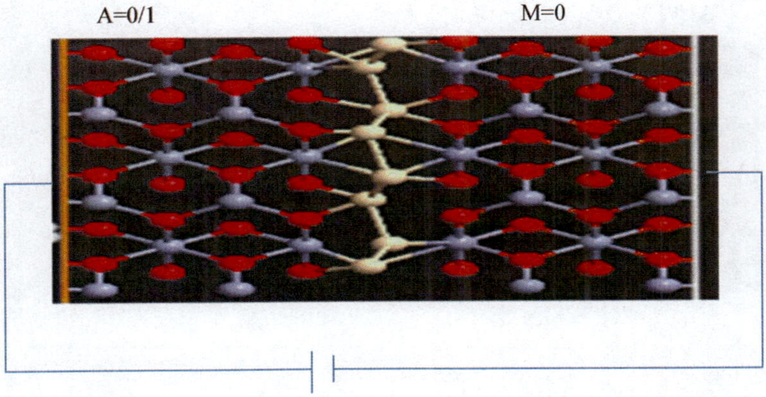

Fig. 6 Diagram for the realization of NOT logic

of the device is represented by the resistance of the device. Small resistance belongs to logic 1, and large resistance belongs to logic 0.

4.1 Not Gate

The schematic for NOT gate realization is shown in Fig. 6. The devices right electrode as shown in Fig. 6 is pinned to spin down configuration (logic 0), and the input A is represented by the magnetization of left electrode, logic 0 for spin down configuration and logic 1 for spin up configuration. The output is 1 (low resistance) when input is 0 and output is 0 when input is 1 and thus the realization of NOT gate.

4.2 AND Gate

The diagram for AND gate realization is shown in Fig. 7. The device corresponds to low impedance state, i.e., output is 1 only when both $A = B = 1$ (spin up configuration) which realizes the AND gate operation.

4.3 NAND Gate

The NAND logic realization can be elucidated from Fig. 8. The net resistance seen by the battery is low (logic 1), when any of the two inputs A/B is 0. The net resistance is high (logic 0) when both $A = B = 1$, hence the realization of NAND gate.

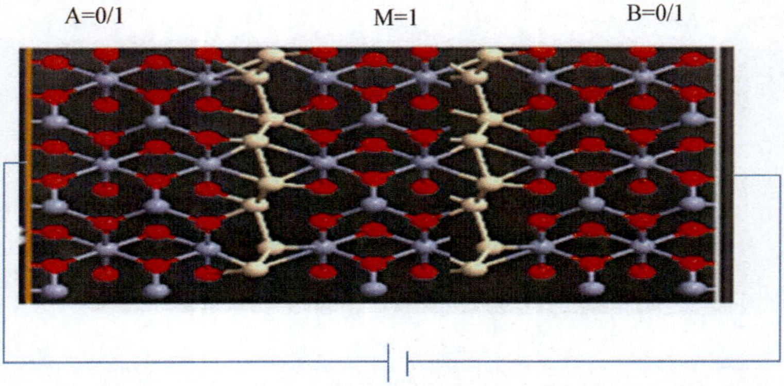

Fig. 7 Schematic diagram for realization of AND gate

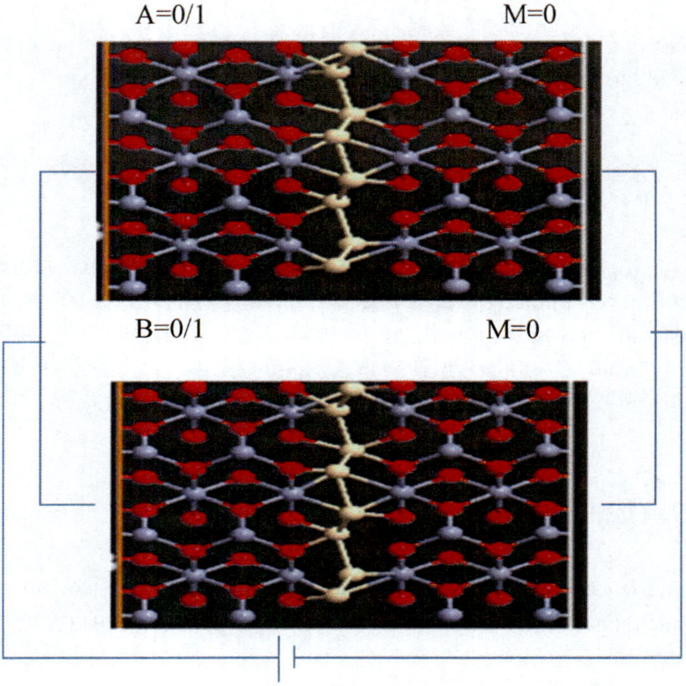

Fig. 8 Schematic diagram for realization of NAND gate

5 Conclusion

In the presented work, we modeled an MTJ consisting of CrO_2 electrodes and out of plane silicene scattering region. The simulation was carried out in ATK software which incorporates non-equilibrium greens function and density function theory for electron transport calculation. The device was simulated to obtain IV characteristics and transmission spectrum. Spin injection efficiency and tunneling magnetoresistance were calculated from the obtained IV characteristics. Furthermore, some logic functions were realized using the modeled device.

References

1. Moore, G.E.: Cramming more components onto integrated circuits. Reprinted from Electronics, volume 38, number 8, April 19, 1965, pp. 114ff, IEEE Solid-State Circ. Soc. Newsl. **11**(3), 33–35 (2006)
2. Zhirnov, V.V., Cavin, R.K., Hutchby, J.A., Bourianoff, G.I.: Limits to binary logic switch scaling-a gedanken model. Proc. IEEE **91**(11), 1934–1939 (2003)
3. Manipatruni, S., Nikonov, D.E., Young, I.A.: Beyond CMOS computing with spin and polarization. Nat. Phys. **14**(4), 338 (2018)
4. Hanyu, T., Endoh, T., Suzuki, D., Koike, H., Ma, Y., Onizawa, N., Natsui, M., Ikeda, S., Ohno, H.: Standby-power-free integrated circuits using MTJ-based VLSI comput-ing. Proc. IEEE **104**(10), 1844–1863 (2016)
5. Ranno, L., Barry, A., Coey, J.M.D.: Production and magnetotransport properties of CrO2 films. J. Appl. Phys. **81**, 5774 (1997)
6. Schwarz, K.H.: CrO_2 predicted as a half-metallic ferromagnet. J. Phys. F: Met. Phys. **16**, L211 (1986)
7. Choudhary, S., Khandate, S.: Implication of hydrogenation in tuning the magnetoresistance of graphene-based magnetic junction. IEEE Trans. Nanotechnol. **18**, 670–675 (2019)
8. Slaughter, J.M.: Materials for magnetoresistive random access memory. Annu. Rev. Mater. Res. **39**, 277–296 (2009)
9. Kent, A.D., Ozyilmaz, B., del Barco, E.: Spin-transfer-induced precessional magnetization reversal. Appl. Phys. Lett. **84**(19), 3897–3899 (2004)
10. Lee, K.J., Redon, O., Dieny, B.: Analytical investigation of spin-transfer dynamics using a perpendicular-to-plane polarizer. Appl. Phys. Lett. **86**(2), 022505(1)–022505(3) (2005)
11. Taniguchi, Y., Miura, Y., Abe, K., Shirai, M.: Theoretical studies on spin-dependent conductance in FePt/MgO/FePt(001) magnetic tunnel junctions. IEEE Trans. Magn. **44**(11), 2585–2588 (2008)
12. Kohn, A., Tal, N., Elkayam, A., Kov_acs, A., Li, D., Wang, S., Ghannadzadeh, S., Hesjedal, T., Ward, R.C.C.: Structure of epitaxial L10-FePt/MgO perpendicular magnetic tunnel junctions. Appl. Phys. Lett. **102**(6), 062403(1)–062403(5) (2013)
13. Ikeda, S., Miura, K., Yamamoto, H., Mizunuma, K., Gan, H.D., Endo, M., Kanai, S., Hayakawa, J., Matsukura, F., Ohno, H.: A perpendicular-anisotropy CoFeB–MgO magnetic tunnel junction. Nat. Mater. **9**(9), 721–724 (2010)
14. Yang, G., Li, D.L., Wang, S.G., Ma, Q.L., Liang, S.H., Wei, H.X., Han, X.F., Hesjedal, T., Ward, R.C.C., Kohn, A., Elkayam, A., Tal, N., Zhang, X.-G.: Effect of interfacial structures on spin dependent tunneling in epitaxial L10-FePt/MgO/FePt perpendicular magnetic tunnel junctions. J. Appl. Phys. **117**(8), 083904(1)–083904(3) (2015)
15. Zheng, H., Zhang, R., Han, H., Liu, C., Yan, Y.: Electric field induced modulation to the magnetic anisotropy of Fe/silicene heterostructures: first-principles study. J. Magn. Magn. Mater. **484**, 172–178 (2019)
16. Brandbyge, M., Mozos, J.L., Ordejon, P., Taylor, J., Stokbro, K.: Density-functional method for nonequilibrium electron transport. Phys. Rev. B **65**, 165401 (2002)

17. Taylor, J., Guo, H., Wang, J.: Ab initio modeling of quantum transport properties of molecular electronic devices. Phys. Rev. B **63**, 245407 (2001)
18. Gani, M., Shah, K.A., Parah, S.A., Misra, P.: Room temperature high Giant Magnetoresistance graphene based spin valve and its application for realization of logic gates. Phys. Lett. A **384**(7), 126171 (2020)

Cloud of Things: A Systematic Review on Issues and Challenges in Integration of Cloud Computing and Internet of Things

Sahilpreet Singh, Arjan Singh, and Vishal Goyal

Abstract Cloud computing and the Internet of things (IoT) are two diverse technologies having complimentary relationship. The IoT generates massive amounts of data, and cloud computing provides a pathway for that data to travel to its destination. In the modern era, by integrating cloud computing and the Internet of things, a new paradigm has been introduced, i.e., cloud of Things. Cloud-based Internet of Things or cloud of things arose as a platform for intelligent use of applications, information in a cost-effective manner. Both technologies help to raise efficiency in the future. But the integration of these two technologies is challenging and bears some key issues. Therefore, this paper provides a brief investigation of cloud of things concept. In this paper, we review the literature about integration, to analyze and discuss the need behind integration in various applications. In the end, we identify some of the issues and challenges for future work in this promising.

Keywords Cloud computing · Internet of things · Cloud of Things

1 Introduction

Today's Internet of things (IoT) is one of the most modern IT buzzwords. The term "Internet of Things" (IoT) was first used by Kevin Ashton, a very renowned British technology pioneer in the year 1999. His ideology was to imagine a smart world where we had computers around us which already knew information that we need to

S. Singh (✉) · V. Goyal
Department of Computer Science, Punjabi University Patiala, Patiala, Punjab, India
e-mail: ersahilpreetsingh@gmail.com

V. Goyal
e-mail: vishal.pup@gmail.com

A. Singh
Department of Mathematics, Punjabi University Patiala, Patiala, Punjab, India
e-mail: arjanpu@gmail.com

P. K. Singh et al. (eds.), *Recent Innovations in Computing*, Lecture Notes in Electrical Engineering 701, https://doi.org/10.1007/978-981-15-8297-4_46

know. He showed a network-oriented structure where all the things and objects shall be interconnected with the support of sensors.

The Internet of things is related as revolution in Internet in contrast with machine-to-machine learning. In beginning, the first version of the Internet was built around the concept to capture and store data created by people only [1]. As the new technologies' evolved, the next version of Internet was developed to store data created by smart things. IoT provides numerous Internet technologies including both wired and wireless communication and further emerging and engaging technologies like embedded systems and micro-electromechanical systems. IoT adds connectivity with people and devices through sensors which help them in generating a free-flowing conversation between man and machine. With the help of advanced technologies like artificial intelligence and machine learning, all the non-living objects or things can be converted into smart things. IoT eventually work when its multiple building blocks simultaneously run and communicate with each other. Beside former technologies, other information systems such as smart buildings and homes, wireless sensor networks, GPS, control systems all supported the IoT [2] (Fig. 1).

1.1 Internet of Things Approach

The primary focus of IoT approach is to merge Internet of things, computer networks, services and people (IoTSP), Internet of energy (IoE), and Internet of media (IoM) in a single information technology concept [3]. Internet of things, service and people (IoTSP) is composed of elements from both physical and as well as Internet world. It keeps people as decision-makers who program or control the process. It includes WSN, RFID technologies, smartphones, databases as well as traditional Internet like Web. In IoTSP, things could be devices, formed a network and other embedded objects that enables the exchange of data. IoE term extends the Internet of things (IoT) to describe a more complex system to the world. It includes machine-to-machine communication, machine-to-people (M2P), and people-to-people (P2P) interaction.

The main focus behind is the integration of IoT with the building block of digital society as well as the digital economy. Internet of things vision revolutionized with the conjunction of having a variety of technologies such as ubiquitous computing, advancements in machine learning rise of wireless communication, and evolution in real-time analytics over the time in recent years. According to IoT expertise, next evolution toward mankind is closer to devices that are not just computers but also appliances of varying nature. The networking giant estimates that by 2020 (2 years from now), the world will have anywhere from 50 to 200 billion connected devices (Fig. 2).

The Internet of things connectivity concept depicts that things and humans can be connected at anytime with anything and anyone, splendidly using whatever path or network and any sort of service.

The visions of Internet of things based on futuristic approach and toward its final architecture are still divergent.

1.2 Cloud Computing

A network-based computing that helps in to store and access data or information over the Internet. It is one of the booming technologies which provides various services that accelerate the computer industry. The major goal of cloud computing is making sure that users should be able to use all the computing resources in very easiest and efficient way by keeping in mind their demand and utilization.

1.2.1 Classification of Cloud [4]

1. **Public Cloud**: As the name describes public cloud is basically open for all systems and services which are and can be accessible to the public. As it delivers a huge number of resources from numerous locations, it is considered as one of the reliable cloud. In any worst case if a resource got failed, a new resource is provided. It has the advantage of low cost of ownership, automated deployments, and scalability.
2. **Private Cloud**: In this, system and services are accessible individually or within an organization. No other user from outside cannot access the cloud. It offers the greatest level of control and security and also offers greater configurability support to any application.
3. **Hybrid Cloud**: The hybrid approach of public and private cloud is known as hybrid cloud. This type of cloud is widely used by organizations. In this scenario, the private cloud is responsible for conducting critical activities and on the other side, public cloud helps in conducting non-critical activities. It is the most control, flexible, and cost-effective cloud.
4. **Community Cloud**: This cloud permits either system or services which can be approachable by variety of numerous organizations. It can be either maintained or operated internally by number of organizations or with the help of third party. On the basis of security, community cloud is much secure in contrast with public cloud and less secure as compared to private cloud.

1.2.2 Cloud Service Models [4]

Cloud models come in three types: Software as a Service (SaaS), Infrastructure as a Service (IaaS), and Platform as a Service (PaaS). Each of the cloud models has their own set of benefits that could serve the needs of various businesses.

1. **Infrastructure as a Service (IaaS)**: One of the most popular and lowest level of cloud service. Earlier name given to it was Hardware as a Service (HaaS). HaaS provides the customer infrastructure with virtual machines and other services. This feature enables the advantage that rather than purchasing a particular server, customers can purchase resources, network equipment. The main benefit here is the independency of the user to choose CPU and memory storage depending

upon its requirement. Microsoft Azure, Amazon EC2, IBM, and many more are IaaS services.

2. **Platform as a service (PaaS)**: PaaS generally provides programming platforms toward all the developers around, where they can do all the programming tasks such as testing, compile, execute, and control all the applications. It is considered as one of the most flexible model that minimizes the organizational cost and can improve the time of development. One of the main advantages of Paas is that developers without worrying about the infrastructure can concentrate on the development and further toward innovation. Services provided by PaaS are Google Apps Engine (GAE), Window Azure, SalesForce.com.
3. **Software as a Service (SaaS)**: "On-Demand Software" is the most favored among all the former services mentioned above. This basically acts as a distributed model in which a variety of applications can be accessed by customers over the Internet by hosting with the help of cloud service provider (CSP). Moreover, it allows organizations to access various work functionalities on a small cost. But the problem is it is totally dependent on the Internet, without the Internet, service is not usable. According to a recent survey, 74% of the cloud workload is Software as a Service (SaaS) workload. Google and Microsoft are the service provider by SaaS.

2 Integration of Cloud and IoT

Both Cloud, as well as IoT, have become closely allied future Internet technologies in which one providing the other a platform for success [1]. They work toward increasing the efficiency of everyday tasks. As we know, technology is rising day by day. The number of connected devices also increases. All these devices contain a huge amount of data or information. So the data or information from these devices needed to store on the cloud. For that purpose, a rental storage space is needed, so that the data can be used in an efficient way. This can be done by integrating both technologies.

2.1 Need for Integration

1. **Storage resources**: IoT basically integrates vast information, producing non-structured, or semi-structured data. On the other hand, cloud is so far considered as a much convenient and cost-effective solution to deal with data produced by IoT [4]. Thus, the integration between cloud and IoT brings a new scenario from which vital opportunities can arise for data processing like data aggregation, dissemination with third parties, and integration.

2. **Computational resources**: One of the prominent drawbacks is on-site data processing of IoT devices. The collected data or information is transmitted through the strong number of nodes present in the system. The numerous data processing capabilities of cloud toward IoT concluded that IOT processing has to be efficiently satisfied and prediction algorithms should be possible and accessible at as less cost as possible.
3. **Communication resources**: IoT and cloud integration provide pervasive ubiquitous in real-world things. The Internet, nowadays, acts as point of queasiness for providing a variety of services as well as distribution toward big data, which can further communicate through dedicated hardware. The major drawback is the cost of communication that quite expensive (Fig. 3 and Table 1).

2.2 Issues in IoT and Cloud Computing Integration [1]

As we know that, IOT and cloud computing are booming technology. Their integration has a great impact on computer technology or network technology. Both technologies are responsible to maintain Quality of Services (QOS). But on the other hand, combining two technologies brings many challenges on the basis of security.

1. **Heterogeneity**: A wide heterogeneity associated with devices, operating systems, platforms, and forces in both technologies one of the big challenges to maintain security from every aspect.
2. **Performance**: For transferring and collecting from IoT devices to cloud, there is a need for high bandwidth. So the challenge is to need of adequate network with a better performance. Moreover, their integration increases the performance and needs of QoS at various levels.
3. **Reliability**: The presence of reliability in both technologies is a challenging task. The need of transmission devices which can access the information can communicate over the cloud network. On cloud, the need for proper storage space and resources that are highly reliable in terms of privacy or security.
4. **Big Data**: Real-time data analysis and security are the major concern of big data. As technology rises day by day, the number of devices in IoT also increases and a survey estimated that possibly a huge estimate of 50 billion devices eventually be networked and connected by the year 2020. The main challenge is to store vast amounts of data on the cloud and to extract important information and regulate patterns from the data. Also, cloud security from various attacks, ransomware, and to find solutions is a big challenge.
5. **Resource Allocation**: When several IoT devices and unexpected things come together asking for resource, then it would be a challenging task. It would be not easy to what kind of resource required for particular IoT. The mapping with resource allocation depends upon the frequency, time, and amount of the sensor or device is being used.
6. **Service Discovery**: On cloud, the responsibility of the cloud manager is to provide or discover new services to the user. We know IoT is so vast that any

object can become any part or can leave at any time. Many IoT nodes are mobile so it would be an issue to update the service according to requirement. So there is a need for a uniform way of service that can track the records and keep updating.

3 Related Work

Khorshed et al. (2011) has presented rule-based learning method to monitor the inside activities on cloud environment. It is been always a challenging task to monitor the attack or malicious activities. So, for the experiment, rule-based learning method has been proposed to identify the activities. To evaluate the result, different classifier algorithms have been tested and concluded that C4.5 classifier rule-based technique is much better than among others.

Patel et al. (2012) have explored the intrusion detection and prevention method in cloud computing. They presented a taxonomy and ideal requirement to design an IDPS. For that, they concentrated on autonomic computing, ontology, risk management, and fuzzy theory. The main goal of the research is to establish trade off relationship and to maximize the security of the system.

Bhattasali et al. (2013) has discussed about the security and trust issues related to cloud of things. Their main focus on to propose a secure, light-weighted cryptography method. For that, they started pairing of communicate among different parties. Proposed mechanism helps in to utilize the benefits between cloud and things.

Zhou et al. (2014) have presented architecture by integrating Internet of things with cloud computing. Their main objective is to accelerate the IoT application on cloud environment. For experimental work, they proposed cloud-based Internet of things platform. Proposed architecture accommodates the cloud models for developing, deploying, and running the IoT applications.

Farooq et al. [5] have analyzed the security issues in Internet of things. They focused on general IoT structure, security goals, and major challenges on each layer. By seeking new possible security solutions, they presented a well-defined architecture for IoT. The proposed framework is responsible to communicate between several IoT devices and provide secure services to end user.

Hodo et al. (2016) have presented threat analysis of IoT using neural network. Their main objective is to focus on threat pattern on IoT network. For experimental work, they used multilayer perceptron classifier algorithm for classifying the attacks. In the end, it is concluded that proposed approach is reliable and monitor the attacks with high accuracy.

Khan et al. (2017) have discussed about IoT security issues, challenges, and blockchain solutions. In their work, they considered popular security issues with IoT-layered architecture, networking protocols, communication, and management. They concluded that blockchain plays a role of a key enabler to solve IoT challenges.

Gupta et al. (2017) have reviewed about various techniques in IoT secure transmission of data. In their research, they identify the right technique for cryptography.

For that purpose, they proposed linguistic technology, semantic reasoning systems, and showed that it is one of the best cryptographic techniques in security purpose.

Agrawal et al. [6] have presented a lightweight approach, that detect DDOS attacks on cloud. Proposed approach experimented on two parameter: they are the traffic flow and the flow count. After working with number of nodes, it has been concluded that, proposed approach was adaptive and detect attacks with high accuracy.

Aldaej et al. (2017) have proposed an intrusion prevention algorithm to enhance the cyber security. Their main focus on to control the bandwidth attacks, which are difficult to detect. The presence of bandwidth attacks on network decreases the performance. The proposed approach helps in to maintain the security, integrity of network from DDOS bandwidth attacks.

Banerjee et al. [2] have observed the problem of integrity of sharing dataset and security. For that purpose, they proposed firmware probing approach for detection and self healing blockchain mechanism for prevention. For experimental research, they work on securities of blockchain mechanisms, applications like Internet of military things, mechanisms of IoT self-healing, approaches toward IoT security, frameworks of Intrusion prevention system, etc.

Wang et al. [7] have discussed about new security challenge to identify algorithm for security of cloud assistance in IoT. In order to collect, store, and access data and for maintaining confidentiality, various methods are discussed. They focused on communication confidentiality between inside objects and edge objects data. For that purpose, they worked on proxy re-encryption-based mechanism, and proposed (CIBPRE) approach abbreviated as conditional-based identity broadcast proxy re-encryption. This approach allows user to confidentially collect and store data on cloud and share among others efficiently.

Wazidet et al. (2018) have discussed about cloud-driven IoT-based environment to show how the objects can route, purify, and swap meaningful data over the Internet. In their work, they focused on big data-based secure authentication schemes for cloud-driven IoT. For experimental work, they proposed method and work on physically secure authentication schemes, cross platform authentication, and granular auditing.

Xiao et al. [8] have reviewed on machine-learning (ML) security solutions in IoT and also investigate the IoT attack model. For that, they worked with supervised learning, unsupervised learning, and reinforcement learning (RL). They concluded that learning-based IoT malware detection-KNN, SVM, Q-learning are shown to be promising protection for the IoT.

Zhou et al. [9] have presented the impacts of newly developed features involved in security as well as privacy. They focused on the majors concerns toward security as well as privacy impacts which can affects the confidentiality, availability of data, and also mentioned the feasible solutions for them as well. They worked on cryptography methods to Insecure Protocols, Privacy Leak, Malware propagation. For experimental work, they have shown security of data would be possible by working on Homomorphic Encryption, Anonymous protocol, and Dynamic Configuration.

Stergiou et al. [10] have focused on gaps based on security and privacy of big data on cloud. They proposed an architecture which improves the security issues like edge node limited processing capability and optimization problem. To overcome this issue, they used deep learning tasks, used different CNN networks, and Python. From all aspects, they have concluded that, cloud computing offers better "green" environment.

Vorakulpipat et al. [11] have focused on IoT security challenges and concerns. For that, they discussed the generations of IoT and challenges related to its security. Their main aim to work on cloud-secure-based data transmission. To minimize the security risks purpose, they proposed a cloud-based-centralized IoT service platforms.

Aggarwal et al. (2018) have presented lightweight intrusion detection method to overcome intrusions. They focused on learning algorithms for monitoring the effectiveness of network security and relatable problems existing in network security such as access control and authentication. For experimental work, they used machine learning concept and proposed a lightweight intrusion detection method algorithm to overcome intrusions.

Gurulakshmi et al. (2018) analyzed and monitored IoT Bots against distributed denial of service attacks with the help of machine learning algorithms. Major goal of their research is toward support vector machine algorithm (SVM) to classify the traffic and abnormal flow in the network traffic. The experimental work with SVM and K-NN classifier in detection resulted out with commendable accuracy.

Hajimirzaei et al. (2018) recommended advanced intrusion detection system (IDS) established on combination of learning algorithm and ant colony algorithm. In their work, learning algorithms identified traffic packets, and training of data is done by ABC algorithm. NSL-KDD data has been approached and fFor designing network Cloud Sim is used.

Jia et al. (2018) have discussed impacts of newly developed features and their relevant solutions on network security as well as privacy. They studied about the various threats on layer of IoT that may affect on internal and external objects. Survey conducted provides reasonable solutions toward network security.

Geetha et al. (2019) proposed a framework for smarter Internet of things devices known as lightweight machine learning-based authentication framework. Traditional-based authentication and biometric template-based authentication approaches were introduced to probe and remove threats that affecting the reliability and efficiency of the system. Simulation method is used and it helps to obtain result with minimal error rate as compared to existing system.

Moon et al. (2019) have presented cloud-edge collaboration framework for IoT data analytics. Their main focus on network dependency, privacy, and delayed response time of IoT data on cloud. For that, they transfer data to edges but still edges are lacking in capacity to store large amount of data. For that purpose, they proposed a new framework structure, that consist of IoT data management module and ML model management module for processing on cloud. It helped to generate the model and actuator operation.

Sharma et al. (2019) presented structure that defines the reaction of attacks in terms of security, reliability, and flexibility. For that purpose, they analyzed or monitor the protection of leakage data, network segmentation, and communication between various devices. For experimental purpose, they proposed Op CloudSec architecture that maximize the identification rate of attacks.

Bojović et al. [12] have implemented a hybrid detection method to detect denial of service attacks. This approach has been simulated under a controlled DDoS based experiment in the real time scenario with two major kind of attacks, i.e., Internet Control message protocol flood attack and Transmission control protocol sync flood attack. To evaluate the results various parameters such as detection rate, recall, precision has been considered.

Wani et al. [13] have analyzed and detect denial of service attacks on cloud using machine learning. To evaluate the performance, various classification algorithms have been used by designing own cloud environment. In the experimental work, the database generated by Snort and was classified by machine learning algorithms support vector machine, random forest, and Naïve Bayes in weka learning tool. Result shows that the SVM algorithm model's has high performance.

Mani et al. (2019) have discussed about security challenges in cloud computing networks. They discuss network security in cloud computing has always been an issue; in their research, they focus on to fill the technological gaps and different approaches need behind it to face the challenge.

Shafi et al. [14] have presented a blockchain preventation method from botnet attacks in Internet of things. They proposed distributed botnet detection for IoT that uses blockchain and software defined networking. For experimental work, minimet emulation tool is used on different topologies. It helps to verify the flow rule and updating of process to prevent from botnet bondage between IoT devices.

4 Current Issues and Challenges in Cloud of Things

Year	Author	Issue	Challenges
2011	Khorsheed et al.	To monitor the attacking activities in cloud environment	To identify the appropriate rue-based algorithm to malicious activities
2012	Distefano et al.	To monitor the resources of cloud models	Need to deal with an ad hoc infrastructure (56)
2012	Patel et al.	To work on various ontology's and fuzzy theories on cloud to maximize the security	For autonomic computing to reduce risk and establish tradeoff relationship
2013	Bhattasali et al.	To examine the security and trust issues in cloud computing	To propose a secure and light-weighted cryptography method for communication purposes
2014	Zhou et al.	To accelerate the IoT application on cloud environment	There is a need to work on the requirement of Internet of things based platform for deploying and running of IoT applications
2014	Aazam et al.	Data Breaching, time framing and uploading	Need of an efficient gateway

(continued)

(continued)

Year	Author	Issue	Challenges
2015	Farooq et al.	To establish communication among each layer of network	Need of secure IoT architecture that helps in to communicate between several IoT devices and provide secure services to end-user
2016	Hodo et al.	To focus on threat analysis of IoT using neural network	Need for appropriate neural classifier attacks to classifying them
2017	Khan et al.	Jamming adversaries, insecure initialization and configuration, buffer reservation attack, RPL routing attack, and insecure interfaces [1]	Measuring signal strength, change of frequencies and locations, Split buffer approach requiring complete transmission of fragments, disallowing weak passwords
2017	Stergiou et al.	For Internet of Things and cloud to recognize the right secure integration technique	Need to work on semantic network process and also on lightweighted techniques [1]
2017	Banerjee et al.	Integrity of sharing dataset and security	Need of better firmware detection and other security-based methods [2]
2017	Rebaah et al.	Insufficient cloud storage, slow processing speed of data, and system security [15]	Need to connect among various devices without human interference
2017	Deogirikar et al.	Finding the most prominent attacks in IoT security	To handle security, need of an efficient-based security mechanism [16]
2018	Xiao et al.	Data privacy, spoofing attacks, intrusions, DoS attacks present on cloud-IoT driven environment	Learning-based IoT malware detection-KNN, SVM, Q-learning [8]
2018	Zhou et al.	Insecure Protocols, Privacy Leak, Malware propagation [9]	Homomorphic encryption, anonymous protocol, dynamic configuration
2018	Stergiou et al.	Edge node limited processing capability, optimization problem [17]	Focus on deep learning tasks, used different CNN networks, Phython
2018	Hajimirzaei et al.	To classify normal and abnormal packets	Need for meta-heuristic methods to detect error rate
2018	Deng et al.	Mobile intrusion detection, fault tolerance and intrusion [18]	Need for more feature selected lightweight intrusion detection method
2018	Moon et al.	Finding the most prominent attacks in IoT security [19]	Need for a security mechanism which handles maximum security
2018	Aggarwal et al.	How to improve the detection mechanism to recognize intrusions [6]	Requirement for better lightweight intrusion detection method

(continued)

(continued)

Year	Author	Issue	Challenges
2019	Bojović et al.	To test DOS attacks in real network [12]	Need for identification method to find various DDOS attacks
2019	Wani et al.	Presence of critical attacks on network compromise the availability [13]	Need of more feature selection model to detect attacks
2019	Bhardwaj et al.	Vulnerabilities in the cloud infrastructure [20]	Need of an efficient DDoS detection and defense mechanisms on cloud platform
2019	Khan et al.	Data entrust, loop hole for misusers and to investigate both supervised, and unsupervised machine learning capability	Need of autoencoders to improve accuracy of result in unsupervised learning [21]
2019	Dutta et al.	Security issue on network layer	Requirement of lightweighted cryptography method [22]
2019	Wu et al.	User privacy, address attacks, access control, secure offloading	Partial state observation, ML-based security, and RL-based security methods
2019	Sahfi et al.	To prevent network from bot forces or botnet attacks [14]	Need for blockchain methodology for different topologies
2019	Taher et al.	To classify network traffic whether it is normal or abnormal	Need of higher capacity server platform [23]

5 Conclusion

The concept of cloud of things which is from the combination of modern cloud technology and IoT represents a big leap ahead in the future Internet. The IoT is becoming an increasingly ubiquitous computing service that requires huge volumes of data storage and processing capabilities. It also has limited capabilities in terms of processing power and storage, also some issues such as security, integrity, privacy, performance, and reliability. As such, integration of cloud and IoT is beneficial to overcome these challenges. This paper demonstrates the need for a cloud-based IoT paradigm and also focused on different application scenarios, challenges, and open research directions that require more attention in the future.

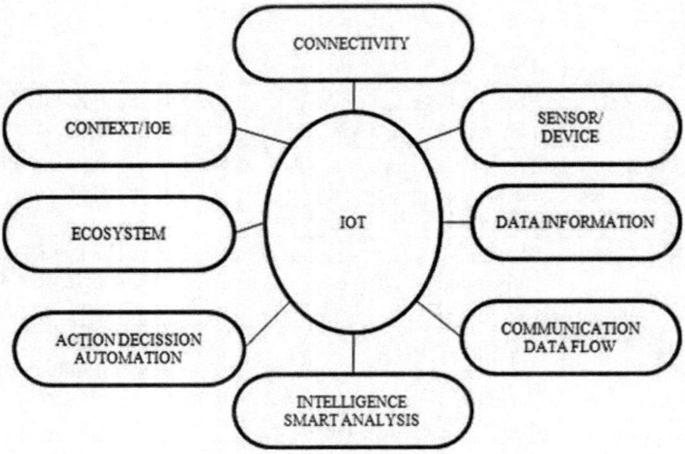

Fig. 1 Basic phenomenon of Internet of things

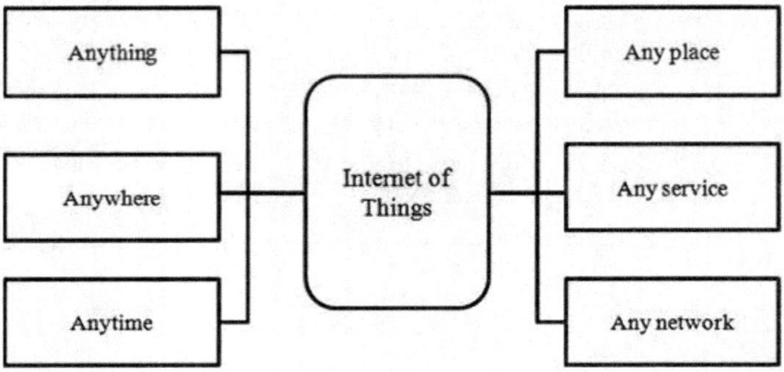

Fig. 2 Connectivity of IoT

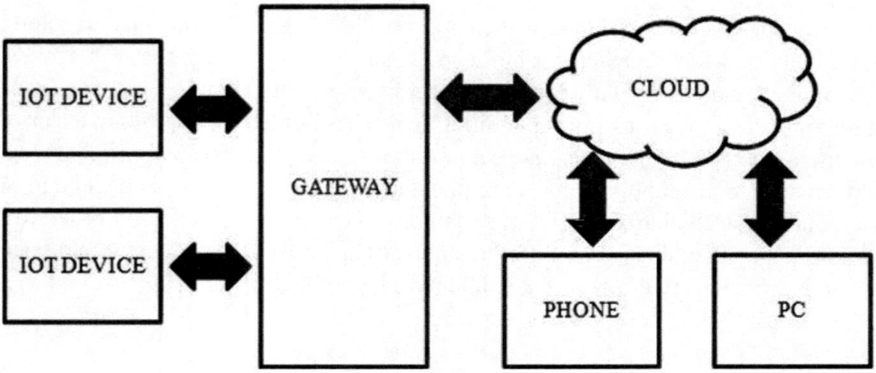

Fig. 3 General architecture of cloud-based IoT

Table 1 Complementary aspects of cloud and IoT

	IoT	Cloud
Displacement	Pervasive	Centralized
Reachability	Limited	Ubiquitous
Component	Real world things	Virtual resources
Computational capabilities	Limited	Virtually unlimited
Storage	Limited	Virtually unlimited

References

1. Stergiou, C., Psannis, K.E., Kim, B.G., Gupta, B.: Secure integration of IoT and cloud computing. Futur. Gener. Comput. Syst. **78**, 964–975 (2016)
2. Banerjee, M., Lee, J., Choo, K.K.R.: A blockchain future to Internet of Things security: a position paper. Dig. Commun. Netw. 217–222 (2017)
3. Gubbi, J., Buyya, R., Marusic, S., Palaniswami, M.: Internet of Things (IoT): a vision, architectural elements, applications and future directions. Futur. Gener. Comput. Syst. **29**(7), 1645–1660 (2013)
4. Cook, A., Robinson, M., Ferrag, M.A., Maglaras, L.A., He, Y., Jones, K., Janicke, H.: Internet of cloud: classification security and privacy issues. In: Cloud Computing for Optimization: Foundations, Applications, and Challenges, pp. 271–301. Springer, Berlin (2018)
5. Farooq, M.U., Waseem, M., Khairi, A., Mazhar, S.: A critical analysis on the security concerns of internet of things (IoT). Int. J. Comput. Appl. 111–117 (2018)
6. Agrawal, N., Tapaswi, S.: A lightweight approach to detect the low/high rate IP spoofed cloud DDoS attacks. In: IEEE 7th International Symposium on Cloud and Service Computing (SC2), pp. 118–123 (2017)
7. Wang, W., Xu, P., Yang, L.T.: Secure data collection, storage and access in cloud-assisted IoT. IEEE Cloud Comput. 77–88 (2018)
8. Xiao, L., Wan, X., Lu, X., Zhang, Y., Wu, D.: IoT security techniques based on machine learning, pp. 277–284 (2018)
9. Zhou, W., Jia, Y., Peng, A., Zhang, Y., Liu, P.: The effect of IoT new features on security and privacy: new threats, existing solutions, and challenges yet to be solved. IEEE Internet of Things J. 367–384 (2018)
10. Stergiou, C., Psannis, K.E., Kim, B.G., Gupta, B.: Secure integration of IoT and cloud computing. Futur. Gener. Comput. Syst. **78**, 964–975 (2018)
11. Vorakulpipat, C., Rattanalerdnusorn, E., Thaenkaew, P., Hai, H.D.: Recent challenges, trends, and concerns related to IoT security: an evolutionary study. In: IEEE 20th International Conference on Advanced Communication Technology (ICACT), pp. 405–410 (2018)
12. Bojović, P.D., Bašičević, I., Ocovaj, S., Popović, M.: A practical approach to detection of distributed denial-of-service attacks using a hybrid detection method. Comput. Electr. Eng. **73**, 84–96 (2019)
13. Wani, A.R., Rana, Q.P., Saxena, U., Pandey, N.: Analysis and detection of DDoS attacks on cloud computing environment using machine learning techniques. In: IEEE Amity International Conference on Artificial Intelligence (AICAI), pp. 870–875 (2019)
14. Shafi, Q., Basit, A.: DDoS botnet prevention using blockchain in software defined Internet of Things. In: IEEE 16th International Bhurban Conference on Applied Sciences and Technology (IBCAST), pp. 624–628 (2019)
15. Rebbah, M., Rebbah, D.E.H., Smail, O.: Intrusion detection in cloud Internet of Things environment. In: IEEE 2017 International Conference on Mathematics and Information Technology (ICMIT), pp. 65–70 (2019)

16. Deogirikar, J., Vidhate, A.: Security attacks in IoT: a survey. In: IEEE International Conference on I-SMAC (IoT in Social, Mobile, Analytics and Cloud) (I-SMAC), 2017, pp. 32–37
17. Aldaej, A.: Enhancing cyber security in modern Internet of Things using intrusion prevention algorithm for IoT. IEEE Access 964–975 (2018)
18. Deng, L., Li, D., Yao, X., Cox, D., Wang, H.: Mobile network intrusion detection for IoT system based on transfer learning algorithm. Cluster Comput. 1–16 (2018)
19. Moon, J., Cho, S., Kum, S., Lee, S.: Cloud-edge collaboration framework for IoT data analytics. In: International Conference on Information and Communication Technology Convergence (ICTC), Oct 2018, pp. 1414–1416. IEEE (2018)
20. Bhardwaj, A., Sharma, A., Mangat, V., Kumar, K., Vig, R.: Experimental analysis of DDoS attacks on OpenStack cloud platform. In: Proceedings of 2nd International Conference on Communication, Computing and Networking, pp. 3–13 (2019)
21. Khan, A.N., Fan, M.Y., Malik, A., Memon, R.A.: Learning from privacy preserved encrypted data on cloud through supervised and unsupervised machine learning. In: IEEE 2019 2nd International Conference on Computing, Mathematics and Engineering Technologies (iCoMET), pp. 1–5 (2019)
22. Dutta, I.K., Ghosh, B., Bayoumi, M.: Lightweight cryptography for Internet of Insecure Things: a survey. In: IEEE 9th Annual Computing and Communication Workshop and Conference (CCWC), pp. 0475–0481 (2019)
23. Taher, K.A., Jisan, B.M.Y., Rahman, M.M.: Network intrusion detection using supervised machine learning technique with feature selection. In: IEEE International Conference on Robotics, Electrical and Signal Processing Techniques (ICREST), pp. 643–646 (2019)
24. Hossain, M.M., Fotouhi, M., Hasan, R.: Towards an analysis of security issues, challenges, and open problems in the internet of things. IEEE World Congress on Services, pp. 21–28 (2015)
25. Goyal, T., Rathi, R., Jain, V.K., Pilli, E.S., Mazumdar, A.P.: Big data handling over cloud for internet of things. Int. J. Inf. Technol. Web Eng. (IJITWE) **13**(2), 37–47 (2018)
26. Xu, S., Yang, G., Mu, Y., Liu, X.: A secure IoT cloud storage system with fine-grained access control and decryption key exposure resistance. Futur. Gener. Comput. Syst. 167–175 (2018)
27. Khan, M.A., Salah, K.: IoT security: review, blockchain solutions, and open challenges. Futur. Gener. Comput. Syst. **82**, 395–411 (2018)
28. Shen, S., Huang, L., Zhou, H., Yu, S., Fan, E., Cao, Q.: Multistage signaling game-based optimal detection strategies for suppressing malware diffusion in fog-cloud-based IoT networks. IEEE Internet of Things J. 1043–1054 (2018)
29. Adat, V., Gupta, B.B.: Security in Internet of Things: issues, challenges, taxonomy, and architecture. Telecommunication Systems **67**(3), 423–441 (2018)
30. John, J., Norman, J.: Major vulnerabilities and their prevention methods in cloud computing. In: Advances in Big Data and Cloud Computing, pp. 11–26. Springer, Singapore (2018)
31. Al-Garadi, M.A., Mohamed, A., Al-Ali, A., Du, X., Guizani, M.: A survey of machine and deep learning methods for internet of things (IoT) security (2018)
32. Aldaej, A.: Enhancing cyber security in modern Internet of things (IoT) using intrusion prevention algorithm for IoT (IPAI). IEEE Access (2019)
33. Li, Z., Yang, Z., Xie, S.: Computing resource trading for edge-cloud-assisted Internet of Things. IEEE Trans. Ind. Inform. (2019)
34. Li, X., Wang, Q., Lan, X., Chen, X., Zhang, N., Chen, D.: Enhancing cloud-based IoT security through trustworthy cloud service: an integration of security and reputation approach. IEEE Access **7**, 9368–9383 (2019)
35. Kumar, Y., Kaul, S., Sood, K.: Effective use of the machine learning approaches on different clouds. 2019 Available at SSRN 3355203
36. Safa, N.S., Maple, C., Haghparast, M., Watson, T., Dianati, M.: An opportunistic resource management model to overcome resource-constraint in the Internet of Things. Concurr. Comput.: Pract. Exp. (2019)
37. Singla, A., Sharma, A.: Physical access system security of IoT devices using machine learning techniques. 2019. Available at SSRN 3356785

38. Stergiou, C., Psannis, K.E., Gupta, B.B., Ishibashi, Y.: Security, privacy & efficiency of sustainable Cloud computing for Big Data & IoT. Sustain. Comput.: Inform. Syst. **19**, 174–184 (2019)
39. Wazid, M., Das, A.K., Hussain, R., Succi, G., Rodrigues, J.J.: Authentication in cloud-driven IoT-based big data environment: survey and outlook. J. Syst. Archit. (2019)
40. Jia, H., Liu, X., Di, X., Qi, H., Cong, L., Li, J., Yang, H.: Security strategy for virtual machine allocation in cloud computing. Procedia Comput. Sci. **147**, 140–144 (2019)
41. Gurulakshmi, K., Nesarani, A.: Analysis of IoT bots against DDOS attack using machine learning algorithm. In: IEEE 2018 2nd International Conference on Trends in Electronics and Informatics (ICOEI), pp. 1052–1057 (2019)
42. Achbarou, O., El Bouanani, S.: Securing cloud computing from different attacks using intrusion detection systems. IJIMAI **4**(3), 61–64 (2019)
43. Kirichek, R., Kulik, V., Koucheryavy, A.: False clouds for Internet of Things and methods of protection. In: IEEE 2016 18th International Conference on Advanced Communication Technology (ICACT), pp. 201–205 (2019)
44. Hassan, W.H.: Current research on Internet of Things (IoT) security: a survey. Comput. Netw. **148**, 283–294 (2019)
45. Nawir, M., Amir, A., Yaakob, N., Lynn, O.B.: Internet of Things (IoT): taxonomy of security attacks. In: IEEE 2016 3rd International Conference on Electronic Design (ICED), pp. 321–326 (2019)
46. Yin, D., Zhang, L., Yang, K.: A DDoS attack detection and mitigation with software-defined Internet of Things framework. IEEE Access **6**, 24694–24705 (2019)
47. Zhang, Y., Li, P., Wang, X.: Intrusion detection for IoT based on improved genetic algorithm and deep belief network. IEEE Access **7**, 31711–31722 (2019)
48. Chaabouni, N., Mosbah, M., Zemmari, A., Sauvignac, C., Faruki, P.: Network intrusion detection for IoT security based on learning techniques. IEEE Commun. Surv. Tutor. (2019)
49. Li, Z., Rios, A.L.G., Xu, G., Trajković, L.: Machine learning techniques for classifying network anomalies and intrusions. In: IEEE 2019 IEEE International Symposium on Circuits and Systems (ISCAS), pp. 1–5 (2019)
50. Chapaneri, R., Shah, S.: A comprehensive survey of machine learning-based network intrusion detection. In: Smart Intelligent Computing and Applications, pp. 345–356 (2019)
51. Praveen, B.V., Mahita, D., Devi, P.R., Sudheshna, A.: Classifying the probe attacks using machine learning techniques in R and Hadoo. Int. J. Appl. Eng. Res. **13**(7), 5175–5178 (2017)
52. Jayasinghe, U., Lee, G.M., Um, T.W., Shi, Q.: Machine learning based trust computational model for IoT services. IEEE Trans. Sustain. Comput. **4**(1), 39–52 (2019)
53. Rath, M., Pati, B.: Security assertion of IoT devices using cloud of things perception. Int. J. Interdiscip. Telecommun. Netw. (IJITN) **11**(4), 17–31 (2019)
54. Distefano, S., Merlino, G., Puliafito, A.: Enabling the cloud of things. In: 2012 Sixth International Conference on Innovative Mobile and Internet Services in Ubiquitous Computing. IEEE (2012)
55. Aazam, M., Huh, E.N., St-Hilaire, M., Lung, C.H., Lambadaris, I.: Cloud of things: integration of IoT with cloud computing. In: Robots and Sensor Clouds, pp. 77–94. Springer, Berlin (2016)

Data Augmentation Using GAN for Parkinson's Disease Prediction

Sukhpal Kaur, Himanshu Aggarwal, and Rinkle Rani

Abstract The disease of Parkinson is a gradual neurodegenerative disorder affecting approximately one million U.S. citizens with nearly sixty thousand new annual clinical health diagnoses [1]. Analysis of voice samples was used to detect Parkinson's disease early (PD) as an efficient tool. The use of deep learning is dependent on the number of samples marked out, which limits the use of deep learning in the smaller sample environment. In this paper, we suggest an approach based on a GAN combined with a deep neural network (DNN). The initial samples were first divided into training and a test range. The GAN learned to generate synthetic sample data to expand the dataset. Last, the synthetic samples are prepared for the DNN classifier. Finally, the classifier testing conducted with the test set, and the indicators confirmed the efficacy of the small sample classification method. Experimental tests have shown greater precision than conventional approaches in the proposed plan. While the classification process appeared to be improved by traditional data increase, an increase of 11:68% was achieved by incorporating GAN-based additions. Moreover, even higher efficiencies can be obtained by combining conventional with GAN-based augmentation schemes.

Keywords Parkinson's disease · Data augmentation · GAN

S. Kaur (✉) · H. Aggarwal
Computer Science and Engineering Department, Punjabi University, Patiala 147002, India
e-mail: sukhpal91@gmail.com

H. Aggarwal
e-mail: himanshu.pup@gmail.com

R. Rani
Department of Computer Science and Engineering, Thapar Institute of Engineering and Technology, Thapar University, Patiala 147004, India
e-mail: raggarwal@thapar.edu

P. K. Singh et al. (eds.), *Recent Innovations in Computing*, Lecture Notes in Electrical Engineering 701, https://doi.org/10.1007/978-981-15-8297-4_47

1 Introduction

Doctor James Parkinson coined Parkinson's disease as "Shaking paralysis" [1] for the very first time in 1817. This is the other common psychiatric disorder in older people, Alzheimer's, [2, 3]. PD is a sort of underlying condition that affects a part of the brain over the years. It induces various symptoms and signs. All symptoms and symbols can be divided into two major groups from a single viewpoint, namely motor and non-engine. Motor symptoms include neurosymptoms and cognitive disorders, problems with sleep, emotions, and autonomic neuropathy (dysautonomia) [4], which affect movement, musculature, and non-motor symptoms. PODs can be seen among several other signs as early as five years before diagnosis [4] by speech disorders. Investigations show that about 90% of PWPs, in particular speech disability [5, 6], have motor problems. Research indicates that vocal impairment of Parkinson's can be characterized by decreased vocal tract volume and language ability, considerably narrowing the pitch range, long pauses, smaller pitch changes, voice strength, and articulation rate. Most researchers, therefore, find automated acoustic analysis a useful method for the non-invasive measurement of PD. Since small datasets typically do not provide enough representative training samples, which could result in overfitting the model, data increase was required to apply deep-seated learning techniques to small training dataset applications. Traditional growth strategies, especially for tasks, are severely restricted where the dataset follows strict standards, as is the case speech samples. Data augmentation using GAN-generated samples has been shown to provide performance improvement for supervised learning tasks. In this paper, we proposed a method of GAN data augmentation for voice classification that uses the prediction uncertainty of the classifier network to determine the optimal GAN samples to augment the training set. The methodology in this paper is an attempt to turn the classical method of statistical learning based on original samples into a deep learning system focused on data enhancement.

2 Proposed Methodology

The data was taken from the UCI machine learning repository [7], a voice study of patients who are stable and Parkinson's diseases. The initial speech samples were divided into an activity set and a test set—the software package for GAN preparation and modification of its hyper-parameters. The trained GAN generator produces synthetic samples and filters these samples by using the discriminator. The synthetic samples were used for the DNN classifier preparation and tested the DNN classifier with the test collection.

2.1 *Generative Adversarial Network*

A GAN consists of two artificial neural networks (ANNs), the generator, and the discriminator, which compete against each other. Firstly, new data are created, and secondly, validity is evaluated. It is the discriminator who assesses the quality of the generator data. As existing feature samples, it collects either from the original dataset or from the generator. It tries to forecast the example source [8]. The pseudo-code of generator and discriminator training described in Algorithm 1 and 2.

2.1.1 Generative Network

We can define the generative network as a function G: $(Z) \rightarrow X$, which has as input data $z \in Z$ and outputs classifier label $x \in X$. The generative network G is a CNN, which takes as input noise and classifier label and outputs the value of the time series for the given time stamp. This transformation is gone through four convolutional transposes (or deconvolution) layers with ReLUs as an activation function and batch normalization at each layer except for the last one. The generative network adjusts its parameter to minimize $\log(1 - D(G(z/t)))$, where $z \in Z$ is the noise vector [9–11].

```
  Algorithm 1.Generator network
  1.Input:-Random noise, Z.□
  2.Output:-Real and fake data.
  3.define_generator(l_dim)
  4.for i in the range do □
  5.model.add(dense(nodes,l_dim=100)//addition  of  random
noise
  6.model.add(Conv2DTranspose(channel,kernel_size,stride,
padding)□
7.model.add(ReLU)           //addition of activation func-
tion
  8.return model.
```

2.1.2 Discriminative Network

The discriminative network implements a function: D: $(X, T) \rightarrow [0; 1]$. This network takes as input real data $x \in X$ or generated data $g \in G$ and gives as output a binary value, deciding whether the data is real or generated. It is composed of two layers of convolution, each followed by a max-pooling layer. In the end, there is a fully connected layer. The discriminative network adjusts its parameter to maximize $\log(D(x/t))$, where x is the time series vector. Leaky ReLU, Adam optimizer, and the use of dropouts selected for this kind of problem are the norm. The input is not an image, so there are no overall layers [11, 12, 16].

```
  Algorithm 2.Discriminator network
  1.Input:-Real and fake data
  2.Output:-class label 1 for real and 0 for fake.
  3.define_discriminator(lmage size)
  4.for i in the range do □
  5.model.add(Conv2D(channel
no,kernel_size,stride,padding)
  6.model.add(LeakyReLU)        //addition  of  activation
function
  7.model.add(dense(no of neurons,activation_fn)    //
fully connected layer
  8.return model
```

Binary cross-entropy losses because the output of a model that has a probability between 0 and 1 is best measured [10, 11]. We checked six different layer and nodal configurations, with all the findings included in the following pages. We evaluated six different configurations. We use a basic network setup in this experiment deliberately to focus on the fundamental proposal to construct synthetic numerical databases. Our neural network consists of 16 inputs in each of the three veiled layers, 10, 20, 10 neurons, and 10 neurons. For 100 and 200 iteration sessions, the network was further trained [12].

2.1.3 Classifier

We use the DNN classifier for both experimental fields to test the results of synthetic training data. The proposed solution is based on a back propagnetization algorithm, three hidden layers with a transfer feature called ReLU, and one with a sigmoid output layer transmission function tangent. The weights and preconditions are chosen randomly initially. The DNN classification is built in Python with Theano Keras' deep learning library. The first hidden layer has 15 neurons, the second hidden layer has 13 neurons, and the third hidden layer has 12 neurons, for the optimum value to be reached. Two neurons are used in the output layer, one for people with Parkinson's disease and one for healthy individuals. Algorithm 3 describes the training method for deep learning classifier. The preparation of the classifier was carried out at the Keras deep learning library after a generation of synthetic samples. Input parameters for the approach that three hidden layers were employed in synthetic voice samples and labeled training data, and the classification model is the output of the proposed method [13–15].

```
Algorithm 3.Deep Learning classifier
1: Input—The synthetic voice samples of healthy and PD
patients (SYsample) and corresponding labels (Ysy ).
2: Output—Classifier for PD prediction.
3: DL = sequential ( )          //Deep learning classi-
fier initialization
4: DL. add (no. of neurons, init, activation function)
           //initialization of Deep learning Model
          layer, neurons and activation function
5: DL = model.fit (SYsample)          //loading of da-
taset
6: Ytest= DL. Predict (Testsample)      // prediction of
class labels
7: Return Ytest.
```

3 Result and Discussion

The Python 2.7.0 programming was performed on one workstation with a microprocessor of the Intel® CoreTM i5-6700 K and 6 GB RAM in all. The analyses were carried out by 196 voice samples of 31 people 23 of whom were affected [16]. The collection of voices was obtained from the Irvin College of California machine learning repository (UCI) [24]. All column of the dataset describes a specific voice estimate for persons and every column of 195 voices registered by those individuals. The only goal of the dataset is to distinguish patients with PD from healthy people based on the default setting level, where 0 is spoken to healthy people, and one speaks to patients with Parkinson's disease. Each of these data is divided into two sub-sets, one for GAN preparation and one for testing (30%). Table 1 includes a description of the datasets.

In both experiments, we generate synthetic data with a similar general GAN architecture. The following parameters are found in GAN architecture: leaky ReLU [17] as a negative tipping feature 0.2, lot size = five, study rates = 0.001, the Adam optimizer [18]. We tried six specific installations, different in the number of layers and the number of nodes in each section, to decide the exhibition of the model whose output is likely to range from 0 to 1. We have tried six specific settings that differ between layers and nodes in each section, and every one of the outcomes remembered for the accompanying areas. In this trial, we purposefully utilize a basic system and design [19], to concentrate on the essential proposition of creating

Table 1 Description of dataset used

Description	Total voice samples	Parkinson's disease patients	Healthy individuals	Sex (male/female)	
Total subjects	196	23	8	44/31 (F)	56/54 (M)

manufactured statistical databases—the condition after that preparing ended. The loss of discrimination requires a loss of 0, so that, combined, actual, and synthetic data can not be remembered. The PD subset had similar misfortune bends in preparation. After the synthetic PD case voice tests have been developed, these samples are utilized for training the DNN classifier. The DNN classifier, at that point, approved with the voice tests of the PD test set. After a progression of examinations on the DNN, the hyper-parameters of the DNN classifier resolved [20–22]. The element of the classifier's information is 22, which is equivalent to the number of PD voice samples. The classification scheme had three hidden layers, each of which had 32 RLUs; the softmax function was [23] used as the output, and the cross-entropy [24, 25] was used as losses. Initially, a deep neural system design [26, 27], which is equipped to accomplish an excellent exhibition on characterizing the two classes (e.g., PD, NC), is chosen. This network educated conventional amplification techniques [28] on the previously cited dataset with (II) and without (I). Sometime later, artificial samples of the training data are introduced to form a composite dataset, again used for classification training. The examination aims to demonstrate that the model prepared with data augmentation using GANs (III) beats the one without (I) (Fig. 1).

Besides, the use of GANs (III) compares with conventional techniques of augmentation (II), while also investigating the results of both forms of extensions (IV). We first found the precision, and reminder is prepared with one fitted on synthetic data by a DNN Classifier on the original data. Shortly, after, synthetic data is applied to the workout data, forming a composite dataset which again uses the use of the conventional enhancement to prepare a specific system with (IV) and without (III).

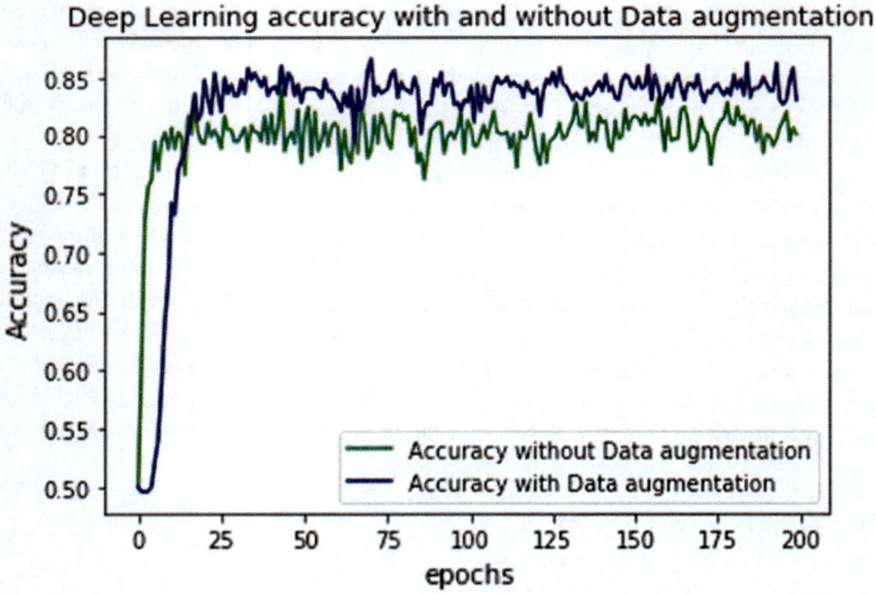

Fig. 1 Deep learning model accuracy with and without data augmentation

We used an alternative number of synthetic samples for the DNN classifier's training to evaluate the influence of the synthetic training sample size on the performance of the DNN classifier. We then tested the three-precision markers, the F-measure, and the G-mean, with real examples. Of the 20 synthetic samples per class, over 20 synthetic samples will be used to train a DNN classifier after each interval. Over 100 synthetic samples are generated each time to prepare the DNN classification, taking over 100 synthetic training samples. The accuracy, F-measure, and G-mean modifications are shown in Fig. 2. By expanding the sample size of synthetic training, precision has grown continuously. The accuracy reached 51.61% with 100 synthetic training samples. The skill becomes more impressive when the sample size of the synthetic training became 1000 (64.52%) when the sample size of the synthetic training arrived in 2000, and the accuracy is 67.74%. However, the expansion of the synthetic sample size did not progress, and efficiency continued to fluctuate by about 67%. The precision was at approximately 70% when the synthetic sample size exceeded 4000. Further, the F-measure's propensity is fundamentally consistent with that of accuracy if the synthetic training sample size has increased, showing that the accuracy of the real samples from each class can be predicted with the general situation.

Working with entirely numeric models enable the GAN to create discrete outcomes. They are likewise quicker to compute and prepare when contrasted with image datasets (Table 2).

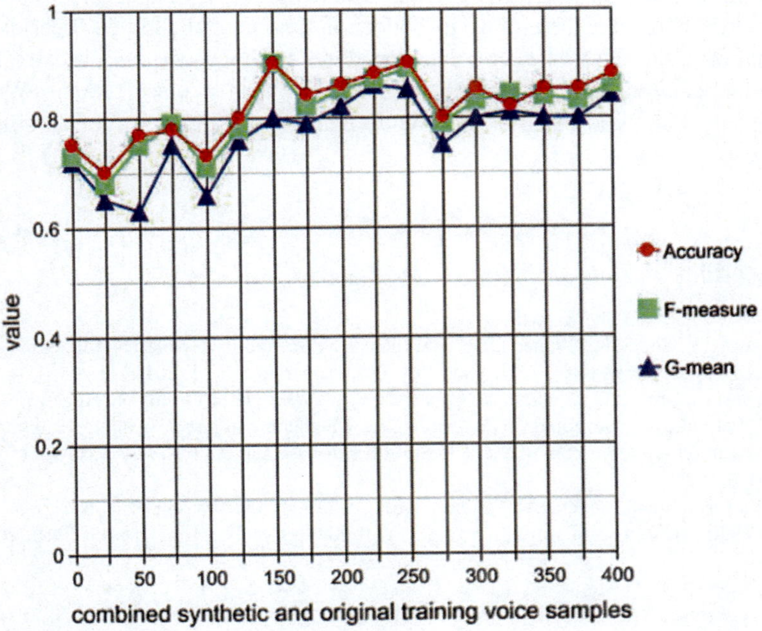

Fig. 2 Performance of DNN classifier over different training synthetic and original voice samples

Table 2 Comparison of the proposed approach with and without data augmentation

Methodology	Accuracy (%)	Specificity (%)	Sensitivity (%)
DNN with data augmentation	88	87.14	87.92
DNN without data augmentation	84.67	83.76	84.13

We are astounded to see that sometimes, the classifier prepared on the synthetic data accomplished preferred outcomes over the one trained in the original data, which proposes that creating synthetic data utilizing GANs can be a decent way to deal with over-fitting. Second, we looked at the classifier's exhibition on an imbalanced dataset that was enlarged by the GAN. In this experiment, the GAN improved the outcomes when contrasted and the first, imbalanced informational index. The deep CNN model utilizing synthetic samples allow the CNN order framework to acquire up to 3% higher accuracy than grouping results using original images. In any case, higher efficiency obtained with more synthetic images.

4 Conclusions and Future Work

In this project, we proposed the use of a GAN to create a synthesis data package for the numerical dataset, to produce a classifier using the hybrid dataset. The classification training using synthetic and original data demonstrated superior precision and consistency to prepare the initial data collection. In a state of conflict, superior to the original data, the GAN synthetic data was done. In this paper, only because of the helpfulness and viability of the GAN for the Parkinson sickness scheme. With the assistance of GANs, the deep learning model performs better in terms of classification accuracy.

References

1. Poewe, W., Seppi, K., Tanner, C.M., Halliday, G.M., Brundin, P., Volkmann, J., Schrag, A.E., Lang, A.E.: Parkinson's Disease. Nat. Rev. Dis. Primers **3**, 17013 (2017)
2. Hopes, L., Grolez, G., Moreau, C., Lopes, R., Ryckewaert, G., Carrière, N., Auger, F., Laloux, C., Petrault, M., Devedjian, J.C., Bordet, R.: Magnetic resonance imaging features of the nigrostriatal system: biomarkers of Parkinson's disease stages? PLoS ONE **11**(4), e0147947 (2016)
3. Pinter, B., Diem Zangerl, A., Wenning, G.K., Scherfler, C., OberaignerW, S.K., Poewe, W.: Mortality in Parkinson's disease: a 38-year follow-up study. Mov. Disord. **30**(2), 266–269 (2015)
4. Lebedev, A.V., Westman, E., Simmons, A., Lebedeva, A., Siepel, F.J., Pereira, J.B., Aarsland, D.: Large-scale resting-state network correlates of cognitive impairment in Parkinson's disease and related dopaminergic deficits. Front. Syst. Neurosci. **8**, 45 (2014)
5. Pringsheim, T., Jette, N., Frolkis, A., Steeves, T.D.: The prevalence of Parkinson's disease: a systematic review and meta-analysis. Mov. Disord. **29**(13), 1583–1590 (2014)

6. Dorsey, E., Constantinescu, R., Thompson, J.P., Biglan, K.M., Holloway, R.G., Kieburtz, K., Marshall, F.J., Ravina, B.M., Schifitto, G., Siderowf, A., Tanner, C.M.: Projected number of people with Parkinson disease in the most populous nations, 2005 through 2030. Neurology **68**(5), 384–386 (2007)
7. Wang, D., Lui, Z., Xu, Y.: Cellular structure image classification with small targeted training samples, pp. 1–7 (2019)
8. Shayan, S., Richard, P., Jian, Z., Joohyun, K., Kisung, L.: Deep generative breast cancer screening and diagnosis, pp. 859–867 (2018). https://doi.org/10.1007/978-3-030-00934-2_95
9. Liu, Y., Zhou, Y., Liu, X., Dong, F., Wang, C., Wang, Z.: Wasserstein GAN-based small-sample augmentation for new-generation artificial intelligence: a case study of cancer-staging data in biology, pp. 156–163 (2019). https://doi.org/10.1016/j.eng.2018.11.018
10. Shams, S., Platania, R., Zhang, J.: Deep Generative Breast Cancer Screening and Diagnosis, pp. 859–867. Springer Nature (2018)
11. Gengxing, W., Wenxiong, K., Qiuxia, W., Zhiyong, W., Junbin, G.: Generative Adversarial Network (GAN) based Data Augmentation for Palmprint Recognition. IEEE (2018)
12. Fatawa, A., Jahardi, M., Ling, S.H.: Efficient diagnosis system for Parkinson's disease using deep belief network, pp. 1324–1330 (2016)
13. Ozcift, A.: SVM feature selection based rotation forest ensemble classifiers to improve computer-aided diagnosis of Parkinson's disease. J. Med. Syst. (2012)
14. Liu, P., Choo, R., Wang, L.: SVM or deep learning, a comparative study on remote sensing image classification. J. Soft Comput. (2016)
15. Kollias, D., Tagaris, A., Stafylopatis, A.: Deep neural architectures for prediction in healthcare. Complex Intell. Syst. **4**, 119–131 (2018)
16. Little, A., McSharry, E., Roberts, S., Costello, D.: Exploiting nonlinear recurrence and fractal scaling properties for voice disorder detection. Biomed. Eng. Online **6**, 23 (2007)
17. Das, R.: A comparison of multiple classification methods for diagnosis of Parkinson's disease. Expert Syst. Appl. **37**, 1568 (2010)
18. Ramani, R.G., Sivagam, G.: Parkinson's disease classification using data mining algorithms. Int. J. Comput. Appl. (0975–8887) **32**, 17 (2011)
19. Khemphila, A., Boonjing, V.: Parkinson's disease classification using neural network and feature selection, world academy of science, engineering, and technology. Int. J. Math., Comput., Phys., Electr. Comput. Eng. **6**, 377 (2012)
20. Rustempasic, I., Can, M.: Diagnosis of Parkinson's disease using principal component analysis and boosting committee machines. Southeast Eur. J. Soft Comput. 102 (2013)
21. Sairam, N., Mandal, I.: New machine-learning algorithms for prediction of Parkinson's disease. Taylor Francis Int. J. Syst. Sci. **45**, 647 (2014)
22. Shahbakhi, M., Taheri, D., Tahami, E.: Speech analysis for diagnosis of Parkinson's disease using genetic algorithm and support vector machine. J. Biomed. Sci. Eng. Sci. Res. **7**, 147 (2014)
23. Suganya, P., Sumathi, C.P.: A novel metaheuristic data mining algorithm for the detection and classification of Parkinson's disease. Indian J. Sci. Technol. **8**, 1 (2015)
24. www.uci repository.com
25. Weng, Y., Zhou, H.: Data augmentation computing model based on generative adversarial network. IEEE Access **7**, 64223–64233 (2019)
26. Sivaranjini, S., Sujatha, C.M.: Deep learning-based diagnosis of Parkinson's disease using convolutional neural network. Multimed. Tools Appl. (2019). https://doi.org/10.1007/s11042-019-7469-8
27. Rustempasic, I., Can, M.: Diagnosis of Parkinson's disease using fuzzy C-means clustering and pattern recognition. Southeast Eur. J. Soft Comput. **42** (2012)
28. Han, Te., Liu, C., Yang, W., Jiang, D.: A novel adversarial learning framework in a deep convolutional neural network for intelligent diagnosis of mechanical faults. Knowl.-Based Syst. (2018). https://doi.org/10.1016/j.knosys.20

Effective Analysis of Tweets Using Hadoop Ecosystem

Ravindra Kumar Singh and Harsh Kumar Verma

Abstract Twitter has gained enough popularity nowadays and collecting people's emotion, opinion, suggestion, feeling, knowledge and current market trends in the form of post on day-by-day basis from different countries, in multiple formats and languages; it is an absolute form of unstructured, rapidly growing million dollar worth data that is difficult to manage and process. This kind of data is mainly referred to as big data. The Hadoop ecosystem evolved around this problem space and offered effective management of this kind of data starting from capturing through processing till workflow management. This research is mainly aimed to provide an effective well-scalable framework to collect, process and analyze tweets using the Hadoop ecosystem. Here, Apache Flume is used to capture and store data in HDFS, Apache Pig and Apache Hive are used for data processing and analysis, and Apache Oozie is used for workflow management and task scheduling. This research also did the performance benchmarking over Hive and Pig on these data to find the recent trends, top influencers and top posts in various data categories. Experimental research concluded that Apache Pig outperformed over Apache Hive in terms of processing time while analytics results were same.

Keywords Social media analytics · Real-time analytics · Big data · Data processing framework · Apache Flume · Apache Hive · Apache Pig · Apache Oozie

R. K. Singh (✉) · H. K. Verma
Dr. B. R. Ambedkar National Institute of Technology, Jalandhar, India
e-mail: ravindra1987singh@gmail.com

H. K. Verma
e-mail: vermah@nitj.ac.in

P. K. Singh et al. (eds.), *Recent Innovations in Computing*, Lecture Notes in Electrical Engineering 701, https://doi.org/10.1007/978-981-15-8297-4_48

1 Introduction

With growing popularity on online portals in our life, the data generation velocity is also increasing rapidly [1]. People are more comfortable in accessing information about their area of interest in their preferred language and format of data on Internets that promotes the data creation and translation in different languages and converts in various multimedia formats for better consumption of the knowledge inside the data. It is going really well for the person to grasp this knowledge, but the challenge behind manual data grasping is the scope limitation, it could not be scaled well, and at the same time it is a costly operation too. So, lots of analytics algorithms [2] came into the picture to collect, process and analyze these data in terms of grasping the knowledge inside the data and report us in a specified form. It contributed significantly well, but rapidly growing data in various forms became challenging to handle and consume these data effectively [3].

On the other side, social media has gained different labels of popularity, and people are crazy about sharing and accessing the information on these portals. Every day something is trolling and trending on social media portals, and people are giving their reactions in their own style even in their own format; someone is sharing video so someone is sharing song, someone sharing screenshot so someone will come up with a URL, but the most important thing is that all these data contain business-critical information [4]. Sentiments about a product, company, person, country, religion, everything matter in the field of analytics. Similarly, people surf the e-commerce portals [5], search various products, read these reviews, add it in a wishlist, compare various products on same portals and sometimes on different portals as well, then buy a few of them and share their feedback, raising complaints and getting an exchange, etc. This process is going forever and gaining popularity like anything; this kind of information along with its associated metadata is worth millions in the business world. This kind of data is not available in a structured format, sometimes they are semi-structured, and sometimes it is unstructured as well.

So, processing and consuming this kind of data in real time is the demand of the industry but it is a tedious job to collect, store and process this elephant size of data. This kind of data is referred to as big data, and Hadoop came to rescue us in dealing with it [6]. Hadoop itself is a distributed file system (HDFS) and a programming paradigm (MapReduce), but it is guided and surrounded by multiple tools or data handling mechanisms. This entire set of things is known as the Hadoop ecosystem. Every component in this ecosystem is very crucial in processing huge volumes of data and plays a different role in the ecosystem. Hadoop ecosystem is very reliable, scalable and configurable in nature, and it could be built with commodity hardware without any additional/special hardware requirements. It is horizontally scalable in zero downtime and performs stably till the petabytes of data; all these advantages are making it so popular in data handling and analytics field and lure the researchers to further research on these topics [7].

This research is mainly aimed to collect, store and analyze real-time tweets in the Hadoop ecosystem in a time-effective manner and to produce an effective and scalable framework to deal tweets in further research. This research is also intended to do the benchmarking on Apache Hive [8] and Apache Pig [9].

2 Background

Tweet's collection [10] and distribution on Hadoop and its analysis approaches have been tried in the past with different use cases, and this literature is helpful in designing the proposed framework. Here in this section, we will specially focus on the literature of fetching tweet streams [7], data distribution on Hadoop, using Apache Flume [11] and Apache Kafka [12] in sinking data in HDFS, Apache Pig and Apache Hive for data analytics tasks done so far. Here, we would cover the literature of current trend analysis [7], finding influencers [13] and top posts on specific areas as well.

2.1 Tweet Collection

Social media analytics, especially on Twitter data, have gained popularity in the research world as well as brought the impact in the industries as well. In this Internet era, data are growing like anything especially on social media platforms, Twitter is one of biggest players of this category, and Twitter posts comprise opinions, feelings, sentiments, reviews, trends, users liking and disliking information, etc. [14]. So, these posts are important to collect and filter the useful information, so for that purpose Twitter has launched its application programming interface to provide data based on given keywords for specific timeline, language with many more filtering criteria along with streaming, and archive data collection mode; these APIs are available in both free and paid modes as per the use case [10]. These APIs have some limitations on the number of tweets per hits, number of hits per minute, number of hits per second as well; these limitations vary on different APIs. These APIs are implemented in different languages as modules for effective use like Tweepy and Twitter4j are implemented for Python and Java, respectively [6]. Sheela [15] and Kumar et al. [16] used Twitter REST API to gather real-time tweets; similarly, Ha et al. [17] used it to acquire historical data from Twitter.

2.2 Tweet Distribution on Hadoop

Storing this huge amount of data and finding insights from data is difficult; thus, Hadoop became handy in this kind of processing [11]. To handle the continuous high velocity of tweets, we need efficient tools for real-time streaming, and Apache Flume is very popular for sinking these posts on Hadoop [8]; below are a few examples of these implementations. Jain et al. [18] and Barskar et al. [9] used Apache Flume to stream data and to process it using Pig; Kumar et al. [16] have used it for collecting large set of Twitter data and its storage on Hadoop and further performing sentiment analysis of these data; similarly, Nadagoud et al. [5] used Apache Flume to collect Twitter data and processing it with Apache Hive to identify the popularity of Flipkart.

2.3 Social Media Analytics Using Hadoop Ecosystem

Social media analytics are nowadays becoming crucial to know the opinions of the user, so a lot of research has been carried out for the same purpose; Chauhan et al. [6], Shang et al. [14], Ennaji et al. [19] and Mullick et al. [20] have used tweets to identify the opinion by using big data technologies. Khade [21] predicted customer behavior using the MapReduce programming model. Cossu et al. [13] detected the real-life influence of people based on their Twitter account. Sheela [15], Kumar et al. [16] and Selvan et al. [22] performed sentiment analysis on Twitter data using Twitter data and for the storage of Twitter data on Hadoop distributed file system. Verma et al. [23] proposed a recommendation system summarizing user reviews, comments, feedback on various topics using the Hadoop framework.

3 Proposed Method

The proposed framework operates on Hadoop ecosystem to effectively draw the analysis on Twitter like complex semi-structured huge amounts of data that is nearly impossible to execute on traditional processing methods [24]. The proposed framework could be explained better in the form of organizing different tools of the Hadoop ecosystem to perform the desired tasks with the help of few config files and scripts to perform specific functions. This framework utilizes Apache Flume to capture/aggregate real-time streamed tweets and distribute it on HDFS with or without using Apache Kafka as per the use case and data volume. Apache Hive and Apache Pig would preprocess and analyze these real-time streamed semi-structured tweets [25]. Apache Oozie could be utilized for scheduling these tasks to run on any specified time or condition. This framework is robust, fault-tolerant and horizontally scalable to bear with a high velocity of data.

A well-organized illustration is given (see Fig. 1) for the architectural overview of the proposed framework.

Above-mentioned blocks of the proposed framework are described well in below sections.

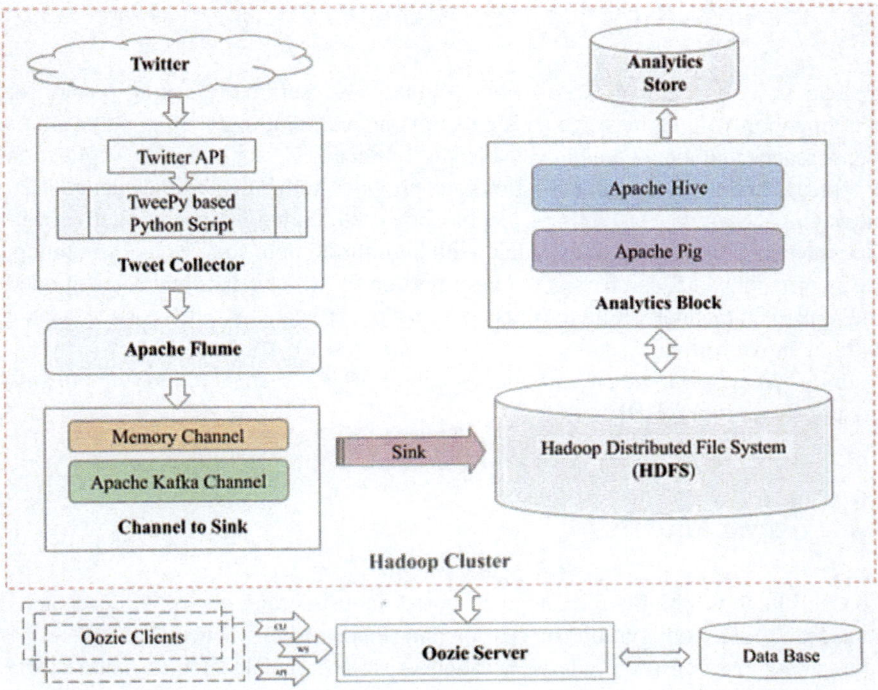

Fig. 1 Proposed framework to capture and analyze tweets with scheduling capabilities

3.1 *Twitter API*

Twitter requires a Twitter application along with its generated consumer key, consumer secret, access token and access token secret in order to fetch the tweets. This application could be generated freely on https://apps.twitter.com/, and there are many tutorials and blogs available to assist in creating Twitter applications and generating required keys. Twitter has restricted the streaming APIs in fetching tweets by limiting the number of tweets in a given time frame.

3.2 *Tweet Collector*

Tweet Collector is a module/script that is utilizing Twitter streaming API to collect tweets matching certain trends or keywords. The most popular modules are Twitter4j for Java users and Tweepy for Python users to stream tweets, they require the Consumer key, Consumer secret, Access token, and Access token secret to interact with Twitter API and fetch the tweets matching certain trends or keywords in JSON data. These tweets could be limited by countries and languages as well.

3.3 Apache Kafka

Apache Kafka is a queue and distributed publish–subscribe messaging system for handling high volume message passing from one endpoint to another. Kafka stores the messages in logical boundaries called as Topic, and Kafka is simply a collection of topics split into one or more partitions and provides both pub–sub and queue-based messaging system in a fast, reliable, stable, persisted, fault tolerance and zero downtime manner [26]. It is capable to deal with high throughput sources like streaming, log aggregators, etc., and persist the messages on the disk in distributed, partitioned and replicated manner within the cluster to ensure zero data loss. It is easily scalable without any downtime.

In this framework, we are utilizing Apache Kafka as a channel in Flume to sink streamed tweets into HDFS.

3.4 Apache Flume

Apache Flume is a highly reliable, distributed, fault-tolerant, horizontally scalable, manageable and configurable service or data ingestion mechanism for collecting aggregating and transporting large amounts of streaming data such as social media data streams, event log, etc., from various sources at very high speed that is difficult to manage on traditional way to a centralized data store as HDFS, HBase, etc., on run time and provides a steady flow of data between them.

In this framework, it rescues us in storing the stream tweets collected from Tweet Collector in JSON format on HDFS. It is configured with two channels: One is Memory Channel for normal traffic of the data, and another one is Kafka Channel for high through systems. Any of them could be utilized as per the use case.

3.5 Hadoop Distributed File System (HDFS)

With growing data velocity, the data size easily outgrows the storage limit of a machine so HDFS came into the pictures to store the data across a network of machines. It is the most reliable file system with a unique design to provide storage for extremely large files (up to petabytes of data) by using commodity hardware with streaming data access patterns (write-once and read-many-times). Machines in the HDFS work in master–slave mode and monitor health using heartbeat signals. It stores data in a distributed manner with multiple replications to provide high availability and fault tolerance [21].

In this framework, HDFS is used to store the stream of tweets for analysis purposes and to maintain its high availability for future use.

3.6 Apache Hive

Apache Hive is an open-source, fast, reliable and scalable data warehouse system built on top of Hadoop used for querying, analyzing and stigmatization of structured and semi-structured large data sets stored in Hadoop files in a much easier manner by using SQL-like language called HiveQL (HQL) which automatically translates SQL-like queries into MapReduce jobs. It also provides file access on various data stores like HDFS and HBase, its tables are defined directly in the HDFS, and it can be portioned and bucketed to improve performance. Hive supports external tables too which make it possible to process data without actually storing in HDFS.

In this framework, semi-structured Twitter data are preprocessed (convert nested JSON data into a structured form that is suitable for analysis), analyzed and summarized by Hive.

3.7 Apache Pig

Apache Pig is an abstraction over MapReduce, used to analyze large sets of data representing them as data flows by using a high-level procedural language called Pig Latin which is enriched with various in-built operators like joins, filters, ordering, etc., in addition to support of various nested data types like tuples, bags and maps. These operators, data types and functions enable extensibility and help to perform any kind of data manipulation operations easily and effectively as compared to MapReduce. Overall, it provides ease of programming and flexibility for user-defined functions in other languages and uses it in Pig Script; all these Pig Scripts are internally converted to MapReduce jobs to perform operations.

In this framework, Twitter data are preprocessed, analyzed and summarized by Pig.

3.8 Apache Oozie

Apache Oozie is an open-source scheduler to manage and execute Hadoop jobs on specified schedules in a distributed environment. It allows combining multiple complex jobs to be run in a sequential and parallel order to achieve a bigger task in a well-planned manner. Oozie is managing and triggering the workflow actions which are using the Hadoop execution engine behind to actually execute the task. Oozie is tightly integrated within Hadoop stack and supporting all kinds of Hadoop jobs including Pig, Hive, Sqoop, Java, Shell scripts, and it is also leveraging the existing Hadoop machinery for load balancing, fail-over, etc., to effectively manage the jobs. Oozie detects and notifies the completion of tasks through callback and polling.

Here in this framework, Apache Oozie is responsible for managing and executing the Oozie Coordinator Jobs to run the analysis block on Apache Pig and Apache Hive on defined intervals.

4 Results and Discussion

This section is providing the experimental parts of the above-mentioned framework, and this includes the system configurations, data collections and usages of the framework for multiple use cases.

4.1 Experimental Setting

This section will provide detailed information about the data collection strategy along with the system configuration to set up this framework.

System Configuration. This research is part of an educational program so a variety of systems were utilized in the collection, storage and processing of the framework but all the benchmarking and result calculation were done on an 8 GB RAM, i5 Processor Ubuntu 14.04 operating system laptop. Hadoop 2.6.5 is installed in pseudo-distributed mode, and on top of that Apache Flume 1.6.0, Apache Kafka 2.2.0, Apache Hive 1.2.3 and Apache Pig 0.16.0 were also installed.

Data Set. This research is done on the reactions of Twitter users on the Citizenship Amendment Act implementation in India. Tweet collection is based on English texts from India only, so the topics, hashtags, events were considered accordingly, and we retrieved approximately 1 million posts for the duration of 11 days from January 1, 2020, to January 11, 2020. Tweet collection was done through Flume using the Tweepy library of Python, and data were sunk directly in HDFS.

4.2 Use Cases of the Framework

There are many use cases for our proposed framework where it boosts the performance and eases the implementation to store and process huge amount of data in relatively less time; below are few use cases implemented on this framework for performance evaluation.

Finding Current Trends on Twitter Data. The trend is a topic or subject of the posts on social media for a particular duration of time; finding such trends requires the processing of the tweets collected over the needed duration. Here in this research, we are considering hashtags of the posts to find the trends. Trends popularity could be examined by many parameters including the number of posts having particular hashtag (nPH) and the number of users involved in sharing the posts of particular hashtag (nUH). In this research, identify the trend score (tScore) calculated with Eq. (1)

$$tScore = nUH \times \sqrt{nPH \div nUH} \tag{1}$$

This research demonstrated the impact of these trends as a word cloud in Fig. 2, and its top 10 trends during the period of research are given in Table 1 along with their tScore. Apart from that, 5 most popular trends with polarity in context were examined for time series analysis on a daily basis, and it is represented in Fig. 3.

Finding Top Posts on Specific Area. The popularity of the post could be examined by multiple parameters including the number of favorites (nF), number of retweets (nRT), number of quoted-tweets (nQT), number of replies (nR), number of the users (nU) interacted with that post, etc. These parameters are combined in some form/formula to calculate and identify the top posts. In this research, post score (pScore) is being calculated with Eq. (2)

$$pScore = \log_{10}(nU \times (2 \times nR + 1.5 \times nQT + 1 \times nRT + 0.5 \times nF) + 1) \tag{2}$$

Top 5 posts during the period of research are given in Table 2 along with their

Fig. 2 Trends representation as a word cloud

Table 1 Top trends and their scores

S. No.	Trend (Hashtag)	tScore	S. No.	Trend (Hashtag)	tScore
1	CAA	82,771	6	CAA_NRC_Protests	18,535
2	IndiaSupportsCAA	31,596	7	CAAProtests	17,379
3	NRC	22,322	8	CAA_NRC_Protest	13,833
4	CAA_NRC	21,403	9	ShaheenBagh	12,320
5	CAA_NRC_NPR	19,725	10	CAA_NRCProtests	12,166

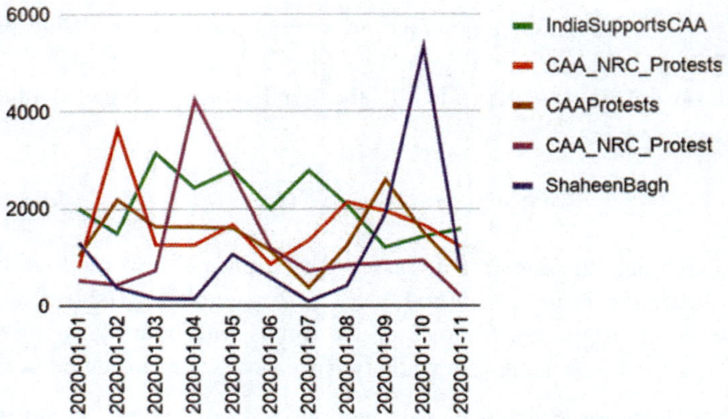

Fig. 3 Time series analysis of trends

Table 2 Top posts and their scores and metadata

S. No.	Post (Tweet ID)	pScore	Post (Tweet ID)	
1	1213850414258380807	9.78	https://t.co/9gzUAVyM8P	RahulGandhi
2	1213057807500529671	9.72	https://t.co/6qD1l50WpS	AmitShah
3	1212663026270138368	9.50	https://t.co/dAoeJ2pftN	PMOIndia
4	1213759380505554944	9.43	https://t.co/tjVkZyXRU8	ArvindKejriwal
5	1213836142828646401	9.37	https://t.co/1uYRhRq6J7	ArvindKejriwal

pScore; apart from that, this research also highlighted the most commonly used words of these posts as a word cloud in Fig. 4.

Finding Top Influencers on Specific Area. An influencer is the user who is being followed by many people, and their ports are being favorite, re-tweeted, quoted-tweeted and replied and viewed by lots of people. In short, those are the people influencing the domain by their activity on social media. Influencing score (iScore)

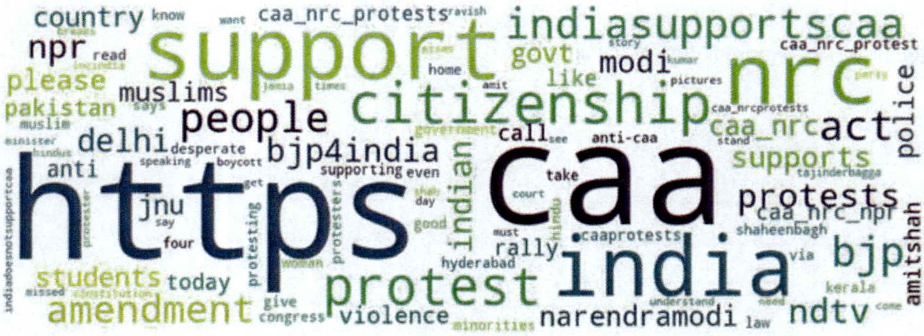

Fig. 4 Most popular word representation as a word cloud

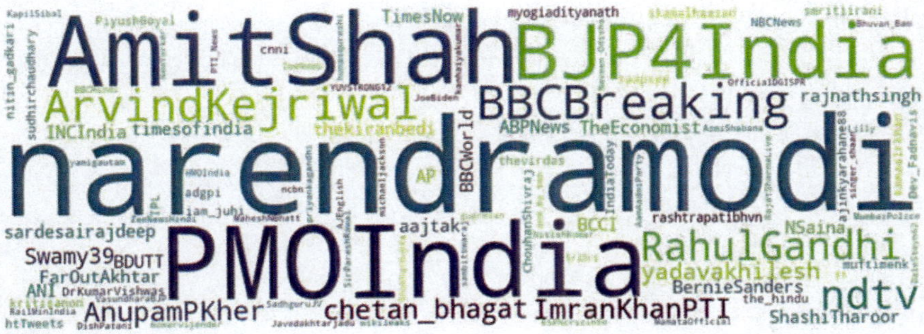

Fig. 5 Influencer's representation as a word cloud

Table 3 Top influencers and their scores

S. No.	Influencer (ScreenName)	iScore	S. No.	Influencer (ScreenName)	iScore
1	narendramodi	179,910	6	ArvindKejriwal	84,064
2	PMOIndia	103,909	7	ndtv	74,069
3	AmitShah	95,940	8	RahulGandhi	68,294
4	BJP4India	94,184	9	ImranKhanPTI	63,361
5	BBCBreaking	91,634	10	chetan_bhagat	61,711

of a user could be calculated by various parameters including the number of posts (nP), number of followers (nFl), number of friends (nFr), average post score (apScore) on their posts, etc. In this research, the influence score of a user is calculated with Eq. (3)

$$iScore = \sqrt{\left(\log_{1000}((nFl + 1) \div (nFr + 1)) + 1\right) \times nFl \times apScore \times \log_{10}(nP + 10)} \tag{3}$$

This research demonstrated the impact of influencers as a word cloud in Fig. 5, and the top 10 influencers participated in this discussion during the period of research are given in Table 3 along with their iScore.

4.3 Performance Evaluation on the Basis of Execution Time

We have used Pig Latin and HiveQL languages to process real-time tweets, depending upon the type of data and use case, developers could choose either to use the Apache Pig or Apache Hive for the analytics purpose. Apache Pig is mainly used by researchers and programmers, whereas Apache Hive is more popular on Extract-Transform-Load (ETL) operations and mainly used by data analysts. In Hive, we

Table 4 Execution time of Apache Pig and Apache Hive

No. of tweets	Task name	Execution time of Apache Hive (in sec)	Execution time of Apache Pig (in sec)
5000	Top influencers	158.20	117.65
5000	Top posts	147.53	104.37
5000	Current trends	140.28	87.34
50,000	Top influencers	507.03	423.73
50,000	Top posts	469.92	396.87
50,000	Current trends	434.63	245.97

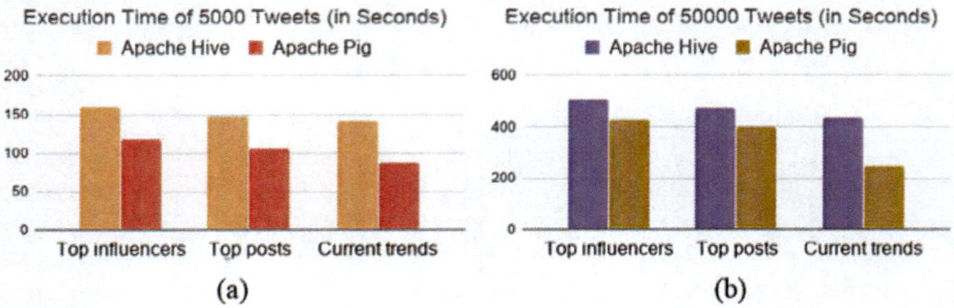

Fig. 6 Execution time of Apache Hive and Apache Pig on different use cases; **a** is the representation of 5000 tweets, and **b** is the representation of 50,000 tweets

need to define the table beforehand and store schema details whereas in Pig there is no dedicated metadata database and schemas.

In this research, we used Apache Hive and Apache Pig; both of them as Tweets are stored in a semi-structured format and performed their benchmarking [26] as well on a batch size of 5000 and 50,000 tweets of the corpus. While the analytics result was same for both of them but the execution of Apache Pig was faster than Apache Hive because the architectural design of Apache Pig supports nested data structures and provides high-level operators for processing semi-structured data, execution times for both of them are given in Table 4 against all three use cases.

In our benchmarking, Apache Pig outperformed the Apache Hive and performed the task in 83.57% on finding top influencers, in 84.45% on finding top posts and in 56.59% on finding current trends. Figure 6 illustrates the benchmarking in 5000 and 50,000 tweets of the politics category.

5 Conclusion

Social media interactions are now a very basic form of communication, expressions of thoughts and putting opinions on approximately all the subjects of our life. Twitter data, e-commerce data, etc., have an extensive dependency on business and politics decisions. In these terms, social media analytics became so popular to collect sentiment analysis, user opinion, etc., but its dependency on efficient mechanism and framework to collect, store and process is inevitable due to the generation of an abundant amount of data on daily basis which makes the mechanism of processing and analyzing of these huge data a supercritical challenge. In recent years, Hadoop proved to be a reliable, scalable, fault-tolerant system to process and manage huge amounts of data. So, this research proposed an efficient framework on the Hadoop ecosystem by integrating Apache Flume, Kafka, Pig, Hive and Oozie to handle the social media data velocity for storage and analytics purposes. Here, Pig Latin and HiveQL came to rescue us to deal with complex MapReduce programs, Apache Flume to smoothen the collection and sinking process and Apache Oozie to manage workflows. This framework was tested on three use cases finding current trends, top influencers and top posts on Twitter data with analytics using Apache Pig and Hive. This research also did benchmarking of execution time for Apache Pig and Apache Hive on 5000 and 5000 batch sizes of tweets, and results concluded that Apache Pig is approximately 20% faster as compared to Apache Hive in executing the mentioned use cases.

6 Future Work

This research is mainly focused on the collection of tweet stream and its distribution on Hadoop with a simple analytics demonstration. This framework could be used to find further analytics on Twitter data, including the determining of new features, sentiment analysis [15], user behavior analysis [20] and so on. Apart from that, this framework could extend the modern analytics tool like Apache Storm and Apache Spark for the analytics process and could include Apache ZooKeeper for coordination and synchronization of various components of the Hadoop ecosystem in this framework.

References

1. Annamoradnejad, I., Habibi, J.: A comprehensive analysis of twitter trending topics. In: 5th International Conference on Web Research (ICWR), Tehran, Iran, IEEE (2019)
2. Fernandes, R., D'Souza, G.L.: Semantic analysis of reviews provided by mobile web services using rule based and supervised machine learning techniques. Int. J. Appl. Eng. Res. (2017)
3. Shrote, K.R., Deorankar, A.V.: Review based service recommendation for big data. In: 2nd International Conference on Advances in Electrical, Electronics, Information, Communication and Bio-Informatics (AEEICB). Chennai, India, IEEE (2016)
4. Yadav, K., Pandey, M., Rautaray, S.S.: Feedback analysis using big data tools. In: International Conference on ICT in Business Industry & Government (ICTBIG), Indore, India: IEEE (2016)
5. Nadagoud, S., Naik, K.: Market sentiment analysis for popularity of Flipkart. Int. J. Adv. Res. Comput. Eng. Technol. (IJARCET) (2015)

6. Chauhan, V., Shukla, A.: Sentimental analysis of social networks using MapReduce and big data technologies. Int. J. Comput. Sci. Netw. (2017)
7. Rodrigues, A.P., Chiplunkar, N.N., Robnik-Šikonja, M.: Real-time Twitter data analysis using Hadoop ecosystem. Cogent Eng. (2018)
8. Bhardwaj, A., Kumar, A.: Big data emerging technologies: a case study with analyzing twitter data using apache hive. In: 2nd International Conference on Recent Advances in Engineering & Computational Sciences (RAECS), Chandigarh, India, IEEE (2015)
9. Barskar, A., Phulre, A.: Opinion mining of twitter data using Hadoop and Apache Pig. Int. J. Comput. Appl. (2017)
10. Twitter Developers Homepage, https://developer.twitter.com/en/apps. Last accessed 2020/02/10
11. Sangeeta: Twitter data analysis using Flume & Hive on Hadoop frame work. Int. J. Recent Adv. Eng. Technol. (2016)
12. Vohra, D.: Apache Kafka. In: Practical Hadoop Ecosystem. Apress, Berkeley, CA (2016)
13. Cossu, J.V., Nicolas Dugué, N., Labatut, V.: Detecting real-world influence through Twitter. In: Second European Network Intelligence Conference, IEEE (2015).
14. Shang, S., Shi, M., Shang, W., Hong, Z.: Research on public opinion based on big data. In: 14th International Conference on Computer and Information Science (ICIS), Las Vegas, NV, USA, IEEE (2015)
15. Sheela, L.J.: A review of sentiment analysis in twitter data using Hadoop. Int. J. Database Theory Appl. (2016)
16. Kumar, M., Bala, A.: Analyzing twitter sentiments through big data. In: 3rd International Conference on Computing for Sustainable Global Development, New Delhi, India, IEEE (2016)
17. Ha, I., Back, B., Ahn, B.: MapReduce functions to analyze sentiment information from social big data. Int. J. Distrib. Sens. Netw. (2015)
18. Jain, A., Bhatnagar, V.: Crime data analysis using pig with Hadoop. Procedia Comput. Sci. (2016)
19. Ennaji, F.Z., Fazziki, A.E., Sadgal, M., Benslimane, D.: Social intelligence framework: Extracting and analyzing opinions for social CRM. In: 12th International Conference of Computer Systems and Applications (AICCSA), IEEE/ACS, Marrakech, Morocco, IEEE (2015)
20. Mullick, A., Goyal, P., Ganguly, N., Gupta, M.: Harnessing Twitter for answering opinion. IEEE Trans. Comput. Soc. Syst. (2018)
21. Khade, A.A.: Performing customer behavior analysis using big data analytics. Procedia Comput. Sci. (2016)
22. Selvan, L.G.S., Moh, T.S.: A framework for fast-feedback opinion mining on Twitter data streams. In: International Conference on Collaboration Technologies and Systems (CTS), Atlanta, GA, USA, IEEE (2015)
23. Verma, J.P., Patel, B., Patel, A.: Big data analysis: recommendation system with Hadoop framework. In: International Conference on Computational Intelligence & Communication Technology (CICT), Ghaziabad, India, IEEE (2015)

24. Uzunkaya, C., Ensari, T., Kavurucu, Y.: Hadoop ecosystem and its analysis on Tweets. In: World Conference on Technology, Innovation and Entrepreneurship, Procedia-Social and Behavioral Sciences. Elsevier (2015).
25. Wiatr, R., Słota, R., Kitowski, J.: Optimising Kafka for stream processing in latency sensitive systems. Procedia Comput. Sci. (2018)
26. Basha, K.J., Balamurugan, M.: A review on Hive and Pig. Int. J. Adv. Res. Basic Eng. Sci. Technol. (2017)

Artificial Intelligence Politicking and Human Rights Violations in UK's Democracy: A Critical Appraisal of the Brexit Referendum

Ikedianchi Ayodele Power Wogu, Sanjay Misra, Oluwakemi Deborah Udoh, Benedict C. Agoha, Muyiwa Adeniyi Sholarin, and Ravin Ahuja

Abstract Following the testimonies of Shaimaire Sanni about the negative wanton use of artificial intelligence (AI) politicking approaches by the Vote-leave group during the 2016 Brexit referendum, the decision by Great Britain (GB) to leave the European Union (EU) had stirred up heated controversies about what would have really been the outcome of the Brexit deal if the Vote-leave group had not cheated with AI politicking systems. Hence, the act of cheating via this platform and the violation of Brexit spending regulations, human rights activists (HRA) like Sanni and Wylie believed, delegitimize the results of the votes obtained for Brexit and for UK's institutions of democracy. Others argue that the allegations raised against the Brexit referendum process justify the agitations for a second Brexit referendum by a section of UK citizens. The Marxian alienation theory and Derrida's critical and analytical method for evaluating qualitative data and arguments gathered on the subject matter of the paper were adopted, with the view to ascertaining the degree of AI politicking approaches that altered the results of UK's Brexit referendum. Marilyn's *ex-post facto* research method was also utilized for interrogating the integrity of UK's democracy in the light of the allegations raised against it. The study observed that most of the

I. A. P. Wogu
Rhema University Nigeria, Aba, Abia State, Nigeria
e-mail: ike.wogu@rhemauniversity.edu.ng

S. Misra (✉) · O. D. Udoh · B. C. Agoha · M. A. Sholarin
Covenant University, Ota, Ogun State, Nigeria
e-mail: sanjay.misra@covenantuniversity.edu.ng

O. D. Udoh
e-mail: oluwakemi.udoh@covenantuniversity.edu.ng

B. C. Agoha
e-mail: ben.agoha@covenantuniversity.edu.ng

M. A. Sholarin
e-mail: solarinadeniyi@gmail.com

R. Ahuja
Shri Viswakarma Skill University, Gurgaon, India
e-mail: ravinahujadce@gmail.com

P. K. Singh et al. (eds.), *Recent Innovations in Computing*, Lecture Notes in Electrical Engineering 701, https://doi.org/10.1007/978-981-15-8297-4_49

allegations raised against UK's Brexit referendum process had merits to their claims, thus justifying their request for a fresh referendum. A positive implementation of AI politicking methods from ethical perspectives was recommended against the current reckless methods adopted by political campaigners.

Keywords Artificial intelligence politicking · AI ad campaigns · Brexit referendum · European Union · Great Britain · Human rights violations · Human rights activists · UK's democracy

1 Introduction

Following the testimony of Christopher Wylie, the Cambridge Analytica (CA) whistle-blower, who allegedly exposed the reckless degree of cheating believed to have taken place during the 2016 US presidential elections when Donald Trump's campaign team hired the service of CA, an AI ad firm commercialized and used its advanced AI technology to manipulate the votes of the US electorates, in the favour of their candidate and the Republican Party. Since then, there have been a series of controversies over the reliability and integrity of the election results collated from the 2016 US presidential election. In view of this, scholars like [1–4] have questioned the sanctity and credibility of future US election. By extrapolation, it also raises issues about the legitimacy of Donald Trump's Presidency in the White House.

In the same vein, the testimony of Shahmir Sanni, the Brexit whistle-blower who brought to light the alleged cases of the violation ranging from the breach of certain electoral spending regulations, to the violation of certain privacy and fundamental human rights of UK citizens, which was allegedly believed, interfered with the eventual results obtained during the 2016 Brexit referendum processes [5]. Consequently, there are those who believe that the outcome of the Brexit referendum votes—which saw the group that voted in favour that Britain should leave the EU winning—would have turned out differently were it not for the last minute negative AI politicking approach which the Vote-leave group deployed via AI ad campaigns from firms known as AggregateIQ (AIQ) [5]. Following the revelation made by the Brexit whistle-blowing, a section of UK citizens called for a fresh EU referendum, on the basis that the first one conducted was allegedly marred [5] by what scholars like [6] referred to as 'certain degrees of criminality'. To corroborate this fact, Sanni had in an interview [7, 8] argued that this very act of cheating undermines the very core feature that defines the British character. In Sanni's words, 'Not cheating is the core of what it means to be British' [9]. Tampering with this culture would annihilate the very foundation on which the British culture is founded upon.

The problematic: The following issues constitute the specific problems which prompted the writing of the paper.

1. That the votes cast during the Brexit referendum of 2016 in the UK were largely flawed and hence not a true picture of the wishes of UK citizens.

2. That the engagement of certain AI data ad firms by the Vote-leave group, during the Brexit campaign and election processes, grossly violated the spending and privacy/human rights regulations of UK citizens.
3. That the values and virtues that define UK's Democratic Institution were grossly distorted and undermined by the activities of the AI data ad firms employed for the Brexit campaign by the Vote-leave group.

Aims and objectives of the paper: In the light of the above issues raised, the paper is poised to do the following:

1. interrogate and evaluate the merits inherent in the claims that the results of the 2016 Brexit referendum votes were largely flawed,
2. determine the validity and extent to which the Vote-leave group used the services of AI Data Firms to supposedly infringe on the privacy and rights of UK citizens,
3. evaluate the plausibility that the adoption of AI politicking methods was inimical to the values and virtues that define what the institution of democracy stands for among UK citizens.

Theoretical perspectives and methodology: The paper adopts the Marxist alienation theory [10] because it offers convenient platforms for interrogating the degree of human rights violations at play in UK's polity. Creswell's mixed research method was adopted for the edge, and it offers researchers [11] evaluating and analysing qualitative data and arguments in the social sciences. Since the study relied mostly on data already gathered from previous studies, Marilyn's ex-*post facto* research method was adopted [7] to facilitate and justify the methods adopted. Derrida's deconstructive and reconstructive method [8] of analysing deeper meanings of concepts and arguments was also utilized for this study.

2 The Brexit Referendum and the European Union (EU)

The word Brexit is a fusion of two: **Br**itain and **exis**t from where 'Brexit' emerged [12]. The words generally indicate the move of the UK as a polity to separate herself from the union of 28 members states which constitutes the European Union [13]. The decisive move of Britain to divorce herself from a long-term marriage from the EU on Thursday, the 23rd of June 2016, sent a rather shocking signal to the political climate of world politics. The vote to leave the EU was substantiated with over 17 million vote, representing 51.9% of the citizens that participated in the process, against the over 16 million votes collated, representing 48% of UK's citizens who wished to remain in the EU, all from a little over 33 million voters who turned out for the referendum. However, the process recorded 46 Million registered voters for the referendum [13]. As it stands, the vote to leave the EU is expected to fully take effect from 11.00 pm on the 29th of March 2019 [12]. Both parties have since provisionally agreed on the divorce settlements.

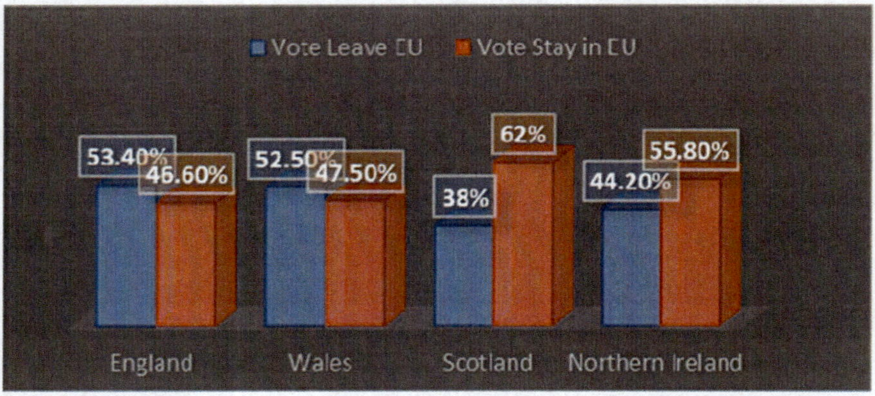

Fig. 1 An analysis of votes of the main actors in UK's referendum of 23rd June 2016

Figure 1 contains the analysis of data and the votes cast by major key players in UK's politics. The diagram revealed that the votes collated in England favoured the Brexit move by 53.4–46.6%. The votes collated in Wales revealed that 52.5% of votes were collated for those who voted for Brexit and 47.5% for those who voted against it. In Scotland and Northern Ireland, however, the case was different. With a vote of 62–38% and 55.8–44.2%, both Scotland and Northern Ireland opposed the idea to leave the EU [12].

3 Artificial Intelligence Politicking in the Twenty-First Century

There is overwhelming consensus that AI has become undoubtedly a global phenomenon regardless of who emerges as the number one leader in AI research and technology [3, 14–17]. Scholars like [1, 2] have argued that one of the sectors that have massively taken advantage of many opportunities and services which AI platforms offer to every sector of life is 'the political sector' [1, 18]. Roy and Rajam among others scholars, for instance, observed that humans and their counterparts (AI machines) have since the past decade—as a result of rising participation and involvement of man in political matters—put mechanisms in place to facilitate the accurate prediction of outcomes of elections for a while now [1]. Other scholars like [19] in their various studies affirmed that high-level machine intelligence (HLMI) systems, and their human counterparts have long achieved the feat of using HLMI systems to enhance political activities. Indeed, the advantage of AI lies in its ability to predict the future of politics and various other states of affairs in different sectors of life's endeavours [1]. This feat notwithstanding, matters arising from the proliferations of AI technology for political campaign purposes have raised several questions among scholars about the credibility and sanctity of politically organized elections and voting processes [20].

3.1 HLMI Systems and Agents of AI Politicking in Twenty-First-Century UK

As mentioned in the above context, studies [2, 3, 18] reveal that one of the sectors which have massively engaged the benefits and services which AI provides is the political sector. Previously, political candidates running for office tend to grope in the dark since there were no tools to help them know and understand the political climate and temperament of electorates. Hence, more often than not, they depended on their instincts and calculated guesses and insights regarding what decision and steps to take vis-à-vis their campaign strategies for running for office, etc. But the advent of innovations in HLMI systems aided by deep learning, big data analytics, machine learning and artificial intelligent systems [19, 20] often rooted in statistical techniques which made it possible for campaign experts to discover and to analyse specific patterns in the culture and behaviour of electorates, long before the day of election. This they do with a view to first collect and process electorates' data for the purpose of understanding their preferences in advance and to know how best to get into their inner mind and conscience to influence them to do otherwise regarding the choice of candidates to vote for when and where the need arises. The studies above revealed that modern AI approaches are now being recklessly deployed during election campaign periods in America.

These abilities in twenty-first-century HLMI systems have already been proven to be possible and effective for campaigning, predicting and manipulating election outcomes. Analysts and experts in America's political arena, for instance, have successfully used it to predict which United States Congress Bills were passed into law. By adopting HLMI systems, one is able to simply evaluating 'the algorithmic assessments of the content of the bill in question and other essential variables like: What time of the year the bill is being proposed? What number of sponsors the bill has had in the past and the exact time and season the bill is being passed to members of congress? [19]. Certain AI ad firms and organizations have since specialized in the art of using these HLMI systems for political campaigns and for manipulating the votes of the electorates in a given polity. The authors of this paper refer to AI data ad firms and organizations that commercialize these AI ad platforms as 'agents of AI politicking'.

Examples of agents that have excelled in the art of tactically collecting sensitive and private data which were later used in most cases to manipulate the psyche of electorates via special ad campaigns [19] include: Cambridge Analytica, Strategic Communication Laboratories (SCL), Group Limited and Global Science Research

(GSR) Limited. These companies, scholars argue, are typical examples of AI politicking agent firms who more often than not pay little or no attention to constituted moral laws in the line of their business [21, 22].

3.2 AI Politicking in UK's Democracy

AI ad firms like Cambridge Analytica and AggregateIQ [9], under the directive of billionaire financier, Robert Mercer—a major key player in bank-rolling Donald Trump's Presidential Campaign—was reported to have sponsored the campaign strategy of using AI ad firm at their disposal to provide expert and technical advice to the Vote-leave campaign group, more especially, on how to target and swing the votes of UK citizens via private and personal data generated from Facebook and other social media platforms without the consent and due authorization of the owners of these data [20]. The service provided by Robert Mercer was observed to have attracted the sum of £625,000. A cost for service paid by the Vote-leave group to an independent referendum campaign organization is known as Beleave [9]. This donation was alleged not to have been officially declared to the referendum electoral commission since such spending exceeded the normal campaign spending limits placed on all the campaign groups [23].

Consequently, the undeclared spending by the Vote-leave group—which was believed to be spent on the technology that influenced the outcome of the Brexit referendum—amounts to what Shahmir Sanni referred to as 'cheating'. The act of cheating, Sanni opines, interfered with what it meant to be British [6, 9]. 'Not to cheat', he submits, 'is the foundation on which UK's democracy was founded upon' [6]. To discover then that the Vote-leave group, a group to which Sanni initially belonged to, could cheat in the above-stated manner was disheartening and unacceptable to Sanni and his fellow human rights activists. In an interview with 'Good Morning Britain' (GMB), Sanni was asked whether the £625,000 spent, over the £7 million originally budgeted for individual campaign group spending, made any significant impact, and Sanni opined that: 'The Vote-leave group won by a 2% margin and that £625,000 was more than enough an amount for getting hundreds of millions, or even billions of impressions on the adverts transmitted during these periods. It's a digital campaign which means millions were seeing the content again and again'. Hence, it should have had a remarkable impact on the results produced at the end of the day.

3.3 *Human Rights Violations in UK's Twenty-First-Century Democracy*

One of the core reasons for the loud outcry against the outcome of the Brexit referendum results was that the citizens of UK discovered that there was a reckless and unlawful use of their private and individual data by the likes of AIQ to manipulating the minds of the electorates, thereby altering the results of the votes that were eventually cast during the Brexit referendum.

AIQ, on the other hand, relied on the service of the parent company and data ad firm known as Cambridge Analytica (CA) based in America. The whistle-blowing efforts of Christopher Wylie revealed that, as much as $1 million was paid to CA to acquire data from unsuspecting Internet users on online platforms like Facebook, Twitter, Instagram etc., this move, Sanni and Christopher argued, amounts to the gross violations of the rights and the privacy of UK citizens, hence justifying the outcry from a section of UK citizens seeking redress and a fresh referendum [6]. It is the act of cheating that these HRA argued had overwhelmed the idea and sanctity of UK's democracy [6]. The consequences of the outcome of the Brexit referendum results exerted feelings that can best be likened to the four classes of the Marxian alienation theory [10]. More on this would be further discussed in the concluding section of this paper.

4 Further Discussions and Summary of Findings

The revelations following Sanni's testimony about the intrigues that took place between the Vote-leave and the Beleave groups, as contained in the 54 paged published dossier by Sanni, raised some very important questions worthy of note. Where it is proven that the last minute spending of £625,000 donation made by the Vote-leave group through the Bealeve group, to AIQ, was what resulted to the decision to leave the EU, the following pertinent questions arise:

1. Is the outcome of the referendum enough reason and justification for the loud outcry by the UK citizens?
2. Is the violation of their fundamental human rights of UK citizens also enough justification for seeking a fresh and second Brexit referendum?

4.1 *Privacy and Human Rights Issues in UK's Democracy*

Evidence from the study conducted so far and the evidence from the testimonies of Sanni tends to portray that the activities of AI politicking agents had to an extent, subverted the democratic processes and institutions of states in the (UK), and hence, the question of the sanctity of future electoral and democratic processes has been raised in the state [23]. Would UK citizens be able to hold future credible elections in the presence of the proliferation of negative AI politicking campaign approaches, which had now taken over a greater part of UK's democratic campaign processes, especially when independent campaign groups wilfully violate regulations stipulated to guide the conduct that governs the hallowed processes of UK's democracy?

There is also the question of the degree of the human rights of UK citizens perceived to have been violated when the private data of individuals were used for election profiteering purposes without due consent or authorization given by the owners of these data. The fact that the said data were also used against the individuals in question by the supposed Believe group, through the AI ad firm contracted for this purpose, in the opinion of most scholars [2, 24], is the most troubling of the issues raised in this section.

This perhaps further explains the fear and feelings of alienation expressed by a section of UK citizens who wondered if they were really in control of their own private actions and the will to cast their votes to candidates of their choice, since AI politicking agents now used negative AI politicking approaches to take over their minds and decisions of the electorates to do otherwise, especially regarding matters of politics. In view of this, UK citizens cannot help but wonder if the results of the votes cast during the referendum would have been different from what it presently is. Would things have turned out differently from what they presently are? Would the Vote-stay group have won the referendum votes? The thoughts of the possible answers to these questions and the feeling of alienation which the citizens now suffer from, it is believed, would hunt British citizens for a long while.

4.2 Summary of Findings

The following specific finding where identified from the study:

1. The study conducted on the first objective revealed that, despite the denials made by Facebook that the 87 million data belonging to Facebook users collected by CA were not part of the data that was used to manipulate the votes eventually recorded for the Brexit deal, there was overwhelming evidence that falsified the claims.
2. The study conducted for the second objective revealed overwhelming evidence, such as: the 54 page dossier published by Sanni which confirmed that the last minute payment of £625,000 and engagement of the AIQ, who deployed negative AI politicking strategies, were indeed the last straw that delivered the winner to the Vote-leave group.
3. Analysis conducted on the third objective revealed that most scholars [4, 23, 24] were convinced that the negative AI politicking strategies used during the said referendum process subverted and in most cases annihilated the institution of democracy in the UK. Their inability to freely exercise their will to choose who

to vote also created feelings of alienation among the people since it influenced their ability to participate in UK's hallowed democratic process.

4.2.1 Recommendations

In the light of the objectives and findings made in the study so far, the authors here are poised to make the following recommendations:

1. Though the study identified overwhelming evidence to support the claims that negative AI politicking was adopted and utilized by the Vote-leave group, the resolution to conduct another referendum would further complicate and jeopardize the already poor socio-political and economic image which the UK was already beginning to suffer from as a result of matters arising from its decision to leave the EU. Hence, UK citizens should rather go ahead and salvage the situation by advancing talks with Brussels which has already progressed significantly.
2. By way of curbing the negative AI politicking campaign strategies adopted by political parties during elections, the authors of this paper recommend that certain EU regulations, which have been inactive before now, be re-enforced such that, while AI politicking in governments may be allowed, it will be done in a manner where the privacy and the fundamental human rights of citizens would not be violated or breached. Where there are breaches of any sort, redress, stiff sanctions and appropriate penalties should be meted to earning individuals, bodies, groups or organizations.
3. Government and all concerned agencies must ensure that politicians and all concerned parties ethically and judiciously use these AI politicking platforms in the manner that would not in any way further jeopardize or undermine the rights and privacy of citizens during elections. Where these recommendations are adopted, it would go a long way to revive the already eroding sanctity, virtue and credibility of elections conducted in UK's democratic institution.

4.2.2 Contribution to Knowledge

The following are the specific contributions which this paper is poised to make to the body of knowledge and to mankind generally:

1. This paper for the first time provides the perspective of an independent study conducted on the Brexit/EU referendum politics, with a view to identify the role and effect of negative AI politicking activities in UK's democracy.
2. The paper identifies that while technology via HLMI systems could be very beneficial for political campaign experts, users of this platforms are prone to become reckless to the point of abusing the platform and infringing on people's fundamental rights and privacy.
3. Where UK citizens succeed in seeing through this complex divorce process from the EU, amides uncertainties and no clear prospects of what the outcome might be, this study would serve as a reference point for researchers and governments

who would want to understand the processes involve and the part played by AI politicking strategies.

5 Conclusion

AI politicking via HLMI systems has become the way of administering twenty-first-century political campaigns and political processes. Nations and organizations who fail to flow in this direction would have themselves to blame. However, while the benefits inherent in AI politicking approaches cannot be over emphasized in this dispensation, this paper is quick to note that where appropriate regulations and ethical foundations are not properly instituted in a polity, as it is the case in the USA and the UK, users and agents of AI politicking platforms are prone to become reckless leading to the negative use of AI politicking strategies leading to the breach of ethical, moral and basic fundamental human rights as was the case for the Vote-leave group who hired the services of AIQ to embark on activities the paper considered illegal and unbecoming for systems and institution which places high regard for democracy.

References

1. Roy, S., Rajam, K.: How AI will decide your future Prime Minister. Your Story. An online publication of Your Story. Retrieved from 20 July 2017. https://yourstory.com/2017/12/artificial-intelligence-politics/
2. Wogu, I.A.P., Sanjay Misra, J., Assibong, P.A., Adewumi, A., Maskeliunas, R., Damasevicius, R..: A critical review of the politics of artificial intelligent machines, alienation and the existential risk threat to America's labour force. In: Gervasi, O., et al. (eds.) Computational Science and Its Applications—ICCSA 2018. Lecture Notes in Computer Science, vol 10963. Springer, Cham. https://doi.org/10.1007/978-3-319-95171-3_18
3. Wogu, I.A.P., Misra, S., Assibong, P.A., Ogiri, S.O., Maskeliunas, R., Damasevicius, R.: Superintelligent machine operations in 21st century manufacturing industries: a boost or doom to political and human development? In: Proceedings of International Conference on Towards Extensible and Adaptable Methods in Computing (TEAMC) 2018, 26–28 Mar 2018. Netaji Subhas Institute of Technology, New Delhi, India (2018). https://drive.google.com/file/d/1bhdcUA0EJBhDTbyMw2Y0mSVkXWlCf8W4/view
4. Wogu, I.A.P., Misra, S., Assibong, P.A., Olu-Owolabi, E.F., Maskeliunas, R., Damasevicius, R.: Artificial intelligence, smart classrooms and online education in the 21st century: implications for human development. J. Case Inf. Technol. (JCIT). Forth Coming Publications of IGI Global (2018)
5. Daly. H.: Good Morning Britain (GMB): 'did it make ANY difference?' Vote Leave whistle-blower savaged in brutal interview (2018). Retrieved on the 20 of June, 2018 from https://www.youtube.com/watch?v=H5-2eITPxo4
6. Eato, G.: We need to take back control: Brexit whistleblower Shahmir Sanni on why there must be a new EU referendum. In Newstateman, an online publication Retrieved on the 20 of June 2018, from https://www.newstatesman.com/politics/uk/2018/03/we-need-take-back-control-brexit-whistleblower-shahmir-sanni-why-there-must-be

7. Marilyn, K.: Ex-post facto research: dissertation and scholarly research, Recipes for success. Seattle, WA: Dissertation Success LLC (2013). https://www.dissertationrecipes.com/wp-content/uploads/2011/04/Ex-Post-Facto-research.pdf.
8. Balkin, J.M.: Deconstructive practice and legal theory. Yale L. J. **96**(15) (1987)
9. Cadwalladr, C., Silver, M., Khalili, M., Phillips, C., Jenkins, A., Search, J., Whipham, S., Rivers, O.: The Brexit whistleblower: 'not cheating is the core of what it means to be British'—video. A Video publication of the Guardian News online. Retrieved on the 20 of June 2018 from: https://www.theguardian.com/politics/video/2018/mar/24/the-brexit-whistleblower-not-cheating-is-the-core-of-what-it-means-to-be-british-video
10. Cox. J.: An introduction to Marx's theory of alienation. (79) 5 International Socialism: Quarterly Journal of the Socialist Workers Party (Britain) Published July 1998 Copyright © International Socialism (1998)
11. Creswell, J.W.: Research Design: Qualitative and Quantitative Approaches. SAGE, Thousand Oaks (2003)
12. Hunt, A., Wheeler, B.: Brexit: all you need to know about the UK leaving the EU. BBC News Online (2018). Retrieved 16 July, 2018. https://www.bbc.com/news/uk-politics-32810887
13. Akabogu, N.N.: Matters arising on Brexit. An online publication of the Sun News Papers (2018). Retrieved on the 16 July 2018 from https://sunnewsonline.com/matters-arising-on-brexit/
14. Lilley, J.: The UK's AI Future Post-Brexit? An online publication of Tech UK (2017). Retrieved on 16 July 2018. https://www.techuk.org/insights/opinions/item/10712-guest-blog-jeremy-lilley-the-uk-s-ai-future-post-brexit
15. Wogu, I.A.P., Olu-Owolabi, F.E., Assibong, P.A., Apeh, H.A., Agoha, B.C., Sholarin, M.A., Elegbeleye, A., Igbokwe, D.: Artificial intelligence, alienation and ontological problems of other minds: a critical investigation into the future of man and machines. In: Proceedings of the IEEE International Conference on Computing, Networking and Informatics (ICCNI 2017). 29–31 Oct 2017, Covenant University, Ota. (2017). https://doi.org/10.1109/ICCNI.2017.8123792
16. Grace, K., Salvatier, J., Dafoe, A., Zhang, B., Evans, O.: When will AI exceed human performance? Evidence from AI experts. arXiv:1705.08807v2 [cs.AI] 30 May 2017. arXiv:1705.08807v2 [cs.AI].
17. Drum, K.: You will lose your job to a robot—sooner than you think! Mother Jones. Online Blogger. Retrieved on the 26 Feb 2018. https://www.motherjones.com/politics/2017/10/you-will-lose-your-job-to-a-robot-and-sooner-than-you-think/
18. Wogu, I.A.P., Misra, S., Olu-Owolabi, F.E., Assibong, P.A., Udoh, O.D., Ogiri, S.O., Maskeliunas, R., Damasevicius, R.: Artificial intelligence, artificial teachers and the fate of learners in the 21st century education sector: implications for theory and practice. Int. J. Pure Appl. Math. **119**(16), 2245–2259 (2018). ISSN: 13114–3395. (Online version). https://www.acadapublic.ecu/hub/. https://acadpubl.eu/2018-119-16/2/232.pdf
19. Polonski, V.: How artificial intelligence conquered democracy: who is really running elections this days. The conversation. An Online publication (2018). Retrieved from https://theconversation.com/how-artificial-intelligence-conquered-democracy-77675
20. Solon, O.: Cambridge Analytica whistleblower says Bannon wanted to suppress voters. The Guardian Online. (2018). Retrieved from 13 June 2018. https://www.theguardian.com/uk-news/2018/may/16/steve-bannon-cambridge-analytica-whistleblower-suppress-voters-testimony
21. Tripathi, T.: Facebook and Cambridge Analytica—where lies privacy? Online publications retrieved on the 19th June, (2018). From: https://www.ihrb.org/focus-areas/information-communication-technology/commentary-facebook-cambridge-analytica.
22. Remnick, D.: Cambridge Analytica and a moral reckoning in Silicon Valley. An online publications of the New Yorker. Retrieved on the 19 June 2018. https://www.newyorker.com/magazine/2018/04/02/cambridge-analytica-and-a-moral-reckoning-in-silicon-valley

23. Banks, A.: Artificial intelligence was key to Brexit campaign. Politic in Descrier (2018). Retrieved from 16 July 2018. Artificial intelligence was key to Brexit campaign. https://descrier.co.uk/politics/artificial-intelligence-key-brexit-campaign-success/
24. Scott, M.: Cambridge Analytica helped 'cheat' Brexit vote and US election, claims whistleblower (2018). Retrieved from POLITICO on Line. https://www.politico.eu/article/cambridge-analytica-chris-wylie-brexit-trump-britain-data-protection-privacy-facebook/

Design of 7 GHz Microstrip Patch Antenna for Satellite IoT- and IoE-Based Devices

Manvinder Sharma, Bikramjit Sharma, Anuj Kumar Gupta, and Bhim Sain Singla

Abstract The next industrial revolution brings machine to machine (M2M) connectivity, Internet of Things (IoT), and Internet of Everything (IoE), and this will take governments, businesses, and people interaction with each other to new level. The global satellite IoT and M2M device market will reach 5.96 million by 2020. Both machine and human can communicate sense and trigger via IoT-based frameworks over a large or remote geographical area using satellite communication. The signal can be sensed at remote location (sea, air, or other unconnected location) and can be uplinked to satellite and can be provided to a central control station. IoT devices should be compact in size so that they can be mounted easily. Microstrip Patch Antenna (MPA) is having advantage of compact design, light weight, easy fabrication method, and low profile over the conventional antennas. Due to their planar structure, microstrip patch antenna is widely used in wireless communication, satellite communication, and in many areas where electromagnetic waves are used. In this paper, inset-fed Microstrip Patch Antenna (MPA) is designed and analyzed for 7 GHz frequency band for satellite communication. The electric field norm plot, radiation pattern are analyzed. Directivity is approximately 12.016 dB, and return loss (S11) calculated is −20.5 dB, and front to back ratio is calculated as 19 dB.

Keywords Microstrip patch antenna · IoT · IoE · Sattelite communication · SIoT

M. Sharma · A. K. Gupta
Chandigarh Group of Colleges, Landran, Punjab, India
e-mail: manvinder.sharma@gmail.com

A. K. Gupta
e-mail: anuj.coecse@cgc.edu.in

B. Sharma
Thapar Institute of Engineering & Technology, Patiala, Punjab, India
e-mail: bikramjit@thapar.edu

B. S. Singla (✉)
College of Engineering and Management, Punjabi University, Patiala, India
e-mail: bhim.pup@gmail.com

P. K. Singh et al. (eds.), *Recent Innovations in Computing*, Lecture Notes
in Electrical Engineering 701, https://doi.org/10.1007/978-981-15-8297-4_50

1 Introduction

In better than IoT which only provide machine to machine (M2M) communication, IoE includes machine to people (M2P) and technology supported people to people (P2P) communication. Rather than only communicating, network intelligence, machine learning, and artificial intelligence are used in IoE. For the proper implementation of Internet of Things (IoT) and Internet of Everything (IoE) ecosystem over a large area, satellite services have inimitable and required characteristics. First, IoT-based device which is mostly used even now a days in ATM machine. For the services "smart application," satellite technology can provide variety of frequencies, speeds, and orbits [1–4]. With the global broadband connectivity via satellite networks satellite IoT (SIoT) provides IoT solutions in remote locations which cannot be reasonable accessed due to cost of terrain. Figure 1 shows satellite-based IoT system. The satellite-based devices can also be deployed in sea, air, or any other unconnected location [5–6]. With the versatility of bandwidth in satellite communication, e.g., narrowband, broadband, and broadcast ability, the different users of IoT and IoE can select required bandwidth and characteristics according to its need. In the industrial IoT ecosystem, satellite technology has already proven its reliability. Figure 2 shows smart city IoT and IoE devices connectivity with satellite [7]. In

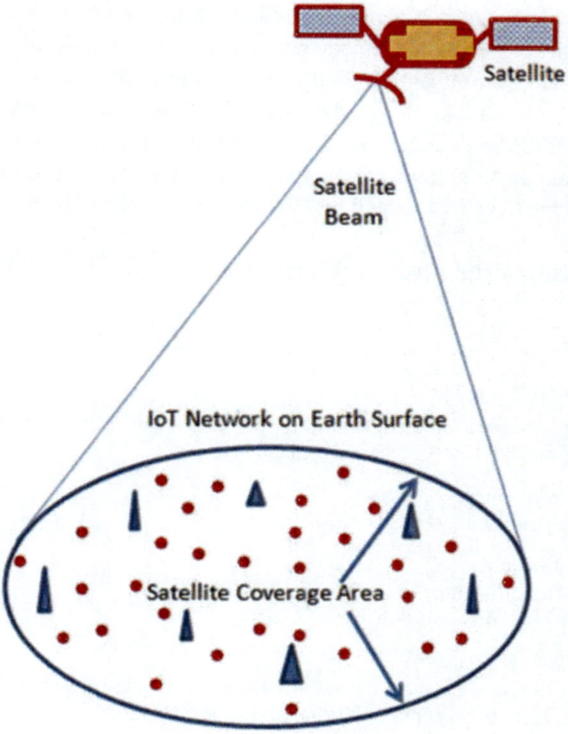

Fig. 1 Satellite-based IoT system (SIoT)

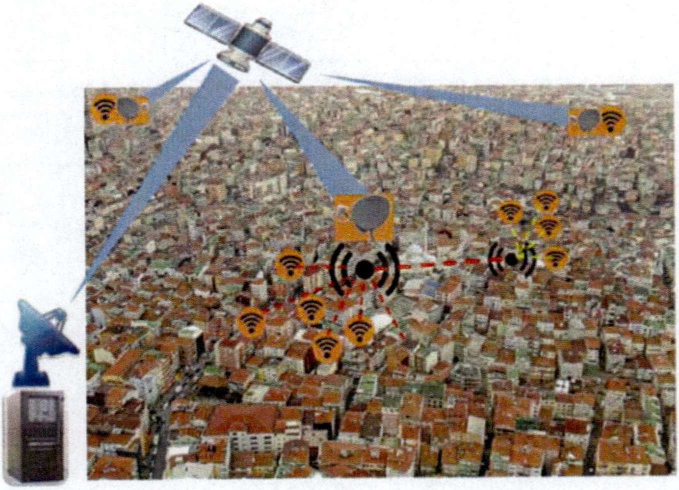

Fig. 2 Connecting smart city IoT and IoE devices to satellite

2016, more than 2.8 million were connected through satellite IoT and will reach 5.96 million by 2020 and 20 billion by 2025. The devices include disaster monitoring, oil and gas, infrastructure, environmental monitoring, military support and aviation. These devices can be reliably operated with satellite services at remote locations or anywhere in the world [8].

Generally, there is three layer IoT architecture model which has client-side (IoT device layer) mid layer of operators on server side (IoT gateway layer) and pathway layer for connecting clients (IoT platform layer). Four-layer IoT architecture is shown in Fig. 3. Using the fundamental three layer model, the requirements are addressed in four-layer model. The four-layer IoT architecture has first layer as sensors and actuators, second as data acquisition system and network gateways, third is edge IT, and fourth layer is data center and cloud. In the first layer, "things" are sensors which convert outer world data into information for analysis and actuators intervene physical reality like switching on/off of other connected device [9–11]. The second layer of Internet gateways and data acquisition system takes input from sensor and connect sensor network to aggregate output. Internet gateways work through Wifi, WLAN, etc., and this stage processes big amount of collected data from sensors and compresses it to optimal size for processing [12]. In the third layer edge IT, preprocessing and enhanced analytics are done. With the use of machine learning and visualization technologies, the architecture can be made more intelligent. In the fourth layer, the data which is analyzed and is stored on cloud and is sent further to communicate other IoT device [13].

IoT and IoE ecosystem requires resilient, ubiquitous, and seamless connectivity all the times to run efficiently. With the SIoT and SIoE system, satellite operators require antennas for communication which are efficient, small in size (to be connected with sensors), and cheap to make satellite IoT and IoE easier [14–15].

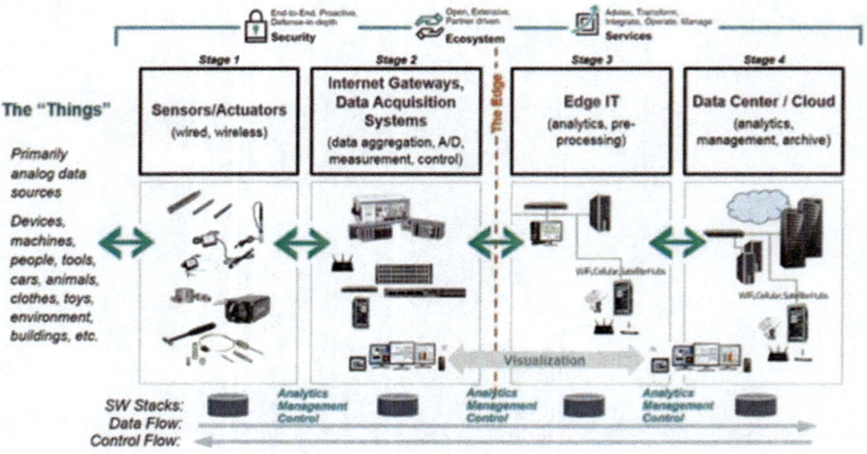

Fig. 3 Four-layer architecture of IoT systems

Microstrip patch antennas is used mostly due to its low-profile conformal design, inexpensive design, easy fabrication, and versatility in design in terms of realization. It can provide good directional patterns according to applications. Figure 4 shows the structure of microstrip patch antenna, the antenna is made up of conducting ground plane and dielectric above it, and the conductive patch is placed over the substrate. Generally, the patch is of the material which is conducting in nature such as copper and can be of any metal and can take any shape [16]. The design of patch is fed with different feeding methods like quarter-wavelength transmission line, coaxial cable or probe feed, coupled feed, inset feed, and aperture feed. The disadvantages of feeding methods are that there is undesirable impedance mismatch with conventional 50 Ω line [17]. To minimize, the mismatch quarter-wavelength transformer

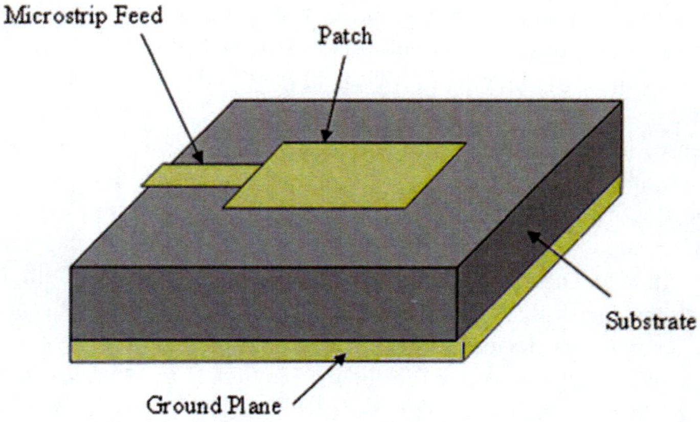

Fig. 4 Structure of microstrip patch antenna

can be employed between 50 Ω line and microstrip feed but this approach also have disadvantages that it increases the size of antenna; hence, overall design size is increased [18–19]. The impedance is higher than 50 Ω if it is fed from edges and if it is fed from center, then the impedance is lower. Thus, the optimal feed point is between edge and center [20–23].

The current at the end of the patch is low, and as we move toward the center, the current increases. If the patch is fed near the center, there is reduction in the impedance.

2 Design Equations of Microstrip Patch Antenna

2.1 *Design of Inset Feed*

Due to sinusoidal distribution in current as shown in Fig. 5, moving along distance *R*, from the ends, the value of current is increased by

$$I = \cos\pi\left(\pi \times \frac{R}{L}\right) \tag{1}$$

If the wavelength is 2*L*, the phase difference is given by

$$\phi = \left(\pi \times \frac{R}{L}\right) \tag{2}$$

As the current increases, there is decrease in the magnitude of voltage using $Z = V/I$. The input impedance is given as

$$Z_m(R) = \cos^2\left(\frac{\pi R}{L}\right)Z_m(0) \tag{3}$$

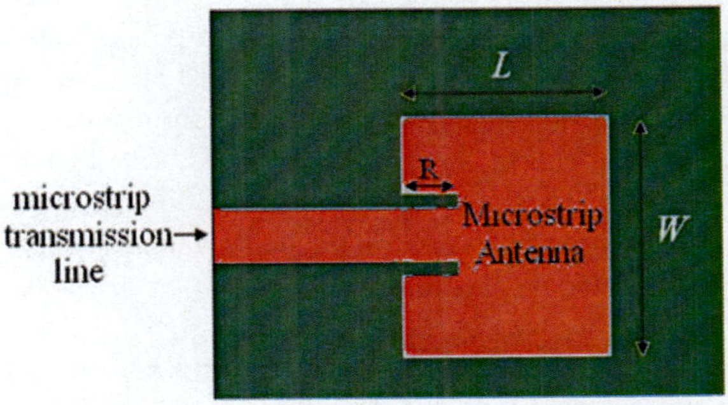

Fig. 5 Inset-fed microstrip antenna

where $Z_{in}(0)$ is the impedance if fed from end.

2.2 Design of Microstrip

The width of patch antenna can be calculated with [14]

$$W = \frac{v_0}{2f_r}\sqrt{\frac{2}{\varepsilon_r + 1}} \tag{4}$$

The length of antenna can be calculated with [15]

$$L = \frac{v_0}{2f_r\sqrt{\varepsilon_{reff}}} - 2\Delta L \tag{5}$$

and

$$\varepsilon_{reff} = \frac{\varepsilon_r + 1}{2} + \frac{\varepsilon_r - 1}{2}\left[1 + 12\frac{h}{W}\right]^{-1/2} \tag{6}$$

$$\Delta L = 0.412h\frac{(\varepsilon_{reff} + 0.3)\left(\frac{W}{h} + 0.264\right)}{(\varepsilon_{reff} - 0.258)\left(\frac{W}{h} + 0.8\right)} \tag{7}$$

3 Modeling and Analysis of Proposed Design

The design is modeled using the equations in software environment for analysis. Electromagnetic frequency domain was used to model the design with 7 GHz frequency which was applied on the lumped port. The analysis of design was done.

Table 1 shows the design values used for model. The model design is shown in Fig. 6. The design structure of microstrip patch antenna is shown in Fig. 7.

Tetrahedral meshing is done. There are five elements per wavelength in the meshing. In meshing, 155,109 tetrahedron values, 22,940 prisms, 15,596 triangles 1140 quad, 1002 edge elements, and 52 vertex elements are used. Inside design of patch maximum element size of 6.6620 is taken with curvature factor of 0.6. The meshed structure is shown in Fig. 8.

Table 1 Description of model

Description	Value (cm)
Substrate thickness	0.1524
50 Ω line width	0.32
Patch width	5.3
Patch length	5.2
Tuning stub width	2
Tuning stub length	1.2
Substrate width	5.3
Substrate length	5.2

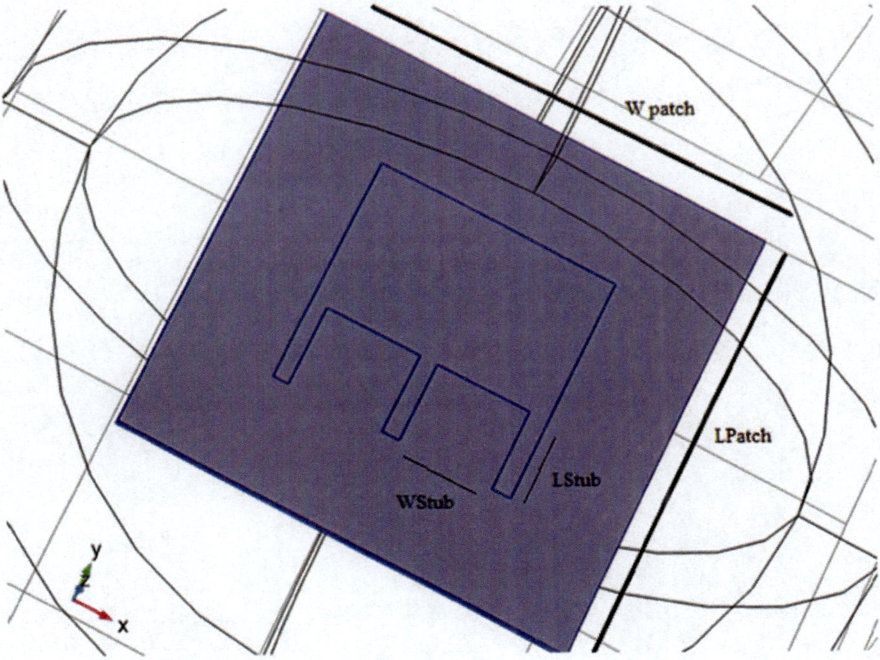

Fig. 6 Dimensions of patch antenna with inset feed

4 Results and Discussion

The modeling and simulation of design are done on 4 × 2.60 GHz processor speed. The simulation used physical memory of 3.6 GB. The frequency 7 GHz is fed to lumped port of antenna as discussed which is in center of patch. The Fig. 9 shows electric field distribution plot and shows the current distribution over the patch.

Figure 10 shows the 2D radiation pattern in H-plane and E-plane. The radiation patterns show directive beam due to ground plane which blocks radiation toward

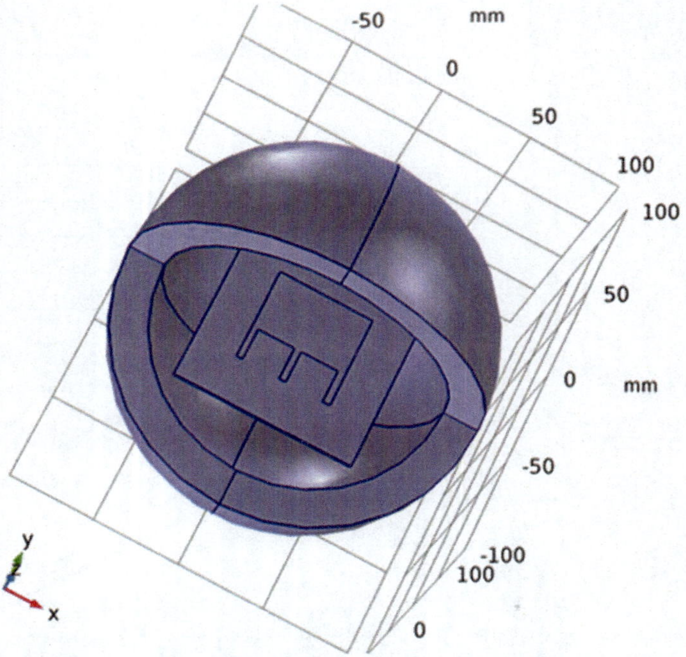

Fig. 7 Structure of microstrip patch antenna

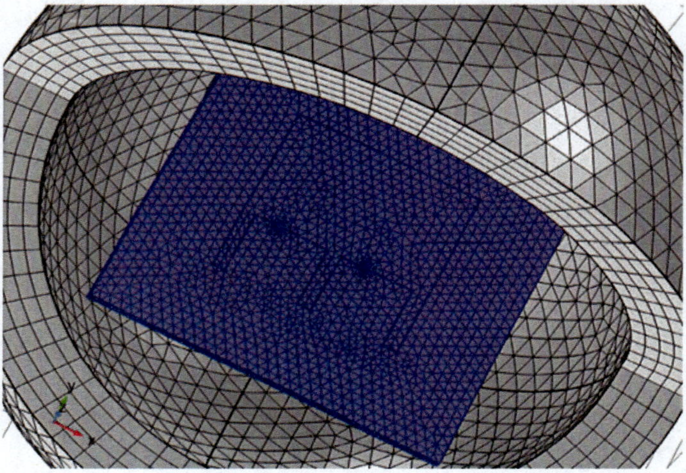

Fig. 8 Meshing of design

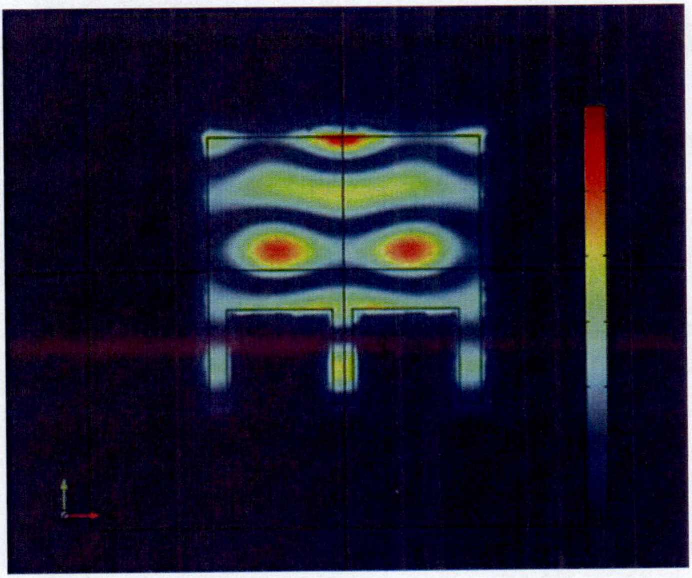

Fig. 9 Electric field distribution plot for 7 GHz

Fig. 10 2D far-field radiation plot

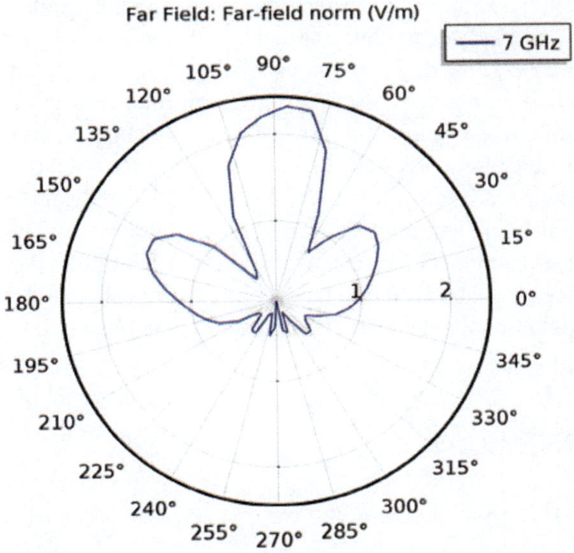

back side. The calculated antenna directivity is 12.016 dB, and the front to back ratio in radiation pattern is more than 19 dB. It can be observed from the radiation pattern that the antenna is radiating like directional antenna which is good for satellite communication. The calculated S_{11} parameter is -20.5 dB which is much better than desired -10 dB. The 3D radiation pattern is shown in Fig. 11.

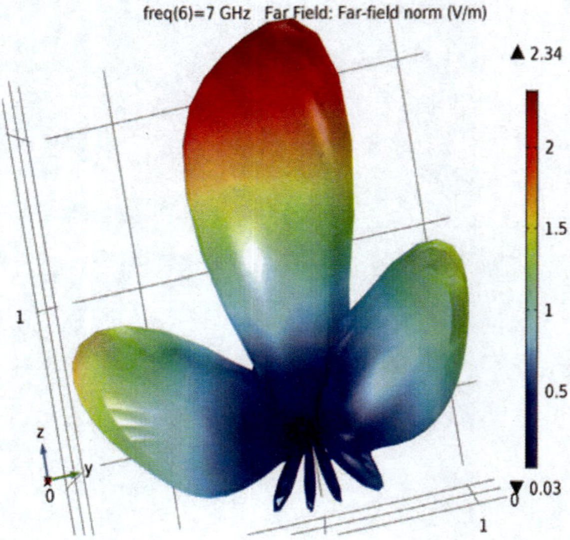

Fig. 11 3D far-field radiation plot

5 Conclusions

With several new applications emerging for satellite-based IoT and IoE devices which include tracking, agriculture, location-based services, healthcare, and security services, the devices "things" will be connected to Internet. For the all time connectivity of remotely located "things" to Internet satellite communication plays a vital role. In this paper, four-layer IoT architecture system is presented, and a microstrip patch antenna is designed and modeled for 7 GHz satellite communication. With the advantages of compact size, inexpensive design and low profile, the microstrip patch antenna can be mounted on "things" easily. The designed antenna is analyzed for parameters like electric field intensity, directivity, front to back ratio, insertion loss S_{11}, and radiation far-filed plot were calculated for 7 GHz. The design showed directional antenna radiation pattern which is good for satellite communication.

References

1. Gineste, M., Deleu, T., Cohen, M., Chuberre, N., Saravanan, V., Frascolla, V., Mueck, M., Strinati, E.C., Dutkiewicz, E.: Narrowband IoT service provision to 5G user equipment via a satellite component. In: 2017 IEEE Globecom Workshops (GC Wkshps), pp. 1–4. IEEE (2017)
2. Yang, Z., Yue, Y., Yang, Y., Peng, Y., Wang, X., Liu, W.: Study and application on the architecture and key technologies for IOT. In: 2011 International Conference on Multimedia Technology, pp. 747–751. IEEE (2011)
3. Sohraby, K., Minoli, D., Occhiogrosso, B., Wang, W.: A review of wireless and satellite-based m2m/iot services in support of smart grids. Mob. Netw. Appl. **23**(4), 881–895 (2018)
4. Minoli, D.: Innovations in Satellite Communication and Satellite Technology. Wiley, New York (2015)

5. Kozlov, D., Veijalainen, J., Ali, Y.: Security and privacy threats in IoT architectures. In: Proceedings of the 7th International Conference on Body Area Networks, pp. 256–262. ICST (Institute for Computer Sciences, Social-Informatics and Telecommunications Engineering) (2012)
6. Ramnath, S., Javali, A., Narang, B., Mishra, P., Routray, S.K.: IoT based localization and tracking. In: 2017 International Conference on IoT and Application (ICIOT), pp. 1–4. IEEE (2017)
7. Miraz, M.H., Ali, M., Excell, P.S., Picking, R.: A review on Internet of Things (IoT), Internet of everything (IoE) and Internet of nano things (IoNT). In: 2015 Internet Technologies and Applications (ITA), pp. 219–224. IEEE (2015)
8. Kaur, S.P., Sharma, M.: Radially optimized zone-divided energy-aware wireless sensor networks (WSN) protocol using BA (bat algorithm). IETE J. Res. **61**(2), 170–179 (2015)
9. Desai, P., Sheth, A., Anantharam, P.: Semantic gateway as a service architecture for iot interoperability. In: 2015 IEEE International Conference on Mobile Services, pp. 313–319. IEEE (2015)
10. Ren, Ju., Guo, H., Chugui, Xu., Zhang, Y.: Serving at the edge: a scalable IoT architecture based on transparent computing. IEEE Netw. **31**(5), 96–105 (2017)
11. Lloret, J., Tomas, J., Canovas, A., Parra, L.: An integrated IoT architecture for smart metering. IEEE Commun. Mag. **54**(12), 50–57 (2016)
12. Li, He., Ota, K., Dong, M.: Learning IoT in edge: deep learning for the Internet of Things with edge computing. IEEE Netw. **32**(1), 96–101 (2018)
13. Tang, J., Sun, D., Liu, S., Gaudiot, J.L.: Enabling deep learning on IoT devices. Computer **50**(10), 92–96 (2017)
14. Ziouvelou, X., Alexandrou, P., Angelopoulos, C.M., Evangelatos, O., Fernandes, J., Loumis, N., McGroarty, F., et al.: Crowd-driven IoT/IoE ecosystems: a multidimensional approach. In: Beyond the Internet of Things, pp. 341–375. Springer, Cham (2017)
15. Sharma, M., Singh, H.: SIW based Leaky wave antenna with Semi C-shaped slots and its Modeling, Design and parametric considerations for different materials of Dielectric. In: 2018 Fifth International Conference on Parallel, Distributed and Grid Computing (PDGC), pp. 252–258. IEEE (2018)
16. Shackelford, A.K., Lee, K.-F., Luk, K.M.: Design of small-size wide-bandwidth microstrip-patch antennas. IEEE Antennas Propag. Mag. **45**(1), 75–83 (2003)
17. Ang, B.K., Chung, B.K.: A wideband E-shaped microstrip patch antenna for 5–6 GHz wireless communications. Prog. Electromagn. Res. **75**, 397–407 (2007)
18. Menzel, W., Grabherr, W.: A microstrip patch antenna with coplanar feed line (2008)
19. Mak, C.L., Luk, K.M., Lee, K.F., Chow, Y.L.: Experimental study of a microstrip patch antenna with an L-shaped probe. IEEE Trans. Antennas Propag. **48**(5), 777–783 (2000)
20. Khosla, D., Kaur, A.: Design of hybrid compression model using DWTDCT-HUFFMAN algorithms for compression of bit stream. Int. J. Eng. Res. Technol. (IJERT) **1**, 2278–3018 (2012)
21. Singh, S., Goyal, S., Sharma, M., Kakkar, R.: Waveguide diplexer design and implementation in communication systems
22. Sharma, M., Singh, H.: Design and analysis of substrate integrated waveguide for high frequency applications. Recent Trends Program. Lang. **6**(1), 1–5 (2019)
23. Sharma, M., Singh, S., Khosla, D., Goyal, S., Gupta, A.: Waveguide diplexer: design and analysis for 5G communication. In: 2018 Fifth International Conference on Parallel, Distributed and Grid Computing (PDGC), pp. 586–590. IEEE (2018)

MWAMLB: Modified Weighted Active Load Balancing Algorithm

Bhagyalakshmi and Deepti Malhotra

Abstract With the fast growth of technology and users, cloud computing has become an important IT paradigm where the resources are available online and on fly. Cloud computing is known for handling large amount of storage and computation data. In the cloud environment, the distinguishing feature of easy availability of resources makes their management a challenging task. One of the most important tasks is the balancing of the load among different virtual machines which in turn leads to proper utilization of resources and good response time. Many researchers have addressed the problem of resource provisioning, but the proactive approach has been gaining a lot of attention in recent years. The resource provisioning can be achieved either by allocating the resources judiciously or by predicting the demand in advance. The traditional methods make use of random selection of virtual machines(VMs) for load balancing. In this research work, a Modified Weighted Active Load Balancing framework (MWAMLB) has been offered with the emergence of cloud computing. The main objective of the MWAMLB framework is to improve the response time of the VM by selecting the virtual machine with maximum weight (W). The weight factor is being calculated on the basis of the availability of RAM, bandwidth and MIPS. The MWAMLB framework have been proposed, implemented and validated in this research paper.

Keywords Cloud computing · Virtual machine · Cloud data center · Load distribution · Resource allocation

---

Bhagyalakshmi (✉) · D. Malhotra
Department of Computer Science & IT, Central University of Jammu, Jammu, India
e-mail: bhagyalakshmi.magotra@gmail.com

D. Malhotra
e-mail: deepti.csit@cujammu.ac.in

P. K. Singh et al. (eds.), *Recent Innovations in Computing*, Lecture Notes in Electrical Engineering 701, https://doi.org/10.1007/978-981-15-8297-4_51

1 Introduction

The cloud refers to an environment that has been developed for remote provisioning of resources to the users that need them. In such a type of computing, the resources are located on a common remote server and are shared among the users rather than owing them personally on personal devices or servers. Irrespective of the fact that cloud computing has shown tremendous growth in the past decade and has proved to be a boon for the IT sector, there are still certain problems that need to be taken care of. Because of the dynamic needs of the users across different geographical locations, these resources need to be managed so as to meet the service-level agreements. Resource management is an umbrella term that covers all the aspects of cloud resources like their characteristics and usage. In cloud computing, resource management is purely based on the allocation of resources where resources are provided to the users by the providers on the pay-per-use basis. There is always a need of effective allocation algorithm which utilizes the resources effectively. Proactive resource provisioning has been gaining a lot of importance these days. Efforts have been made by the researchers to put forward various such schemes to distribute the load evenly among the machines because overloading leads to the degradation of the performance of servers. This challenging task of load balancing has been categorized into two broad categories, namely static load balancing and dynamic load balancing. The examples of the former one are round robin, shortest job first, weighted round robin, etc., whereas the latter one include throttled, ant colony, particle swarm and various other genetic, prediction and heuristic approaches. The effective utilization of resources and management of load across multiple machines can be therefore achieved using the schemes that either use the proactive approach for resource allocation or by predicting the future load.

In this research paper, much optimized load balancing scheme has been proposed to schedule the tasks in the cloud environment. The remaining section of paper is organized as follows: Sect. 2 presents various resource allocation schemes being discussed by researchers. Comparative analysis of the existing techniques for the resource allocation is given in Sect. 3. The proposed MWAMLB framework has been illustrated and discussed in Sect. 4. Section 4 shows experimental setup, Sect. 5 shows simulation result and analysis, and Sect. 6 concludes the paper.

2 Background and Related Work

There have been various research benefactions whose nerve center is based upon allocation of resources on demand basis and overcoming problems like limitation of resources, certain environmental constraints, etc. [1] has proposed the two phase techniques for monitoring and balancing the load of resources in the data center. Graph theory has been used for the implementation of these two phases. The cloud data center has been represented in the form of a graph in the first phase, where the concept of the minimum dominating set has been used for tracking the load in the network. In the second phase, live virtual machine migrations are carried out taking into consideration the system and network parameters for balancing the load. An algorithm named *system and traffic-aware live VM migration for load balancing* (ST-LVM-LB) algorithm has been proposed, and the comparison of the same has been done with the existing works. The advantage of the proposed algorithm is that it improvises the service delivery performance of the cloud data center using the concept of graphs. Depending upon the job characteristics, [2] have mainly focused on the allocation of VM to the user. The jobs with high deadline are considered to be low priority, whereas the jobs with low deadline are considered to be of high priority. The major criteria fulfilling the execution of tasks are such that the low priority jobs should not hamper the execution of high priority jobs and the high priority jobs should be given preference over the low priority jobs. In the proposed scheme, the higher priority jobs are given preference over the low priority jobs. Whenever a high priority job arrives at the data center, no new machine is created for its execution. However, a low priority job is suspended and preference is given to the high priority newly arrived job. The suspended low priority job gets its slot for execution only after the execution of all high priority jobs or if any of the machines becomes free. The proposed scheme reduces the overhead of creation of the machines. In [3], the authors have introduced a scheduling algorithm where the containers are allocated to the appropriate servers with the aim of achieving the balanced load in the system. For the scheduling of tasks being constrained by certain deadlines, an algorithm has been proposed by [4] that takes care of the cost as well as maintains the QoS of the network. The proposed algorithm combines the features of two queuing policies, namely first come first serve (FCFS) and earliest deadline first (EDF). Results show that the algorithm outperforms the existing schemes in terms of achieving deadlines with the deduction in transfer cost also. Li et al. [5] have proposed two algorithms, namely dynamic cloud list scheduling and Dynamic Cloud Min-Min Scheduling, taking into consideration the heavy workload along with the conflict over resources. The authors have divided the tasks into two categories: advanced reservation list and best effort list. The tasks in the Advanced Reservation are given priority over tasks in the Best Effort list. The latter one are preempted on occurrence of any task from the Advanced Reservation list. The results show that the proposed scheme proves to be more energy efficient and also improves the utilization of the system. Reference [6] presented two scheduling algorithms which are dependent upon two major factors of execution of tasks. The first being the requirement of the task in terms of processing and the second is the capacity of the resource to fulfill the needed amount. The experiments are carried out using the CloudSim toolkit, and the results showed that the proposed scheme works better than the existing techniques while handling heavy loads in the system and provided better resource utilization. Host load prediction also plays a significant role for improving VM allocation and utilization in cloud computing. In the past few years, many researchers have focused on the prediction of workload among the hosts. The consumption of resources in cloud environment keeps on changing which makes it quite difficult to anticipate the future needs. In [7], the authors have proposed two algorithms for the prediction of dynamic workload in cloud environment. One of the algorithms is based constraint programming, while other uses the concept of neural networks. The first step takes as input the traces of historical data and groups them into clusters with the help of k-means clustering algorithm. These grouped data are then used for making predictions. The constraint programming is used for the optimization of heuristics, whereas neural networks help in making better predictions. The two programming constructs complement each other making the system more efficient. A deep belief network (DBN)-based approach [8] has been used to predict the requests for resources in the cloud. The researchers introduced the analysis of variance and orthogonal experimental design techniques into learning parameter of DBN. The approach has been used for both long-term and short-term predictions. Google cloud trace has been used to trace the resource requests. The preprocessing of the requests has been done with the help of the differential transformation, normalization, autoregression (AR) and autocorrelation tests. Each RBM from bottom to top is trained to form a DBN that is then fine-tuned using back-propagation (BP) to minimize its loss function. In [9], the authors have proposed an ensemble prediction algorithm. The proposed cloud resource model is based on learning automata (LA) theory that combines different prediction models and also examines the importance of each constituent model according to its performance. The proposed algorithm executes by predicting the average of all the values predicted by the individual model. The proposed model tends to predict the values more suitably as compared to individual predictive models. The authors have proposed a combination of median absolute deviation and Markov chain and named it as Median Absolute Deviation Markov Chain Host Detection algorithm (MadMCHD) [10]. The median absolute deviation is used to find out the current utilization of resources, whereas the Markov chains are used to predict the future utilizations with the help of the current values.

3 Comparative Analysis of Various Techniques and Approaches of Resource Allocation

Different approaches of resource allocation given by the researchers is given in Table 1.

4 MWAMLB_Modified Weighted Active Load Balancing Algorithm

A scheme that takes into consideration the heterogeneity of the system has been proposed. MWAMLB policy calculates the weight of each virtual machine based upon the remaining RAM, bandwidth, number of processors and speed of each processor with the help of Eq. 1.

Table 1 Study of various techniques/approaches used in resource allocation

S. No	Author	Approaches/techniques used	Results
1	Devi et al. [1]	The concept of minimum dominating sets using graph theory has been incorporated for load balancing	Service delivery performance has been improved
2	Sarawathi et al. [2]	Priority-based resource allocation has been demonstrated. Low deadline jobs are given priority over high deadline jobs	Overhead of VM creation is marginally reduced
3	Bossche et al. [4]	Authors have combined the features of earliest deadline first and first come first serve policies	Data transfer cost has been reduced while improving the transfer time
4	Li et al. [5]	The tasks are divided into two lists: advanced reservation (AR) and best effort (BE). Jobs in list AR are given priority over jobs in BE list	System utilization is improved and energy wastage is reduced
5	Sindhu et al. [6]	The proposed algorithm takes into consideration two major factors: the processing need of the task and the ability of the resource to fulfill this need	Better resource utilization with heavy workloads
6	Madi wamba et al. [7]	K-means clustering, constraint programming and neural networks have been used	Results show optimized heuristics and better predictions
7	Asghar et al. [9]	Learning automata has been used to predict the workload values	Better prediction as compared to traditional methods of prediction
8	Melhem et al. [10]	Combination of Markov chains and median absolute value has been proposed	Provides better prediction of future resource utilization
9	Barati et al. [11]	Tuned support vector regression combining the features of genetic algorithm and particle swarm optimization has been used to predict the load	Improved results in terms of overall system metrics
10	Zhong et al. [12]	Authors have combined the concept of weighted wavelet support vector machine and support vector machine for load prediction and distribution	Much efficient system in terms of energy is achieved

Table 2 Average response time in milliseconds

S. No	No. of VMs	RT using MWAMLB	RT using AMLB
1	50	215.79	220.81
2	100	223.75	228.82
3	200	234.62	241.69
4	500	347.15	367.47
5	700	439.42	488.72

4.1 MWAMLB Framework

The framework for the proposed algorithm is shown in Fig. 1.

4.1.1. Components used in framework

(1) Data control center comprises virtual machines, index table and the allocation table. An index table keeps the record of the weight of each VM calculated with the help of Eq. 1. And an allocation table updates the parameters (RAM, BW and MIPS) of the VM upon each allocation and de-allocation.
(2) The load balancer component is responsible for finding the virtual machine with maximum weight.

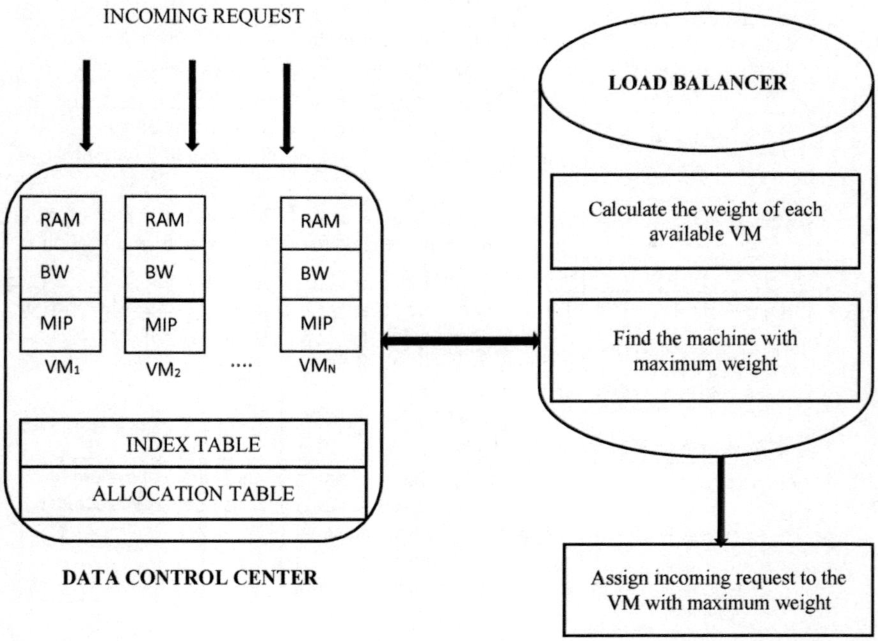

Fig. 1 Framework for modified weighted active monitoring load balancing algorithm

4.2 Proposed Strategy

Calculate the weight W of each virtual machine using the following formula:

$$W = W_1[x_1 + x_2 + x_3 + \cdots] + W_2[y_1 + y_2 + y_3 + \cdots] + W_3[z_1 + z_2 + z_3 + \cdots]; \quad (z_1 = \mathrm{ps}_{\mathrm{pe1}}, z_2 = ps_{\mathrm{pe2}}, \ldots) \quad (1)$$

where W_1, W_2, W_3 are the predefined weights of corresponding system parameter. The weights have been decided on the basis of the generalization of the parameter. More general the parameter, larger the weight. Following this approach, the weights have been assigned as $W_1 = 0.4$, $W_2 = 0.3$, $W_3 = 0.3$ such that their summation becomes one. $\times 1$, $\times 2$ are available RAM, y_1, y_2 are available bandwidth, z_1, z_2 are the processing speeds ($\mathrm{ps}_{\mathrm{pe1}}$ is the processing speed of processing element 1).

4.3 Flowchart

The flow of activities occurring in the execution of MWAMLB has been illustrated in Fig. 2.

4.4 MWAMLB Algorithm

Input:
Number of incoming requests (cloudlets) $r_1, r_2, r_3, r_4, \ldots, r_\mathrm{n}$.
Available virtual machines $\mathrm{VM}_1, \mathrm{VM}_2, \mathrm{VM}_3, \mathrm{VM}_4, \ldots, \mathrm{VM}_m$.

Output:

All incoming requests $r_1, r_2, r_3, r_4, \ldots, r_\mathrm{n}$ are allocated to available VM with highest weight value among the available virtual machines $\mathrm{VM}_1, \mathrm{VM}_2, \mathrm{VM}_3, \mathrm{VM}_4, \ldots \mathrm{VM}_m$.

1. Initially, all the virtual machines are available. The proposed algorithm maintains two tables.
2. On receiving a request, data control center first parses the allocation table to find out the VMs that can accommodate the incoming request and then for these VMs, checks the index table to find out the virtual machine with highest weight. If more than one machine has same highest weight, then the first available machine in the index table is selected for execution.
3. The id of this identified VM is sent to DCC.
4. A request is sent to the VM by DCC for the execution of the incoming request.
5. DCC notifies MWAMLB for new allocation.
6. Tables are updated.

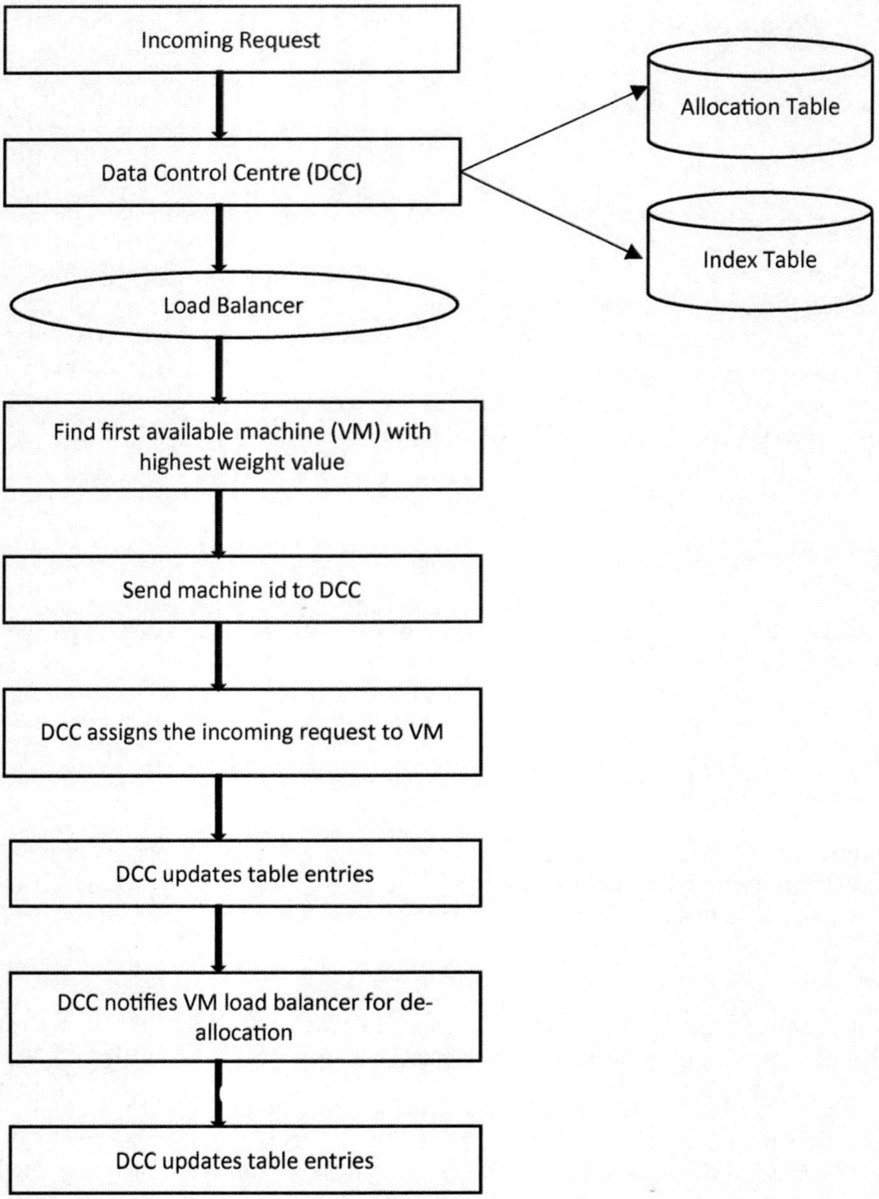

Fig. 2 Flowchart of the proposed weighted algorithm (MWAMLB)

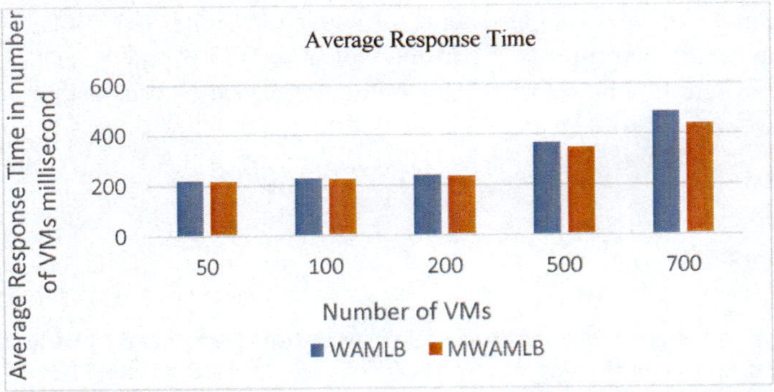

Fig. 3 Average response time

7. DCC notifies MWAMLB for deallocation after getting the response from VM about the completion of the request.
8. MWAMLB updates the allocation Table.
9. Continue from step 2 for the next request.

5 Simulation and Results

In order to evaluate the efficiency of the proposed load balancing algorithm, the experiments were accomplished into five test case groups using the CloudSim simulator. The test cases maintain the heterogeneity of the distributed environment by increasing the number of available VMs. For simulation purpose, VMs ranging from 50 to 700 in number have been considered, all having different sizes in terms of RAM, different bandwidths and different processors having different speeds. Simulation results show that the proposed algorithm has better response time as compared to the existing algorithm. Table 2 shows the overall average response time (RT) in millisecond (ms). Graphical representation of the performance is shown in Fig. 3.

6 Conclusion

The drawback of the existing scheme is that the total capacity of the VM is taken into consideration. However, in heterogeneous environments, the availability of resources keeps on changing with allocations and deallocations. MWAMLB focuses on effective utilization and allocation of VMs by assigning them to incoming requests on the basis of calculated weight, using available, RAM, BW and MIPs. The parameters are individually multiplied to the pre-decided weights, and the summed-up value provides the average calculated weight value. The machine with highest calculated weight is the assigned to the incoming request. The proposed scheme has also been compared with the existing scheme through CloudSim. The results show better response time with our proposed algorithm indicating improvement over the existing algorithm. In future, a new strategy for prediction based load balancing technique will be proposed.

References

1. Devi, R.K.: Load monitoring and system-traffic-aware live VM migration-based load balancing in cloud data center using graph theoretic solutions. Cluster Comput. pp. 1–16 (2018). https://doi.org/10.1007/s10586-018-2303-z.
2. Saraswathi, A.T., Kalaashri, Y.R.A., Padmavathi, S.: Dynamic resource allocation scheme in cloud computing. Procedia Comput. Sci. **47**, 30–36 (2015). https://doi.org/10.1016/J.PROCS.2015.03.180
3. Amani, A., Zamanifar, K.: Improving the time of live migration virtual machine by optimized algorithm scheduler credit. In: Proceedings 4th International Conference on Computer and Knowledge Engineering ICCKE 2014, pp. 346–351, 2014. https://doi.org/10.1109/ICCKE.2014.6993374.
4. Van Den Bossche, R., Vanmechelen, K., Broeckhove, J.: Online cost-efficient scheduling of deadline-constrained workloads on hybrid clouds. Futur. Gener. Comput. Syst. **29**(4), 973–985 (2013). https://doi.org/10.1016/j.future.2012.12.012
5. Li, J., Qiu, M., Ming, Z., Quan, G., Qin, X., Gu, Z.: Online optimization for scheduling preemptable tasks on IaaS cloud systems. J. Parall. Distrib. Comput. **72**(5), 666–677 (2012). https://doi.org/10.1016/j.jpdc.2012.02.002
6. Sindhu, S., Mukherjee, S.: Efficient Task Scheduling Algorithms for Cloud Computing Environment, pp. 79–83. Springer, Berlin, Heidelberg (2011)
7. Madi-wamba, G., Li, Y., Beldiceanu, N., Menaud, J.: Cloud workload prediction and generation models. (2017). https://doi.org/10.1109/SBAC-PAD.2017.19.
8. Zhang, W., et al.: Resource requests prediction in the cloud computing environment with a deep belief network. Softw.—Pract. Exp. **47**(3), 473–488 (2017). https://doi.org/10.1002/spe.2426
9. Asghar, A., Arani, M.G.: A learning automata-based ensemble resource usage prediction algorithm for cloud computing environment. Futur. Gener. Comput. Syst. (2017). https://doi.org/10.1016/j.future.2017.09.049
10. Melhem, S.B., Agarwal, A., Member, S.: Markov prediction model for host load detection and VM placement in live migration. **6** (2018)
11. Barati, M., Sharifian, S.: A hybrid heuristic-based tuned support vector regression model for cloud load prediction. J. Supercomput. **71**(11), 4235–4259 (2015). https://doi.org/10.1007/s11227-015-1520-y
12. Zhong, W., Zhuang, Y., Sun, J., Gu, J.: A load prediction model for cloud computing using PSO-based weighted wavelet support vector machine. Appl. Intell. **48**(11), 4072–4083 (2018). https://doi.org/10.1007/s10489-018-1194-2

Security and Privacy

Security Issues in Internet of Things: Principles, Challenges, Taxonomy

Manik Gupta, Shaily Jain, and R. B. Patel

Abstract The Internet of Things (IoT) has great potential to change the fundamental way of interacting with technology in daily life, and for ease, it also observes and records user preferences that challenge privacy in another way. IoT devices are suspended to extensive usage even more than mobile phones and attain more access to private and secured data. With the growth of connected devices, mobile security is already a challenge, so perspective challenges for IoT connected devices must be much greater than considered at present and can be primarily categorized into safety, security and privacy. Rigorous development of security techniques should be an essential process toward the foundation of strong IoT systems to achieve and retain user trust. The survey in this paper reviewed and analyzed security principles, attacks and countermeasures at different layers of IoT-layered architecture, considering the bottlenecks of IoT systems.

Keywords Internet of things · Security · Privacy · Attacks · Countermeasures · Threats · Challenges

M. Gupta (✉) · S. Jain
Chitkara University School of Engineering & Technology, Chitkara University, Himachal Pradesh, India
e-mail: manik.gupta@chitkarauniversity.edu.in

S. Jain
e-mail: shaily.jain@chitkarauniversity.edu.in

R. B. Patel
Department of Computer Science & Engineering, Chandigarh College of Engineering & Technology, Chandigarh, India
e-mail: drpatelrb@gmail.com

P. K. Singh et al. (eds.), *Recent Innovations in Computing*, Lecture Notes in Electrical Engineering 701, https://doi.org/10.1007/978-981-15-8297-4_52

1 Introduction

The term "Internet of Things" first proposed by Kevin Ashton in 1999 [1] refers to all the things, commonly referred to as daily-life objects or devices with some intelligence to carry certain applications and connected with the Internet to communicate with other devices. IoT can be summarized as a network of interconnected heterogeneous entities that may be present in different parts of the world and can be located with the Internet. In fact, IoT is an extension of the existing Internet that connects smart physical devices rather than humans. The Internet of Things (IoT) is revolutionizing human lives in almost every aspect of life as much as possible. As part of the research, the IoT phenomenon is still in nonage, but their applications are growing rapidly to comfort human life as the available solutions are creating a connected environment with devices that are easy to port, easy to implant and easy to wear. IoT is build up with a huge number of devices ranging from pint-sized chips to colossal-sized machines, connected with wireless technology, and their scope and rate of expansion are really astonishing. Besides domestic applications, IoT technology is marking its great presence in industrial automation and upgradation with a huge impact in human life by encompassing machines, processes, data as well as individuals, thus better renamed as the Internet of Everything (IoE). The IT infrastructure around the world is ready to support around 50 billion devices and things, irrespective of their size and applications that are affecting humans very rapidly with an ecosystem of information extracted from devices all around [2]. IoT is marking its presence in various applications like smart homes, smart cities, wearable devices, smart farming, M2M, etc. [3].

Generally, Internet works over traditional TCP/IP for network communication; however, IoT already contains billions of devices that raise much more traffic with the need of greater storage [4] besides other major issues like security and privacy [5, 6], scalability, reliability, quality of service, interoperability, etc. To cater these issues, three-layered architecture of IoT [7] was considered as best over five-layered and seven-layered architecture due to low overhead. Here, each layer merges the subsequent layer matching their functionality domain and is shown in Fig. 1.

Perception layer or device layer is the lowest layer of IoT architecture and consists of perception and physical devices and GPS, WSN, RFID, RSN, MEMS NEMS

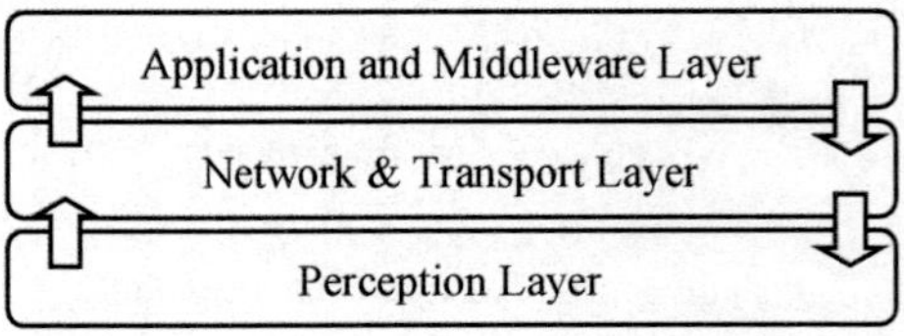

Fig. 1 Three-layered architecture of IoT

and/or sensing devices. It involves perception and controlling objects, gathering and identifying device-specific information from objects or the environment around, which is then transformed into a digital setup for transferring to the network and transport layer to process the information securely. The threats in terms of security at this layer mostly strike from outside and are at node level or replace the device software to act the devices as per their wish and gather network information; perception layer supports various security services to the object devices involved [8, 9].

Network and Transport layer can be wired or wireless, and collectively known as the transmission layer that transfers data securely for information processing from the perception layer's physical devices to the application and middleware layer. This layer supports different combinations of heterogeneous networks and can be divided into local area networks, core networks or access networks. The protrusive security concerns in this layer are the authentication and integrity of network data. Since this layer carries a large amount of data so, this layer is highly prone to attacks which may cause network congestion and provides ubiquitous access to the perception layer information [10].

IoT supports the bridge between different types of devices it connects and communicates among pairs sharing similar service types. *Application and middleware layer* accepts information from the network layer performs actions like ubiquitous computing, decision making and finally store this information in the database. Further, this object information is managed globally by various applications like smart home, smart farming, e-health, etc. The distribution speed of risks over the devices is more prone than the Internet with IoT devices [8].

2 Principles for Secure Communication in IoT

A large number of information security systems have been investigated by [11] and proposed comprehensive security principles in terms of security and assurance referred to as IAS (Information, Assurance, Security) octave.

Confidentiality (C) ensures access to sensitive information should be restricted to authorized users only which may be in the form of assets like machines, internal objects, services as well as humans. This principle strongly follows the management of data and its classification policy as it states that assess to assets and information should be assessable only on the basis of requirement so that the information which is restricted to a particular view should now have a public view and assessable to every asset of the network.

Integrity (I) ensures the maintenance of trustworthiness, accuracy and consistency of data throughout its lifecycle, i.e., the data or information should not have tampered while they are stored in devices or floating from source to destination. Security mechanisms like data encryption, checksums, hashing, etc., ensure the integrity of data. A very common way to maintain integrity of data is one-way-hash, which involves calculation of hash for particular set of data before transmission which is supposed to float along with original message and at receiver's end if computed hash doesn't match the hash received reflects data loss during transition of message in the network thus ensuring end-to-end security in IoT.

Availability (A) ensures the availability of all the resources and data to all the IoT devices which need them. To achieve this, it involves the maintenance of hardware involved in IoT devices, software patching and network optimization. However, availability can be defeated by natural calamities and disasters, to maintain the processes redundancy; fault tolerance, failover, high-availability clusters and RAID systems can help in maintaining the availability.

Non-repudiation (NR) ensures that none of the nodes in the network can deny packet or message transmission, which it has previously sent. It is a process in which all the incidents and non-incident of any event in the network can be validated by IoT.

Privacy (P) ensures that IoT systems should follow privacy policies, rules and allow network nodes to control the sensitive data stored with them and to support context-specific anonymity. The data should be carefully protected by all the nodes in the network, which are involved in the management of data as well as the acquisition of data.

Audibility (AU) ensures performing firm monitoring on any actions in an IoT system which appear to focus more on vulnerability assessment and security configuration management.

Accountability (AC) ensures the acceptability of IoT users and the fitness of technologies deployed in the network. The IoT nodes maintain accounts of users taking charge of the nodes and raise specific accountability challenges related to security, privacy, safety, etc. due to the autonomous, physical and pervasive nature of the nodes.

Trustworthiness (TW) ensures trust among third-party devices by proving their identity with minimum requirements of security, safety, privacy, resilience and reliability among IoT systems.

3 Security Challenges in IoT

The IoT paradigm focuses on a wide range of devices ranging from micro-embedded chips to high-end servers; each of them faces a variety of security issues. Implementing conventional security techniques on IoT nodes is not that easy as they face resource constraints. A number of security attacks need to be fixed at different layers of IoT; the following section summarizes major security threats in IoT systems.

3.1 Tag Cloning

Tags are generally installed on several things that are deployed in open access areas, where the things can be accessed physically, and the data can be sniffed and altered by an adversary or can be compromised by replicating the tags so that the reader cannot differentiate it with the original [12].

3.2 Brute Force Attack

A brute force attack carries all possible combinations of key sets from the key pool using automated software in order to find the correct encryption key. However, with the increase in computing power and growth in success rate to capture the keys using brute force attacks, many other attacks successfully started reducing the search space and retrieving keys.

3.3 Eavesdropping

Eavesdropping or sniffing involves interception of communication between two points of the network by sniffing or recording data packets and thus stealing information among them by using cryptographic analysis tools. Since in IoT, most of the devices are left unattended and communication between the connected devices is through the Internet; hence, IoT devices are more vulnerable to such attacks. In RFID systems, since most of the RFID tags lag in encryption due to memory constraints, so eavesdropping is generally launched during the data transmission between the tag and readers. However, in the case of near-field communication systems, data communication between nodes takes place in close proximity, and NFCs generally lag in security. Here, an attacker can intrude in close proximity or even use a powerful antenna to intercept the data values floating in the communication channel [13].

A very common form of eavesdropping that specifically targets at network layer rather than the perception layer is a man-in-the-middle attack. It also takes control of the network communication between different nodes of the network and malicious nodes. It can forge their identities to communicate normally and acquire more access in the network [14] to modify and relay the data to other malicious nodes in the network.

3.4 Routing Attack

The routing protocol for low power and lossy networks commonly referred to as RPL [15] is considered as a standard routing protocol for the Internet of Things [16]. It supports one-to-one, one-to-many and many-to-one communication, thus composed of one node as a sink node and it is based on a destination-oriented directed acyclic graph [17]. RPL strongly follows a destination-oriented directed acyclic graph (DODAG) topology, which supports unidirectional traffic toward the root node and bidirectional traffic among devices. Each node in the network contains its respective IDs, the ID of the parent node, IDs of neighbor nodes and its rank, which represents its position concerning the root node as well as other nodes in the network. This rank increases downward from the root toward node and vice versa. The major attacks that encounter RPL are sinkhole, selective forwarding, sybil, wormhole, blackhole, identity and hello flooding attack.

3.5 Flooding Attack

Flooding attacks can be performed either internally or externally in a network and are responsible for increasing the volume of traffic in the network to make resources unavailable and exhaust them in worst cases that may be in the form of reaching other devices or the links. "Hello" flooding attack is one of the major attacks in this category. Here, devices possessing to join the IoT network send a broadcast packet to existing devices in the network, known as "hello" packet with high-power antenna and represent itself as a neighbor device to all other devices in the network thus creating a fake one-hop network which does not need encryption breaking [18]. Besides this, in the case of RPL networks, an attacker can saturate the RPL nodes and reset the trickle timer of neighboring nodes by broadcast or unicast the DODAG Informational Solicitation (DIS) message and accepts DODAG Information Object (DIO) message in return. Reference [19] studied the consequences of flooding attacks and analyzed a significant increase in the control message overhead without affecting the delivery ratio.

3.6 Wormhole Attack

Wormhole attack affects both network traffic flow and network topology. This attack can be launched by transmitting all the packets, by creating a private channel or tunnel between two attackers in the network [20]. There are primarily two types of wormhole attacks—using packet encapsulation and using packet relay. In the packet encapsulation technique, a malicious node sends packets that are encapsulated in the payload to neighboring nodes in the network, and on other end, another malicious node can get the packet from the payload and transit it again in the network. In the packet relay technique, a malicious node relays a packet between different legitimate nodes in the network to convince them that they are neighbors.

3.7 Replay Attack

Replay attacks also known as playback attacks are classified as an intermediate level of IoT attack in which malicious node pretends to be a legitimate node and detects transmission of data to further carry its retransmission or delay in the network thereby fooling the receivers and thus hindering the packet re-assembly at receiver nodes' end. This may focus to affect the processing of legitimate packet transmission in the network and depletion of resources in the network, rebooting of other nodes connected in malicious nodes' communication cycle and/or buffer overflows to increase latency in the network [21, 22].

3.8 Spoofing and Sybil Attack

Sybil attack in IOTs enforces malicious nodes to forge their identities to disseminate spam phishing and degrade the functionality of the network [23]. Malicious nodes may successfully launch spoofing attacks in IoT by forging the MAC or IP address of any legitimate node in the network and thereby claiming itself a legitimate node, thus gaining the illegal access to the IoT network to launch more advanced attacks like DOS attacks or man in the middle attack [24]. A huge number of security mechanisms rely on the identity of the nodes however in perception layer, a sybil node may provide fake MAC value with a motive to deplete network resources and as a result, legitimate nodes may be restricted to access network resources; this is referred as low-level sybil attack [25]. A solution proposed by [26, 27] is to measure the signal strength and channel estimation. Similar to low-level sybil attacks, an intermediate-level sybil attack focuses on the network layer to violate the data privacy by forging the node's identity and vitiate the network performance by advertising malware and launching phishing attacks [28]. References [29–32] suggest some solutions like random traversing throughout the network graph, user behavior analysis and maintaining clear discrimination lists among trusted and non-trusted users.

3.9 Sinkhole Attack

Networks in IoT which act more in hostile areas are usually left unattended with devices having limited battery backup, computational capability and range of communication. They are usually more prone to sinkhole attack in which an adversary compromises a node in the network and based on the routing metrics, it will start establishing communication with neighboring nodes. At this point, the adversary node will launch an attack; some attacks that are generally launched in sinkhole are routing drop, spoofing attack and selective forwarding attack. Moreover, sinkholes can also forge information send to the cluster heads and/or base stations. Some of the anomaly-based solutions for sinkhole attack were [33] proposed an RSSI solution that involves an extra monitor node in the network, [34] can detect sinkhole attacks based on LQI, i.e., liquid strength indicator where rate of detection increases when number of detector node increases, [35] proposed a message digest algorithm to achieve data authentication and integrity [35] also proposed a statistical-based algorithm to identify the malicious node in less time with a low false-positive rate and detect sinkhole attack.

3.10 Denial of Service and Sleep Deprivation Attack

Sensing a hostile environment with battery-operated devices like in wireless sensor networks makes the network more vulnerable to attacks like battery drainage as the replacement of batteries is not possible in such environments due to the absence of infrastructure. One of the most devastating attacks of such kind is sleep deprivation where adversaries try to maximize the power consumption of nodes in the network to minimize the wakeup span of the nodes. Such attacks are considered low-level attacks and are possible by increasing useless traffic toward a node in the 6loWPAN environment so that they completely deplete their batteries and stops working anymore thus further leading to denial-of-service (dos) attacks. The solutions proposed by [36] are to apply a multi-level intrusion detection system, and [37] builds architecture on top of bits network framework to detects more complicated 6loWPAN-related DOS attacks.

3.11 Insecure Software, Firmware and Interfaces

The interfaces used to access IoT services are based upon cloud, web or mobile interfaces that are highly prone to attacks and thus may affect data privacy. Besides interfaces, vulnerabilities may be caused by insecure firmware or software; thus, their updates need to be carried out securely.

3.12 *CoAP and Middleware Security*

The application layer containing web transfer protocols like the Constrained Application Protocol (CoAP) is also prone to attacks [38–40]. To provide end-to-end security in constrained devices, CoAP follows a specific message format defined in RFC-7252 [41] and uses DTLS bindings with several security modes. RFC-7252-based CoAP messages require encryption for secure communication, whereas multicast support in CoAP needs authentication and key management. The middleware in the IoT paradigm used by different environments, and interfaces must be secured enough to provide communication among various heterogeneous entities.

4 Taxonomy of Security Threats and Countermeasures in IoT

Vulnerabilities can be exploited at all the layers of IoT security architecture, and they can be based upon physical components, connectivity, communication, data, operating system, software or firmware-level attacks. The countermeasures for the security threats address these vulnerabilities to achieve security at different layers. This section focuses on the taxonomy of various possible security threats and their countermeasures at different layers of IoT security. Figure 2 and Table 1 highlight the comparative analysis of various security threats, their implications, level of attack, occurrence in architecture model, security principles compromised and their countermeasures.

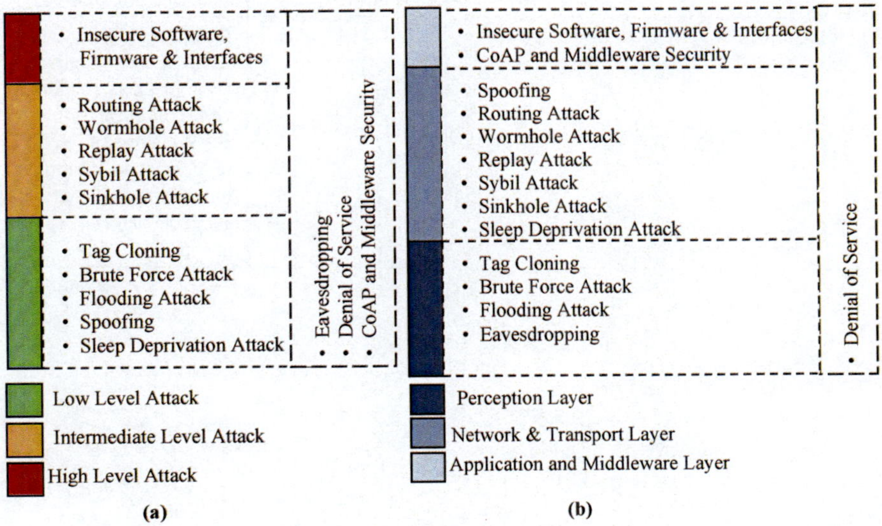

Fig. 2 **a** Measurement of level of attacks, **b** layer affected by attacks

Table 1 Affected security principles and their countermeasures

Security threats	Implication	Security principle affected	Security countermeasures
Tag cloning	Privacy violation	All	Tag isolation, distance estimation, tag blocking, kill/sleep command, implementing authentication techniques, hash-based schemes, encryption techniques [42], OTP synchronization between tag and backend
Brute force attack	Privacy violation	ALL	Cryptography techniques, firmware security update
Eaves-dropping	Privacy violation, man-in-the-middle Attacks	C, NR, P, I	Authentication techniques, secured channel, RFID communication channel encryption [43], pre-installation of network key on devices, mutual authentication and Tamper Detection
Routing attack	Man-in-the-middle attacks, eavesdropping	C, I, A, AC, NR, P	Distributed hash tables, storing and tracking identities of each instances of nodes in RPL, signature-based authentication [44–46]
Flooding attack	Denial of service and disruption	AU, C, I, AC, NR, P, A	Computing signal strength, packet delivery ration, packets encoding with error correction codes, change of frequency and locations, firewall implementation and deep packet inception [47–50]

(continued)

Table 1 (continued)

Security threats	Implication	Security principle affected	Security countermeasures
Wormhole attack	Denial of service	C, I, AC, NR, P	Trust level management, key management, signal strength measurement, geographic information binding and graph traversal [35, 51–59]
Replay attack	Denial of service and disruption	C, I, AC, NR, P	Time-stamping, hash chain-based verification of fragments [60, 61
Spoofing	Denial of service and network disruption	All	Channel estimation, measuring signal strength, authentication and encryption techniques, Message authentication, Filtering, SSL authentication [62]
Sybil attack	Spamming, unreliable broadcast, privacy violation, Byzantine faults	C, I, AC, NR, P	Classification-based sybil detection (BCSD), user behavior analysis, trusted and untrusted user list maintenance, random walk on social graphs [28–32, 63], Douceur's approach –Trusted certification
Sinkhole attack	Denial of service	A, C, I	Parent fail-over, IDS solution, generating identity certificates and rank authentication techniques [52, 64], message digest algorithm [65]
Denial of service	Increase in traffic, energy drain	A, AC, AU, NR, P	Suspicious devices list maintenance, access control lists, policies provided by providers [66–68],], Load balancing [69]

(continued)

Table 1 (continued)

Security threats	Implication	Security principle affected	Security countermeasures
Sleep deprivation attack	Energy consumption	P, I, C, AU, TW, NR	Content chaining approach, Round Robin scheme, multi-layer-based intrusion detection, split buffer approach, target IPv6 defense movement in 6LoWPAN [22, 36], random vote
Insecure software, firmware and interfaces	Privacy violation, network disruption, denial of service	C, P, I, TW	Regular device updates, file encryption using acceptable encryption techniques, file transmission via encrypted connection, secured update server
CoAP and middleware security	Network disruption, privacy violation, denial of service	C, P, TW	VIRTUS Middleware [70], security policies, secure middleware for embedded peer-to-peer systems (SMEPP), lightweight DTLS, TLS-DTLS mapping, HTTP-CoAP mapping, TLS-DTLS tunneling, message filtration using 6LBR, service layer M2M security [38–40, 70–74]]

5 Conclusion

In this survey, principles for secure communication in IoT are covered along with the security challenges and attacks at the layered architecture of IoT applications. We presented several security threats along with their implication and countermeasures related to the layers of IoT architecture and affected security principles. These threats are also measured at a certain level of impact over the IoT applications. IoT can provide every possible solution and raise more human dependency in the future if major security concerns like authentication, trust management, access control, privacy, confidentiality, key management and end-to-end security are fixed at both hardware and software levels. Several countermeasures are presented in this paper to handle various security challenges, but still a wide spectrum of security issues needs more attention toward smart device hardening and detection. In the future, more focus should be diverted toward the development and implementation of standardized security solutions at the production level of IoT devices; on the other hand, end-users should also need to understand the objectives and risks around the device, while interconnecting. The survey is expected to be a useful resource for more secure IoT applications in the future.

References

1. Ashton, K.: That 'Internet of Things' thing. RFID J. **22**(7), 97–114 (2009)
2. Bodkhe, U., Mehta, D., Tanwar, S., Bhattacharya, P., Singh, P.K., Hong, W.: A survey on decentralized consensus mechanisms for cyber physical systems. IEEE Access **8**, 54371–54401 (2020)
3. Sharma, A., Sharma, R.: A review of applications, approaches, and challenges in Internet of Things (IoT). In: Proceedings of ICRIC 2019, pp. 257–269, Springer (2020).
4. Tan, L., Wang, N.: Future Internet: The Internet of Things. In: 3rd International Conference On Advanced Computer Theory & Engineering, vol. 5, pp. V5–376. IEEE (2010).
5. Gan, G., Lu, Z., Jiang, J.: Internet of Things security analysis. In: International Conference On Internet Technology And Applications, pp. 1–4. IEEE (2011).
6. Rastogi, N., Singh, S.K. Singh, P.K.: Privacy and security issues in big data: through Indian prospective. In: 3rd International Conference on Internet of Things: Smart Innovation and Usages, pp. 1–11, Bhimtal (2018).
7. Liu, L., Lai, S.: ALOHA-based anti-collision algorithms used in RFID system. In: International Conference On Wireless Communications, Networking And Mobile Computing, pp. 1–4. IEEE (2006).
8. Suo, H., Wan, J., Zou, C. Liu, J.: Security In the Internet of Things: a review. In: International Conference On Computer Science And Electronics Engineering, vol. 3, pp. 648–651. IEEE (2012).
9. Xiaohui, X.: Study on security problems and key technologies of the Internet of Things. In: International Conference On Computational And Information Sciences, pp. 407–410. IEEE (2013).
10. Zhang, L., Wang, Z.: Integration of RFID Into wireless sensor networks: architectures, opportunities and challenging problems. In: International Conference On Grid And Cooperative Computing Workshops, pp. 463–469. IEEE (2006).

11. Cherdantseva, Y., Hilton, J.: A reference model of information assurance & security. In: International Conference On Availability, Reliability And Security, pp. 546–555. IEEE (2013).
12. Burmester, M., De Medeiros, B.: RFID security: attacks, countermeasures and challenges. In: RFID Academic Convocation, The RFID Journal Conference (2007).
13. Thorat, N.B., Sreevardhan, C.: Survey on security threats and solutions for near field communication. Int. J. Res. Eng. Technol. **3**(12), 291–295. IJRET (2014).
14. Padhy, R.P., Patra, M.R., Satapathy, S.C.: Cloud computing: security issues and research challenges. Int. J. Comput. Sci. Inf. Technol. Secur. (IJCSITS) **1**(2), 136–146 (2011)
15. Wallgren, L., Raza, S., Voigt, T.: Routing attacks and countermeasures In the RPL-based Internet of Things. Int. J. Distrib. Sens. Netw. **9**(8), 794326 (2013)
16. Winter, T., Thubert, P., Brandt, A., Hui, J., Kelsey, R., Levis, P., Pister, K., Struik, R., Vasseur, J.P., Alexander, R.: RPL: IPv6 routing protocol for low-power and lossy networks, No. RFC 6550 (2012).
17. Asim, M., Iqbal, W.: IoT operating systems and security challenges. Int. J. Comput. Sci. Inf. Secur. **14**(7), 314 (2016).
18. Sen, J.: Security in wireless sensor networks. Wirel. Sens. Netw.: Curr. Status Fut. Trends 407 (2012).
19. Le, A., Loo, J., Luo, Y., Lasebae, A.: The impacts of internal threats towards routing protocol for low power and Lossy network performance. In: IEEE Symposium on Computers and Communications, pp. 000789–000794. IEEE (2013).
20. Mayzaud, A., Badonnel, R., Chrisment, I.: A taxonomy of attacks in RPL-based Internet of Things. Int. J. Netw. Secur. **18**(3), 459–473 (2016)
21. Kim, H.: Protection against packet fragmentation attacks At 6LoWPAN adaptation layer. In: International Conference on Convergence and Hybrid Information Technology, pp. 796–801. IEEE (2008).
22. Hummen, R., Hiller, J., Wirtz, H., Henze, M., Shafagh, H., Wehrle, K.: 6LoWPAN fragmentation attacks & mitigation mechanisms. In: Proceedings of The 6th ACM Conference on Security & Privacy In Wireless & Mobile Networks, pp. 55–66. ACM (2013).
23. Kumar, R., Chauhan, N., Kumar, P., Chand, N., Khan, A.U.: Privacy aware prevention of Sybil attack in vehicular ad hoc networks. In: Singh, P., Bhargava, B., Paprzycki, M., Kaushal, N., Hong, W.C. (eds.) Handbook of Wireless Sensor Networks: Issues and Challenges in Current Scenario's, vol. 1132, pp. 364–380. Advances in Intelligent Systems and Computing. Springer, Cham (2020)
24. Zeng, K., Govindan, K., Mohapatra, P.: Non-cryptographic authentication and identification in wireless networks. Netw. Secur. **1**, 3 (2010)
25. Xiao, L., Greenstein, L.J., Mandayam, N.B., Trappe, W.: Channel-based detection of Sybil attacks in wireless networks. IEEE Trans. Inf. Forensics Secur. **4**(3), 492–503 (2009)
26. Demirbas, M., & Song, Y.: An RSSI-Based Scheme For Sybil Attack Detection In Wireless Sensor Networks. In: International Symposium on a World of Wireless, Mobile and Multimedia Networks, pp. 5. IEEE (2006).
27. Li, Q., Trappe, W.: Light-weight detection of spoofing attacks in wireless networks. In: IEEE International Conference on Mobile Ad Hoc and Sensor Systems, pp. 845–851. IEEE (2006).
28. Zhang, K., Liang, X., Lu, R., Shen, X.: Sybil attacks and their defenses in the Internet of Things. IEEE Internet Things J. **1**(5), 372–383 (2014)
29. Alvisi, L., Clement, A., Epasto, A., Lattanzi, S., Panconesi, A.: SOK: The evolution of Sybil defense via social networks. In: IEEE Symposium on Security And Privacy, pp. 382–396. IEEE (2013).
30. Cao, Q., & Yang, X.: SybilFence: improving social-graph-based Sybil defenses with user negative feedback. In: arXiv preprint at arXiv:1304.3819. (2013).
31. Mohaisen, A., Hopper, N., Kim, Y.: Keep your friends close: incorporating trust into social network-based Sybil defenses. In: Proceedings IEEE INFOCOM, pp. 1943–1951. IEEE (2011).
32. Quercia, D., Hailes, S.: Sybil attacks against mobile users: friends and foes to the rescue. In: Proceedings IEEE INFOCOM, pp. 1–5. IEEE (2010).

33. Tumrongwittayapak, C., Varakulsiripunth, R.: Detecting sinkhole attacks in wireless sensor networks. In: ICCAS-SICE (pp. 1966–1971). IEEE (2009).
34. Choi, B.G., Cho, E.J., Kim, J.H., Hong, C.S., Kim, J.H.: A sinkhole attack detection mechanism for LQI based mesh routing in WSN. In: International Conference on Information Networking, pp. 1–5. IEEE (2009).
35. Sharmila, S., Umamaheswari, G.: Detection of sinkhole attack in wireless sensor networks using message digest algorithms. In: International Conference on Process Automation, Control and Computing (pp. 1–6). IEEE (2011).
36. Bhattasali, T., Chaki, R.: A survey of recent intrusion detection systems for wireless sensor network. In: International Conference on Network Security and Applications, pp. 268–280. Springer, Berlin, Heidelberg (2011).
37. Kasinathan, P., Pastrone, C., Spirito, M.A., Vinkovits, M.: Denial-of-service detection in 6LoWPAN based Internet of Things. In: International Conference on Wireless and Mobile Computing, Networking and Communications (WiMob), pp. 600–607. IEEE (2013).
38. Brachmann, M., Keoh, S.L., Morchon, O.G., Kumar, S.S.: End-to-end transport security in the IP-based Internet of Things. In: International Conference on Computer Communications and Networks, pp. 1–5. IEEE (2012).
39. Granjal, J., Monteiro, E., Silva, J.S.: Application-layer security for the WoT: extending CoAP to support end-to-end message security for internet-integrated sensing applications. In: International Conference on Wired/Wireless Internet Communication, pp. 140–153. Springer, Berlin, Heidelberg (2013).
40. Sethi, M., Arkko, J., Keränen, A.: End-to-end security for sleepy smart object networks. In: 37th Annual IEEE Conference on Local Computer Networks-Workshops, pp. 964–972. IEEE (2012).
41. Shelby, Z., Hartke, K., Bormann, C.: The constrained application protocol (CoAP). URL: https://tools.ietf.org/html/rfc7252 (2014).
42. Khedr, W.I.: SRFID: a hash-based security scheme for low cost RFID systems. Egypt. Inf. J. **14**(1), 89–98 (2013)
43. Mitrokotsa, A., Rieback, M.R., Tanenbaum, A.S.: Classifying RFID attacks and defenses. Inf. Syst. Frontiers **12**(5), 491–505 (2010)
44. Pongle, P., Chavan, G.: A survey: attacks on RPL and 6LoWPAN in IoT. In: International conference on pervasive computing ICPC, pp. 1–6. IEEE (2015).
45. Dvir, A., Buttyan, L.: VeRA-version number and rank authentication In RPL. In: IEEE 8th International Conference on Mobile Ad-Hoc and Sensor Systems, pp. 709–714. IEEE (2011).
46. Le, A., Loo, J., Lasebae, A., Vinel, A., Chen, Y., Chai, M.: The impact of rank attack on network topology of routing protocol for low-power and Lossy networks. IEEE Sens. J. **13**(10), 3685–3692 (2013)
47. What is a UDP flood—ddos attack glossary—incapsula. URL https://www.incapsula.com/ddos/attack-glossary/udp-flood.html. Last accessed 2020/2/27.
48. Xu, W., Trappe, W., Zhang, Y., Wood, T.: The feasibility of launching and detecting jamming attacks in wireless networks. In: Proceedings of the 6th ACM International Symposium On Mobile Ad Hoc Networking and Computing, pp. 46–57. ACM (2005).
49. Noubir, G., Lin, G.: Low-power DoS attacks in data wireless LANs and countermeasures. ACM SIGMOBILE: Mob. Comput. Commun. Rev. **7**(3), 29–30 (2003)
50. Xu, W., Wood, T., Trappe, W., Zhang, Y.: Channel surfing and spatial retreats: defenses against wireless denial of service. In: Proceedings of the 3rd ACM Workshop on Wireless security, pp. 80–89. ACM (2004).
51. Jang, J., Kwon, T., Song, J.: A time-based key management protocol for wireless sensor networks. In: International Conference on Information Security Practice and Experience, pp. 314–328. Springer, Berlin, Heidelberg (2007).
52. Weekly, K., Pister, K.: Evaluating sinkhole defense techniques in RPL networks. In: IEEE International Conference on Network Protocols (pp. 1–6). IEEE (2012).
53. Ahmed, F., Ko, Y.B.: Mitigation of black hole attacks in routing protocol for low power and Lossy networks. Secur. Commun. Netw. **9**(18), 5143–5154 (2016)

54. Wazid, M., Das, A.K., Kumari, S., Khan, M.K.: Design of sinkhole node detection mechanism for hierarchical wireless sensor networks. Secur. Commun. Netw. **9**(17), 4596–4614 (2016)
55. Krontiris, I., Dimitriou, T., Giannetsos, T., Mpasoukos, M.: Intrusion detection of sinkhole attacks in wireless sensor networks. In: International Symposium on Algorithms And Experiments For Sensor Systems, Wireless Networks And Distributed Robotics, pp. 150–161. Springer, Berlin, Heidelberg (2007)
56. Raju, I., Parwekar, P.: Detection of sinkhole attack in wireless sensor network. In: Proceedings of Second International Conference on Computer and Communication Technologies, 3, 629–636, Springer (2015).
57. Ngai, E.C., Liu, J., Lyu, M.R.: On the intruder detection for sinkhole attack in wireless sensor networks. In: IEEE International Conference on Communications, vol. 8, pp. 3383–3389. IEEE (2006).
58. Poovendran, R., Lazos, L.: A graph theoretic framework for preventing the wormhole attack in wireless ad hoc networks. Wirel. Netw. **13**(1), 27–59 (2007)
59. Salehi, S. A., Razzaque, M. A., Naraei, P., Farrokhtala, A.: Detection of sinkhole attack in wireless sensor networks. In 2013 IEEE International Conference on Space Science and Communication, pp. 361–365. IEEE (2013).
60. Xiao, Q., Boulet, C., Gibbons, T.: RFID security issues in military supply chains. In: Second International Conference on Availability, Reliability and Security, pp. 599–605. IEEE (2007).
61. Vidgren, N., Haataja, K., Patino-Andres, J.L., Ramirez-Sanchis, J.J., Toivanen, P.: Security threats in ZigBee-enabled systems: vulnerability evaluation, practical experiments, countermeasures, and lessons learned. In: Hawaii International Conference on System Sciences, pp. 5132–5138. IEEE (2013).
62. Tay, H. J., Tan, J., & Narasimhan, P.: A survey of security vulnerabilities in Bluetooth low energy beacons. Carnegie Mellon University Parallel Data Lab Technical Report CMU-PDL-16-109 (2016).
63. Wang, G., Mohanlal, M., Wilson, C., Wang, X., Metzger, M., Zheng, H., Zhao, B. Y.: Social turing tests: crowdsourcing Sybil detection. In: arXiv preprint arXiv:1205.3856. (2012).
64. Raza, S., Wallgren, L., Voigt, T.: SVELTE: real-time intrusion detection in The Internet of Things. Ad Hoc Netw. **11**(8), 2661–2674 (2013)
65. Kibirige, G.W., & Sanga, C.: A survey on detection of sinkhole attack in wireless sensor network. arXiv preprint arXiv:1505.01941 (2015).
66. Haataja, K.: Bluetooth network vulnerability to disclosure, integrity and Denial-of-service attacks. In: Proceedings of the Annual Finnish Data Processing Week at the University of Petrozavodsk, vol. 7, 63–103 (2005)
67. Ongtang, M., McLaughlin, S., Enck, W., McDaniel, P.: Semantically rich application centric security in Android. Secur. Commun. Netw. **5**(6), 658–673 (2012)
68. Chandna, S., Singh, R., Akhtar, F.: Data Scavenging threat in cloud computing. Int. J. Adv. Comput. Sci. Cloud Comput. **2**(2), 17–22 (2014)
69. Gupta, M., Gopalakrishnan, G., Sharman, R.: Countermeasures against Distributed Denial of Service. School of Management State University of New York Buffalo, NY (2017)
70. Conzon, D., Bolognesi, T., Brizzi, P., Lotito, A., Tomasi, R., Spirito, M.A.: The Virtus middleware: an XMPP based architecture for secure IoT communications. In: International Conference on Computer Communications and Networks, pp. 1–6. IEEE (2012).
71. Brachmann, M., Garcia-Mochon, O., Keoh, S.L., Kumar, S.S.: Security considerations around end-to-end security in the IP-based Internet of Things. In: Workshop on Smart Object Security, In Conjunction With IETF83, Paris, France (2012).
72. A Gómez-Goiri P Orduña J Diego D López-De-Ipiña 2014 OTSOPACK: lightweight semantic framework for interoperable ambient intelligence applications Comput. Hum. Behav. 30 460 467

73. Liu, C.H., Yang, B., Liu, T.: Efficient naming, addressing and profile services in Internet-of-Things sensory environments. Ad Hoc Netw. **18**, 85–101 (2014)
74. Ferreira, H.G.C., de Sousa, R.T., de Deus, F.E.G., Canedo, E.D.: Proposal of a secure, deployable and transparent middleware for Internet of Things. In: 9th Iberian Conference on Information Systems And Technologies (CISTI), pp. 1–4. IEEE (2014).

A Review of Anomaly Detection Techniques Using Computer Vision

Vandana Mohindru and Shafali Singla

Abstract Obtaining videos for surveillance purpose or to use them for future predictions is a challenging task as a video has a large number of image frames displayed in a sequence, and modeling every frame is not possible, so various methods are used for building an intelligent vision system, which is used for obtaining videos and also in video anomaly detection. This paper provides an overview of research directions for different types of anomalies and also tells about different techniques in machine learning for managing the problem of anomaly detection in videos and images using computer vision. Computer vision is used to make computers capable of extracting information from digital images or videos. It trains computers to interpret and understand the visual world. When machines detect errors, abnormal or unnatural behavior of datasets, it is called anomaly detection. In this paper, anomaly detection techniques and anomaly detection in datasets using computer vision are classified accordingly.

Keywords Anomaly detection · Outliers · Computer vision · Artificial intelligence · Feature extraction · Classification

1 Introduction

Information that can be processed, interpreted and studied only by humans is called human-readable information, for example, image, videos and text. Machine-readable (or structured data) information is that which cannot be easily interpreted by humans and only computer programs can process it. Now, artificial intelligence is playing a very important role in making machines capable that they can understand and interpret every type of human-readable information (unstructured data) like videos,

V. Mohindru (✉) · S. Singla
Department of Computer Science & Engineering, College of Engineering, Chandigarh Group of Colleges, Mohali, Punjab, India
e-mail: vandanamohindru@gmail.com

S. Singla
e-mail: shafalisinghal23@gmail.com

P. K. Singh et al. (eds.), *Recent Innovations in Computing*, Lecture Notes in Electrical Engineering 701, https://doi.org/10.1007/978-981-15-8297-4_53

images, written data and even human sentiments also, so in artificial intelligence [1], there are algorithms (or we can say programs) that train machines which make them as smart as human, and in this process, we need sample data to train them. We have to make sure that the training data is correct and errorless, so we use anomaly detection [2] to find the errors in data and remove them. Anomaly detection is a method to detect abnormal or unusual behavior of data in any dataset. Along with this, as we know that while working in the real-time environment, a machine can detect a lot of inputs for which it is not trained because the real world is full of unpredicted events, and it is not practically possible for training the machine for every possibility, so the unexpected inputs are considered as anomalies, and these anomalies can decrease the efficiency and accuracy of the machine; that is why it becomes very important to detect these anomalies (unusual inputs) and remove them. This helps a lot in increasing the decision-making power and efficiency of the machine.

Anomaly detection methods are used in almost every area like in monitoring network traffic for suspicious activities (intrusion detection [3]), to prevent money or property from being obtained through the unauthorized access in banking and insurance (fraud detection [4]), in monitoring a system and identifying when a fault occurred (fault detection [5]), to maintain a system inoperable condition (system health monitoring [6]), in sensor networks event detection, in detecting ecosystem disturbances [7] like earthquakes, storms, droughts, etc., using machine sensors, and also, it is used for datasets to remove anomalous data from it.

Anomaly detection is very popular in surveillance and anomalous event detection [8] by using computer vision also. Computer vision is used to gain a good understanding of images and videos. The manual labeling of abnormal events for every image and video which is captured from the surveillance cameras present in crowded and public places is inefficient and time-consuming. Therefore, systems that can detect anomalies are urgently needed for surveillance [9] purpose. In this, images and videos are firstly converted in readable form like an array of data or table of data, etc., and then, these datasets are used as an input to the anomaly detection algorithms, and then, abnormal data is detected. In intelligent video surveillance, anomaly detection is very difficult, but some advancement has been done in anomaly detection and selection. The main problem is that in most of the surveillance videos, the identity of the anomaly is indefinite. In general, we can say that the events that are expressively diverse from normal events are anomalies. But sometimes an event anomalous in one scene might not be anomalous in second, because the regular events in the second scene may contain the anomalous event of the first scene. Hence, anomalies are of inadequate dimensions, and resemblances must be excellently modeled because sometimes, the training data of the existing datasets is not capable of effective anomaly measuring and comprises only standard events while the data to be proved comprises both normal and abnormal events. In this paper, we are discussing various anomaly detection methods; we can use according to the training and testing datasets in detail.

The rest of the paper is organized as follows: Sect. 2 presents the details of anomalies and anomalies detection. Section 3 explains the comparison of different types of anomaly detection techniques. Section 4 discusses the anomaly detection using computer vision. Section 5 finally presents the conclusion and future scope of the work.

2 Anomalies and Anomalies Detection

Exceptions or outliers are the data objects that do not have normal behavior in a dataset. The process of finding and processing exceptions in the data set is called anomaly detection. The anomaly detection can be:

1. Supervised;
2. Unsupervised;
3. Semi-supervised.

In supervised anomaly detection [10] technique, datasets that have data along with labels have been used. These labels tell whether the data is "normal" or "abnormal." The abnormal data is determined as an anomaly.

In unsupervised anomaly detection [2] technique, datasets that do not have labels for data have been used for detecting anomalies under the assumption that most of the events in the dataset are normal and the event which is incompatible with the other events is considered an anomaly.

In semi-supervised anomaly detection [11] techniques, the datasets are neither fully labeled nor fully unlabeled that means some data of the data is labeled and some are unlabeled; that is why it is called semi-supervised anomaly detection method.

Anomalies can be categorized (Fig. 1) into three types as follows:

Point Anomalies—If only a single data value has irregular behavior in comparison with the rest of the data, then it is called point anomaly (e.g., purchase with large transaction value).

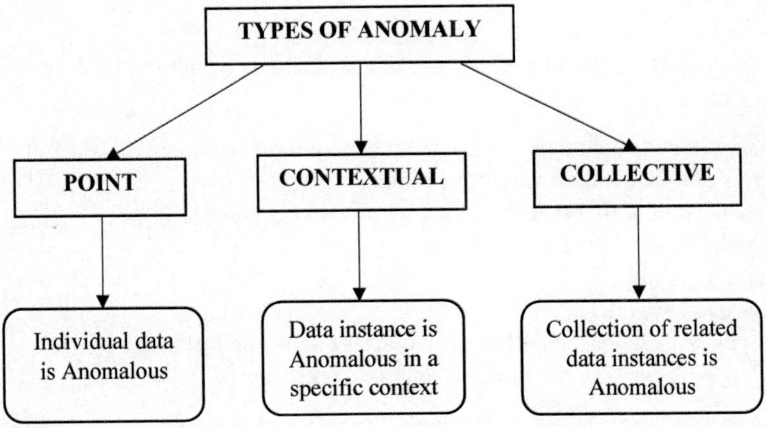

Fig. 1 Classification of anomalies

Contextual Anomalies—If a data value has irregular behavior in comparison with the rest of the data in a specific context, but not otherwise, then it is called contextual anomaly (anomaly if occur at a certain time or a certain region, e.g. large spike at the middle of the night).

Collective Anomalies—If not single data values but collections of related data values have irregular behavior in comparison with the rest of the data, then it is called a collective anomaly. They have two variations.

1. Events in unexpected order (ordered, e.g., breaking rhythm in ECG)
2. Unexpected value combinations (unordered, e.g., buying a large number of expensive items).

The anomaly detection with the help of computer vision is used for detecting anomalous behavior in images and videos. Anomalies in videos are broadly defined as events that are unusual and signify irregular behavior. Manual anomaly detection in long video sequences is a very meticulous task that often requires more manpower because there is a very less possibility of occurring unusual or abnormal events in long video sequences, e.g., surveillance footage. Real-world anomalies have complications and distinctiveness, and that is why it is difficult to list all of the possible abnormal behaviors of data. Therefore, the anomaly detection algorithm must not only dependent on the knowledge about the events. In other words, anomaly detection should be done with minimum supervision. That is why there is the concept of computer vision in video and image anomaly detection, as it uses an algorithm to have an understanding of the images. Consequently, anomaly detection has broad applications in many different areas, including surveillance, intrusion detection, health monitoring and event detection.

3 Comparison of Different Types of Anomaly Detection Techniques

Several anomaly detection techniques have been proposed in the literature. Some of the popular

- Density-based techniques like k-nearest neighbor, local outlier factor, isolation forests and many more variations of this concept.
- Tensor-based outlier detection for high-dimensional data;
- One-class support vector machines;
- Neural networks;
- Bayesian networks;
- Cluster analysis-based outlier detection like k-mean clustering;
- Fuzzy logic-based outlier detection like fuzzy c-means.

The performance of different methods depends a lot on the dataset and parameters, and while comparing with various datasets and parameters, some methods have

Table 1 Comparison of the different types of anomaly detection techniques

Technique	Methodology	Advantages
K-nearest neighbor [12]	Assign the anomaly score to data instances relative to their neighbors and perform anomaly detection based on anomaly scores	Very easy to understand for building models, that involve non- standard data types, such as text
Neural network [2]	Predict different user's behavior in systems and have the capability to solve many problems detected by rule-based approaches	A neural network learns and does not need to be programmed again and again It can be implemented in any application
Decision tree [13]	Main components: nodes (feature attribute), arcs (feature value) and leaves (category), and it classifies data points by starting from the root node and traversing until the leaf node	Simple, requires little data preparation, able to handle both numerical and categorical data, robust
Support vector machine [14]	It can be used for both regression and classification; it finds a hyperplane that separates different data classes	Find optimal separating hyperplane, can deal with high-dimensional data, and usually work very well
Self-organizing map [15]	A single-layered neural network with of neurons arranged which calculates the similarity between the input and its weights	Simple unsupervised clustering algorithm that works with the nonlinear dataset and can work on high-dimensional data and reduce its dimensionality
K-means [16]	divides the data into different clusters of similar data according to the value of k	The necessity of specifying and also, it is very noise sensitive
Fuzzy c-means [17]	Clustering method, in which one data instance can belong to two or more clusters based on a fuzzy function	Have to define cluster number, and are sensitive to the initial assignment of centroids

little advantages over another. And out of all anomaly detection techniques, some techniques are more popular which are given below in Table 1.

4 Anomaly Detection Using Computer Vision

As we know computer vision process includes methods for processing, analyzing, and understanding digital images. Also, this process involves the extraction of high dimensional data from the real world in order to produce information in the form of decisions. In this, we feed images as input and get important information about the image as output. In the case of videos, analyzing videos and getting information from it is a lengthy process as we know that video is a sequence of images to form a moving picture, and to analyze a video, we have to analyze that sequence of the image which needs a lot of processing.

The computer vision methods generally did not work on raw images, but they use images processed by the image processor that is used to extract only useful information from the image, and these enhanced images are more appropriate for computer vision tasks. The computer vision gives us information about the broad category of objects in the image, the type of a given object, key points for the object, pixels belong to the object, etc. Many popular computer vision applications [18] are:

- Object classification;
- Object identification;
- Object verification;
- Object detection;
- Object landmark detection;
- Object segmentation;
- Object recognition.

The objects that present in an image require localization and classification, and the object detection methods classify images and detect objects by drawing a line around each object which is present in the image. All these applications help a lot in getting in-depth knowledge about images and videos. Despite all these applications, computer vision plays a very important role in anomaly detection in images and videos, which means detect and identify those areas of the region whose pattern or behaviors do not conform to expect in the datasets of images and videos. This application is very useful in surveillance [19], health monitoring [20, 21], traffic analysis [22, 23], self-driven cars, smart homes, action classification, etc.

The image and video anomaly detection by computer vision follows the procedure (Fig. 2):

1. Data collection;
2. Preprocessing;
3. Feature extraction;

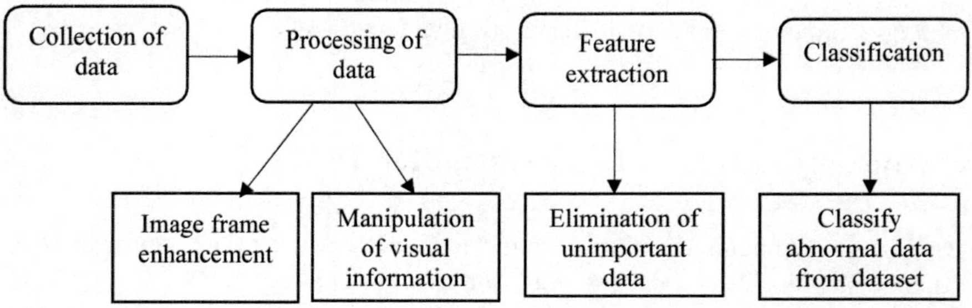

Fig. 2 Computer vision process

4. Classification.

Data collection—The data is collected in the form of images and videos. In the case of the video, the object movement is recorded by multiple cameras to avoid hiding objects which is an unwanted situation that occurs when the region of interest is blocked or masked by another object.

Pre-processing—After the collection of data, the data is processed. This video is split into video frames (in case of the video anomaly detection) and image frames are enhanced, and manipulation and interpretation of visual information are done.

Feature extraction—Feature extraction stands for selecting and abstracting the most important and useful data from the dataset. It is performed by abstracting unimportant and unnecessary data from the dataset.

Classification—Classifiers try to find dissimilarity between positive and negative examples. For the classification purpose, we use different methods like

1. Linear classifiers: logistic regression, Naive Bayes classifier;
2. Nearest neighbor;
3. Support vector machines;
4. Decision trees;
5. Random forest;
6. Neural networks, etc.

In the classification method, classifiers should be trained by sample data before classifications. The sample data should contain objects and background images for image anomaly detection.

Anomaly detection techniques have many applications such as in surveillance cameras (to detect abnormal human and object behavior), traffic analysis, health monitoring (to monitor patient health conditions), in many industries to detect defected products and so on.

In this paper, we discussed various anomaly detection techniques which can be used on datasets, but these techniques have some drawbacks also like all these techniques are very complex and hard to implement, need a lot of storage space, very time-consuming, and also, energy consumption of these techniques is very high. For example, although self-organizing map is easy to implement but it is very time-consuming, and the neural network is feasible for every application but needs high-level training for implementation on data. On the other side, k-means is less complex but needs training to specify k whereas decision trees can handle both numerical and categorical data but have a very complex structure. The SVM works very well but requires lots of memory whereas k-nearest neighbor is very easy to understand but is very expensive.

5 Conclusion

In this paper, we have revisited important anomaly detection techniques and computer vision-based survey papers. Then, we explored various types of anomalies. After that, we presented a comparative analysis of various anomaly detection techniques followed by knowledge of the contribution of computer vision in anomaly detection. Various methods we presented for the anomaly detection scheme can be regarded as a starting point for using a variation autoencoder to detect anomalies. Further research directions will contain joining the input with richer temporal and contextual information and joining the feature extraction with the final anomaly decision. Moreover, we can understand from the deep neural network frameworks for object detection and classification tasks to design more sophisticated frameworks, to represent the multiple patterns from the input video. The current techniques are just focusing on anomaly detection in every application rather than making anomaly detection more efficient. So, energy-efficient, less complex, cost-effective and less time-consuming analysis for the presence of anomalies is the current focus of the researchers.

References

1. Russell, S., Dewey, D., Tegmark, M.: Research priorities for robust and beneficial artificial intelligence. Ai Magaz. **36**(4), 105–114 (2015)
2. Kwon, D., Kim, H., Kim, J., Suh, S.C., Kim, I., Kim, K.J.: A survey of deep learning-based network anomaly detection. Cluster Comput. 1–13 (2017)
3. Haq, N.F., Onik, A.R., Hridoy, M.A.K., Rafni, M., Muhammad Shah, F., Farid, D.M.: Application of machine learning approaches in intrusion detection system: a survey. Int. J. Adv. Res. Artif. Int. (IJARAI) **4**(3) (2015)
4. Lopez-Rojas, E.A., Axelsson, S.: A review of computer simulation for fraud detection research in financial datasets. In: 2016 Future Technologies Conference (FTC), pp. 932–935. IEEE (2016)
5. Capozzoli, A., Lauro, F., Khan, I.: Fault detection analysis using data mining techniques for a cluster of smart office buildings. Expert Syst. Appl. **42**(9), 4324–4338 (2015)
6. Haghi, M., Thurow, K., Stoll, R.: Wearable devices in medical internet of things: scientific research and commercially available devices. Healthcare Inf. Res. **23**(1), 4–15 (2017)
7. Seidl, R., Spies, T.A., Peterson, D.L., Stephens, S.L., Hicke, J.A.: Searching for resilience: addressing the impacts of changing disturbance regimes on forest ecosystem services. J. Appl. Ecol. **53**(1), 120–129 (2016)
8. Nguyen, T.H., Grishman, R.: Event detection and domain adaptation with convolutional neural networks. In: Proceedings of the 53rd Annual Meeting of the Association for Computational Linguistics and the 7th International Joint Conference on Natural Language Processing (Volume 2: Short Papers), pp. 365–371 (2015)
9. Ibrahim, S.W.: A comprehensive review on intelligent surveillance systems. Commun. Sci. Technol. **1**(1) (2016)
10. Ahmed, M., Mahmood, A.N., Hu, J.: A survey of network anomaly detection techniques. J. Network Comput. Appl. **60**, 19–31 (2016)
11. Song, H., Jiang, Z., Men, A., Yang, B.: A hybrid semi-supervised anomaly detection model for high-dimensional data. Comput. Int. Neurosci. **2017** (2017)

12. Khan, J.A., Jain, N.: Improving intrusion detection system based on KNN and KNN-DS with detection of U2R, R2L attack for network probe attack detection. Int. J. Sci. Res. Sci. Eng. Technol. **2**(5), 209–212 (2016)
13. Kevric, J., Jukic, S., Subasi, A.: An effective combining classifier approach using tree algorithms for network intrusion detection. Neural Comput. Appl. **28**(1), 1051–1058 (2017)
14. Erfani, S.M., Rajasegarar, S., Karunasekera, S., Leckie, C.: High-dimensional and large-scale anomaly detection using a linear one-class SVM with deep learning. Pattern Recogn. **58**, 121–134 (2016)
15. Liu, J., Chen, S., Zhou, Z., Wu, T.: An anomaly detection algorithm of cloud platform based on self-organizing maps. Math. Prob. Eng. **2016** (2016)
16. Karami, A., Guerrero-Zapata, M.: A fuzzy anomaly detection system based on hybrid PSO-Kmeans algorithm in content-centric networks. Neurocomputing **149**, 1253–1269 (2015)
17. Nayak, J., Naik, B., Behera, H.S.: Fuzzy C-means (FCM) clustering algorithm: a decade review from 2000 to 2014. In: Computational Intelligence in Data Mining-Volume 2, pp. 133–149. Springer, New Delhi (2015)
18. Naik, N., Kominers, S.D., Raskar, R., Glaeser, E.L., Hidalgo, C.A.: Computer vision uncovers predictors of physical urban change. Proc. Natl. Acad. Sci. **114**(29), 7571–7576 (2017)
19. Liu, Y., Yu, H., Gong, C., Chen, Y.: A real time expert system for anomaly detection of aerators based on computer vision and surveillance cameras. J. Vis. Commun. Image Represent. **68**, 102767 (2020)
20. Singh, P.K., Kar, A.K., Singh, Y., Kolekar, M.H., Tanwar, S.: Proceedings of ICRIC 2019, Recent Innovations in Computing, 2020. Lecture Notes in Electrical Engineering, Springer: Cham, Switzerland, vol. 597, pp. 3–920
21. Bao, Y., Tang, Z., Li, H., Zhang, Y.: Computer vision and deep learning–based data anomaly detection method for structural health monitoring. Struct. Health Monitoring **18**(2), 401–421 (2019)
22. Singh, P.K., Bhargava, B.K., Paprzycki, M., Kaushal, N.C., Hong, W.C.: Handbook of wireless sensor networks: issues and challenges in current scenario's, advances in intelligent systems and computing, Springer: Cham, Switzerland, vol. 1132, pp. 155–437 (2020)
23. Kumaran, S.K., Dogra, D.P. Roy, P.P.: Anomaly detection in road traffic using visual surveillance: a survey. 2019 arXiv preprint arXiv:1901.08292

Security Attacks in Internet of Things: A Review

Vandana Mohindru and Anjali Garg

Abstract As time flows by, the quality and quantity of technology keep on increasing in an individual's life. Every coming day either a new technology is being discovered or an already existing popular technology is being made more and more efficient. One such technology on which lots of research is being done is IoT. IoT is one of the major technologies used these days and continuous research is being done over it to fill in the loopholes. IoT has a traditional layered architecture in which each layer has its great importance. As believed that the future will be linked with IoT all around, rigorous research on security issues of IoT is being done. As a thing with great perfection or use surely has some flaws too, in the same way, IoT too had various types of security issues associated with it. This paper deals with defining the architecture of IoT in detail. The security goals and the issues found in IoT are also explained. An analysis of some of the security attacks in IoT is done to have a closer look at the effects of attacks on the network. Various types of attacks are present in IoT among which some are easily detectable and can be prevented, but some are very difficult to detect and prevent. The paper helps in knowing about the different aspects of security issues and security attacks. Also, some future work related to research areas has been mentioned in the paper.

Keywords Security · Attacks · Internet of Things (IoT) · Sensor · Authentication · Confidentiality · Integrity · Malicious node

1 Introduction

In today's era, technology is everywhere. We often hear or read news about new technology coming into existence and gaining popularity. The concept of most of

V. Mohindru (✉) · A. Garg
Department of Computer Science & Engineering, College of Engineering, Chandigarh Group of Colleges, Mohali, Punjab, India
e-mail: vandanamohindru@gmail.com

A. Garg
e-mail: anjaligarg1999@gmail.com

P. K. Singh et al. (eds.), *Recent Innovations in Computing*, Lecture Notes in Electrical Engineering 701, https://doi.org/10.1007/978-981-15-8297-4_54

the technologies prevailed since the 19th or eighteenth century. But as the demand is less and the pace of development is slow, those technologies weren't acknowledged much at that time. But with the progress in time and needs, the same technologies started rising drastically. One such technology is IoT whose concept did exist in the late 1830s but didn't catch people's eye then [1, 2]. But in today's world IoT has become the most popular and demanding technologies by gaining popularity at a very large scale.

IoT refers to the Internet of Things. The Internet of Things can be understood as a large number of devices or things connected to the internet for sharing and collecting data all by themselves by using the internet without any intervention of humans. In technical terms, IoT is a system in which interrelated computing devices, digital and mechanical l machines, objects, people, or animals provided with unique identifiers and quality to move data over a network without expecting human-to-human or human-to-computer interaction are connected [3, 4]. According to McKinsey, the definition of IoT can be stated as: "Sensors and actuators embedded in physical objects are linked through wired and wireless networks, often using the same Internet Protocol (IP) that connects the Internet." The most popular application areas of IoT are wearable, smart cities, smart grids, industrial internet, connected cars, connected health, smart retail, smart farming, etc. IoT has its roots stretched out in almost every field of technology.

Since the 1970s, IoT was known with the name "embedded internet" or "pervasive computing". The term Internet of Things was invented in 1999 to promote RFID technology. Hence IoT is a very old term, as old as used 20 years back. Kevin Ashton was the first person to use the term "IoT" in 1999 during his work at Procter & Gamble. The internet is the hottest new trend in 1999, it was significant to call his presentation "Internet of Things". In the summer of 2010, the concept of IoT gained some popularity [1, 5, 6]. The late Mark Weiser (at the time chief scientist at the XEROX Palo Alto Research Center) once quoted: "The most profound technologies are those that disappear. They weave themselves into the fabric of everyday life until they are indistinguishable from it." [7] Going by Mark Weiser's statement, it is clearly stated that the dedicated devices will be no more seen one day whereas the information processing capabilities are going to rule all the surroundings. With the help of integrated information processing capacity, smart capabilities in industrial products can be seen. Electronic identities will be queried remotely or may have sensors embedded to detect the physical changes around them [8, 9]. All the static objects will be converted to dynamic ones. IoT will make the machines capable of interacting with each other without human intervention.

IoT [10, 11] helps in making intelligent devices around us to perform daily tasks and chores. The terms used in the relevance of IoT are smart homes, smart cities, smart transportation and infrastructure, smart agriculture, etc. IoT use can range from personal to enterprise environments. IoT also plays an important role in building up social relationships and lets users interact with their surrounding environment. In the transportation domain, smart roads, smart cars, and smart traffic lights serve very convenient and safe transportation facilities. In enterprises and industries, IoT helps in better finance, banking, marketing, etc. activities. It helps in maintaining monitoring agricultural activities through the use of various sensors.

The rest of the paper is organized as follows: Sect. 2 presents the IoT architecture. Section 3 explains the security goals. Section 4 discusses IoT security issues. Section 5 explains the various security attacks in IoT. Section 6 gives the analysis of security attacks in IoT. Section 7 finally presents the conclusion and future scope of the work.

2 IoT Architecture

IoT has a three-layered architecture in general. In the architecture, each layer has its definition based on the functions the particular layer does and the type of devices that are used in that layer. The number of layers in IoT is based on the opinions of different research groups across the world [12]. Generally, IoT works on 3 tier architecture. The three layers of IoT are named as Perception, Network, and Application layer. All the layers have different enabling technologies and their features. Each layer possesses its security issues. The three-layered architecture of IoT can be defined as given in Fig. 1 [13].

The three layers of IoT can be explained as follows:

Fig. 1 Architecture of IoT

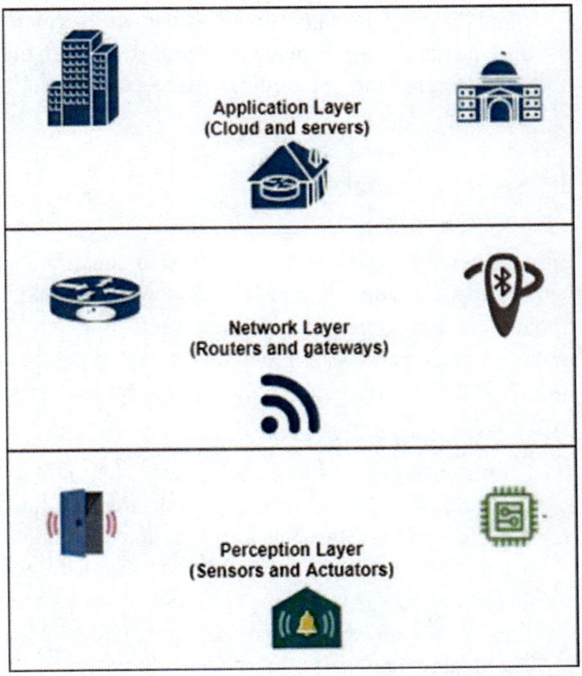

1. **Perception Layer**: It is also known as the "physical layer" or "sensors layer "as it is concerned with the physical devices [12]. This layer performs the most important work of IoT, i.e., collection of information. The information is collected with the help of different devices like smart cards, RFID tag reader and sensor networks, etc. [13]. The information from the environment is obtained using actuators and sensors [12]. Uncovering service and device identification is the main work of the perception layer. By using communication technology, devices of this layer should be able to connect either directly or indirectly via the Internet [14, 15]. Also, each device should have a unique tag to connect successfully to the network. The data is detected, collected, processed, and transmitted to the next layer, i.e., network layer.
2. **Network Layer**: Routing data and transmission of it to the different IoT hubs is the main task of the network layer over the Internet. As suggested by the name, this layer includes communication channels, network interfaces, information maintenance, network management, and intelligent processing. Here the technologies used by different network devices are WiFi, LTE, Bluetooth, 3G, ZigBee, etc. [12]. This layer uses wireless sensors to send information to any information processing system that may be present within the system or outside the system. Limited processing and computing power help the small-sized sensors to have low electricity consumption. The data received is processed, wirelessly transmitted, and presented to the end-user [15].
3. **Application Layer**: This layer is a service-oriented layer in which the type of services ensured among the connected devices is the same [16]. It provides capabilities of storage as well as data collection as it is capable of storing data into the database. Outside communication from the system of the devices is done via different kinds of applications as per the user's needs [17, 18].

3 Security Goals

Enforcement of security goals is a must to construct a secure communication framework. Ensuring proper identity authentication mechanisms, providing confidentiality to data, etc. are some of the primary security goals of IoT. The three main goals to be kept in mind are—confidentiality, integrity, and availability. In case of a breach in any of these serious issues in the system are caused. The security goals can be explained as follows [2, 19, 20]:

1. **Confidentiality**: It should be ensured that only authorized users have access to data and also data is secured. Users may be internal objects, human, external objects, and machines and services. It can be taken as an ability to give users the privacy of sensitive information via different mechanisms. The revelation to the unauthorized party can be prevented. It can be accessed by permitted users only. This makes the exchanged messages understandable for the intended entities. The data management mechanisms to be applied, the person responsible for the

management, and assurance of data is protected throughout should be well versed to the end-user of IoT.

2. **Integrity**: The accuracy of data should be ensured as data in IoT is exchanged between many devices. By maintaining end-to-end security, integrity can be imposed in an IoT network. Firewalls and protocols are used to manage data traffic. Due to the characteristic nature of low computational power, the security at endpoints cannot be guaranteed. It ensures that no tampering has been done on data and that the right sender is sending the data.
3. **Availability**: The data should be available to the users whenever needed. In disastrous conditions too, the availability of data should be ensured. The prevention of bottleneck situations that lead to the prevention of information flow is ensured. It ensures the reachability and availability of devices and services when needed for achieving the expectations of IoT. This requirement is targeted by the Denial of service attack.
4. **Authentication**: The object present in the network should be capable of clearly identifying and authenticating other objects also that are present in the network. It checks whether the entities are the same as what they claim to be actually. It is targeted by the masquerade attack or an impersonation attack. A mechanism to authenticate entities mutually in every interaction is needed.

4 IoT Security Issues

Each layer of IoT architecture has some security issues associated with it. These security issues make the layer vulnerable to different types of security attacks [20, 21, 22]. The security issues can be classified as shown in Fig. 2.

The three main categories of security issues are:

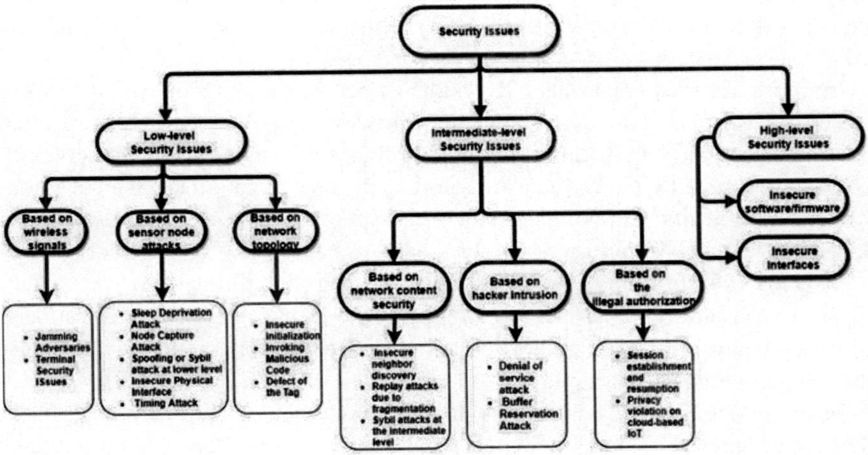

Fig. 2 Security issues in IoT

- Low-level security issues (At perception Layer)
- Intermediate -level security issues (At network Layer)
- High-level security issues (At application Layer).

4.1 Low-Level Security Issues

It is concerned with the physical and data link layers of communication as well as hardware-level security issues. This level consists of all the issues that take place at the perception layer of IoT architecture and hinder the privacy and confidentiality of nodes and data in it. At the perception layer, we divide the security issues based on—wireless signals, sensor node attacks, and the network topology of the IoT network [23]. The issues lying under the low level of IoT architecture are described below:

Based on Wireless Signals

Jamming adversaries: In this attack, the networks are dropped by emitting radio frequency signals without following any specific protocol. This interference severely impacts the operations of the network and leads to unpredictable behavior of the system.

Terminal Security Issues: Large terminals are needed to represent real-time collected data to the users. These terminals are vulnerable to security attacks which may cause leakage of confidential information, tampering, terminal virus, copying, and other issues.

Based on Sensor Node Attacks

Sleep Deprivation Attack: The sensor nodes are made awoke most of the time due to which their battery life is depleted. This is the sleep deprivation attack done on energy-constrained devices.

Node capture Attack: In this, confidentiality is attacked as the attacker takes over the node and captures all the data and information related to that node.

Spoofing or Sybil attack at a low level: In this, a malicious Sybil node gets into the network and uses fake identities which ultimately degrade the functionality of IoT.

Insecure Physical Interface: Here the proper functioning of IoT devices is hindered. The tools for testing/debugging, software access through physical interfaces, and poor physical security may lead to compromising nodes in the network.

Timing Attack: In this, the attacker tries to gain the encryption key from the node by analyzing the time required to perform encryption.

Based on Network Topology

Insecure initialization: A secure initialization and configuring of IoT done at the physical layer ensure the proper functioning of the whole system. But the vulnerability of the physical layer causes the whole system not to communicate well. Hence physical communication should also be secured.

Invoking Malicious Code: In this, malicious code or program is spread in the network which easily affects the wireless and sensor network. The determination of malicious code and recovery from it is hard to achieve.

Defect of the Tag: The attacker has easy access to the information on the tag and can illegally access the RFID system because the limited cost of the tag doesn't provide enough security to protect the network from the attacker. Any rewritable tag when once accessed by the attacker can be easily copied, decoded, or fabricated by him.

4.2 Intermediate Level Security Issues

It is concerned with the communication, routing, and session management that takes place at the transport and network layers of IoT. At the network layer, we divide the security issues based on—network content security, hacker intrusion, and illegal authorization [24]. The issues related to the low level of IoT architecture are described below:

Based on Network Content Security

Insecure neighbor discovery: To share data from the source to destination, neighbor discovery is done before data transmission. The neighbor discovery packets are used for this purpose. No proper verification of these packets leads to Denial of service attack as no proper identification and verification of the neighbors is carried out.

Replay attacks due to fragmentation: The IPv6 packet is fragmented for the devices conforming to the IEEE 802.15.4 standard as the frame size of these devices is small. The packets are then reconstructed at the 6LoWPAN layer which may cause the depletion of resources, overflow buffer, and rebooting of devices. The duplicate packets that are sent by the malicious nodes hinder the reassembly of packets and hence the processing of other legitimate packets is also hindered.

Sybil attacks at the intermediate level: Fake identities are used by Sybil nodes to carry out communication in the network which ultimately leads to spamming, disseminating malware, or the launch of phishing attacks.

Based on Hacker Intrusion

***Denial of service attack*:** In this, the required node is denied of the services from the server as the attacker keeps the server busy which ultimately leads to denial of service to the required node.

Buffer Reservation Attack: In this, the attacker exploits the buffer space for reassembly of packets of a receiver node by sending incomplete packets to it. Now as the buffer space is already filled with the incomplete packets so the actual fragmented packets are denied and hence the receiver node doesn't receive the packet.

Based on the Illegal Authorization

Session establishment and resumption: The session is hijacked with forged messages which lead to the denial-of-service. The attacking node impersonates itself as the victim node and continues the communication between the two nodes. The nodes communicating can require re-transmission of messages through the alteration of sequence numbers.

Privacy violation on cloud-based IoT: In this, different attacks violating the identity and location privacy are launched in the cloud. Also, a malicious cloud server can be used for the deployment of IoT which may access the confidential information being shared in the network.

4.3 High-Level Security Issues

These issues are concerned with the applications that are run on the IoT network [25]. Some of the security issues occurring here are:

Insecure interfaces—The interfaces used through web, mobile, and cloud for accessing the IoT services are very vulnerable to the attacks hindering the data privacy.

Insecure software/firmware—It consists of vulnerabilities caused by insecure software or firmware. The proper testing of code written in JSON, XML, SQLi, and XSS are to be done. Updating software/firmware should be done securely.

5 Security Attacks in IoT

Security attacks are the attacks that hinder the security of the different layers of IoT. These attacks can cause a security breach and data breach from the network [2, 26]. The security attacks in IoT can be classified as follows:

5.1 Based on IoT Assets

It has further classifications into it as follows:

Based on Device Property

Low-end device class attack: In it, the attack is made using IoT devices with identical abilities and configurations of the IoT devices of a native network. For example, a malicious wearable device containing malicious applications can get unauthorized access to the smart TV of smart homes and launch various threatening attacks.

High-end device class attack: In it, more powerful devices as compared to the devices of native networks are used to get access to the network and lunch attacks.

Based on Adversary Location

Internal Attack: In this attack, the attacker is either in the close proximity of the devices or in the same network. He uses either his malicious device or legitimate device is compromised.

External Attack: In this attack, the attacker is not a part of the native network. He gets unauthorized access to the devices and resources of the network or compromises with one of the trusted devices of the network.

Based on Access Level

Active Attacks: When the attacker causes some malicious activities inside the network to disrupt the usual functionality of IoT devices are known as an ac.

Passive Attacks: When the attacker gathers the information about the network by imitating an authorized device of the network and not interrupting the communication of the devices of the network is known as passive attacks.

Based on Strategy of Attack

Physical attack: Physical attacks are the attacks that cause physical damage to the device or change its configurations and properties. Example-malicious code injection.

Logical Attack: Logical attacks are the attacks that might launch some kind of attack which would make the devices dysfunctional rather than giving physical damage to the devices. Example- an attack on the communication channel.

Based on Level of Information Damage: They focus on damaging the floating data in the network to disrupt the network or compromising the information. Some of the in-transit attacks are:

Interruption: Other than the interruptions like service shutdowns. Attacks like DoS are used for causing resource exhaustion and therefore resulting in the unavailability of some services. Here the implementation of disaster recovery mechanisms is important.

Man-in-the-Middle attack: Here, the communication is intercepted which is taking place between two nodes. Two parties are made to think that they are communicating with each other securely but actually, the attacker is communicating with them.

Eavesdropping: Here the attacker can listen to the private communication's information. It poses a threat to message confidentiality and RFID devices are most susceptible to it.

Alteration: In this, unauthorized access is taken on the information and then tampers the information to create confusion and misleads into the network. It is an attack on the message integrity of the network.

Fabrication: Add-ons to the data or activity is done by the attacker which doesn't have any existence. The main purpose of this attack is to create confusion among the parties and threatening the message genuineness by using external or internal resources.

Message Replay: In this, the parties are made to mislead or confused that are not time-aware and included in the communication. In this, the message freshness is harmed.

Based on the Host

User compromise: With the use of an unsporting maneuver, the users are entrapped and made to expose their security credentials. A secured credentials transfer must be needed in this.

Software compromise: The vulnerabilities of system software running on the node are taken advantage of. One such method is of pushing a device into an exhaustion state by using resource buffer overflow.

Hardware compromise: The hardware is tempered in it. Here the attacker steals the embedded credentials such as data, keys, or program code of the device.

5.2 Based on the Area of Attack

Physical Attacks: These attacks are done on the physical layer of the IoT architecture which consists of the hardware devices. For this, the attacker needs to be present in the network or stay close to it. They harm the functionality and the lifetime of the attacked hardware device. Some of the examples are:

Node Tempering: In this, the sensor node can be either completely replaced physically or a part of its hardware can be replaced. Also, a node can be electronically interrogated to alter sensitive information by gaining access.

Malicious Node Injection: In this, a new malicious node can be deployed in the system between any two nodes and then control all the data flow and the operation of the nodes. Another name of this is the Man-in-the-Middle attack.

Social Engineering: In it, the manipulation of users is done. The manipulation is done by the attacker either to extract some private information or to perform some actions which serve his own goals. Physical interaction of the attacker with the users is a must for the achievement of his goals. Due to this reason, it is kept under physical attacks.

Sleep Deprivation Attack: To extend the battery life of the nodes, they are powered with replaceable batteries and programming to follow the sleep routines is done. In this attack, the attacker does not allow the nodes to follow the sleep routines which ultimately makes the node stay awake and hence shutting down the nodes early.

RF Interference on RFIDs: An RFID tag is easily prone to Denial-of-service attack by including noise signals in the radio frequency signals used by RFIDs for communication. And hence the communication is disturbed due to the noise signals presence.

Network Attacks: Network attacks are those in which the attack is done on the network layer of the IoT architecture. The attacker doesn't need to be part of the network or be close to it. Some of the examples of these attacks are:

Traffic Analysis Attacks: The attacker can pull out the confidential information or any data which is flowing from the RFID technologies because of their wireless nature. Before the employment of attack, firstly the attacker gathers some information about the network. Applications such as port scanning application, packet sniffer application, etc. help in making it happen.

Sinkhole Attack: In this, all the traffic from WSN nodes is lured which ultimately creates a sinkhole. In this attack, the confidentiality of data is breached and also the services to the network are denied as all the packets are dropped instead of passing the packets to the destination node.

Denial of Service Attack: In this, a network can be bombarded with more and more traffic into the IoT network which can lead to a successful denial of service attack. The destination node is made busy by broadcasting lots of data via the bombarded traffic which results in a denial of the source node for sending data.

RFID cloning: In this, data fetching from the victim node and copying it to another RFID tag can lead to RFID cloning. But in this, the id of the RFID tag remains identical. This helps to distinguish between the actual and copied tag.

Software Attacks: One of the main sources of vulnerabilities in security in the computerized system is software attacks. They have their hand in exploiting the system by using spyware, Trojan horse programs, malicious scripts, and worms. This causes tampering of data, harm to the devices, information stealing, and service denies. Some of the examples of software attacks are:

Phishing Attacks: In this, first, the authentication credentials are taken. Then confidential data is gained. The confidential data is obtained with the help of phishing websites or infected emails.

Malicious Scripts: The user controlling the gateway is fooled into running active-x scripts which are executable. This leads to the complete shutdown of the system or data theft.

Virus, Worms, Spyware, Trojan horse, and Aware: the malicious software infects the system in a variety of outcomes such as tampering data, denial of service, or stealing information.

Encryption Attacks: These are related to encryption scheme breakage that has been used in the IoT system. Some of the encryption attacks are:

Side-Channel Attacks: The attacker can retrieve the key used for encryption and decryption of data by using particular techniques like power analysis, fault analysis, timing analysis, or electromagnetic analysis.

Cryptoanalysis Attacks: In this, the plaintext or ciphertext is captured to find the encryption key and break the system's encryption scheme. Examples of such types of cryptoanalysis attacks are Chosen-plaintext attack, Known-plaintext attack, Ciphertext only attack, and Chosen-Ciphertext attack.

Man in the Middle Attack: An adversary positions itself between the two nodes which are communicating with each other and during the challenge-response scenario exchange the keys for establishing a secure communication channel. The adversary performs the key exchange with the two nodes separately and hence decrypt/encrypt the data coming from the nodes. While the two nodes think that they are communicating with each other.

6 Analysis of Security Attacks in IoT

In IoT, different types of attacks are found. For some attacks, proper solutions have been discovered while some still research is going on. It has been found that some of the attacks can be easily prevented and detected. But some attacks do exist which are very difficult to detect because of various constraints or the nature of the network. In this paper, some of the attacks have been discussed in the following table in terms of whether it is an active or passive attack, which layer of IoT architecture is harmed, what can be the location of the attacker while attacking the network, how the attack is brought into action and whether the attack can be prevented or not. Analysis of some of the security attacks has been done in Table 1.

From the above table, we can easily analyze what all effects various attacks have on the network. The detection of sinkhole attacks can be done with the help of rule-based detection, an anomaly-based detection or by key management detection. The most common method of detecting DoS attacks is by using ACL counters. We use a network guard device for detecting and blocking the incoming and/or outgoing packets to know if the traffic of the network is being analyzed or not. Some attacks like DoS cannot be prevented easily but the further spread can be stopped [27]. In the coming future, the DDoS attacks will be very common and easy to implement [28]. All the traditional attack detecting techniques are rule-based. In contrast to the traditional detection techniques, a machine learning and deep learning-based detection technique has been found which works more efficiently. Machine learning can help in detecting any intruders into the normal activity of the system. IDS are anomaly-based which makes use of machine learning techniques for detecting attacks in real-time in IoT networks. As per [29, 30], Intrusion Detection Systems (IDSs) acts as a great countermeasure for many of the attacks taking place in IoT. But still, more efficient IDSs are being searched which should be lightweight and also undoubtedly provide very high protection.

7 Conclusion

However, IoT is one of the most emerging technologies of today's time. Hence, the security of IoT devices is the main concern as in coming future it is believed that the world will be driven by smart devices. Going by the architecture, IoT has a traditional network architecture that has its flaws. Various attacks are possible that pose a threat to the security of the devices present in IoT architecture. More and more research needs to be done to find some of the more capable techniques which can help in overcoming the barriers which make detection of various attacks very difficult. Various attacks hinder various security goals of IoT. The security goals of IoT need to be preserved so that the devices are kept secure. This paper discusses some of the common security issues and attacks. Moving from a rule-based technique to an anomaly-based technique for attack detection is a considerable step towards the easy detection of attacks and proposing a solution for them. The researchers can work on designing a new architecture that can overcome the maximum vulnerabilities and loopholes of the present architecture and hence make the network more and more secure. Also, more efficient attack detection techniques should be researched so that the attacks are easily detected and dealt with.

Table 1 Analysis of attacks in IoT

Name of Attack	Type of attack	Layer attacked	Location of the attacker	Based on	Security goal hindered	Can it be prevented?
Sinkhole Attack	Active attack	Network layer	External	Routing	Availability, Confidentiality	Yes if the node can be authenticated
Denial of Service Attack	Active attack	Network layer	Internal	Flooding the target with traffic	Availability	Not likely but prevention can be of being a medium of attack for another system
Traffic Analysis attack	Passive attack	Network layer	External	Keeping a check on the node's interaction	Confidentiality	Yes by using Traffic-flow security
Sleep Deprivation Attack	Active attack	Physical Layer	Internal	Malicious node make requests to victim often to keep it awake	The lifetime of the victim node reduced drastically	Yes by using clustering
Malicious Node Injection	Active attack	Physical layer	External and internal both	Inserting malicious node	Availability	Yes if replication attack can be prevented
Phishing Attack	Passive attack	Application layer	Internal	Malicious link or attachment	Confidentiality	Yes by using anti-spyware and firewall settings
Node Tempering	Active attack	Physical layer	Internal	Physical replacement of the complete node or part of it	Integrity, Confidentiality	Yes by using the strong and lengthy encryption key
Man in the Middle attack	Active attack	Network layer	Internal	Suspicious Node	Confidentiality, Integrity	Yes by using firewalls and security protocols
Worm attack	Passive attack	Application, Physical	Internal	Malicious Code	Availability, Integrity, Authenticity	Yes if we do not allow suspicious files and sites to access our system

References

1. Hassan, W.H.: Current research on Internet of Things (IoT) security: a survey. Comput. Netw. **148**, 283–294 (2019)
2. Mohindru, V., Singh, Y., Bhatt, R.: Prevention of Node Clone Attack: secure and energy efficient algorithms in WSN (2020)
3. Singh, P.K., Bhargava, B.K., Paprzycki, M., Kaushal, N.C., Hong, W.C.: Handbook of wireless sensor networks: issues and challenges in current scenario's. Advances in Intelligent Systems and Computing, Vol. 1132, pp. 155–437. Springer Cham, Switzerland (2020)
4. Tanwar, S., Thakkar, K., Thakor, R., Singh, P.K.: M-Tesla-based security assessment in wireless sensor network. Proc. Comput. Sci. **132**, 1154–1162 (2018)
5. Deogirikar, J., Vidhate, A.: Security attacks in IoT: a survey. In 2017 International Conference on I-SMAC (IoT in Social, Mobile, Analytics, and Cloud) (I-SMAC), IEEE, pp. 32–37 (2017)
6. Singh, P.K., Kar, A.K., Singh, Y., Kolekar, M.H., Tanwar, S.: Proceedings of ICRIC 2019, Recent Innovations in Computing. Lecture Notes in Electrical Engineering, Vol. 597, pp. 3–920. Springer, Cham, Switzerland (2020)
7. Sengupta, J., Ruj, S., Bit, S.D.: A Comprehensive survey on attacks, security issues and blockchain solutions for IoT and IIoT. J. Netw. Comput. Appl. **149**, 102481 (2020)
8. Tahaei, H., Afifi, F., Asemi, A., Zaki, F., Anuar, N.B.: The rise of traffic classification in IoT networks: a survey. J. Netw. Comput. Appl. **154**, 102538 (2020)
9. Singh, P.K., Bhargava, B.K., Paprzycki, M., Kaushal, N.C., Hong, W.C.: Handbook of wireless sensor networks: issues and challenges in current scenario's. In: Advances in Intelligent Systems and Computing, Vol. 1132, pp. 155–437. Springer, Cham, Switzerland (2020)
10. Mohindru, V., Bhatt, R., Singh, Y.: Reauthentication scheme for mobile wireless sensor networks. Sustain. Comput. Informatics Syst. **23**, 158–166 (2019)
11. Matta, P., Pant, B.: Internet of things: genesis, challenges and applications. J. Eng. Sci. Technol. **14**(3), 1717–1750 (2019)
12. Mahmoud, R., Yousuf, T., Aloul, F., Zualkernan, I.: Internet of things (IoT) security: current status, challenges, and prospective measures. In: 2015 10th International Conference for Internet Technology and Secured Transactions (ICITST), pp. 336–341. IEEE (2015)
13. Singh, P.K., Pawłowski, W., Tanwar, S., Kumar, N., Rodrigues, Joel, J.P.C.: Proceedings of First International Conference on Computing, Communications, and Cyber-Security (IC4S 2019). Lecture Notes in Networks and Systems, Vol. 121, pp. 3–917. Springer, Cham, Switzerland (2020)
14. Mahmoud, R., Yousuf, T., Aloul, F., Zualkernan, I.: Internet of things (IoT) security: current status, challenges and prospective measures. In: 2015 10th International Conference for Internet Technology and Secured Transactions (ICITST), pp. 336–341. IEEE (2015)
15. Mohindru, V., Singh, Y., Bhatt, R.: Hybrid cryptography algorithm for securing wireless sensor networks from Node Clone Attack. Recent Adv. Electrical Electronic Eng. (Formerly Recent Patents on Electrical & Electronic Engineering) **13**(2), 251–259 (2020a)
16. Bhabad, M.A., Bagade, S.T.: Internet of things: architecture, security issues, and countermeasures. Int. J. Comput. Appl. **125**(14)
17. Andrea, I., Chrysostomou, C., Hadjichristofi, G.: Internet of Things: security vulnerabilities and challenges. In: 2015 IEEE Symposium on Computers and Communication (ISCC), pp. 180–187. IEEE (2015)
18. Mohindru, V., Singh, Y., Bhatt, R.: Securing wireless sensor networks from node clone attack: a lightweight message authentication algorithm. Int. J. Inf. Comput. Secur. **12**(2–3), 217–233 (2020b)
19. Khan, R., Khan, S.U., Zaheer, R., Khan, S.: Future internet: the Internet of Things architecture, possible applications, and key challenges. In: FIT, pp. 257–260 (2012)
20. Gubbi, J., Buyya, R., Marusic, S., Palaniswami, M.: Internet of Things (IoT): a vision, architectural elements, and future directions. Future Gener. Comput. Syst. **29**(7), 1645–1660 (2013)

21. Mahmud, H.M., Fotouhi, M., Hasan, R.: Towards an analysis of security issues, challenges, and open problems in the internet of things. In: 2015 IEEE World Congress on Services, pp. 21–28. IEEE (2015)
22. Khan, M.A., Salah, K.: IoT security: review, blockchain solutions, and open challenges. Future Gener. Comput. Syst. **82**, 395–411 (2018)
23. Ferrag, M.A., Shu, L., Yang, X., Derhab, A., Maglaras, L.: Security and privacy for green IoT-based agriculture: review, blockchain solutions, and challenges. IEEE Access **8**, 32031–32053 (2020)
24. Alaba, F.A., Othman, M., Hashem, I.A.T., Alotaibi, F.: Internet of Things security: a survey. J. Netw. Comput. Appl. **88**, 10–28 (2017)
25. Dabbagh, M., Rayes, A.: Internet of Things security and privacy. In: Internet of Things from Hype to Reality, pp. 211–238. Springer Cham (2019)
26. Jain, A., Sharma, B., Gupta, P.: Internet of Things: architecture, security goals, and challenges—a survey. Int. J. Innov. Res. Sci. Eng. **2**(4), 154–163 (2016)
27. Mohindru, V., Singh, Y.: Node authentication algorithm for securing static wireless sensor networks from node clone attack. Int. J. Inf. Comput. Secur. **10**(2–3), 129–148 (2018)
28. Nortan Homepage. https://us.norton.com/internetsecurity-iot-5-predictions-for-the-future-of-iot.html. Last accessed 2020/02/20
29. Khan, Z.A., Herrmann, P.: Recent advancements in intrusion detection systems for the internet of things. Secur. Commun. Netw. Volume 2019, Article ID 4301409, 19 pages (2019). https://doi.org/10.1155/2019/4301409
30. Thamilarasu, G., Chawla, S.: Towards deep-learning-driven intrusion detection for the internet of things. Sensors **19**(9), 1977 (2019)

Texture Feature Technique for Security of Indian Currency

Snehlata and Vipin Saxena

Abstract In recent years, security of currency has gain importance in the field of research. With the advent of digital technology, color printer, and color scanner are the cheapest way for counterfeiter to produce fake currency. Feature extraction is the most important technique in paper currency recognition. According to reviewer, texture feature plays an important role for paper currency detection. Texture feature is generally a statistical-based approach and in the present work, a new model is proposed for paper currency detection. The presented model is computing the texture properties like Gray Level Co-occurrence Matrix (GLCM) of Rs. 500 for real and fake currency. The Principle Component Analysis (PCA) is used for reduction of higher dimension of images. The proposed work provides better results with the collaboration of PCA and GLCM. The texture properties have been used and GLCM measured the variation in intensity at pixel of interest of the currency. The computed results have been presented in the form of table and graphs.

Keywords Currency · Security · Texture feature · Gray-level · GLCM

1 Introduction

In the digital era of technology, it is very difficult to identify fake paper currency. Illegal replication of original currency is known as counterfeiting. The Government of India (GOI) has used the term fake currency in place of counterfeit. This is a very challenging issue to identify the fake currency. Reserve Bank of India (RBI) is the only body that takes sole responsibility to print paper currency in India. Extracting security features from Indian currency is most important for accuracy and robustness of the automated system. Features extraction in the form of colour, shape, texture or context are the most important techniques of image processing and pattern recognition. Although, many techniques are available in the literature to find out the pattern of paper currency, but feature extraction is the most common technique

Snehlata (✉) · V. Saxena
Babasaheb Bhimrao Ambedkar University, Lucknow, India
e-mail: loginsneha91@gmail.com

P. K. Singh et al. (eds.), *Recent Innovations in Computing*, Lecture Notes in Electrical Engineering 701, https://doi.org/10.1007/978-981-15-8297-4_55

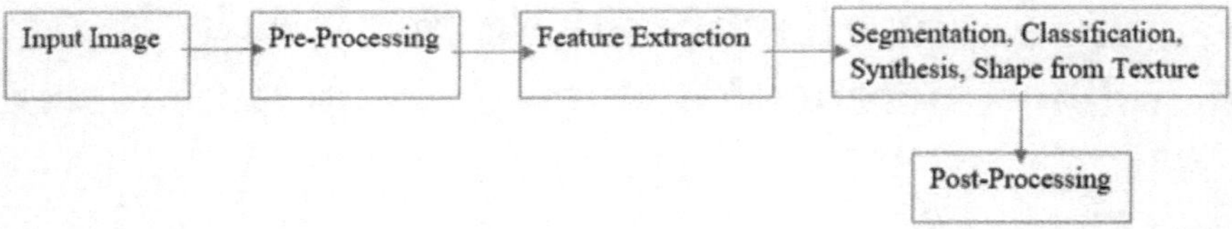

Fig. 1 Representation of texture feature analysis

in the form of colour features, transform features, shape features, texture features, edge and boundary features. From the literature of the currency identification, it is observed that there is no work on the newly developed paper currency by RBI but some little information in the literature are available for the Mexican banknotes, Euro, US Dollar related to the countries namely Australia, Cyprus, Sri Lankan and Bangladesh. Further, it is found from the literature that the banknotes of each country are modelled by extracting the very important feature that depends on the nationality of currency. Therefore, the contribution of systematic approach is to recognize the currency by texture feature analysis. In the following, Fig. 1, the traditional way to image analysis in image processing is represented.

In the present paper, Sect. 2 describes a state of art of work on the texture feature analysis, Sect. 3 describes the feature extraction methods, Sect. 4 is related to the result of implementation and finally Sect. 5 describes the discussion and concluding remarks of the work.

2 State of Art of Texture Feature Analysis of Paper Currency

In the texture feature analysis, authors have to first identify the texture features on the basis of Pattern and analysis the same by various approaches. In this reference, P. Mohanaiah et al. [15] have described the GLCM approach for image texture feature extraction. In the work, it is presented an application of GLCM to extract the second-order statistical texture feature analysis for estimation of the image using Xilinx FPGA. The computed results present the texture feature analyses with high accuracy in less computation time and it is efficiently used for real-time pattern recognition application. Alnowaini et al. [2] proposed a new automated system that identifies the Yemeni paper currency real or fake based on image processing technology with machine learning algorithms. This proposed framework has different five steps such as currency acquisition, preprocessing, feature extraction, classification and verification. The system has found to detect the currency real or fake in respect of accuracy and fast processing time. Chavan et al. [5] designed an automatic coin dispensing machine to resolve this issue. There have three modules such as detection of a genuine currency note, determining the value of the note and dispensing of equivalent coins. Here, the genuine currency analyses through security thread with the help of histogram of currency captured by Ultraviolet (UV) light. The robust system has 92.12% accuracy for detection of denomination of the currency, 100% accuracy for dispensing of equivalent coins, and 90.07% accuracy for the genuineness of currency. Kekre et al. [13] have explained a method for image retrieval using texture feature extraction from GLCM, Linde-Buzo-Gray (LBG) and Kekre's Proportionate Error (KPE) algorithms. Further, Chao Tong et al. (2017) proposed a new algorithm to differentiate the new and old currency based on five level statistical parameters of texture of currency in grayscale image, gray level-gradient co-occurrence matrices and multi-DAG-SVM classifier. The experimental outcomes demonstrate the algorithm with high speed and greater efficiency at the classification level. Hamza and Al-Assadi [9] also described a new method of image improvement on optimal texture feature extracted from GLCM using genetic algorithm which finds ideal texture features separated from GLCM dependent on the fitness function. Yan et al. [22] developed a prototype that identifies the currency. There are two highlight vectors which comprised of colour features and texture features. Lot of research work is available on currency being grouped through a Feedforward Neural Network (FNN) and also measured difference among sample and counterfeit banknote. Garcia-Lamont et al. [8] addressed the use of the artificial vision to recognition of Mexican currencies which may be grouped by colour and texture features with respect to the RGB space and the local binary patterns. This mechanism displays the characterization outcome for the Mexican currencies and the planned strategy may be enforced to identify the currencies of different countries by the use of colours to recognize the groups. Snehlata and Saxena [19] have proposed a case study of Indian currency identification system which is based on currently launched Rs. 500 and Rs. 2000 through object-oriented approach. Arya and Sasikumar [4] have proposed Fake currency notes are increasing day by day, in order to overcome this, design a very helpful and efficient system to detect the fake currency. For detecting the fake currency note is done by counting the number of interruptions in the thread line. For predicting the note is real or fake on the basis of a number of interruptions. If the number of interruption is zero, if it is real note otherwise it is fake. And also we calculate the entropy of the currency notes for the efficient detection of a fake currency note. MATLAB software is used to detect fake currency note. Ahmadi et al. [1] explained a method that is based on PCA which increases the reliability of recognition system and proposed system recognizes the six different types of bills of US dollars. First, image is captured by Line sensor, then, image is extracted with main feature and decreased the dimension of the image using PCA technique. It is used Linear Vector Quantization (LVQ) network and outcomes show that the dependability has been incremented up to 95% when the PCA component and LVQ vectors are taken correctly to recognize the currencies for Australian blind people. Hinwood et al. [11] have proposed a Money Talker that electronically recognize the different colours and texture of Australian currencies. First, currency is captured by the system then light is passed through currency and respective sensors and detect the different ranges of values depending on the colour of the currency. Takeda et al. [20] described a method to recognize Thai banknotes with slab values that are stored in digital form as input of Neural Network (NN) for recognition process. Aoba et al. [3] presented Euro banknotes recognition system which uses the two types of NN's. Frosini et al. [7] explained the neural-based recognition technique which is used in banknotes machines. Snehlata and Saxena [18] described the efficient technique for currency detection by the use of PCA for providing the better way to currency identification. Generally, PCA is used for dimension reduction. The eigen-values and eigen-vectors provide a better discrimination of the fake and real currency that provides a better results in the form of graph and table. Priyal has proposed [6] proposed an efficient Image processing technique that has to extract various features from the U.S. currency. The features of currency are dimensions, colour written letters, security thread, texture, federal reserve indicator, serial number, and various other details that help to increase the accuracy of fake currency detection. The proposed framework has used machine learning and deep neural network to extract the more feature more and improve accuracy.

3 An Overview of Feature Extraction

This section briefly elaborates about digital Image processing, pattern recognition and feature extraction. Feature extraction is the most essential chunk of pattern recognition that deals with the dimensionality reduction. To deal with enormous information that contains redundant data, the feature extraction method is used to effectively reduce the data size without losing important and relevant information. In general, feature contains information about colour, shape, surface or context. The kind of literature has suggested lot of techniques for feature extraction from images, some are listed in Fig. 2.

3.1 GLCM

GLCM is a technique for extraction of texture features and according to, Guiying Li (2012), GLCM is defined as the texture of continual arrangement of information having the structure of regular intervals. Texture study means size, shape, density, arrangement, surface characteristics and appearance of an object and also of its elementary part. As texture contains vital information regarding any object, so that texture feature extraction leads to the key function in the Image processing field.

Singh et al. [16] proposed a novel method of object recognition using point feature technique for colour images. The SURF algorithm used to find the object detection for unique feature matches of colour images. This approach of detection has robustly find object between colored cluttered images. Singh et al. [17] proposed a method suggests to uses K-means and EK means clustering methods on various tumor images. The characteristics are observed based on geometrical features. Images were trained by advanced machine learning algorithms like Artificial Neural Network (ANN) and Support Vector Machine (SVM) where it divides the tumor into the malignant

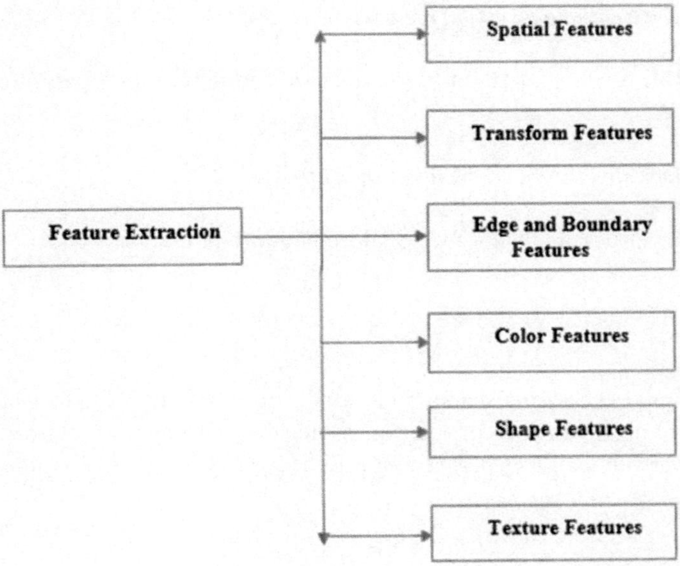

Fig. 2 Various types of feature extraction

type and due to segmentation it gives tumorous part. That system used characteristic extraction by further segmenting GLCM characteristics into Haralick features.

Neville et al. (2003) demonstrated various texture feature extraction methods and characterized as model-based and transform information, statistical, structural methods. Among all listed methods, statistical based feature extraction method is prominent for texture feature extraction that incorporates first-order statistics, second-order statistics and higher-order statistics. In the present scenario, the co-occurrence matrix and texture feature are most popularly used in the area of image identification which was suggested by Haralick in early 1973 [12] that is based on second-order statistical features. Suggestions regarding texture feature contain the following two steps:

- Firstly, the computation of the co-occurrence matrix;
- Based on computation of co-occurrence matrix, texture features are calculated.

This idea is quite useful in Image analysis for currency identification. After optimizing the fourteen texture features, the idea of Haralick (1973), Andrea Baraldi and Flavio Parmiggiani (1995) proposed a new approach of the five statistical parameters energy, entropy, contrast, homogeneity and correlation. The explanation is as follows:

3.1.1 *M, N:Coefficient of Co-Occurrence Matrix;*

3.1.2 *X (M, N):Element in the Co-Occurrence Matrix at the Coordinates M, N;*

3.1.3 *K:Dimension of Co-Occurrence Matrix.*

The five statistical parameters are explained below in brief:

(a) Energy

The performance can be measures in terms of energy (E) which can be measured as the extent of pixel pair repetitions. It measures the uniformity of an image. When pixels are very similar, the energy value will be large. The energy is defined as:

$$E = \sqrt{\sum_{m=0}^{K-1}\sum_{n=0}^{K-1} X^2(m,n)} \tag{1}$$

(b) Entropy

The Entropy (Ent) is the proportion of arbitrariness that is utilized to describe the surface of the information picture. It's worth will be maximum when all the components of the co-occurrence matrix are the equivalent. It is defined below:

$$\text{Ent} = \sum_{m=0}^{K-1}\sum_{n=0}^{K-1} X(m,n)(-\text{In}(X(m,n))) \tag{2}$$

(c) Contrast

The Contrast (Con) is characterized in Eq. (3), which is a proportion of power of a pixel and its neighbour over the picture. In the visual impression of a genuine world, differentiate is deflect mined by the distinction in colour and brightness of the object and different objects with-in a similar field of view.

$$\text{Con} = \sum_{m=0}^{K-1}\sum_{n=0}^{K-1} (m-n)^2 X(m,n) \tag{3}$$

(d) Homogeneity

Homogeneity returns a value that measures the closeness of the distribution of elements in GLCM to the GLCM diagonal. The value of energy is 1 for a diagonal GLCM and having the range as [0 1]

$$\text{Hom} = \sum_{m,n} \frac{X(m,n)}{1+|m-n|} \tag{4}$$

(e) Correlation

Correlation returns a measure of how correlated a pixel into its neighbour over the whole image. It's range is [−1 1] and defined below:

$$\text{Corr} = \sum_{m,n} \frac{(m-\mu m)(n-\mu n)X(m,n)}{\sigma_m \sigma_n} \tag{5}$$

3.2 Principle Component Analysis

According to the data hypothesis, extract the important feature of currency image for identification and compare one currency to another currency feature to a database. A basic way to deal with extricating the data contained in an image of a currency is to by one way or another catch the variety in a collection of currency image, autonomous of results of features and utilize this data to computed and analyse the single currency image [15]. Snehlata and Saxena [18] briefly discussed that Principle Component Analysis (PCA) is a method of recognition based on the feature of texture that features are expressed in the term of resemblance and dissimilarity. Once the pattern is matched in data then it can be compressed.

The transformation function as given below:

$$M = YW \tag{6}$$

where.
Y denotes $a * b$ matrix of a observations on b variables;
M denotes $a * b$ of a values for each of b components;
W denotes $b * b$ of coefficient defining the linear transformation.
There are included two phases: Training and Testing.

The Training phase contains the following stages:

(1) Generate the currency database;

(2) Features are extracted from the currency;
(3) Calculate Mean;
(4) Subtract mean from each currency image, known as mean aligned currency;
(5) Calculate covariance of the mean aligned currency;
(6) Calculate Eigenvalues and Eigenvectors;
(7) Identify Eigenvectors and find principle Eigenvectors;
(8) Find the direction of Data.

The Testing phase contains the following stages:

(1) For testing, a test currency image, make it a column vector;
(2) Subtract mean of currency to this test currency;
(3) Calculate the distance.

4 Results and Discussion

Let us first describe the characteristics of Rs. 500 Indian paper currency which is depicted in Fig. 3.

The presented currency has the following silent features listed below:

1. See through register
2. Latent image
3. Denomination numerical in Devanagari
4. Mahatma Gandhi's portrait in centre facing to right
5. Windowed security thread
6. Guarantee clause
7. Portrait and electrotype watermark
8. Number panel with numerals growing from small to big on the top left side and bottom right side
9. Denomination in numerals with rupee symbol in colour changing ink (green to blue) on the bottom right
10. Asoka pillar emblem on the right
11. Circle with Rs. 500 in raised print on the right
12. Five bleed lines on left and right in raised print
13. Year of printing of the note on left
14. Swachh Bharat logo with the slogan
15. Language panel towards the centre
16. Red Fort with Indian flag
17. Denomination numeral in Devanagari on right.

The image database used in this study served as the main feature extraction of the texture using GLCM. The GLCM method incorporates the averaging of coefficient matrix having the direction of 0°, 45°, 90° and 135° [10]. The above-mentioned six features of GLCM used in this research like are Energy, Entropy, Contrast, Homogeneity and Correlation. The values of these features are depicted as the training and

Fig. 3 Representation of Rs. 500 Indian paper currency

Table 1 Characteristics of GLCM database training

Energy	Entropy	Contrast	Homogeneity	Correlation
0.20645	1.35718	0.83730	0.811055	0.749905

test database from our keenly observed image data set. Table 1 shows the value of the characteristics of GLCM used as input to database training.

4.1 Classification and Identification System Testing

The determination of value k as in our cases 1, 3 and 5 is carried out with the help of PCA classification process for fixed amount of training data. The right value of k is totally depending upon the level of closeness and accuracy of the identification

Table 2 Accuracy of the identification result

Banknote	k-values	Accuracy (%)
Rs. 500	k = 1	100
	k = 3	90
	k = 5	80

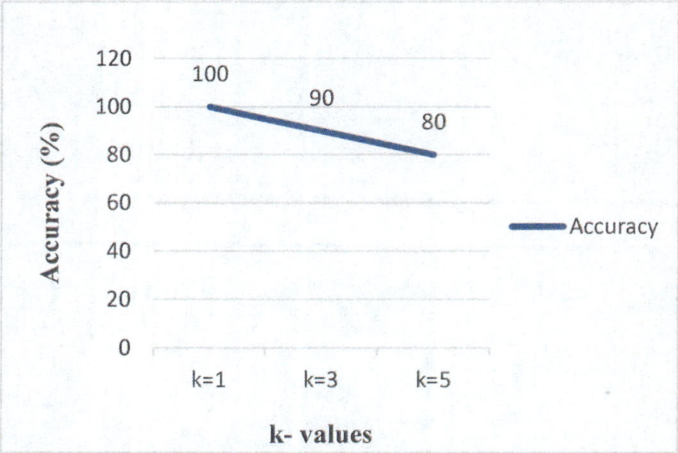

Fig.4 The Graph between *k*-values and Accuracy

process. Hence as a result of the identification process to determine the value of k testing process are observed with the certain accuracy as depicted in Table 2.

The proposed technique is implemented using MATLAB to get real currency from the database. Based on *k*-value obtained different accuracy to currency detection, graph plot between *k*-value and accuracy in Fig. 4.

5 Conclusion

The texture pattern of the genuine Rs. 500 currency is exceptionally powerful in the process information securing can influence the estimation of surface element extraction utilizing the GLCM technique. The estimation of k in the arrangement input utilizing the PCA technique can decide the exactness of the characterization procedure and ID of the realness of rupee currency. The after-effects of the recognizable proof of the authenticity of the rupee tried right now an accuracy of 100% when the value of $k = 1$, accuracy of 90% when the estimation of $k = 3$ and accuracy of 80% when the estimation of $k = 5$ with the arrangement of the measure of 20 training data and 10 testing data. The accuracy of the value for $k = 1$ is the most elevated of accuracy value. It is on the grounds that when an estimation of $k = 1$ is the default value research can be inferred that the more noteworthy of the separation technique that has 1 nearest neighbour of that information worth or it very well may be considered that information possesses. Right now be reasoned that the more prominent the estimation of k given to the framework during the arrangement and identification procedure will cause the lower level of accuracy produced from the system.

References

1. Ahmadi, A., Omatu, S., Kosaka, T., Fujinaka, T.: A reliable method for classification of bank notes using artificial neural networks. Artif. Life Robot. **8**(2), 133–139 (2004)
2. Alnowaini, G., Alabsi, A., Ali, H.: Yemeni paper currency detection system. In: 2019 First International Conference of Intelligent Computing and Engineering (ICOICE), IEEE, pp. 1–7 (2019)
3. Aoba, M., Kikuchi, T., Takefuji, Y.: Euro banknote recognition system using a three-layered perceptron and RBF networks. Trans. Math. Modeling Appl. **44**(SIG7 (TOM 8)), 99–109 (2003)
4. Arya, S., Sasikumar, M.: Fake currency detection. In: 2019 International Conference on Recent Advances in Energy-Efficient Computing and Communication (ICRAECC), IEEE, pp. 1–4 (2019)
5. Chavan, S.S., Fernandes, C., Dumane, P.R., Varma, S.L.: Design and Implementation of Automatic Coin Dispensing Machine. In: ICCCE 2019, pp. 379–385. Springer, Singapore (2020)
6. Doshi, P.: Currency Feature Extraction Using Image Processing Techniques (2020)
7. Frosini, A., Gori, M., Priami, P.: A neural network-based model for paper currency recognition and verification. IEEE Trans. Neural Netw. **7**(6), 1482–1490 (1996)
8. García-Lamont, F., Cervantes, J., López, A.: Recognition of Mexican banknotes via their color and texture features. Expert Syst. Appl. **39**(10), 9651–9660 (2012)
9. Hamza, R.M., Al-Assadi, T.A.: Genetic Algorithm to Find Optimal GLCM Features. Department of Computer Science, College of Information Technology (2012)
10. Hardani, D.N.K., Luthfianto, T., Tamam, M.T.: Identify the authenticity of Rupiah currency using K Nearest Neighbor (K-NN) algorithm. Jurnal Ilmiah Teknik Elektro Komputer dan Informatika (JITEKI) **5**(1), 1–7 (2019)
11. Hinwood, A., Preston, P., Suaning, G.J., et al.: Bank note recognition for the vision impaired. Australas. Phys. Eng. Sci. Med. **29**, 229 (2006)
12. https://shodhganga.inflibnet.ac.in/bitstream/10603/24460/9/09_chapter4.pdf. Last accessed 2020/01/21
13. Kekre, H.B., Thepade, S.D., Sarode, T.K., Suryawanshi, V.: Image retrieval using texture features extracted from GLCM, LBG and KPE. Int. J. Comput. Theor. Eng. **2**(5), 695 (2010)
14. Lamsal, S., Shakya, A.: Counterfeit paper banknote identification based on color and texture. In: Proceedings of the IOE Graduate Conference, pp. 160–168 (2015)
15. Mohanaiah, P., Sathyanarayana, P., GuruKumar, L.: Image texture feature extraction using GLCM approach. Int. J. Sci. Res. Publ. **3**(5), 1–5 (2013)
16. Singh, P.K., Kar, A.K., Singh, Y., Kolekar, M.H., Tanwar, S.: Proceedings of ICRIC 2019. In: Recent Innovations in Computing. Lecture Notes in Electrical Engineering, Vol. 597, pp. 3–920. Springer, Cham, Switzerland (2020)
17. Singh, P.K., Pawłowski, W., Tanwar, S., Kumar, N., Rodrigues, J.J.P.C.: Proceedings of First International Conference on Computing, Communications, and Cyber-Security (IC4S 2019). Lecture Notes in Networks and Systems, Vol. 121, pp. 3–917. Springer, Cham, Switzerland (2020)
18. Snehlata, S., Saxena, V.: An efficient technique for detection of fake currency. Int. J. Recent Technol. Eng. (IJRTE). Vol. 8, Issue 3, ISSN: 2277–3878 (2019)

19. Snehlata, S., Saxena, V.: Identification of fake currency: a case study of Indian scenario. Int. J. Adv. Res. Comput. Sci. **8**(3) (2017)
20. Takeda, F., Sakoobunthu, L., Satou, H.: Thai banknote recognition using neural network and continues learning by DSP unit. Lect. Notes Artif. Intell. **2773**, 1169–1177 (2003)
21. Turk, M., Pentland, A.: Face recognition using Eigenfaces. In: Proceedings of 1991 IEEE Computer Society Conference on Computer Vision and Pattern Recognition, pp. 586–587 (1991)
22. Yan, W.Q., Chambers, J., Garhwal, A: An empirical approach for currency identification. Multimedia Tools Appl. **74**(13), 4723–4733 (2015)

Optimal Unit Commitment for Secure Operation of Solar Energy Integrated Smart Grid

Aniket Agarwal and Kirti Pal

Abstract Currently, the majority of the world's electricity demand is met by thermal power generation stations that run purely on traditional fossil fuels. Utilities rely mainly on these sources to spend huge revenue to meet the ever-increasing demand for electricity. This motivates utilities to manage their generation most cost-effectively according to the load demand. Thus, an optimum generation allocation among the various power generating units can save considerable fuel inputs and expenses. Extending this optimization technique to decide which of these units would participate in the optimum allocation could theoretically save a greater amount of fuel costs. In other words, the determination of whether the device has to be ON/OFF is important. This is termed as unit commitment (UC). As for deciding the optimal generation dispatch for the minimum cost, a complete optimal power flow (OPF) is run over the UC time horizon for each hour's commitment. Conventional OPF solves all constraints such as fixed bus voltage limits, line power flows, transformer tap positions, etc., for optimum dispatch adjustment, resulting in a secure solution. In this paper, with the integration of solar thermal power plant, the total generation cost is reduced. The suggested method is tested by applying it to the standard IEEE 14 bus test system. The findings of the UC, demonstrated in both the presence and absence of STPP, show the method's efficiency. We propose a mathematical programming-based approach with alternative current optimal power flow (ACOPF) network constraints, to optimize the unit commitment problem.

Keywords Unit commitment · AC optimal power flow · Solar thermal power plant (STPP) · IEEE 14 bus test system

A. Agarwal (✉) · K. Pal
Department of Electrical Engineering, Gautam Buddha University, Gautam Buddh Nagar, Greater Noida, UP 201312, India
e-mail: aniagarwal1697@gmail.com

K. Pal
e-mail: kirti.pal@gbu.ac.in

P. K. Singh et al. (eds.), *Recent Innovations in Computing*, Lecture Notes in Electrical Engineering 701, https://doi.org/10.1007/978-981-15-8297-4_56

1 Introduction

The most important problem with electric power generation scheduling is unit commitment. Unit commitment means coordinating short-term unit generation to meet load demand forecasting. The problem of unit commitment is a complex problem of optimization. It has a variable integer as well as a constant variable [1]. The unit commitment (UC) function in the power system refers to the question of optimization to evaluate the ON/OFF states of generating units that reduce operating costs for a given time horizon [2]. The committed units must meet the system's forecast demand and spinning reserve requirement at the minimum operating cost, subject to a wide set of operating restrictions.

In practice as well as in the state-of-the-art research on deterministic UC models, what has varied most significantly are the assumptions underlying security (i.e., contingency [3]), network, and operational constraints for alternating current (AC) power systems. Improving Garver's seminal UC formulation [4], many prominent methods to date have centered on mixed-integer linear programming (MILP) with tighter convex hull representations of non-network constraints of thermal unit operations [5–10]. These UC formulations have usually been expanded to include a linear (DC) representation of the network, either with (e.g., [11, 12]) or without (e.g., [13]) called actual power losses. The resulting commitment schedules neglect the constraints of reactive power dispatch and AC power flow, which must subsequently be compensated for via corrective and generally ad hoc processes [14, 15].

UC is an important optimization task in modern power system day-to-day service planning. UC schedules are generally set a day ahead. Accurate load forecasting offers the hourly load demand for UC problems. The only optimizing criterion in the determination of the UC schedule is the cost of generation that must be minimized during the planning period while meeting all system constraints resulting from the physical capabilities of the generating unit and the network configuration of the transmission system. The recent scenario of generating electricity is the installation of power generators of large size. This makes the network of power systems more complex and tends to pollute the environment. Therefore, we need a better approach to determining the commitment schedule for the economic–emission unit.

The UC problem is a mix of two sub-problems. One determines the units to be committed, and the other focuses on the amount of generation that each of these committed units will produce. Generating units exhibit various operating efficiencies and performance characteristics that reflect on the inputs required. Thus, the generation cost also depends on the output amount of each committed unit, apart from the choice of generating units. Therefore, two stages of a UC problem are solved that are economic dispatch and optimal power flow. With the rapid development of the world economy and the depletion of fossil fuels, significant attention has been given to renewable energy sources (RES) in present society, as characteristic of non-contamination and inexhaustible [16]. By introducing renewable energy sources, we can reduce the total fuel cost and execution time.

In this paper, we have discussed about UC and its importance. The formulation of the UC schedule in the presence of renewable energy resource becomes tedious, which is done by integration of solar in existing system to reduce total generation cost. The proposed methodology is implemented on IEEE 14 bus system.

2 Unit Commitment and Its Importance

Economic operation results in increasing operational efficiencies, thereby minimizing costs per kilowatt-hour. Total power system load varies at any time, generally higher during the daytime and early evening when industrial loads are heavy, lights are on, and so on, and lower in the late evening and early morning when the majority of the population sleeps. The option to turn ON enough units and leave them online is therefore not viable due to the costs involved so that the variable load demand is met at all times. This occasionally causes some of the units to operate close to their minimum capacity, leading to lower system efficiency and increased economics. Therefore, in order to optimize device operation, units must be shut down as the load goes down and must be brought online as it goes up again [1].

Electrical utilities must schedule their generation in advance to meet this varying demand, as to which generators are to be started and when to synchronize them into the network and the sequence where working systems have to be shut down. The decision-making process is well-known as 'unit commitment.' The word 'commit' refers to making a computer 'turn ON' a unit.

Therefore, the unit commitment issue is to plan the ON and OFF hours of the generating units with the overall minimum cost while maintaining the operating restrictions of the unit, such as peak up/down periods, ramp rate limits, and maximum and minimum power generation limits. The main component of the costs incurred in generation is the cost of fuel supply per hour for all generators, while maintenance costs only contribute to a small extent. This assessment of the fuel costs is more relevant to thermal and nuclear power plants, which is not the case for hydrostations where the energy is derived from the storage of water in dams built for irrigation purposes and is apparently free. Savings in fuel costs can be achieved by proper load allocation among the engaged units.

3 Unit Commitment with Renewable Resources in Power System

With the prolonged and excessive use, it has become clear that supplies of fossil fuel are increasingly depleting and that the age of fossil fuel is slowly coming to a close. This is particularly true of coal, oil, and natural gas resources. The large-scale use of fossil fuels has caused detrimental environmental effects over the years. The result of this is the movement toward global warming which is of great concern for the future of human life. In pursuit of alternative sources, the situation has caused the whole world to meet the energy needs.

Renewable energy sources, such as solar, wind, tidal, and geothermal, were the alternatives that attracted the attention. It is a success to extract energy from the solar and wind to a considerable level. The research focuses on the use of solar thermal energy for electricity generation here. To model a solar integrated power system near a realistic scenario, a solar thermal power plant is installed into the existing power grid.

4 Problem Formulation

4.1 Unit Commitment Model with ACOPF Constraints

Here, a unit commitment model with ACOPF constraints is formulated, also known as UC + ACOPF, in which the goal is to reduce the total cost of generation required to meet the requirements of load with reserve.

Objective function: The sum of all committed generator fuel costs must be minimized.

$$F_T = F_1 + F_2 + \cdots + F_{ng} \tag{1}$$

$$= \sum_{i=1}^{ng} F_i(Pg_i) \tag{2}$$

$$= \sum_{i=1}^{ng} (a_i Pg_i^2 + b_i Pg_i + c_i) \tag{3}$$

Subjected to OPF constraints:-

1. Power balance in the network.
2. Unit generation limits.
3. Limits on load bus voltage magnitudes.
4. Limits on transmission line flows, transformer tap settings, and phase shifter angles.

   (a) Active and reactive power balance in the network:

$$Pg_i - Pd_i - P_i = 0 \quad i = 1, 2, \ldots N \tag{4}$$

$$Qg_i - Qd_i - Q_i = 0 \quad i = 1, 2, \ldots Nb \tag{5}$$

where Pg_i and Qg_i represent active and reactive power generation, P_i and Q_i represent active and reactive power injections at bus i, Pd_i and Qd_i represent active and reactive power demands at bus i, N is total number of buses, and Nb is the total number of load buses in the system.

(b) Limits on active and reactive power generations on all generator buses:

$$Pg_{i,\min} \leq Pg_i \leq Pg_{i,\max} \quad i = 1, 2, \ldots ng \tag{6}$$

$$Qg_{i,\min} \leq Qg_i \leq Qg_{i,\max} \quad i = 1, 2, \ldots ng \tag{7}$$

(c) Limits on voltage magnitudes and phase angles on all load buses:

$$V_{i,\min} \leq V_i \leq V_{i,\max} \; i = 1, 2, \ldots Nb \tag{8}$$

$$\delta_{i,\min} \leq \delta_i \leq \delta_{i,\max} \tag{9}$$

(d) Limits on line flows can be expressed either in MW, amperes, or MVA, if it is expressed in
MW then:

$$P_{ij,\min} \leq P_{ij} \leq P_{ij,\max} \; i = 1, 2, \ldots Nl \tag{10}$$

where P_{ij} is the active power flow between buses i and j. $P_{ij,\min}$,$P_{ij,\max}$ are the corresponding minimum and maximum limits, and Nl is the total number of transmission lines.

4.2 *IEEE 14 Bus Test System*

Figure 1 shows the network containing five generator buses, nine load buses, and twenty transmission lines. OPF is performed as a sub-problem to assess the optimum cost of generating for the commitment of the hour.

5 Result Analysis

The proposed methodology is implemented on the standard IEEE 14 bus test system. The results of simulation studies are depicted in three cases.

Case 1: Implementation of UC-ACOPF in standard IEEE 14 bus test system
Here, UC-ACOPF is scheduled for 24 h to show the effectiveness of the proposed methodology. A load profile is developed for 24 h to keep in mind that during office

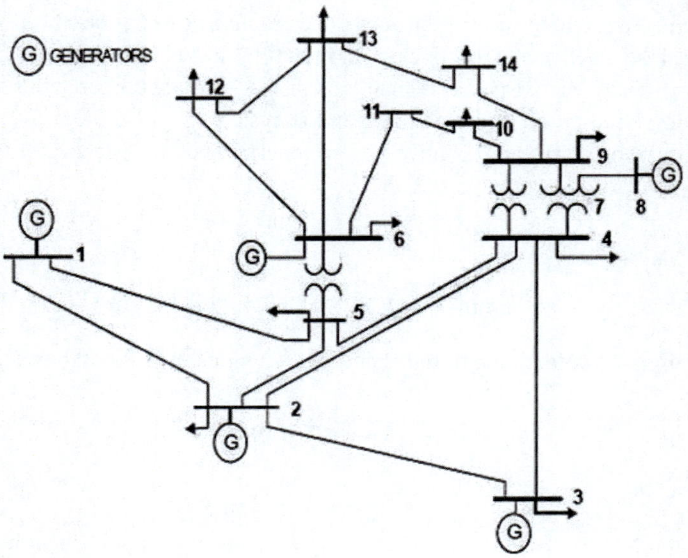

Fig. 1 Single line diagram of IEEE 14 bus system

hour load is high while in the morning, lunchtime, and late night load is low. Figure 2 shows the load profile for a scheduled day. The load variation is visible in Fig. 2. Hour 1 is started from the midnight.

Table 1 shows the 24 h UC schedule for standard IEEE 14 bus test data for load profile shown in Fig. 2. Table 1 clearly shows hourly load demand, the total power generated, unit status, power output of the individual generator, and total cost of UC schedule with hourly production costs.

Unit status 1 indicates generator is ‘ON,’ and 0 indicates generator is ‘OFF.’ It is clear from the table that during peak load hours total generator cost is very high as uneconomical generators 3 and 8 are ON during peak load hours. To reduce the cost of conventional generator during peak hours, here we introduce solar thermal power plant (STPP) in the standard IEEE 14 bus system. The optimal location of STPP is an issue that is discussed in case 2.

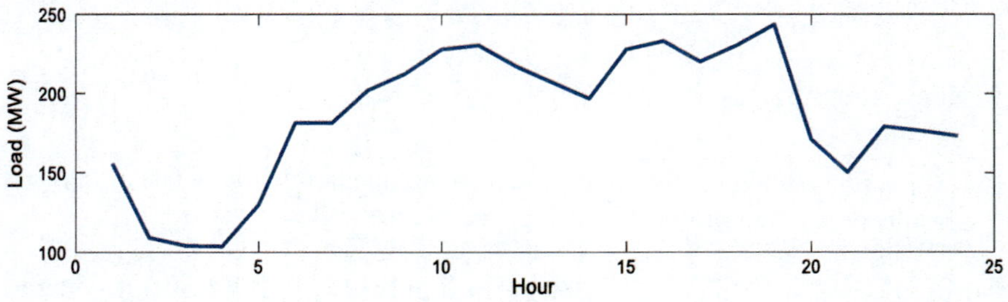

Fig. 2 Load profile for 24 h

Table 1 UC for IEEE 14 bus test system without STPP integrated

Time (h)	Load (MW)	Unit status					Power output (MW)					Total power (MW)	Cost ($/h)
		1	2	3	6	8	1	2	3	6	8		
1	155.40	1	1	0	0	0	134.91	25.16	0	0	0	160.07	4142
2	109.00	1	1	0	0	0	93.78	17.32	0	0	0	111.1	2675
3	104.00	1	1	0	0	0	89.25	16.46	0	0	0	105.71	2524
4	103.60	1	1	0	0	0	89.25	16.46	0	0	0	105.71	2524
5	129.50	1	1	0	0	0	111.98	20.77	0	0	0	132.75	3302
6	181.30	1	1	0	0	0	158.04	29.64	0	0	0	187.68	5047
7	181.30	1	1	0	0	0	158.04	29.64	0	0	0	187.68	5047
8	202.02	1	1	0	0	0	176.69	33.29	0	0	0	209.98	5819
9	212.38	1	1	1	0	0	186.23	35.16	0.14	0	0	221.53	6234
10	**227.92**	**1**	**1**	**1**	**0**	**0**	**189.78**	**35.84**	**11.32**	**0**	**0**	**236.94**	**6836**
11	**230.51**	**1**	**1**	**1**	**0**	**0**	**190.34**	**35.95**	**13.27**	**0**	**0**	**239.56**	**6939**
12	217.56	1	1	1	0	0	187.52	35.41	3.54	0	0	226.47	6426
13	207.20	1	1	0	0	0	181.37	34.21	0	0	0	215.58	6019
14	196.84	1	1	0	0	0	172.02	32.37	0	0	0	204.39	5622
15	**227.92**	**1**	**1**	**1**	**0**	**0**	**189.78**	**35.84**	**11.32**	**0**	**0**	**236.94**	**6836**
16	**233.10**	**1**	**1**	**1**	**0**	**0**	**190.91**	**36.06**	**15.21**	**0**	**0**	**242.18**	**7042**
17	220.15	1	1	1	0	0	188.08	35.52	5.48	0	0	229.08	6528
18	**230.51**	**1**	**1**	**1**	**0**	**0**	**190.34**	**35.95**	**13.27**	**0**	**0**	**239.56**	**6939**
19	**243.46**	**1**	**1**	**1**	**0**	**1**	**192.79**	**36.42**	**21.98**	**0**	**1.47**	**252.66**	**7457**
20	170.94	1	1	0	0	0	148.77	27.84	0	0	0	176.61	4678
21	150.22	1	1	0	0	0	130.31	24.28	0	0	0	154.59	3969
22	179.00	1	1	0	0	0	155.72	29.19	0	0	0	184.91	4954
23	176.30	1	1	0	0	0	153.40	28.74	0	0	0	221.74	4861
24	173.10	1	1	0	0	0	151.09	28.29	0	0	0	179.38	4769
Overall cost = 1,27,189.00 ($/day)													

Bold values represent the unit and peak load

Case 2: Optimal location of STPP

In order to minimize the conventional generator cost, STPP is integrated into the standard IEEE 14 bus system. STPP has to be added as an extra plant into the existing system. To identify the optimal location for the integration of STPP, load at each bus is increased by 10% and then by 20% to observe the impact on consumer cost. Table 2 shows that under the base case or when load is increased by 10% and 20% in all three cases the cost is higher at bus numbers 13 and 14. To minimize the cost and to minimize the use of conventional generator, two STPPs with capacity of 10 MW each are integrated at buses 13 and 14.

Table 2 Load impact on cost

Bus No.	Base case (259 MW)	With 10% increase in load (284.9 MW)	With 20% increase in load (310.8 MW)
	Cost ($/h)	Cost ($/h)	Cost ($/h)
1	36.724	36.925	37.061
2	38.360	38.590	38.754
3	40.575	40.789	40.970
4	40.190	40.437	40.603
5	39.661	39.917	40.068
6	39.734	40.030	40.160
7	40.172	40.394	40.570
8	40.170	40.390	40.564
9	40.166	40.380	40.564
10	40.318	40.567	40.767
11	40.155	40.438	40.618
12	40.379	40.742	40.942
13	**40.575**	**40.944**	**41.168**
14	**41.197**	**41.573**	**41.873**

Case 3: Implementation of UC-ACOPF in standard IEEE 14 bus test system with STPP

As the optimal location is identified in case 2, add the STPP in bus nos. 13 and 14 of the standard IEEE 14 bus test system. As the STPP can work in daytime, only the following schedule for generating units is assumed:

1. From 12.00 a.m. to 08.00 a.m., only the conventional thermal units have to cater to the load.
2. From 08.00 a.m. to 6.00 p.m., conventional units and solar thermal unit cater the load.
3. From 6.00 p.m. to 11.00 p.m., again only conventional units cater to the load.

It is clear from the time frame that during daytime or office peak hours, STPP will remain ON. During the absence or low output periods of solar generation, load demand has to be catered by the other units reliably without violating any operational constraints. Thus, the choice of generating a unit that is to be substituted by the modeled STPP is required to be judicious. Without any loss of generality, the lowest capacity unit in the test system is chosen for the same.

Table 3 shows the UC schedule for 24 h along with unit status, total generation, and hourly generation cost in the presence of STPP. During peak load hours such as for hours 10, 11, 15, 16, 18, and 19, total generation cost is considerably reduced while using STPP as compared with total generation cost without STPP as shown in Table 1. From Tables 1 and 3, the overall cost is calculated in the presence and

Table 3 UC for IEEE 14 bus test system with STPP

Time (h)	Load (MW)	Unit status							Power output (MW)							Total power (MW)	Cost ($/h)
		1	2	3	6	8	13	14	1	2	3	6	8	13	14		
1	155.40	1	1	0	0	0	0	0	134.91	25.16	0	0	0	0	0	160.07	4142
2	109.00	1	1	0	0	0	0	0	93.78	17.32	0	0	0	0	0	111.1	2675
3	104.00	1	1	0	0	0	0	0	89.25	16.46	0	0	0	0	0	105.71	2524
4	103.60	1	1	0	0	0	0	0	89.25	16.46	0	0	0	0	0	105.71	2524
5	129.50	1	1	0	0	0	0	0	111.98	20.77	0	0	0	0	0	132.75	3302
6	181.30	1	1	0	0	0	0	0	158.04	29.64	0	0	0	0	0	187.68	5047
7	181.30	1	1	0	0	0	0	0	158.04	29.64	0	0	0	0	0	187.68	5047
8	202.02	1	1	0	0	0	0	0	176.69	33.29	0	0	0	0	0	209.98	5819
9	212.38	1	1	0	0	0	1	1	167.96	31.62	0	0	0	10	10	219.58	5455
10	**227.92**	**1**	**1**	**0**	**0**	**0**	**1**	**1**	**181.97**	**34.37**	**0**	**0**	**0**	**10**	**10**	**236.34**	**6046**
11	**230.51**	**1**	**1**	**0**	**0**	**0**	**1**	**1**	**184.30**	**34.83**	**0**	**0**	**0**	**10**	**10**	**239.13**	**6146**
12	217.56	1	1	0	0	0	1	1	172.62	32.54	0	0	0	10	10	225.16	5649
13	207.20	1	1	0	0	0	1	1	163.31	30.71	0	0	0	10	10	214.02	5263
14	196.84	1	1	0	0	0	1	1	154.02	28.90	0	0	0	**10**	**10**	202.92	4887
15	**227.92**	**1**	**1**	**0**	**0**	**0**	**1**	**1**	**182.66**	**34.51**	**0**	**0**	**0**	**10**	**10**	**237.17**	**6076**
16	**233.10**	**1**	**1**	**1**	**0**	**0**	**1**	**1**	**185.41**	**35.05**	**1.36**	**0**	**0**	10	10	**241.82**	**6249**
17	220.15	1	1	0	0	0	1	1	174.96	32.99	0	0	0	10	10	227.95	5747
18	**230.51**	**1**	**1**	**1**	**0**	**0**	**1**	**1**	**184.30**	**34.83**	**0.02**	**0**	**0**	**10**	**10**	**239.15**	**6147**
19	**243.46**	**1**	**1**	**1**	**0**	**0**	**1**	**1**	**187.66**	**35.48**	**9.10**	**0**	**0**	10		**252.44**	**6657**
20	170.94	1	1	0	0	0	0	0	148.77	27.84	0	0	0	0	0	176.61	4678

(continued)

Table 3 (continued)

Time (h)	Load (MW)	Unit status							Power output (MW)							Total power (MW)	Cost ($/h)
		1	2	3	6	8	13	14	1	2	3	6	8	13	14		
21	150.22	1	1	0	0	0	0	0	130.31	24.28	0	0	0	0	0	154.59	3969
22	179.00	1	1	0	0	0	0	0	155.72	29.19	0	0	0	0	0	184.91	4954
23	176.30	1	1	0	0	0	0	0	153.40	28.74	0	0	0	0	0	182.14	4861
24	173.10	1	1	0	0	0	0	0	151.09	28.29	0	0	0	0	0	179.38	4769
Overall cost = 1,18,633.0 ($/day)																	

Bold values represent the unit and peak load

absence of STPP and the difference is 8556.00 \$/h. Thus, we have calculated the saving in fuel cost.

6 Conclusion

Unit commitment is a complex problem of optimization in power systems. A prior knowledge of the units to be committed among the available to meet the anticipated load demand not only reduces the generation expense, but also allows system operators to operate smoothly. But, with the introduction of renewable energy sources into the power grid the UC issue becomes more tedious as it reduces the total generation cost and fuel cost. An overall optimal schedule is found only by the optimal hourly commitments of units and dispatching the load among them economically. In this paper, the unit commitment problem is formulated with constraints involved in finding an optimal schedule. Results for standard IEEE 14 bus test systems are demonstrated. Unit commitment is scheduled for 24 h for IEEE 14 bus system in the presence and absence of STPP in order to reduce the overall generation cost. Results show the efficiency of proposed methodologies in evaluating UC schedules for the safe operation of an integrated smart power grid with solar energy.

References

1. Kumar, D., Kumar, A., Yadav, L.K.: Unit commitment of thermal power plant in integration with wind and solar plant using genetic algorithm. Int. J. Eng. Res. Technol. **7** (2014)
2. Zhao, B., Guo, C.X., Bai, B.R., Cao, Y.J.: An improved particle swarm optimization algorithm for unit commitment. Int. J. Electr. Power and Energy Syst. **24** (2006)
3. Stott, B., Alsaç, O.: Optimal power flow: Basic requirements for real-life problems and their solutions (2012)
4. Garver, L.L.: Power generation scheduling by integer programming—development and theory. Trans. Amer. Inst. Elect. Eng. Power App. Syst. 730–734 (1962)
5. Rajan, D., Takriti, S.: Minimum up/down polytopes of the unit commitment problem with start-up costs. IBM, Res. Rep. RC23628 (2005)
6. Carrión, M., Arroyo, J.: A computationally efficient mixed-integer linear formulation for the thermal unit commitment problem. IEEE Trans. Power Syst. 69–77 (2011)
7. Ostrowski, J., Anjos, M.F., Vannelli, A.: Tight mixed integer linear programming formulations for the unit commitment problem. IEEE Trans. Power Syst. 39–46 (2012)
8. Morales-España, G., Atorre, J.M., Ramos, A.: Tight and compact MILP formulation for the thermal unit commitment problem. IEEE Trans. Power Syst., 4897–4908 (2013)
9. Atakan, S., Lulli, G., Sen, S.: An improved MIP formulation for the unit commitment problem (2015)
10. Jabr, R.A.: Tight polyhedral approximation for mixed-integer linear programming unit commitment formulations. IET Gener. Transm. Distrib. 1104–1111 (2012)
11. Morales, J.M., Conejo, A.J., Pérez-Ruiz, J.: Economic valuation of reserves in power systems with high penetration of wind power. IEEE Trans. Power Syst. 900–910 (2009)
12. Wu, L., Shahidehpour, M.: Accelerating benders decomposition for network-constrained unit commitment problems. Energy Syst. **1**(3):339–376 (2010)

13. Feizollahi, M.J., Costley, M., Ahmed, S., Grijalva, S.: Large-scale decentralized unit commitment. Electr. Power Energy Syst 97–106 (2015)
14. Padhy, N.: Unit commitment—a bibliographical survey. IEEE Trans. Power Syst. 1196–1205 (2004)
15. Bhardwaj, A., Kamboj, K., Shukla, V., Singh, B., Khurana, P.: Unit commitment in electrical power system—a literature review. In: Proceeding IEEE PEOCO, Melaka, Malaysia (2012)
16. Xiu, L., Kang, Z., Huang, P.: Unit commitment using improved adjustable robust optimization for large-scale new energy power stations (2019)

Project Management Method-Based Cryptographic Algorithm Employing IC Engine Transmission Ratio and Simple Interest Formula

Rajdeep Chowdhury, Smriti Kumari, and Sukhwant Kumar

Abstract In this rationalized epoch, the orb is gyrating around technological progression with encouraging upshot. Nevertheless, it is decent being expectant, though circumstances urge for cynicism too. Being conscious of the fact that constructive facets of technology should not be a dispiriting aspect, rather emphasis should be toward curbing continuing cybercrimes, diurnal data stealth, malicious intrusions, etc. Computer systems are frequently harmed by network security. Employment of cryptography and cryptographic modus operandi is ensured to overawe such issues. The paper emphasizes utilization of cipher text in determining the critical path employing project management method. Project management method demonstrates all the activities requisite to accomplish a task in the finest feasible manner. The time though which it consumes sets out all the individual activities, partaking in crafting a colossal project. The apposite order has to be upheld as one activity could transpire only if other activity is accomplished, whereas in other cases, some activities transpire concurrently. The most convenient path forms the critical path. The proposed work is principally an amalgamation of project management method along with IC Engine Transmission Ratio and Simple Interest Formula, developed with a purpose of data safekeeping, which is an indispensable facet for apiece organizational accentuation.

Keywords Cybercrime · Network security · Project management method · Critical path · EST · IC engine · Transmission ratio · Simple interest

R. Chowdhury (✉)
Chinsurah, Hooghly, West Bengal 712101, India
e-mail: dujon18@yahoo.co.in

S. Kumari
Institute of Management Studies, Banaras Hindu University, Varanasi, Uttar Pradesh 221005, India
e-mail: smriti.kumariimbhu@gmail.com

S. Kumar
Department of Mechanical Engineering, JIS College of Engineering, Kalyani, Nadia, West Bengal 741235, India
e-mail: 00saket08@gmail.com

P. K. Singh et al. (eds.), *Recent Innovations in Computing*, Lecture Notes in Electrical Engineering 701, https://doi.org/10.1007/978-981-15-8297-4_57

1 Introduction

In the contemporary epoch, fresh technological tools are being hurled in the market diurnally [1]. With ample master minds emergent apiece day, there is also copious prospect of cybercrimes. Cybercrime is essentially a delinquency which comprises of computer and network. The data is either pilfered or viewed by another person, without the owner's knowledge. The term cybercrime typically threatens an individual or a group of individuals. It becomes a foremost predicament when it seizes confidential and private information [2].

Cryptography is a substantial solution to all such quandary at bay. It is a procedure for communication safekeeping or in a wider sense it is a mode for crafting and analyzing protocols which averts any unauthorized individual from accessing the data [2–9].

The paper illustrates amalgamation of project mnagement method with IC Engine Transmission Ratio and Simple Interest Formula. Project management method determines the critical path method. It is a procedure which is scheduled for a cluster of project actions. A critical path is established by classifying the protracted of reliant actions. Furthermore, it is also employed for measuring the total time requisite for accomplishing the activities from start to finish [10]. It also embraces simple interest formula which is an effortless as well as hasty mode of evaluating interest claimed on a credit. The same is determined by intensifying the daily interest proportion by the principal amount in contrast of numeral days that elapses amid expenses.

The study focuses on ensuring computer systems free from delinquencies by keeping apiece data secured. It endows with a diverse way of safekeeping data. It is the most expedient procedure which employs couple of distinct disciples beneath the identical caption [1].

2 Proposed Work

The section elucidates exclusively with the flowchart of the proposed work, followed by key engendering, encryption modus operandi and decryption modus operandi (Fig. 1).

Elucidating with Exemplification
Consider @12Smriti as plain text.

2.1 Key Engendering

Step 1: Consider @12Smriti as plain text. Plain text would be inserted in the table and ASCII value would be written adjacent to the text (Table 1).

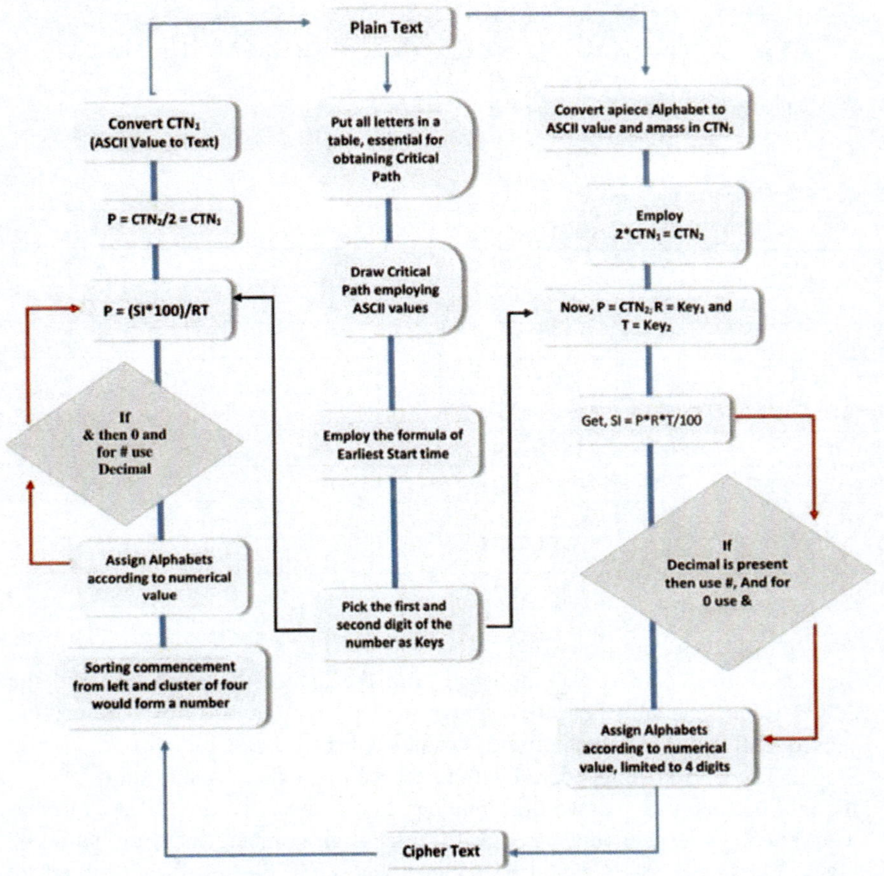

Fig. 1 Flowchart of the proposed work

Table 1 Key engendering step 1

Plain text	Activity	Immediate predators	ASCII value (path)
@12Smriti	@	Nil	64
	1	@	49
	2	1	50
	S	2	83
	M	S	109
	R	m	114
	I	r	105
	T	i	116
	I	t	105

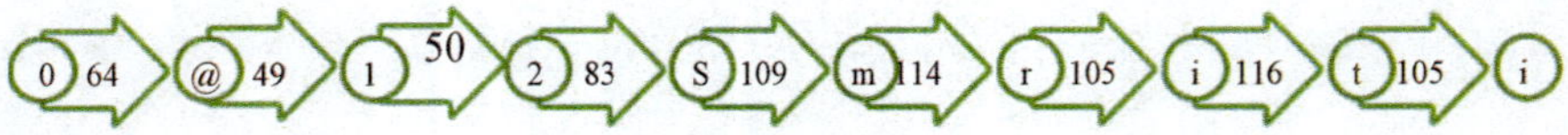

Fig. 2 Critical path image of the text

Table 2 Key engendering step 2

Activity	Earliest start time (EST)
0	0
1, (@)	0 + 64 = 64
2, (1)	64 + 49 = 113
3, (2)	64 + 49 + 50 = 163
4, (S)	64 + 49 + 50 + 83 = 246
5, (m)	64 + 49 + 50 + 83 + 109 = 355
6, (r)	64 + 49 + 50 + 83 + 109 + 114 = 469
7, (i)	64 + 49 + 50 + 83 + 60 + 114 + 105 = 574
8, (t)	64 + 49 + 50 + 83 + 60 + 114 + 105 + 116 = 690
9, (i)	64 + 49 + 50 + 83 + 60 + 114 + 105 + 116 + 105 = 795

Step 2: Critical path is accomplished with the aid of the plain text and the ASCII value. The table beneath displays the activity of apiece ASCII value, and subsequently, the earliest start time is calculated (Fig. 2 and Table 2).

Step 3: The keys are engendered with reference to the earliest start time.

It could be seen that for two-digit number, first and second digit of the number would be keys respectively, whereas for three-digit number, first digit would be a key and second digit would be another key while the third digit would be eliminated (Table 3).

Table 3 Key engendering final step

Earliest start time (EST)	Key1	Key2
0	0	0
64	6	4
113	1	1
163	1	6
246	2	4
355	3	5
469	4	6
574	5	7
690	6	9
795	7	9

2.2 Encryption Modus Operandi

Step 1: Consider the ASCII value as number of teeth in camshaft (Table 4). The formula from IC engine is consequently employed:

$$\text{No. Crankshaft's Teeth/No. Camshaft's Teeth} = 2$$

Step 2: The crankshaft teeth number is considered as the principal. Principal along with the two keys, assumed as rate and time, respectively, is therefore employed for calculating the simple interest (Table 5).

Table 4 Encryption initial process

ASCII value (camshaft teeth number) CTN^1	Crankshaft teeth number CTN^2 = (2 * camshaft teeth number)
64	128
49	98
50	100
83	166
109	218
114	228
105	210
116	232
105	210

Table 5 Encryption modus operandi step 2

P	R	T	SI = $P * R * T/100$
128	6	4	30.72
98	1	1	00.98
100	1	6	06.00
166	2	4	13.28
218	3	5	32.70
228	4	6	54.72
210	5	7	73.50
232	6	9	125.28
210	7	9	132.30

Employing Formula,

$$SI = \{P * R * T\}/100$$

Now,

$$\begin{aligned} CTN^2 &= P(\text{Principal}) \\ \text{Key1} &= R(\text{Rate of Interest}) \\ \text{Key2} &= T(\text{Time in Years}) \end{aligned}$$

Step 3: The alphabetic coding is taken from the cognitive slice which aids in formulating the cipher text. The diverse variables are taken for the value of decimal and zero. They are indicated as follows:

*For Decimal value # and for 0---&

* The digit should always be taken such as the total number of digits equals to 4, where '0' could be employed prior and after the digit in that number to make it as aforementioned (Table 6).

2.3 Decryption Modus Operandi

Decryption modus operandi is essentially reverse methodology with an exception that cipher text sorting is realized in a cluster of five, commencing from left. Nevertheless, the cipher text is sorted.

3 Result Analysis

In this section, the proposed work is assessed and compared with existing works on cryptography and values are analyzed in a tabular form and further displayed with the aid of graph. Executable text files pertaining to encryption time and decryption time are compared and exhibited in Table 7 (Fig. 3).

4 Conclusion

The comprehensive analysis of the subsequent modus operandi displays the amalgamation of project management method and cryptography, thereby ushering prominence toward the necessity of interdisciplinary research. Data safekeeping is not

Table 6 Cipher text engendering final step

SI = $P * R * T/100$	Alphabetic code	Cipher text
30.72	C&#GB	**C&#GB&&#IH&F#&&AC#BHCB#G&ED#GBGC#E&ABE#BACB#C**
00.98	&&#IH	
06.00	&F#&&	
13.28	AC#BH	
32.70	CB#G&	
54.72	ED#GB	
73.50	GC#E&	
125.2	ABE#B	
132.3	ACB#C	

Table 7 Comparison chart for .txt files

SI. No.	Source file	Source file	PMMBCA (in mm. s)		AES (in mm. s)		TDES (in mm. s)	
	Name	Size (in bytes)						
			Enc.	Dec.	Enc.	Dec.	Enc.	Dec
1	File1.txt	2565	13	16	16	0	0	16
2	File2.txt	8282	16	0	62	0	0	16
3	File3.txt	26,585	31	13	328	0	16	0
4	File4.txt	52,852	36	32	36	13	16	0
5	File5.txt	82,825	61	32	333	13	13	16
6	File6.txt	157,848	31	29	31	32	31	31
7	File7.txt	343,587	108	71	47	32	125	141
8	File8.txt	737,157	143	94	62	48	250	152
9	File9.txt	782,732	152	158	89	63	159	215
10	File10.txt	1,375,453	214	159	89	89	291	329
11	File11.txt	1,737,050	253	299	94	94	344	360
12	File12.txt	2,107,551	328	359	109	125	438	453
13	File13.txt	2,770,747	501	463	158	235	562	641
14	File14.txt	3,284,377	522	539	140	156	815	865
15	File15.txt	3,785,411	612	585	298	313	953	1109

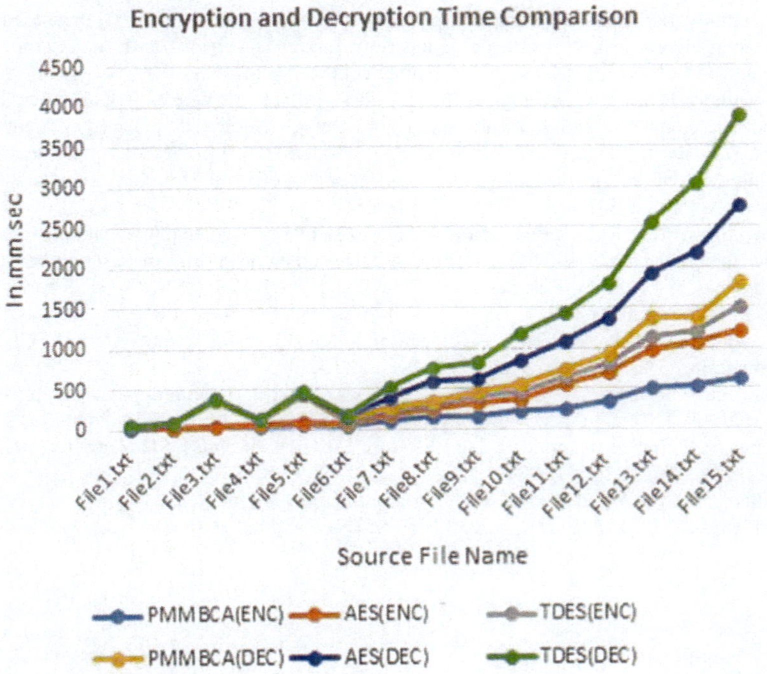

Fig. 3 Time lapse comparison graph for encryption and decryption

only crucial but the most significant facet of data clandestineness. The proposed work could be employed in diverse facets for data safekeeping and as means of ensuring fact safety. Furthermore, it could act as a comparative and analytical tool for future researchers during literature review, stressing on interesting findings, thereby furnishing optimum upshot. Exhaustive design and apt implementation of the proposed cryptographic algorithm is indispensable as the same could be integrated in diverse stratums of organizational repositories.

References

1. Kahate, A.: Cryptography and Network Security. Published by the Tata McGraw-Hill Publishing Company Limited, New Delhi (2008). ISBN (10)–0-07-064823-9, ISBN (13)–978-007-06-4823-4
2. Anciaux, N., Bouganim, L., Pucheral, P.: Data confidentiality: to which extent cryptography and secured hardware can help. Ann. Telecommun. **61**(3–4), 01–20 (2006)
3. Chowdhury, R., Roy, O., Datta, S., Dasgupta, S.: Virtual data warehouse model employing crypto–math modus operandi and intelligent sensor algorithm for cosseted transference and output augmentation. In: Knowledge Computing and Its Applications, pp. 111–129. Springer, Singapore (2018), ISBN (O)–978-981-10-6680-1, ISBN (P)–978-981-10-6679-5

4. Chowdhury, R., Datta, S., Dasgupta, S., De, M.: Implementation of central dogma based cryptographic algorithm in data warehouse for performance enhancement. Int. J. Adv. Comput. Sci. Appl. **6**(11), 29–34 (2015). ISSN (O)–2156 5570, ISSN (P)–2158 107X
5. Chowdhury, R., Dey, S.K., Datta, S., Shaw, S.: Design and implementation of proposed drawer model based data warehouse architecture incorporating DNA translation cryptographic algorithm for security enhancement. In: Proceedings of International Conference on Contemporary Computing and Informatics, IC3I 2014, Mysore, pp. 55–60. Published and Archived in IEEE Digital Xplore, (2014). ISBN–978-1-4799-6629-5
6. Chowdhury, R., Chatterjee, P., Mitra, P., Roy, O.: Design and implementation of security mechanism for data warehouse performance enhancement using two tier user authentication techniques. Int. J. Innov. Res. Sci. Eng. Technol. **3**(6), 165–172. ISSN (O)–2319 8753, ISSN (P)–2347 6710 (2014)
7. Chowdhury, R., Saha, A., Dutta, A.: Logarithmic function based cryptosystem [LFC]. Int. J. Comput. Inf. Syst. **2**(4), 70–76 (2011). ISSN–2229 5208
8. Chowdhury, R., Saha, A., Biswas, P., Dutta, A.: Matrix and mutation based cryptosystem [MMC]. Int. J. Comput. Sci. Netw. Secur. **11**(3), 7–14 (2011). ISSN–1738 7906
9. Mamta, P.A.: Image encryption using RSA with 2 bit rotation. Int. J. Res. Dev. Technol. **5**(7), 154–158 (2016). ISSN (O)–2349 3585
10. Bishnoi, N.: Critical path method (CPM): a coordinating tool. Int. Res. J. Manage. Sci. Technol. **9**(1), 459–467 (2018). ISSN (O)–2250 1959, ISSN (P)–2348 9367

Design of Reversible Gate-Based Fingerprint Authentication System in Quantum-Dot Cellular Automata for Secure Nanocomputing

Suhaib Ahmed, Soha Maqbool Bhat, and Seok-Bum Ko

Abstract The issues faced by CMOS technology in the nanoregime has led to the research of other possible technologies which can operate with same functionalities, however, with higher speed and lower power dissipation. One such technology is quantum-dot cellular automata (QCA). In this paper, QCA and reversible logic have been combined to design a 2 × 2 Feynman reversible gate-based fingerprint authentication system (FSA). An 8 × 8 size input fingerprint image is compared with the images present in the database and upon successful match, the FSA gives an output of logic '1' to confirm the match. Based on the performance analysis, it is shown that the proposed design achieves performance improvement of up to 89.05% compared to the previously reported design with respect to various parameters such as cell count, area, quantum cost, etc.

Keywords Reversible computing · Reversible gate · Quantum computing · Fingerprint authentication · Biometric system

1 Introduction

The miniaturization of conventional silicon CMOS transistors has been the driving force behind the rapid growth of semiconductor industry because it helped to increase the chip density, reduce power dissipation and also to increase the switching speed of integrated circuits. However, in the deep nanometer regime, a quantum effect comes

S. Ahmed (✉)
Department of ECE, BGSB University, Rajouri, India
e-mail: sabatt@outlook.com

S. M. Bhat
School of ECE, SMVD University, Katra, India
e-mail: soha.bhat@outlook.com

S.-B. Ko
Department of Electrical and Computer Engineering, University of Saskatchewan, Saskatoon, Canada
e-mail: seokbum.ko@usask.ca

P. K. Singh et al. (eds.), *Recent Innovations in Computing*, Lecture Notes in Electrical Engineering 701, https://doi.org/10.1007/978-981-15-8297-4_58

into the picture and plays a major role in the transistor operation. Other problems in designing circuits with deep nanometer transistor are high power consumption and leakage current along with electron migration [1–3]. Various alternatives to remedy the problems being faced by CMOS technology are being investigated by International Technology and Roadmap for Semiconductors (ITRS) [3, 4]. Given the current and future demand of digital systems, emerging nanotechnologies are a good market to invest in [2, 5, 6]. In this context, quantum-dot cellular automata (QCA) is one such new technology that is being developed for digital logic designs while offering the solutions to various issues faced by CMOS technology in the nanoregime [4, 7]. QCA can be developed by using molecules, and hence, the scalability of QCA is much better compared to traditional silicon CMOS transistor scaling.

1.1 QCA Nanotechnology

One of the most promising alternative technology being explored and investigated by the researchers is the nanotechnology based on quantum-dot cellular automata. The concept of QCA was first proposed by C. S. Lent [1] in 1993. QCA is developed based on the concept of cells, and the interaction between the charges in the cells is sufficient perform computation and information transformation [4, 7]. Another advantage of QCA-based electronic circuit design is that for the interconnection of cells, wires are not required. The primary component in a QCA circuit is a QCA cell, shown in Fig. 1. It consists of four quantum dots or wells wherein only two electrons are localized. These quantum dots are separated by quantum tunnel junctions. Coulombic interactions between the cells are responsible for the electron flow. Due to the coulombic repulsion, the electrons reside only at antipodal sites, i.e., diagonally thereby achieving minimum repulsion [8–10]. These electron-residing positions result in two polarization states, viz. binary ‘0’ and binary ‘1’. The QCA cell polarization states are shown in Fig. 2. Compared to CMOS technology, there is relatively low energy dissipation since there is negligible energy dissipation during the propagation and state transition in QCA [3, 11, 12].

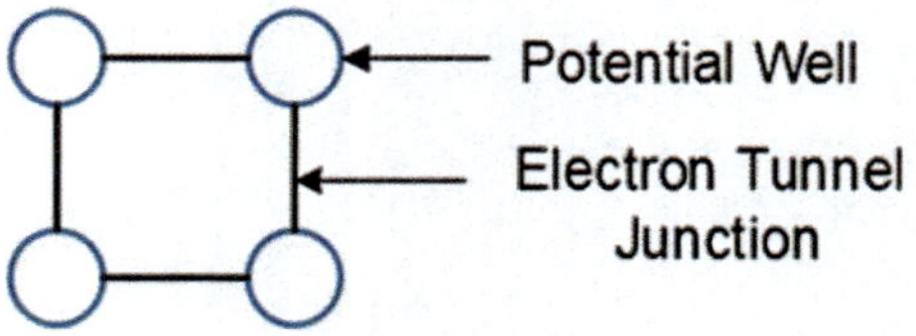

Fig. 1 QCA cell

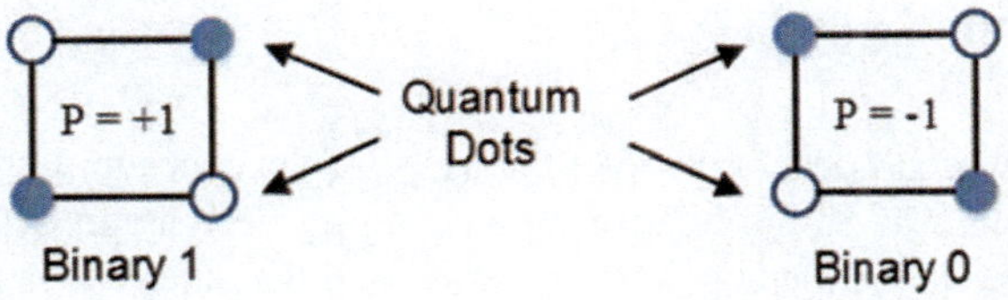

Fig. 2 QCA cell polarization

1.2 Clocking in QCA

Unlike CMOS circuits, where we have only two states in clocking which are 'LOW' and 'HIGH', QCA circuits have four-phase clock in which every phase has a 90° phase difference between them. At a particular instant, a particular clock phase enhances or reduces the barrier potential; controlling the quantum tunneling of electrons to other dots. This makes data transmission possible by a phenomenon of pipelining. The QCA-based circuits are provided with clock signals, and these clock signals control the information flow between the QCA cells. This clocking consists of four phases, viz. switch, hold, release and relax. The QCA cells are initially in an unpolarized state thereby having low potential. The polarization of a QCA cell is mainly influenced by the neighboring cell polarization during the switch phase and the potential vitality of the electrons begin to rise. Further, no state change occurs as the electron attains highest potential vitality toward the end of the switch phase. In the hold phase, cell holds the previous state and potential vitality is maintained high. In the release and relax phase, the potential vitality of the electrons begins to decrease and eventually achieving null polarization again [4, 6, 7]. The different clock phases are shown in Fig. 3.

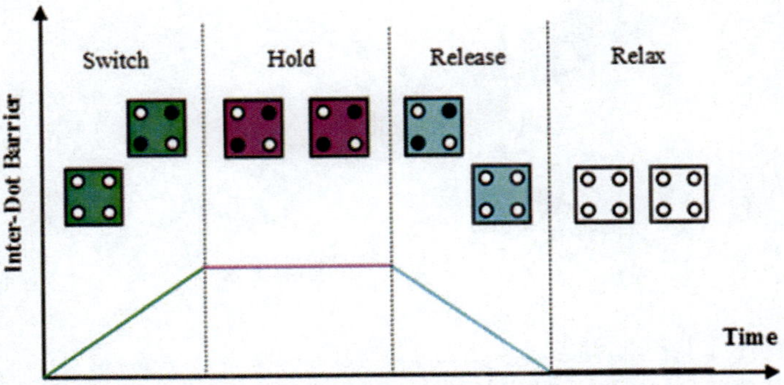

Fig. 3 Graphical representations of various clock phases in QCA

1.3 Crossovers in QCA

Many times, the circuit design involves the usage of the concept of crossover of the QCA wires in order to design relatively less complex designs. In QCA, crossovers are of two types. One is the coplanar crossover, and the other one is the multilayer crossover. The coplanar crossover lies in the single plane wherein simple and rotated cells are used to crossover the wires as there is no interaction between the cells or alternate clock zones are used for crossovers, as depicted in Fig. 4 [13–17]. The other one is the multilayer crossover in which more than one layers are involved, as shown in Fig. 5. The designing of multilayer crossover is complicated, but it is preferred over coplanar crossover to reduce the cell count and area of the circuit [1, 18–20]. However, in context to fabrication feasibility, coplanar is preferred.

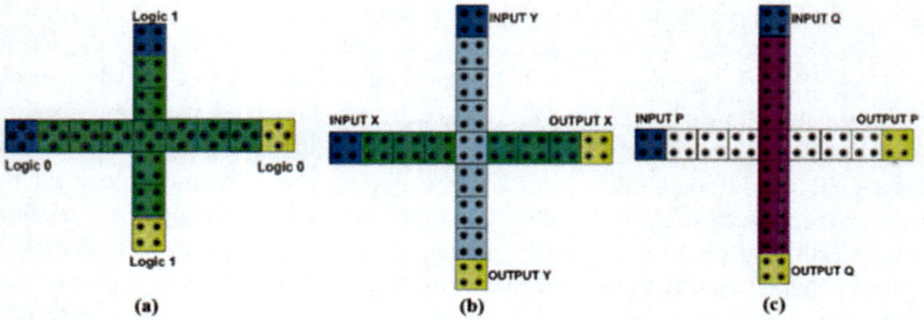

Fig. 4 Illustration of coplanar wire crossing **a** using rotated cells, **b** using clock 0 and clock 2, **c** using clock 1 and clock 3 in QCA [21]

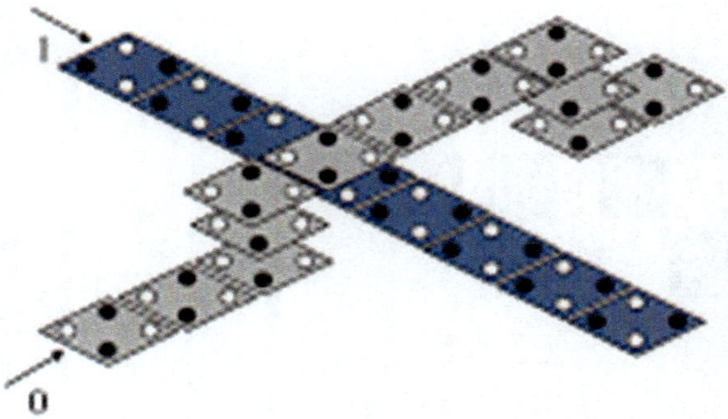

Fig. 5 Illustration of multilayer crossover in QCA [22]

Table 1 Some reversible gates

Gate size	Gate
2 × 2	Feynman [23]
3 × 3	Fredkin [24], RG-QCA [3], SSG-QCA [6], IMG [25], NNG [26], Peres [27], TR [28], QCA1 [29]
4 × 4	HNG [30], MRLG [31], RAM [32], HNFG [33], PFAG [34]

1.4 Reversible Logic

A logic is reversible in nature when the outputs and inputs are equal in number and if there is a one-to-one mapping between the inputs and outputs which leads to a different combination of output vector for each set combination of input vector. Till date various reversible logic gates of different sizes have been proposed, some of which are shown in Table 1.

2 Fingerprint Authentication System (FAS)

The digital logic-based fingerprint authentication system (FAS), shown in Fig. 6, is used to match the input fingerprint image with existing images available in the database. If match exists, the user is authenticated. In digital logic-based systems, all bits of the input image are checked sequentially with the images in the database and when the match if found the corresponding output bit is set to logic '1'. If at the end of matching, all the bits of output are '1' then match exists. This phenomenon can be understood from Table 2.

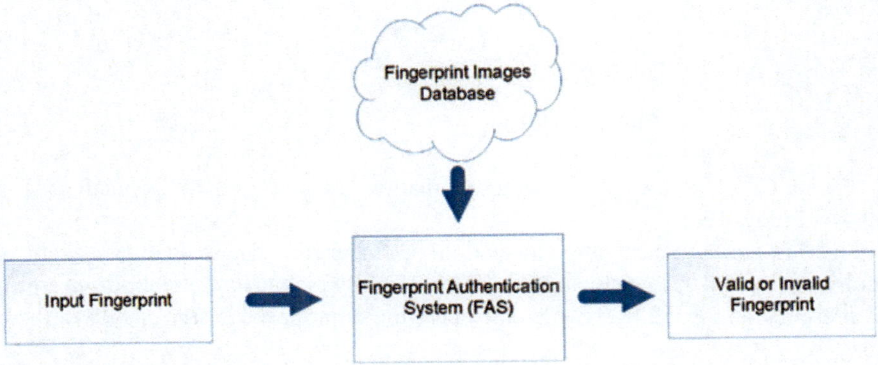

Fig. 6 Block diagram representation of proposed IMG gate

Table 2 Truth table of fingerprint authentication system (FAS)

Input *A*	Input *B*	Output *Y*
'0'	'0'	'1'
'0'	'1'	'0'
'1'	'0'	'0'
'1'	'1'	'1'

Upon closely inspecting Table 2, it is observed that the output function of the FAS is nothing but a XNOR gate with output equation given as:

$$Y = \overline{A.B} + A.B \tag{1}$$

The entire algorithm can be visualized using the flowchart presented in Fig. 7.

3 QCA Implementation of FAS

The FAS can be designed using reversible gates in QCA as follows:

A. Feynman Gate [23]:

One fingerprint authenticator (FPA) was proposed in [35], as shown in Fig. 8, using a 2 × 2 Feynman Gate.

Authors in [35] modified the input *A* of the Feynman gate to *A*′ to operate it as FPA as shown in Fig. 9.

This FPA was proposed in QCA using the concept of 3-input majority voter equations given below:

$$\text{Gar} = \overline{X} \tag{2}$$

$$\text{FPA}_{\text{out}} = M\left(M\left(\overline{X}, \overline{Y}, 0\right), M(X, Y, 0), 1\right) \tag{3}$$

The QCA implementation of these equations as proposed by Debnath [35] is shown in Fig. 10.

In this paper, a new implementation of fingerprint authentication system is proposed using 2 × 2 Feynman gate as shown in Fig. 11, and its simulation waveform is presented in Fig. 12 from which the operation of proposed design can be verified.

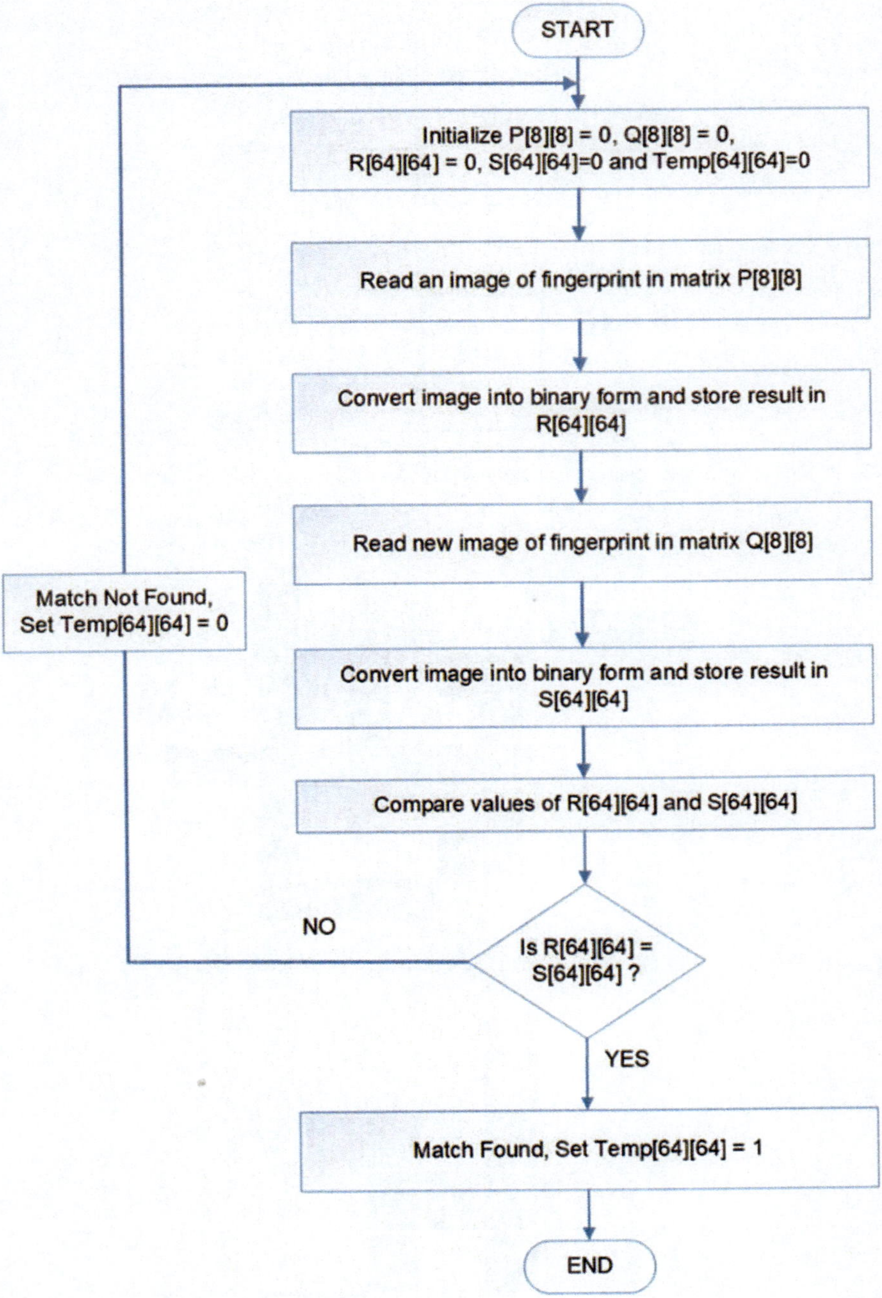

Fig. 7 Flowchart of fingerprint authentication system (FAS)

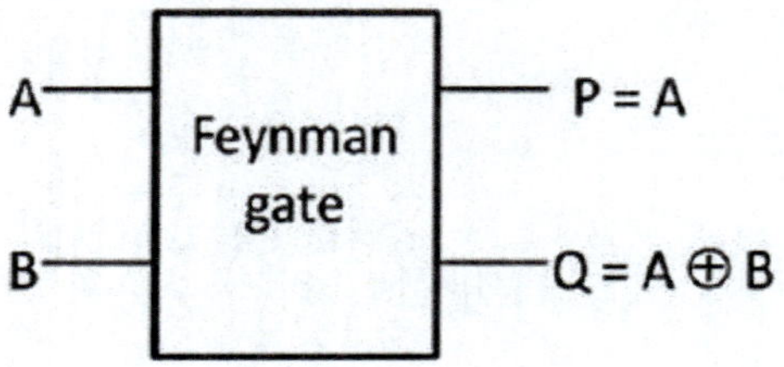

Fig. 8 2 × 2 Feynman gate

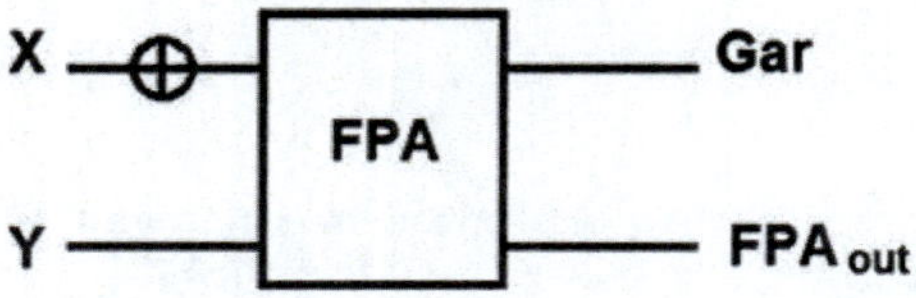

Fig. 9 FPA using 2 × 2 Feynman gate proposed in [35]

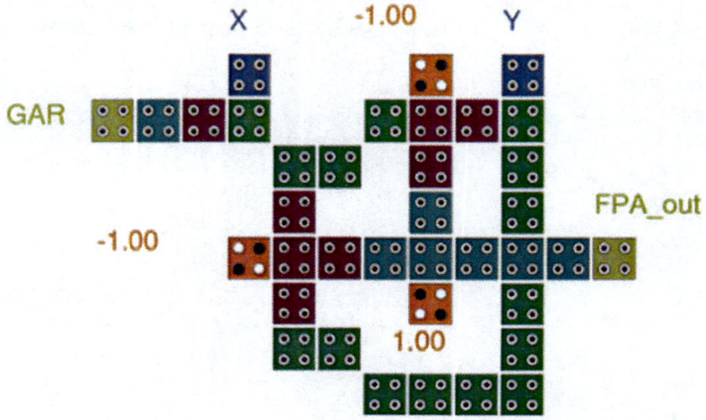

Fig. 10 FPA using 2 × 2 Feynman gate in QCA proposed by Debnath [35]

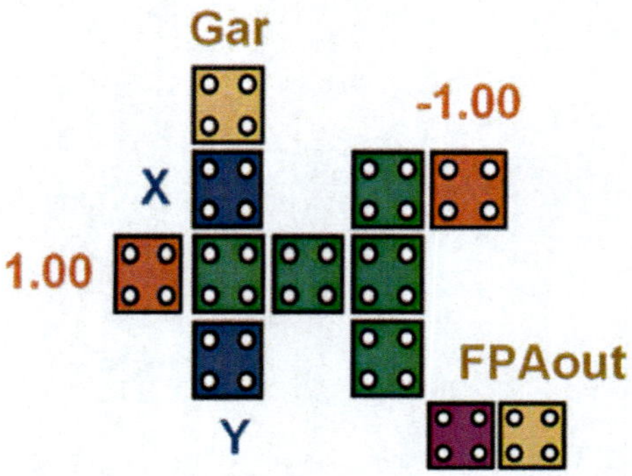

Fig. 11 Proposed QCA implementation (design-1) of FPA using 2 × 2 Feynman gate

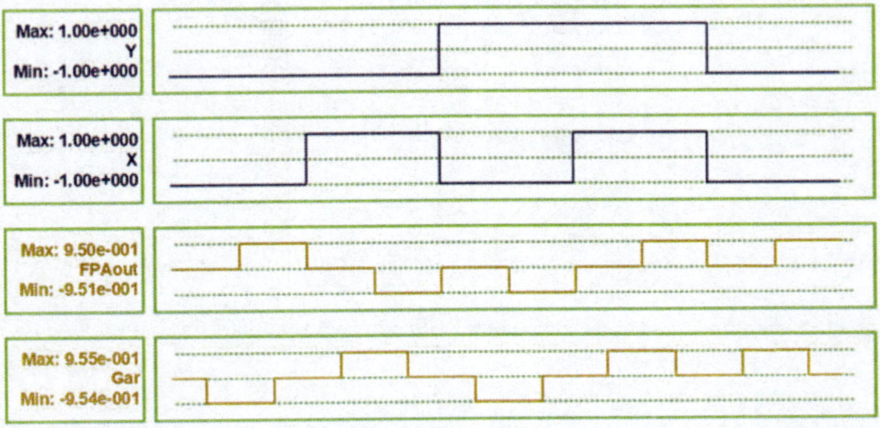

Fig. 12 Simulation waveform of proposed design of FPA

4 Energy Estimation Analysis

The energy dissipation analysis of the proposed 1-bit comparator circuit has been done using a probabilistic modeling QCAPro tool [36]. This tool uses Hartree–Fock mean-field approach approximation for power dissipation analysis as illustrated in [37, 38].

$$H = \begin{bmatrix} \frac{-E_k}{2}\sum_i C_i f_{i,j} & -\gamma \\ -\gamma & \frac{E_k}{2}\sum_i C_i f_{i,j} \end{bmatrix} = \begin{bmatrix} \frac{-E_k}{2}(C_{j-1}+C_{j+1}) & -\gamma \\ -\gamma & \frac{E_k}{2}(C_{j-1}+C_{j+1}) \end{bmatrix} \tag{4}$$

The power dissipation of a single QCA cell is estimated using upper bound power dissipation model as:

$$P_{\text{diss}} = \frac{E_{\text{diss}}}{T_{\text{cc}}}\langle\frac{\hbar}{2T_{\text{cc}}}\vec{\Gamma}_+ \times \left[-\frac{\vec{\Gamma}_+}{\left|\vec{\Gamma}_+\right|}\tanh\left(\frac{\hbar\left|\vec{\Gamma}_+\right|}{k_B T}\right) + \frac{\vec{\Gamma}_-}{\left|\vec{\Gamma}_-\right|}\tanh\left(\frac{\hbar\left|\vec{\Gamma}_-\right|}{k_B T}\right)\right]\rangle \tag{5}$$

Figure 13 shows the power dissipation maps of proposed FAS at temperature of 2 K and tunneling energy levels of 0.5 E_k, whereas Table 3 shows the dissipation analysis for different simulation conditions.

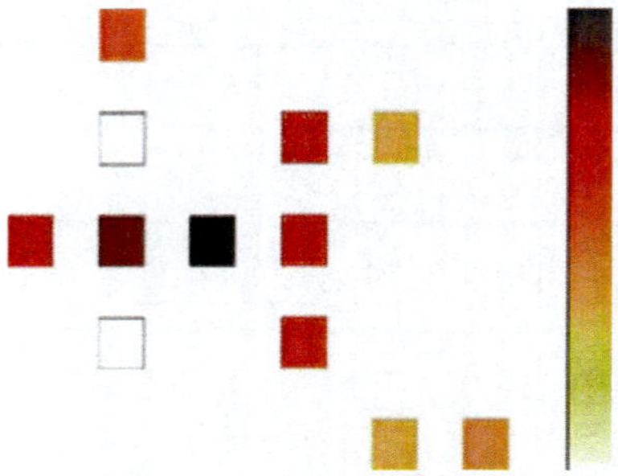

Fig. 13 Energy dissipation map of proposed FAS design at 2 K temp and 0.5 E_k energy level

Table 3 Energy dissipation of proposed FAS design

Energy dissipation type	Tunneling energy level		
	0.5 E_k	1.0 E_k	1.5 E_k
Avg. leakage (eV)	0.00438	0.01143	0.01902
Avg. switching (eV)	0.00951	0.00794	0.00665
Total (eV)	0.01389	0.01937	0.02567

5 Performance Comparison

The performance of the proposed reversible FAS design is evaluated by computing the various parameters such as cell count, total area utilized by the design, latency or delay (which is equal to the number of clock cycles utilized in the design to get the desired outputs), and quantum cost (which is equal to the area delay product). The performance comparison with existing design in literature is presented in Table 4.

It is evident that the proposed design is highly area and cost efficient, thereby making it suitable for biometric systems requiring low power consumption.

Table 4 Comparison of proposed fingerprint authentication systems (FAS) design

Parameter	[35]	Proposed design	% Age improvement
Cell count	37	12	67.57%
Cell area (μm^2)	0.0148	0.00389	73.72%
Total area (μm^2)	0.0352	0.01156	67.16%
Latency	0.75	0.25	66.67%
Area delay product	0.0264	0.00289	89.05%

6 Conclusion

A new 2 × 2 Feynman gate-based reversible fingerprint authentication system (FAS) has been proposed in quantum-dot cellular automata (QCA) for nanosecure applications. In view of the performance evaluation, it is seen that the proposed design of fingerprint authentication system (FPA) is efficient and accomplishes improvement up to 89.05% in terms of quantum cost and up to 73.72% in terms of cell area for implementing the reversible logic-based fingerprint authentication system.

References

1. Lent, C.S., Tougaw, P.D., Porod, W., Bernstein, G.H.: Quantum cellular automata. Nanotechnology **4**, 49 (1993)
2. Smith, C.G.: Computation without current. Science **284**, 274–274 (1999)
3. Bilal, B., Ahmed, S., Kakkar, V.: Modular adder designs using optimal reversible and fault tolerant gates in field-coupled QCA nanocomputing. Int. J. Theor. Phys. **57**(5), 1356–1375 (2018)
4. Bilal, B., Ahmed, S., Kakkar, V.: An insight into beyond CMOS next generation computing using quantum-dot cellular automata nanotechnology. Int. J. Eng. Manuf. **8**(1), 25–37 (2018)
5. Tougaw, P.D., Lent, C.S.: Logical devices implemented using quantum cellular automata. J. Appl. Phys. **75**, 1818–1825 (1994)
6. Bhat, S.M., Ahmed, S.: Design of ultra-efficient reversible gate based 1-bit full adder in QCA with power dissipation analysis. Int. J. Theor. Phys. **58**(12), 4042–4063 (2019)
7. Bilal, B., Ahmed, S., Kakkar, V.: Quantum dot cellular automata: a new paradigm for digital design. Int. J. Nanoelectron. Mater. **11**(1), 87–98 (2018)
8. Frost, S.E., Rodrigues, A.F., Janiszewski, A.W., Rausch, R.T., Kogge, P.M.: Memory in motion: a study of storage structures in QCA. In: Proceedings of First Workshop on Non-Silicon Computing (2002)
9. Niemier, M.T., Kogge, P.M.: Logic in wire: using quantum dots to implement a microprocessor. In: Proceedings of 6th International Conference on Electronics, Circuits and Systems, pp. 1211–1215. IEEE (1999)
10. Amlani, I., Orlov, A.O., Toth, G., Bernstein, G.H., Lent, C.S., Snider, G.L.: Digital logic gate using quantum-dot cellular automata. Science **284**, 289–291 (1999)
11. Ahmad, F., Ahmed, S., Kakkar, V., Bhat, G.M., Bahar, A.N., Wani, S.: Modular design of ultra-efficient reversible full adder-subtractor in QCA with power dissipation analysis. Int. J. Theor. Phys. **57**(9), 2863–2880 (2018)
12. Bilal, B., Ahmed, S., Kakkar, V.: Multifunction reversible logic gate: logic synthesis and design implementation in QCA. In: Proceedings of International Conference on Computing, Communication and Automation (ICCCA), 5–6 May 2017, pp. 1385–1390. IEEE (2017)
13. Sang Sefidi, M., Abedi, D., Moradian, M.: Design a collector with more reliability against defects during manufacturing in nanometer technology QCA. J. Softw. Eng. Appl. **6**(6), 304–312 (2013)
14. Gin, A., Tougaw, P.D., Williams, S.: An alternative geometry for quantum-dot cellular automata. J. Appl. Phys. **85**(12), 8281–8286 (1999)
15. Devadoss, R., Paul, K., Balakrishnan, M.: Coplanar QCA crossovers. Electron. Lett. **45**(24), 1234–1235 (2009)
16. Arjmand, M.M., Soryani, M., Navi, K.: Coplanar wire crossing in quantum cellular automata using a ternary cell. IET Circ. Dev. Syst. **7**(5), 263–272 (2013)

17. Dysart, T.J., Kogge, P.M.: Probabilistic analysis of a molecular quantum-dot cellular automata adder. In: Proceedings of International Symposium on Defect and Fault-Tolerance in VLSI Systems, pp. 478–486. IEEE (2007)
18. Qi, H., Sharma, S., Li, Z., Snider, G.L., Orlov, A.O., Lent, C.S., Fehlner, T.P.: Molecular quantum cellular automata cells: electric field driven switching of a silicon surface bound array of vertically oriented two-dot molecular quantum cellular automata. J. Am. Chem. Soc. **125**(49), 15250–15259 (2003)
19. Antonelli, D.A., Chen, D.Z., Dysart, T.J., Hu, X.S., Kahng, A.B., Kogger, P.M., Murphy, R.C., Niernier, M.T.: Quantum-dot cellular automata (QCA) circuit partitioning: problem modeling and solutions. In: Proceedings of 41st Design Automation Conference (DAC), 7–11 July 2004, pp. 363–368. IEEE (2004)
20. Lent, C.S., Tougaw, P.D., Porod, W.: Bistable saturation in coupled quantum dots for quantum cellular automata. Appl. Phys. Lett. **62**, 714–716 (1993)
21. Mallaiah, A., Swamy, G.N., Padmapriya, K.: 1-bit and 2-bit comparator designs and analysis for quantum-dot cellular automata. Nanosyst. Phys. Chem. Math. **8**(6), 709–716 (2017)
22. Gassoumi, I., Touil, L., Ouni, B., Mtibaa, A.: An efficient design of CORDIC in quantum dot cellular automata technology. Int. J. Electron. **106**(12), 2039–2056 (2019)
23. Feynman, R.: Quantum mechanical computers. Found. Phys. **16**(6), 507–531 (1986)
24. Fredkin, E., Toffoli, T.: Conservative logic. Int. J. Theor. Phys. **21**, 219–253 (1982)
25. Manzoor, I., Nafees, N., Baba, M.I., Bhat, S.M., Puri, V., Ahmed, S.: Logic design and modeling of an ultra-efficient 3 × 3 reversible gate for nanoscale applications. In: Algorithm for Intelligent Systems, pp. 1433–1442. Springer, Berlin, Heidelberg (2020)
26. Nafees, N., Manzoor, I., Baba, M.I., Bhat, S.M., Puri, V., Ahmed, S.: Modeling and logic synthesis of multifunctional and universal 3 × 3 reversible gate for nanoscale applications. In: Algorithm for Intelligent Systems, pp. 1423–1432. Springer, Berlin, Heidelberg (2020)
27. Peres, A.: Reversible logic and quantum computers. Phys. Rev. A **32**, 3266–3276 (1985)
28. Saravanan, S., Vennila, I., Mohanram, S.: Design and implementation of an efficient reversible comparator using TR gate. Circ. Syst. **7**, 2578–2592 (2016)
29. Ma, X., Huang, J., Metra, C., Lombardi, F.: Reversible and testable circuits for molecular QCA design. In: Emerging Nanotechnologies, pp. 157–202, Springer, Berlin, Heidelberg (2008)
30. Biswas, A.K., Hasan, M.M., Chowdhury, A.R., Babu, H.M.H.: Efficient approaches for designing reversible binary coded decimal adders. Microelectron. J. **39**, 1693–1703 (2008)
31. Kumar, V., Dhawan, D.: Design of reversible adder subtractor using multifunction reversible logic gate (MRLG). Int. J. Adv. Comput. Electron. Eng. **2**, 5–11 (2016)
32. Rangaraju, H., Suresh, A.B., Muralidhara, K.: Design and optimization of reversible multiplier circuit. Int. J. Comput. Appl. **52**(10), 44–50 (2012)
33. Haghparast, M., Navi, K.A.: Novel reversible BCD adder for nanotechnology based systems. Am. J. Appl. Sci. **5**, 282–288 (2008)
34. Islam, M.S., Rahman, M.M., Begum, Z., Hafiz, M.Z.: Low cost quantum realization of reversible multiplier circuit. Inf. Technol. J. **8**(2), 208–213 (2009)
35. Debnath, B., Das, J.C., De, D.: Fingerprint authentication using QCA technology. In: Proceedings of conference on devices for integrated circuit (DevIC), 23–24 Mar 2017, Kalyani, India, pp. 125–130. IEEE (2017)
36. Srivastava, S., Asthana, A., Bhanja, S., Sarkar, S.: QCAPro-an error-power estimation tool for QCA circuit design. In: Proceedings of International symposium of circuits and systems (ISCAS), pp. 2377–2380. IEEE (2011)
37. Timler, J., Lent, C.S.: Power gain and dissipation in quantum-dot cellular automata. J. Appl. Phys. **91**, 823–831 (2002)
38. Srivastava, S., Sarkar, S., Bhanja, S.: Estimation of upper bound of power dissipation in QCA circuits. IEEE Trans. Nanotechnol. **8**, 116–127 (2009)

A Survey on Blockchain Technologies and Its Consensus Algorithms

Rahul Katarya and Vinay Kumar Vats

Abstract The evolution and development in blockchain technologies have attracted both research academia and industries. A typical blockchain stores data in a permanent and immutable way in form of blocks connecting, forming a chain of data. The whole system is made decentralized so that anyone connected to the network can verify the data, defining its P2P distributed nature. Blockchain has many components among which the core component is consensus protocol. This protocol is responsible for the security and performance of the blockchain. The consensus protocol introduced by Nakamoto in Bitcoin led the foundation stone for more innovative alternative consensus mechanisms. In this paper, we will conduct a systematic review of blockchain technology and its consensus algorithms and further analyze them based on some essential features and factors.

Keywords Blockchain · Bitcoin · Consensus algorithms

1 Introduction

S. Haber and W. S. Stornetta developed a technique for time stamping a digital document [1] in 1991, which is similar to the technique used by the present-day blockchain system. In the past decade, the blockchain technology became quite popular after the introduction of Nakamoto's Bitcoin [2] in the industry. The digital payment system was greatly impacted by Bitcoin. It is envisaged that a wide range of financial service industries as well as many non-financial sectors will be revolutionized by blockchain technology and applications based upon it. The Bitcoin blockchain keeps a record of all transactions occurred in the network from the first block created until the last block generated after every ten minutes. It is a decentralized system which means

R. Katarya · V. K. Vats (✉)
Department of Computer Science and Engineering, Delhi Technological University, Shahbad Daulatpur, New Delhi, Delhi 110042, India
e-mail: veenu.kumarvats@gmail.com

R. Katarya
e-mail: rahulkatarya@dtu.ac.in

P. K. Singh et al. (eds.), *Recent Innovations in Computing*, Lecture Notes in Electrical Engineering 701, https://doi.org/10.1007/978-981-15-8297-4_59

that there is no central authority which controls the whole system. The system is distributed among all the network nodes in the system. It is secured by its design. Being decentralized, it also does not assume trust between the network nodes. The data structure adopted by blockchain system uses hashes, which stores the hash of the previous block in the hash of the next block generated creating a chain of blocks. This chain assures that the data stored on the block cannot be changed. Due to this property of blockchain, it is considered as the backbone of upcoming peer-to-peer, decentralized and open-source access technologies. Some also compare to emerging Web 3.0 structure of future Internet.

But, while applying these, blockchain technologies come with some hidden and popular problems and challenges. Between these problems, the major one faced by a decentralized system is to reach network consensus. The decentralized nodes have to validate a transaction and need to agree on a common validation. This is done to ensure that the system remains identical for all nodes. In this type of system where nodes are distributed, every node of the system can act a host and a server and they exchange data between each other to establish a consensus in the network.

Although this technique of checking the transactions through different nodes in the network causes confusion between them as if every node tries simultaneously to transmit a new block. To prevent this situation, the "Consensus Algorithms" [3] must be implemented on the system and network nodes to manage processes for updating chains.

A modern-day blockchain deploys a range of cryptographic techniques [4] for data management and cryptographically equates on-chain identities of the users for the transactions they make using their tokens. This property of blockchain ensures proof of asset (i.e., token) flow by transfer authentication and also proof of asset ownership. Also, the whole system preserves a random order of the transaction records by chaining the record subsets cryptographically to their chronic predecessors in the structure of data "blocks."

2 Background

The blockchain technology has three features which describe its nature and make it unique. These features are.

2.1 Decentralized

The blockchain is a P2P distributed network system which provides secured operation without the need for a third-party intervention. This exclusion of third party relieves the transaction cost used by the networks for designing centralized ledgers and governing body. The whole database is shared with all nodes participating in the network. A centralized system can be described as a central node or authority which has all the data and functionalities with it, so for any operation, other node has to request data from that central node only. The information is not handled by a single entity in a decentralized system. That means, the knowledge is shared by everyone within the network. If you decided to communicate with your friend in a decentralized network, then you can do so directly without going through a third party. This removal of third party removes the transaction cost one needs to pay for each transaction in a centralized system.

2.2 *Immutable*

Immutability is defined as the ability of a blockchain system to remain unchanged, which means the chain cannot be altered. This means that once something enters the blockchain, it cannot be tampered. For the assurance of this, blockchain uses cryptographic hash functions. Hashing's technique refers to taking the length of an input string and sending out a fixed-length output. One can relate hashing to encryption where an input of random length gives an output of a fixed-length ciphertext. If it is altered, then it can be easily detected. This property of blockchain is ensured by using hashes. Every next block in the chain contains a unique hash of the previous block, and when combined, they create the unique hash for the present block. Due to this, hashes becomes the identifiers of blocks for exchange and addressing.

Your Hash: 8cf9f1209f18a3accf611b8fb830c164
Your String: This is new system
Your Hash: 0f7af0a31393c0d0fbdc5163ffbe5223
Your String: This system is new

With the help of the above example, one can see that changing a single word in a the sentence or a single alphabet in a word will change the encrypted hash of the input drastically. This property is called avalanche effect. Even if you make a small adjustment to your data, the changes reflected in the hash will be enormous.

2.3 *Transparency*

The transparency in a blockchain can be seen as all the blockchain nodes have full copy and access to the blockchain, and they all can see transactions of each others. A person's identity in a blockchain network is hidden via complex cryptographic algorithms and represented in the network only by their public address.

Table 1 Comparison between different types of blockchain

Type	Public	Private	Consortium
Participation	Anyone	Specific organization	In between organizations
Directed By	Anyone	Organization itself	Selected authority
Efficiency	Low	High	High
User node identity	Pseudonymous	Revealed	Some nodes are revealed and some are pseudonymous
Network size	Large	Small	Medium
Example	Bitcoin, Ethereum [6, 7]	Blockchain of a specific department of an organization	Hyperledger [8]

So, I one view a person's transaction history, in a Bitcoin wallet, you will not see

"Merry sent 1 BTC"
Rather, you will see
"1LF1bhsFLkBppp9vpFYEmvwT2TbyCt8Mnz sent 1 BTC".

So, while the real identity of the individual [5] is secure, you will still see all the transactions that their public address has done.

Types of Blockchain (Table 1):

3 Working of Blockchain

The first transaction is deposited in the first block referred to as the genesis block [9]. A transaction which occurs on the system requires two phases of verification before it is added to the chain (Fig. 1):

1. Verification which is carried out in the same chain by some other nodes.
2. Sender's digital signature verification achieved by public and private key cryptography.

The proper transaction was annexed to the block, and the copy of this block is sent for verification by every node in the system. Carrying out verification of block this way causes confusion between the nodes if each node wants to update its verified block to the blockchain. So, this condition can be prevented by an agreement which must be reached between nodes, by applying a consensus algorithm, to decide which block is to be appended and by which node.

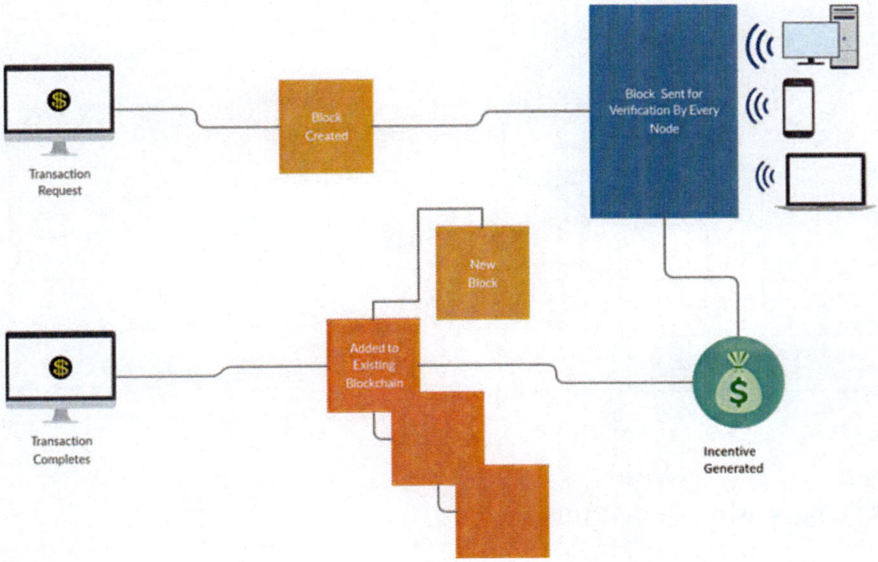

Fig. 1 Working of blockchain storing transaction history

Problems faced by the previous blockchain strategies:

There are two major problems which were faced in blockchain development previously. But most of the present-day blockchain's networks are immune to these two problems.

1. **Double-Spending Problem**

The problem arises when two separate transactions involve the same digital coin and try to spend it twice over. The reason behind this is that it takes a certain amount of time in between transaction request and its completion, and the completion here refers to the transaction which is updated to the blockchain, for example, any Bitcoin transaction takes ten minutes for processing. In this time gap, the same digital coin can be spent as many time [10].

2. **Byzantine Generals Problem (BGP)**

In a distributed P2P system, transactions are accepted by the network nodes after achieving consensus on the transaction validity. This becomes a security issue if an attacker tries to exploit the system by manipulating network nodes to verify a fake transaction [11] and temper the blockchain. So that's why consequently achieving a consensus state by untrustworthy nodes in blockchain networks is an important issue.

Nevertheless, if the intruder acquires 51% of the network resources, these issues would effectively compromise the Blockchain network. The consensus algorithm, therefore, has to be carefully designed to avoid such an attack.

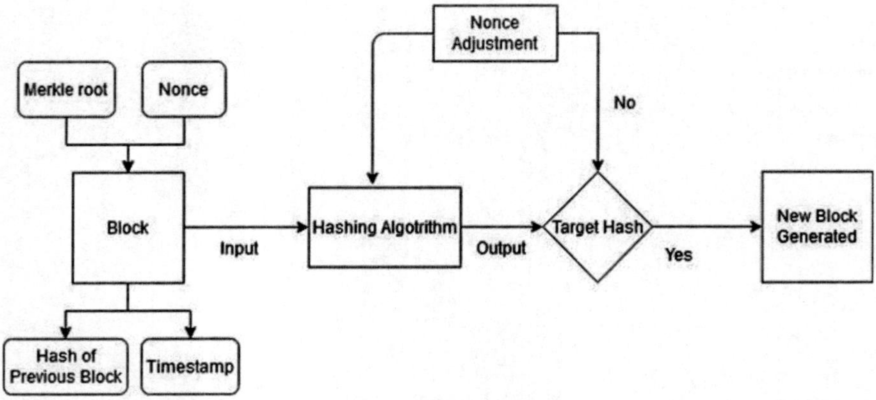

Fig. 2 Working: proof-of-work algorithm

4 Consensus Algorithms

Consensus algorithms would allow the next block to be added to the chain to be the only version of the truth and prevent attackers from infecting the chain [12]. The consensus algorithms are divided into two types. One of them uses proof of node and the other one uses voting between nodes to achieve common consensus.

4.1 Proof-Based Consensus Algorithms

4.1.1 Proof of Work (PoW)

The proof-of-work [2] mechanism was recommended by Satoshi Nakamoto for the administration of Bitcoin network. His primary goal for developing this mechanism was to resolve the double-spending problem. This technique first time used the term mining. In PoW, a complex cryptographic problem is attached to the new transaction block which is ready to be added in the blockchain, the first node which solves this problem adds this block to the blockchain and an incentive for its mining of block given to the node. Besides its features of securing the system PoW faces, many drawbacks like the calculation power take a lot of intensity and handling (Fig. 2).

4.1.2 Proof of Stake (PoS)

PoS evolved as a substitute to PoW which works on the principle of the economic stake of a validator in the blockchain network. Stake here refers to the currency number that a network participant owns in the form of tokens. The transaction validator must possess the necessary measure of tokens for a particular time before

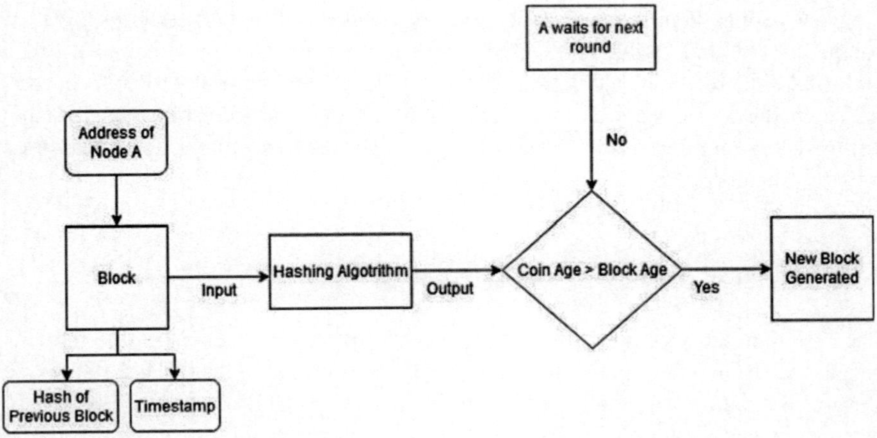

Fig. 3 Working: proof-of-stake algorithm

taking part in the transaction approval procedure. The main reason behind PoS development was to eliminate the energy and computing power loss [13] faced due to establishing PoW. But due to this stake holding issue, PoS is very much prone to the 51% assault, which means that the one holding more than 51% of stakes in the network will be chosen as the ledger with PoS protocol. To avoid this problem, fines are forced for validating any bogus confirmation and verification [11]. The PoS also has some future perspectives discussed in [14] which can be useful with some upgradations (Fig. 3).

4.1.3 Delegated Proof of Stake (DPoS)

DPoS [15] can be seen as the upgraded version of PoS protocol where an election is performed between all the stakeholder nodes and the chosen one becomes delegates. These chosen delegates become in charge of the consensus process, while in the PoS mechanism, majority of the network nodes participate in the consensus process, which becomes a problem if the network size is increasing. Thus, DPoS solves this problem by electing delegates through a continuous approval voting system. Whenever a delegate decreases its participation in the transaction approval mechanism, then it will be rejected and new a delegate replaces it. DPoS is thus adopted by many cryptocurrencies like EOS [16], BitShares [17], etc.

4.2 Voting-Based Consensus Algorithms

Opposite to evidence-based consensus protocols, these consensus protocols require the hubs' identifiable proof that will be involved in the review proceedings before the work begins. In turn, the exchange will be checked by all network hubs together. As the hub will talk to others to agree to add another block to their chain. Casting a ballot-based agreement calculations functions like the techniques of tolerating faults revised in the correct state. Thus, certain conditions like nodes crashing or subverting will resist certain algorithms. Voting-based consensus algorithms can be classified into two categories [3].

4.2.1 Practical Byzantine Fault Tolerance (pBFT)

The pBFT initiative was proposed to reduce the Byzantine weakness that remains because the BGP is dissatisfied with consensus. The purpose of the pBFT is to provide a useful Byzantine state machine replication to survive the Byzantine deficiency. To be superior and allow low overhead, this formula is advanced to work in an offbeat setting. The majority option has come to experience different stages right now. The calculation of pBFT forms the exchange of demands at speed, yet the overhead resulting from correspondence limits their adaptability. Moreover, the character of the framework is probably going to be undermined as a result of how the calculation is structured. Likewise, it is hard to use this calculation in the open condition because of the prerequisite of distinguishing the number of hubs before begin working. Besides, this calculation does not create a prize for mining [6].

4.2.2 Raft

The approach was introduced in the framework where the innovative political decision mechanism relies on randomized clocks to deal with the replicated logs forces solid administration. This measure increases the level of security by consistent changes for the server participation. Raft cannot accommodate malicious nodes but can reach fault tolerance of up to 50% [3].

5 Discussion and Analysis

In this section, we discuss and examine the majority of aspects that affect consensus algorithms.

5.1 Method for Managing User Node Identity

In public blockchains, as we have discussed that the identity of users are not revealed. So for these public blockchains where network nodes can openly connect to a network, algorithms such as PoW, PoS and DPoS are more reliable and efficient, whereas the consensus algorithms like pBFT and Raft need to test the node identity before participating in the voting process. Thus, these algorithms are ideal for private or cooperative blockchain networks where most of the nodes are private and revealed and less number of public nodes.

5.2 Mining Efficiency

A very crucial point in judging a consensus algorithm for blockchain is the mining process employed. Much is influenced by the power consumption, efficiency [18], reliability and speed (transactions completed per second). The proof-of-work mechanism involves solving a complex mathematical problem that enhances security although it requires a great deal of electrical and computational resources. Due to which, the PoW algorithm becomes less available and least used nowadays, whereas in pBFT, the voting process requires the exchange of many broadcast messages between networks nodes which also, in the end, causes network overhead.

5.3 Power Consumption

Without any doubt, PoW is the most power-consuming algorithm used in blockchain technology. A huge amount of electric energy is required for solving the complex mathematical puzzle, which includes successive calculations to obtain the desired hash value of the generated new block. The transaction speed of PoW transactions per second is 3–7 which is very less and limits the PoW potential. In PoS and DPoS, they all have identical weaknesses like the volume of calculations is smaller, which makes them use less power than PoW, while only stakeholders in the network receive block reward, which in the end leads to a significant reduction in coin liquidity. As far as the voting-based techniques are considered, they do not perform any mining process due to which energy consumption is very low in pBFT and Raft.

5.4 Incentive Mechanism

In a public blockchain network, there are many advantages to miners or the nodes participating in the consensus management. The resources, time, computational and electrical energy they spend for verifying a transaction they get an incentive for their service. In private blockchain networks, the mining operations depend upon business capital and the connected nodes are the organization's private nodes only. Due to this, they do not require an incentive.

5.5 *Performance and Scalability*

A blockchain's efficiency varies with the network's increase in size. Like if the network size increases, the efficiency decreases because the speed of verification for a transaction decreases, and this reduces the device throughput. The speed of creation of the block is also decreasing. All of PoW, PoS and DPoS are scaled well. Although they do not have very high efficiency (transaction per second), there are few ways to improve the scalability [19]. For example, the Bitcoin network adopts a lightning network to equip transactions off-chain to increase the scalability. As pBFT is applicable for only a small number of high-performance network nodes, the scalability of pBFT is reduced.

5.6 *Choosing Miners from the Network*

Blockchain transactions are secured and reviewed by mining. Blocks are secured by blockchain miners in the ledgers, and they form a chain linked to each other. Miners must add a legitimate block to the existing blockchain. In PoW, the node with the highest computing power will be able to solve the cryptographic puzzle quickly, and it will be chosen as the miner. Both PoS and DPoS choose the validator according to the stake of the nodes. A node with a higher stake has higher chances of being picked as the block validator. pBFT uses mathematical methods to choose the block validators, whereas Raft uses randomized timers to achieve the same.

Comparison (Table 2):

6 Conclusion

Blockchain is an innovative technology having potential industrial applications which facilitate safe transactions without a central authority's intervention. While the use of blockchain technology is still in its early stages, it is based on commonly accepted and established principles of cryptography.

The consensus protocol is the guarantee that networks with blockchain will operate stably. With the aid of consensus algorithms, the nodes decide on the state of a transaction. We implemented some common and realistic algorithms for the process of blockchain consensus and found their strengths, limitations and implementation through study and comparison. An effective consensus protocol is one which takes into consideration not only good fault tolerance but also the best possible implementation.

Table 2 Comparison between different consensus algorithms

Type	Proof-based consensus algorithms			Voting-based consensus algorithms	
Factors	PoW	PoS	DPoS	pBFT	Raft
Node identity management	Public	Public	Public	Private	Private
Mining efficiency	Low	High	High	Medium	Medium
Power consumption	High	Low	Low	Negligible	Negligible
Scalability	Good	Great	Great	Bad	Bad
Block reward incentive	Only for first miner	No	Only for elected witnesses	No	No
Transaction fees incentive	Yes	Yes	Yes	No	No
Byzantine fault tolerance (%)	50	50	50	33	50
Double spending	Yes	No	No	No	Not confirm
Miners chosen based on	Computing Hash	Stake	Stake	Mathematically	Randomized timers
Examples	Smart contracts [22]	Crypto currency [4, 14], Smart contracts	BitShares [17]	Smart contracts, distributed systems [20]	General applications [21]

References

1. Haber, S., Stornetta, W.S.: How to time-stamp a digital document. J. Cryptol. **3**(2), 99–111 (1991)
2. Nakamoto, S.: Bitcoin: a peer-to-peer electronic cash system (2008). https://bitcoin.org/bitcoin.pdf. Last accessed 2019/12/20
3. Nguyen, G.T., Kim, K.: A survey about consensus algorithms used in blockchain. J. Inf. Process. Syst. **14**(1), 101–128 (2018)
4. Lecture Notes in Computer Science, 2019. Advances in Cryptology—CRYPTO. Springer, Berlin (2019)
5. Lecture Notes in Computer Science: Data Privacy Management, Cryptocurrencies and Blockchain Technology, Springer, Berlin (2019)
6. Woord, G.: Ethereum: A Secure Decentralised Generalised Transaction Ledger (2014)
7. Wood, G.: Ethereum: a secure decentralised generalised transaction ledger. https://ethereum.github.io/yellowpaper/paper.pdf. Last accessed 2019/12/22
8. Sukhwani, H., Martinez, J.M., Chang, X., et al.: Performance modeling of PBFT consensus process for permissioned blockchain network (Hyperledger Fabric). In: Reliable Distributed Systems, pp. 253–255. IEEE (2017)
9. Singhal, B., Dhameja, G., Panda, P.S.: Beginning Blockchain. Apress, Berkeley, CA (2018)
10. Bradbury, D.: The problem with Bitcoin, Comput. Fraud Secur. **2013**(11), 5–8 (2013)
11. Lamport, L., Shostak, R., Pease, M.: The byzantine generals problem. ACM Trans. Program. Lang. Syst. **4**(3), 382–401 (1982)

12. Ouattara, H.F., Ahmat, D., Ouédraogo, F.T., Bissyandé, T.F., Sié, O.: Blockchain consensus protocols. In: Odumuyiwa, V., Adegboyega, O., Uwadia, C. (eds.) e-Infrastructure and e-Services for Developing Countries. AFRICOMM 2017. Lecture Notes of the Institute for Computer Sciences, Social Informatics and Telecommunications Engineering, vol. 250. Springer, Cham (2018)
13. Bartoletti, M., Lande, S., Podda, A.S.: A proof-of-stake protocol for consensus on bitcoin subchains. S. Afr. J. Anim. Sci. **36**(5, Suppl 1), 568–584 (2017)
14. Nguyen, C.T., Hoang, D.T., Nguyen, D.N., Niyato, D., Nguyen, H.T., Dutkiewicz, E.: Proof-of-stake consensus mechanisms for future blockchain networks: fundamentals, applications and opportunities. IEEE Access **7**, 85727–85745 (2019)
15. Snider, M., Samani, K., Jain, T.: Delegated proof of stake: features & tradeoffs. Multicoin Cap. 1–19 (2018)
16. "EOS.IOTechnicalWhitepaperv2" (2018) https://github.com/EOSIO/Documentation/blob/master/TechnicalWhitePaper.md. Last accessed 2019/12/28
17. BitShares 2.0—Industrial-grade decentralized (DPoS) eco-system on blockchain. https://bitshares.org/. Last accessed 2020/1/18
18. Eklund, P.W., Beck, R.: Factors that impact blockchain scalability. In: Proceedings of the 11th International Conference on Management of Digital EcoSystems (MEDES '19). Association for Computing Machinery, New York, NY, USA, pp. 126–133 (2019)
19. Chauhan, A., Malviya, O.P., Verma, M., Mor, T.S.: Blockchain and scalability. In: 2018 IEEE International Conference on Software Quality, Reliability and Security Companion (QRS-C), Lisbon, pp. 122–128 (2018)
20. Vukolić, M.: The quest for scalable blockchain fabric: Proof-of-work vs. BFT replication. In: International Workshop on Open Problems in Network Security, pp. 112–125. Springer, Berlin (2016)
21. Ongaro, D., Ousterhout, J.: In Search of an Understandable Consensus Algorithm. In: 2014 USENIX Annual Technical Conference (2014)
22. Alharby, M., van Moorsel, A.: Blockchain based smart contracts: a systematic mapping study. Comput. Sci. Inf. Technol. (CS IT) 125–140 (2017)

Digital India

Toward Prediction of Student's Guardian in the Secondary Schools for the Real Time

Chaman Verma, Veronika Stoffová, Zoltán Illés, and Deepak Kumar

Abstract To ease of school management for the identification of student's protector during his or her schooling and evaluate the performance of a student, the guardian predictive models are presented with the help of machine learning algorithms. For this, standard secondary dataset of two secondary schools was considered from Portugal belonging to the language course. The initial dataset consisted of 649 instances and 33 features. These features belong to the student's academic, demography, family and personal features. The guardian feature has been considered as a class variable and others (significant) features assumed as predictors. In the orange platform, three machine learning algorithms, support vector machine (SVM), random forest (RF) and neural network (NN), were used with three testing techniques. On one hand, the SVM computed the highest prediction probabilities of 0.996 for other class and another hand, the NN gives the largest prediction probabilities such as 0.906 for father class and 0.889 for mother class. The NN attained the most guardian prediction accuracy of 89% and outperformed others. Also, leave-one-out method significantly enhanced the prediction accuracy of each learner except the SVM. Also, it proved the NN learner slower with prediction time (23 s) and makes the SVM as faster with time (14 s). This study may not only helpful to the school management but also support the social administration of the district or state. Using the model, it must be significant to predict the care-taker of the student.

C. Verma (✉) · Z. Illés
Eötvös Loránd University, Budapest, Hungary
e-mail: chaman@inf.elte.hu

Z. Illés
e-mail: illes@inf.elte.hu

V. Stoffová
Trnava University, Trnava, Slovakia
e-mail: NikaStoffova@seznam.cz

D. Kumar
Guru Kashi University, Punjab, India
e-mail: dr.d.k.mehta81@gmail.com

P. K. Singh et al. (eds.), *Recent Innovations in Computing*, Lecture Notes in Electrical Engineering 701, https://doi.org/10.1007/978-981-15-8297-4_60

Keywords Machine learning · Guardian prediction · School student · Classification

1 Introduction

Machine learning is a domain under the shelter of artificial intelligence. It has been using in numerous fields of research such as medical, computer vision, banking fraud identification, robotics, IoT and others. One of the important types of machine learning is supervised learning which emphasizes the prediction of target class labels based on the predictors class labels under the supervisor of learner with specified rules. The application of machine learning is trending to apply the predictive models on the educational datasets. In addition to a student performance of computer students [1] and M.Sc. students [2], the attitude [3] and technical awareness levels [4] were predicted with supervised machine learning algorithms. It has been also found that placement possibilities of students predicted well based on their historical records supported the job placement in the institutions [5, 6]. A smart approach to automatic the real-time gender predictive model was presented [7]. Also, the few significant genders predicted models with most accuracy have been proposed with the machine learning techniques [8–10]. Based on the residence of educators like locality [11–13] state [14], country [15], nationality [16], machine learning spotted most significant features with algorithms. Also, the age-group of students toward ICT [17] and level of study (bachelor and master) [18] reflecting binary classification problem has been resolved with the help of machine learning. The present and growth of trending technology have been also predicted with feature mapping on machine learning algorithms [19]. The identification of educational institutions based on the technical features was presented [20]. Freshly, in the Portuguese secondary school, the enthusiasm of students has been predicted based on various parameters [21]. The residency of Indian faculties has been predicted with various learner algorithms [22]. Further, many researchers worked on different problems using machine learning [23–25]. The five major sections depict the formation of the present paper. Section 1 belongs to the basic theory and concerned related work to this paper. Section 2 about the approach to be research methodology. Section 3 keeps the discussion theory about the experiments performed. Section 4 compares the important performance measure to compare each model, and Sect. 5 concludes the major findings of the work done.

2 Research Approach

2.1 Dataset Definition

The student performance dataset has 13 categorical, and 16 numeric and 04 nominal features. The categorical and nominal features have been numerically encoded. Table

Table 1 Dataset features

Demographic (5)	Family (7)	Personal (8)	Academic (13)
School	Famsize	Freetime	Nursery, higher
Sex	Medu	Gout	Absences, paid
Age	Fedu	Dalc	Famsup, schoolsup
Address	Mjob	Walc	Failures, Studytime
Guardian	Fjob	Health	Traveltime, reason
	Famrel Psize	Romantic	g1, g2, g3
		Internet	
		Activities	

1 depicts the four major categories of 33 features. It can be seen that the demographic parameter has five features that reflect the gender, age, school, locality, and guardian. Under the family, parent's education and job, family size and relation were found. For the personal parameter, student's health, free-time, outing, romantic, drinking frequency (daily and weekly), etc., were considered. In the canopy of the academic parameter, student's nursery schooling, failures, absences, grades, family and school support and others have been considered. All the features are considered as predictors, and the guardian is set to the class or response variable.

2.2 Dataset Preprocess

The student performance dataset has 17 categorical and 16 numeric features that are available on the UCI repository. Because of the mixed scale of features, normalization applied with a scale of 0–1 to implement experiments. Figure 1a initially, the guardian class has 455 instances of the mother, 153 instances of father and 41 of others. Figure 1b firstly, the instances belonging to others are separated and a total of instances are 608 and minority oversampling (smote) has been performed to enhance the father's instances and Fig. 1c displays the balance of father and mother class with a total of 910 instances. Later, 41 other class instances have been added into the balanced set of father and mother class depicted in Fig. 1d. Again, smote has been applied to make more virtual smote points with $k = 5$ nearest neighbor from the training dataset, and Fig. 1e shows the complete balancing of three classes and now 1365 instances are ready to train and test using classifiers.

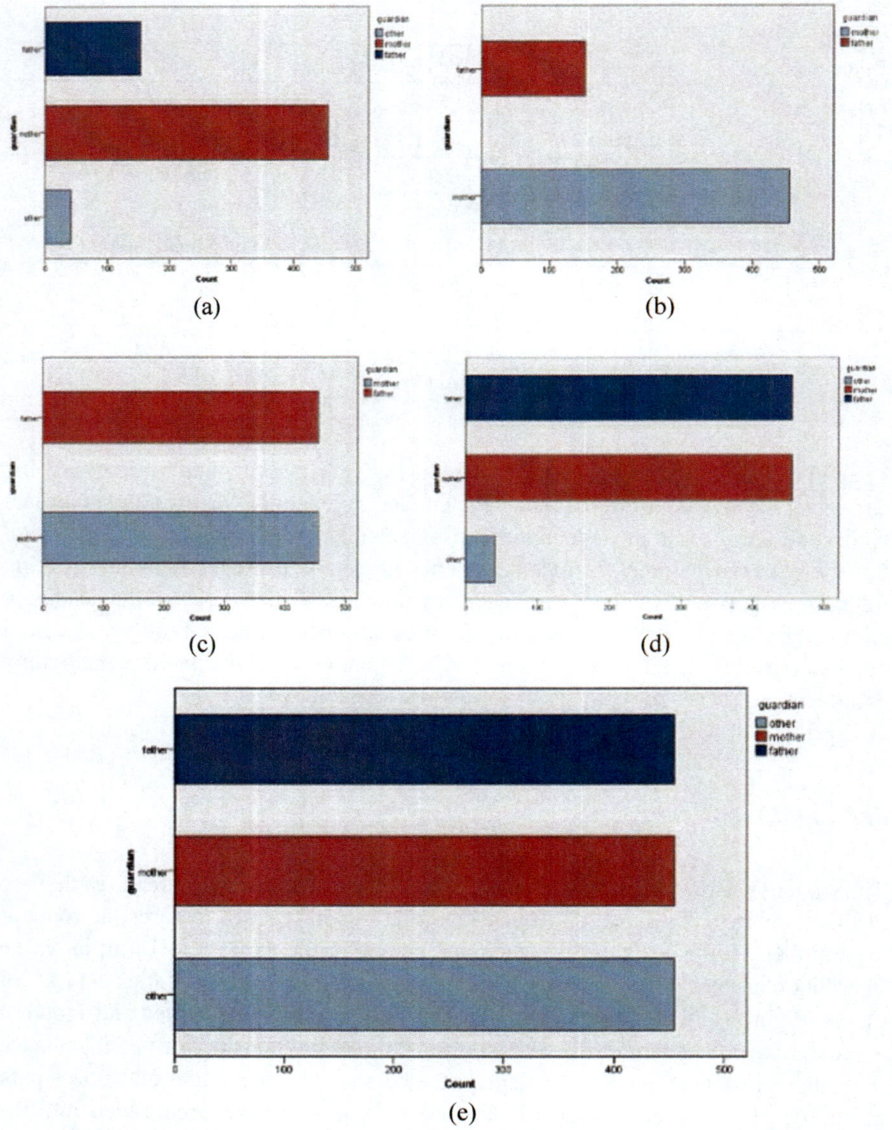

Fig. 1 **a** Initial dataset, **b** father and mother dataset, **c** balanced father and mother dataset, **d** merged balanced dataset with other, **e** full-balanced dataset

2.3 Algorithms and Tool

To analyze the dataset and deploy the models, the orange machine learning tool is used. To make each class balanced, the smote algorithm is used in the IBM SPSS modeler. To predict the student's guardian based on various features, three supervised learners applied. In orange, the attributes are loaded in the File widget. A *column widget* is used to set the guardian as the target attribute. The *test and score* widget used to make and compare the power of three learners with three testing methods. A *confusion matrix* widget is used to produce comparative matrices of the learner. A *calibration plot* displays probabilities of guardian prediction. The RF's parameters: number of trees = 10 and subset splitting >5 are used. A neural network with a multi-layer perceptron with backpropagation is used. The 100 neurons per hidden layers with rectified linear unit function (ReLu) activation function are used. A stochastic gradient-based optimizer as a weight solver method is considered. The alpha: L2 penalty as a regularization set to 0.00010. The highest iteration to train is set to 200. In the *SVM type*, the general C parameter is used, the cost parameter is 1, the *Numerical Tolerance* is 0.0010, *the Iteration limit* is 100. The *kernel* function is RBF which is a radial basis function kernel. The *regression loss* is 0.10.

2.4 Performance Measures

Following performance key metrics are applied and compared appropriately to test the predictive power of each model collectively using widgets.

1. Accuracy: It presents a total right prediction in terms of percentage.
2. Confusion matrix: It depicts the predicted versus actual instances.
3. Lift Curve: It plots the collective order of probability versus a joint number of TP rates.
4. Calibration Curve: It plots the contest between the prediction of the model's probability versus probability of actual class.
5. F1: It is a harmonic mean of recall and precision values of classification.
6. AUC: It is the area under the curve of the ROC curve.
7. Precision: It is the quantity of true positive records classified as positive.
8. Recall: It is the number of true positives among the whole positive records.
9. 1-Specificity: It is a false positive rate (FP rate) of records.
10. Test time: It is a time taken by the learner to identify the unseen record.

3 Experiments and Discussion

This section discusses the results of three various experiments performed in the machine learning tool Orange. The stabilized and preprocessed dataset has been trained and tested using with leave one out, hold out and *k*-fold methods. Firstly, dataset partitions achieved with training ratios of 50:50 which means that the 50% dataset records have been trained with testing the 50% records randomly. Secondly, a stratified cross-validation technique with standard split value $k = 10$ is applied which makes 10 equal subsets of the main dataset. Each learner algorithm has been tested with onefold (a group of records) at a time and the resultant tested dataset again tested with second fold and this process repeated for all the tenfold. Sometimes, it is called partially cross-validation. During the third experiment, we performed a full cross-validation method also known as leave one out which makes each record get a chance to become a test sample, and the rest of all are considered as a train set.

Figure 2 compares the accuracies of the student's guardian prediction about various testing techniques. The uppermost accuracy is 89% which is attained by NN with full cross-validation testing where $k = 1365$ and the SVM have the least prediction accuracy of 79.9% with the same testing approach. It has been also observed that fully cross-validation methods significantly enhanced the prediction accuracy except for the SVM. It seems that the holdout methods best suited to the SVM. The RF algorithm has an identical prediction accuracy of 81.4% with a k-fold and training ratio. However, the NN proved outperformer algorithm as compared to the SVM and the RF in the prediction of the guardian of the student.

Figure 3a shows the confusion matrix of NN, where the highest 1215 instances (i)

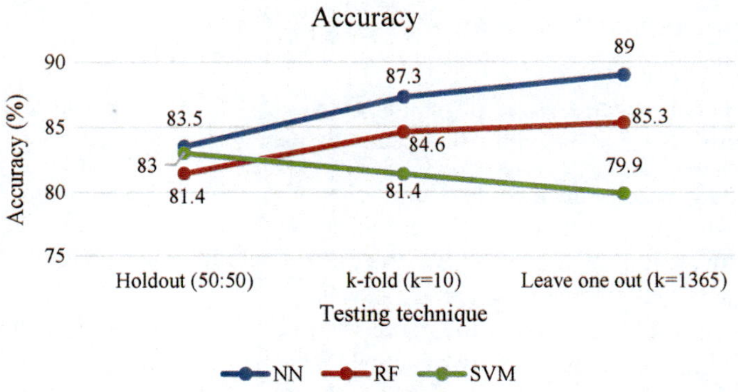

Fig. 2 Guardian prediction accuracy comparison

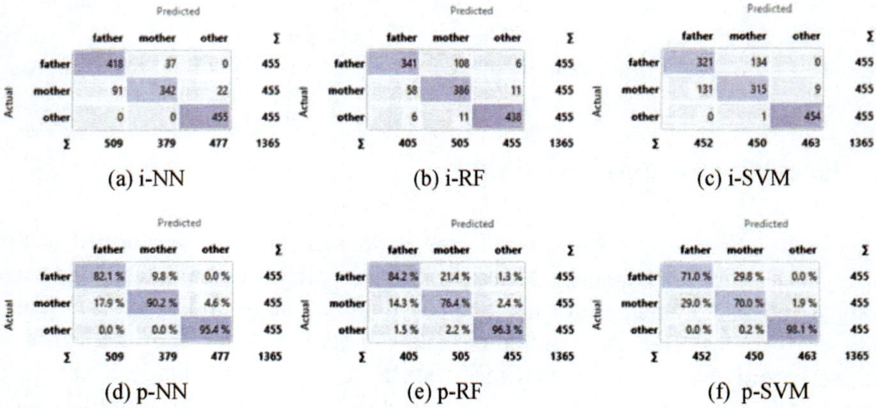

Fig. 3 Guardian prediction accuracy comparison

Table 2 Train versus test time at leave one out

Dataset	Training time (s)	Testing time (s)
NN	4687	23
RF	129	14
SVM	565	17

are classified accurately and 1165 instances correctly identified with the RF in Fig. 3b. The least accurate prediction of 1090 instances can be seen in Fig. 3c. Hence, the NN predicted maximum instances as compared to the RF and the SVM. Figure 3d depicts that in the NN classification, the highest proportion of 95.4% accurate prediction is computed in other class and the father class has the least 82.1% amount of right predicted instances. In Fig. 3e, for the RF, the mother class has the least proportion (p) of 76.4% right prediction, and the most proportion of 96.3% is measured in other classes. In Fig. 3d, no major difference has been found in both classes named mother and father considering the almost same proportion of 70 and 71%.

4 Performances Comparison

4.1 Time

The training time is the total time taken to train the model, and testing time is actual prediction time to predict the guardian-based unseen record inputting the learner. Table 2 keeps the two types of time in seconds (s) for each learner applied. If we compare the time, then NN is found slower than others in both training and testing time. The RF predictive model is found faster than the ANN and the SVM.

4.2 Probability

To compare the prediction probabilities of each learner with the real class probabilities with varying cutoff points, the calibration curve seems appropriate. The two rugs (stripped horizontal bar) signifies the right and wrong probabilities of learners. The leftmost side of the lowest rug keeps the low probability of the guardian class, and the right-side portion stores the erroneously allocated high probabilities. We set the threshold (p) is equal to 0.5 which spotted with a centered black dotted line. In both Fig. 4a, b, the NN learners have the highest prediction probability to identify the father (0.906) and mother (0.889) as compared to the SVM and the RF learners. In Fig. 4c, the SVM learner shows the maximum probability of 0.996 to predict the other class and the least probability of 0.977 computed by the RF.

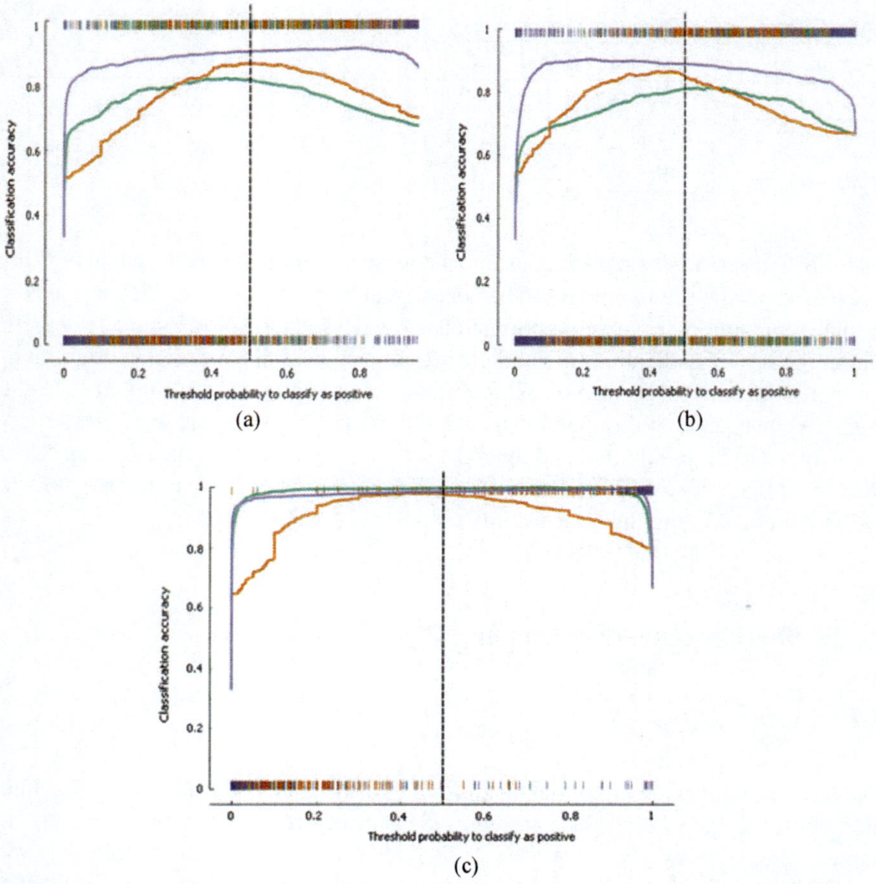

Fig. 4 Calibration curve of class, **a** father, **b** mother, **c** other

4.3 Performance Metrics

The *test and score* widget also important to provide others most important measures of classification. Table 3 stores the more results of first experiment provided with training ratios at 50:50 and compares each with specific learner algorithm. Almost

Table 3 Metric with stratified holdout 50:50

Dataset	F1	Precision	Recall	Specificity	AUC
NN	0.831	0.834	0.835	0.917	0.935
RF	0.812	0.813	0.813	0.906	0.936
SVM	0.829	0.828	0.830	0.915	0.943

Table 4 Metric with K-fold

Dataset	F1	Precision	Recall	Specificity	AUC
NN	0.870	0.874	0.873	0.936	0.957
RF	0.854	0.856	0.854	0.927	0.952
SVM	0.813	0.813	0.814	0.907	0.933

Table 5 Metric with leave one out

Dataset	F1	Precision	Recall	Specificity	AUC
NN	0.888	0.892	0.890	0.945	0.963
RF	0.849	0.851	0.848	0.924	0.950
SVM	0.798	0.797	0.799	0.899	0.932

all key-metrics are found maximum of NN learner as compared to the RF and the SVM.

Table 4 displays metrics of the second experiment held with k-fold with $k = 10$. On the one hand, the all measures belonging to the NN and the RF are significantly increased but on the another hand, SVM's metrics get reduced at tenfold.

Table 5 keeps the metrics of the third experiment held with full k-fold with $k = 1365$ where we tested each record as a test set with training data. Fortunately, again important increment is observed in the performance matrices of the RF and the NN learners except the SVM.

5 Conclusion

To identify the guardian of the student based on four major important parameters, machine learning played an energetic role. Major three experiments were executed with three testing techniques. The first experiment revealed that the NN and the SVM learner have the most prediction accuracy of 83.5% and 83%, respectively. Later, in the second experiment, the k-fold method enhanced the accuracy of each learner. At tenfold, the uppermost accuracy of 87.3% has been computed with an NN learner. In the third experiment, full cross-validation perfectly enhanced the proceeding accuracies of NN by 1.7% and the RF by 0.7%. The findings of all testing methods proved that the NN learner outperformed others. The NN learners have gained a maximum prediction probability of 90.6% to predict the father class and the probability 89% to recognize the mother class. Also, the SVM learner depicted the maximum probability of 99.6% to predict the other class. The NN and the RF have accurately predicted a total of 1215 and 1165 instances, respectively. It is also found that the NN classification has the highest proportion of 95.4% accurate prediction which has been computed in 23 s. From the accuracy prospective, the NN learner is proved outperformer and from the viewpoint of prediction time, the RF learner is a fast learner as compared to others.

Acknowledgements The first author's work was support by Hungarian Government and European Social Funds under the project "Talent Management in Autonomous Vehicle Control Technologies (EFOP-3.6.3-VEKOP-16-2017-00001)".

References

1. Fernández, D.B., Gil, D., Mora, S.L.: Application of machine learning in predicting performance for computer engineering students: a case study. Sustain. J. 1–18 (2019). https://doi.org/10.3390/su11102833
2. Koutina, M., Kermanidis, K.L.: Predicting postgraduate student's performance using machine learning techniques. In: IFIP Advances in Information and Communication Technology, vol. 364 (2011). https://doi.org/10.1007/978-3-642-23960-1_20
3. Verma, C., Illés, Z.: Attitude prediction towards ICT and mobile technology for the real-time: an experimental study using machine learning. In: The 15th International Scientific Conference eLearning and Software for Education, pp. 247–254, Romania (2019). https://doi.org/10.12753/2066-026X-19-171
4. Verma, C., Stoffová, V., Illés, Z.: Prediction of students' awareness level towards ICT and mobile technology in Indian and Hungarian University for the real-time: preliminary results. Heliyon **5**(6), 1–7 (2019a). https://doi.org/10.1016/j.heliyon.2019.e01806
5. Harinath, S., Prasad, A., Suma, H.S., Suraksha, A., Mathew, T.: Student placement prediction using machine learning. Int. Res. J. Eng. Technol. **6**(4), 4577–4579 (2019)
6. Manvitha, P., Swaroopa, N.: Campus placement prediction using supervised machine learning techniques. Int. J. Appl. Eng. Res. **14**(9), 2188–2191 (2019)
7. Bathla, Y., Verma, C., Kumar, N.: Smart approach for real time gender prediction of European School's principal using machine learning. In: Proceeding of ICRIC 2019, Lecture Notes in Electrical Engineering (LNEE), pp. 159–175 Springer, Berlin (2019). https://doi.org/10.1007/978-3-030-29407-6_14
8. Verma, C., Tarawneh, A.S., Stoffová, V., Illés, Z., Dahiya, S.: Gender prediction of the European school's teachers using machine learning: preliminary results. In: Proceeding of 8th IEEE International Advance Computing Conference, pp. 213–220, IEEE (2018). https://doi.org/10.1109/IADCC.2018.8692100
9. Verma, C., Stoffová, V., Illés, Z.: An ensemble approach to identifying the student gender towards information and communication technology awareness in european schools using machine learning. Int. J. Eng. Technol. **7**(4), 3392–3396 (2018)
10. Verma, C., Illés, Z., Stoffová, V.: Gender prediction of Indian and Hungarian students towards ICT and mobile technology for the real-time. Int. J. Innov. Technol. Explor. Eng. **8**(9S3), 1260–1264 (2019)
11. Verma, C., Stoffová, V., Illés, Z.: Ensemble methods to predict the locality scope of Indian and Hungarian students for the real time. In: Proceeding of ICACIE-2019, Advances in Intelligent Systems and Computing, pp. 1–13, Springer, Berlin (2020). (in press)
12. Verma, C., Stoffová, V., Illés, Z.: Real-time prediction of student's locality towards information communication and mobile technology: preliminary results. Int. J. Recent Technol. Eng. **8**(1), 580–585 (2019b)
13. Verma, C., Illés, Z., Stoffová, V.: Real-time classification of national and international students for ICT and mobile technology: an experimental study on Indian and Hungarian university. J. Phys.: Conf. Ser. **14032**, 1–8 (2020). https://doi.org/10.1088/1742-6596/1432/1/012091

14. Verma, C., Stoffová, V., Illés, Z.: Feature selection to identify the residence state of teachers for the real-time. In: IEEE International Conference on Intelligent Engineering and Management, pp. 1–6, Accepted, London (2020)
15. Verma, C., Stoffová, V., Illés, Z.: Prediction of residence country of student towards information, communication and mobile technology for real-time: preliminary results. Procedia Comput. Sci. **167C**, 224–234 (2020). In: Proceedings of ICCIDS-2019. Elsevier. https://doi.org/10.1016/j.procs.2020.03.213
16. Verma, C., Tarawneh, A.S., Illés, Z., Stoffová, V., Singh, M.: National identity predictive models for the real time prediction of European school's students: preliminary results. In: IEEE International Conference on Automation, Computational and Technology Management, pp. 418–423, IEEE (2019). https://doi.org/10.1109/ICACTM.2019.8776842
17. Verma, C., Stoffová, V., Illés, Z.: Age group predictive models for the real time prediction of the university students using machine learning: preliminary results. In: 2019 IEEE Third International Conference on Electrical, Computer and Communication, pp. 1–7 (2019). https://doi.org/10.1109/ICECCT.2019.8869136
18. Verma, C., Illés, Z., Stoffová, V.: Study level prediction of Indian and Hungarian students towards ICT and Mobile Technology for the real-time. In: IEEE International Conference on Computation, Automation and Knowledge Management, pp. 219–223, UAE (2020). (in press). https://doi.org/10.1109/ICCAKM46823.2020.9051551
19. Verma, C., Illés, Z., Stoffová, V.: Real-time prediction of development and availability of ICT and mobile technology in Indian and Hungarian university. In: Proceeding of ICRIC 2019, Lecture Notes in Electrical Engineering (LNEE), pp. 605–615. Springer, Berlin (2020). https://doi.org/10.1007/978-3-030-29407-6_43
20. Verma, C., Illés, Z., Stoffová, V., Singh, M.: ICT and mobile technology features predicting the university of Indian and Hungarian student for the real-time. In: IEEE System Modeling & Advancement in Research Trends, pp. 85–90, IEEE (2020). https://doi.org/10.1109/SMART46866.2019.9117414
21. Singh, M., Verma, C., Kumar, R., Juneja, P.: Towards enthusiasm prediction of Portuguese school's students towards higher education in real time. In: IEEE International Conference on Computation, Automation and Knowledge Management, pp. 427–431. IEEE (2020). https://doi.org/10.1109/ICCAKM46823.2020.9051459
22. Verma, C., Illés, Z., Stoffová, V.: Predictive modeling to predict the residency of teachers using machine learning for the real-time. In: Proceeding of FTNCT- 2019, Communications in Computer and Information Science (CCIS), pp. 592–601, Springer, Berlin (2020). https://doi.org/10.1007/978-981-15-4451-4_47
23. Singh, P.K., Kar, A.K., Singh, Y., Kolekar, M.H., Tanwar, S., Proceedings of ICRIC 2019, Recent Innovations in Computing, 2020, Lecture Notes in Electrical Engineering, vol. 597, pp. 3–920. Springer, Cham. https://doi.org/10.1007/978-3-030-29407-6
24. Singh, P.K., Bhargava, B.K., Paprzycki, M., Kaushal, N.C., Hong, W.C.: Handbook of wireless sensor networks: issues and challenges in current scenario's. Adv. Intell. Syst. Comput. **1132**, 155–437 (2020). https://doi.org/10.1007/978-3-030-40305-8
25. Tanwar, S., Bhatia, Q., Patel, P., Kumari, A., Singh, P.K., Hong, W.: Machine learning adoption in blockchain-based smart applications: the challenges, and a way forward. IEEE Access **8**, 474–488 (2020). https://doi.org/10.1109/ACCESS.2019.2961372

A Review on Enhanced Techniques for Multimodal Fake News Detection

Vidhu Tanwar and Kapil Sharma

Abstract This paper is a review of enhanced techniques for detecting the multimodal fake news. It helps to develop an insight into the characterization of a news story with different content types and its influence among the readers. We review different techniques on machine learning and deep learning with its merits and demerits. The paper is concluded with the open research challenges that can assist the upcoming researchers.

Keywords Fake news detection · Multimodal · Social medium

1 Introduction

Dissemination and detection of fake news of social media have become a major challenge for the past several years [1]. Due to the popularity gained by social media, millions of data are generated on a daily basis. Likewise, different devices are also associated with social media that led to an increased presence of data. Different social platforms like Twitter, YouTube, Facebook and so on will act as great information source to the media content developers. It shares the real-time data which helped for promoting the contents. These promotions may be spread as false news at some time constraints. Henceforth, social media has become a powerful tool for different sorts of journalism like sports, healthcare and political. Fig. 1 presents the overall workflow of the false news detection systems in social media [2].

V. Tanwar (✉) · K. Sharma
Delhi Technological University, Delhi 110042, India
e-mail: tanwar.vidhu22@gmail.com

K. Sharma
e-mail: kapil@ieee.org

P. K. Singh et al. (eds.), *Recent Innovations in Computing*, Lecture Notes in Electrical Engineering 701, https://doi.org/10.1007/978-981-15-8297-4_61

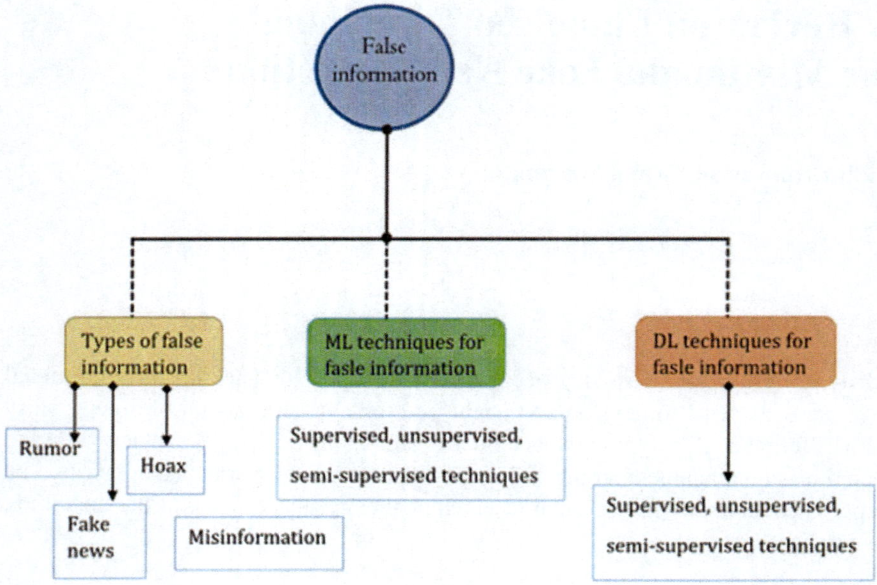

Fig. 1 False news detection system [1]

1.1 Types of Data in News

Since social media content is used, the content (or) data may be in the form of text, video and audio. Each type of data spread false information in four forms, namely [3].

1. Rumors: It is a piece of information where the information is insufficient with fewer trust factors. Mostly, it is created by panic toward public opinions of different credibility.
2. Fake news: In order to gain the attention of the readers, the false news are written and published.
3. Misinformation: Some inappropriate information is communicated as if it is true content.
4. Hoax: Most of the political news were hoaxed by journalists, so as to gain the attention of the news.

1.2 Types of Fake News

The general categorization of the fake news is classified as follows [4]:

1. Visual-based features: It includes graphical representation of the fake news like images, video and so on.

2. User-based features: Several fake accounts were created by social users. Thus, the target variables are included in age, groups, etc.
3. Post-based features: It appears in social media contents like tweets, memes, etc.
4. Network-based features: It describes the associativity between users and networks.
5. Propagation-based features: It describes how false sentences are written and published.
6. Knowledge-based features: It provides false content of unresolved issues.

1.3 Types of Dataset

Different sorts of datasets are available for the researchers. Here, commonly used datasets were given as follows [5]:

1. MediaEval dataset
2. Weibo Dataset
3. Kaggle dataset
4. Sina Weibo dataset
5. FakeNewsNet
6. r/ Fakeddit.

The rest of the paper is organized as follows: Sect. 2 presents reviews of existing studies; Sect. 3 presents the comparative analysis of existing studies; Sect. 4 presents the future challenges and scope and finally concludes in Sect. 5.

2 Review of Existing Studies

This section presents a review of the different studies that have developed various techniques for detecting fake news. In Zafarani and Zhou [6], the authors have surveyed different detection methods and opportunities. The study was covered by four perspectives, how the false knowledge proceeds, writing styles, propagation patterns and the credibility of the creators and the spreaders. Most of the fake news is not analyzed in the aspects of complex patterns and the network traffic data patterns. The author in Mahid et al. [7] presented a survey on different detection techniques of fake news on social media. Different aspects like content-based, social context-based and hybrid-based methods were analyzed under different influential factors of the users. The review was done for gaining knowledge in terms of interpretation, extraction and analysis of the detection mechanisms. In Jan, [8] provide an insight of clustering of tweets as spam or non-spam.

In Zhang et al. [9], the authors presented detector models using neural network systems. Some deceptive words in social media networks have greatly influenced social users. Due to the weakened communication coordination among diverse news articles, the extraction of latent features from text information is not analyzed. From the aspects of data mining techniques, the detection of false news in social media was explored by Shu et al. [10] They stated the importance of user engagements in social media to make determined infrastructure models. False news spreaders are not able to prevent for futuristic events. Cognition security (CogSec) [11] presented to explore the potential impacts of fake news by decision-making systems. Human-content cognition and social influence and opinion diffusion were explored on PolitiFact datasets. Open challenges are human potentiality model on fake news, social contagion and diffusion models, earlier detection models and the debunking of fake news.

In Gupta et al. [12], the authors presented the Community Infused Matrix-Tensor Coupled Factorization (CIMT) which detected the false news by the presence of echo chambers and mouse-click events. News Cohort Analysis and Collaborative News Recommendation were the two methodologies used for experimental purpose. Performance measures such as precision, recall and F1 score were analyzed on two datasets like Buzzfeed and PolitiFact and achieved an accuracy of 0.93406 (Buzzfeed) and 0.90626 (PolitiFact). Simpler detection process employed on large-scale profile information yet the complex fake news patterns are ignored. Online text may subject to different false news due to public opinion. Therefore, recurrent neural networks (RNN) was designed by organizing the contents. It was applied in LIAR datasets on different RNN models like LSTM, GRU and Vanilla. Compared to prior techniques, increased accuracy was achieved by LSTM (0.2166), Vanilla (0.215) and GRU (0.217).

The author in Cușmaliuc et al. [13] explained random forest, SVM and Naive Bayes algorithms in identifying the fake news, even for re-tweeted data. It was applied on Facebook and Twitter datasets. The system has achieved an accuracy of 92.43% (Naive Bayes); 95% (SVM) and 95.93% (random forests). Some incorrect classes have reduced the effects of feature mapping and higher error rate. In Li et al. [14], the authors discussed the rumor detection of social media from different aspects of datasets. It is observed that multitask learning models have performed better than the other learning models in the aspects of user credibility and source credibility.

Multimodal variational autoencoder (MVAE) was suggested by Khattar et al. [15] that detects the fake news via learning the probabilistic latent variable models. Most of the complex patterns are ignored, which was improved by multimodal representations. Experimental results have shown that the improvement of 6% in accuracy and 5% in F1 score. Some characteristics of the users are not explored under neural network architectures. A multimodal framework was developed by Singhal et al. [16] which exploited textual and visual features. Different language models like BERT were combined with VGG-19 pre-trained architecture on Imagenet datasets. The suggested model has achieved an accuracy of 77.77% (Twitter) and 89.23% (Weibo). Limited hidden layers are taken for analytic purpose. Zhou et al. [17] provided a similarity between text and images features which helps to identify fake news.

3 Comparative Analysis

Finally, a comparative table is developed based on the studies discussed above. The studies are compared based on the objectives, techniques, results obtained, the demerits of further research (Table 1).

4 Research Challenges and Future Scope

The open research challenges are

1. Datasets in multimodal: Most of the public repositories contain variants of fake news. It is also an open challenge for focusing the research objects that covers all news data types.
2. Verification models on multimodal data: Different linguistic models were designed for detecting [8] the false news. Visual-based features are difficult to recognize false news.
3. Source verification: Origin of the fake news is not explored by any researchers.
4. Credibility assessment: Chain of false news under the same (or) different authors is not studied from the aspects of propagation and knowledge-based features.

5 Conclusion

Presence of fake news has been around over several years with the intervention of social media and the journalisms. This paper aims to present an insight into false news characteristics, and its techniques designed are reviewed. Given the challenges related to detecting the false news have made the researchers, to understand the fundamentals of those origins of fake news. The comparative analysis will help the upcoming researchers of open challenges in this field.

Table 1 Comparative analysis of fake news detection methods

References	Objectives	Techniques	Obtained results with merits	Demerits
Cao et al. [18]	To detect false news in social media using fusion approaches	Here, false news was detected for both image and text data. Different semantics approaches like LSTM, GRU, TextCNN, Bert, MVNN and att-RNN	Performance metrics such as accuracy, precision, recall and F1. The accuracy obtained as 0.864 (LSTM); 0.857 (GRU); 0.851 (TextCNN); 0.867 (BERT); 0.728 (pre-trained VGG19); 0.759 (fine-tuned VGG19); 0.805 (MVNN); 0.876 (early fusion); 0.846 (late fusion) and 0.852 (attRNN)	Limited benchmark datasets
Kaur et al. [19]	To detect fake news of social media by ensuring the verification process via multi-voting model	Term frequency-inverse document frequency; count-vectorizer and hash-vectorizer were used as a feature extraction process. Then, passive aggressive (PA), logistic regression (LR), linear support vector (LSV) and linear SVM were used for classification purpose	Suggested models were applied on three datasets, namely, News Trends, Kaggle and Reuters. Performance metrics such as accuracy, precision, recall, F1 score and specificity. Multi-voting model is the novel approach employed. The news trends datasets have achieved an accuracy of 94.5 (Tf-IDF); 93.6 (CV); 87.1 (HV). Kaggle datasets have achieved an accuracy of 98.9 (Tf-IDF); 98.7 (CV); 95.8 (HV). Likewise, Reuters datasets have achieved an accuracy of 97.2 (Tf-IDF); 96.5 (CV); 90.2 (HV)	Though the system has improved the detection accuracy, the efficiency of the detection classifier is not explored If the input size increases, then the system lowered the efficiency of the classifier

(continued)

Table 1 (continued)

References	Objectives	Techniques	Obtained results with merits	Demerits
Orlov and Litvak [20]	To develop a user behavior model on detecting the false news on Twitter	An unsupervised approach was employed here. Clustering and frequent itemset mining were used for constructing the classifiers	The suggested classifier was studied in military airstrikes in Syria in Sep. 2017. For eight clusters, the system has achieved 100% precision	Lack of geolocation prediction and analysis
Jwa et al. [21]	To detect automatic fake news by improving pre-training classifiers	Bidirectional Encoder Representations from Transformers (BERT)	CNN and daily mail datasets were used for analytic purpose. Performance measures analyzed are precision, recall and f1 score. Compared to prior algorithms, 0.14 F1 score was improved	Data imbalance issue arises, when the authenticity of the data is altered
Zhou, et al. [22]	To detect fake news earlier by theory-based approaches	News content analyzed at four levels, namely, lexicon-level, syntax-level, semantic-level and discourse-level. Then, a supervised approach was framed to classify the contents	Semi-supervised classifiers such as SVM, random forest and XG Boost were used for study purpose. PolitiFact & Buzzfeed datasets were for experimental purpose. The suggested model has achieved 0.892 (accuracy); 0.877 (precision); 0.908 (recall) & 0.892 (F1-score) for PolitiFact dataset. Likewise, Buzzfeed dataset has helped for achieving 0.879 (accuracy); 0.85 (precision); 0.902 (recall) & 0.879 (F1-score)	Interpretation of the data and its relationships are not effectively approached Some complex news data are ignored for study purpose

(continued)

Table 1 (continued)

References	Objectives	Techniques	Obtained results with merits	Demerits
Balwant [23]	To detect the fake news of different sources of social media	Improved part of speech (POS) bidirectional LSTM and convolutional neural networks	Liar-Liar datasets were used on this hybrid model LSTM and CNN and achieved an accuracy of 42.2% with gain 3.3%	Some learning patterns of the news content are difficult to formalize the hidden layers
Volkova et al. [24]	To detect fake news on different multimodals deceptive systems	Different neural networks architecture was used for study purpose. AdaBoost and NN models were explored	The class with the highest incorrect prediction in this manner is disinformation (40.08% of tweets) followed by the conspiracy (39.13%) and propaganda (37.45%). The least incorrectly predicted class is satire (0.72%), then hoax (2.19%), verified (5.55%), and clickbait (11.26%). Between all collections, about 31.5% of tweets fool all of our models in this way	The false-positive rate is higher in the combination of disinformation and propaganda posts
Parikh and Atrey [25]	To study about the media-rich fake news detection models	Surveyed about the characterization of a news story of different c Parikh and Atrey [25] Ontent types	This paper has provided better insights into fake news detection systems	
Zhang et al. [9]	To detect the fake news and also developed as detecting tool	A novel FakeDetector model was suggested by deep diffusive neural networks models which extracted latent features	Compared to prior algorithms, the suggested model has achieved an accuracy of 14.5% higher Availability of the content made the classifiers for all heterogeneous data models	Though it has taken multiple inputs, the extraction of the relevant features is still in underdevelopmental process

(continued)

Table 1 (continued)

References	Objectives	Techniques	Obtained results with merits	Demerits
Wang et al. [26]	To detect the fake news, even for newly arrived events	Event adversarial neural networks that composed of three components, namely, multimodal feature extractor, the fake news detector and the event discriminator. Features-related text, visual and shared data were analyzed for pre-training the classifiers	Twitter and Weibo datasets were used for experimental purpose. Performance measures such as accuracy, precision and f1-score. Twitter datasets have achieved 0.715 (accuracy); 0.822 (precision); 0.638 (recall) and 0.719 (F1 measure)	Some features were not available for different semantic measures
Jin et al. [27]	To develop a multimodal fusion on different microblogs	Recurrent neural networks with an attention mechanism (Att-RNN)	Weibo and Twitter datasets have analyzed for all social context features. Weibo dataset helped to achieve 0.788 accuracies, whereas 0.682 achieved by Twitter datasets	Some complex data patterns are not analyzed

References

1. Habib, A., Asghar, M.Z., Khan, A., Hsbib, A., Khan, A.: False information detection in online content and its role in decision making: a systematic literature review. Soc. Network Anal. Mining **9**(1), 50 (2019)
2. Guo, B., Ding, Y., Sun, Y., Ma, S., Li, K.: The Mass, Fake news, and cognition security. ArXiv preprint arXiv:1907.07759. (2019)
3. Tandoc Jr., E.C., Wei, L.Z., Richard, L.: Defining "fake news" a typology of scholarly definitions. Dig Journalism **6**(2), 137–153 (2018)
4. Victoria, L., Rubin, C.Y., Conroy, N.J.: Deception detection for news: three types of fakes. Proc Assoc Inf Sci Technol **52**(1), 1–4 (2015)
5. Ruchansky, N., Seo, S., Liu, Y.: Csi: A hybrid deep model for fake news detection. In: Proceedings of the 2017 ACM on Conference on Information and Knowledge Management, pp. 797–806 (2017)
6. Zafarani, R., Zhou, X.: Fake news: a survey of research, detection methods, and opportunities. ArXiv preprint arXiv:1812.00315 (2018)
7. Mahid, Z.I., Selvakumar M., Karuppayah, S.: Fake news on social media: brief review on detection techniques. In: 2018 Fourth International Conference on Advances in Computing, Communication & Automation (ICACCA), pp. 1–5. IEEE (2018)
8. Jan, T.G.: Clustering of tweets: a novel approach to label the unlabelled tweets. In: Proceedings of ICRIC 2019, pp. 671–685. Springer, Cham (2020)
9. Zhang, J., Cui, L., Fu, Y., Gouza, F.B.: Fake news detection with deep diffusive network model. ArXiv preprint arXiv:1805.08751 (2018)
10. Shu, K., Sliva, A., Suhang, W., Jiliang, T., Huan, L.: Fake news detection on social media: A data mining perspective. ACM SIGKDD Explor. Newsl **19**(1), 22–36 (2017)
11. Guo, B., Yasan, D., Lina, Y., Yunji, L., Zhiwen, Y.: The future of misinformation detection: new perspectives and trends. ArXiv preprint arXiv:1909.03654 (2019)
12. Gupta, S., Thirukovalluru, R., Sinha, M., Mannarswamy, S.: CIMTDetect: a community infused matrix-tensor coupled factorization based method for fake news detection. In: 2018 IEEE/ACM International Conference on Advances in Social Networks Analysis and Mining (ASONAM), pp. 278–281. IEEE (2018)
13. Cuşmaliuc, C.-G., Coca, L.-G., Iftene, A.: Identifying fake news on twitter using naive bayes, SVM and random forest distributed algorithms. In: Proceedings of The 13th Edition of the International Conference on Linguistic Resources and Tools for Processing Romanian Language (ConsILR-2018). ISSN, pp. 177–188 (2018) (1843)
14. Li, Q., Zhang, Q., Si, L., Liu, Y.: Rumor detection on social media: datasets, methods and opportunities. ArXiv preprint arXiv:1911.07199 (2019)
15. Khattar, D., Singh, G.J., Manish, G., Vasudeva, V.: Mvae: multimodal variational autoencoder for fake news detection. In: The World Wide Web Conference, pp. 2915–2921 (2019)
16. Singhal, S., Shah, R.R., Chakraborty, T., Kumaraguru, P., Satoh, S.: SpotFake: a multimodal framework for fake news detection. In: 2019 IEEE Fifth International Conference on Multimedia Big Data (BigMM), pp. 39–47. IEEE (2019)
17. Zhou, X., Wu, J., Zafarani, R.: SAFE: similarity-aware multi-modal fake news detection. ArXiv preprint arXiv:2003.04981 (2020)
18. Cao, J., Sheng, Q., Qi, P., Zhong, L., Wang, Y.: False news detection on social media. ArXiv preprint arXiv:1908.10818 (2019)
19. Kaur, S., Kumar, P., Kumaraguru, P.: Automating fake news detection system using multi-level voting model. Soft Comput. 1–21 (2019)
20. Orlov, M., Litvak, M.: Using behavior and text analysis to detect propagandists and misinformers on twitter. In: Annual International Symposium on Information Management and Big Data, pp. 67–74. Springer, Cham (2018)
21. Jwa, H., Dongsuk, O., Park, K., Ka, J.M.: exBAKE: automatic fake news detection model based on bidirectional encoder representations from transformers (BERT). Appl. Sci. **9**(19), 4062 (2019)

22. Zhou, X., Jain, A., Phoha, V.V., Rez: Fake news early detection: a theory-driven model. ArXiv preprint arXiv:1904.11679 (2019)
23. Balwant, M.K.: Bidirectional LSTM Based on POS tags and CNN architecture for fake news detection. In: 2019 10th International Conference on Computing, Communication and Networking Technologies (ICCCNT), pp. 1–6. IEEE (2019)
24. Volkova, S., Ayton, E., Arendt, D.L., Huang, Z.: Explaining multimodal deceptive news prediction models. In: Proceedings of the International AAAI Conference on Web and Social Media, vol. 13, no. 01, pp. 659–662 (2019)
25. Parikh, S.B., Atrey, P.K.: Media-rich fake news detection: a survey. In: 2018 IEEE Conference on Multimedia Information Processing and Retrieval (MIPR), pp. 436–441. IEEE (2018)
26. Wang, Y., Fenglong, M., Zhiwei, J., Ye, Y., Guangxu, X., Kishlay, J., Su, L., Jing, G.: Eann: Event adversarial neural networks for multi-modal fake news detection. In: Proceedings of the 24th ACM SIGKDD International Conference on Knowledge Discovery & Data Mining, pp. 849–857 (2018)
27. Jin, Z., Juan, C., Han, G., Yongdong, Z., Jiebo, L.: Multimodal fusion with recurrent neural networks for rumor detection on microblogs. In: Proceedings of the 25th ACM International Conference on Multimedia, pp. 795–816 (2017)

An Intelligent Student Hostel Allocation System Based on Web Applications

Ambrose Azeeta, Sanjay Misra, Modupe Odusami, Onyepunuka Ugochukwu Peter, and Ravin Ahuja

Abstract Student hostel management system is a software programme designed to manage the activities of allocating students to a hostel and other activities involved in managing the students in the hostel. Managing the student's hostel allocation is a complex task. This study develops an algorithm and techniques for automatic student hostel allocation system based on Web applications that would allocate student to the hall and room-based certain constraints. System is faced with lot of challenges. The system will use MySQL for the manipulation and storage of data, PHP to create dynamic Web pages and Sublime Text as the integrate development environment/text editor for HTML/CSS and PHP. The proposed algorithm and techniques for automatic student hostel allocation system will perform the task of allocating student to hostel rooms in a timely fashion based on the defined constraints. The algorithms and techniques were implemented and validated. Result shows that the proposed model provides information about the room occupancy at any given time, enabling the management in making decisions to improve condition of living in hostels and grants the hostel management team the statistics they need on the hostel.

Keywords Algorithm · Hostel management · Hostel security · Sudents management · Web application

---

A. Azeeta · S. Misra (✉) · M. Odusami · O. U. Peter
Covenant University, Ota, Ogun State, Nigeria
e-mail: sanjy.misra@covenantuniversity.edu.ng

A. Azeeta
e-mail: ambrose.azeta@covenantuniversity.edu.ng

M. Odusami
e-mail: modupe.odusami@covenantuniversity.edu.ng

O. U. Peter
e-mail: Onyepunuka.Ugochukwupeter@covenantuniversity.edu.ng

R. Ahuja
Shri Vishwakarma Skill University, Gurgaon, India
e-mail: ravinahujadce@gmail.com

P. K. Singh et al. (eds.), *Recent Innovations in Computing*, Lecture Notes in Electrical Engineering 701, https://doi.org/10.1007/978-981-15-8297-4_62

1 Introduction

Web-based applications are seen as applications that are distributed over a network on a Web browser which can be found on different mobile phones, smartphones, tablets, desktop, laptops, etc. [1]. It is a software that uses the client server architecture where the client-side logic and the user interface run on a Web browser. Some of the common capacities of that can be found on a Web application are online sales, Webmail, push notifications and many more different capacities using the Internet and mobile network which can be accessed anywhere and anytime with great flexibility and ease [2, 3]. A student is primarily a person enlisted in a school or other instructive establishment who goes to classes in a course to accomplish the suitable level of authority of a subject under the direction of an educator. In the more extensive sense, an understudy is any individual who puts forth a concentrated effort to the serious scholarly commitment with some issue important to first-rate it as a major aspect of some purposeful undertaking in which such authority is crucial or definitive [4]. One of the efficient ways to manage the sudents sucessfully is the provision of good housing.

There are a lot of issues with the allocation process that could be easily avoided. Examples of such issues are students being allocated to room which are in a bad state, students who are not physically enabled to use stairs up to the fourth floor and students allocated roommates who are bullies, students allocated a room far from their course mates, thereby limiting information of activities that would take place in class. The identification of the drawbacks of the existing system leads to the development of a computerized system that would be compatible with the existing system, user-friendly and more graphical user interface (GUI) oriented.

The motivation for this research is hinged on the need to make hostel allocation system a means of providing a clearer understanding of room habitable and provide necessary information, thereby enabling the management in making decisions. Many approaches have been developed to solve HSAP, rule-based model [5, 6] heuristics model [7–9] which may search optimal solutions within reasonable computational time but the solution tend to stuck in local optimal and thus may not be effective for domain-specific knowledge [10] metaheuristic model [11–13]. In this paper, we propose algorithms and techniques that is a Web-based application that would make use of MySQL Database for the database manipulation, PHP for the application tier it receives input from the users and sends the information needed to the data tier, HTML/CSS for the front end and Sublime Text as the integrated development environment. This study aims at providing an automated system that allocates student to hostels using decision tree algorithm university. The rest of the paper is sectioned as follows: the next section discusses the related works. Section 3 presents the step-by-step methods and algorithms of the proposed model. Section 4 presents the implementation of the model, followed by results analysis and discussion. Section 5 concludes and suggests future areas of research.

2 Related Works

Olantunji [5] introduced a model for hostel accomodation system based on fuzzy inference in decision making which was implemented using server side scripting language (PHP) in conjunction with MySQL being used as the relational database and Apache serving as the Web server. This model used fuzzy logic rather than Boolean logic in making decision. Ajibola [14] utilized heuristics approach to solve hostel space assignment problem (HSAP) in higher establishments of learning. SAP is a combinatorial streamlining issue that includes the dissemination of spaces accessible among an arrangement of meriting substances (rooms, bed spaces, and office spaces and so forth.), with the goal that the accessible spaces are ideally used and agreed to the given arrangement of imperatives. Hill climbing (HC), simulated tempering (SA), tabu search (TS), late acceptance hill climbing (LAHC) and GA were connected to circulate the understudies at all the three levels of distribution. At each level, a correlation of the calculations is displayed. Results acquired with the datasets additionally show the suitability of applying tried activities explore strategies effectively to tackle new occurrence of SAP. Authors in [15] employed genetic algorithm-based optimization model for urban land use allocation by considering multi-objective function for planners. The model simultaneously maximizes land prices and reduces incompatibility among adjacent land uses for an area. The authors concluded that more heuristics approach could be considered for future. Yip et al. [16] introduced an integrated reservation optimization and operations management system to enhance the management of hospital nurse quarters based on certain constraints and objective functions. The system allowed an overall better management of the Nurses' Quarters. However, the system is not flexible owing to the fact that it does not allow the choices. Adewumi and Ali [17] presented a multi-stage genetic algorithm to solve HSAP that is multi-stage in nature. Result shows that the use of heuristics can be leverage on for automated hostel allocation and the use of metaheuristics and their variants for HSAP were recommended for future research. Obit et al. [10] utilized stochastic algorithms on hostel space allocation problem where the background problems are related with hard constraints and soft constraints. İn order to enhace the quality of the proposed system, Great Deluge with linear and nonlinear decay rate and simulated annealing with linear reduction are used. Result shows that simulated annealing with linear reduction temperature performs better. Authors in [18] presented a how ICT impacts on management of Students' Records in University Administration. Patnaik [19] utilized a Web-based application targeted to all sectors of management thereby providing information within a college or university. Based on their level of access, a student or a staff can download or upload file from the database. The system can be divided into six modules (admin, student, login/signup, payment, room allotment, complaint) and each module can be developed independently. Authors in [20] discussed the use of Web application/Web-based information for hotel reservation system which allows the hotels to manage their sales rate, available services and marketing to be available to sales channels. In some other related works, authors in [21] proposed a framework for allocation of work among the software engineers distributed globally.

3 Model Approach

The approaches utilized to carry out the proposed algorithm are discussed in this section.

3.1 Functional Requirements of the Proposed System

The functional requirements of the student hostel allocation system include:

1. Users (Applicants) must be able to register, login and log out of the system.
2. Users (Applicants) must be able to view their allocation status.
3. The system must be able to allocate rooms to student based on the criteria provided to enhance co-existence between roommates.
4. Users (Admin) must be able to upload student details to the system.
5. Users (Admin) must be able to view students details and hostel allocation arrangement.
6. Users (Admin) must be able to generate allocation report for the various hall containing the list of students in each room.
7. Users (Admin) must be able to view multiple complaints that have been logged into the system.

3.2 Non-functional Requirements of the Proposed System

The non-functional requirements of the student hostel allocation system include:

1. Accuracy: The system must allocate room accurately based on the criteria's used in the decision tree algorithm.
2. Excellent human computer interface: The system must be user-friendly providing optimal satisfaction to the user with easy navigations in and out of different parts of the system as well as ease of use to the user.
3. Speed: The system must respond rapidly to users request and provide the appropriate result.
4. Security: The system data should be protected against unauthorized users by using appropriate measure (authentication and password encryption).

The proposed system model is represented using graphical notations to illustrate how the system works which is mostly based on the notations in the Unified Modelling Language (UML). The (UML) Unified Modelling Language is a broadly useful visual displaying framework that gives traditions used by a software engineer to indicate, imagine, build and archive the result of a system. The proposed student hostel allocation system utilizes the following UML: use case diagram, sequence diagram, activity diagram and class diagram. The use case diagrams utilized for the proposed system are depicted in Figs. 1 and 2.

From Figs. 1 and 2, the use case diagrams specify the operations that the admin in charge of the system and the students are permitted to perform respectively. Tables 1, 2, 3, 4, 5, 6, 7, 8, 9 and 10 show various use case scenerio that can be used in the system design.

From Table 1, the student login use case scenario is detailed and explained.

From Table 2, the view room concise details are described.

From Table 3, the upload students details use case scenario is described.

From Table 4, the add session use case scenario is explained.

From Table 5, the view room status use case scenario is explicitly described. It indicates certain parameters and conditions necessary for an admin to view the conditions of a room and also to see the total number of allocated rooms.

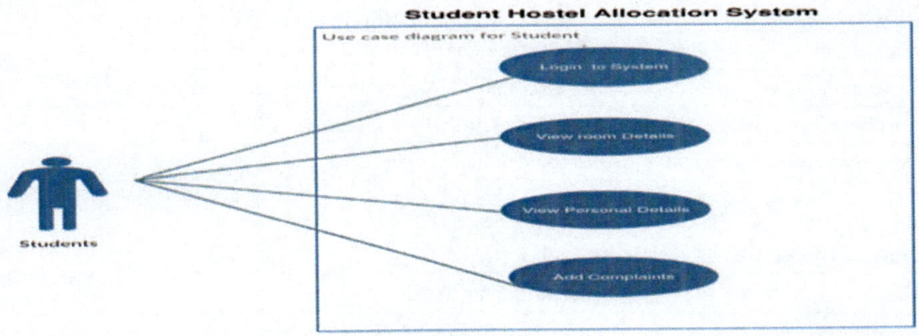

Fig. 1 Use case diagram for student

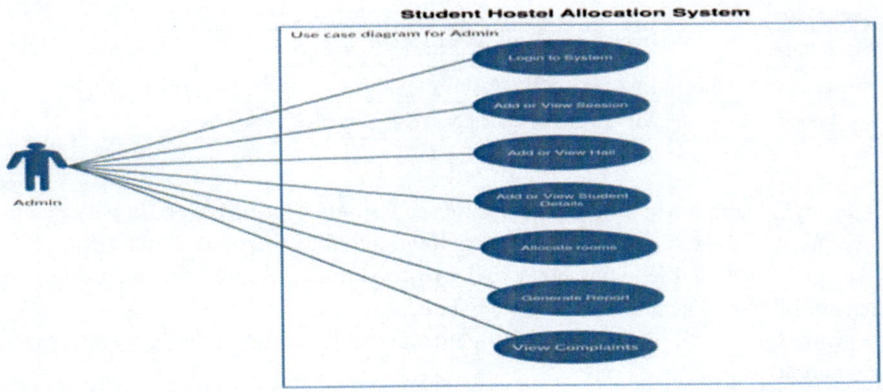

Fig. 2 Use case diagram for admin

Table 1 Login use case scenarios

Use case 1	Students login to the system
Brief description	The students are to login to the system using their login details
Actor(s)	User (students)
Level	Students use case
Parameters	Students login details
Pre-conditions	User (student) must provide the correct login details
Post-conditions (success end)	User has successfully logged in to the system
Post-conditions (failed end)	Provided incorrect details user not found

Table 2 View room details use case scenarios

Use case 2	View personal room details
Brief description	The student can view his/her roommates as well as his room number
Actor(s)	User (students)
Level	A simple view room details use case
Parameters	Matriculation number
Pre-condition(s)	The user must be logged in to view job opening
Post-conditions (success end)	User successfully views room details
Post-conditions (failed end)	No room assigned to user

Table 3 Upload students details use case scenario

Use case 3	Upload students details
Brief description	To successfully upload the details of students
Actor(s)	User (Admin)
Level	A simple upload students details use case
Parameters	The required details needed for a student to be allocated rooms
Pre-conditions	All necessary parameters must be filled before upload
Post-conditions (success end)	User successfully uploads the student information
Post-conditions (failed end)	Student already exist, incomplete form filled

From Table 6, the admin login use case scenario is described. Certain parameters and conditions necessary for an admin to login into the system are explained.

From Table 7, the parameters and pre-conditions required for an admin to successfully view a students record are stated.

From Table 8, the admin can only generate results based on certain parameters and conditions (Table 11).

Table 4 Add session use case scenario

Use case 4	Add session
Brief description	Add a new session to the system before allocation begins
Actor(s)	User (Admin)
Level	A simple add session use case
Parameters	The name of the new session to be added
Pre-conditions	The admin must be logged in to add session and session must no be in existence
Post-conditions (success end)	User successfully adds session
Post-conditions (failed end)	Failed to add session, session already exist

Table 5 View room status case scenario

Use case 5	View room status
Brief description	The admin can view the conditions of a room and can also see those allocated the room
Actor(s)	User (Admin)
Level	A simple view room status use case
Parameters	Room number
Pre-conditions	The admin must be logged in
Post-conditions (success end)	The admin successfully views the room status
Post-conditions (failed end)	Room number does not exist

Table 6 Admin login use case scenario

Use case 6	Admin login to system
Brief description	The admin logs in to the system using their login details
Actor(s)	User (Admin)
Level	Admin login use case
Parameters	Admins' login details
Pre-conditions	The admin must have the correct details
Post-conditions (success end)	Admin successfully logs into the system
Post-conditions (failed end)	Incorrect user details user not found

Table 7 View students use case scenario

Use case 7	View students
Brief description	The admin can view all the students in the database to be allocated rooms and their information
Actor(s)	User (Admin)
Level	A simple view student use case
Parameters	Student name or matric number
Pre-conditions	The user must be an admin and must be correctly logged in

Table 8 Generate report use case scenario

Use case 8	Generate report
Brief description	The admin can generate a report of the allocation of the students
Actor(s)	User (Admin)
Level	This is a simple generate report use case
Parameters	Based on the session and semester
Pre-conditions	The admin must be correctly logged in
Post-conditions (success end)	Admin generates a report successfully from the system
Post-conditions (failed end)	Failed to generate report

Table 9 Student lodge complaints use case scenario

Use case 9	Student lodge complaints
Brief description	The student can make several complaints on the system
Actor(s)	User (student)
Level	A simple student adds complaints use case
Pre-conditions	The user must be logged in as a student
Post-conditions (success end)	User adds complaint successfully
Post-conditions (failed end)	Failed to add complaints

Table 10 Admin view complaints use case scenario

Use case 10	Admin view complaints
Brief description	The admin can view, edit the various complaints made by students
Actor(s)	User (admin)
Level	A simple admin view complaint use case
Parameters	The time of the complaint
Pre-conditions	The user must be correctly logged in as admin

Table 11 Admin allocates room use case scenario

Use case 11	Admin allocates rooms
Brief description	The admin allocates rooms to students based on the criteria's checked
Actor(s)	User (Admin)
Level	Admin allocates rooms use case
Parameters	Criteria's for which the allocation is based
Pre-conditions	User must select the criteria for allocation
Post-conditions (success end)	Admin successfully allocates room
Post-conditions (failed end)	Allocation failed

3.3 Web Application Platform of the Proposed System

The proposed system utilized a three–tier architecture consisting of the client tier, the application tier and the data tier. The three tiers are designed using HTML and CSS., PHP, and MySql, respectively. The brief description of the hall database for the proposed system is depicted in Table 12.

Brief description of the room and hall database is displayed in Tables 13 and 14, respectively.

Table 12 Allocate_hall database

Name	Type (length)	Description
Allocate_hall_id	Int (11)	
Level	Int (11)	This stands for the level who has been allocated the hall
Sex	Varchar (1000)	This stands for the gender who the hall has been located which has only two variable male and female
Hall_id	Int (11)	This is the id for the hall that was allocated
Semester_id	Int (11)	This stands for the id of the semester the allocation took place

Table 13 Allocate_room database table

Name	Type (length)	Description
Allocate_room_id	Int (11)	This is the allocation id for a room that has been allocated
Allocate_hall_id	Int (11)	This is the allocation id for the hall that has been allocated
Student_id	Int (11)	This is the student id for the student that has been allocated a room
Hall_id	Int (11)	This is the standard id of the halls
Room_id	Int (11)	This is the standard id for a room in a hall
Level	Varchar (1000)	The level of the student that has been allocated a room
Semester_id	Int (11)	The is the standard id for the semester allocation took place
Status	Int (11)	This is the status of the room which was allocated usually in 0 s (bad) and 1 s (good)

Table 14 Halls database table

Name	Type (length)	Description
Hall_id	Int (11)	This is the standard id of the halls generated once added to the database
Hall_name	Varchar (1000)	This is the name of the hall that is in the database
Gender	Varchar (1000)	This stands for the gender that the hall can be allocated to
Status	Int (11)	This is the status of the hall usually in 0 s (bad) and 1 s (good)

Table 15 Sudent database

Name	Type (length)	Description
Student_id	Int (11)	This represents the id given to the student once added to the database
Name	Varchar (1000)	This is the full name of the student
Level	Varchar (1000)	This represents the level the student is currently in
Mat_no	Varchar (1000)	This represents the student matriculation number
Programme	Varchar (1000)	This represents the course of study of the student
GPA	Varchar (1000)	This represents the grade point average of the student
Bully_record	Varchar (1000)	This represent whether a student has a bully record or not. It is represented in form of 0 s (no) and 1 s (yes)
Personality	Varchar (1000)	This represents the student personality type for allocation
Disability	Varchar (1000)	This represent whether a student has physical disability or not. It is represented in form of 0 s (no) and 1 s (yes)
Sex	Varchar (1000)	This represents the gender of the student in the database
Password	Varchar (1000)	This is the password the student uses to login to the system
Status	Int (11) *Database table*	This represents whether the students have been allocated a room

A brief description of student databbase is depicted in Table 15.

4 Results and Discussion

The system modules and how it operates are discuss below:

The login page (Admin) is the first page that opens to an admin as the admin logs on to the website. This page is used to receive and confirm the administrator credentials before granting access to the system. Figure 3 depicts a screenshot of the login page. The administrator dashboard shows the possible operations the admin can perform on the system along with some basic information such as the number of students, number of halls and number of rooms. A screenshot of the administrator dashboard is shown in Fig. 4. The view student page displays the list of students in the database in a tabular format and allows the administrator to export the list in file format of his/her choice. Figure 5 shows the screenshot of the view student page.

The student personal info page displays the personal information of students in the database, e.g. name, matriculation number, bully record and personality type. A screenshot of the student personal info page is shown in Fig. 6. The allocate hall page is the page where the admin selects which hall is allocated to a particular gender as depicted in Fig. 6.

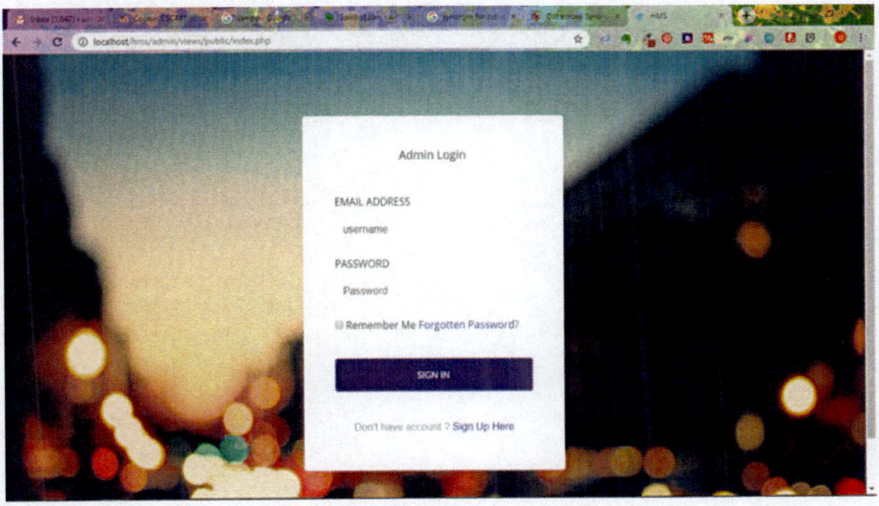

Fig. 3 Showing the admin home page

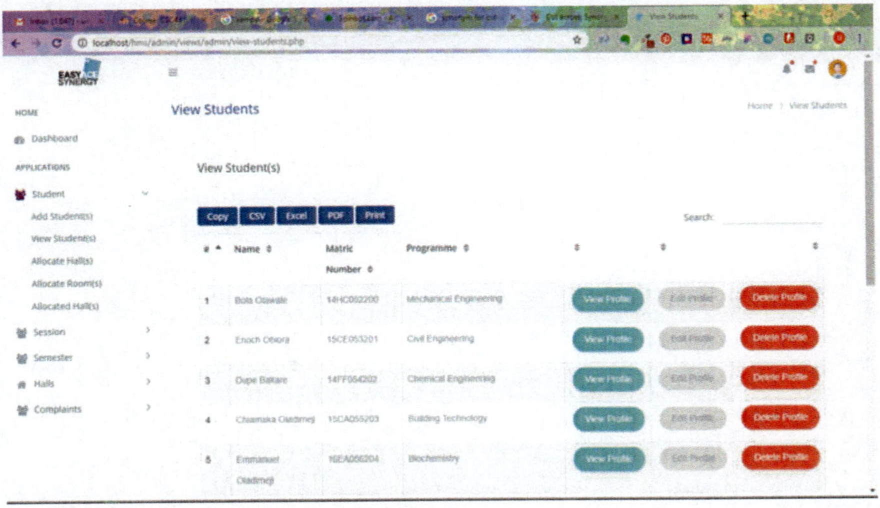

Fig. 4 Showing administrator dashboard

5 Conclusion and Future Work

In this study, a Web application for hostel allocation using decision tree algorithm was implemented using PHP for the backend, CSS for the front end and MYSQL for the database and data manipulation. The application was designed and developed using an integrated development environment called Sublime Text, visual paradigm in the design of the unified modelling language diagrams like activity diagram, sequence

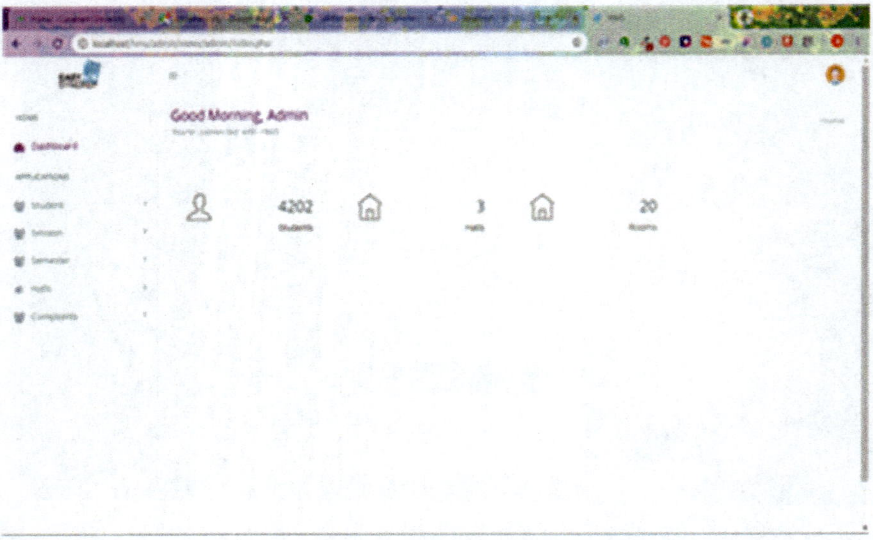

Fig. 5 View students page

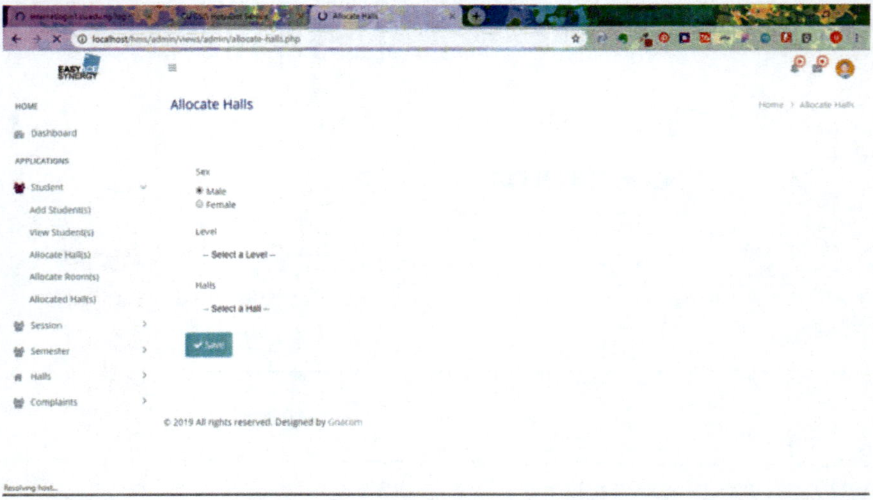

Fig. 6 Allocate hall page

diagram and use case diagram. The database server used in the implementation of the system was XAMPP. The system is not only able to provide a standard framework for hostel management with robust database, but also stores every information that relates to hostel and update such information.

References

1. Holl, K., Elberzhager, F.: Mobile application quality assurance. In: Advances in Computers, vol. 112, pp. 1–77. Elsevier (2019)
2. Roberts, M.L., Zahay, D.: Internet Marketing: Integrating Online and Offline Strategies. Cengage Learning (2012)
3. Varia, J., Mathew, S.: Overview of Amazon Web Services. Amazon Web Services, pp.1–22 (2014)
4. George, J.M., Jones, G.R.: Understanding and Managing Organizational behavior (2012)
5. Olatunji, K.A., et al.: Fuzzy—based accommodation allocation system. Int. J. Eng. Sci. Res. Technol. **4**(8) (2015) ISSN: 2277-9655
6. Bolaji, A.L.A., Michael, I., Shola, P.B.: Optimization of office-space allocation problem using artificial bee colony algorithm. In: International Conference on Swarm Intelligence pp. 337–346. Springer, Cham (2017)
7. Wei, Y., Xue, B., Ning, Z., Zhicheng, B.: The research of intelligent storage space allocation for exported containers based on rule base. In: Proceeding of the 2012 International Conference of Modern Computer Science and Applications, pp. 421–425. Springer, Berlin (2013)
8. Rodrigues, E., Gaspar, A.R., Gomes, Á.: An evolutionary strategy enhanced with a local search technique for the space allocation problem in architecture, Part 1: Methodology. Comput. Aided Des. **45**(5), 887–897 (2013)
9. Gajjar, H.K., Adil, G.K.: A dynamic programming heuristic for retail shelf space allocation problem. Asia-Pacific J. Oper. Res. **28**(02), 183–199 (2011)
10. Obit, J.H., Junn, K.Y., Alfred, R., Bolongkikit, J., Sheng, O.Y.: An investigation towards hostel space allocation problem with stochastic algorithms. In: Computational Science and Technology, pp. 227–236. Springer, Singapore (2019)
11. Zeng, M., Cheng, W., Guo, P.: Modelling and metaheuristic for gantry crane scheduling and storage space allocation problem in railway container terminals. Discrete Dyn. Nat. Soc. **2017** (2017)
12. Da Silva, G.C., Bahiense, L., Ochi, L.S., Boaventura-Netto, P.O.: The dynamic space allocation problem: applying hybrid GRASP and Tabu search metaheuristics. Comput. Oper. Res. **39**(3), 671–677 (2012)
13. Chang, Y., Zhu, X.: A novel two-stage heuristic for solving storage space allocation problems in rail-water intermodal container terminals. Symmetry **11**(10), 1229 (2019)
14. Ajibola, A.S., 2013. Studies of Heuristics for Hostel Space Allocation Problem. Doctoral dissertation
15. Haque, A., Asami, Y.: Optimizing urban land use allocation for planners and real estate developers. Comput. Environ. Urban Syst. **46**, 57–69 (2014)
16. Yip, K., Huang, K., Chang, S., Chui, E.: A mathematical optimization model for efficient management of nurses' quarters in a teaching and referral hospital in Hong Kong. Oper. Res. Health Care **8**, 1–8 (2016)
17. Adewumi, A.O., Ali, M.M.: A multi-level genetic algorithm for a multi-stage space allocation problem. Math. Comput. Model. **51**(1–2), 109–126 (2010)
18. Egoeze, F., Misra, S., Maskeliūnas, R., Damaševičius, R.: Impact of ICT on universities administrative services and management of students' records. Int. J. Human Capital Inf. Technol. Prof. (IJHCITP) **9**(2), 1–15 (2018)
19. Patnaik, S., Kumari Singh, K., Ranjan, R., Kumari, N.: College management system. Int. Res. J. Eng. Technol. (IRJET) **3**(05)
20. Bemile, R., Achampong, A., Danquah, E.: Online hotel reservation system. Int. J. Innovative Sci. Eng. Technol. **1**(9), 2014 (2014)
21. Ruano-Mayoral, M., Casado-Lumbreras, C., Garbarino-Alberti, H., Misra, S.: Methodological framework for the allocation of work packages in global software development. J. Software: Evol. Proc. **26**(5), 476–487 (2014)

Artificial Intelligence in the Energy World—Getting the Act Together

Vasundhra Gupta and Rajiv Bali

Abstract The chance of growing a framework that may "think" has captivated individuals on account since authentic cases. Man-made consciousness (computer-based intelligence) frameworks contain two preeminent districts, proficient/master frameworks (ES), and engineered/counterfeit neural systems (ANNs). The significant goal of this paper is to show how engineered insight methods may play a basic capacity in demonstrating and forecast of the exhibition of inexhaustible power frameworks. The paper plots know-how of how expert structures and neural systems perform by utilizing way of giving an assortment of issues inside the remarkable controls of inexhaustible force designing. The different utilizations of expert structures and neural systems are given in a topical rather than a sequential or some other request. Results introduced on this paper are declaration to the limit of manufactured knowledge as a plan instrument in heaps of locales of sustainable power engineering.

Keywords Artificial intelligence (AI) · Energy consumption · Forecasting · Renewable energy

1 Introduction

Energy consumption is among one of the important subjects of energy systems. Energy intake came beneath the eye after the strength disaster in nineteen seventies [1]. Also, it is proven that electricity intake throughout the world is all of sudden increasing [2]. Therefore, every of us tries to use as less energy as viable in their use as distinctive areas from building to farms, from industrial manner to automobiles [3]. As power comes from three unique resources like fossil fuels, renewable, and nuclear assets [4], it wants a lot try to preserve monitoring of electricity intake of these types in exceptional area. However, by using doing so, we are capable of count on the amount of power, that is consumed in unique regions and try and make plans, specialized for a specific usage and place. For all energy kinds noted

V. Gupta (✉) · R. Bali
Government College of Engineering and Technology Jammu, Jammu, India
e-mail: Vasundhragupta09@gmail.com

P. K. Singh et al. (eds.), *Recent Innovations in Computing*, Lecture Notes in Electrical Engineering 701, https://doi.org/10.1007/978-981-15-8297-4_63

above, estimating the utilization is beneficial for choice and coverage makers. By knowing how a bargain energy may be used for his or her technique or art work, they will be able to consider a few modifications in them to lessen the quantity electricity utilization. Predicting future electricity utilization every in short-term and long-term manner will assist us even to understand, that in which type energy is used broadly speaking and try to trade the fashion, as it is miles happened within the contemporary years for fossil fuels and now we have renewable strength. The quantity of electricity applied in special regions is inspired by way of various factors which includes water, wind, and temperature. Having a couple of factors, predicting the electricity consumption is a complicated trouble [5]. Nowadays, ML models are being applied in first rate areas due to the fact they may be beneficial and the way ML (machine learning) works is like a function which fantastic maps the input facts to output. Machine gaining knowledge of fashions can produce prediction for power intake with excessive accuracy. So, they may be used by governments to implement energy-saving suggestions. They additionally can be used to expect the destiny use of various sorts of power like electricity or herbal fuel [6]. This research artwork has been finished in the prediction of various electricity kind usage. Predictions may be done on using strength in a particular system in business enterprise [7] or the general amount of electricity used in a rustic [7]. This looks at attempts to have a look at the latest research related to modeling and estimating of power consumption in unique place.

2 Artificial Intelligence (AI)

Artificial intelligence (AI) has been around since the mid-1950s, but with a recent boom in massive computational power, big data, and algorithms, AI has seen tremendous growth and is shaping the daily functions of the utility industry. With energy industry trends, such as renewables and the industrial IoT, AI technologies can help utilities capture these opportunities. From the perspective of power consulting at ABB, AI can be implemented with daily operations, grid safety, reliability, and resilience (Fig. 1).

3 Artificial Neural Networks

The idea of neural system assessment has been found almost 50 years inside the past, anyway it is far best inside a definitive twenty years that applications programming program has been advanced to manage sensible issues. The records and rule of neural systems have been characterized in a major amount of distributed writing and will now not be covered on this paper aside from a totally short appraisal of strategies neural systems work [3]. ANNs are well for a couple of obligations while ailing in some others. In particular, they will be right for commitments concerning deficient

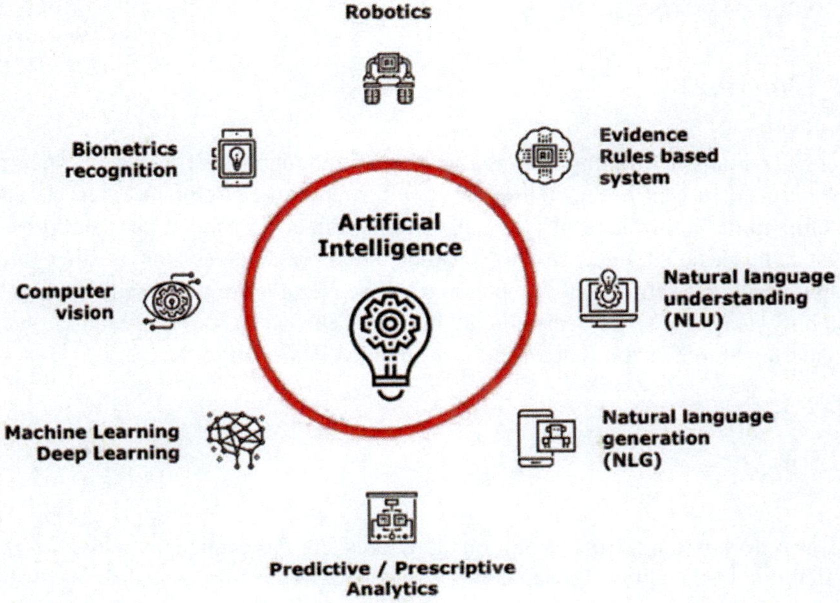

Fig. 1 Main components of AI

records units, fluffy or fragmented data, and for perceptibly perplexing and not well-portrayed difficulties, wherein individuals by and large settle on an intuitional establishment. They can concentrate from models and are fit for address non-straight difficulties. Besides they show heartiness and deficiency tolerance [5]. The obligations that ANNs cannot deal with effectively are the ones requiring over the top exactness and accuracy as in typical feel and arithmeticians had been done strongly in different fields of science, designing, medication, financial aspects, nervous system science, and masses of others. A portion of the most extreme basic ones are; in example, sound and discourse acknowledgment, inside the investigation of electromyography and standout clinical marks, in the recognizable proof of naval force objectives and inside the ID of explosives in traveler bags [4]. They have furthermore being used in atmosphere and commercial center demeanors determining, inside the expectation of mineral investigation sites, in electric controlled and warm burden forecast, in versatile and mechanical oversee and masses of others. Neural systems are utilized for machine oversee because of the reality they are ready to build prescient styles of the strategy from multidimensional records routinely collected from sensors.

4 Energy Forecasting

4.1 *Nnergix*

Nnergix is a statistics mining and Internet-based strength forecasting utility. Nnergix agency makes use of both satellite for pc information from climate forecasts and system gaining knowledge of (ML) algorithms trained to analyze the enterprise's statistics to make extra accurate forecasting. The excessive-decision weather forecasting is generated from satellite pictures. These pictures are then used to generate both huge-scale and small-scale climate fashions. The ML algorithms examine such statistics and predicts the state of the surroundings in a specific area [3].

4.2 *Xcel*

Xcel is enforcing an AI that ambitions at addressing the challenges related to the unreliability of the climate-based power assets such as sun and wind. This software can inform whether the electricity source will range in energy (that is influenced through the varying climate). Xcel is utilized in getting access to climate reports with higher accuracy and properly specific. The company gives the opportunity to hire more precautions when harnessing and maintaining the generated power. To provide this type of unique record, the AI gadget mines facts from nearby satellite for pc reviews, wind firms, and climate stations [2]. The ML set of rules that drives the machine is trained to become aware of styles the use of those information units to expect the era of strength.

4.3 *PowerScout*

This application makes use of AI to version capability savings on utility expenses using enterprise records. PowerScout leverages facts analytics to identify "smart domestic improvement assignment", and this is based on particular functions and electricity utilization within the domestic of clients. Basically, the AI acts as a market advisor by using providing pointers to assist customers in making knowledgeable choices when buying renewable power technologies for their houses. By 2017, this system had overseen the installation of solar capability that is almost sufficient to energy 250,000 homes [6]. **Some of PowerScout partners are the United States Department of energy and Google.**

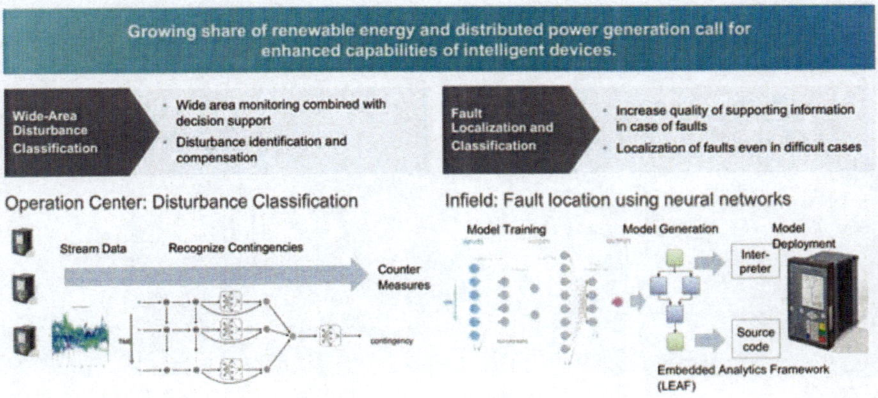

Fig. 2 Artificial intelligence deep learning example. Example: Online decision support for power grids

5 Optimizing Renewables

With increasingly more renewables being delivered to the grid, AI will help utilities to optimize their electricity resources, in particular renewables, through matching supply and call for. According to BP's Energy Outlook, by using 2040, renewables will make up over 20% of general. Worldwide power technology even as hydro and nuclear will make up approximately 10% and 8%, respectively. This worldwide technology blend might be the most assorted the arena has ever seen. Additionally, this boom in renewables is principal in developed and developing countries, which include OECD countries, China, and India [4]. To assist, manipulate the diverse era risk and lessen curtailment from renewables (Fig. 2).

AI will decide the best source of energy at any given point in time. It can also manage DERs, like wind and battery storage, and connect electricity customers to their preferred energy source. Not only will the grid become more optimized, but utilities are able to create a more personalized experience for their consumers (Fig. 3).

6 Conclusion

From the above problem portrayals, you will have the option to see that both expert structures and ANNs were actualized in a gigantic scope of fields for demonstrating and forecast in sustainable quality structures. What is required for setting up such structures is insights that speaks to the past history and execution of a genuine contraption and choice of suitable expert framework or neural network rendition. Without a doubt the quantity of projects gave directly here is neither whole nor comprehensive yet only an example of bundles that show the helpfulness of man-made intelligence strategies. AI styles like each other estimation methods have relative favors

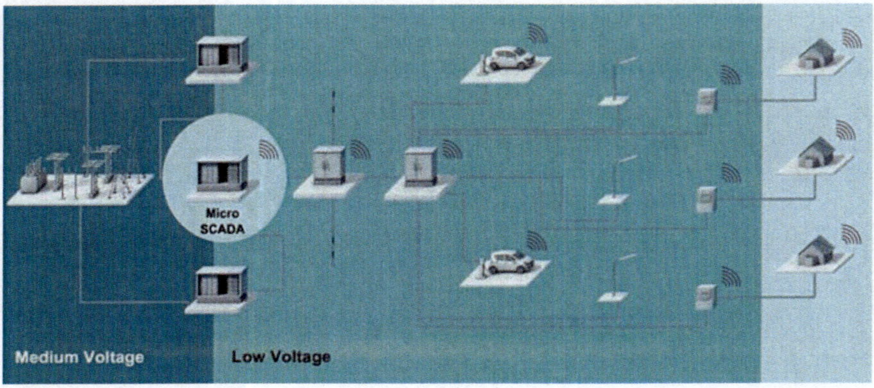

Fig. 3 Local artificial intelligence in distribution grids

and inconveniences. There are no arrangements as to while this exact technique is proper for programming. In the perspective on masterful manifestations gave here its miles acknowledged that reproduced insight gives an elective framework which need not be barely cared about. The vitality business is out of the blue changing over as it faces disturbance from decarbonization, decentralization, and digitalization. Artificial intelligence can help utilities effectively enhance the framework and hold unwavering quality and versatility. Gigantic computational vitality, the blast of huge realities, and better calculations have pushed simulated intelligence time than take care of a few issues in each industry. Also, artificial intelligence streamlines inexhaustible resources at the framework and might improve unwavering quality and strength. It moreover gives an open door for utilities to make a customized client appreciate.

References

1. Ni, Z.: A survey of the development and application of artificial intelligence technology. Coal. Mine Mach. (02), 4–7 (2009)
2. Yu, Z.: Artificial intelligence technology development overview. J. Nanjing Univ. Inf. Sci. Technol. **9**(03), 297–304 (2017)
3. Li, B., Gao, Z.: Application analysis and prospect of artificial intelligence technology in smart grid. Electr. Power **50**(12), 136–140 (2017)
4. Bengio, Y.: Learning deep architectures for AI[J]. Found. Trends in Mach. Learn. (2009)
5. Wen, S.: Application of Computer Science in Smart Grid. China High Technol. Enterprises **21**, 47–49 (2016)
6. Zhu, C., Yang, J., Chen, S., Luo, Z.: Demand response technology based on artificial intelligence theory in the context of smart grid[J]. Shaanxi Electr. Power **43**(07), 63–69 (2015)
7. Rigas, E.S., Ramchurn, S.D., Bassiliades, N.: Managing electric vehicles in the smart grid using artificial intelligence: a survey. IEEE Trans. Int. Trans. Sys. **16**(4), 1619–1635 (2015)

Non-functional Requirements Engineering Questionnaire: Novel Visions and Review of Literature

Naina Handa, Anil Sharma, and Amardeep Gupta

Abstract The aim of this research study is to evaluate the non-functional requirements (NFRs) of educational websites from the usability perspective. The online questionnaire is used for gathering and analyzing the NFRs. The non-functional questionnaires are filled by the 52 software developers. The questionnaire contains different questions related to different factors of ISO 25010. The different NFRs that are related to the usability are considered accessibility, orphan pages, irritating elements, placement and content of sitemap, website content updating, download time, hyperlink description, design consistency and compatibility of website with different web browsers to mention a few. The different views that are given by software developers on the basis of these questions are discussed. We have also added the suggestions column in questionnaire.

Keywords Elicitation · Questionnaire · Requirements

1 Introduction

Several studies have been conducted to investigate the significance of non-functional requirements during the initial stages of software development [1]. Unmasking non-functional requirements (NFRs) such as consistency parameters, functionality specifications and software design limitations are critical in identifying technical alternatives for device from an early design point [2]. Extraction of NFRs from requirement documents is needed for the development of quality software product. If this

N. Handa (✉) · A. Sharma
Lovely Professional University, Phagwara, Punjab, India
e-mail: naina.41500146@lpu.co.in

A. Sharma
e-mail: anil.19656@lpu.co.in

A. Gupta
DAV College, Dasuha, Punjab, India
e-mail: dramardeepgupta@gmail.com

P. K. Singh et al. (eds.), *Recent Innovations in Computing*, Lecture Notes in Electrical Engineering 701, https://doi.org/10.1007/978-981-15-8297-4_64

procedure is automatic, the process will be effective, the human effort, time and mental fatigue involved in defining specific needs from a large number of criteria in a text [3]. Thus, the effective elicitation of NFRs plays a vital role in the process of software development. Non-functional specifications are known as utilities such as flexibility, modifiability, durability, portability, scalability, maintenance adaptability and a few terms for customizability. Implicit requirements or expected from the product are non-functional requirements [4]. These are expected attributes so-called quality attributes. NFRs derive the technical architecture. This elaborates on system performance characteristics. NFRs need to be given equal priority to FRs, as NFRs can be the determining factor in a project's success or failure [5].

The customer and developer focused most of their time on designing practical specifications in both standard and agile software development methods. Non-functional requirements are regarded as second-class requirements, typically overlooked until the end of the development cycle. Often these are hidden, overshadowed and thus neglected or forgotten [6]. These are difficult to develop and test for modeling. While non-functional requirements are largely ignored or retrofitted late in software development, software repair in later stages may result in lower quality. The understanding of NFRs is very important for application in the process of software development. The development of better methods for eliciting NFRs is therefore of interest [7].

2 NFR...Why?

NFRs are important for consideration in the early stages of the life cycle of system development. It identifies the technology selection, hardware allocation and standards adopted during software development. IT project deficiency is often related to design process defects [8]. In a review of the US Air Force program, for example, it was found that more than 40% of the uncovered errors could be traced back to requirement errors [9]. It was also reported that failures in finding and fixing specifications account for 70–85% of IT project rework costs. A very difficult part of specifications architecture is dealing with non-functional requirements and their management. The failure to take proper account of NFRs will result to the most difficult and expensive errors that can only be corrected once a system is completed [10].

It is classified as one of the ten most major risks in requirement engineering [11]. Failure to address NFRs during the design phase may result in a failure of the software product, even if it meets all functional needs. Client or customer consider what they want the device to do naturally and do not care about the money, portability, reliability, protection or efficiency they require. Yet non-functional specifications are very important to manage and should be understood in development because they impact database preference, programming language, operating system, etc. The non-functional criteria are the restrictions on the facilities or functionality provided by the program, e.g., time limits, implementation cycle constraints, etc. This explains how it will be achieved by the machine and not what it will be doing [12]. Non-functional

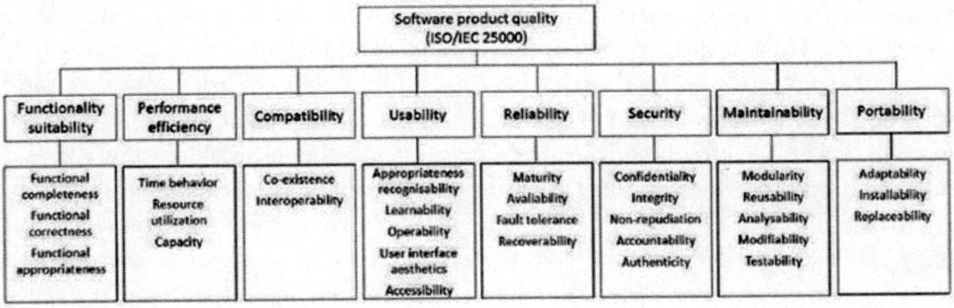

Fig. 1 Software product quality

specifications are typically very complex and arbitrary. These are usually informally defined, frequently inconsistent, difficult to execute during development.

3 Standard Parameters: ISO 25010

One key success factor for successful software projects is requirements engineering with its underlying process and methodology. The functional and non-functional requirements are elicit and defined in this phase. They form the basis for the software project as a whole [13]. The use of scenario in agile projects has evolved as an appropriate way of describing the requirements. From a user's point of view, scenarios depict requirements. Their high quality is very critical, as the entire software project is focused on the specifications [14]. For that purpose, the IEEE created a set of requirement consistency features. We are outlined in the IEEE recommended practice for IEEE Standard 830-1998 device criteria specifications and its counterpart, ISO/IEC/IEEE 29148 [15]. Nonetheless, current scenario-based architecture criteria methodologies do not specifically consider these qualitative characteristics. ISO 25010 has specified quality characteristics and subcharacteristics like functional suitability, usability, compatibility, etc. and their subcharacteristics [16] (Fig. 1).

4 Usability

It is one of the important quality factors and focused on UI and UX. UI defines the user interface and UX defines the user experience. It must be considered while software development. Requirement engineering is very important for the various domains such as business, education, industry, entertainment, etc. As a result, concerns are growing over how websites are developed and the degree of quality delivered. Developing a website should be progressed through several design guidelines to ensure that the website can fulfill the aims and objectives it wants to achieve [17]. Additionally, the website of an organization is a gateway to their information, products and services. Ideally, it should be a representation of the interests of the consumers it represents. However, the creation of websites also relies on infrastructure, organizational structure or business artifacts [18]. Unfortunately, website design is often motivated not by user needs but by technology, organizational structure or business objectives. Nevertheless, website owners and creators have increasingly begun understanding and addressing the issue of usability in recent years [19]. Usability is considered as one of the key factors that determine a website's performance. As defined by ISO 9241-11, it is "the extent to which specified users can use a product to achieve specified objectives with efficiency, effectiveness and satisfaction in a specified context of use [20]. Many evaluation techniques were introduced to determine website usability and make changes to website design. Several methods are present in literature; some methods are experts in addressing, while others are geared at user. Such methods include.

Walkthroughs including one or a group of evaluators who review a user interface through a series of tasks to determine website comprehensibility and learning facilities [21]. Questionnaires are used to retrieve document and collect information to assess user satisfaction with the usability of the website [22]. In general, any website should satisfy the needs of its different stakeholders [23]. Users of the educational website mainly deal with the critical problem of NFRs like performance, security, portability and usability is main concern.

5 Usability Elicitation Techniques

Elicitation of NFRs is very important in the early stages. Elicitation methodology relies heavily on brainstorming, questionnaires, checklists and design. These are used to capture stakeholder feedback. The NFRs catalog helps researchers define criteria for consistency and reduce disagreements. Detection techniques detect the early aspects of low-level design and code from various requirement specification documents. The elicitation methods are used to define the NFRs requires a lot of user interaction, so different non-functional criteria need to be defined and retrieved automatically.

6 Literature Review

Evaluations of the usability of websites have been conducted over the years and for many domains.

Kopczynska et al. explored the value of e-commerce-based NFR templates for novice requirements elicitor. By using 41 industrial projects, the authors simulated the creation of catalog of NFR models and also assessed the maintenance effort required for catalogs. The experiment showed that templates perform better than

Adoc's requirement elicitation approach and also improved the quality of NFR extraction [5].

Mishra et al. explored a method called situation requirement method system (SRMS) was developed which is a web-based application. The proposed technique is validated by make a use of questionnaire. The test showed tool's utility [24].

Ganiyu et al. designed a new method for evaluating the accessibility of University of Istanbul's department websites from usability perspective. The tool comprised of two components; the first part can be used by users who visit and answer questions about the website, and the second part has been developed for the use of administrators where they can handle the usability evaluation. The measurement was based on two environments: the conventional laboratory environment and the Internet [25].

Kirakowski assessed user satisfaction of 5 websites using a questionnaire method. For the evaluation, the authors developed a new questionnaire (called WAMMI). The reader is directed to for more information about the WAMMI (https://www.ucc.ie/hfrg/questionnaires/wammi). The questionnaire indicated that user satisfaction appraisal leads to the successful creation of websites [26, 27].

The research by Chiew and Salim centered on the creation of a web-based application (called WEBUSE) consisting of 24 questions to determine website usability. The report the tool generates indicates how good the website is in terms of usability. The researchers suggest that WEBUSE is ideal for testing websites of all sorts and for any field. The tool will help webmasters develop their websites based on the feedback the users of the expected websites receive [28].

Adepoju et al. evaluated the usability of website of their University. The researchers used quantifiable test metrics, such as navigation times, to determine the usefulness of site. The study showed that the website is suffering from several issues including website-specific terminology, unorganized link patterns and poor evaluation recommendations proposed by the authors to improve and unify the website of the university [29].

Corry et al. performed an assessment of accessibility of an existing website of Midwestern Universities. An experiment to restructure the information of current website was conducted; a prototype was designed and tested against the existing website. Usability was based on the ability of respondents (such as pupils, guardians and faculty) to find the answers to a set of questions quickly and accurately. While the study did work well, the usability measurement methodologies were limited to the total time of the assignment and the amount of user errors [30].

Taj et al. focused on characterizing requirements in FRs and NFRs. The writers also developed a model to help in the phase of requirement elicitation and classification. The requirements are gathered using a method of crowdsourcing and various stakeholders have participated in the process of elicitation. For the classification model, the Naïve Bayes and the decision tree used. The case study was performed to show the model's competence [31].

Portugal et al. focused on using unstructured data to mines the NFRs. The developers developed NFRFinder, a semi-automatic tool. At first, it was applied to structured text and used as a performance metric recall and precision. NFRFinder has delivered promising results in a structured way and showed that classification of

NFRs is influenced by the context and the stakeholders involved in classification [33].

7 Usability-Based Questionnaire

In this study, the online questionnaire-based method has used. The google forms have been used for designing the questionnaire for elicit the NFRs from software developers. The questionnaire is focused on the educational websites and particularly on the usability perspective of NFRs. Questionnaire has identified the different aspects like accessibility, download time, watermarks, readability and navigation or side maps related to educational websites. The questionnaire has distributed into two segments. The first segment represents the UX, i.e., the user experience interaction and the second segment represent the UI, i.e., user interface. The questionnaire includes the 15 questions. The questions were divided into three categories like user interface design, navigation and readability. An extensive literature review has done and find out the different usability attributes which need to be consider while developing the project. The different usability questions have measured on the scales.

8 Results and Discussion

A total of 52 software developers have filled the questionnaire. Most of them are having good experience in software development field and exposer of non-functional requirements. The 20% of developers who have filled the questionnaire are particularly deal with NFRs in their field.

It provides a description of the results obtained from the process of assessment using the questionnaire. Figure 2 shows the distribution of the responses for the question regarding the importance of website accessibility over the four scales like

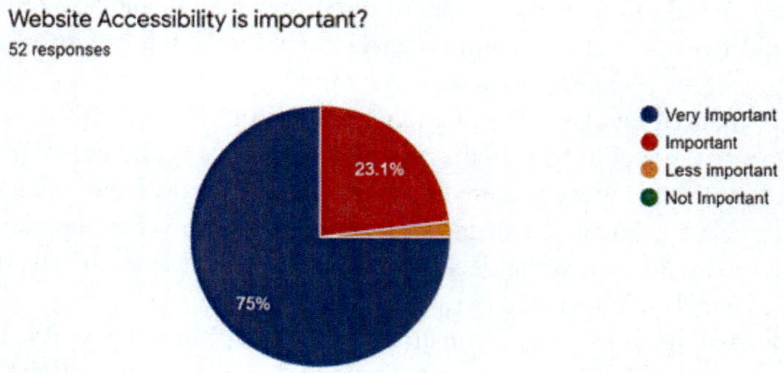

Fig. 2 Website accessibility significance

very important, important, less important and not important. According the four item scale, the result indicates that the 75% software developers have considered the website accessibility is very important. In comparison, about 23% of respondents were considered it important and about 2% considered it as less important. So accessibility is important NFR to be considered while developing the software.

Figure 3 showed that the 48% respondents are annoyed with the irritating elements on web page and only 7% are not considering it.

Figure 4 depicts that the 65% software developers respond that the low downtime is required for downloading a web page is very important, 20% respond as maybe and 15% considered it as not important at all.

Figure 5 has shown that the 69% agreed for the requirement of compatibility of website with different web browser, 14% respond as no need of compatibility and 17% answered as maybe.

Figure 6 depicts the result of question regarding the helpfulness of placement and side map. The 51% respondents are agreed to it, 23% are strongly agreed about the helpfulness of side map and only 2% are disagree for the helpfulness of it. Figure 7

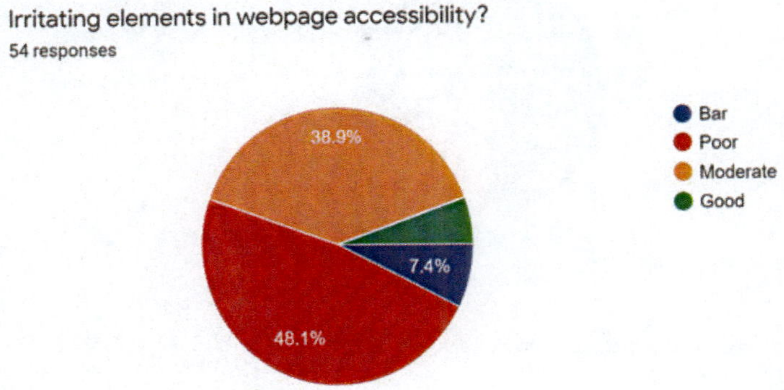

Fig. 3 Irritating elements in accessibility

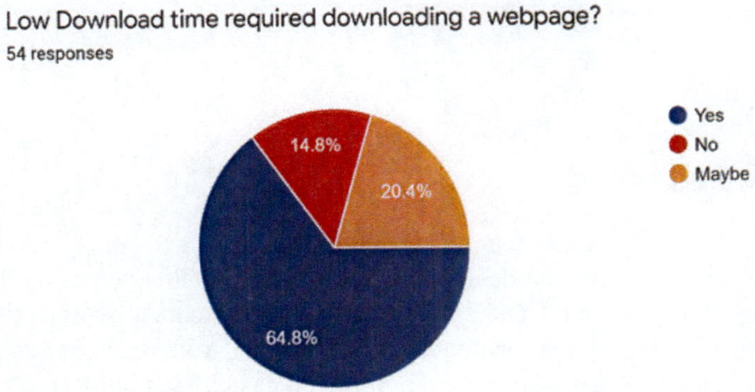

Fig. 4 Low download time in downloading

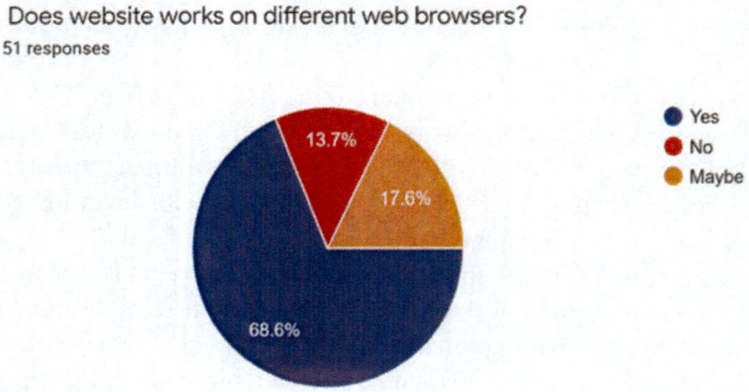

Fig. 5 Compatibility with different web browsers

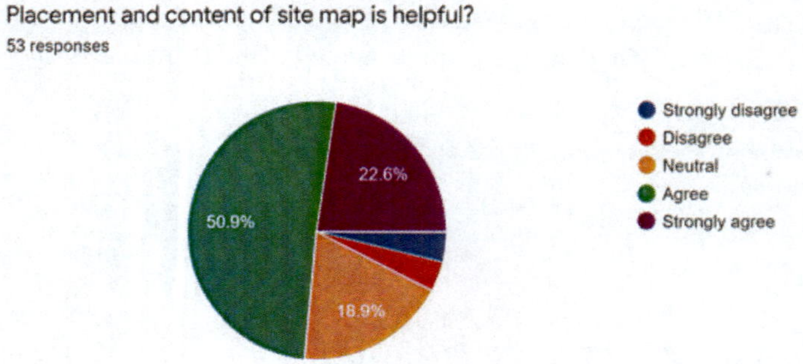

Fig. 6 Helpfulness of site map

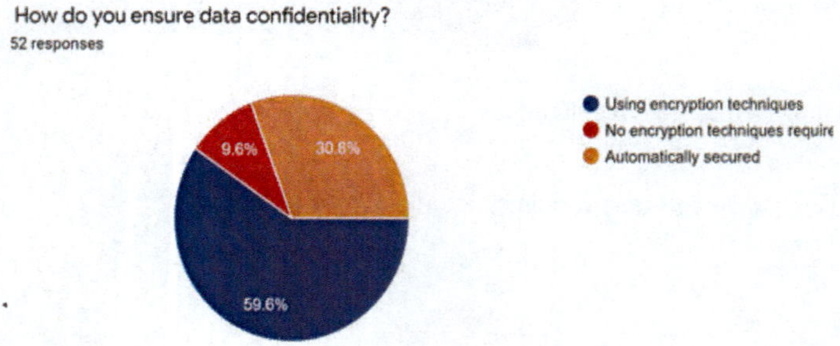

Fig. 7 Data confidentiality

depicts the results of data confidentiality 60% responds that the encryption techniques are required for data confidentiality, 30% said it is automatically secured and only 9% said that no encryption technique has required. It can be interpreted that the encryption techniques are quit important for the data confidentiality.

9 Conclusion and Future Prospects

Quality requirements are of prime significance in systems and stakeholders cannot compromise with quality. Considering the response of 52 developers with regard to usability. The ideas for extending and improving questionnaire are given. The questionnaire has two sections UI and UX which focuses on different aspects like accessibility, download time, irritating elements on web page, etc. Designing an online tool for usability in particular and NFR in general is a future scope to the current work.

References

1. Kopczyńska, S., Ochodek, M., Nawrocki, J.: On importance of non-functional requirements in agile software projects—a survey. In Integrating Research and Practice in Software Engineering pp. 145–158. Springer, Cham (2020)
2. Khan, F., Jan, S.R., Tahir, M., Khan, S., Ullah, F.: Survey: dealing non-functional requirements at architecture level. VFAST Trans. Software Eng. **9**(2), 7–13 (2016)
3. Bagga, P., Joshi, A., Hans, R.: QoS based web service selection and multi-criteria decision making methods. Int. J. Interact. Multimedia Artif. Int. **5**(4), 113 (2019)
4. Eckhardt, J., Vogelsang, A., Fernández, D. M.: Are "non-functional" requirements really non-functional? An investigation of non-functional requirements in practice. In: Proceedings of the 38th International Conference on Software Engineering, pp. 832–842 (2016)
5. Kopczyńska, S., Nawrocki, J., Ochodek, M.: When NFR templates pay back? A study on evolution of catalog of NFR templates. In: International Conference on Product-Focused Software Process Improvement, pp. 145–160. Springer, Cham (2019)
6. Ameller, D., Franch, X., Gómez, C., Martínez-Fernández, S., Araújo, J., Biffl, S., ... Muccini, H.: Dealing with non-functional requirements in model-driven development: a survey. IEEE Trans. Softw. Eng. (2019)
7. Afreen, N., Khatoon, A., Sadiq, M.: A taxonomy of software's non-functional requirements. In Proceedings of the Second International Conference on Computer and Communication Technologies, pp. 47–53. Springer, New Delhi (2016)
8. Behutiye, W., Karhapää, P., Costal, D., Oivo, M., Franch, X.: Non-functional requirements documentation in agile software development: challenges and solution proposal. In: International Conference on Product-Focused Software Process Improvement, pp. 515–522. Springer, Cham (2017)
9. Finkelstein, A., Dowell, J.: A comedy of errors: the London ambulance service case study. In: Proceedings of the 8th International Workshop on Software Specification and Design, pp. 2–4. IEEE (1996)
10. Chung, L., do Prado Leite, J.C.S.: On non-functional requirements in software engineering. In: Conceptual Modeling: Foundations and Applications pp. 363–379. Springer, Berlin, Heidelberg (2009)
11. Jung, H.T., Lee, G.H.: A systematic software development process for non-functional requirements. In: 2010 International Conference on Information and Communication Technology Convergence (ICTC), pp. 431–436. IEEE (2010)
12. Ullah, S., Iqbal, M., Khan, A.M.: A survey on issues in non-functional requirements elicitation. In: International Conference on Computer Networks and Information Technology, pp. 333–340. IEEE (2011)

13. Mouzakitis, S., Tsapelas, G., Pelekis, S., Ntanopoulos, S., Askounis, D., Osinga, S., Athanasiadis, I.N.: Investigation of common big data analytics and decision-making requirements across diverse precision agriculture and livestock farming use cases. In: International symposium on environmental software systems, pp. 139–150. Springer, Cham (2020)
14. Oriol, M., Seppänen, P., Behutiye, W., Farré, C., Kozik, R., Martínez-Fernández, S., ... Choras, M.: Data-driven elicitation of quality requirements in agile companies. In: International Conference on the Quality of Information and Communications Technology, pp. 49–63. Springer, Cham (2019)
15. Labidi, T., Sakhrawi, Z., Sellami, A., Mtibaa, A.: An Ontology-based approach for preventing incompatibility problems of quality requirements during cloud SLA establishment. In: International Conference on Computational Collective Intelligence, pp. 663–675. Springer, Cham (2019)
16. Anugrah, S., Putra, A.E.: Analisis Kualitas ISO 25010 Aplikasi artificial intelligence troubleshooting Komputer dengan FURPS. e-Tech: Jurnal Ilmiah Teknologi Pendidikan, **6**(2) (2019)
17. Nichols, E., Olmsted-Hawala, E., Holland, T., Riemer, A.A.: Usability testing online questionnaires: experiences at the US census bureau. Adv. Questionnaire Des. Dev. Eval. Test. 315–348 (2020)
18. García-Peñalvo, F.J., Vázquez-Ingelmo, A., García-Holgado, A., Seoane-Pardo, A.M.: Analyzing the usability of the WYRED Platform with undergraduate students to improve its features. Univ. Access Inf. Soc. **18**(3), 455–468 (2019)
19. Blanchard, E.: Usability: low tech, high security. Doctoral dissertation, Université Sorbonne Paris Cité (2019)
20. Arthana, I.K.R., Pradnyana, I.M.A., Dantes, G.R.: Usability testing on website wadaya based on ISO 9241-11. In: Journal of Physics: Conference Series, vol. 1165, no. 1, p. 012012. IOP Publishing (2019)
21. Bağcı, S. (2019). Assesssing the walkability principles: The case study of Mehmetçik Boulevard (Master's thesis, Izmir Institute of Technology).
22. Gouveia, V.V., Ribeiro, M.G.C., de Aquino, T.A.A., Loureto, G.D.L., Nascimento, B.S., Rezende, A.T.: Gratitude Questionnarie (GQ-6): evidence of construct validity in Brazil. Curr. Psychol. 1–9 (2019)
23. Nong, Z., Gainsbury, S.: Website design features: exploring how social cues present in the online environment may impact risk taking. Hum. Behav. Emerg. Technol. **2**(1), 39–49 (2020)
24. Mishra, D., Aydin, S., Mishra, A., Ostrovska, S.: Knowledge management in requirement elicitation: Situational methods view. Comput. Stand. Int. **56**, 49–61 (2018)
25. Ganiyu, A.A., Mishra, A., Elijah, J., Gana, U.M.: The Importance of Usability of a Website. IUP J. Inf. Technol. **13**(3), 27–35 (2017)
26. Kirakowski, J.: Questionnaires in usability engineering, a List of Frequently Asked Questions, Web-site compiled by: Jurek Human Factors Research Group, Cork, Ireland (2006)
27. Kirakowski, J.: Questionnaires in usability engineering: a list of frequently asked questions. Human Factors Research Group, Cork, Ireland, 15 (2000)
28. Chiew, T.K., Salim, S.S.: Webuse: Website usability evaluation tool. Malays. J. Comput. Sci. **16**(1), 47–57 (2003)
29. Adepoju, S.A., Oyefolahan, I.O., Abdullahi, M.B., Mohammed, A.A.: A survey of research trends on university websites' usability evaluation. i-Manager's J. Inf. Technol. **8**(2), 11 (2019)
30. Corry, M.D., Frick, T.W., Hansen, L.: User-centered design and usability testing of a web site: an illustrative case study. Educ. Tech. Res. Dev. **45**(4), 65–76 (1997)
31. Taj, S., Arain, Q., Memon, I., Zubedi, A.: To apply data mining for classification of crowd sourced software requirements. In: Proceedings of the 2019 8th International Conference on Software and Information Engineering, pp. 42–46 (2019)

Data Ingestion and Analysis Framework for Geoscience Data

Niti Shah, Smita Agrawal, and Parita Oza

Abstract Big earth data analytics is an emerging field since environmental sciences are probably going to profit by its different systems supporting the handling of the enormous measure of earth observation data, gained and produced through perceptions. It additionally benefits by giving enormous stockpiling and registering capacities. Be that as it may, big earth data analytics requires explicitly planned instruments to show specificities as far as significance of the geospatial data, intricacy of handling, and wide heterogeneity of information models and arrangements [1]. Data ingestion and analysis framework for geoscience data is the study and implementation of extracting data on the system and processing it for change detection and to increase the interoperability with the help of analytical frameworks which aims at facilitating the understanding of the data in a systematic manner. In this paper, we address the challenges and opportunities in the climate data through the climate data toolbox for MATLAB [2] and how it can be beneficial to resolve various climate-change-related analytical difficulties.

Keywords Earth data · Data ingestion · Hadoop · Earth observation data · Climate data

1 Introduction

Big data is the terminology which describes excessively large and complex data sets to be worked with applying some generally accessible tools and techniques. An advanced analytic technique that operates on big data sets is big data analytics.

N. Shah (✉) · S. Agrawal · P. Oza
CE Department, Institute of Technology, Nirma University, Ahmedabad, India
e-mail: 17MCA041@nirmauni.ac.in

S. Agrawal
e-mail: smita.agrawal@nirmauni.ac.in

P. Oza
e-mail: parita.oza@nirmauni.ac.in

P. K. Singh et al. (eds.), *Recent Innovations in Computing*, Lecture Notes in Electrical Engineering 701, https://doi.org/10.1007/978-981-15-8297-4_65

Henceforth, big data analytics principally comprises of two things, big data and analytics. Big data analytics have formed one of the most reflective trends in business intelligence today. Big data is commonly defined based on the size of the data. Volume matters, yet there are other significant features of big data analytics, that is data variety and velocity. The three Vs that is volume, variety, and velocity of big data establish an inclusive definition, which clarifies the myth that big data is only about volume of the data [1].

Individually, the three Vs have its particular out-turn for analytics. From the volume outlook, the overrun of input data is the foremost aspect which we need to address because it may lead to paralysis of the data analytics directly. From the aspect of velocity, streaming or real-time data leads to the issue of huge amount of data approaching into the system within a short-term period but system inclusive of devices may be incapacitate to hold the volume of input data. From the view point of variety, the input data may have dissimilar types or have insufficient data in such case how to hold that data is another concern for the operators [3, 4].

Many of us may see big data as a source for collecting more useful information through large volume of data. In contrast, the truth is that collecting more data do not certainly mean getting more useful and relevant information. The data may contain abnormal and ambiguous data. For illustration, consider an individual account may have multiple users or a user that may have multiple accounts which can lead to inaccurate outcomes. Hence, data analytics is coming up with numerous new issues, such as storage, security, privacy, quality of data, and fault tolerance [5]. To overcome such issues, most of the key applications of future generation for the distributed and the parallel systems are in big data analytics.

Climate science is one such big data domain application that is undergoing unprecedented development. Some of the major climatology big data challenges are easily understood. The large data mines for such applications are swiftly increasing in volume. For such large data sets, the issue to move that data arises and it cannot easily be moved as an alternative, certain analytical procedures must be followed to migrate it to where the data exist in. Data often exist on platforms with fluctuating network and computational capabilities [6].

Large amount of data surges the significance of the provenance management, discovery, and metadata. Henceforth, a need for resource balancing, fast networks, the ability to respond quickly to customer demands and intermediation, is created for analytic products and migrating codes within a growing network of computational resources and storage. Moreover, the unforeseen uses for climatology information requires superior agility in deploying and building applications [7, 8].

While addressing these challenges, it becomes essential to take into consideration that the climatology contains two aspects that is, social practice and technical practice. There exist certain established procedures for analyzing, sharing, and creating scientific data sets. The paper presented addresses some of climatology big-data-related issues including data ingestion and data preprocessing, also the technical resolutions that are being developed to improve the data analytics, accessibility and publication [9].

2 Data Ingestion

Data ingestion is the process of obtaining and importing data for immediate use or storage in a database. To ingest, something is to take something in or absorb something. Ingestion of data can be performed in two ways that is in batches or in real time. In data obtained in batches, the data is gathered and imported at regular period of time in distinct chunks. While, when the data is collected and imported in the real-time data ingestion, each time data is collected and imported as it is received from the source. The ideal data ingestion procedure initiates by ordering the data set sources, authenticating the individual data files, and then directing data to the particular destination.

Batch processing is the most commonly used data ingestion methodology. In batch processing, the ingestion layer collects the source data at regular intervals and groups it accordingly and then sends it to the destination system. The data is grouped based on a simple schedule, the activation of certain conditions or any logical ordering. When it is not essential to use the real-time data ingestion, the batch processing is used for data ingestion as it is more affordable and easy to use as compared to real-time streaming of data. Moreover, the real-time streaming do not involve the grouping as the data is sourced, manipulated, and loaded immediately on creation or recognition by the data ingestion layer. As it requires the systems to constantly monitor the sources and input new data, it is more expensive to use. Nevertheless, it may be useful and more appropriate for the analytics of the system data that needs continuous refreshed information [10].

However, data ingestion comes with its own challenges. The first and one of the major challenge is that it is slow. The data has grown so much large in volume and has become more diverse and complex over the time which makes old methods like ETL tools incapable of keeping up with the volume of data and the modern data sources as they are not fast enough. The other major challenge is that it is highly complex to work with that because with the explosion of new data sources like sensors, smart phones, and other connected devices, the organizations are having difficulty to get the desired value from the input data, due to the complexity of schema mismatches and detecting and removing errors in the data.

Another challenge is that it is expensive. Numerous factors collaborate to make data ingestion expensive. One such factor is the infrastructure that is required for the different data sources. Also, the proprietary tools are very expensive to maintain over time. Moreover, it is also costly to maintain a group of experts to support the ingestion. The risk factor also leads to expense when the real money is lost when the business decisions are not taken quickly. Another major aspect challenged is security issues. While moving the data, security is always an issue as during ingestion data is often staged at various steps, which makes it difficult throughout the process to meet compliance standards [11].

3 Analysis Framework

Analytical frameworks are designed to support logical thinking of the analyst, in a systematic manner. The analytical frameworks are often presented visually and aim to facilitate and guide sense making and understanding. Analytical frameworks guides and supports the analysis, storage, and collection of information by identifying and categorizing the key products and analytical outcomes at each step of the analysis. We can easily find a way to organize what information to collect and how to analyze the same. It also improves the way of analyzing some common gaps and deficiencies that have particular outcomes and also identify the risks related to the geographic location of intervention.

With the shortcoming of currently available tools and techniques adopted by variety of industries, big data analytics come with a comparatively newer standards to analyze the datasets. Extracting relevant and useful information from large pool of datasets is big data analytics. It includes the methodologies to uncover the unknown patterns and correlations, which can be further used to make effective decisions and to give powerful insights. For the data generated every day, the methodologies are required for the volume, variety, and velocity. For the system to identify the faulty behavior and failures, monitoring becomes extremely crucial [12].

4 Geoscience Data

Geospatial or geoscience data refers to scattered data sets surpassing the volume of the current figuring frameworks. A noteworthy segment of huge information is really geoscience information, and the volume of the information is developing quickly at any rate by 20% consistently. Geospatial information has consistently been huge information. In nowadays, huge information investigation for geospatial information is accepting impressive thoughtfulness regarding enable clients to dissect enormous measures of geospatial information. Geospatial huge information normally alludes to spatial informational indexes surpassing limit of the current processing frameworks. Along with this exponential increment of geospatial enormous information, the ability of elite registering is being required significantly than any time in recent memory, for demonstrating and recreation of geospatially empowered substance [13]. The geospatial dataset used for research and implementation is obtained from the Indian Space Research Organization (ISRO) Web-based utility called BHUVAN.

Link: https://bhuvan-app3.nrsc.gov.in/data/download/index.php (Table 1).

Table 1 Analysis frameworks used in the various applications of GeoScience data [1, 11–15]

Geoscience problem statement	Related to	Objective	Framework
Utilizing cloud computing to address big geospatial data challenges	Climate studies, geospatial knowledge mining. Land cover simulation, and dust storm modeling	To provide cloud computing as a utility service for addressing different processing needs with (a) on demand services (b) pooled resources (c) elasticity (d) broad band access and (e) measured services	Hadoop distributed file system spark
A framework for processing large scale geospatial and remote sensing data in map reduce Environment	Management and data processing of spatial and remote sensing data in distributed environment	To provide extensibility and adaptability to enable previously implemented algorithms and existing toolkits to be easily adapted to distributed execution without major effect	Apache hadoop
GeoSpark a cluster computing framework for processing large-scale spatial data	GeoSpark an in-memory cluster computing framework for processing large-scale spatial data	To provide a geometrical operations library that accesses spatial resilient distributed data sets (RDD) to perform basic geometrical operations (e.g., overlap, intersect)	GeoSpark
Big earth data: a comprehensive analysis of visualization analytics issues	Comprehensive review of the technology and terminology within the big earth data problem space	(a) Presents exam pies of state-of-the-art project sin each major branch of big earth data research, (b) Current issues within big earth data research are highlighted and potential future solutions identified	Hadoop FS
Geospatial big data: challenges and opportunities	Various challenges and opportunities which geospatial big data brought us. including fuel and time saving, revenue increase, urban planning, and healthcare	(a) Introduce new emerging platforms for sharing the collected geospatial big data and for tracking human mobility via mobile devices, (b) Do research on interactive analytics of real-time or dynamic data	Hadoop, Hive

(continued)

Table 1 (continued)

Geoscience problem statement	Related to	Objective	Framework
Big data analytics for earth sciences: the earth server approach	Big earth data analytics	(a) To provide a solution for coverage type datasets, built around a high performance array database technology. (b) The adoption and enhancement of standards for service interaction	Map reduce

5 Climate Data Toolbox

As a very much tried, all around reported stage for interdisciplinary coordinated efforts, the climate data toolbox for MATLAB expects to diminish time consumed composing low level code, that let scientists center around material science instead of coding and energize increasingly adequate code sharing.

Climate science data is profoundly integrative commonly, so to know the interactions between earth forms characteristically permit the utilization of logical programming that can work over the controls of earth science. Climate data toolbox for MATLAB contains in excess of 100 capacities that length the significant atmosphere-related orders of earth science. The toolbox empowers modernized, totally scriptable work processes that are instinctive to compose and simple to share. It incorporates capacities to assess vulnerability, perform framework tasks, compute atmosphere records, and create normal information shows.

CDT contains more than 100 all around recorded capacities intended to help clients at every venture of logical investigation, from bringing in and handling information to plotting and deciphering outcomes. The capacities are proposed to rationalize work processes and confirm that clients never feel aground at any progression of the examination. As needs be, the sorts of capacities in CDT length the array from straightforward utilities, to capacities for summed up measurable investigation [2], [16].

5.1 CDT Contents

Climatology: The climatology function gives typical values of a variable as it varies throughout the year. The output of this function includes the overall mean. It does not have any trend through time. Its only variability is that with the seasons. CDT has another function called season. The only difference is that the output of climatology includes the mean of the variable, whereas the output of the season will always have a 0 mean value (Fig. 1).

Monthly: The monthly function calculates statistics of a variable for specified months of the year. For example, will the sea surface temperature in March of next year be about the same as it was in March of last year, or is there a tremendous amount of variability between Marches? Or What is the average springtime Antarctic sea ice extent? From these examples, we can see that it helps in loading the time series and plot it to get a sense of analytical questions asked (Fig. 2).

Polyfitw: The polyfitw function computes weighted polyfits. Calculates the unweighted polynomial fit of x versus y to the nth order, exactly like polyfit. Specifies weights to apply to each y value. Returns the structure and centering/scaling values mu for use in polyval (Fig. 3).

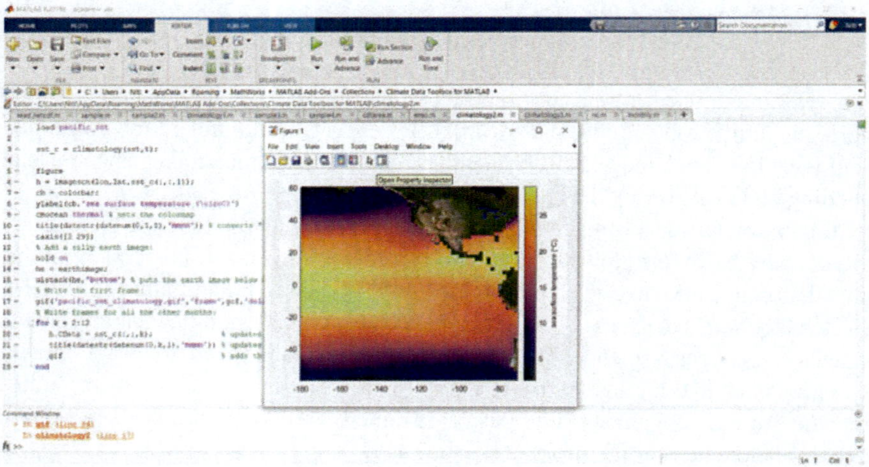

Fig. 1 Gridded data of sea surface temperatures for each month of the typical year. It makes a gif by plotting the frames [2]

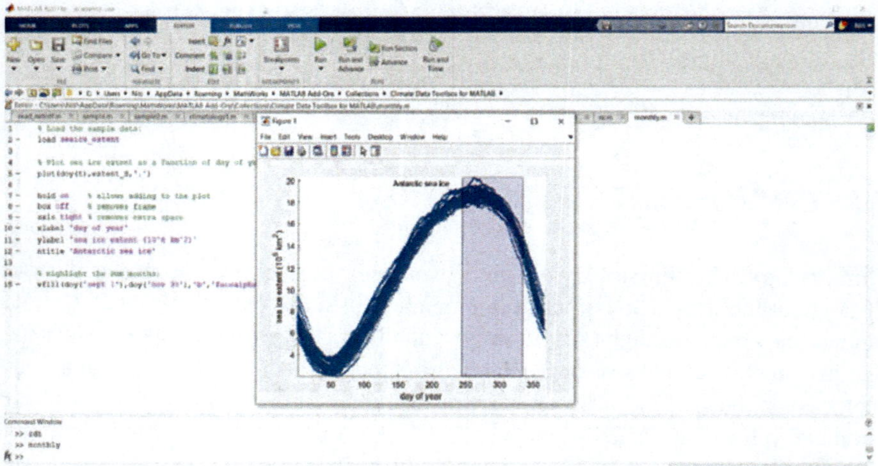

Fig. 2 Daily sea ice extent from a particular year to present and it is plotted as a function of day of year [2]

Polyplot: This function plots a polynomial fit to scattered x, y data. It adds a linear trend line or other polynomial fit to a data plot. It places a least-squares linear trend line through scattered x, y data and specifies the degree n of the polynomial fit to the x, y data (Fig. 4).

PoltpSD: This function plots a power spectral density of a time series using the periodogram function. It plots a power spectrum of 1D array y at sampling frequency

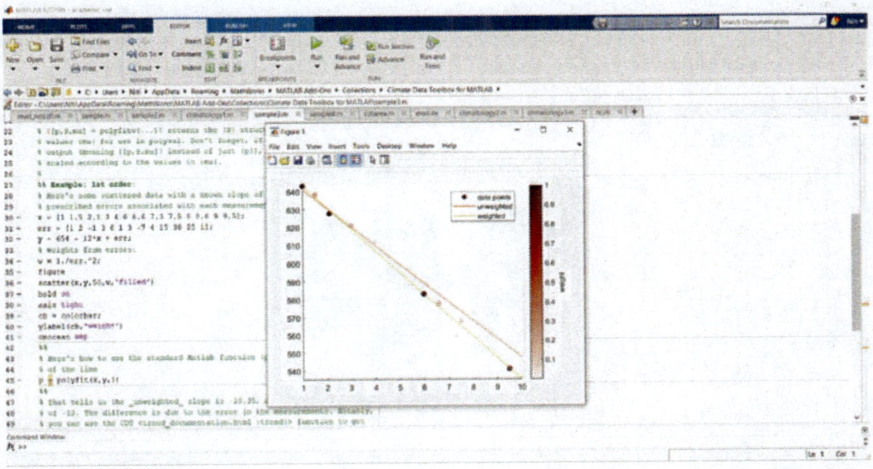

Fig. 3 Standard MATLAB function polyfit to find the unweighted slope of the line and the difference due to the error in the measurements [2]

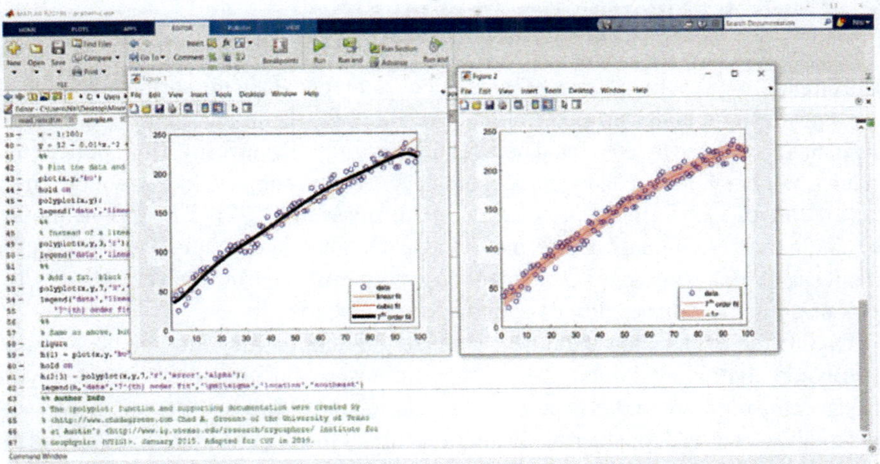

Fig. 4 Linear trend line with black seventh-order polynomial fit and ±1 standard deviation of error lines [2]

Fs using the periodogram function. Sampling frequency Fs must be a scalar. It plots a power spectrum of y referenced to an independent variable *x* (Fig. 5).

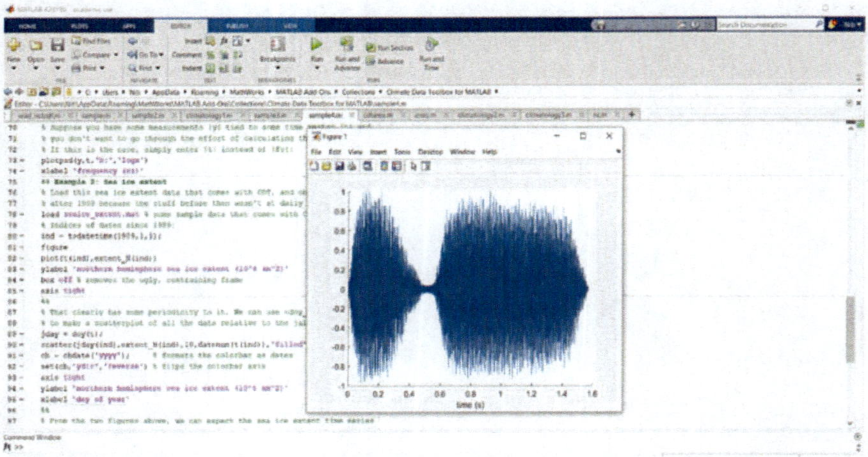

Fig. 5 Power spectrum of the train signal by using the inbuilt train whistle example and plotting the time series for context [2]

6 Research Opportunities and Future Scope

The climatology is still in its early stage of development when the sustainable aspect is connected to it, considering the already existing studies on Internet of Things and big data. It is noticed that the focal point of huge information investigation application in environmental change is by all accounts uneven, and subjects like waste or reuse the executives with significant potential are dismissed [17–22]. The reasons might be an absence of comparing information assets, an absence of conservative benefit and research financing, poor correspondence between gatherings of specialists with various abilities, vulnerability data, and so forth (Fig. 6).

One of the exploration patterns recognized among the ongoing applications is distributed computing, which gives a superior answer for huge information stockpiling, transmittingm and computational prerequisites. With the growth of Internet of Things, it is worth considering the architecture of the platform that empowers real

Fig. 6 Future opportunities in climate change data to achieve sustainability through IoT that is Internet of Things

time, efficient, storage of large datasets and in memory cloud computing, utilizing the advanced techniques by integrating it in the future research to attain sustainability [8].

7 Conclusion

Earth observations are likely bound to keep on developing apace. Fortunately, progressions in the field of analytics algorithms, especially in the machine learning area, offer great prospects for keeping up with the massive data growth. In this paper, we discussed the challenges and future opportunities which geoscience and climatology brought us. This study reviews a variety of Data Ingestion tools and methodologies for analyzing the observations made from the various outcomes used for handling geospatial big data.

Having examined the current and future advancements of GeoScience Big Data visualization analysis, it is clear that the volume, variety, and velocity of GeoScience data being gathered is developing quickly and we have initiated to explore how can we utilize this data toward development. One of the biggest challenge concerning geospatial data is climate change data which is of storing and processing this data. In this study, we have addressed the challenges and future opportunities in the direction of climate change data. To aid our observations and to make a precisive analysis, we have used climate data toolbox (CDT).

CDT as a toolbox is exceptional in that it is integrative, yet altogether and academically recorded. CDT gives specialized entrance ramps to enable new clients to start their examination and calculated scaffolds to interface course reading hypothesis to genuine information investigation. The tools in CDT address the issues of atmosphere researchers at each phase of investigation, from bringing in and breaking down information to showing results. The functions benefit the users to keep their attention on physics rather than coding, and CDT linguistic structure is straightforward, instinctive, and speedy to learn. Examinations achieved in CDT are completely scriptable and direct, which collectively, we expect will empower simple coordinated efforts, an expansion in code distribution, and a higher level of replicability for science overall [23].

References

1. Baumann, P., et al.: Big data analytics for earth sciences: the earth server approach. Int. J. Digi. Earth **9**, 1–27 (2015). https://doi.org/10.1080/17538947.2014.1003106
2. https://www.mathworks.com/matlabcentral/fileexchange/70338-climate-data-toolbox-for-matlab
3. Russom, P.: Big data analytics. Big Data Analytics, **38**

4. Verma, J.P., Agrawal, S., Patel, B., Patel, A.: Big data analytics: challenges and applications for text, audio, video, and social media data, international journal on soft computing. Arti. Intel. Appl. (IJSCAI) **5**(1), 41–51 (2016). https://doi.org/10.5121/ijscai.2016.5105
5. Tsai, C.-W., et al.: Big data analytics: a survey. J. Big Data **2**(1), 21. https://doi.org/10.1186/s40537-015-0030-3
6. Trends in Big Data Analytics. J. Parallel Distrib. Comput. **74**(7):2561–2573. https://doi.org/10.1016/j.jpdc.2014.01.003
7. Agrawal, S., Patel, A.: A study on graph storage database of Nosql. Int. J. Soft Comput. Artif. Int. Appl. (IJSCAI) **5**(1), 33–39 (2016). https://doi.org/10.5121/ijscai.2016.5104. URL http://aircconline.com/ijscai/V5N1/5116ijscai04.pdf
8. Masani, K.I., Oza, P., Agrawal, S.: Predictive maintenance and monitoring of industrial machine using machine learning. Scalable Comput. Pract. Experience **20**(4), 663–668 (2019)
9. Schnase, J.L., et al.: Big data challenges in climate science, 11. Data ingestion: the first step to a sound data strategy. stitch resource. Stitch https://www.stitchdata.com/resources/data-ingestion/. Accessed 6 Nov 2019
10. Schnase, J.L., et al.: Big data challenges in climate science, 11. Data Ingestion: the first step to a sound data strategy stitch resource. https://www.stitchdata.com/resources/data-ingestion/. Accessed 6 Nov 2019
11. (PDF) Big data analytics framework for improved decision making. https://www.researchgate.net/publication/273818434_Big_Data_Analytics_Framework_for_Improved_Decision_Making. Accessed 6 Nov 2019
12. Yang, C., et al.: Utilizing cloud computing to address big geospatial data challenges. Comput. Environ. Urban Syst. Geospatial Cloud Comput. Big Data **61**, 120–128 (2017). https://doi.org/10.1016/j.compenvurbsys.2016.10.010
13. Giachetta, R.: A framework for processing large scale geospatial and remote sensing data in mapreduce environment. Comput. Graph. **49**, 37–46 (2015). https://doi.org/10.1016/j.cag.2015.03.003
14. Merritt, P., et al.: Big earth data: a comprehensive analysis of visualization analytics issues. Big Earth Data **2**(4), 321–350 (2018). https://doi.org/10.1080/20964471.2019.1576260
15. Lee, J.-G., Kang, M.: Geospatial big data: challenges and opportunities. Big Data Res. **2**(2), 74–81 (2015). https://doi.org/10.1016/j.bdr.2015.01.003
16. Yu, J., Wu, J., Sarwat, M.: GeoSpark: a cluster computing framework for processing large-scale spatial data. In: Proceedings of the 23rd SIGSPATIAL International Conference on Advances In Geographic Information Systems, SIGSPATIAL'15, pp. 1–4. Association for Computing Machinery, Eattle, Washington (2015). https://doi.org/10.1145/2820783.2820860
17. Desai, K., Devulapalli, V., Agrawal, S., Kathiria, P.: Patel, A.: Web crawler: review of different types of web crawler, its issues, applications and research opportunities. Int. J. Adv. Res. Comput. Sci. **8**(3) (2017)
18. Agrawal, S., Verma, J.P., Mahidhariya, B., Patel, N., Patel, A.: Survey on mongodb: an open-source document database. Int. J. Adv. Res. Eng. Technol. **1**(2), 4 (2015)
19. Yadav, S., Verma, J., Agrawal, S.: SUTRON: IoT-based industrial/home security and automation system to compete the smarter world. Int. J. Appl. Res. Inf. Technol. Comput. **8**(2), 193–198 (2017)
20. Desai, R., Gandhi, A., Agrawal, S., Kathiria, P., Oza, P.: Iot-based home automation with smart fan and ac using nodemcu. In: Proceedings of ICRIC 2019, Springer, 2020, pp. 197–207
21. Agrawal, S., Patel, A.: Clustering algorithm for community detection in complex network: a comprehensive review. Recent Adv. Comput. Sci. Commun. **13**(1), 1–8 (2020). https://doi.org/10.2174/2213275912666190710183635. http://www.eurekaselect.com/node/173402/article
22. Agrawal, S.S., Patel, A.: CSG cluster: A collaborative similarity based graph clustering for community detection in complex networks. Int. J. Eng. Adv. Technol. **8**(5), 1682–1687 (2019)
23. The Climate Data Toolbox for MATLAB—Greene—2019—Geochemistry, Geophysics, Geosystems—Wiley Online Library. https://agupubs.onlinelibrary.wiley.com/doi/abs/10.1029/2019GC008392. Accessed 29 Feb 2020

Design of Waste Heat Recovery System for Green Environment

Shruti Jain, Pramod Kumar, and Meenakshi Sood

Abstract The global energy problem due to population growth, industrialization, and depletion of natural resources and environmental issues is a cause of concern. A lot of active research on energy harvesting from waste heat from various sources is being done. The prime concern is to not only reduce the wastage of energy resources but also recycle the energy resources. This will also mitigate the carbon footprints of the environment. A high potential for harvesting the energy lies in the energy obtained from household chores. The energy recycled from waste energy can be made sufficient enough required for microelectronic devices. In this research work, a novel module is designed to generate energy by utilizing the heat obtained from household cooking stoves or chullahs. The designed module employs a thermoelectric energy system coupled with heat sink, DC-DC converter, and booster circuits. Three different modules were designed by modifying different key parameters, and the performance of an energy harvesting system has been evaluated. The results obtained are in good agreement with the simulation results and capable of charging a mobile device.

Keywords Thermoelectric generator · Energy harvesting system · Renewable energy source · Green environment

S. Jain · P. Kumar
Department of ECE, Jaypee University of Information Technology, Solan, Himachal Pradesh 173234, India
e-mail: jain.shruti15@gmail.com

P. Kumar
e-mail: promod.kumar@juit.ac.in

M. Sood (✉)
Department of CDC, NITTTR, Chandigarh 160019, India
e-mail: meenakshi@nitttrchd.ac.in

P. K. Singh et al. (eds.), *Recent Innovations in Computing*, Lecture Notes in Electrical Engineering 701, https://doi.org/10.1007/978-981-15-8297-4_66

1 Introduction

A renewable energy source has emerged all over the world. The energy harvesting system has become a popular terminology in both industrial and academics because traditional power generation resources like fossil fuels or nuclear fission are costly or facing a global crisis [1, 2]. There are various sources of energy harvesting systems like electromagnetic energy from communication and broadcast, temperature gradient from the combustion engine [3–5], and motion from human movement [6]. Recently, areas of research interests consist of waste heat recovery [7–9], pyroelectric energy harvesting, and piezoelectric energy harvesting. The recent areas of energy harvesting system are capable of generating sufficient power that can drive low-power electronic devices.

As per the World Health Organization (WHO) statistics, 3 billion people cook in open fire by burning different biomass like animal dung, coal, wood, and crop waste. About 4 million people die from illness due to household air pollution by cooking with different biomass fuels [10, 11]. The incomplete combustion of biomass fuels leads to dense soot formation that is highly hazardous for small children or women [12, 13]. In developing countries, biomass energy leads to about 90% of the total rural supplies. Environmental degradation has increased over the last several decades [14, 15]. Rapid industrialization, vehicular emissions, and urbanization are some major causes of air pollution. Sulfur dioxide (SO_2), suspended particulate matter (SPM), and respirable particulate matter (RPM) are some damaging effects on human health due to air pollutants. Among all, SPM is considered a leading cause that affects health. After China and the USA, India is third among the countries with the highest CO_2 emissions. As per WHO statistics, air pollution leads to 0.6–1.4% of disease in developing countries.

This paper aims at developing a portable system waste heat recovery system (WHRS) that can be used for low-power applications. The advancement of thermoelectric generators (TEG) has led to extensive research in the field of greener sources of energy and focus on to convert the waste heat into useful electrical energy. The novelty of this paper lies in the utilization of the waste heat and converts it into voltage/ power which can be used in many applications. The designed prototype has been evaluated for three different modules to achieve better results. All the results were simulated and experimentally verified for different stove systems at different loads. The aim and motivation of this paper is to encourage environmentally and eco-friendly self-sustainable local households.

Our specific objectives are as follows:

1. Designing a model for testing electrical power output across various loads.
2. Interfacing of the material with an energy harvesting circuit.
3. Testing of the prototype with traditional Indian cooking furnace.
4. Harvest the energy for final application.

In Sect. 2, the proposed methodology is discussed. Section 3 explains the results and discussion of the implementation of different module followed by conclusion and future work.

2 Methodology

The world is moving toward an era where there is a decrease in energy resources and an increase in pollution (exhaust heat from vehicles/ companies). In recent years, the main concern of environmental issues of emissions and the limited energy resources has resulted in extensive research into novel technologies of generating electrical power [16, 17]. The non-renewable resources are getting exhausted day by day, and a lot of work is being done and proposed renewable energy resources. In this research, paper authors have proposed a waste heat recovery system (WHRS) prototype which can use waste heat from the cooking system for voltage generation. TE materials motivate the development of TEG. TEG is a suitable energy source among other various sources, especially where other sources are not resulting in better results.

The proposed methodology consists of four modules namely cooking system, TEG system, control system, and load system (shown in Fig. 1). As biomass fuel burns in Chullha or stove, it leads to useful energy as well as heat that gets wasted in environment. This waste heat is used and converted to electric energy using TEG. The energy flows in the control system which is designed to harvest max possible power that can be stored in a battery which further may be used for low-power electronics applications.

Cooking System: There are various types of Chullhas used in rural areas with varying hearth height and hearth diameter. Depending on these dimensions, different Chullhas were designed [18].

TEG System: Thomas Johann Seebeck discovered that thermal gradient formed between two dissimilar conductors produces a voltage. The flow of charge carrier between hot and cold region in turn creates a voltage difference. The ideal TEG materials have higher value of electrical conductivity, with low thermal conductivity, and high Seebeck coefficient. Low thermal conductivity is necessary to have high gradient between the junctions [8].

Control System: This system consists of DC-DC converter and booster circuit. The DC voltage converts to AC first then steps up the AC voltage. Later, this AC voltage is fed in to the power management circuit which converts back to DC voltage [19].

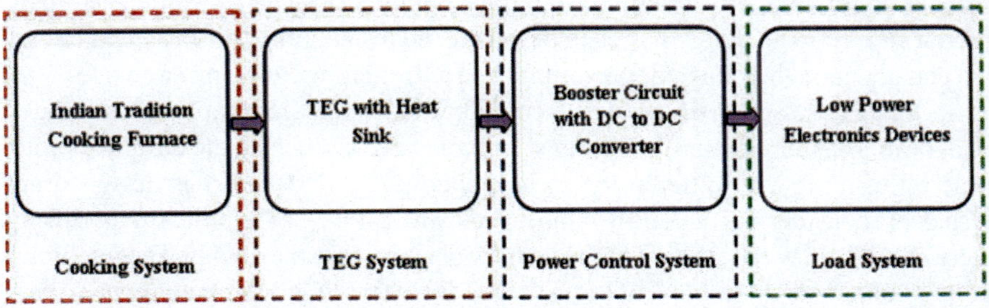

Fig. 1 Waste heat recovery system (WHRS)

Load System: The amplified DC is used for storage in battery or any other low-power electronic devices.

The generator could generate enough energy to power devices like mobile phone, chargeable lights, and other such similar low-power applications. WHRS is a portable, affordable, and long-lasting generator that works in all climatic conditions. The outcome of this system would lead to reductions in fuel consumption. The heat which is wasted out and pollutes the environment can be used by common man.

3 Results and Discussion

Waste heat generated from Indian cooking furnace can be used to convert into the energy that can be stored and reused further to charge low-power devices. This is a step toward a greener source of energy as it does not hinder the normal working of the chullhas. The waste heat from any source can be utilized to develop different modules for practical applications.

The developed design is a simple one and can be used in all households in every type of chullahs. The power output and efficiency are dependent on the temperature of the cold and hot side and also the temperature difference between them. To obtain higher efficiency and power, the temperature difference should be maximum. Components evaluation and selection (such as high ZT value, good stability, thermoelectric module, non-toxicity, good contact property, low cost) play a prominent role. Moreover, for better performance, minimization of thermal resistance between module and heat sinks is required. This is achieved by using a good flat surface, high-temperature thermal grease, and uniform clamping pressure. The power management system is also essential to obtain the maximum power of TEG. The internal resistance of TEG must be comparable to the effective resistance of the battery. Several performance metrics as cost, reliability, and power are taken into account for selecting the suitable best alternative. Different design parameters like reliability, cost, and power are required for the selection of the best alternative.

In this paper, the circuitry has been designed that can convert the waste heat to the electrical energy and tested across various loads. Also, the power conditioning circuit has been designed specifically for traditional Indian cooking furnace and tested to achieve continuous power output. Additional circuitry that can convert and store the produced electrical energy has been designed and tested. From the designed circuit, we can get enough voltage or power that helps in charging low-power devices.

In this paper, authors have designed and tested three different types of models under different temperature conditions. Authors have also tested the designed prototype by using it with different types of chulhas. ***The first model*** is designed and tested by considering the spirit lamp as shown in Fig. 2. Empirical experiments were conducted with several TEGs and their placement in the circuit. The maximum voltage obtained by single TEG is 237 mV, for two TEGs', the maximum voltage obtained is 352 mV, while when using 4 TEGs, 485 mV is achieved when employed

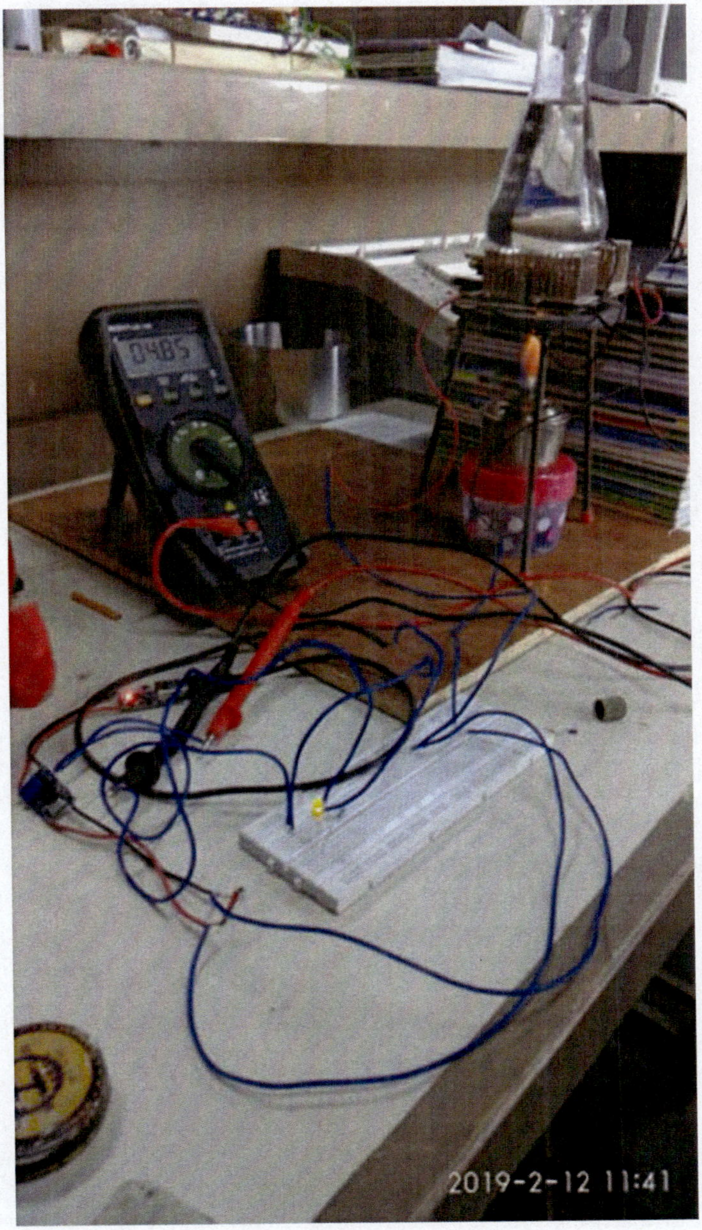

Fig. 2 Prototype model using spirit lamp in the laboratory

Fig. 3 Prototype model using copper plates on conventional chullah

with boiling water heat source. When this voltage is further fed to the booster circuit, a voltage of 3.5 V is obtained.

To improve the voltage and its application, a novel type of ***the second model*** with two copper heat-conducting flat plates (shown in Fig. 3) installed oppositely either of sides of chulhas is designed to integrate a relatively large number of TE modules. The TEGs are connected to the copper plate with thermal grease; the voltage generated by a single TEG is nearly 500 mV. This voltage is fed to the DC-DC converter LTC3108 yielding load current of 150 mA when connected with a 50-Ω load that is used to charge a capacitor. To increase the voltage, TEG should have a proper temperature difference. Authors have used a module with water to keep hot and cold plates at a particular temperature (as shown in Fig. 4) to result in a better current value that helps in charging a device.

To improve the current, a new model has been proposed. Four steel rods (two in series and two in parallel) with TEG's are placed as shown in Fig. 5.

Figures 5 and 6 represent the experimentations under different temperature condition. To maintain the temperature difference of TEGs, an aluminium-based model is designed (as shown in Fig. 7). This model results in 4 V, 250 mA current with four TEGs which is sufficient to charge a low-power EC devices.

Table 1 tabulates the voltage and current values for our three proposed models. It is interpreted that model 3 results in maximum voltage and current values.

The final proposed model has been designed for Indian cooking furnace (as shown in Fig. 8) and tested to achieve continuous power output.

Testing of prototype has been done with different types of chullhas and different types of loads. Different models are designed and tested for achieving highest efficiency for charging.

Fig. 4 Conventional chullah installed with prototype model

Fig. 5 Designed structure for WHRS

4 Conclusion

In this work, waste heat is utilized which could be extracted from chullhas employed in every household in rural areas. The TEG-based energy harvesting circuit presents an efficient method of generating electricity from waste heat. The source of powering

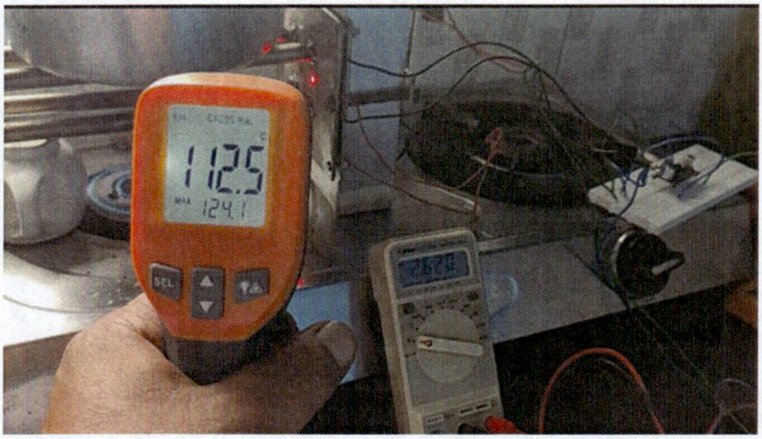

Fig. 6 Proposed WHRS in the laboratory set-up depicting V and I ratings

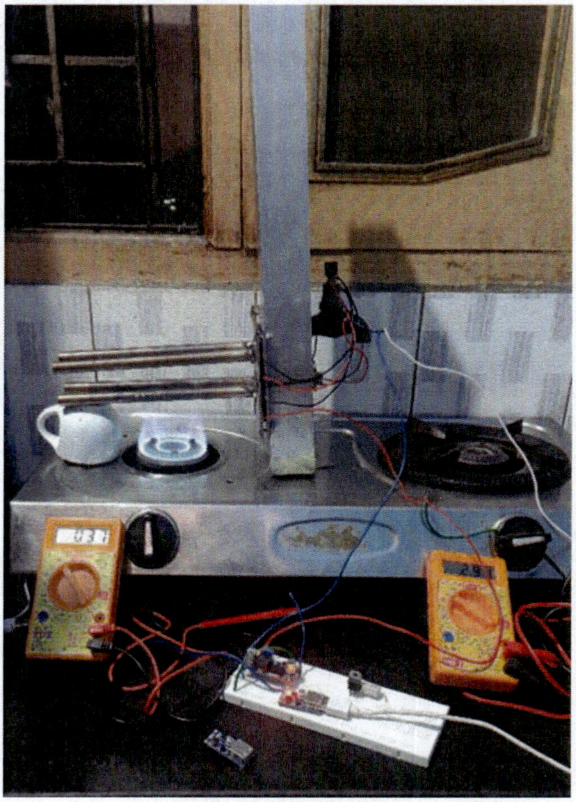

Fig. 7 Final model with booster circuit for WHRS system

Table 1 Electrical parameters for proposed models

Proposed models		Voltage (in V) (4 TEG's)	Current (mA)
Model 1	Using spirit lamp	485 mV	100
Model 2	Using two copper plates	500 mV	150
Model 3	Using four steel rods	2 V	250

Fig. 8 Mobile charging by designed module employing the proposed WHRS

the device is inexhaustible, and the energy harvested from this circuit will be helpful in supplementing the increasing demand for electricity. The waste heat obtained from primary source has low efficiency and, to enhance the energy content, energy harvesting and booster circuits are designed for utilizing the recovered energy.

1. In this paper, different energy harvesting modules were designed and tested for different types of chullhas and load.
2. By designing the proposed energy harvesting system, we were able to utilize the waste heat from chullhas and use it for low-power microelectronic devices.

This is a promising technology for utilizing the waste heat without polluting the environment of households. In the future, we hope that technology will help to replace dependence on environmentally unfriendly batteries and over-reliance on using power outlets to charge small devices.

Acknowledgements This work is funded vide project no. SCSTE/F(8)-1/2017-5085-5086 by Himachal Pradesh Council for Science, Technology and Environment, Himachal Pradesh, India.

References

1. Sood, M., Kashyap, V, Jain, S.: Energy harvesting system from household waste heat employing thermoelectric generator. In: 3rd International Conference on Advanced Informatics for Computing Research (ICAICR-2019), pp 3–12, Solan, India, (2019)
2. Singh, K.R., Sharma, O., Kathuria, M., Jain, S.: Design of energy harvesting generators from waste heat of home chimney using thermocouples" Project report
3. Liu, C.K., Han, W.K., Chun H.: Thermoelectric waste heat recovery for automotive. In: Microsystems, Packaging, Assembly and Circuits Technology Conference (IMPACT), 9th International (2014)
4. Zhang, X., Chan, C.C., Li W.: An automotive thermoelectric energy system with parallel configuration for engine waste heat recovery. In: International Conference on Electrical Machines and Systems (ICEMS), pp. 1–6 (2011)
5. Hsu. C.-T., Yao, D.-J., Ye, K.-J., Yu, B.: Renewable energy of waste heat recovery system for automobiles. J. Renew. Sustain. Energy. **2**(1) (2010)
6. Carmo, J.P., et al.: Thermoelectric micro convertor for energy harvesting system. IEEE Trans. Ind. Electron. **57**(3), 861–867 (2010)
7. Prashantha, K., Wango, S.: Smart power generation from waste heat from thermoelectric generator. Int. J. Mech. Prod. Eng. 45–49 (2016)
8. Kashyap, J.S., Sood, V.M.: Optimizing booster circuits for thermo electric generators to utilize waste heat. In: International Conference on Sustainable Computing in Science, Technology and Management, Amity University Rajasthan, pp 36–46. Jaipur, India (2019)
9. Zhang, X., et al. The match of output power and conversion efficiency of thermoelectric generation technology for vehicle exhaust waste heat. In: Eighth International Conference on Measuring Technology and Mechatronics Automation. pp. 805–810 (2016)
10. Champier, D., et al.: Thermoelectric power generation from biomass cook stove. Energy **35**(2), 935–942 (2010)
11. Mishra, R., Jain, S., Durgaprasad, C., Sahu, S.: Vibration energy harvesting using drum harvesters. Int. J. Appl. Eng. Res. **10**(14), 34995–35001 (2015)
12. Jain, S., Kashyap, V., Sood, M.: Design and analysis of thermoelectric energy harvesting module for recovery of household waste heat. In: 2nd International Conference on Recent Innovations in Computing (ICRIC-2019), Central University of Jammu, J & K (2019)
13. Mishra, R., Jain, S., Thakur, B., Verma, Y.P., Durgaprasad, C.: Performance analysis of piezoelectric drum transducers as shoe-based energy harvesters. Int. J. Electron. Lett. **5**(4), 1–15 (2016)
14. Kutt, L., Lehtonen, H.: Automotive waste heat harvesting for electricity generation using thermoelectric system-an overview. In: IEEE International Conference on Power Engineering, Energy and Electrical Drives, pp. 55–61 (2015)
15. Sornek, K., et al.: The development of a thermoelectric power generator dedicated to stove-fireplaces with heat accumulation system. Energy Convers. Manage. 1–9 (2016)
16. Sultana, A., et al.: A pyroelectric generator as a self powered temperature sensor for sustainable thermal energy harvesting from waste heat and human body heat. Appl. Energy **221**, 299–307 (2018)
17. Hoque, M.N., Rahman, M.S., Nahar, N.: Development of portable traditional triple mouth Chula. J. Bangladesh Agric. Univ. **8**(1), 141–145 (2010)
18. Gao, H.B., et al.: Development of stove powered thermoelectric generator: a Review. Appl. Therm. Eng. **96**, 1–42 (2015)
19. Xie, W., et al.: A maximum power point tracking controller for thermoelectric generator. In: Proceedings of the 36th Chinese Control Conference, pp. 9079–9084 (2017)

Template Attacks and Protection in Multi-biometric System: A Systematic Review

Syed Umaya Anayat and Arvind Selwal

Abstract In the modern era of computing, the automatic authentication using human biometrical traits is being rapidly gaining popularity. The biometrical infrastructure installations are facing a variety of challenges ranging from limitations in uni-biometrics and attacks by imposters. Multi-biometric systems are being deployed to overcome the limitations of uni-biometric counterparts. The key information of all the enrolled users in biometric systems is stored centrally in the template database. The attacks on the template database in these systems may either completely result in failure or degradation in the performance of the system. This paper presents a systematic review of various template attack and their countermeasures in multi-biometric systems. The survey reveals that information fusion, compatibility of templates and size of templates are still an open challenge for researchers. Furthermore, the need is to design more robust and efficient template security schemes for multi-biometric system which meets the standard characteristics of ideal schemes.

Keywords Template security · Attacks · Multi-biometric systems · Fingerprint

1 Introduction

In the traditional systems to protect private data or to reserve the access rights, people have been using passwords, PINs, etc. But it has some limitations, due to which there came a need for biometrics. Biometrics is described as a branch of science used for the identification of a person by using their physical, chemical or behavioral qualities [1]. A biometric system consists of four main modules, such as **sensor module, which** captures the raw biometric information of a person. **Feature extraction module** extracts the unique traits from the biometric data. For example, the orientations and

S. U. Anayat (✉) · A. Selwal
Department of Computer Science and Information Technology, Central University of Jammu, Samba, Jammu and Kashmir 181143, India
e-mail: syedumayandrabi@gmail.com

A. Selwal
e-mail: arvind.cuj@gmail.com

P. K. Singh et al. (eds.), *Recent Innovations in Computing*, Lecture Notes in Electrical Engineering 701, https://doi.org/10.1007/978-981-15-8297-4_67

position of the minutia points in a finger are extracted as features for the fingerprint image. **Matching and decision-making module**: Matching module uses the classifier to contrast the extracted characteristic set with the stored template to produce the score value. The score value is used by the decision module for acquiring or denying an individual. **System database module** is used to store the biometric information [1, 2].

The biometric systems may work in two different modes, namely identification mode or verification mode. In case of identification mode, the user is identified by comparing his input with the templates that are already stored in the database, whereas in the verification mode, the user's identity is verified against the claimed identity and checked whether the user is genuine or not [3, 4].

Generally, the biometric systems are being deployed in various application areas, such as **commercial**, for example, ATM, distance learning and PDA; **government**, for example, social security, border and airport security; **forensic**, for example, corpse identification, parenthood determination and criminal investigation [1].

The remainder of the paper is organized as follows: Section 2 briefly defines multi-biometric system and fusion levels. In Sect. 3, we provide an overview of biometric template attacks. In Sect. 4, we give an outline of template protection schemes. Research challenges, opportunities and conclusion are defined in the last section.

2 The State-of-the-Art of Multi-biometric Systems

The biometric framework has been divided into two categories depending upon the number of traits used, namely **uni-modal and multi-modal system** in which the system uses a single biometric trait of the individual for identification and verification [5]. There are certain limitations associated with these systems such as intra-class variation, spoofing attack, failure to enroll and inter-class variation [4, 6]. To overcome these limitations, the multi-biometric system was introduced. These systems are capable of using two or more modalities to identify an individual [5, 7]. The accuracy performance of a biometric system can be increased by employing the multi-biometric system rather than the uni-biometric system [8].

The fusion plays a major role in multi-biometrics. It is carried out at five different levels as shown in Fig. 1.

(i) Sensor-level fusion: At this level, raw data is generated from different sensors, and this data is combined to produce new raw fused information. There are three scenarios where sensor-level fusion can be applied, such as single-sensor multi-sample, multi-modal and multi-sensor. (ii) Feature-level fusion: In this scheme, a single feature vector is obtained from the different feature vectors which are generated from different biometric systems. The two different scenarios where feature-level fusion can be applied are multi-modal and multi-algorithm. (iii) Score-level fusion: In this method, a single match score is obtained from combining multiple scores generated from the different or same modalities. The scenarios where score-level

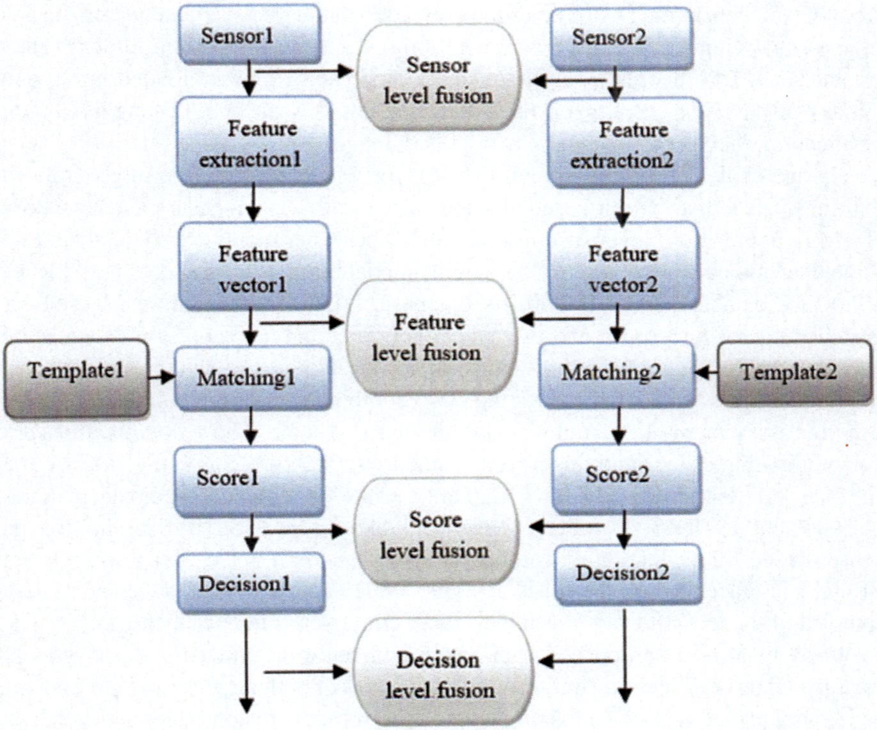

Fig. 1 Fusion levels in biometric system (adopted from [9])

fusion is applied involves multi-modal and multi-sample systems. (iv) Decision-level fusion: In this scheme, a single decision can be obtained by combining the different information generated from the multiple-decision modules. The methods used in decision level fusion include 'AND', 'OR', majority voting and weighted majority voting. (v) Rank-level fusion: In this mechanism, multiple ranks are combined to produce a single rank which is used to identify the individual [7, 9].

Kabir et al. [10] proposed a matcher performance-based (MPB) fusion scheme. They used modalities such as fingerprint, palm print and ear print. They perform fusion at two levels, i.e., feature level and score level, for improving the complete identification efficiency. In feature-level fusion, the encoded characteristics of two traits with high EER are firstly fused. In score-level fusion, the modality which has the lowest EER and score acquired from the feature level fusion are combined. The proposed normalization technique is referred to as overlap extrema variation anchored min-max (OEVBAMM). The confidence-based weighting techniques were used for generating FRR @0.5 FA, and EER, GAR @0.5 FAR.

Arteaga-falconi et al. [11] proposed an ECG matcher that used an SVM classifier with the RBF radial basis function kernel. For fingerprint, they used the MINDTCT algorithm as minutiae extractor and the BOZORTH3 algorithm as minutiae matcher. They fused the biometric results at the decision level. They used fingerprint samples from 73 subjects in the DBI type of the FVC2006 database. Every subject has seven images involving six images for verification and one image for enrollment. They selected 73 ECG records from the database, and each signal is divided into seven chunks of data; for enrollment, they used one chunk with 60 s in length, and for verification, they used six chunks with 40 s in length.

Sharma et al. [12] proposed a unified scheme which is composed of generalized extrema value distribution-based normalization and Dezert–Smarandache theory (DSMT)-based score-level fusion technique. The experiments were conducted on three databases, namely NIST BSSR1 multi-modal biometric score database, FRGC V2.0 face database and LG4000 iris database. NIST BSSR1 dataset is used for various units of fingerprint and two face algorithms and having resemblance value vectors of 6000 images for 6000 subjects. They used half of values for evaluation of training variables and a half for testing. Two combinations of three modalities were used for assessment in the multi-modal biometric scheme. These combinations are (i) combination of face and fingerprint biometric modalities from NIST BSSR1 and (ii) face and iris traits from FRGC V2.0 and LG 4000 datasets. Resemblance values of 3000 samples for 3000 subjects have been taken from NIST BSSR1, half of the score is used for evaluation of training variables, and half is used for testing. From FRGC V2.0 face and LG 4000 iris datasets, 19,079 samples of the face and 19,079 samples of the left iris from 46 subjects were employed for experiment.

Waisy et al. [13] proposed an effective and real-time multi-biometric system established on deep learning description for samples of both the right and left irises of an individual, and rank-level fusion method is used for the fusion. The proposed deep learning system is called IrisConvNet established on the combination of the CNN and Softmax classifier, and without domain knowledge, particular characteristics were extracted from iris images and classified it into one of N classes. The efficiency and durability of the proposed methods were tested on three different databases: SDUMLA-HMT, CASIA-iris-V3 interval and IITD iris.

Bifari et al. [14] presented a weighted score matching algorithm for a multi-modal biometric system established on fingerprint and hand geometry. In this work, authors used five databases called fingerprint verification competition for testing fingerprint, and each database contains 800 samples, i.e., 100 different people and 8 samples for each. They also used COEP (College of Engineering Pune-41,005) palm print database for testing hand geometry, and this database contains 100 persons. The experimental outcome produced the average EER of the multi-modal system was 3.27%, while EER of the fingerprint was 8.86% and EER of hand geometry was 8.89%.

Conti et al. [15] proposed an approach that performed fingerprint matching using the segmented regions (ROIs) surrounding fingerprint singularity points, and iris preprocessing is used to check the circular region surrounding the iris. They used log-Gabor algorithm-based codifier for encoding both fingerprint and iris characteristics to generate a consolidated homogeneous template. Moreover, for similarity index computation, Hamming distance was applied on the fused template. Different datasets from FVC2002 DB2 fingerprint database and BATH iris database were used for testing the multi-modal biometric system. Results of the first test operated on ten users are FAR = 0% and FRR = 5.71% although the results of the tests conducted on the FVC2002 DB2A and BATH databases are FAR = 0% and FRR = 7.28% ÷ 9.7% (Table 1).

Table 1 A comparative summary of different biometric and fusion types

Method	Year	Types of biometric and fusion	Author	Performance measures and database
Frequency-based approach	2010	Multi-modal biometric system and feature-level fusion	Conti et al.	FAR = 0%, FRR = 5.71%, EER = 2.36, On FVC2002 DB2B and BATH databases FAR = 0%, FRR = 7.28÷9.7%, EER = 3.17÷5.76 on FVC2002 DB2A and BATH databases
Weighted sum rule	2017	Multi-modal biometric system and score-level fusion	Bifari et al.	EER = 3.27% and FVC and COEP databases
Ranking-level fusion method	2017	Multi-biometric system and rank-level fusion	Al et al.	Identification rate = 100% and recognition time < 1 s per person. And SDUMLA-HMT, CASIA-Iris V3 Interval and IITD iris databases
SVM classifier and minutiae extractor	2018	Bi-modal and decision-level fusion	Arteaga-falconi et al.	EER = 0.46% and FVC2006
Generalized extreme value (GEV) distribution and Dezert–Smarandache theory	2018	Multi-biometric system and score-level fusion	Sharma et al.	NIST BSSR LG4000 and FRGC V2.0 databases
Overlap extrema variation-based anchored min-max-, and confidence-based weighting method	2019	Multi-biometric system and feature- and score-level fusion	Kabir et al.	EER = 0.47%,GAR@0.5% FAR = 99.73, FRR@0.5%FAR = 0.27%, and processing time 1.1 s per image and FVC2002-DB1-A, COEP and AMI databases

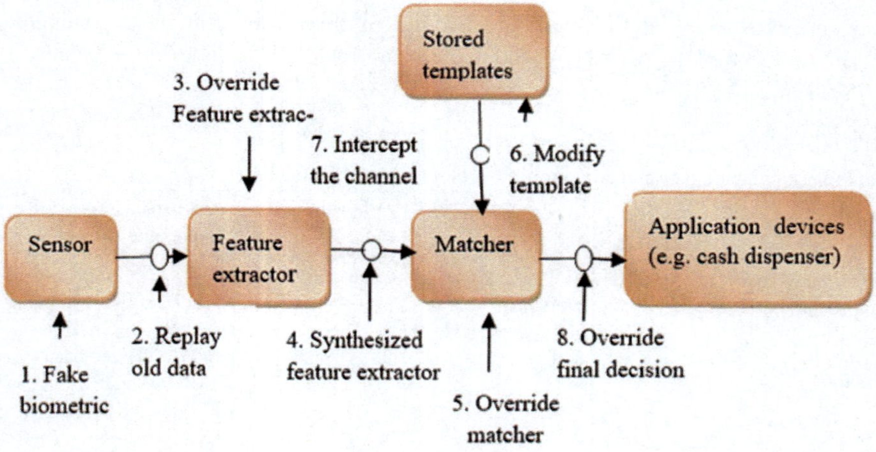

Fig. 2 Attack points in a biometric system (adopted from [16])

3 Biometric System Attacks

The biometric systems are vulnerable to certain kinds of attacks. There are eight attack points in the typical biometric system where attackers can breach the security as shown in Fig. 2.

In this study, we only focus on 'Type 6 attack,' which is an intrusion on the database where the biometric information of the users is stored [3].

4 Background

In this section, we present the review of existing template security schemes in the literature.

To model a supreme biometric template security system, the following four principles are required to be satisfied:

Diversity: To ensuring the user's privacy, the cross-matching across the database should be prevented by generating distinctive secure templates from the same source.

Revocability: One should be capable to revoke the stolen or modified template, and unique template can be generated from the native template. **Security**: One should not obtain the original biometric template from the secure template. **Performance**: The biometric template security scheme should maintain the identification achievement of the biometric system.

The template security schemes are generally classified into two categories, namely biometric cryptosystem and cancelable biometric [17] (Fig. 3).

Biometric cryptosystem: The aim of designing the biometric cryptosystem is to securing a cryptographic pivotal to a biometric or producing a biometric key from

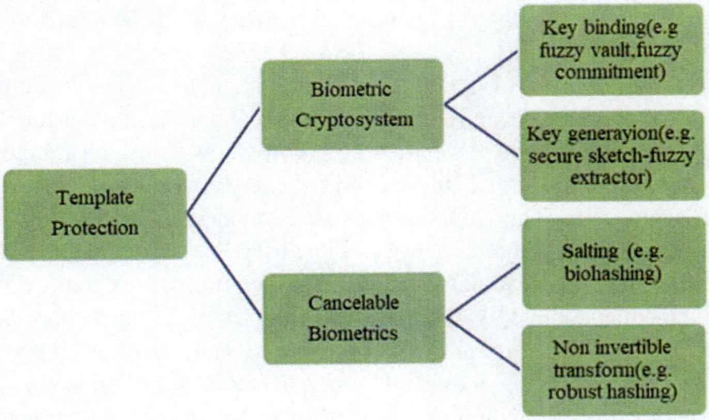

Fig. 3 A taxonomy of template security schemes (adopted from [16])

biometric characteristics. Biometric cryptosystems have needed some general data (called as helper data) for producing key or recapturing key. The helper does not acknowledge important information about the original biometric template [18]. The biometric cryptosystem is also called a helper-data-based method. The helper data is used for the extraction of crypto keys from the query biometric trait during matching [16]. The authenticating key validities are indirectly used for biometric comparisons. The result of the authentication mechanism is either a key or failure message as shown in Fig. 4 [18].

A biometric cryptosystem is further classified into two categories, namely key binding and key generation systems. In key binding schemes, cryptographic keys are not dependent on biometric features. The original data can be obtained from the secured template only by those attackers who know the private key. In key generation cryptosystem, recovery of the key string is not easy as data are not stored directly [19]. The biometric template is only the one used, from which the helper data can

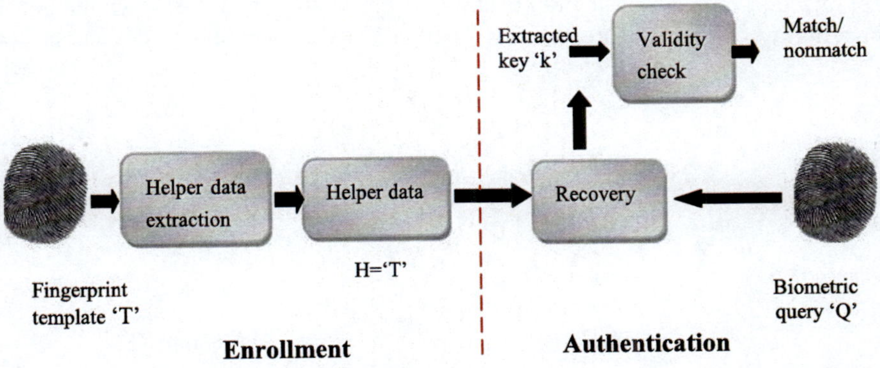

Fig. 4 Enrollment and authentication mechanism in biometric cryptosystem (adopted from [16])

be acquired. The cryptographic key is achieved from the helper information and biometric characteristics [18].

Elrefaei et al. [20] proposed a biometric cryptosystem using a fuzzy commitment scheme. They gathered the gait templates for the machine vision sensor. The local ternary pattern (LTP) is used for extracting gait features from gait images as LTP upper and LTP lower and then calculating the gait energy samples for upper and lower LTP as GEIU and GEIL. These two samples are combined using 2D joint histogram, and joined images have a high dimension. They used PCA to reduce the dimension and make the image compatible with the fuzzy commitment scheme. The Bose–Chaudhuri–Hocquenghem code (BCH) is used for key encoding in enrollment and for decoding in the verification phase. They take nine samples for each person in the CMU MOBO database and then used eight samples for the enrollment phase and one sample for verification phase. For the enrollment phase range of the feature, a vector is 65,536 * 200 later generating a joint histogram and now applying PCA to decrease the size of the characteristic vector to 199 * 200. They take five samples for each person in CASIA A database and then four samples used for the enrollment phase and one sample for verification phase. For enrollment phase, size of the characteristic vector is 65,536 * 76 after generating a joint histogram and now applying PCA to decrease the size of the feature vector to 75 * 76. The FAR = 0% and FRR = 4% in CMU database for a fast and slow walk. Moreover, FAR = 0% and FRR = 0% in CASIA A database for 45-degree direction.

Cancelable biometrics (CB) or Feature transformation: In this approach, biometric template (T) is converted into transformed template (f(T; K)) by applying a transformation function (f) and only the transformed template is reserved into the database as shown in Fig. 5. The matching is done between the transformed template and the transformed query (converted by using the same transformation function (f) to query features (Q)) [16].

Beom et al. [21] proposed partial local structure (PLS) a different alignment-free minutia-based description that is used to construct a binary cancelable fingerprint template with low-efficiency loss. They contour a different transformation method established on the permutated randomized non-negative least square (PR-NNLS) optimization problem on the top of the PLS descriptor which fascinates the non-invertibility of the template security. They used five databases for evaluating the

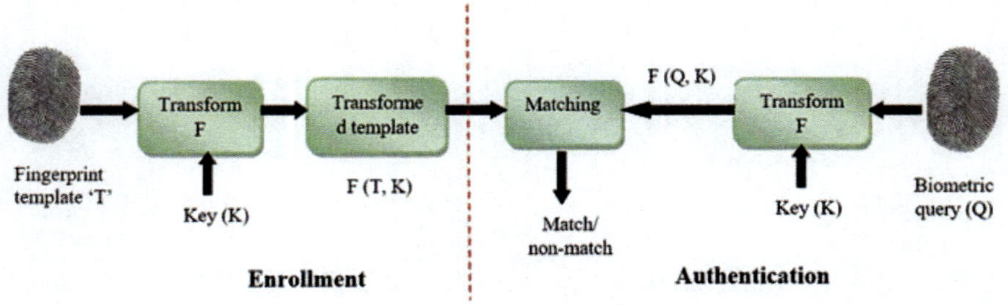

Fig. 5 Fingerprint biometric template after feature transformation (adopted from [16])

proposed method. Moreover, they used FVC2004DB1 for establishing MD and PCA projection matrix. In all dataset, 100 subjects with eight images per subject are prompt for minutia extraction. VeriFinger 6.2 is used, and for experimentation, they have used two distinct protocols, namely original FVC and the 1vs 1 protocol. The system achieves cancellability, non-invertibility, unlinkability and prevents from privacy and security attacks.

There are two types of cancelable biometrics based on aspects of transformation function (f), namely biometric salting and non-invertible transforms. In the case of biometric salting, the transformation function (f) which is invertible is used to the biometric template for obtaining the transformed template. The primary biometric information can be reconstructed by the opponent from the transformed template if transform specification is compromised and the accomplishment of the system gets strayed [18]. As the key is distinct to the person, different templates can be generating for the same biometric information. The compromised template can be easily revoked and changed with the template achieved by using alternate user-specific key [16]. In case of non-invertible transformation, the biometric template is transformed by using the non-invertible transformation function (f). The features of applied non-invertible transforms are changed, to produced updatable templates. The benefit of this approach is that, when the transforms are comprised, the opponent cannot generate all biometric information [18].

Gomez-barrero et al. [22] proposed a general framework based on bloom filters for biometric and multi-biometric template protection. The weighted feature-level fusion for protecting templates was proposed to enhance verification efficiency along with the level of solitary protection presented. The irreversibility and unlinkability of the system are accomplished by using this approach. The irreversibility of the templates is increased because OR operation is performed in this approach. Multi-hypothesis sequential probability ratio test (MSPRT) is only evaluated for the protected template. Experiments are carried out on different databases for development and testing stages.

Gomez-barrero et al. [23] proposed a framework for multi-biometric template protection based on homomorphic probabilistic encryption for enhancing the performance and accomplishing more protected and privacy conserving systems. They carry out two distinct measures and contrast these in terms of verification performance, irreversibility, unlinkability and computational complexity. They evaluate and describe distinct models for three fusion levels; feature-level fusion has achieved better performance and produced a solitary template for every subject. The experiments were performed out on the fingerprint and online signature using BiosecureID multi-modal database. They also achieved irreversibility and unlinkability. In the feature level, EER = 0.12% was obtained.

Selwal et al. [24] proposed three multi-modal frameworks based upon fingerprint and hand geometry biometric modalities. In the case of the multi-modal framework1 feature, templates are generated by using two different feature extractors. The fusion has occurred at feature level, and the fused template is reserved into the database. In the case of multi-modal framework2, two distinct feature templates are extracted, no fusion takes place and templates are stored in different databases. In the case of multi-modal framework3, various feature extractors are used for both the biometric traits and generate four different feature templates. These templates are combined by using rank-level fusion. They used the fuzzy analytic hierarchy process (FAHP) technique and identified five variables affecting the design of multi-modal. According to the fuzzy weight matrix, the weight of the template security is 0.3940 that means it is the highest essential parameter, and next is template size overhead with weight 0.342. By considering all the three proposed system frameworks, framework1 has an ideal design with a result of 0.72796.

Wild et al. [25] investigated that spoofing and recognition are jointly important for each other. They also assess the 1-median filter as a different multi-biometric fusion method consolidating liveness and identification scores. They employed three base classifiers, namely regularized LR, single-layer perceptron and SVM. Three databases were used, i.e., LivDet 2013 crossmatch: contains 4500 samples of 99 users; Idiap Replay-Attack: 1300 clips of 50 users with 320 * 240 pixels resolution; and CASIA antispoofing face: 600 clips of 50 users with 640 * 480 pixels resolution. The EER of this 1-media filter is between 0.47 and 1.81% for $m = 1 - 5$, i.e., EER is low as compared to other rule and d-prime is high from 3.18 to 3.08. They proposed a bootstrap aggregating classifier for antispoofing and show accuracy = 84% on challenging database LivDet 2013crossmatch dataset.

Chin et al. [26] presented a three-stage hybrid biometric template protection method for a secure multi-biometric system, namely (i) feature-level fusion used for obtaining a unified template, and the fingerprint and palm print biometric systems are combined at the feature level; (ii) random tiling (RT)—the revocability and diversity are achieved by abstracting distinctive and random attributes from the combined features by employing a user-determined key; and (iii) equal–probable 2N discretization (eq 2N)—ultimate bit–string template is to produce from the consolidated feature vector confer to the allocated index. They used three fingerprints and two palm print databases, and 100 subjects and 8 samples for each participant are present in all databases. The revocability is achieved by using various user-determined keys; templates of bit–string generated are also distinct from each other. They used a short-time Fourier transform analysis for enhancing the sample aspect of fingerprint and Gaussian low-pass filter for smoothing the aspect of the palm print samples. Moreover, they adopted a set of Gabor filters along with eight distinct angles and used them to filter the sub-bands generated by two-dimensional discrete wavelet transform (2DDWT) (Table 2).

5 Research Challenges and Opportunities

After going through the literature, it is clear that designing an efficient template protection scheme for multi-biometric systems is still an open challenge. Moreover, designing a template protection scheme which satisfies all the ideal characteristics of a template security scheme is a typical opportunity for researchers. Following are the major research challenges which need to be addressed in future.

Table 2 A comparative overview of various template protection schemes

Method	Year	Types of biometric and fusion type	Author	Performance measures and database
Three-stage hybrid template protection method	2013	Multi-biometric system and feature-level fusion	Chin et al.	–
1-median filter	2016	Multi-modal and score-level fusion	Wild et al.	EER = 0.47–1.81% and d-prime = 3.18–3.08 and LivDet 2013 cross Match, Idiap Replay-Attack and CASIA antispoofing face
Fuzzy analytic hierarchy process	2016	Multi-modal and three levels of fusion	Selwal et al.	Overall value = 0.72796
Homomorphic encryption	2017	Multi-biometric and three levels of fusion	Gomez-barrero et al.	EER = 0.12% and BiosecureID database
Bloom filters	2018	Multi-biometric system and feature-level fusion	Gomez-barrero et al.	–

1. Designing the more secure system which is less vulnerable to attacks. For example, in uni-biometric systems, the sensor is more vulnerable to attacks like attack on the face or fingerprint biometric system using fake artifacts.
2. Dimensionality reduction: For example, the dimension of a fingerprint is n*3, and dimension of the iris is fixed, i.e., 512 bytes.
3. Deciding the level where fusion can occur so that the performance and accuracy of the system must be improved.
4. The selection of the protection schemes for ensuring the template protection of the system.
5. Selection of multiple sources (e.g., multi-modal, multi-algorithm, etc.) to overcome the limitations of the uni-biometric system.
6. Designing adaptive and dynamic fusion system: For example, in online learning, data can be regularly changed so we have to design such fusion method to regularly progress overtime to receive changes in system specification along with modification in the data distribution.
7. The conflict between systems must be resolved. For example, in identification mode, different modalities may produce a completely different list of ranked identities or in verification mode, one half of classifier should confirm the claimed identity, another half disprove the claimed identity, so for such cases, it is necessary to have a principled way to produce the decision.

The multi-modal systems are developed with highly secured traits like fingerprint and iris multi-modal system for improving recognition rate and lower error rate.

Moreover, for the security of the biometric system, multi-factor authentication methods needed to be developed.

6 Conclusions

At present, the world needs very secure and protected authentication systems. When security is the primary concern, then the multi-biometric system is used although it takes much more time than the uni-biometric system. This research work mainly focuses on template attacks and their countermeasures in multi-biometric systems. The biometric templates that are stored in a database are mostly attacked by the attackers. The template biometric schemes are used to secure the template, but their selection depends on application scenario and specification. We can achieve a high-protection multi-biometric system by using suitable normalization and fusion techniques.

References

1. Jain, A.K., Flynn, P. Ross, A.A.: Handbook of biometrics handbook of biometrics (2007)
2. Jain, A.K., Ross, A., Prabhakar, S.: An introduction to biometric recognition. **14**, 4–20 (2004)
3. Jain, R., Kant, C.: Attacks on biometric systems : an overview * correspondence info **01**, 283–288 (2015)
4. Jain, A.K., Nandakumar, K., Ross, A.: 50 years of biometric research : accomplishments, challenges, and opportunities, **79**, 80–105 (2016)
5. Oloyede, M.O., Member, S., Hancke, G.P.: Unimodal and multimodal biometric sensing systems : a review. **3536**, (2016)
6. Jain, A., Nandakumar, K., Ross, A.: Score normalization in multimodal biometric systems, **38**, 2270–2285 (2005)
7. Kumar, S., Modak, S., Jha, V.K.: Multibiometric fusion strategy and its applications: a review. Inf. Fusion **49**, 174–204 (2019)
8. Martiri, E.A.: Feature fusion scheme for multi-biometric template protection. In: IFAC Proceedings Volumes, vol. 46 IFAC (2013)
9. Ali, M.M.H., Gaikwad, A.T.: Multimodal biometrics enhancement recognition system based on fusion of fingerprint and PalmPrint : a review (2016)
10. Kabir, W., Member, S., Ahmad, M.O.: A multi-biometric system based on feature and score level fusions. IEEE Access **7**, 59437–59450 (2019)
11. Arteaga-falconi, J.S., Osman, H., Al & Saddik, A.El.: ECG and fingerprint bimodal authentication. Sustain. Cities Soc. **40**, 274–283 (2018)
12. Sharma, R., Das, S. Joshi, P.: Score-level fusion using generalized extreme value distribution and DSmT, for multi- biometric systems, pp. 1–8 (2018)
13. Al, A.S., et al.: Short Paper a multi-biometric iris recognition system based on a deep learning approach. Pattern Anal. Appl. (2017)
14. Bifari, E.N., Elrefaei, L.A.: Aweighted score matching algorithm for a multi-modal biometric system based on **3**, 1–24 (2017)

15. Conti, V., Militello, C., Sorbello, F., Vitabile, S.: A Frequency-based approach for features fusion in fingerprint and iris multimodal biometric identification systems **40**, 384–395 (2010)
16. Jain, A.K., Nandakumar, K., Nagar, A.: Biometric template security. **2008** (2008)
17. Rafiq, S., Selwal, A.: Template security in Iris recognition systems: research challenges and opportunities. In: Singh P., Kar A., Singh Y., Kolekar M., Tanwar S. (eds.), Proceedings of ICRIC 2019. Lecture Notes in Electrical Engineering, vol. 597. Springer, Cham (2020). https://doi.org/10.1007/978-3-030-29407-6_55.
18. Rathgeb, C., Uhl, A.: A survey on biometric cryptosystems and cancelable biometrics, pp. 1–25 (2011)
19. Müftüoğlu, Z., Yildirim, T.: ScienceDirect Comparative analysis of crypto systems using biometric key comparative of crypto systems using biometric key research on analysis the innovation of protecting intangible cultural heritage in the "Internet Era *, Tülay Plus ". Procedia Comput. Sci. **154**, 327–331 (2019)
20. Elrefaei, L.A., Al-mohammadi, A.M.: Machine vision gait-based biometric cryptosystem using a fuzzy commitment scheme. J. King Saud Univ. Comput. Inf., Sci (2019)
21. Beom, J., Kim, J., Kim, I., Teoh, A.B.J.: Cancelable fingerprint template design with randomized non-negative least squares. Pattern Recognit. **91**, 245–260 (2019)
22. Gomez-barrero, M., et al.: Multi-biometric template protection based on bloom fi lters. Inf. Fusion **42**, 37–50 (2018)
23. Gomez-barrero, M., Maiorana, E., Galbally, J., Campisi, P., Fierrez, J.: Multi-biometric template protection based on Homomorphic Encryption. Pattern Recognit. **67**, 149–163 (2017)
24. Selwal, A., Kumar, S.: Template security analysis of multimodal biometric frameworks based on fingerprint and hand geometry. Perspect. Sci. **8**, 705–708 (2016)
25. Wild, P., Radu, P., Chen, L., Ferryman, J.: Robust multimodal face and fingerprint fusion in the presence of spoofing attacks. Pattern Recognit. **50**, 17–25 (2016)
26. Chin, Y.J., Ong, T.S., Teoh, A.B.J., Goh, K.O.M.: Integrated biometrics template protection technique based on fingerprint and palmprint feature-level fusion. Inf. Fusion. (2013)

Predictive Analysis for User Mobility Using Geospatial Data

Jai Prakash Verma, Sudeep Tanwar, Archies Desai, Poojan Khatri, and Zdzislaw Polkowski

Abstract Extremely usage of smart wearable devices such as smartphones and smartwatches which contain various sensors for location detection such as Wi-Fi, LTE, GPS and motion detection such as accelerometer, it has become easier to obtain user mobility data. Today communication systems are becoming more popular due to the developments in communication technologies. There are various services provided which also help to access the data such as video, audio, images from which we can be used to grab the information or pattern of user mobility. The user mobility where user's movements and locations can be predicted using various methods and algorithms. It can be predicted through data mining, machine learning, and deep learning algorithms where user's data are fetched from the communication system. A comparative data mining model base on DBSCAN and RNN-LSTM was proposed for predicting the user's future location-based information predicted from the last locations reported. Mobility prediction based on the transition matrix prediction is done from cell to cell and calculated with the help of the previous inter-cell movement.

Keywords User mobility · Predictive analytic · Machine learning · Deep learning · Geospatial data

J. P. Verma (✉) · S. Tanwar · A. Desai · P. Khatri
Institute of Technology, Nirma University, Ahmedabad, Gujarat, India
e-mail: jaiprakash.verma@nirmauni.ac.in

S. Tanwar
e-mail: sudeep.tanwar@nirmauni.ac.in

A. Desai
e-mail: 16BIT021@nirmauni.ac.in

P. Khatri
e-mail: 16BIT027@nirmauni.ac.in

Z. Polkowski
Wroclaw University of Economics and Business, Wroclaw, Poland
e-mail: zdzislaw.polkowski@ue.wroc.pl

P. K. Singh et al. (eds.), *Recent Innovations in Computing*, Lecture Notes in Electrical Engineering 701, https://doi.org/10.1007/978-981-15-8297-4_68

1 Introduction

User mobility detection has become a job for daily life where mobility is predicted from wearable devices like smartphones, watches, etc. In this paper, a data analytic approach is used for prediction of the user future location based on their geospatial data generated from socially connect networks [19]. User mobility detection is widely used nowadays for detecting the location, tracing the path, for detection the ratio of population, mobile users, an internet user, etc. thus mobility prediction is used in various domains [5]. This data can be used for various applications ranging in scope from a convention hall (party management) to city-wide (city planning), and spanning multiple cities (disaster management) [10, 16, 17].

User mobility detection and prediction is an important research area. Basically user mobility means the user movement. A record of this user movement is taken and then by the use of various techniques the future mobility of the user can be detected [1]. This can be useful in many circumstances which we will encounter in the next sections During the progression of this work, we will design a machine learning-based data analysis model trained in such a way that given the input of the previous locations visited by the user, wherein the location is taken in the general form of latitude and longitude; the next location of the user can be predicted accordingly [9, 21].

As per Fig. 1 a total number of smartphone users has risen to an estimated 2.71 billion and the total number of internet users has risen to 3.9 billion as illustrated in the diagram above. These users collectively generate a huge amount of data which when used correctly can be extremely useful for many types of applications [5, 20]. Figure 1 is showing a quick fact that is collected from google trends, it shows how the interest of people for internet uses increases the data volume which can be used for user mobility detection and prediction.

Though there are high variations in the mobility prediction there also exhibit the structural pattern due to geographic and social constraints where the user's locations based on their social relationship. It predicts a group of similar places of friends or people visited in the past. But though there are some unanswered questions due to difficulty in obtaining large scale human mobility data [1, 14]. Due to the rapid emerges of the new technologies and expansion of the networks there are new possibilities for understanding human behavior. Smart devices such as smartwatches or smartphones identify its environment and follow every action based on built-in sensors. Those sensors are also used in areas such as health, sports and general user monitoring where the transportation system is very well adapted to this type of system [6, 9]. The routing of vehicles can be predicted through the smart device sensors such as an accelerometer, magnetometer, and gyroscope.

The main objective of the study here is to learn various methods of Machine Learning [15, 18]. With the use of these methods, we work on the data analysis model to provide a solution to the above-mentioned problem. The data in real life is too large and hence handling them can be difficult and requires more resources for

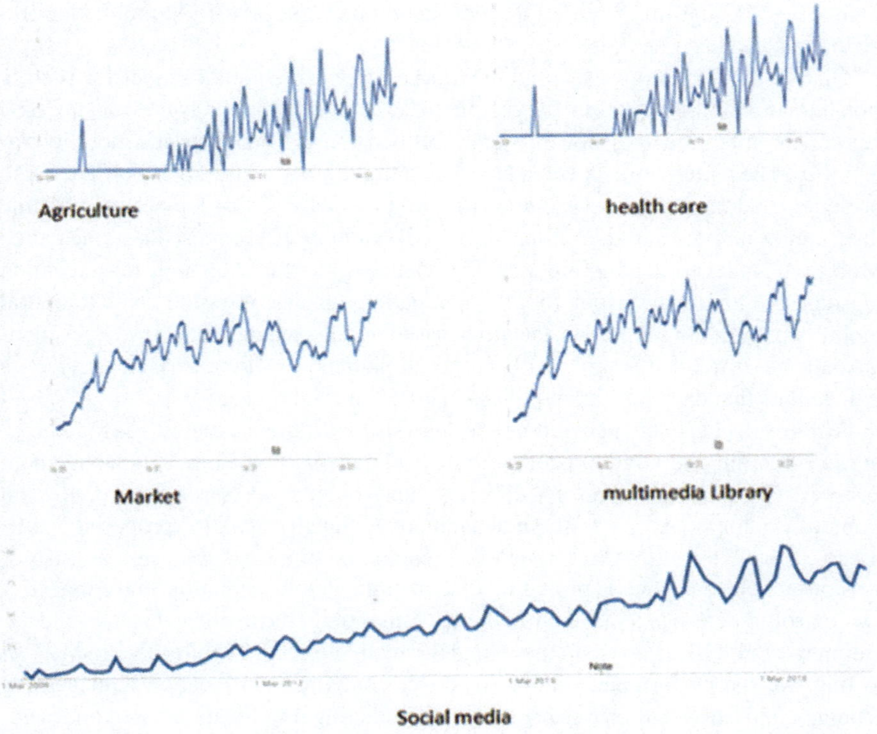

Fig. 1 Internet users in the world during March 2004-March 2019 in different domains

utilization. So for this project, we keep the objective is to just test the project for a small group of people whose mobility is taken into consideration.

The remaining part of the paper is organized as per the following way. Section 2 shows the related work done by different researchers in the area of user mobility detection and prediction. Section 3 presents the proposed research work with proposed system architecture. The methodologies applied to achieve defined objectives are presented in Sect. 4. Section 5 discussed the results and the conclusion of this paper is discussed in Sect. 6.

2 Related Work

Much research has been conducted in the field of user mobility detection and prediction in many application areas. We enumerate some of it as follows. Zhao et al. [23], proposed and presented a user mobility prediction model based on supervised learning approaches. They excrement with Nokia Mobile Challenge Data (MDC) dataset with three ML algorithms decision trees, neural network prediction/multilayer precision, and Bayes predictor.

Qiao et al. [12] proposed a hybrid Markov based prediction model for prediction human mobility. Dataset was generated by capturing mobile patterns using LTE network architecture. It discovers user mobility patterns from the user's mobility trajectories. The pattern is discovered based on the similarity of user mobility. They also presented that the prediction accuracy to be improved by adding the occurrence time distribution into the variable-order Markov. Bayir et al. [2] proposed a frame called Mobility Profiler for the mobile user. The framework is based on high-level location log data required for various mobile applications. It discovers the spatiotemporal mobility pattern of mobile user locations based on selected datasets. The experimental analysis shows that a total of 15% of a cell phone user's time is spent on average in locations that each appears with less than 1% of total time.

Attila et al. [11] published a complete analysis for different data sources for emerging information and communication system based smart cities traffic management systems. This paper also presents different data storage models as well as prediction frameworks for smart city traffic management. Cerqueira et al. [4] proposed Multi-user Session Control (MUSC) to allow fixed and mobile users to access multi-user sessions ubiquitously while providing QoS mapping, QoS adaptation, and connectivity control in heterogeneous environments with mobile receivers and static senders. Quintero et al. [13] proposed a new model called Seamless Mobile IPv6 (SMIPv6) to improve the performance of the handover component in location management schemes. This model improves handover by predicting user location based on Users' Mobility Profiles. The overall goals of SMIPv6 are to reduce both handover latency and signaling loads generated during the location update process.

Celik et al. [3] sought to discover socially similar users, and they did so based on frequently visited or socially important locations of social media users instead of all locations that users visited. The authors also proposed a new interest measure, which is based on Levenshtein distance, to quantify user similarity based on their socially important locations and two algorithms were developed using the proposed method and interest measure. Xu et al. [22] proposed an analytical framework by coupling large scale mobile phone and urban socioeconomic datasets to better understand human mobility patterns and their relationships with travelers' socioeconomic status (SES). The study was conducted for the cities of Boston and Singapore. As can be seen, the research is not limited to only scientific applications but can and does extend to social applications. This goes to show the breadth of the application and research area.

3 Proposed Research Work

For user mobility detection, we need to follow the steps for the prediction of the user movement that is based on the patterns and the mined rules (see Fig. 2). Pattern mining is used to recognize the pattern from the dataset on the trace id of the user. It

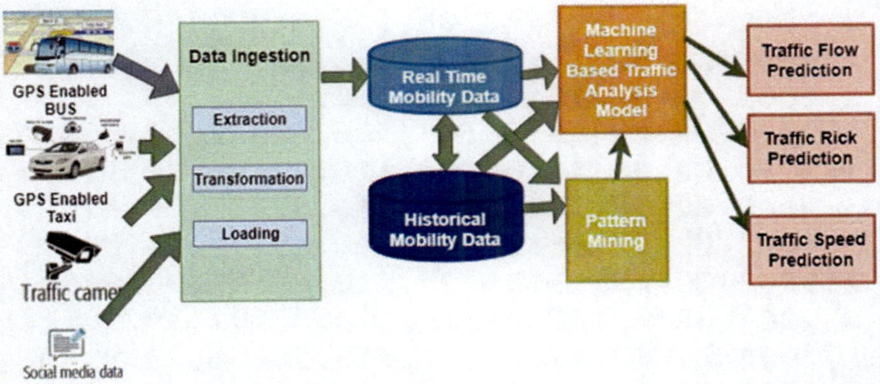

Fig. 2 System architecture for proposed research work

can be predicted the future location of the user from the pattern which is generated. From the pattern which is generated by the pattern mining through applying the mining methods. Some patterns are generated according to the patterns rules are to be followed which is used for detection of the user, that is the cell the user is residing. After applying the mining method to the dataset and patterns are evaluated from which mining's rules are being applied for the prediction of the user location where the distance of the user previous location and the present location is being calculated and the distance between the cells is being calculated to detect the location of the user in future.

4 Methodology and Algorithm

4.1 Dataset Selection

For the experimental analysis, we have selected to use the Nokia Mobile Data Challenge (MDC) dataset [8]. As this dataset comes from a well-established company like Nokia and a prestigious competition like Mobile Data Challenge we can be assured of its quality. This dataset has four features: ID, Date-Time, Latitude, and Longitude (refer Fig. 3). Where in the datasets folder there are different chunks of the data for the particular id. Similarly, we have data of particular id which is of 10,357 unique id and its datasets.

```
1,2008-02-02 15:36:08,116.51172,39.92123
1,2008-02-02 15:46:08,116.51135,39.93883
1,2008-02-02 15:46:08,116.51135,39.93883
1,2008-02-02 15:56:08,116.51627,39.91034
1,2008-02-02 16:06:08,116.47186,39.91248
1,2008-02-02 16:16:08,116.47217,39.92498
1,2008-02-02 16:26:08,116.47179,39.90718
1,2008-02-02 16:36:08,116.45617,39.90531
1,2008-02-02 17:00:24,116.47191,39.90577
1,2008-02-02 17:10:24,116.50661,39.9145
1,2008-02-02 20:30:34,116.49625,39.9146
1,2008-02-02 20:40:33,116.50962,39.91071
1,2008-02-02 20:50:33,116.52231,39.91588
```

Fig. 3 Format for MDC (nokia mobile data challenge) dataset

4.2 Concepts and Methodology

The problem of user mobility prediction comes under the general umbrella of time-series prediction problems. In time series prediction, the model needs to consider all the examples that came before the current example and then make the prediction i.e. when predicting the output of nth example, all the examples from 1 to n-1 impact the output in different magnitude. For this, we need an algorithm that can "remember" the 1 to n-1 examples. We cannot use traditional neural networks as these only work when examples are independent of each other i.e. output of every example is dependent only on the inputs of the current example. Traditional neural networks don't have a "memory" so to speak. There are a few algorithms like RNN and LSTM which can be used for proposed research work in this paper [7]. We are proposing to go with LSTM as these are shown to be generally better in most applications, which will also improvise the accuracy of prediction. The diagram below shows how recurrent neural networks operate in general (see Fig. 4). A recurrent neural network looks at input Xt, performs computation and produces output ht. This is like any other neural network model or any other algorithm in general but the step that differentiates the recurrent neural network is that it saves the computation that was performed for getting the output, it remembers the "thoughts" that led to forming the current output.

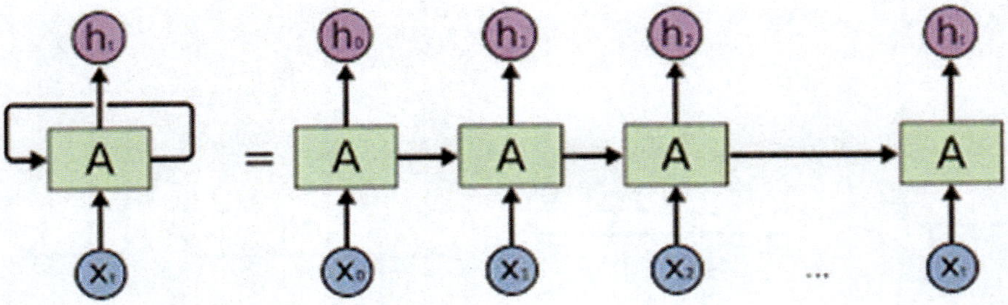

Fig. 4 Long short term memory execution flow

4.3 Working Model Analysis

Figure 5 is showing a more detailed diagram showing the core computation that happens in each cell. The problem with standard RNNs is that they don't work when long term dependencies are required i.e. it cannot look too far back into the past. Also, there is no way of differentiating useful information from not useful information. This is problematic and leads to wrong predictions. LSTMs are specially equipped to address this problem as they are more sophisticated and most importantly allow differentiation between useful and not useful information. LSTMs can be broken down into three internal structures: Input gate, Output gate and Forget gate.

Forget gate (see Fig. 6a): Perhaps the most important part of LSTMs, forget gate is what differentiate LSTMs from standard RNNs. Forget gate gives LSTMs the ability to remember the context while forgetting the previous subject which means that when a new example is given to LSTMs they "forget" the last input and output but "remember" what computations led to obtaining the output.

Input gate (see Fig. 6b): The input gate decides what information to input to the cell. The input gate analysis the newly available information decides what is useful and what is not, normalizes the useful information and inputs it to the cell.

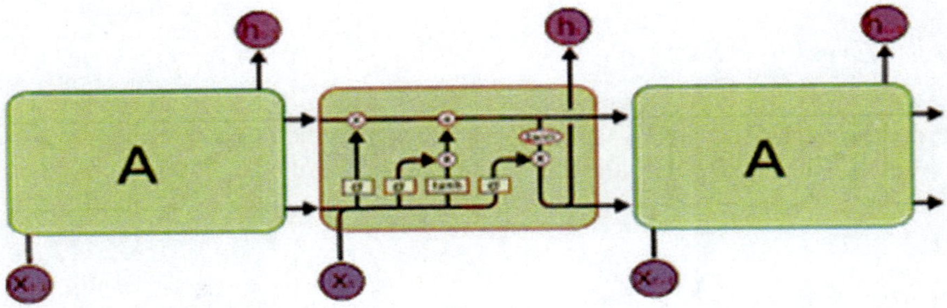

Fig. 5 Detailed diagram for core computation

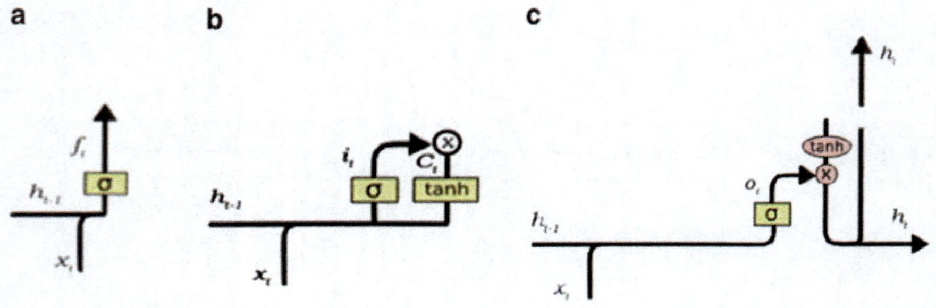

Fig. 6 **a** Forget Gate. **b** Input gate. **c** Output gate

Output gate (see Fig. 6c): Not all the information that is available in the cell has to be output. The main job of the output gate is to decide what to output and when scaling the output back to the original scale and conveying what led to the output to the next cell.

Here the basic idea is to predict the user position based on the latitude and longitude. Since we do not have any class labels, therefore, we cannot classify and predict the classes. Hence clustering techniques can be used to help us. As per DBSCAN clustering, the data samples are taken into consideration and plotted on a 2-dimensional graph. These points have a maximum distance threshold. If the distance between these points becomes greater than the maximum distance threshold, then these two points are said to be in the different clusters. Then according to our data, we create clusters and assign cluster labels to the corresponding points that belong to that particular cluster. This is how we perform clustering. After performing clustering the cluster labels are ready with ourselves. Hence now we can work it like classification by taking a training and test data sample. The training data sample is used to train the model and then we use different algorithms to predict the outcome. Here we can use 4 types of algorithms: K-Nearest Neighbors, Logistic Regression, Decision Trees, and Naíve Bayes. With the help of these models, we can see whether the predicted data belonged to the cluster label it was assigned or not.

4.4 Execution Steps

Pseudo-code 1 is defined as the execution steps required to achieve the proposed objective using the methodology defined in the working model section.

Algorithm 1: Execution Steps

Input: Dataset D, n independent variable xi=x1,x2,.., xn
Output: dependent output variable y
Step1: Using RNN and LSTMs: Import all the necessary libraries
Step2: Create a function for creating dataset from file data
Step3: Read data from file:
- As LSTMs are sensitive to scale, normalise the values of dataset
- Divide dataset into training set and testing set
- Reshape into X=t and Y=t+1
- Reshape input to be [samples, time steps, features]

Step4: Creating and fitting the LSTM model
Step5: Make predictions
- Convert normalised predictions to original scale
- Find error for training and testing data
- Shift predictions on training data for plotting
- Shift predictions on testing data for plotting

5 Results and Discussion

Plot original data and predicted data (blue-original data, orange-predictions of test data, yellow-predictions on training data). Using various Clustering methods: Let us begin with the implementation of our model using clustering. Here we begin first by importing the necessary libraries required for the model. We then import the dataset and fill in the missing values by the mean of the respective columns (here latitude and longitude). Then for doing clustering using DBSCAN (see Figs. 7 and Figure 8), we require the max distance after which any two points will be considered in a different group.

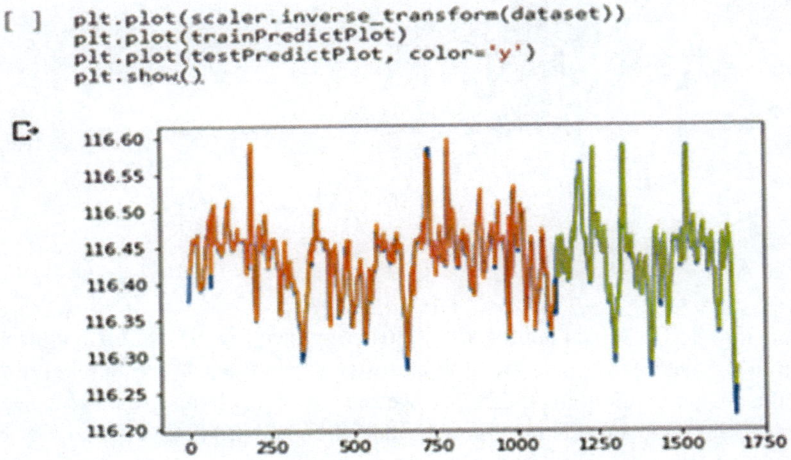

Fig. 7 DBSCAN based prediction

```
sns.countplot(datamin['cluster_labels'],label="Count")
plt.show()
```

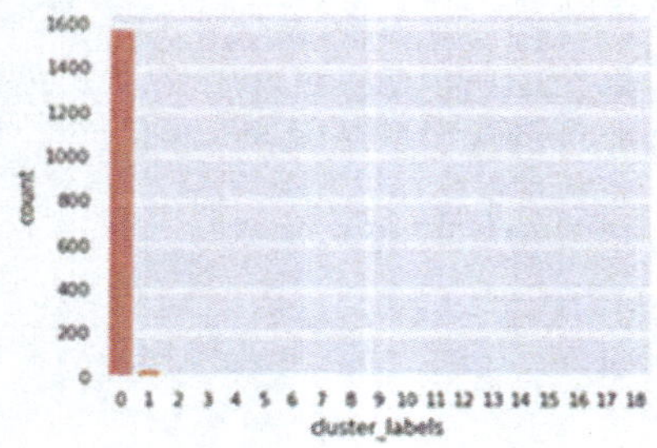


```
feature_names=['lat','long']
X=datamin[feature_names]
y=datamin['cluster_labels']
```

```
from sklearn.model_selection import train_test_split
X_train, X_test, y_train, y_test = train_test_split(X, y, random_state=0)
from sklearn.preprocessing import MinMaxScaler
scaler = MinMaxScaler()
X_train = scaler.fit_transform(X_train)
X_test = scaler.transform(X_test)
```

Fig. 8 DBSCAN based cluster information

```
[28] from sklearn.cluster import KMeans
     kmeans = KMeans(n_clusters=19)
     kmeans.fit(Z)
     y_kmeans = kmeans.predict(Z)
```

```
plt.scatter(Z.iloc[:, 0], Z.iloc[:, 1], c=y_kmeans, s=50, cmap='viridis')
centers = kmeans.cluster_centers_
plt.scatter(centers[:, 0], centers[:, 1], c='black', s=200, alpha=0.5);
```

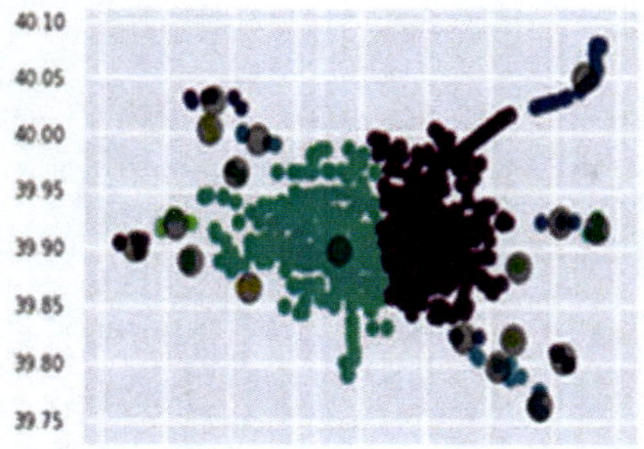

Fig. 9 Machine learning based learning model

Then using the DBSCAN approach for clustering (see Fig. 9), we form clusters and then in the dataset variable we assign cluster labels to each one belonging to the specific cluster. After doing clustering, we see how many items are there in each cluster label. As the dataset is user mobility detection and prediction, a majority of the time the user movement is very slow or sometimes still. As a result, there are

Table 1 Confusion matrix

S. No.	Precision	Recall	F1-score	Support
1.	0.99	1.00	0.99	393
2.	1.00	1.00	1.00	11
3.	0.00	0.00	0.00	1
4.	0.00	0.00	0.00	2
5.	0.50	1.00	0.67	2
6.	1.00	1.00	1.00	2
7.	1.00	1.00	1.00	1
8.	0.00	0.00	0.00	2
9.	0.00	0.00	0.00	1
10.	0.50	1.00	0.67	1
11.	0.00	0.00	0.00	1
12.	1.00	1.00	1.00	2

many points belonging to the same cluster. This shows that the user movement is mostly similar but varies only sometimes. Then we also divide our data into train and test data wherein the X contains the latitude and longitude information and y contains the cluster labels information.

After doing this we apply various classification models like logistic regression, decision tree, etc. to find out whether the predicted point belongs to the same cluster label or different cluster label. Their accuracy are shown below. Here we also create the confusion matrix (Refer Table 1).

And now by using plot functions we give a figurative idea of what each cluster looks like, where the black dots represent the center of their cluster. Thus this is how we implement clustering to predict future locations.

6 Conclusion and Future Work

With the advent of time, the number of mobiles has increased and as a result, the user movement has become of concern. We are proposing to use Predictive Analytic as an approach to deal with these types of data. First, we apply Machine Learning using the data of user movement, which consists of latitude and longitude at specific date and time, then we are proposing to apply clustering-based learning for mobility prediction. We also apply a different method of clustering for prediction. At the moment, having seen the various methods of how we can predict the user movement in the future, many more lie ahead in the future. Hence the scope of work here is of great importance. As a future enhancement of this work, Deep Learning Algorithms like RNN and LSTM will be implemented to increase the accuracy of our predictions.

Acknowledgements The experimentation analysis was done by the fifth-semester students (Archies Desai and Poojan Khatri) as part of their mini-project.

References

1. Ashbrook, D., Starner, T.: Using gps to learn significant locations and predict movement across multiple users. personal and ubiquitous computing 7(5), 275-286. Personal and Ubiquitous Computing **7**, 275–286 (2003). https://doi.org/10.1007/s00779-003-0240-0
2. Bayir, M.A., Demirbas, M., Eagle, N.: Mobility profiler: a framework for discovering mobility profiles of cell phone users. Pervasive Mob. Comput. **6**(4), 435 – 454 (2010) (Human Behavior in Ubiquitous Environments: Modeling of Human Mobility Patterns)
3. Celik, M., Dokuz, A.S.: Discovering socially similar users in social media datasets based on their socially important locations. Inf. Proc. Manage. **54**(6), 1154–1168 (2018)
4. Cerqueira, E., Veloso, L., Neto, A., Curado, M., Monteiro, E., Mendes, P.: Mobility management for multi-user sessions in next generation wireless systems. Comput. Commun. **31**(5), 915–934 (2008) (Mobility Management and Wireless Access)
5. Eagle, N., (Sandy) Pentland, A.: Reality mining: sensing complex social systems. Personal Ubiquitous Comput. **10**(4), 255–268 (2006)
6. Gong, J., Huang, Y., Chow, P.I., Fua, K., Gerber, M.S., Teachman, B.A., Barnes, L.E.: Understanding behavioral dynamics of social anxiety among college students through smartphone sensors. Inf. Fusion **49**, 57–68 (2019)
7. Kaushal, N.C., Paprzycki, M., Bhargava, B.K., Singh, P.K., Hong, W.C.: Ditzinger: handbook of wireless sensor networks: issues and challenges in current scenarios. Springer International Publishing (2020)
8. Laurila, J.K., Gatica-Perez, D., Aad, I., Blom, J., Bornet, O., Do, T.M.T., Dousse, O., Eberle, J., Miettinen, M.: From big smartphone data to worldwide research: the mobile data challenge. Pervasive Mob. Comput. **9**(6), 752–771 (2013) (Mobile Data Challenge)
9. Lu, X., Bengtsson, L., Holme, P.: Predictability of population displacement after the 2010 haiti earthquake. Proc. National Acad. Sci. **109**(29), 11576–11581 (2012). https://doi.org/10.1073/pnas.1203882109, https://www.pnas.org/content/109/29/11576
10. Marmasse, N., Schmandt, C.: A user-centered location model. Personal Ubiquitous Comput. **6**(5–6), 318–321 (2002). https://doi.org/10.1007/s007790200035, https://doi.org/10.1007/s007790200035
11. Nagy, A., Simon, V.: Survey on traffic prediction in smart cities. Pervasive Mob. Comput. (2018). https://doi.org/10.1016/j.pmcj.2018.07.004
12. Qiao, Y., Si, Z., Zhang, Y., Abdesslem, F., Zhang, X., Yang, J.: A hybrid markov-based model for human mobility prediction. Neurocomputing **278** (2017). https://doi.org/10.1016/j.neucom.2017.05.101
13. Quintero, A., Pierre, S., Alaoui, L.: A mobility management model based on users' mobility profiles for ipv6 networks. Comput. Commun. **30**(1), 66–80 (2006)
14. Rodríguez, L., Palanca, J., del Val, E., Rebollo, M.: Analyzing urban mobility paths based on users' activity in social networks. Future Gener. Comput. Syst. **102**, 333–346 (2020)
15. Singh, P.K.: Futuristic Trends in Network and Communication Technologies: First International Conference, FTNCT 2018, Solan, India, February 9–10, 2018: revised selected papers. Springer (2019)
16. Singh, P.K.: Proceedings of First International Conference on Computing, Communications, and Cyber-Security (IC4S 2019). Springer (2020)
17. Singh, P.K., Kar, A.K., Singh, Y., Kolekar, M.H., Tanwar, S.: Proceedings of ICRIC 2019: Recent Innovations in Computing. Springer (2020)
18. Singh, P.K., Sood, S., Kumar, Y., Paprzycki, M., Pljonkin, A., Hong, W.C.: Futuristic Trends in Networks and Computing Technologies: Second International Conference, FTNCT 2019, Chandigarh, India, November 22–23, 2019, Revised Selected Papers. Springer (2020)

19. Verma, J.P., Mankad, S.H., Garg, S.: A graph based analysis of user mobility for a smart city project. In: Next Generation Computing Technologies on Computational Intelligence. pp. 140–151. Springer Singapore, Singapore (2019)
20. Verma, J.P., Patel, A.: Evaluation of unsupervised learning based extractive text summarization technique for large scale review and feedback data. Ind. J. Sci. Technol. **10**(17), (2017)
21. Verma, J.P., Patel, B., Patel, A.: Article: Web mining: opinion and feedback analysis for educational institutions. Int. J. Comput. Appl. **84**(6), 17–22 (2013) (Full text available)
22. Xu, Y., Belyi, A., Bojic, I., Ratti, C.: Human mobility and socioeconomic status: analysis of Singapore and Boston. Comput. Environ. Urban Syst. **72**, 51–67 (2018)
23. Zhao, Z., Karimzadeh, M., Gerber, F., Braun, T.: Mobile crowd location prediction with hybrid features using ensemble learning. Future Gener. Comput. Syst. (2018)

A Literature Review of Critical Success Factors in Agile Testing Method of Software Development

Abhishek Srivastava, Deepti Mehrotra, P. K. Kapur, and Anu G. Aggarwal

Abstract Software development process is a collaborative effort to fulfil multidimensional product requirements. The development methodology plays an important role to meet the expectations of all the stakeholders especially the client. So, agile teams are constituted having experts level of knowledge and versatility in software development. The team tests the software in errands and takes regular feedback to develop a final product. For this, researchers have suggested numerous approaches in their research works. This paper is a literature review of those earlier research works suggesting different approaches of software development using agile testing method to increase efficacy of software as compared to traditional methods. The purpose is to list the objectives considered which help in deciding the development approach to be undertaken. It brings forth the related agile attributes which influence the software development task in terms of quality, scope, timeliness and cost effectiveness. Further, a framework has been presented for use by any of the stakeholders involved in agile software development methodology.

Keywords Agile testing · Software development · Attribute · Agile testing · Critical success factor

A. Srivastava (✉) · D. Mehrotra
Amity School of Engineering and Technology, Noida, Uttar Pradesh, India
e-mail: abhishek.sri13@gmail.com

D. Mehrotra
e-mail: mehdeepti@gmail.com

P. K. Kapur
Amity Center for Interdisciplinary Research, Amity University, Noida, Uttar Pradesh, India
e-mail: pkkapur1@gmail.com

A. G. Aggarwal
Department of Operational Research, Faculty of Mathematical Sciences, University of Delhi, Delhi, India
e-mail: anuagg17@gmail.com

P. K. Singh et al. (eds.), *Recent Innovations in Computing*, Lecture Notes in Electrical Engineering 701, https://doi.org/10.1007/978-981-15-8297-4_69

1 Introduction

Software development has been evolving to become more collaborative and have human approach. It is due to continuous challenges faced by the software engineers at the time of project delivery. It has been observed that there are several instances of software rejection or project delays or even product desertion due to the product ineffectiveness despite of thorough efforts of all the teams. This grew requirement of a collaborative team known as agile team (AT). This team works towards a clear path of developing flawless, quality driven end product through articulate communication between the client and the development team. Agile teams or Dev teams comprise software developers teamed with testers. Agile testing is not a separate phase in it but a stage within the same team [1]. One of team member acts as an agile tester (AT) to test the software at first and bring improvements with time as per a high-level plan on case to case basis [2]. Within the agile team, ATs are responsible for finding defects in software, communicate with the software developers in their terminology from the perspective of client's requirements. Though ATs may not have coding skills, it needs to be highly dynamic in their working [2]. The entire functionality of an agile team is managed by cross-functional teams through self-organization and is requirements and solution-driven [3].

It is notable that each iteration gives fully functional software which could be released. There are some widely used agile methodologies of software development like Extreme Programming (XP), Lean Development (LD), Test-Driven Development (TDD), Scrum, Feature-Driven Development (FDD) and Dynamic Systems Development Method (DSDM) [4]. These methodologies have several iterations and incremental development patterns within the agile framework for software development within the stipulated time frame. Else the further improvement plan is considered as a backlog instead of extending the project delivery date [5].

Agile software development follows a set of four core values and has twelve principles, which are termed as Agile Manifesto. Software developers follow the Agile Manifesto values and principles as it helps in meeting the client requirements through continuous dialogue to build and deliver at a faster pace. The software development life cycle is focused on working as a team, collaborative efforts and process smoothness within some limitations as per agile development and testing team principles [4]. It makes the software developers powerful, as they become part of team creating more business value through simplification of design but having better technical features [6]. Researchers also referred agile methodology as consisting of process ease with less formalized structured way of process [7]. This also brings a brought more flexibility concept in work frame along with bringing more rapid change due to the deftness and leanness of the team which led to reduction in response time [8].

However, this brought in few challenges also for the testers when compared to plan-driven process. As an agile tester, some people faced challenges working under scrum framework, as identified by the researcher. The faced were like less cooperation, lack of trustworthiness and loyalty to the company and follow-up on regulatory governance [2]. One of the agile testers also faced problem in communication, taking team opinion, and automation related challenges while executing a rewrite project on legacy system. Therefore, during this manoeuvre, collaboration and internal communication can be major challenges for ATs to work seamlessly [9]. Due to collaborative nature of work in agile teams, ATs also bear added challenges like team integration, integrated system testing, enhancing developers' skills and developing high-level testing strategy [9]. Knowledge discourse on the domain and agile framework must be ensured to keep from further complications.

Further, challenges faced by the agile team have been discussed along with their associated approach.

2 Literature Review

The agile team often faces the challenge of deciding the most appropriate approach to develop the required product in the stipulated time and also find best or any possible solutions to the problems in the software. Sometimes, if a suitable solution could not be found, the projects are abandoned even. This is due to the fact that the agile development process of the software that keeps the process evolving always with the perspective of business [10]. As the technology is continuously changing and also a clearer idea of the product developed is obtained after iterations demo, so the development methodology also changes. This agility has given numerous software development approaches such as Extreme Programming, Crystal, SCRUM, Lean and Kanban.

In general software development work in larger organizations, smaller teams work on specific part of the product. This keeps the team focused and gives less control to avoid much collaboration. However, to introduce any new feature to the product needs more collaborative effort. So, in agile methodology, a self-organizing team is arranged, which decides on the team accessibility during the development and post release access also. Therefore, this team is power-driven and capable of coordinating themselves smoothly. This affects the productivity of the team positively and elevates the software development process. This gives the team an opportunity to improve on their professional expertise. This skill enhancement motivates them too [11].

However, it has been observed that considerable number of organizations still plan the software development process according to a presaged plan. In such scenario, these teams are shifted to agile teams later, which is quite an interest to many develop new learnings, as the presaged plan for overcoming the procedural difficulties during development process without hindrance is being studied by many researchers as this process has witnessed lot of dead ends which forces the company to discontinue the development project [12]. This often demotivates the team with lot of questions overshadowing any minimal benefits such as the cost of transition. Also, Smite et al. [11] found in his research that many software companies are exploring the avenues for global sourcing for software development which means the teams anyways will confront challenges of collaborating between teams that are far apart in their demographics and temperament. It brings another set of challenges leading to agile approach of software development.

This motivated the researcher to further delve into the software development process using agile approach and explore the success and critical factors. This paper is focused on finding the knowledge contributions in agile in relation to the attributes and how does the team decide the best matching approach for software development.

2.1 *Critical Success Factor for Agile Teams*

Critical success factor is introduced as an approach which detects names and evaluates an organization's performance [13, 14]. The researchers considered critical success factor as factors that ensure accomplishment, satisfaction and bring motivation for individuals with a healthy competitive work environment. But it is limited to few domains only. These factors are relevant project management techniques with a mix of software development approach and business strategies [15]. Some researchers also consider critical success factors as variety of dimensions like product development cycle, strategic planning, project management techniques, forecasting, estimation, resource management, etc. [16].

Also, it is necessary to learn about possible problems to be ready beforehand, as have been shared below.

2.2 *Pitfalls in Agile Teams*

Agile development is also about learning from the pitfalls or failures which other ATs faced. So, problem searching is another way of agile development methodology, based on experience of other specific projects which are sufficient to be generalized for other projects as learnings [17]. Researchers have listed about the mistakes committed and some misunderstandings due to the multidimensional approach of agile projects [18]. Management challenges handled during the development process were listed [19, 20] from the perspective of technology, processes and available resources for transitioning the general development process into agile development process.

2.3 *Work Dimensions for Agile Teams to Drive Success*

2.3.1 Organizational

To keep the agile development process outcome driven, organizations focus on equipping the team with all sort of resources, trainings, cultural changes like collective rewards, acknowledgements, etc. Team building exercises help in developing a motivated task force [21].

2.3.2 People

Agile teams interact with all stakeholders especially amongst themselves or with others teams, client and vendors too. This impacts their performance directly [22]. Therefore, it becomes necessary that more transparency in work processes and communications is practiced. Team meetings and interpersonal skills are considered to have better team work. Conflict management and better coordination play important role in such coordination tasks.

2.3.3 Process

Process following agile approach requires the following project management techniques in same manner keeping team communications very critical. Regular team meetings, following working schedule, communicating and following client requirements regularly makes the process agile positively.

2.3.4 Technical

The agile team members have to be well versed with the programming languages and must have gained sufficient experience through previous engagements. This is very much required to keep pace within the team. Merely length of experience of a software developer does not work in agile. The products developed using the required set of skills indicate decent software expertise to be part of AT teams [23].

2.3.5 Project

Software development is also part of the project management but in IT domain. It involves project planning, design thinking, strategic planning and finally process of software development starts. Similar risks are involved in this as well being the environment is highly dynamic, with numerous teams working together on it subparts. Table 1 shows the attributes considered for agile development of the software.

Table 1 Agile attributes

Attribute name	Attribute description	References
Continuous integration	This attribute is used when development process is arranged as per the developer's knowledge about certain functions and thus functional description of requirements It integrates the work of the all developers to reduce the process complexity, as multiple developers work on same project. The integrated work is checked for errors or bugs	[24–27]
Test environment availability	This attribute is focused on defects which exist in the software or may be highlighted later The developers focus on making software defect free. The development cycle has four methods to make software organized and includes prevention and removal of existing defects. Also, defect tolerance and its forecasting are within scope of this attribute	[28, 29]
Test-driven development	In this attribute, development is focused on the software requirements and client specifications. Therefore, software test run is conducted, and if there is any change or fault found, new code is written again to iterate the testing process	[29, 30]
Adequate documentation	This attribute focuses on documenting the development path to simplify the process of agile development. Any updates made are documented side by side for reference at any given time and help timely delivery of project	[31, 32]
Feedback mechanism	Here, the agile team developers focus on feedback iterations to help achieved the desired product. This helps the team to put productive efforts to run loops which re-examine at every step during the software release	[33, 34]
Availability of test data	This attribute of agile software development is to store the complete testing data for any time reference by the team. A storage or entire data back option of test data is must to support the agile testing team with all possible test environments tried. It is also important as one may require to restore the test environment	[35]

(continued)

Table 1 (continued)

Attribute name	Attribute description	References
Supporting tools	Continuous integration tools like team management tools which help in modelling the agile system or help in creating and following suitable patterns of software development by keeping them from any possible deviations from the necessary requirements accomplishments. Such tools are deliberately designed depending on the work settings and required development tracking system during all iterations and testing	[35–38]
Regular stand-up meetings	Daily crisp meeting is a usual way of bringing all in the team on same platform and affirm focus on the requirement sets. It helps each member of team to be on same page and share the work progression and discuss problems faced within time	[35]
Clearly defined requirement	As the requirements document is updated several times, it must have clearly stated requirements for continuous reference by the agile team	[35]
Automated test cases	It refers to automatic testing of the agile software as and when any updates are made to avoid any software bug in functionality. Regression testing or black box testing is a method of automated testing	[39, 40]
Agile method used-repeated	In this, the entire agile development process is bifurcated into errands which are the basically augmentations of the software. Each errand is run repeatedly, and the process is handled by the deputed agile team for the specified errand which may be focused on enhancing its code or prerequisites or positioning or any component testing or for responsiveness Scrum, Kanban, XP are the most popular methods used in this attribute	[41, 42]
Customer involvement	Customer involvement is needed in every sprint cycle so that the feedback can be incorporated in real time and customer feedback can be taken to increase the efficiency and improve the quality of the product	[29, 34]

(continued)

Table 1 (continued)

Attribute name	Attribute description	References
Defect management process	This defines the daily checks and maintenance action to eliminate any complications in the development progression	[29, 37]
Agile testing mindset	It refers to basic and repeated testing of each module of the code as fast as possible to know the best possible functionality	[31, 34]
Root cause analysis (RCA)	It is the attribute in which the focus is on removing the root cause of any problem to reduce the probability of the problem recurrence	[43]
Mutation testing (MT)	The focus of system is to test the code in cycles to ensure thorough testing of the code and avoid any false sense of correctness of code	[44]
Cyclic testing and lean approach	The attribute keeps the project as per the timelines decided, keeping with the quality enhancement accomplishments. The overall time consumed must fall within the overall project guidelines	[45]
Process action	This attribute engages a process action team (PAT) to devise a project strategy to complete the required tasks. The strategy is taken as the task completion/progress path by the agile software development team	[46]
IT governance strategy	This suggests to devise a strategy to implement the final software post-delivery. An effective strategy would be executed to ensure business excellence and achieve required outcomes	[47]
Organizational change	This attribute ensures smooth transition to bring the organizational change from one process to the fresh software. This would be in keeping with the principles of change management for harmonizing the changes	[48]
Effective risk mitigation	During the transition process, the risk management attribute proves its importance as the change progresses. All possible risks impacting the system have to be monitored and managed through a well-designed plan which can mitigate the risks to any level. Proper indicative charts and supportive documents would give clear status of each sprint of risk exposure	[49]

(continued)

Table 1 (continued)

Attribute name	Attribute description	References
Tangible outcomes and feedback	This is an essential attribute of any software development process. The strong feedback system helps in regular refinement of the end product both in quality adjoined with performance. Also, a joint meeting for feedback discussion brings both client and the ATs to know what best is achievable ensuring business excellence	[10]

The entire study has helped the researcher in categorizing the attributes of agile software development teams into five categories on the basis of which the development strategy is planned with consideration to certain suitable attributes which have been shown in the Fig. 1.

Figure 1 shows the classification of all agile attributes chosen keeping in view the prime focus on the suitable approach. If the development approach is associated to source control, then it is driven by strong planning variable represents the set of factors associated to source control. The associated system attributes are test estimates, automated tests, clarity about requirements and agile Method used.

If the development approach is associated with technique being used, it means, the approach focuses on execution of the agile development process. Thus, the development team follows continuous integration with completion of required documentation with proper backup of overall test data created while the several iterations and defect management processes.

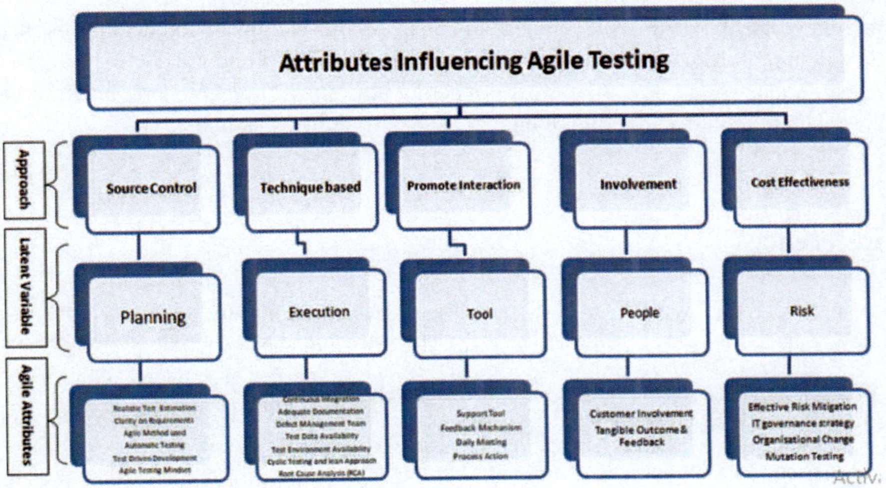

Fig. 1 Five categories of agile software development

If the agile team's approach is mainly interaction-driven within or outside the team, then the team develops tools that act as supporting tools for interaction and developing strong feedback mechanism and have daily stand-up meeting with all team members.

However, if the team wants more involvement from stakeholder and team functioning is people-driven, then main attributes of the development process are client involvement, whole team tests and availability of testing data.

Also, if the overall strategy is mainly for business excellence mainly with perspective of cost factor, then the agile team develops its strategy oriented towards taking minimum risk through attributes like effective risk mitigation, IT governance strategy, organizational change management and mutation testing.

3 Conclusion

In this literature review, we have listed the possible attributes of software development process which would be helpful for the agile team to decide the approach of strategizing the software development process. The critical success factors as well as pitfalls can be checked by looking into the possible dimensions of the development process to drive success. The agile teams are exposed to all stake holders requirements and feedback on regular basis to refine the final product. These teams work in highly dynamic environment to meet the requirements of the client. Therefore, a category chart will be an instant way to decide the possible approach towards product development as per the client requirements and matching work dimension based on approach oriented towards source control, technique, people interaction, involvement or cost. Each approach has its specific latent variable that will give the team its related attributes to focus on deriving a successful plan of action accomplishing the client requirements. However, further research shall be conducted to experiential research also to lay down the most impacting attributes for each of the undertaken approach potentially in terms of quality, scope, timeliness and cost.

References

1. Penmetsa, J.R.: Agile testing. In: Mohanty et al. (eds.) Trends in Software Testing (2017)
2. Crispin, L., Gregory, J.: Agile testing: A Practical Guide for Testers and Agile Teams. Pearson Education (2008)
3. Beck, K. et al.: Manifesto for Agile Software Development (2001)
4. Padmini, K.V.J., Bandara, H.M.N.D., Perera, I.: Use of software metrics in agile software development process. In: Proceeding Moratuwa Engineering Research Conference (2015)
5. Itkonen, J., Rautiainen, K., Lassenius: Towards understanding quality assurance in agile software development. In: Proceeding of International Conference on Agility Helsinki (2005)
6. Henderson-Sellers, B., Serour, M.K.: Creating a dual-agility method: the value of method engineering. J. Database Manage. **16**, 1–23 (2005)
7. Cockburn, A.: Agile Software Development: The Cooperative Game. Addison-Wesley (2007)

8. Erickson, J., Lyytinen, K., Siau, K.: Agile modeling, agile software development, and extreme programming. J. Database Manage. **16**, 88–100 (2005)
9. Gupta, R.K., Manikreddy, P., Abhinandan, G.V.: Challenges in adapting agile testing in a legacy product. In: Proceeding of 11th IEEE International Conference on Global Software Engineering (2016)
10. Williams, L., Cockburn, A.: Agile software development: it's about feedback and change. Computer **36**, 39–43 (2003)
11. Smite, D., Moe, N.B., Agerfalk, P.J.: Agility Across Time and Space, pp. 333–337. Springer-Verlag, Berlin, Heidelberg (2010)
12. Moe, N.B., Dingsoyr, T., Dyba, T.: A teamwork model for understanding an agile team: a case study of a scrum project. Inf. Softw. Technol. **52**, 480–491 (2010)
13. Bullen, C.V., Rockhart, J.F.: A primer on critical success factors (Working Paper No. 69), Massachusetts Institute of Technology, Sloan School of Management, Center for Information Systems Research, Cambridge, Massachusetts (1981)
14. Rockhart, J.F., Crescenzi, A.D:. Engaging top management in information technology. Sloan Manage. Rev. **25**(4), 3–16 (1984)
15. Bytheway, A.J.: Successful software projects and how to achieve them. IEEE Softw. **16**(3), 15–17 (1999)
16. Bosghossian, Z.J.: An investigation into the critical success factors of software development process, time, and quality, Ph.D. Thesis, Pepperdine University, Malibu, California (2002)
17. Cohn, M., Ford, D.: Introducing an agile process to an organization. Computer **36**(6), 74–78 (2003)
18. Larman, C.: Scaling Lean and Agile: Large. Multisite or Offshore Delivery (2011)
19. Boehm, B., Turner, R.: Management challenges to implement agile processes in traditional development organizations. IEEE Softw. **22**(5), 30–39 (2005)
20. Nerur, S., Mahapatra, R.K., Mangalaraj, G.: Challenges of migrating to agile methodologies. Commun. ACM **48**(5), 72–78 (2005)
21. Shore, J., Warden, S.: The Art of Agile Development. O'Reilly Media Inc, Beijing Sebastopol, CA (2008)
22. Cao, L., Mohan, K., Xu, P., Ramesh, B.A.: Framework for adapting agile development methodologies. Eur. J. Inf. Syst. **18**, 332–343 (2009)
23. Doran, H.D.: Agile knowledge management in practice. In: Melnik, G., Holz, H. (eds.) Advances in Learning Software Organizations, Proceedings, pp. 137–143(2004)
24. Perry, W.E.: Effective methods for software testing, 3rd edn. Wiley, Indianapolis, IN (2006)
25. Schaefer, A., Reichenbach, M., Fey, D.: Continuous Integration and Automation for DevOps, IAENG Transactions on Engineering Technologies, pp. 345–358. Springer, Netherlands (2013)
26. Stillwell, M., Coutinho, J.G.: A DevOps approach to integration of software components in an EU research project. QUDOS (2015)
27. Zampetti, F., Vassallo, C., Panichella, S., et al.: An empirical characterization of bad practices in continuous integration. Empir Softw. Eng. **25**, 1095–1135 (2020)
28. Artač, M., Borovšak, T., Di Nitto, E., Guerriero, M., Tamburri, D.A.: Model-driven continuous deployment for quality DevOps. In: 2nd International Workshop on Quality-Aware DevOps. ACM, pp. 40–41. New York (2016)
29. Patwardhan, A., Kidd J., Urena, T., Rajgopalan, A.: Embracing agile methodology during DevOps developer internship program (2016)
30. Sommerville, I.: Integrated requirements engineering: a tutorial. IEEE Softw. **22**, 16–23 (2005)
31. Mullaguru, S.N.: Changing scenario of testing paradigms using DevOps–a comparative study with classical models. Glob. J. Comput. Sci. Technol. **15**(2) (2015)
32. Saaty, T.L.: Decision making with the analytic hierarchy process. Int. J. Ser. Sci. **1**(1), 83–98 (2008)
33. Erich, F., Amrit, C., Daneva, M.: A mapping study on cooperation between information system development and operations. In: International Conference on Product-Focused Software Process Improvement, pp. 277–280. Springer International Publishing (2014)

34. Mohamed, S.I.: DevOps maturity calculator DOMC—value oriented approach. Int. J. Eng. Res. Sci. **2**(2):25–35 (2016)
35. Shihab, E., Bird, C., Zimmermann, T.: The effect of branching strategies on software quality. In: ACM-IEEE International Symposium on Empirical Software Engineering and Measurement, pp. 301–310. IEEE (2012)
36. Appleton, B., Steve, B., Ralph, C., Robert, O.: Streamed lines: branching patterns for parallel software development. In: Fifth Annual Conference on Pattern Languages of Programs, Monticello (1998)
37. Hsieh, C.Y., Chen, C.T.: Patterns for continuous integration builds in cross-platform agile software development. J. Inf. Sci. Eng. **31**(3), 897–924 (2015)
38. Larman, C., Basili, V.: Iterative and incremental development: a brief history. Computer **36**, 47–56 (2003)
39. Binkley, D.: The application of program slicing to regression testing. Inf. Softw. Technol. **40**(11–12), 583–594 (1998)
40. Ghafari, M., Ghezzi, C., Rubinov, K.: Automatically identifying focal methods under test in unit test cases. In: IEEE 15th International Working Conference on Source Code Analysis and Manipulation (SCAM), Bremen, pp. 61–70 (2015)
41. Lwakatare, L.E., Kuvaja, P., Oivo, M.: Dimensions of DevOps. In: International Conference on Agile Software Development, Springer International Publishing, pp. 212–217, Finland (2015)
42. Sacks, M.: Devops principles for successful web sites. Pro Website Development and Operations, Apress, pp. 1–14 (2012)
43. Hall, M.A.: Root cause analysis: a tool for closer supply chain integration in construction. In: Akintoye, A. (Ed.), 17th Annual ARCOM Conference, vol. 1, pp. 929–938, Salford (2001)
44. Jia, Y., Harman, M.: An analysis and survey of the development of mutation testing. IEEE Trans. Software Eng. **37**(5), 649–678 (2011)
45. Nidagundi, P., Novickis, L.: Introduction to lean canvas transformation models and metrics in software testing. Appl. Comput. Syst. **19**(1), 30–36 (2016)
46. Salo, O., Abrahamsson, P.: Integrating agile software development and software process improvement: a longitudinal case study. In: International Symposium on Empirical Software Engineering, p. 10. Noosa Heads, Qld (2005)
47. Qumer A.: Defining an Integrated agile governance for large agile software development environments. In: Concas, G., Damiani, E., Scotto, M., Succi, G. (eds.) Agile processes in software engineering and extreme programming. XP 2007. Lecture Notes in Computer Science, vol. 4536. Springer, Heidelberg (2007)
48. Pikkarainen, M., Salo, O., Kuusela, R., et al.: Strengths and barriers behind the successful agile deployment—insights from the three software intensive companies in Finland. Empir Softw. Eng. **17**, 675–702 (2012)
49. Siddique, L., Hussein, B.A.: Practical insight about risk management process in agile software projects In Norway. In: IEEE International Technology Management Conference, pp. 1–4. Chicago (2014)

Author Index

P. K. Singh et al. (eds.), *Recent Innovations in Computing*, Lecture Notes in Electrical Engineering 701, https://doi.org/10.1007/978-981-15-8297-4